Lymphocyte Signalling

Lymphocyte Signalling
Mechanisms, Subversion and Manipulation

Edited by

Margaret M. Harnett

University of Glasgow, UK

and

Kevin P. Rigley

The Edward Jenner Institute for Vaccine Research, UK

JOHN WILEY & SONS

Chichester · New York · Weinheim · Brisbane · Singapore · Toronto

Other Wiley Editorial Offices

John Wiley & Sons, Inc., 605 Third Avenue,
New York, NY 10158–0012, USA

VCH Verlagsgesellschaft mbh, Pappelallee 3,
D-69469 Weinheim, Germany

Jacaranda Wiley Ltd, 33 Park Road, Milton,
Queensland 4064, Australia

John Wiley & Sons (Asia) Pte Ltd, 2 Clementi Loop #02-01,
Jin Xing Distripark, Singapore 129809

John Wiley & Sons (Canada) Ltd, 22 Worcester Road,
Rexdale, Ontario M9W 1L1, Canada

Library of Congress Cataloging-in-Publication Data
Lymphocyte signalling : mechanisms, subversion, and manipulation / edited by Margaret M. Harnett and Kevin P. Rigley.
    p.   cm.
  Includes bibliographical references.
  ISBN 0-471-95903-0 (hbk. : alk. paper)
  1. Lymphocytes. 2. Cellular signal transduction. 3. Lymphocytes–Receptors. 4. Lymphocyte transformation.
5. Immunopathology. I. Harnett, Margaret M.   II. Rigley, Kevin P.
  QR185.8. L9L868   1996   96–19084
616.07′9—dc20   CIP

British Library Cataloguing in Publication Data
A catalogue record for this book is available from the British Library
ISBN 0-471-95903-0

Typeset in 10/11.5 pt Times from the author's disks by Puretech (India) Ltd
Printed and bound in Great Britain by Bookcraft, Midsomer Norton, Somerset
This book is printed on acid-free paper responsibly manufactured from sustainable forestation, for which at least two trees are
planted for each one used for paper production.

Contents

Contributors

D.R. Alexander *The T Cell Laboratory, Department of Immunology, The Babraham Institute, Cambridge CB2 4AT, UK*

A. Arnaiz-Villena *IMMUNOLOGIA, Hospital Universitario '12 de Octubre', Universidad Complutense, Carretera Andalucia, 28041 Madrid, Spain*

J. Ashwell *Laboratory of Immune Cell Biology, National Cancer Institute, National Institutes of Health, Bethesda MD 20892–1152, USA*

J-P. Aubry *Glaxo IMB, 14 Chemin des Aulx, Plan-les-Ouates, CH-1228 Geneva, Switzerland*

J-Y Bonnefoy *Glaxo IMB, 14 Chemin des Aulx, Plan-les-Ouates, CH-1228 Geneva, Switzerland*

B.L. Brown *Institute of Endocrinology, University of Sheffield Medical School, Beech Hill Road, Sheffield S10 2RX, UK*

E.L. Burlinson *Division of Biochemistry and Molecular Biology, Institute of Biomedical and Life Sciences, University of Glasgow, Glasgow G12 8QQ, UK*

R.E. Callard *Cellular Immunology, Institute of Child Health, 30 Guilford Street, London WC1 4EN, UK*

J. Cambier *National Jewish Center for Immunology and Respiratory Medicine, 1400 Jackson Street, Denver CO 80208, USA*

A. M. Carbone *National Jewish Center for Immunology and Respiratory Medicine, 1400 Jackson Street, Denver CO 80208, USA*

R.H. Carter *Departments of Medicine and Microbiology, University of Alabama at Birmingham, Birmingham AL 35294, USA*

P.A. Clarke *Molecular Immunology, Institute of Child Health, 30 Guilford Street, London WC1 4EN, UK*

A. Correll *IMMUNOLOGIA Hospital Universitario '12 de Octubre', Universidad Complutense, Carretera Andalucia, 28041 Madrid, Spain*

W. Cushley, *Division of Biochemistry and Molecular Biology, Institute of Biomedical and Life Sciences, University of Glasgow, Glasgow G12 8QQ, UK*

P.R.M. Dobson *Institute of Cancer Studies, University of Sheffield Medical School, Beech Hill Road, Sheffield S10 2RX, UK*

C. Edmead *Institute of Rheumatic Diseases, Trim Bridge, Bath BA1 1HD, UK*

C.A. Evans *Department of Biochemistry and Applied Molecular Biology, Leukaemia Research Fund Cellular Development Unit, UMIST, Manchester M60 1QD, UK*

S. Ezhevsky *Immunology Department, American Red Cross, Holland Laboratory, 15601 Crabbs Branch Way, Rockville, MD 20855, USA*

D. Fearon *Wellcome Trust Immunology Unit, University of Cambridge, School of Clinical Medicine, Cambridge, CB2 2SP, UK*

C.R. Gallego *IMMUNOLOGIA, Hospital Universitario '12 de Octubre', Universidad Complutense, Carretera Andalucia, 28041 Madrid, Spain*

J.F. Gauchet *Glaxo IMB, 14 Chemin des Aulx, Plan-les-Ouabes, CH-1228 Geneva, Switzerland*

D.J. Gilfoyle *Department of Biochemistry, School of Biological Sciences, University of Sussex, Brighton BN1 9QG, UK*

P. Graber *Glaxo IMB, 14 Chemin des Aulx, Plan-les-Ouates, CH-1228 Geneva, Switzerland*

D. Gray *Department of Immunology, RPMS, Hammersmith Hospital, Du Cane Road, London W12 ONN, UK*

Margaret M. Harnett *Department of Immunology, University of Glasgow, Glasgow G12 8QQ, UK*

W. Harnett *Department of Immunology, University of Strathclyde, Glasgow G4 0NR, UK*

C.M. Heyworth *CRC Department of Experimental Haematology, Paterson Institute for Cancer Research, Christie Hospital NHS Trust, Manchester M20 9BY, UK*

S. Hinschelwood *Molecular Immunology Unit, Institute of Child Health, 30 Guilford Street, London WC1N 1EH, UK*

M. Houslay *Division of Biochemistry and Molecular Biology, Institute of Biomedical and Life Sciences, University of Glasgow, Glasgow G12 8QQ, UK*

M. Howard *DNAX Research Institute, 901 California Avenue, Palo Alto CA 94304–1104, USA*

D.R. Jones *Instituto de Investigaciones Biomedicas, 28029 Madrid, Spain*

C. H. June *Immune Cell Biology Program, Naval Medical Research Institute, Bethesda MD 20889, USA*

J.E. Kay *Department of Biochemistry, School of Biological Sciences, University of Sussex, Brighton BN1 9QG, UK*

C. Kinnon *Molecular Immunology Unit, Institute of Child Health, 30 Guilford Street, London WC1N 1EH, UK*

S. Lecoanet-Henchoz *Glaxo IMB, 14 Chemin des Aulx, Plan-les-Ouates, CH-1228 Geneva, Switzerland*

R.C. Lovering *Molecular Immunology Unit, Institute of Child Health, 30 Guilford Street, London WC1N 1EH, UK*

B. Maddox *Immunology Department, American Red Cross, Holland Laboratory, 15601 Crabbs Branch Way, Rockville, MD 20855, USA*

D.J. Matthews *Cellular Immunology, Institute of Child Health, 30 Guilford Street, London WC1 4EN, UK*

J.J. Murphy *Infection and Immunity Group, Division of Life Sciences, King's College London, London W8 7AH, UK*

J.J. Newton *Infection and Immunity Group, Division of Life Sciences King's College London, London W8 7AH, UK*

J.D. Norton *CRC Department of Gene Regulation, Paterson Institute for Cancer Research, Christie Hospital NHS Trust, Manchester M20 9BY, UK*

Keith Nye *The Medical College of Saint Bartholomew's Hospital, Department of Immunology, 38 Little Britain, West Smithfield, London EC1A 7BE, UK*

B.W. Ozanne *CRC Beatson Laboratories, Garscube Estate, Switchback Road, Glasgow G61 1BD, UK*

L. Pao *National Jewish Center for Immunology and Respiratory Medicine, 1400 Jackson Street, Denver CO 80208 USA*

Mike Parkhouse *BBSRC Institute for Animal Health, Pirbright Laboratory, Ash Road, Pirbright, Woking GU24 0NF, UK*

R. Parry *Department of Pharmacology, School of Pharmacy, University of Bath, Bath BA2 7AY, UK*

Kevin P. Rigley *The Edward Jenner Institute for Vaccine Research, Compton, Berks, RG20 7NN, UK.*

C.M. Roifman *Division of Immunology and Allergy, The Hospital for Sick Children, 555 University Avenue, Toronto, Ontario, Canada M5G 1X8*

D.M. Sansom *Institute of Rheumatic Diseases, Trim Bridge, Bath BA1 1HD, UK*

D.W. Scott *Immunology Department, American Red Cross, Holland Laboratory, 15601 Crabbs Branch Way, Rockville MD 20855, USA*

N. Sharfe *Division of Immunology and Allergy, The Hospital for Sick Children, 555 University Avenue, Toronto, Ontario, Canada M5G 1X8*

Y. Shi *Immunology Department, American Red Cross, Holland Laboratory, 15601 Crabbs Branch Way, Rockville MD 20855, USA*

J.N. Siegel   *Immune Cell Biology Program, Naval Medical Research Institute, Bethesda MD 20889, USA*

D.A. Taylor-Fishwick   *Immune Cell Biology Program, Naval Medical Research Institute, Bethesda MD 20889, USA*

S. Ward   *Department of Pharmacology, School of Pharmacy, University of Bath, Bath BA2 7AY, UK*

K. Washart   *Immunology Department, American Red Cross, Holland Laboratory, 15601 Crabbs Branch Way, Rockville MD 20855*

T. H. Watts   *Department of Immunology, University of Toronto, Toronto, Ontario, Canada*

A.D. Whetton   *Department of Biochemistry and Applied Molecular Biology, Leukaemia Research Fund Cellular Development Unit, UMIST, Manchester M60 1QD, UK*

Y. Yang   *Laboratory of Immune Cell Biology, National Cancer Institute, National Institutes of Health, Bethesda MD 20892–1152, USA*

X.R. Yao   *Immunology Department, American Red Cross, Holland Laboratory, 15601 Crabbs Branch Way, Rockville MD 20855, USA*

Preface

The last decade has witnessed explosive progress in our understanding of the molecular mechanisms underlying lymphocyte development, proliferation and apoptosis. In particular, major advances in identifying the roles of tyrosine kinases in coupling the antigen and cytokine receptors to the Ras-MAPkinase and Stat/transcription factor pathways have provided a framework for elucidating the precise signal transduction events emanating from the plasma membrane to the nucleus. Moreover, elucidation of the genetic and/or signalling defects underlying major immunodeficiencies has not only underscored the pivotal roles that these signalling pathways play in lymphocyte biology but has also identified key potential targets of therapeutic intervention and immunomodulation.

Lymphocyte Signalling is divided into four major sections. *Part I: Signalling Via the Antigen Receptors on B and T Cells—From the Receptor to the Nucleus* provides a state-of-the-art overview of the structural–functional relationship of these ITAM motif-antigen receptors which now provide a paradigm for recruitment and activation of tyrosine kinase-dependent signalling elements during cell activation. *Part II: Accessory Molecule Signalling in B and T Cells* illustrates the recent, rapid progress made in understanding the role of coreceptor-signals in modulating the immune response and identifying novel mechanisms of receptor crosstalk and manipulation. Moreover, these systems clearly indicate the cellular potential for rewiring and integration of differential signalling cassettes at distinct maturation states of lymphocyte development. This theme is developed further in *Part III: Life or Death: Signalling Throughout B and T Cell Development* which deals with the identification, by biochemical, transgenic or gene knockout studies, of key signals involved in determining lymphocyte development, proliferation or apotosis and, thus, sheds insight into the molecular mechanisms underlying positive and negative selection of B and T cells and the generation of immunological tolerance. Finally, *Part IV: Disease and Disruption of Lymphocyte Signalling—Immunopharmacology and Potential Sites of Therapeutic Intervention* highlights how lesions in lymphocyte signalling subverts the normal immune response and leads to the generation of primary immunodeficiencies (such as HLA, hyper IgM syndrome and SCIDX), immune deficiencies resulting from infection (viral, bacterial, fungal and parasitic) and defects resulting from immune system dysfunction (such as cancer, atopy or autoimmunity). Integration of our knowledge of signalling defects resulting in immune dysfunction with the elucidation of B and T cell signal transduction pathways has dramatically advanced our understanding of the biological roles of particular signalling elements and has also identified key potential targets of rational drug design for therapeutic intervention and immunomodulation of lymphocyte responses. Thus, the future perspectives highlighted by the authors throughout *Lymphocyte Signalling* should appeal to, and stimulate, immune signallers, whether their interests lie in basic B and T cell signalling, immune dysfunction or immunotherapy.

Margaret M. Harnett
Kevin P. Rigley

Part I

Signalling Via the Antigen Receptors on B and T cells—From the Receptor to the Nucleus

Antigen Receptor Structure and Signaling in B Cells

Lily Pao, Amy M. Carbone and John C. Cambier

National Jewish Center for Immunology and Respiratory Medicine, Denver, Colorado, USA

INTRODUCTION

B lymphocytes express cell surface immunoglobulins that function to capture and internalize antigens for processing and presentation to T cells. These receptors also function in transmembrane transduction of signals that lead to a variety of cellular responses. The outcome of this signal transduction is partly dictated by the structure for the antigens. For instance, multimeric antigens complexed with carbohydrates induce resting B cells to proliferate in a T cell independent manner. On the other hand, antigens which are proteinaceous and paucivalent prime B cells to collaborate with T cells but do not induce proliferation. Priming involves increased expression of surface MHC class II, CD80 (B7.2), and presumably other molecules. While MHC class II is recognized by T cell antigen receptors, provided it contains the appropriate peptide, CD80 functions as a T cell activating ligand which interacts with CD28 and CTLA4. Antigen receptor signaling also causes a transcription dependent change in the mode of MHC class II signal transduction, such that CD4 and T cell antigen receptor (TCR) ligation of these molecules activate B cells. In the absence of T cell helper signals, antigen stimulation of B cells can lead to cell anergy and premature cell death by apoptosis. Finally, it is noteworthy that antigen receptor signaling may determine whether antigen presenting B cells deliver an activating or tolerizing signal to T cells. Hence, the quality and quantity of receptor signaling in B cells plays an important role in both the T and B cell immune response.

B cell binding to antigens in different structural contexts is also an important determinant in B cell development and differentiation. Immature B cell binding to cell-bound, i.e. polyvalent, antigens leads to deletion by cell death, or *de facto* deletion by receptor editing, which results in altered specificity. In contrast, immature B cells are rendered anergic by encounters with antigens in soluble form. Germinal center B cells, although they are selected for clonal expansion by antigens, also apoptose following exposure to antigen in the proper context. These examples further illustrate the pleiotropy of B cell responses to antigen receptor ligation.

How crosslinking of the same receptors leads to such diverse biological responses is unknown. Different responses are in part a function of the quality and/or quantity of antigen receptor signals, but they must also be determined by other factors, such as coreceptor ligation or differential cellular expression of signal transduction intermediaries and regulatory molecules. Recent studies have identified several membrane immunoglobulin associated molecules, including Igα and Igβ, that have been shown to be the primary signal transducers for membrane immunoglobulin (mIg). Coreceptors that modulate antigen induced signal-transduction include CD45, CD22, FcγRIIB1, and the CD19/CD21 complex. These cell surface molecules may function by modulating the biochemical events in signal propagation. In this chapter, we summarize the recent progress made in elucidating the structure and function of the B cell–antigen receptor complex, and discuss a possible mechanism for signal propagation. The basis of signal modification by coreceptors will also be discussed.

Lymphocyte Signalling: Mechanisms, Subversion and Manipulation. Edited by M. M. Harnett and K. P. Rigley © 1997 John Wiley & Sons Ltd.

ANTIGEN RECEPTOR STRUCTURE

The B cell–antigen receptor complex, or BCR, is composed of a membrane form of immunoglobulin (mIg) noncovalently associated with disulfide linked Igα (CD79a) and Igβ (CD79b) heterodimers [1–3]. All five classes of immunoglobulin, including IgM, IgD, IgG, IgA and IgE, can occur as receptors. The structure of mIgM, which serves as a prototype, is depicted in Figure 1.1. Membrane immunoglobulin consists of homodimeric heavy chains that are each covalently associated with a light chain, forming a symmetrical four chain structure with two antigen binding sites. The five isotypes mentioned above are defined by their heavy chain constant region sequences. The membrane-bound forms of Igs differ from their secreted counterparts in containing a juxtamembrane spacer, membrane spanning region, and a cytoplasmic domain C-terminal to their last constant region domain. Based on the Kyte and Doolittle algorithm [4], the cytoplasmic tails of the most common mIg isotypes, mIgM and mIgD, are composed of only three amino acids, KVK. This is in contrast to other mIg isotypes, which exhibit more extended cytoplasmic domains of up to 28 residues (Figure 1.2). However, the Klein, Kanehisa and Delisi algorithm [5] predicts that the cytoplasmic tails of mIgM and mIgD are composed of eleven, rather than three, residues, potentially providing sufficient structure to interact with cytoplasmic effectors. The precise length of the cytoplasmic tail and importance of this region of mIg in BCR function is unclear (see below).

The B cell specific BCR components Igα and Igβ are encoded by the mb1 and B29 genes, respectively, and are members of the immunoglobulin supergene family [6, 7]. These molecules form disulfide linked heterodimers and are expressed on the cell surface as type I glycoproteins. Igα has an extracellular domain of 109 amino acids and a cytoplasmic domain of 61 amino acids. In contrast, Igβ has a larger extracellular domain, consisting of 129 amino acids, but a smaller cytoplasmic domain of 48 amino acids. The membrane spanning region of both is predicted to be 22 residues. The extracellular portions of both contain conserved residues found in other members of the Ig supergene family, and an additional cysteine that presumably forms the interchain disulfide bond between the two molecules [8]. Association of these molecules with mIg is required for surface expression of BCR [1, 2], although mIgG$_{2a}$, mIgG$_{2b}$, and mIgD can be expressed on the cell surface in their absence [9, 10]. Interestingly, mIgM-and mIgD-associated Igα molecules are differentially glycosylated [11]. In addition, a C-terminally truncated version of Igβ, termed Igγ, has also been identified in murine intermediate and low density splenic B cells and bone marrow cells [12]. Alternatively spliced RNA products of mb-1 and B29 transcripts, encoding the extracellular domain deleted forms of Igα and Igβ, have also been found in a variety of human B cells and B cell lines [13].

Igα and Igβ heterodimers are intimately involved in BCR signal transduction. A conserved motif, termed *I*mmunoreceptor *T*yrosine-based *A*ctivation *M*otif (ITAM), is present within their cytoplasmic domains [14, 15]. As indicated in Figure 1.3, this motif, D/E-X$_7$-D/E-XX-YXXL-X$_7$-YXXL/I, is also present in the cytoplasmic domains of other members of the *M*ultichain *I*mmune *R*ecognition *R*eceptor family (MIRRs) [16], which includes BCR, TCR, FcεRI, FcγRI, FcγRIIA and C, FcγRIIIA, and FcαR. The importance of ITAMs in MIRR signaling was underscored by the demonstration that the ITAM contains sufficient structural information for activation of signaling pathways in lymphocytes [17–21], and that mutations in either of the conserved tyrosines in the motif disables receptor signaling [17, 18, 22–25].

Additional molecules reportedly associate with mIgM and mIgD. Most notable among these is a group of ubiquitously expressed cytoplasmic proteins, called BCR-associated proteins or BAPs [26, 27]. BAPs exhibit isotype specific association, such that BAP32, 37 and 41 (numbers refer to their apparent molecular weights $\times 10^{-3}$) associate noncovalently with mIgM, and BAP29 and 31 preferentially associate with mIgD. BAP32 has been identified as a homolog of rat prohibitin; all other BAPs are uncloned. Although BAPs are associated with mIgM and mIgD, they do not appear in the plasma membrane, suggesting that they do not function in transmembrane signaling. Analysis of the CD5$^+$ and CD5$^-$ B cells also showed two proteins of 54 kDa and 67 kDa that could specifically be coprecipitated with Igα from the CD5$^+$ popula-

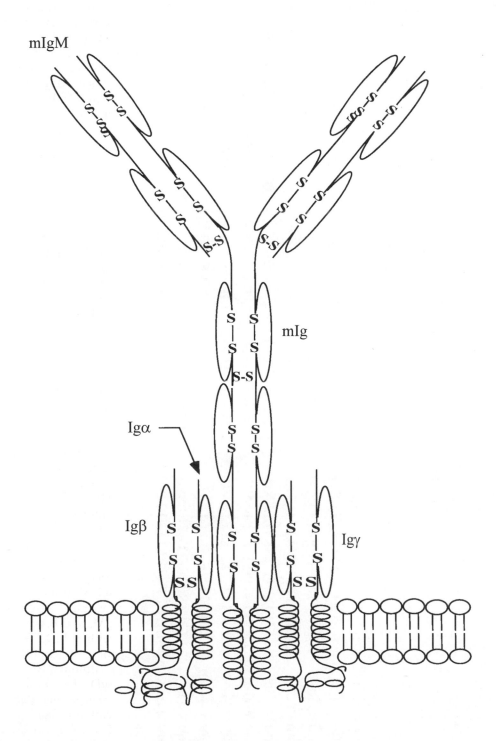

Figure 1.1 Structure of mIgM. As shown, mIgM is composed of a mIg and disulfide linked heterodimeric Igα and Igβ. Alternatively, mIg could also associate with disulfide linked Igα and Igγ, instead of Igβ

	Transmembrane Domain		Cytoplasmic Tail
mouse μm	NLWTTASTFIVLFLLSLF	(YSTTVTLF)	KVK
human μm	NLWATASTFIVLFLLSLF	(YSTTVTLF)	KVK
mouse δm	GLWPTMCTFVALFLLTLL	(YSGFVTFI)	KVK
human δm	SLWTTLSTFVALFILTLL	(YSGIVTFI)	KVK
mouse γ3m	GLWTTITIFISLFLLSVC	(YSASVTLF)	KVKWIFSSVVQVKQTAIPDYRNMIGQGA
mouse γ1m	GLWTTITIFISLFLLSVC	(YSAAVTLF)	KVKWIFSSVVELKQTLVPEYKNMIGQAP
mouse γ2bm	GLWTTITIFISLFLLSVC	(YSASVTLF)	KVKWIFSSVVELKQKISPDYGNMIGQGA
mouse γ2am	GLWTTITIFISLFLLSVC	(YSASVTLF)	KVKWIFSSVVELKQTISPDYRNMIGQGA
mouse εm	ELWTSICVFITLFLLSVS	(YGATVTVL)	KVKWVLSTPMQDTPQTFQDYANILQTRA
mouse αm	SLWPTTVTFLTLFLLSLF	(YSTALTVT)	TVRGPFGSKEVPQY

Figure 1.2 Sequence comparison between mouse and human transmembrane and cytoplasmic domains of different isotypes of immunoglobulins. The transmembrane region shown is based on the prediction by Kyte and Doolittle algorithm [4]. However, based on the Klein, Kanehisa and Delisi algorithm [5], the transmembrane region also includes sequence within the parentheses. Sequence information is derived from Reth [8] and Rogers and Wall [225]

Figure 1.3 Distribution and sequence of ITAMs. A comparison between mouse and human ITAMs found in the cytoplasmic domains of various members of MIRRs.

The figure is a multiple sequence alignment (rotated 90° on the page) of ITAM motifs. Reading each entry N→C, the rows are:

Consensus:	D/E · · · · – – – – · · · – – – – · · Y · E · L/I · · · Y · E · L/H
m MB-1	D M A T · P D D D – – – – D Y E L · N C S M E Y D L H
h MB-1	D A G K · A D D D – – – – D Y E N · C S A M E Y D H
m B29	D G K R · G A K A – – – – Q Y M G · T T E T S Y S L
h B29	D S R R · K A Q D – – – – D M G H · T T E E T Y D H
m TCR-ζ/ηa	E T A N · L Q D P N – – – N Q L Y N E L N L R K A V E Y D V L
m TCR-ζ/ηb	K Q R R · R N P Q – – – M Q E L Y Q P L S D A E S S Y E H
m TCR-ζc	E R R G · R K I – – – D G Q L Y D T D K D T T D S S Y A L
m TCR-ηc	E R R G · K K – – – F D R Y D R R K K P R D L G R L
m CD3-γ	D K Q T · L N – D D P F Q F D D G Q L Y D Q D L S H Y E S Q L
m CD3-δ	E V Q A · A N – D D Q E P Q L Y D T E R D V P S H P Y S L
m CD3-ε	N K E R · R P P – D D T P A N P D Y E P R N K L G N N T L
c T11.15	D R Q N · – – – – L D Q L Y L V A N D G Q N G P D L
m FcεRI-γ	A I A S · S D E N I M – A D A V Y A R T H N V P R A Y E T L
m FcεRI-β	E L E S · K K V P T T H – D R L Y R R D H V S T E S L Y I L
h FcγRIIa	E T N D · Y E T V P P T P A D G G Y T G N P K G K N P P M Y L
BLV gp30a	P E I S · L T P K P K T – D H P S A D D K Q H E S I P Y L
BLV gp30b	D Y Q A · L P S P A S P – H I S P V S E H L S H Y N I
EBV LMP2	D P Y W · G N G D R – D H S D W T G D Q D G Q S G Y T G H
EBV EBNA2	– – – – · – – – – P T M G L H I G – F H Y V L L

tion [28]. p67 was identified as CD5 itself, and the identity of p54 remains to be determined. While the role of CD5 in signaling is not clear, Lankester *et al.* [28] demonstrated that CD5 coprecipiated with BCR could be phosphorylated *in vitro*; in addition, BCR ligation also led to tyrosine phosphorylation of CD45 *in vivo*. Several other unknown proteins have also been reported to associate with mIgs. For instance, a 52 kDa surface molecule phosphorylated following phorbol myristate acetate stimulation was shown to coprecipitate with Igα [29]. Yellen-Shaw and Monroe [30] described a 56 kDa cytoplasmic protein which reportedly coprecipitated with mIgM in mature B cells. However, recent experiments suggest this may actually be an IgG heavy chain (J. Monroe, unpublished observation). Finally, Igarashi *et al.* [31] described a monoclonal antibody (SIG1) generated by immunizing rats with receptor complexes isolated from a digitonin lysate of WEHI-231 B lymphoma cells. This antibody immunoprecipitates a cell surface molecule of 160 kDa (p. 160), which showed increased associated kinase activity upon BCR ligation in mature B cells. In immature B cells, however, BCR stimulation caused a transient loss of p160 associated kinase activity. This p160 molecule may be involved in determining the differential BCR signaling in mature and immature B cells, but functions of this and other mIg associated proteins require further characterization.

BCR SIGNAL TRANSDUCTION

Crosslinking of the BCR initiates transduction of signals that have as their earliest known consequence the activation of protein tyrosine phosphorylation [32, 33]. Proteins phosphorylated include mIg associated Igα and Igβ heterodimers [34], Src-family protein tyrosine kinases (PTKs) Lyn, Fyn, Lck, Blk and Fgr [35–39], and other cytoplasmic tyrosine kinases, including Syk and Btk [40–43]. While Src-family kinases are associated with "resting" receptors at apparently low stoichiometry, receptor activation results in recruitment of additional kinase molecules to the receptor complex. BCR mediated activation of tyrosine kinases appears to be temporally regulated, with the Src-family kinase activity reaching maximum within seconds, followed by Btk activity

which peaks at about five minutes. Syk activity was found to increase later than others, reaching a maximum at 10 to 60 minutes [44]. Protein tyrosine phosphatase IC, PTP1C(or SHP1), has also been shown to be part of the resting receptor complex, and may function in keeping Igα and Igβ in the dephosphorylated state in resting receptors [45]. Following initial activation of protein tyrosine kinases, the signal transduction cascade seems to diverge into three parallel but potentially cross-regulatory pathways. These include the phospholipase C (PLC) pathway, the p21^{ras} pathway, and the phosphatidylinositol-3 kinase (PI3-k) pathway. As depicted in Figure 1.4, these pathways are characterized by activation of several distinct effector molecules. Briefly, the PLC pathway is activated by PLCγ tyrosine phosphorylation and activation, leading to breakdown of phosphatidylinositol biphosphate (PIP$_2$) and generation of inositol 1,4,5-triphosphate (IP$_3$) and diacylglycerol (DAG). The p21^{ras} pathway involves BCR induced tyrosine phosphorylation of several putative regulators of p21^{ras} activity, including GTPase activating protein (GAP), p120 rasGAP, guanine nucleotide exchange factor (GEF), Vav, and a linker molecule, Shc. Lastly, the PI3-k pathway is represented by PI3-k activation. These pathways and their respective effector molecules are further discussed below.

As mentioned, one of the signaling pathways activated upon BCR ligation involves PLC activation. Activated PLC hydrolyzes PIP$_2$ to generate two products, IP$_3$ and DAG, which function as second messengers in mediating calcium mobilization and protein kinase C activation (see review by Berridge [46]). Phosphoinositide (PI) turnover was among the earliest BCR signaling events defined. In 1975, Maino *et al.* [47] demonstrated an anti-Ig induced increase in incorporation of ^{32}P into the PI pool of B cells. Consistent with this, Coggeshall and Cambier [48] showed a rapid increase in phosphatidic acid generation preceding PI synthesis by B cells responding to anti-Ig treatment. In addition, direct assessment of BCR mediated phosphoinositide degradation, yielding IP$_3$ and DAG, was carried out by Bijsterbosch *et al.* [49], providing more evidence for the involvement of a PLC in BCR signal transduction. Ransom *et al.* [50] further linked BCR induced IP$_3$ to calcium mobilization following receptor crosslinking. These

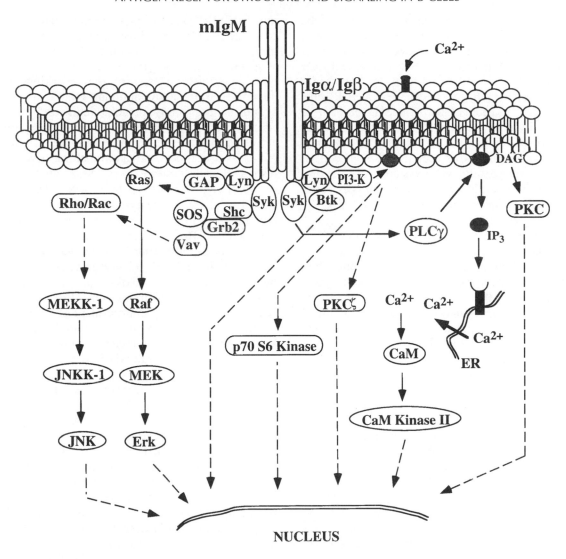

Figure 1.4 BCR signal transduction pathways. A schematic representation of the parallel cascades initiated from the BCR to the nucleus

early studies of BCR mediated signaling events were appreciated when more recent reports of BCR mediated tyrosine phosphorylation and activation of PLCγ1 and PLCγ2 [51, 52] provided a mechanism for BCR initiated PIP₂ hydrolysis. In B lymphocytes, PLCγ2 appears to be the predominant PLC isoform [52], and its enzymatic activity is probably increased by tyrosine phosphorylation, as shown for PLCγ1 [53]. The tyrosine kinase responsible for phosphorylation of PLCγ

is most likely to be Syk. A recent study using a Syk knockout in chicken B cells suggests Syk kinase activity is required for phosphorylation of PLCγ2 and generation of IP₃ [54]. Aggregation of chimeric CD16-Syk molecules also leads to tyrosine phosphorylation of PLCγ and calcium mobilization, apparently independent of other PTKs, indicating that Syk activation is sufficient for this response [55]. Direct involvement of Syk in the activation of PLCγ1 is further suggested by their *in vivo* interaction [56], which is likely to be mediated by binding of the Src homology 2 (SH2) domain of PLCγ1 to Syk [57]. However, Lyn may also participate in Syk mediated PLCγ tyrosine phosphoryation. Syk and Lyn reportedly associate *in vivo* [56, 58], and in activated B cells, Syk is likely to interact with the SH2 domain of Lyn [58]. Lyn may also regulate Syk activation, as demonstrated by Syk phosphorylation when coexpressed with Lyn in Cos cells [59]. This is consistent with the results of Iwashima *et al.* that functional Src-family kinase expression is essential for TCR mediated activation of Syk/ZAP-70 family kinases [60].

The biologic significance of PI hydrolysis in B cells remains unclear, but it is implicated in antigen mediated induction of several genes, including *c-myc*, *c-fos*, and *egr-1* [61–64]. While both PKC activation and calcium mobilization seem to be required for *c-fos* induction [64, 65], induction of *egr-1* and *c-myc* appears to require only PKC activation [64, 66]. Furthermore, BCR ligation also leads to phosphorylation of Ets-1, which is dependent on calcium mobilization, but not PKC activation [67].

Another signaling pathway initiated by BCR ligation involves activation of $p21^{ras}$ [68, 69], linking BCR to the MAP kinase pathway. Several putative regulators of Ras activity are implicated in this process by virtue of their tyrosine phosphorylation upon BCR ligation. One protein known to interact with $p21^{ras}$ is the Ras GTPase-activating protein, rasGAP [70]. GAP functions as a negative regulator of $p21^{ras}$ by stimulating the intrinsic GTPase activity of $p21^{ras}$, promoting the conversion of $p21^{ras}$ to an inactive GDP-bound state. BCR ligation leads to increased tyrosine phosphoryation of GAP, and association of GAP with tyrosine phosphorylated p190 and p62 [71]. p190 functions as a GAP for the small GTP binding protein, Rho [72] and its association with ras-GAP may provide a "bridge", linking BCR signal transduction to regulation of the Rho/Rac family of small GTP binding proteins. p62, which has homology to a glycine-rich putative hnRNP, GRP33, exhibits detectable RNA binding activity, but its biologic function remains unknown [73]. Association of p190 molecules with GAP in B cells may have a negligible role in regulating GAP activity, as Lazarus *et al.* [69] showed antigen induced $p21^{ras}$ activation in mature B cells with concomitant tyrosine phosphorylation and inhibition of rasGAP activity that is independent of p190 association. Several reports suggest that the tyrosine kinase(s) responsible for linking BCR ligation to $p21^{ras}$ activation via rasGAP are members of the Src-family. rasGAP was found in association with Fyn, Lyn and Yes in thrombin activated platelets [74]. *In vitro* binding experiments also mapped association of rasGAP from resting B cell lysates to the *N*-terminal unique region of Lyn [75]. Studies using recombinant proteins further demonstrated the ability of Lck to tyrosine phosphory-late rasGAP *in vitro* [76].

In addition to rasGAP, BCR ligation leads to tyrosine phosphorylation and activation of Vav [77, 78]. Vav reportedly has Ras guanine nucleotide exchange activity (GEF) [78, 79], and it may contribute to most of the Ras GTP/GDP exchange activity in B cells [78]. Pretreatment of B cells with herbimycin A blocked BCR mediated increase in Vav enzymatic activity, suggesting the importance of Vav tyrosine phosphorylation in regulation of its activity [78]. However, the role of Vav as a GEF for Ras has been controversial. Its sequence homology to human Dbl, Bcr and yeast Cdc24 suggests that it may function as a GEF for the Rho/Rac family instead of Ras [80, 81]. In addition, transforming activities and phenotypes of Vav differ from that of Ras [82]. Nevertheless, BCR mediated proliferation appears to be dependent on Vav expression, as revealed by two independently derived Vav knockout mice [83, 84]. Vav may also play a significant role in B cell development, especially in the generation of the $CD5^{+}$ population, since these cells are absent in the peritoneal cavity of Vav deficient mice.

Another Ras GEF, SOS, has also been implicated in BCR mediated Ras activation. BCR ligation induces Shc tyrosine phosphorylation and formation of complexes that contain Shc, Grb2, SOS-1, and several other tyrosine phosphorylated proteins [85–87]. Activated BCR also recruits Shc to the receptor complex by a SH2–phosphotyrosine interaction [88]. This recruitment is likely to promote Shc phosphorylation by virtue of its apposition to Lyn and Syk in the receptor complex. Both Lyn and Syk may be required for BCR mediated phosphorylation of Shc [89]. Recent demonstration of the role of Sos in epidermal growth-factor-triggered Ras activation suggests a mechanism by which recruitment of Sos to the plasma membrane may be sufficient to cause Ras activation [90]. Hence, receptor association and phosphorylation of Shc may mediate Ras activation simply by recruiting Grb2/Sos complexes to the receptor complex, where Ras is localized [91].

The relative contribution of rasGAP, Vav and Sos in BCR mediated $p21^{ras}$ activation is unknown. Events downstream of $p21^{ras}$ activation in B cells are even less clear. BCR ligation leads to activation of Raf [92, 93], MEK [93] and MAP kinase [93, 94]. Using dominant–negative mutants of $p21^{ras}$ and Raf-1, McMahon and Monroe [95] demonstrated that BCR mediated induction of egr-1 is dependent on activation of the $p21^{ras}$ pathway. Finally, antigen induced activation of the $p21^{ras}$ pathway in B cells may also lead to activation of other intermediary or related enzymes in the MAPK and JNK cascades as indicated in Figure 1.4.

A third signaling cascade initiated upon BCR crosslinking is the PI3-k pathway. This kinase phosphorylates the inositol ring of phosphoinositides at the D-3 position, and was first implicated in B cell signal transduction by virtue of its association with Lyn [96] and increased activity immunoprecipitable with anti-phosphotyrosine antibodies following BCR stimulation [96, 97]. Mutations in a variety of other receptors and oncoproteins that eliminate activation of PI3-k have been shown to compromise certain biological responses, including cell growth [98], chemotaxis [99, 100], receptor down-regulation [101], and basophil degranulation [102]. The downstream targets of PI3-k are poorly defined, however, products of PI3-k, such as $PI-3,4-P_2$ and $PI-3,4,5-P_3$, reportedly activate $PKC\zeta$ and calcium independent PKCs δ, ϵ and η in vitro [103, 104]. Among the isoforms mentioned, $PKC\delta$ and ζ are expressed in B cells [105]. Activated $PKC\zeta$ leads to phosphorylation and inactivation of $I\kappa B$, which plays a critical role in the regulation of transcription factor $NF\kappa B$ [106, 107]. In addition, PI3-k has also been implicated in activation of p70/p85 S6 kinase [108, 109].

BCR mediated activation of PI3-k has been more closely examined. Pleiman et al. [110] demonstrated that the Lyn and PI3-k association described by Yamanashi et al. (96) occurs via binding of the Lyn SH3 domain to a proline rich sequence in the p85 subunit of PI3-k. This binding also results in activation of PI3-k in vitro, and is required for receptor mediated activation of PI3-k [110]. PI3-k may additionally be regulated by Ras, via a pathway distinct from Raf, as shown in another system [111]. Moreover, PI3-k is not the only lipid kinase involved in receptor signaling in B cells. Report of an increased level of phosphatidylinositol tetrakisphosphate following receptor stimulation by Gold and Aebersold [112] further implicates activation of a PI4-k in BCR signal transduction.

While most of the molecules involved in immediate aspects of BCR signal transducation seem to impact on one or more of the signaling pathways described above, the role of several other tyrosine kinase substrates, including $p120^{cbl}$ and HS1, are less defined. The protein product of c-cbl proto-oncogene was implicated in antigen receptor signaling by being one of the major proteins tyrosine phosphorylated upon TCR and BCR stimulation [113, 114]. When tyrosine phosphorylated, Cbl can associate with GST fusion proteins containing SH2 domains of Fyn, Lck, Blk, GAP, and $PLC\gamma1$ [113]. Donovan et al. [113] also identified Cbl as the 120 kDa protein constitutively associated with the N-terminal SH3 domain of Grb2 [115, 116] and SH3 domains of Fyn and Lck [117]. These SH3 domains are likely to interact with the C-terminal proline-rich sequence within Cbl. Cbl has several PXXP motifs, defined by Yu et al. [118] to be potential SH3 binding sites. Association of Cbl with Grb2 and Grb2-associated Sos, and p36 and p75 phosphoproteins, suggests that Cbl may function in the $p21^{ras}$ pathway. In addition to Grb2, Cbl also interacts with another linker molecule, Nck, in vitro and in vivo [119], but the significance of this association is not clear.

Compared to Cbl, HS1 has been better characterized. HS1 is a 75 kDa protein that becomes tyrosine phosphorylated and associates with Lyn following BCR stimulation in human B cells [120]. Tyrosine phosphorylated HS1 was shown to bind to SH2 domains of Lyn, Blk and Fyn [120, 121]; however, a recently cloned HS1 homolog in murine T cells was shown to interact with Lck via the kinase's SH3 domain [122]. HS1 seems to play a role in apoptosis. Fukuda et al. [123] showed that two variant cell lines of WEHI-231 that were resistant to anti-IgM induced apoptosis exhibited reduced levels of HS1 protein expression. The apoptotic response could be restored in one cell line by HS1 transfection, suggesting HS1 may be one of the components that control apoptosis. The function of HS1 was also assessed by generation of HS1 knockout mice. In these mice, development of the lymphoid cells appears normal; however, proliferative responses driven by crosslinking of BCR or TCR are impaired [124]. When HS1 deficient mice were crossed to mice bearing transgenes encoding male H-Y antigen specific TCR, the resulting mice also showed defective negative selection during thymocyte development [124]. These data suggest that HS1 may be important for deletion of developing self-reactive T and B cells as well as positive selection and activation.

In summary, BCR signal transduction involves an initial activation of protein tyrosine kinases, followed by a divergence of parallel signal propagation pathways, which subsequently converge in the nucleus. Many of the intermediary events in BCR signal transduction remain poorly defined. Understanding of the basis of differential cell integration of signals which arise from binding to structurally distinct antigens must await a more thorough understanding of basic BCR signaling processes.

MEMBRANE IMMUNOGLOBULIN STRUCTURE–FUNCTION RELATIONSHIPS

Studies of B cell antigen receptor structure–function relationships have focused mainly on mIgM containing receptor complexes. Mutational analyses of the mIgM heavy chain (μm) have defined specific regions that are required for receptor transport, signal transduction and antigen presentation. In the absence of Igα/Igβ heterodimers, mIg is retained in the endoplasmic reticulum (ER). Initial truncation and substitution experiments suggested that mIgM contains ER retention sequences that are located in the heavy chain C_H4 and the transmembrane spanning regions. These two domains are also sufficient to confer cell surface expression of receptors by virtue of mIg association with Igα/Igβ [2, 125]. The transmembrane

spanning region defined based on the prediction by the Kyte and Doolittle algorithm [4] contains two distinct polar regions, $T_{572}TAST$ and $Y_{587}STTVT$ in mouse, or $A_{572}TAST$ and $Y_{587}SATTVT$ in human, which are potential sites of protein–protein interaction (Figure 1.5, underlined residues). The sequences that constitute these regions are also well conserved between mouse and human, differing only at position 572, which is a threonine in mouse but an alanine in human (Figure 1.5). Mutations in either of these two polar patches, such as changing $T_{572}TAST_{576}$ to VVAAV, or $Y_{587}S_{588}$ to VV, are sufficient to disrupt ER retention of mIg and allow surface expression in the absence of Igα/Igβ association [125–127].

The ability of YS to VV mutant to overcome retention in the ER has been examined more closely in relation to calnexin, an ER chaperone [128]. While both wildtype and mutant μm can bind calnexin, mutant μm does not require Igα expression to facilitate the release of the heavy chain from calnexin. Thus, calnexin provides a possible architectural editing mechanism for assembly of the receptor complex. However, it is not clear how the YS to VV mutant escapes this regulation.

Mutations in regions outside of μm transmembrane region reportedly also disrupt ER retention. Cherayil *et al.* [129] have shown that deletion of just the C_H1 domain allows surface expression of IgM in nonlymphoid cells. This suggests the presence of ER retention signal(s) in C_H1 domain, deletion of which eliminates the need for Igα and Igβ. This C_H1 domain was previously shown to bind to another ER chaperone, BiP, implicated in ER retention of mIg [130]. Finally, mutant mIgM, which lacks the transmembrane spanning region and associates with the plasma membrane via a phosphatidylinositol-glycan [131], does not require Igα and Igβ for surface expression [132]. Thus, remarkable as it may seen, it appears that three or perhaps four retention signals are present in μm, and disruption of any one of these allows surface expression of mIg.

Based on the discussion above, participation of Igα and Igβ in receptor transport is clear. However, the mechanism by which Igα and Igβ may participate in assembly and release the receptors retained in the ER by chaperones, such as calnexin and BiP, requires further examination.

Wildtype

	μm transmembrane region(Tm)	cytoplasmic tail
Human(h):	NLW<u>ATAST</u>FIVLFLLSLF<u>YSTTVT</u>LF	KVK
Murine(m):	NLW<u>TTAST</u>FIVLFLLSLF<u>YSTTVT</u>LF	KVK
	569	597

Mutants	Sequence	tail	Calcium Mobilization (Stimulus)	Antigen Presentation	Igα/β Association
			(PC/Ab1-2)		
mTm:I-Aα/26	**TVVCALGLSVGLVGIVVGTIFIIQGL**	KVK	-	-	?
mTm:I-Aα/17	**TVVCALGLSVGLVGIVV**FYSTTVTLF	KVK	-	-	?
mTm:I-Aα/8	**TVVCALGL**FIVLFLLSLFYSTTVTLF	KVK	+	-	+
			(PC/Ab)		
mTm:TT/AA	NLW**AA**ASTFIVLFLLSLFYSTTVTLF	KVK	+	?	+
mTm:TAAST /AAAAA	NLW**AAAAA**FIVLFLLSLFYSTTVTLF	KVK	+	?	+
mTm:MutA	NLW**VV**A**VV**FIVLFLLSLF**FAVV**V**V**LF	KVK	?	+	+
mTm:MutB	**NMATVAVLVVLGAAIVTGAVVAFVM**	KVK	?	-	-
			(PC)		
hTm:YS/VV	NLWATASTFIVLFLLSLF**VV**TTVTLF	KVK	-	-	-
hTm:ST/VV	NLWATA**VV**FIVLFLLSLFYSTTVTLF	KVK	+	+	+
hTm:YS/FA	NLWATASTFIVLFLLSLF**FA**TTVTLF	KVK	+	-	?
hTm:S/A	NLWATASTFIVLFLLSLFY**A**TTVTLF	KVK	+	+	?
hTm:Y/F	NLWATASTFIVLFLLSLF**F**STTVTLF	KVK	+	-	+
			(GAM)		
mTm:YS/FA	NLWTTASTFIVLFLLSLF**FA**TTVTLF	KVK	+	?	+
mTm:YS/VV	NLWTTASTFIVLFLLSLF**VV**TTVTLF	KVK	+/-	?	+/-
			(GAM) (b-7-6) (NP)		
mTm:S/A	NLWTTASTFIVLFLL**A**LFYSTTVTLF	KVK	+ + -	?	+
mTm:Y/F	NLWTTASTFIVLFLLSLF**F**STTVTLF	KVK	+ - -	?	+
mTm:T/V	NLWTTASTFIVLFLLSLFYSTTV**V**LF	KVK	+ - -	?	+
mTm:K/I	NLWTTASTFIVLFLLSLFYSTTVTLF	KV**I**	+ + -	?	+

Figure 1.5 A summary of μm transmembrane mutations and the ability of the mutated IgM to mediate calcium mobilization and antigen presentation. Association of mutated mIg with Igα/β is also indicated. The polar patches are underlined in the wildtype mouse and human sequences. Mutations within the mouse (m) or human (h) transmembrane domain (Tm) are in bold and underlined. Receptor crosslinking reagents, including antigens PC (phosphorylcholine) and NP (nitrophenyl), and antibodies, Ab1–2, an anti-idiotypic monoclonal antibody (mAb), b. 7–6, a rat anti-mouse IgM mAb, and GAM, goat anti-mouse IgM, are also indicated. Information compiled and obtained from Grupp *et al.* [126], Michnoff *et al.* [137], Neuberger *et al.* [226], Parikh *et al.* [134, 135], Pleiman *et al.* [138], Shaw *et al.* [141], and Stevens *et al.* [127]

Mutational analyses of μm transmembrane region have also revealed residues that are critical for signal transduction and antigen presentation. The μm transmembrane domain was first implicated in receptor signaling by the finding that substitution of this region with that of I-Aα prevented receptor mediated signaling in CH33 cells [133]. Similarly, replacement of the *N*-terminal 17 residues of the μm transmembrane sequence with corresponding I-Aα residues abrogated receptor mediated calcium mobilization [134, 135]. However, if *N*-terminal 8 residues, instead of 17, of

μm transmembrane region were replaced with I-Aα sequence, the receptors retained their ability to mobilize calcium [135]. Point mutations made by Patel and Neuberger [136] further identified specific residues that are critical for signal transduction. A summary of several reported μm mutants and their effect on cellular responses in B cell transfectants is shown in Figure 1.5. An important aspect of BCR signaling revealed by analyses of the μm transmembrane mutants is that most of the signaling property of BCR can be attributed to the mIg associated Igα and Igβ. For example, human mIgM bearing $Y_{587}S_{588}$ to VV mutations described by Shaw et al., which could not mobilize calcium and lacked associated kinase activity in vitro, were expressed on the cell surface in the absence of Igα and Igβ in murine B cells [126]. Michnoff et al. [137] similarly reported correlation between the association of Igα/β with mIgM bearing I-Aα transmembrane sequence and the ability of such receptors to initiate protein tyrosine phosphorylation and calcium mobilization. In one case, when association of Igα/Igβ with $Y_{587}S$ to VV mutated mouse mIg could be weakly detected, the receptors also exhibited partial ability to signal upon stimulation with Ig crosslinking antibodies [127]. Restoration of signaling ability of mutated receptors achieved by attachment of the cytoplasmic domains of Igα or Igβ to mutated μm [24, 25] further supports the idea that Igα and Igβ are the signaling components of the BCR. However, other evidence suggests that mIgs may possess signaling function in addition to mere association with Igα and Igβ. Pleiman et al. [138] showed defective antigen induced mutant mIgM signaling even when the μm was apparently normally associated with Igα/Igβ. As summarized in Figure 1.5, Y_{587} to F and T_{592} to V mutations abolished both antigen, nitrophenyl (NP), and monoclonal anti-μ antibody, b-7-6, induced protein tyrosine phosphorylation and calcium mobilization. In contrast, S_{584} to A and K_{597} to I mutations only affected antigen induced signaling [138]. In all cases, the effect of the point mutations could be overcome by using a polyclonal goat anti-mouse IgM antibodies (GAM, Figure 1.5) as a stimulus. These results suggest that the association with Igα and Igβ is not always sufficient to sustain the ability of mIgM to signal. Perhaps, while mutations do not abolish Igα and Igβ association, they weaken the

interaction with μm such that the mutant receptors can only transduce signals under the condition of more extensive crosslinking. This possibility is unlikely, since in this study the association of Igα/β with Y_{587} to F mutated mIg survived the rigors of the coprecipitation procedure. Alternatively, mutations may change μm-Igα/β orientation or disrupt μm association with other molecules, which may assist in Igα and Igβ mediated signaling. Based on Klein, Kanehesia and Delisi algorithm [5], $Y_{587}STTVTLF_{594}$ is predicted to be cytoplasmic rather than transmembrane (see Figure 1.2). Hence, these residues may be associated with effector(s) involved for triggering the signaling cascades. This is supported by the findings that mIgs exhibit serine/threonine kinase activity [24] and may associate with cytoplasmic PTK, Syk, independently of Igα/Igβ [139].

While the cytoplasmic tails of IgM and IgD may be too short for interaction with other molecules, the cytoplasmic domains of other isotypes are considerably larger (see Figure 1.2) and may potentially be involved in receptor signaling. Kim et al. [140] have constructed a chimeric molecule which consists of the extracellular and transmembrane domains of CD8 and the cytoplasmic region of IgG$_{2a}$. This chimera failed to elicit calcium mobilization and protein tyrosine phosphorylation upon crosslinking, suggesting that the cytoplasmic domain of IgG$_{2a}$ lacks structural features capable of initiating signal transduction. However, the possibility that the cytoplasmic domains of this and other isotypes may carry out isotype specific functions remains open.

As shown in Figure 1.5, several μm mutations also revealed residues that are necessary for antigen presentation. Replacement of the N-terminal eight residues of the μm transmembrane sequence with that of the I-Aα sequence can effectively abrogate antigen presentation function [135]. Interestingly, if mutations in this N-terminal region of μm are only limited to two residues, with $S_{576}T_{577}$ altered to VV, then the mutated mIgM exhibits wildtype antigen presentation function. In addition, antigen presentation function could also be disrupted by mutating $Y_{587}S_{588}$ to VV or FA, or Y_{587} to F [141]. Antigen presentation function of the receptors appears to be independent of the Igα/β association and their signaling ability. With the exception of the receptors bearing $Y_{587}S_{588}$ to VV mutations, all other above-mentioned mutants were capable of calcium mobilization upon antigen stimulation, and were shown to associated with Igα and Igβ at the cell surface [126, 137]. Consistent with this, chimeric mIgs containing the YS to VV mutations fused to either Igβ or Igα still could not mediate antigen presentation, despite a restored signaling capability [142]. In contrast, Patel and Neuberger [136] demonstrated that the cytoplasmic domain of Igβ was sufficient to restore antigen presentation activity of a mutated mIgM (see Figure 1.5, mTm : MutB), presumably by allowing more efficient antigen internalization. Yet, this is apparently independent of Igβ signaling function, since the ITAM tyrosines of Igβ are dispensable for restoration of the antigen presentation function of the mutant receptor.

Different mechanisms may be responsible for the observed defect of mIgM mutations on antigen presentation. One may speculate that the hydroxyl group of Y_{587} is involved in binding of the molecule(s) critical for mediating antigen presentation function in B cells. Mutations may also prevent endocytosis of the receptors or cause improper trafficking of the internalized antigen–mIg complexes. The latter two possibilities were examined by Mitchell et al. [142]. These authors showed that while antigen endocytosis and peptide degradation were normal in cells expressing the Y_{587} to F or Y_{587}S to VV transmembrane mutants, the defect in antigen presentation is likely to be the result of improper trafficking of the antigens, such that the antigenic peptides could not associate with MHC class II molecules. The relationship between μm sequence and antigen presentation function of B cells is obviously more complex than a simple association with Igα and Igβ. Proper orientation of the receptor components and perhaps other accessory molecules may also be involved.

Igα AND Igβ STRUCTURE–FUNCTION RELATIONSHIPS

While extensive mutational analyses have identified the specific residues in the transmembrane domain of mIgM that are critical for association with Igα/Igβ heterodimers, identification of the residues in Igα and/or Igβ that are required for BCR assembly has not been undertaken. Transfection experiments revealed that mouse Igα is more capable of transporting murine mIgM on to the cell surface than human Igα in plasmacytomas [1]. Since murine and human Igα have identical transmembrane residues, this result suggests the more divergent extracellular domain of Igα contributes to association with mIg. Reth [8] has proposed a model of μm–Igα/Igβ complex formation, based on the arrangement of the transmembrane domains of these components in α-helices. The contact between the μm dimer and Igα/β heterodimer is proposed to be mediated by Igα, which, unlike Igβ, has a transmembrane region containing polar amino acids on both sides of the α-helix. In addition, since mIg exhibits bilateral symmetry, each mIg is presumably associated with two Igα/β heterodimers, resulting in a complex organized as Igβ/Igα–mIg–Igα/Igβ. In this arrangement, the polar amino acids of each component can also be orientated to face each other, exposing the hydrophobic residues to the lipid bilayer. It should be noted, however, that points of mIg and Igα/Igβ association are not limited to the transmembrane domain; removal of the C_H4 region of μm also prevents complex formation with Igα and Igβ [2, 125]. Perhaps this domain interacts with the extracellular Ig-like domain of Igα as suggested earlier.

As discussed in the previous section, Igα and Igβ appear to play a central role in BCR signal transduction. To study the localization of signaling functions within Igα and Igβ, several groups have constructed chimeric molecules by fusing cytoplasmic domains of Igα and Igβ with inert transmembrane and extracellular domains of other receptors [21, 25, 140, 143, 144]. These chimeras were shown to be competent to transduce signals, mediating activation of both receptor proximal and distal events, including protein tyrosine phosphorylation, calcium mobilization and IL-2 secretion. In some cases, Igα chimeras were shown to generate more robust protein tyrosine phosphorylation and calcium mobilization signals than Igβ [25, 140, 143]. Choquet et al. [143] further reported that Igα has a unique ability to trigger IL-2 secretion and mediate calcium mobilization responses that are qualitatively distinct from those triggered through Igβ.

Studies using chimeric receptors have demonstrated that Igα and Igβ signal transduction is mediated in large part by the ITAMs, which are present in a single copy in each of their cytoplasmic domains. Amino acid sequences of ITAMs of Igα, Igβ, and other members of MIRR family are summarized in Figure 1.3. This motif was first recognized by Reth [14], and its significance in antigen receptor signaling was only later revealed by a series of studies initially involving CD3 components, TCRζ and FcεRIγ chains. By using chimeric receptors, cytoplasmic domains of FcεRIγ and TCRζ were shown to couple the receptors to several proximal and distal signaling events, including protein tyrosine kinase activation, calcium mobilization, cellular cytotoxicity, IL-2 production and basophil degranulation [145–147]. Subsequent mutational analyses further defined the residues within TCRζ and CD3ε cytoplasmic domains that are necessary for signal transduction [17–19]. In particular, the two ITAM tyrosines as well as the residue at the Y+3 position were identified as critical for ITAMs to signal [17, 18]. Similar studies conducted using Igα and Igβ sequences demonstrated that the ITAM alone is sufficient to confer signaling functions upon Igα and Igβ [21]. Consistent with TCR studies, the two motif tyrosines and the residue at position Y+3 are indispensable for ITAMs of Igα and Igβ to function [22–25, 144].

The importance of each Igα ITAM tyrosine and its phosphorylation in BCR signal transduction was more closely examined by Flaswinkel and Reth [22] in J558Lμm cells, which express mIgM and Igβ. In this model system, manipulated Igα was transfected into the cells to reconstitute BCR expression, and BCR mediated protein tyrosine phosphorylation was assessed. These authors showed that phosphorylation of Igβ and Igα was ablated by mutating the first conserved tyrosine in Igα (Y182). In contrast, mutating the second motif tyrosine in Igα (Y193) only slightly diminished tyrosine phosphorylation of Igα and Igβ. This indicates that Y182 is the major in vivo Igα phosphorylation site. In addition, reduced Igβ tyrosine phosphorylation consequent to Igα mutation suggests that phosphorylation and function of Igβ may partly be dependent upon the Igα-ITAM in BCR. Additional analysis of the same Igα mutations in chimeric receptors showed that mutation of either ITAM tyrosine is sufficiently disruptive to prevent receptor induced tyrosine phosphorylation and calcium mobilization responses. Experiments carried out by others similarly showed that mutation of Igβ-ITAM tyrosines could destroy the ability of Igβ to signal [24, 25]. These data suggest that both tyrosines of ITAMs are required for signaling, although only one is significantly phosphorylated. This may reflect some phosphorylation independent function of the second tyrosine. Alternatively, only the subpopulation of receptors which is phosphorylated at both tyrosines may be active in signal transduction.

Association of BCR with protein tyrosine kinases has been well documented [35, 36, 41, 139]. Lin and Justement [148] showed that Igα and Igβ heterodimers are responsible for coupling the resting BCR to Src-family kinases. In vitro binding experiments allowed localization of the binding site of these kinases to the ITAM of Igα [149]. Since non-phosphorylated Igα cytoplasmic tails were utilized in the study, this binding presumably reflects kinase association with resting receptors. The specificity of association between Src-family kinases and non-phosphorylated Igα-ITAM appears to be determined by the sequence DCSM located between the two motif tyrosines (see Figure 1.3). Substituting Igβ residues QTAT at the equivalent positions with DCSM is sufficient to confer binding of Src-family kinases to Igβ [150]. The precise structural limits of the binding site for Src-family kinases in the non-phosphorylated Igα-ITAM are not known. Clearly, the site requires DCSM, but additional Igα-ITAM sequence is presumably also important. In addition to Src-family kinases, Clark et al. [149] reported association of Igα-ITAM with an unidentified molecule, p38, and association of Igβ-ITAM with p40 and p42. These and other identified molecules, such as p52 discussed previously, may also be important in signal transduction.

ITAM phosphorylation is critical for signal transduction. Work by Songyang et al. [151, 152] suggests that pYxxI and pYxxL found in ITAMs are recognition sites for Src family and Syk/ZAP-70 tyrosine kinase SH2 domains. Consistent with this, Src-family kinases and Syk are rapidly recruited to the activated BCR by binding to phosphorylated ITAMs via their SH2 domains [21, 153–155]. High affinity binding of Syk to phos-

ZAP 70 crystal structure??

phorylated ITAMs requires both functional SH2 domains [154]. This is consistent with earlier studies which showed that mutating either ITAM tyrosines or the residue at Y+3 position can effectively abort signal transduction [22, 24, 25, 144]. Interestingly, maximal binding of Src-family kinases to ITAMs appears to require both tyrosines be phosphorylated [150], yet these kinases contain only one SH2 domain. This suggests that binding of ITAM phosphotyrosines to SH2 domains of Src-family kinases may involve a complex regulatory mechanism. Perhaps phosphorylation of one ITAM tyrosine affects availability of the other phosphotyrosine for binding to the SH2 domain.

Phosphorylation of ITAMs is apparently also important in recruiting downstream effector molecules to activated receptors. For example, following BCR ligation, Shc, an adaptor molecule implicated in $p21^{ras}$ activation, is phosphorylated, recruited to the plasma membrane, and binds to phosphorylated ITAMs via its SH2 domain [85, 87, 88]. *In vitro* binding experiments have revealed association of doubly phosphorylated Igα-ITAMs with additional SH2 domain containing effectors, e.g. PI3-k, rasGAP, and PLCγ1 ([156] and unpublished observation). However, whether these associations are direct or indirect requires further study.

The importance of doubly phosphorylated ITAMs extends beyond their acting as a scaffold to recruit and organize kinases and effectors. Binding of Src-family and Syk kinases to doubly tyrosine phosphorylated ITAMs also results in enzyme activation *in vitro* [150, 157, 158] and in a permeabilized cell system [156]. The molecular basis of this mode of kinase activation is unknown. Binding of these kinases to doubly phosphorylated ITAMs may allosterically enhance the kinase activity, via an allosteric mechanism or by dimerization, leading to autophosphorylation and full activation. Particularly, in the case of Src-family kinases, ITAMs may "organize" the kinase such that transphosphorylation can occur, resulting in kinase activation.

In summary, phosphorylated ITAMs not only recruit kinases and effectors, but they also activate associated kinases in the Src and Syk/ZAP-70 families, leading to signal propagation.

MOLECULAR BASIS OF COR ACCESSORY MOLECULE MODULATION OF ANTIGEN RECEPTOR SIGNAL TRANSDUCTION

Other integral membrane proteins modulate the signaling triggered by BCR crosslinking. The structural and modulatory effects of CD45, FcγRIIB, CD22 and the CD19/21 complex are described below.

CD45 is a member of the receptor type phosphotyrosine phosphatase family, with relative molecular mass that ranges from 180 to 220 kDa, depending upon splicing and glycosylation. It is also expressed selectively on cells of the hematopoietic lineage [159]. Early studies using monoclonal antibodies implicated CD45 in regulation of B cell function. Yakura *et al.* [160] demonstrated that a monoclonal anti-CD45 antibody could selectively inhibit generation of plaque forming cells to TI-type 2 antigens, but not TI-type 1 antigens in spleen cell cultures. LPS-induced polyclonal IgG secretion was also significantly reduced by treatment with anti-CD45 antibody [161, 162]. This inhibitory effect was evident in reduced steady-state level of Cγ mRNAs in LPS treated B cells in the presence of anti-CD45 antibody [163]. Mittler *et al.* [164] further showed inhibition of anti-IgM mediated B cell proliferation by several different anti-CD45 monoclonal antibodies. These antibodies seemed to exert their inhibitory effect at an early stage of B cell activation, since addition of antibodies to B cell cultures at later time points had little or no effect on anti-IgM mediated proliferative responses [164]. Additional evidence suggesting participation of CD45 in BCR signal transduction was shown by studies in which ligation of CD45 by antibodies could reduce anti-IgM, CD19 or Bgp95 mediated calcium mobilization [165, 166].

More definitive studies on the role of CD45 in B cell signal transduction are based on the use of CD45 deficient cell lines and knockout mice. Justement *et al.* [167] showed that crosslinking of IgM on the surface of a CD45 negative plasmacytoma cell line, J558Lμm3, failed to mobilize calcium. This response was restored by reconstituting expression of CD45, suggesting a requirement for CD45 in BCR mediated phospholipase activation and phosphoinositide hydrolysis. Similarly, BCR induced $p21^{ras}$ activation and cell proliferation were found to be dependent on expression of CD45 [168, 169], indicating that CD45 is a positive regulator of BCR coupling to the $p21^{ras}$ pathway. In addition, elimination of CD45 expression in the BAL-17 mature B cell line also results in absence of anti-IgM mediated growth arrest, but it has little effect on lgM mediated apoptosis [170]. While BCR mediated calcium mobilization, $p21^{ras}$ activation, and growth arrest appear to be dependent on CD45 expression, antigen induced protein tyrosine phosphorylation is evident in the CD45 negative variants of J558L myeloma and BAL-17 B cell lymphoma [170–172b]. This suggests that partial signaling, including activation of a subset of protein tyrosine kinase(s) as well as certain biologic responses, can occur in the absence of CD45.

Analysis of a CD45 deficient variant of the "immature" B cell lymphoma, WEHI-231, yielded some contrasting results. Ogimoto et al. [173] demonstrated that in CD45 negative WEHI-231, tyrosine phosphorylation of cellular substrates was constitutive, and could not be enhanced by BCR stimulation. In addition, BCR mediated growth arrest and DNA fragmentation were also more pronounced in the absence of CD45, suggesting CD45 may negatively regulate BCR signaling in immature B cells. The discrepant results from the studies of these CD45 deficient B cell lines may reflect different CD45 function in BCR signal transduction at different stages of B cell development, or they may be due to technical differences.

The molecular mechanism by which CD45 mediates its effect on BCR signal transduction is not clear. One possibility is that CD45 regulates activation of the Src-family kinases. This possibility was first addressed in experiments in which it was shown that CD45 could dephosphorylate Lck and Fyn at their conserved C-terminal negative regulatory site in vitro, enhancing kinase activity [174–176]. Consistent with this, in CD45 negative T cells these kinases were shown to be phosphorylated in vivo at this site at high stoichiometry [177–179], and they could also be co-immunoprecipitated or colocalized with CD45 [176, 180, 181]. In addition, Lck and Fyn in several CD45 negative T cells exhibited decreased kinase activity, compared to their CD45 positive counterparts [179, 182, 183].

Despite mounting evidence correlating absence of CD45 to hyperphosphorylation of Lck and Fyn at C-terminal negative regulatory sites and decreased kinase activity, Burns et al. [184] showed in three T cell lines that absence of CD45 leads to an increase in Lck and Fyn kinase activity. Lck and Fyn in these CD45 negative cells are also hyperphosphorylated at their negative regulatory sites. Thus, regulation of the Src-family kinases by CD45 may be more complicated than a simple dephosphorylation of the negative regulatory phosphotyrosine to mediate kinase activation.

In resting B cells, CD45 reportedly associates with Lyn and Igα/β heterodimers, suggesting Lyn and Igα/β may be substrates of CD45 [185]. This finding is in agreement with increased phosphorylation of Igα and Igβ after sequestration of CD45 in splenic B cells [186a] and elevated basal Igα, Igβ, and Lyn tyrosine phosphorylation of a CD45 negative variant of the J558Lμm3 cell line compared to its CD45 positive form [186b]. The effect of CD45 on activity of Src-family kinases was further examined in the J558Lμm3 system [186b]. In a CD45 negative variant of J558Lμm3, Lyn hyperphosphorylation was mapped to Y508, the kinase's negative regulatory site, and the kinase was found to be inactive. These data suggest CD45 may specifically regulate activation of Src-family kinases in B cells by dephosphorylating the negative regulatory phosphotyrosine residues. Importantly, this kinase was not recruited to the receptor and could not be activated following receptor crosslinking, despite Igα and Igβ phosphorylation. CD45 determines the ability of Src-family kinases to participate in receptor signaling; interestingly, receptor mediated activation of a non-Src-family kinase, such as Syk, which lacks a negative regulatory tyrosine, was not abrogated in the absence of CD45 expression [186b].

The overall effect of CD45 in BCR signal transduction may indeed be dependent on the maturation stage of B cells. In CD45 exon-6-deficient mice, the absence of CD45 expression does not seem to affect the BCR mediated signaling needed for B cell development, since B cells develop to the immature stage. However, the majority of peripheral B cells from these mice do not progress beyond this stage [187]. These 'immature' peripheral B cells also fail to proliferate upon anti-lgM stimulation [169]. Thus, while these cells cannot fully mature and respond to antigen, differential maturation stage-dependent functions of CD45 may have allowed such B cells to develop, at least partially, in the absence of CD45. Alternatively, compensatory mechanisms, such as overexpression of Src-family kinases, may also allow B cell development in the absence of CD45. By shifting the kinase–phosphatase equilibrium, this may provide sufficient Src-family kinase which is not phosphorylated at the negative regulatory site to provide for effective signaling, bypassing the normal CD45 requirement. Likewise, such cells may also overexpress other protein tyrosine phosphatases, such as Syp and PTP1C, or underexpress Csk, etc., to restore sufficiently the balance between kinase and phosphatase activity in cells to allow selection.

Another receptor known to modulate BCR signal transduction is CD22. CD22 is a B cell specific glycoprotein of about 135 kDa [188, 189], which may actually be part of the BCR itself, as indicated by its coprecipitation with Syk and BCR in digitonin lysates [190, 191]. Anti-CD22 monoclonal antibodies augment BCR mediated responses and lower the threshold of signals required for B cell activation [192]. Anti-Ig mediated calcium mobilization and proliferation of tonsillar B cells were also found to require expression of CD22, although direct crosslinking of CD22 failed to induce calcium mobilization [193]. Tyrosine phosphorylation of CD22 following BCR ligation has also been demonstrated [191, 194]. Both Lyn and Syk kinases were reportedly recruited to tyrosine phosphorylated CD22 following BCR ligation [195]. This recruitment is presumably via binding of the kinases' SH2 domains to the phosphotyrosines within the cytoplasmic domain of CD22. In this respect, CD22 may function as a positive regulator by bringing more kinases into the vicinity of activated receptor complexes.

More recently, association of tyrosine phosphorylated CD22 with PTP1C, a tyrosine phosphatase, has been demonstrated [196, 197]. Doody et al. [197] showed that tyrosine phosphorylated CD22 could recruit and activate PTP1C. Since sequestering CD22 away from BCR results in reduced threshold of anti-IgM induced proliferation, CD22 may be primarily a negative regulator of BCR signal transduction by recruiting PTP1C into the activated receptor complex to terminate signal transduction. How CD22 may deliver a positive and a negative signal to BCR signal transduction is unclear; recruitment of effectors may follow a specific temporal program in which PTP1C arrives later than the kinases.

Another receptor known to associate with PTP1C is FcγRIIB1 [198]. Coligation of BCR and FcγRIIB1 by immune complexes results in inhibition of antibody production [199, 200]. This inhibitory effect occurs at the level of BCR signaling, since coligation of FcγRIIB and BCR leads to premature termination of IP$_3$ generation and extracellular calcium influx [201–203]. Deletional analysis demonstrated that a 13 amino acid motif in the cytoplasmic domain of FcγRIIB1 is required for the inhibitory function [204]. Later studies showed that the 13 amino acid motif could mediate the effect, and that phosphorylation of Y309 within the motif is critical for its function [205]. D'Ambrosio et al. [198] further demonstrated that the effector molecule responsible for this inhibitory function is PTP1C. Coligation of BCR and FcγRIIB1 leads to tyrosine phosphorylation of Y309 and recruitment of PTP1C via interaction between phosphotyrosine and the C-terminal SH2 domain of PTP1C. This association also results in activation of PTP1C phosphatase activity. Thus, FcγRIIB1 may mediate negative signaling both by bringing the phosphatase into the activated receptor complex and by activating the phosphatase.

The CD19/CD21 complex also modulates BCR signal transduction. CD21, the type 2 complement receptor (CR2), is a receptor for various ligands, including complement components iC3b, C3dg and C3d fragments, and Epstein–Barr virus [206, 207]. In contrast, the ligand for CD19, a member of the Ig superfamily [208], is unknown. TAPA-1 (CD81) and Leu-13 are also found in the CD19/21 complex, and may contribute to the signaling properties of CD19/21 [209–211]. Earlier studies using monoclonal antibodies against CD19 or ligands to CD21 suggest synergy between the complex and BCR in B cell activation [209, 212, 213]. However, antibody binding to CD19 can also have an inhibitory effect on anti-Ig induction of B cell proliferation [214]. Indeed, the effect of CD19 ligation on B cells seems to depend on the stimulatory conditions and the response assessed [215]. Carter and Fearon [216] have demonstrated that CD19 functions by lowering the threshold for antigen required to stimulate proliferation nearly 100-fold. In addition, the CD19/21 complex may also be involved in enhancing the response of B cells to antigens by interacting with components of the classical complement pathway [217]. Functions of CD19 were more recently examined in two independently derived CD19 knockout mice [218, 219]. Both CD19 knockout mice showed normal B cell development in the bone marrow, but reduced numbers of peritoneal B cells, and defective responses to TD antigens. The impaired humoral response was characterized by the reduced production of antigen-specific antibodies. Rickert *et al.* [218] further examined the response of their CD19$^{-/-}$ cells to TD antigens and showed that the impaired antibody production was accompanied by an absence of high affinity antibodies and a lack of germinal center formation, but the kinetics of the response was comparable to the wildtype B cells. The two CD19 deficient mice also showed some subtle differences. For instance, the mice generated by Engel *et al.* [219] have a reduced number of B cells in the periphery, but those generated by Rickert *et al.* have normal numbers. In addition, B cells from the former mice also seemed to be less responsive to a variety of B cell stimuli *in vitro*, but the B cells from the latter mice showed no such defect. Nevertheless, the importance of CD19 in augmenting BCR mediated responses, in particular to TD antigens, is clearly implicated in both CD19 deficient mice. Since B cells from the CD19 knockout mice generated by Rickert *et al.* responded normally to the TI-type 2 antigen, NP-Ficoll, a high degree of BCR crosslinking can apparently override the requirement for CD19.

The idea that CD19 augments BCR mediated signal transduction is further supported by the phenotype of mice overexpressing human CD19 transgene. B cells from these mice are hyperresponsive to mitogens *in vitro*, and despite a greatly reduced number of peripheral B cells, they also have increased levels of serum immunoglobulins [219]. Interestingly, overexpression of CD19 appears to be detrimental to B cell development, since the frequency of immature and mature B cells is greatly reduced in these mice [219]. As suggested by Engel *et al.* [219], this phenomenon may likely be due to increased B cell sensitivity to transmembrane signaling, resulting in lowering the threshold for negative selection, and increased clonal delection of B cells.

CD19 enhancement of BCR signaling may be due to its ability to mediate signal transduction, in addition to its ability to augment BCR avidity and thus crosslinking via association with carrier molecules, e.g. C3dg. In surface Ig negative pre-B cells, crosslinking of CD19 results in protein tyrosine phosphorylation of several cellular substrates, including CD19 itself [220, 221]. Phosphorylated CD19 can also bind to Src-family kinases, including Lck, Lyn, Fyn and Yes [220–222]. This binding is probably mediated by SH2 domains of the kinases [221]. The cytoplasmic domain of CD19 contains nine conserved tyrosine residues, several of which may be tyrosine phosphorylated and serve as SH2 binding sites. The SH2 containing proteins that are predicted to recognize these phosphotyrosine motifs include PI3-k, Abl, 3BP2 and Vav [151, 152]. Tuveson *et al.* [223] have shown coprecipitation of the p85 subunit of PI3-k with CD19, following crosslinking of CD19 or BCR to induce tyrosine phosphorylation of CD19. The amount of p85 coprecipitated correlates with the intensity of CD19 tyrosine phosphorylation, and association of PI3-k to CD19 is dependent on phosphorylation of Y484 and Y515, both of which are part of the YXXM motifs found in the cytoplasmic domain of CD19. PI3-k activity is also co-immunoprecipitated with CD19, following crosslinking of BCR or CD19 alone. In addition, Weng *et al.* [224] have observed tyrosine phosphorylation of Vav and formation of CD19/Vav/PI3-k complexes induced by CD19 ligation. Significantly, CD19 crosslinking leads to MAP kinase and MEK activation [224]. Coligation of BCR and CD19 also augments BCR mediated tyrosine phosphorylation and translocation of Shc/Grb2/Sos complex to the membrane [87]. In conclusion, CD19 may function both as an independent receptor and as a BCR coreceptor by enhancing signaling through the PI3-k and Ras pathways.

A MODEL OF BCR COUPLING TO SIGNAL PROPAGATION MECHANISMS

Based on the data discussed above, we propose the following as a mechanism for BCR signal transduction. In the resting state, several Src-family

PTKs, including Lyn and Fyn, are associated with Igα and Igβ. This binding is via the unique region of the Src-family kinases to the residues between the two ITAM tyrosines. Syk kinase may also be part of the complex by virtue of a direct association with Igα and Igβ, or with mIg. Although associated kinases exhibit a significant basal activity, this receptor complex is "inactive", as BCR associated PTP1C keeps the Igα and Igβ dephosphorylated. CD45 may also function at this stage by providing a pool of receptor associated Src-family PTKs in a non-phosphorylated state. Upon crosslinking of BCR, a series of biochemical events is triggered. The activatable pool of Igα and Igβ associated Src-family kinases transphosphorylate Igα and Igβ on ITAM tyrosines of neighboring receptors and also phosphory late CD19. The outcome of phosphorylation is reorientation of receptor associated Src-family kinases and recruitment of more Src and Syk family kinases. These kinases bind to phosphorylated Igα and Igβ via a SH2-phosphotyrosine interaction and, importantly, this interaction leads to kinase activation. BCR mediated CD19 and CD22 tyrosine phosphorylation also leads to recruitment and activation of more PTKs and lipid kinases to the activated receptor complex.

Following phosphorylation of ITAMs, other effector molecules such Shc, Vav, PLCγ, rasGAP, and PI3-k can be recruited to pITAMs and activated by the adjacent PTKs. For instance, activated Lyn and Syk may phosphorylate Shc bound to the phosphorylated Igα/Igβ, allowing assembly of Shc/Grb2/Sos complex in close proximity to p21^{ras}. Activation of p21^{ras} may be a cooperative effect of inactivation of rasGAP, activation of Vav and translocation of Sos to the membrane. Lyn may also activate PI3-K by binding to the proline rich region within the p85 subunit of PI3-k via its SH3 domain. Phosphorylation and activation of PLCγ is likely to be mediated by activated Syk. Second messengers such as DAG and IP$_3$ generated by activated PLC enzyme further propagate signaling via calcium mobilization and PKC activation.

During an immune response, the BCR signaling process described above is modified by participation of coreceptors. Immune complexes or opsonized bacteria displaying complement C3dg fragments may modulate the humoral response by coengaging CD19/21 and BCR. Crosslinking of CD19 reduces the antigen threshold required to activate B cells through the BCR, by enhancing activation of PI3-k and MAPK pathways. Late in the response, accumulating immune complexes containing IgG coligate BCR and FcγRIIB1 on resting B cells, resulting in recruitment and activation of PTP1C to counteract any activation signal, and maintain the receptors at resting state. This may ultimately enhance affinity maturation and epitope dominance by preventing participation of newly generated antigen specific cells in the ongoing immune response.

FUTURE PROSPECTS

Although great progress has been made, many steps in BCR signal transduction and their relationship to downstream biologic responses remain poorly understood. In addition, studies that attempt to elucidate the molecular mechanism of BCR signaling have focused mainly on activation of PTKs and positive signaling. Recent studies on PTP1C have only begun to advance our knowledge of "negative" signaling, and the importance of this mechanism in the immune regulation. More insights into this aspect of BCR signal transduction and integration of signals from coreceptors will further our understanding of the basis of immunity.

REFERENCES

1. Hombach J, Tsubata T, Leclercq L, Stappert H and Reth M. Molecular components of the B-cell antigen receptor complex of the IgM class. *Nature* 1990; **343**: 760–762.
2. Hombach J, Lottspeich F and Reth M. Identification of the genes encoding the IgM-α and Ig-β components of the IgM antigen receptor complex by amino terminal sequencing. *Eur. J. Immunol.* 1990; **20**: 2765–2799.
3. Campbell KS and Cambier JC. B lymphocyte antigen receptors (mIg) are noncovalently associated with a disulfide-linked, inducibly phosphorylated glycoprotein complex. *EMBO J.* 1990; **9**: 441–448.
4. Kyte J and Doolittle RF. A simple method for displaying the hydropathic character of a protein. *J. Mol. Biol.* 1982; **157**: 105–132.
5. Klein P, Kanehisa M and Delisi C. The detection and classification of membrane-spanning proteins. *Biochem. Biophys. Acta* 1985; **815**: 468–476.

6. Sakaguchi N, Kashiwamura S, Kimoto M, Thalmann P and Melchers F. B lymphocyte lineage-restricted expression of mb-1, a gene with CD3-like structural properties. *EMBO J.* 1988; **7**: 3457–3464.

7. Hermanson GG, Eisenberg D, Kincade PW and Wall R. B29: a member of the immunoglobulin gene superfamily exclusively expressed on B-lineage cells. *Proc. Natl. Acad. Sci. USA* 1988; **85**: 6890–6894.

8. Reth M. Antigen receptors on B lymphocytes. *Ann. Rev. Immunol.* 1992; **10**: 97–121.

9. Weiser P, Riesterer C and Reth M. The internalization of the IgG2a antigen receptor does not require the association with Ig-α and Ig-β but the activation of protein tyrosine kinases does. *Eur. J. Immunol.* 1994; **24**: 665–671.

10. Venkitaraman AR, Williams GT, Dariavach P and Neuberger MS. The B-cell antigen receptor of the five immunoglobulin classes. *Nature* 1991; **352**: 777–781.

11. Campbell KS, Hager EJ and Cambier JC. α-Chains of IgM and IgD antigen receptor complexes are differentially *N*-glycosylated MB-1-related molecules. *J. Immunol.* 1991; **147**(5): 1575–1580.

12. Friedrich A, Campbell KS and Cambier JC. The γ subunit of the B cell antigen-receptor complex is a *C*-terminally truncated product of the B29 gene. *J. Immunol.* 1993; **150**: 2814–2822.

13. Hashimoto S, Chiorazzi N and Gregersen PK. Alternative splicing of CD79a (Ig-α/mb-1) and CD79b (Ig-β/B29) RNA transcripts in human B cells. *Mol. Immunol.* 1995; **32**: 651–659.

14. Reth M. Antigen receptor tail clue. *Nature* 1989; **338**: 383–384.

15. Cambier JC. New nomenclature for the Reth motif (or ARH1/TAM/ARAM/YXXL). *Immunol. Today* 1995; **16**: 110.

16. Keegan AD and Paul WE. Multichain immune recognition receptors: similarities in structure and signaling pathways. *Immunol. Today* 1992; **13**(2): 63–68.

17. Letourneur F and Klausner RD. Activation of T cells by a tyrosine kinase activation domain in the cytoplasmic tail of CD3 ϵ. *Science* 1992; **255**: 79–82.

18. Romeo C, Amiot M, Seed B. Sequence requirements for induction of cytolysis by the T cell antigen/Fc receptor ζ chain. *Cell* 1992; **68**: 889–897.

19. Wegener AMK, Letourneur F, Hoeveler A, Brocker T, Luton F and Malissen B. The T cell receptor/CD3 complex is composed of at least two autonomous transduction modules. *Cell* 1992; **68**: 83–95.

20. Irving BA, Chan AC and Weiss A. Functional characterization of a signal transducing motif present in the T cell antigen receptor ζ chain. *J. Exp. Med.* 1993; **177**: 1093–1103.

21. Law DA, Chan VWF, Datta SK, DeFranco AL. B-cell antigen receptor motifs have redundant signalling capabilities and bind the tyrosine kinases PTK72, Lyn and Fyn. *Curr. Biol.* 1993; **3**: 645–657.

22. Flaswinkel H and Reth M. Dual role of the tyrosine kinase activation motif of Ig-α protein during signal transduction via the B cell antigen receptor. *EMBO J.* 1994; **13**(1): 83–89.

23. Burkhardt AL, Costa T, Misulovin Z, Stealy B, Bolen JB and Nussenzweig MC. Igα and Igβ are functionally homologous to the signaling proteins of the T-cell receptor. *Mol. Cell. Biol.* 1994; **14**: 1095–1103.

24. Williams GT, Peaker CJG, Patel KJ and Neuberger MS. The α/β sheath and its cytoplasmic tyrosines are required for signaling by the B cell antigen receptor but not for capping or serine/threonine-kinase recruitment. *Proc. Natl. Acad. Sci. USA* 1994; **91**: 474–478.

25. Sanchez M, Misulovin Z, Burkhardt AL, Mahajan S, Costa T, Franke R *et al.* Signal transduction by immunoglobulin is mediated through Ig-α and Ig-β. *J. Exp. Med.* 1993; **178**: 1049–1057.

26. Terashima M, Kim K-M, Adachi T, Nielsen PJ, Reth M, Kohler G *et al.* The IgM antigen receptor of B lymphocytes is associated with prohibitin and a prohibitin-related protein. *EMBO J.* 1994; **13**: 3782–3792.

27. Kim K-M, Adachi T, Nielsen PJ, Terashima M, Lamers MC, Kohler G *et al.* Two new proteins preferentially associated with membrane immunoglobulin D. *EMBO J.* 1994; **13**: 3793–3800.

28. Lankester AC, van Schijndel GMW, Cordell JL, van Noesel CJM and van Lier RAW. CD5 is associated with the human B cell antigen receptor complex. *Eur. J. Immunol.* 1994; **24**: 812–816.

29. Kuwahara K, Matsuo T, Nomura J, Igarashi H, Kimoto M, Inui S *et al.* Identification of a 52-kDa molecule (p52) coprecipitated with the Ig receptor-related MB-1 protein that is inducibly phosphorylated by the stimulation with phorbol myristate acetate. *J. Immunol.* 1994; **152**: 2742–2752.

30. Yellen-Shaw AJ and Monroe JG. Developmentally regulated association of a 56-kD member of the surface immunoglobulin M receptor complex. *J. Exp. Med.* 1992; **176**: 129–137.

31. Igarashi H, Kuwahara K, Nomura J, Matsuda A, Kikuchi K, Inui S *et al.* B cell ag receptor mediates different types of signals in the protein kinase activity between immature B cell and mature B cells. *J. Immunol.* 1994; **153**: 2381–2393.

32. Gold MR, Law DA, DeFranco AL. Stimulation of protein tyrosine phosphorylation by the B-lymphocyte antigen receptor. *Nature* 1990; **345**(6278): 810–813.

33. Campbell MA and Sefton BM. Protein tyrosine phosphorylation is induced in murine B lymphocytes in response to stimulation with anti-immunoglobulin. *EMBO J.* 1990; **9**: 2125–2131.

34. Gold MR, Matsuuchi L, Kelly RB and DeFranco AL. Tyrosine phosphorylation of components of the B-cell antigen receptors following receptor crosslinking. *Proc. Natl. Acad. Sci. USA* 1991; **88**: 3436–3440.

35. Yamanashi Y, Kakiuchi T, Mizuguchi J, Yamamoto T and Toyoshima K. Association of B cell antigen receptor with protein tyrosine kinase lyn. *Science* 1991; **251**: 192–194.

36. Burkhardt AL, Brunswick M, Bolen JB and Mond JJ. Anti-immunoglobulin stimulation of B lymphocytes activates src-related protein-tyrosine kinases. *Proc. Natl. Acad. Sci. USA* 1991; **88**: 7410–7414.

37. Campbell M-A and Sefton BM. Association between B-lymphocyte membrane immunoglobulin and multiple mem-

bers of the src family of protein tyrosine kinases. *Mol. Cell. Biol.* 1992; **12**: 2315–2321.

38. Gold MR, Chiu R, Ingham RJ, Saxton TM, van Oosteen I, Watts JD *et al.* Activation and serine phosphorylation of the p56[lck] protein tyrosine kinase in response to antigen receptor cross-linking in B lymphocytes. *J. Immunol.* 1994; **153**: 2369–2380.

39. Wechsler RJ and Monroe JG. *src*-family tyrosine kinase p55[fgr] is expressed in murine splenic B cells and is activated in response to antigen receptor cross-linking. *J. Immunol,* 1995; **154**: 3234–3244.

40. Hutchcroft JE, Harrison ML and Geahlen RL. B lymphocyte activation is accompanied by phosphorylation of a 72 kDa protein-tyrosine kinase. *J. Biol. Chem.* 1991; **266**: 14846–14849.

41. Yamada T, Taniguchi T, Yang C, Yasue S, Saito H and Yamamura H. Association with B-cell antigen receptor with protein-tyrosine kinase p72[syk] and activation by engagement of membrane IgM. *Eur. J. Biochem.* 1993; **213**: 455–459.

42. de Weers M, Brouns GS, Hinshelwood S, Kinnon C, Schuurman RKB, Hendriks RW *et al.* B-cell antigen receptor stimulation activates the human Bruton's tyrosine kinase, which is deficient in X-linked agammaglobulinemia. *J. Biol. Chem.* 1994; **269**: 23857–23860.

43. Aoki Y, Isselbacher KJ and Pillai S. Brunton tyrosine kinase is tyrosine phosphorylated and activated in pre-B lymphocytes and receptor-ligated B cells. *Proc. Natl. Acad. Sci. USA* 1994; **91**: 10606–10609.

44. Saouaf SJ, Mahajan S, Rowley RB, Kut SA, Fargnoli J, Burkhardt AL *et al.* Temporal differences in the activation of three classes of non-transmembrane protein tyrosine kinases following B-cell antigen receptor surface engagement. *Proc. Natl. Acad. Sci. USA* 1994; **91**: 9524–9528.

45. Pani G, Kozlowski M, Cambier JC, Mills GB and Siminovitch KA. Identification of the tyrosine phosphatase PTP1C as a B cell antigen receptor-associated protein involved in the regulation of B cell signaling. *J. Exp. Med.* 1995; **181**: 2077–2084.

46. Berridge MJ. Inositol trisphosphate and calcium signalling. *Nature* 1993; **361**: 315–325.

47. Maino VC, Hayman MJ and Crumpton MJ. Relationship between enhanced turnover of phosphatidylinositol and lymphocyte activation by mitogen. *Biochem. J.* 1975; **146**: 247–252.

48. Coggeshall KM and Cambier JC. B cell activation. VIII. Membrane immunoglobulins transduce signals via activation of phosphatidylinositol hydrolysis. *J. Immunol.* 1984; **133**: 3382–3386.

49. Bijsterbosch MK, Meade JC, Turner GA and Klaus GGB. B lymphocyte receptors and polyphosphoinositide degradation. *Cell* 1985; **41**: 999–1006.

50. Ransom JT, Harris LK and Cambier JC. Anti-Ig induces release of inositol 1,4,5-trisphosphate, which mediates mobilization of intracelular Ca^{++} stores in B lymphocytes. *J. Immunol.* 1986; **137**: 708–714.

51. Carter RH, Park DJ, Rhee SG and Fearon DT. Tyrosine phosphorylation of phospholipase C induced by membrane immunoglobulin in B lymphocytes. *Proc. Natl. Acad. Sci. USA* 1991; **88**: 2745–2749.

52. Coggeshall KM, McHugh JC and Altman A. Predominant expression and activation-induced tyrosine phosphorylation of phospholipase C-γ2 in B lymphocytes. *Proc. Natl. Acad. Sci. USA* 1992; **89**: 5660–5664.

53. Nishibe S, Wahl MI, Hernandez-Sotomayor SMT, Tonks NK, Rhee SG and Carpenter G. Increase of the catalytic activity of phospholipase C-γ1 by tyrosine phosphorylation. *Science* 1990; **250**: 1253–1256.

54. Takata M, Sabe H, Hata A, Inazu T, Himma Y, Nukada T *et al.* Tyrosine kinases Lyn and Syk regulate B cell receptor-coupled Ca^{2+} mobilization through distinct pathways. *EMBO Journal* 1994; **13**: 1341–1349.

55. Kolanus W, Romeo C and Seed B. T cell activation by clustered tryosine kinases. *Cell* 1993; **74**: 171–183.

56. Sidorenko SP, Law C-L, Chandran KA and Clark EA. Human spleen tyrosine kinase p72Syk associates with the Src-family kinase p53/56Lyn and a 120-kDa phosphoprotein. *Proc. Natl. Acad. Sci. USA* 1995; **92**: 359–363.

57. Sillman AL and Monroe JG. Association of p72[syk] with the src homology-2 (SH2) domains of the PLCγ1 in B lymphocytes. *J. Biol. Chem.* 1995; **270**: 11806–11811.

58. Aoki Y, Kim Y-T, Stillwell R, Kim TJ and Pillai S. The SH2 domains of Src family kinases associate with Syk. *J. Biol. Chem.* 1995; **270**: 15658–15663.

59. Kurosaki T, Takata M, Yamanashi Y, Inazu T, Taniguchi T, Yamamoto T *et al.* Syk activation by the src-family tyrosine kinase in the B cell receptor signaling. *J. Exp. Med.* 1994; **179**: 1725–1729.

60. Iwashima M, Irving BA, van Oers N, Chan AC and Weiss A. Sequential interaction of the TCR with two distinct cytoplamic tyrosine kinases. *Science* 1994; **263**: 1136–1139.

61. Kelly K, Cochran B, Stiles CD and Leder P. Cell-specific regulation of the c-myc gene by lymphocyte mitogens and platelet-derived growth factor. *Cell* 1983; **35**: 603–610.

62. Monroe JG and Kass MJ. Molecular events in B cell activation. I. Signals required to stimulate G0 to G1 transition of resting B lymphocytes. *J. Immunol.* 1985; **135**: 1674–1682.

63. Seyfert VL, Sukhatme VP and Monroe JG. Differential expression of a zinc finger-encoded gene in response to positive versus negative signaling though receptor immunoglobulin in murine B lymphocytes. *Mol. Cell. Biol.* 1989; **9**: 2083–2088.

64. Klemsz MJ, Justement LB, Palmer E and Cambier JC. Induction of c-fos and c-myc expression during B cell activation by IL-4 and immunoglobulin binding ligands. *J. Immunol.* 1989; **143**: 1032–1039.

65. Mittelstadt PR and DeFranco AL. Induction of early response genes by cross-linking membrane Ig on B lymphocytes. *J. Immunol.* 1993; **150**: 4822–4832.

66. Seyfert VL, McMahon S, Glenn W, Cao X, Sukhatme VP and Monroe JG. *Egr-1* expression in surface Ig-mediated B cell activation. *J. Immunol.* 1990; **145**: 3647–3653.

67. Fisher CL, Ghysdael J and Cambier JC. Ligation of membrane Ig leads to calcium-mediated phosphorylation of the proto-oncogene product, Ets-1 [published erratum appears in *J. Immunol.* 1991 Sep 15; **147**(6): 2068]. *J. Immunol.* 1991; **146**(6): 1743–1749.

68. Harwood AE and Cambier JC. B cell antigen receptor cross-linking triggers rapid protein kinase C independent activation of p21[ras]. *J. Immunol.* 1993; **151**: 4513–4522.

69. Lazarus AH, Kawauchi K, Rapoport MJ and Delovitch TL. Antigen-induced B lymphocyte activation involves the p21^{ras} and rasGAP signaling pathway. *J. Exp. Med.* 1993; **178**(5): 1765–1769.

70. Trahey M and McCormick F. A cytoplasmic protein stimulates normal N-*ras* p21 GTPase, but does not affect oncogenic mutants. *Science* 1987; **238**: 542–545.

71. Gold MR, Crowley MT, Martin GA, McCormick F and DeFranco AL. Targets of B lymphocyte antigen receptor signal transduction include the p21ras GTPase-activating protein (GAP) and two GAP-associated proteins. *J. Immunol.* 1993; **150**: 377–386.

72. Settleman J, Albright CF, Foster LC and Weinberg RA. Association between GTPase activators for Rho and Ras families. *Nature* 1992; **359**(6391): 153–154.

73. Wong G, Muller O, Clark R, Conroy L, Moran MF, Polakis P *et al.* Molecular cloning and nucleic acid binding properties of the GAP-associated tyrosine phosphoprotein p62. *Cell* 1992; **69**: 551–558.

74. Cichowski K, McCormick F and Brugge JS. p21^{ras}-GAP association with Fyn, Lyn, and Yes in thrombin-activated platelets. *J. Biol. Chem.* 1992; **267**(8): 5025–5028.

75. Pleiman CM, Clark MR, Winitz S, Coggeshall K, Johnson GL and Cambier JC. Mapping of sites on the src-family protein tyrosine kinases p55^{blk}, p59^{fyn} and p56^{lyn} which interact with the effector molecules phospholipase C-γ2, microtubule-associated protein kinase, GTPase-activating protein, and phosphatidylinositol 3-kinase. *Mol. Cell. Biol.* 1993; **13**: 5877–5887.

76. Amrein KE, Flint N, Panholzer B and Burn P. Ras GTPase-activating protein: a substrate and a potential binding protein of the protein-tyrosine kinase p56^{lck}. *Proc. Natl. Acad. Sci. USA* 1992; **89**: 3343–3346.

77. Bustelo XR and Barbacid M. Tyrosine phosphorylation of the vav proto-oncogene product in activated B cells. *Science* 1992; **256**: 1196–1199.

78. Gulbins E, Langlet C, Baier G, Bonnefoy-Berard N, Herbert E, Katzav S *et al.* Tyrosine phosphorylation and activation of VAV GTP/GDP exchange activity in antigen receptor-triggered B cells. *J. Immunol.* 1994; **152**: 2123–2129.

79. Gulbins E, Coggeshall KM, Langlet C, Baier G, Bonnefoy-Berard N, Burn P *et al.* Activation of Ras *in vitro* and in intact fibroblasts by the Vav guanine nucleotide exchange protein. *Mol. Cell. Biol.* 1994; **14**: 906–913.

80. Katzav S, Martin-Zanca D and Barbacid M. *vav*, a novel human oncogene derived from a locus ubiquitously expressed in hematopoietic cells. *EMBO J.* 1989; **8**: 2283–2290.

81. Adams JM, Houston H, Allen J, Lints T and Harvey R. The hematopoietically expressed *vav* proto-oncogene shares homology with the *dbl* GDP-GTP exchange factor, the *bcr* gene and a yeast gene (CDC24) involved in cytoskeletal organization. *Oncogene* 1992; **7**: 611–618.

82. Khosravi-Far R, Chrzanowska-Wodnicka M, Solski PA, Eva A, Burridge K and Der CJ. Dbl and Vav mediate transformation via mitogen-activated protein kinase pathways that are distinct from those activated by oncogenic Ras. *Mol. Cell. Biol.* 1994; **14**: 6848–6857.

83. Zhang R, Alt FW, Davidson L, Orkin SH and Swat W. Defective signalling through the T-and B-cell antigen receptors in lymphoid cells lacking the vav proto-oncogene. *Nature* 1995; **374**: 470–473.

84. Tarakhovsky A, Turner M, Schaal S, Mee PJ, Duddy LP, Rajewsky K *et al.* Defective antigen receptor-mediated proliferation of B and T cells in the absence of Vav. *Nature* 1995; **374**: 467–470.

85. Saxton TM, van Oostveen I, Bowtell D, Aebersold R and Gold MR. B cell antigen receptor cross-linking induces phosphorylation of the p21ras oncoprotein activators SHC and mSOS1 as well as assembly of complexes containing SHC, GRB-2, mSOS1, and a 145 kDa tyrosine phosphorylated protein. *J. Immunol.* 1994; **153**: 623–631.

86. Smit L, de Vries-Smits AMM, Bos JL and Borst J. B cell antigen receptor stimulation induces formation of a Shc–Grb2 complex containing multiple tyrosine-phosphorylated proteins. *J. Biol. Chem.* 1994; **269**: 20209–20212.

87. Lankester AC, van Schijndel GMW, Rood PML, Verhoeven AJ and van Lier RAW. B cell antigen receptor cross-linking induces tyrosine phosphorylation and membrane translocation of a multimeric Shc complex that is augmented by CD19 co-ligation. *Eur. J. Immunol.* 1994; **24**: 2818–2825.

88. D'Ambrosio D, Hippen KL, Cambier JC. Distinct mechanisms mediate SHC association with the activated and resting B cell antigen receptor. *Eur. J. Immunol.* 1996; **26**: 1960-1965.

89. Nagai K, Takata M, Yamamura H and Kurosaki T. Tyrosine phosphorylation of Shc is mediated through Lyn and Syk in B cell receptor signaling. *J. Biol. Chem.* 1995; **270**: 6824–6829.

90. Buday L and Downward J. Epidermal growth factor regulates p21(ras) through the formation of a complex of receptor, Grb2 adapter protein, and Sos nucleotide exchange factor. *Cell* 1993; **73**(3): 611–620.

91. Graziadei L, Raibowol K and Bar-Sagi D. Co-capping of ras proteins with surface immunoglobulins in B lymphocytes. *Nature* 1990; **347**: 396–400.

92. Tamaki T, Kanakura Y, Kuriu A, Ikeda H, Mitsui H, Yagura H *et al.* Surface immunoglobulin-mediated signal transduction involves rapid phosphorylation and activation of the protooncogene product Raf-1 in human B cells. *Cancer Research* 1992; **52**: 566–570.

93. Tordai A, Franklin RA, Patel H, Gardner AM, Johnson GL and Gelfand EW. Cross-linking of surface IgM stimulates the ras/raf-1/MEK/MAPK cascade in human B lymphocytes. *J. Biol. Chem.* 1994; **269**: 7538–7543.

94. Casillas A, Hanekom C, Williams K, Katz R and Nel AE. Stimulation of B-cells via the membrane immunoglobulin receptor or with phorbol myristate 13-acetate induces tyrosine phosphorylation and activation of a 42-kDa microtubule-associated protein-2 kinase. *J. Biol. Chem.* 1991; **266**: 19088–19094.

95. McMahon SB and Monroe JG. Activation of the p21^{ras} pathway couples antigen receptor stimulation to induction of the primary response gene *egr*-1 in B lymphocytes. *J. Exp. Med.* 1995; **181**: 417–422.

96. Yamanashi Y, Fukui Y, Wongsasant W, Kinoshita Y, Ichimori Y, Toyoshima K *et al.* Activation of Src-like

protein-tyrosine kinase Lyn and its association with phosphatidylinositol 3-kinase upon B-cell antigen receptor-mediated signaling. *Proc. Natl. Acad. Sci. USA* 1992; **89**: 1118–1122.

97. Gold MR, Chan VW-F, Turck CW and DeFranco AL. Membrane Ig cross-linking regulates phosphatidylinositol 3-kinase in B lymphocytes. *J. Immunol.* 1992; **148**: 2012–2022.

98. Cantley LC, Auger KR, Carpenter C, Duckworth B, Graziani A, Kapeller R *et al.* Oncogenes and signal transduction. *Cell* 1991; **64**: 281–302.

99. Wennström S, Siegbahn A, Yokote K, Arvidsson A-K, Heldin C-H, Mori S *et al.* Membrane ruffling and chemotaxis transduced by the PDGFβ-receptor require the binding site for phosphatidylinositol 3' kinase. *Oncogene* 1994; **9**: 651–660.

100. Kundra V, Escobedo JA, Kazlauskas A, Kim HK, Rhee SG, Williams LT *et al.* Regulation of chemotaxis by the platelet-derived growth factor receptor-β. *Nature* 1994; **367**: 474–476.

101. Joly M, Kazlauskas A, Fay FS and Corvera S. Disruption of PDGF receptor trafficking by mutation of its Pl-3 kinase binding sites. *Science* 1994; **263**: 684–687.

102. Yano H, Nakanishi S, Kimura K, Hanai N, Saitoh Y, Fukui Y *et al.* Inhibition of histamine secretion by wortmannin through the blockade of phosphatidylinositol 3-kinase in RBL-2H3 cells. *J. Biol. Chem.* 1993; **268**: 25846–25856.

103. Nakanishi H, Brewer KA and Exton JH. Activation of the ζ isozyme of protein kinase C by phosphatidylinositol 3,4,5-trisphosphate. *J. Biol. Chem.* 1993; **268**: 13–16.

104. Toker A, Meyer M, Reddy KK, Falck JR, Aneja R, Aneja S *et al.* Activation of protein kinase C family members by the novel polyphosphoinositides PtdIns-3,4-P$_2$ and PtdIns-3,4,5-P$_3$. *J. Biol. Chem.* 1994; **269**: 32358–32367.

105. Mischak H, Kolch W, Goodnight J, Davidson WF, Rapp U, Rose-Joh S *et al.* Expression of protein kinase C genes in hemopoietic cells is cell-type and B cell differentiation stage specific. *J. Immunol.* 1991; **174**: 3981–3987.

106. Diaz-Meco MT, Dominguez I, Sanz L, Dent P, Lozano J, Municio MM *et al.* ζPKC induces phosphorylation and inactivation of IκB-α *in vitro*. *EMBO J.* 1994; **13**: 2842–2848.

107. Diaz-Meco MT, Berra E, Municio MM, Sanz L, Lozano J, Dominguez I *et al.* A dominant negative protein kinase C ζ subspecies blocks NF-κB activation. *Mol. Cell Biol.* 1993; **13**: 4770–4775.

108. Chung J, Grammer TC, Lemon KP, Kazlauskas A and Blenis J. PDGF-and insulin-dependent pp70^{S6k} activation mediated by phosphatidylinositol-3-OH kinase. *Nature* 1994; **370**: 71–75.

109. Weng Q-P, Andrabi K, Klippel A, Kozlowski MT, Williams LT and Avruch J. Phosphatidylinositol 3-kinase signals activation of p70 S6 kinase *in situ* through site-specific p70 phosphorylation. *Proc. Natl. Acad. Sci. USA* 1995; **92**: 5744–5748.

110. Pleiman CM, Hertz WM and Cambier JC. Activation of phosphatidylinositol-3' kinase by Src-family kinase SH3 domain binding to the p85 subunit. *Science* 1994; **263**: 1609–1612.

111. Rodriguez-Viciana P, Warne PH, Dhang R, Vanhaeseb-roeck B, Gout I, Fry MJ *et al.* Phosphatidylinositol-3-OH kinase as a direct target of Ras. *Nature* 1994; **370**: 527–532.

112. Gold MR and Aebersold R. Both phosphatidylinositol 3-kinase and phosphatidylinositol 4-kinase products are increased by antigen receptor signaling in B cells. *J. Immunol.* 1994; **152**: 42–50.

113. Donovan JA, Wange RL, Langdon WY and Samelson LE. The protein product of the c-cbl protooncogene is the 120-kDa tyrosine-phosphorylated protein in Jurkat cells activated via the T cell antigen receptor. *J. Biol. Chem.* 1994; **269**: 22921–22924.

114. Cory GOC, Lovering RC, Hinshelwood S, MacCarthy-Morrogh L, Levinsky RJ and Kinnon C. The protein product of the c-cbl protooncogene is phosphorylated after B cell receptor stimulation and binds the SH3 domain of Bruton's tyrosine kinase. *J. Exp. Med.* 1995; **182**: 611–615.

115. Reif K, Buday L, Downward J, Cantrell DA. SH3 domains of the adapter molecule Grb2 complex with two proteins in T cells: the guanine nucleotide exchange protein Sos and a 75-kDa protein that is a substrate for T cell antigen receptor-activated tyrosine kinases. *J. Biol. Chem.* 1994; **269**: 14081–14087.

116. Buday L, Egan SE, Viciana PR, Cantrell DA and Downward J. A complex of Grb2 adaptor protein, Sos exchange factor, and a 36-kDa membrane-bound tyrosine phosphoprotein is implicated in Ras activation in T cells. *J. Biol. Chem.* 1994; **269**: 9019–9023.

117. Reedquist KA, Fukazawa T, Druker B, Panchamoorthy G, Shoelson SE and Band H. Rapid T-cell receptor-mediated tyrosine phosphorylation of p120, an Fyn/Lck Src homology 3 domain-binding protein. *Proc. Natl. Acad. Sci. USA* 1994; **91**: 4135–4139.

118. Yu H, Chen JK, Feng S, Dalgarno DC, Brauer AW, Schreiber SL. Structural basis for the binding of proline-rich peptides to SH3 domains. *Cell* 1994; **76**: 933–945.

119. Rivero-Lezcano OM, Sameshima JH, Marcilla A and Robbins KC. Physical association between Src homology 3 elements and the protein product of the c-cbl proto-oncogene. *J. Biol. Chem.* 1994; **269**: 17363–17366.

120. Yamanashi Y, Okada M, Semba T, Yamori T, Umemori H, Tsunasawa S *et al.* Identification of HS1 protein as a major substrate of protein-tyrosine kinase(s) upon B-cell antigen receptor-mediated signaling. *Proc. Natl. Acad. Sci. USA* 1993; **90**: 3631–3635.

121. Baumann G, Maier D, Freuler F, Tschopp C, Baudisch K and Wienands J. *In vitro* characterization of major ligands for Src homology 2 domains derived from protein tyrosine kinases, from the adaptor protein SHC and from GTPase-activating protein in Ramos B cells. *Eur J Immunol* 1994; **24**: 1799–1807.

122. Takemoto Y, Furuta M, Li X-K, Strong-Sparks WJ and Hashimoto Y. LckBP1, a proline-rich protein expressed in haematopoietic lineage cells, directly associates with the SH3 domain of protein tyrosine kinase p56^{lck}. *EMBO J.* 1995; **14**: 3403–3414.

123. Fukuda T, Kitamura D, Taniuchi I, Maekawa Y, Benhamou LE, Sarthou P *et al.* Restoration of surface IgM-mediated apoptosis in an anti-IgM-resistant variant of

WEHI-231 lymphoma cells by HS1, a protein-tyrosine kinase substrate. *Proc. Natl. Acad. Sci. USA* 1995; **92**: 7302–7306.

124. Taniuchi I, Kitamura D, Maekawa Y, Fukuda T, Kishi H and Watanabe T. Antigen-receptor induced clonal expansion and deletion of lymphocytes are impaired in mice lacking HS1 protein, a substrate of the antigen-receptor-coupled tyrosine kinases. *EMBO J.* 1995; **14**: 3664–3678.

125. Williams GT, Venkitaraman A, Gilmore D and Neuberger M. The sequence of the μ transmembrane segment determines the tissue specificity of the transport of immunoglobulin M to the cell surface. *J Exp Med* 1990; **171**: 947–952.

126. Grupp SA, Campbell K, Mitchell RN, Cambier JC and Abbas AK. Signaling-defective mutants of the B lymphocyte antigen receptor fail to associate with Ig-α and Ig-β/γ. *J. Biol. Chem.* 1993; **268**: 25776–25779.

127. Stevens TL, Blum JH, Foy SP, Matsuuchi L and DeFranco AL. A mutation of the μ transmembrane that disrupts endoplasmic reticulum retention. *J. Immunol.* 1994; **152**: 4397–4406.

128. Grupp SA, Mitchell RN, Schreiber KL, McKean DJ and Abbas AK. Molecular mechanisms that control expression of the B lymphocyte antigen receptor complex. *J. Exp. Med.* 1995; **181**: 161–168.

129. Cherayil BJ, MacDonald K, Waneck GL and Pillai H. Surface transport and internalization of the membrane IgM H chain in the absence of the Mb-1 and B29 proteins. *J. Immunol.* 1993; **151**: 11–19.

130. Haas IG and Wabl M. Immunoglobulin heavy chain binding protein. *Nature* 1983; **306**: 387–389.

131. Mitchell RN, Shaw AC, Weaver YK, Leder P and Abbas AK. Cytoplasmic tail deletion converts membrane immunoglobulin to a phosphatidylinositol-linked form lacking signaling and efficient antigen internalization functions. *J. Biol. Chem.* 1991; **266**: 8856–8860.

132. Wienends J and Reth M. Glycosyl–phosphatidylinositol linkage as a mechanism for cell-surface expression of immunoglobulin D. *Nature* 1992; **356**: 246–248.

133. Webb CF, Nakai C and Tucker PW. Immunoglobulin receptor signaling depends on the carboxyl terminus but not the heavy-chain class. *Proc. Natl. Acad. Sci. USA* 1989; **86**: 1977–1981.

134. Parikh VS, Nakai C, Yokota SJ, Bankert RB and Tucker PW. COOH terminus of membrane IgM is essential for an antigen-specific induction of some but not all early activation events in mature B cells. *J. Exp. Med.* 1991; **174**: 1103–1109.

135. Parikh VS, Bishop GA, Liu KJ, Do BT, Ghosh MR, Kim BS *et al.* Differential structure–function requirements of the transmembranal domain of the B-cell antigen receptor. *J. Exp. Med.* 1992; **176**(4): 1025–1031.

136. Patel KJ and Neuberger MS. Antigen presentation by the B cell antigen receptor is driven by the α/β sheath and occurs independently of its cytoplasmic tyrosines. *Cell* 1993; **74**: 939–946.

137. Michnoff CH, Parikh VS, Lelsz DL and Tucker PW. Mutations within the NH_2-terminal transmembrane domain of membrane immunoglobulin (Ig) M alters Igα and Igβ association and signal transduction. *J. Biol. Chem.* 1994; **269**: 24237–24244.

138. Pleiman CM, Chien NC and Cambier JC. Point mutations define a mIgM transmembrane region motif that determines intersubunit signal transduction in the antigen receptor. *J. Immunol.* 1994; **152**: 2837–2844.

139. Hutchcroft JE, Harrison ML and Geahlen RL. Association of the 72-kDa protein-tyrosine kinase PTK72 with the B cell antigen receptor. *J. Biol. Chem.* 1992; **267**: 8613–8619.

140. Kim KM, Alber G, Weiser P and Reth M. Differential signaling through the Ig-α and Ig-β components of the B cell antigen receptor. *Eur. J. Immunol.* 1993; **23**: 911–916.

141. Shaw AC, Mitchell RN, Weaver YK, Campos-Torres J, Abbas AK and Leder P. Mutations of immunoglobulin transmembrane and cytoplasmic domains: effects on intracellular signaling and antigen presentation. *Cell* 1990; **63**: 381–392.

142. Mitchell RN, Barnes KA, Grupp SA, Sanchez M, Misulovin Z, Nussenzweig MC *et al.* Intracellular targeting of antigens internalized by membrane immunoglobulin in B lymphocytes. *J. Exp. Med.* 1995; **181**: 1705–1714.

143. Choquet D, Ku G, Cassard S, Malissen B, Korn H, Fridman WH *et al.* Different patterns of calcium signaling triggered through two components of the B lymphocyte antigen receptor. *J. Biol. Chem.* 1994; **269**: 6491–6497.

144. Taddie JA, Hurley TR, Hardwick BS, Sefton BM. Activation of B-and T-cells by the cytoplasmic domains of the B-cell antigen receptor proteins Ig-α and Ig-β. *J. Biol. Chem.* 1994; **18**: 13529–13535.

145. Irving BA and Weiss A. The cytoplasmic domain of the T cell receptor ζ chain is sufficient to couple to receptor-associated signal transduction pathways. *Cell* 1991; **64**: 891–901.

146. Romeo C and Seed B. Cellular immunity to HIV activated by CD4 fused to T cell or Fc receptor polypeptides. *Cell* 1991; **64**: 1037–1046.

147. Letourneur F and Klausner RD. T-cell and basophil activation through the cytoplasmic tail of T-cell-receptor ζ family proteins. *Proc. Natl. Acad. Sci. USA* 1991; **88**: 8905–8909.

148. Lin J-J and Justement LB. The MB-1/B29 heterodimer couples the B cell antigen receptor to multiple src family protein tyrosine kinases. *J. Immunol.* 1992; **149**: 1548–1555.

149. Clark MR, Campbell KS, Kazlauskas A, Johnson SA, Hertz M, Potter TA *et al.* The B cell antigen receptor complex: association of Ig-α and Ig-β with distinct cytoplasmic effectors. *Science* 1992; **258**: 123–126.

150. Clark MR, Johnson SA, Cambier JC. Analysis of Ig-α – tyrosine kinase interaction reveals two levels of binding specificity and tyrosine phosphorylated Ig-α stimulation of Fyn activity. *EMBO J.* 1994; **13**: 1911–1919.

151. Songyang Z, Shoelson SE, Chaudhuri M, Gish G, Pawson T, Hasert WG *et al.* SH2 domains recognize specific phosphopeptide sequences. *Cell* 1993; **72**: 767–778.

152. Songyang Z, Shoelson SE, McGlade J, Oliver P, Pawson T, Bustelo XR *et al.* Specific motifs recognized by the SH2 domains of Csk, 3BP2, fps/fes, GRB-2, HCP, SHC, Syk, and Vav. *Mol. Cell Biol.* 1994; **14**: 2777–2785.

153. Pleiman CM, Abrams C, Timson-Gauen L, Bedzyk W, Jongstra J, Shaw AS et al. Distinct domains within p53/56^{lyn} and p59^{fyn} bind nonphosphorylated and phosphorylated Ig-α. Proc. Natl. Acad. Sci. USA 1994; **91**: 4268–4272.

154. Kurosaki T, Johnson SA, Pao L, Sada K, Yamamura H and Cambier JC. Role of the syk autophosphorylation site and SH2 domains in B cell antigen receptor signaling. J. Exp. Med. 1995; **182**: 1815-1823.

155. Burg DL, Furlong MT, Harrison ML and Geahlen RL. Interactions of Lyn with the antigen receptor during B cell activation. J. Biol. Chem. 1994; **269**: 28136–28142.

156. Johnson SA, Pleiman CM, Pao L, Schneringer J, Hippen K and Cambier JC. Phosphorylated immunoreceptor signaling motifs (ITAMs) exhibit unique abilities to bind and activate Lyn and Syk tyrosine kinases. J. Immunol. 1995; **155**: 4596-4603.

157. Rowley RB, Burkhardt AL, Chao H-G, Matsueda GR and Bolen JB. Syk protein-tyrosine kinase is regulated by tyrosine-phosphorylated Igα/Igβ immunoreceptor tyrosine activation motif binding and autophosphorylation. J. Biol. Chem. 1995; **270**: 11590–11594.

158. Shiue L, Zoller MJ and Brugge JS. Syk is activated by phosphotyrosine-containing peptides representing the tyrosine- based activation motifs of the high affinity receptor for IgE. J. Biol. Chem. 1995; **270**: 10498–10502.

159. Thomas ML. The leukocyte common antigen family. Ann. Rev. Immunol. 1989; **7**: 339.

160. Yakura H, Shen F-W, Bourcet E and Boyse EA. On the function of Ly-5 in the regulation of antigen-driven B cell differentiation. J. Exp. Med. 1983; **157**: 1077–1088.

161. Yakura H, Kawabata I, Shen F-W and Katagiri M. Selective inhibition of lipopolysaccharide-induced polyclonal IgG response by monoclonal Ly-5 antibody. J. Immunol. 1986; **136**: 2729–2733.

162. Yakura H, Kawabata I, Ashida T and Katagiri M. Differential regulation by Ly-5 and Lyb-2 of IgG production induced by lipopolysaccharide and B cell stimulatory factor-1 (IL-4). J. Immunol. 1988; **141**: 875–880.

163. Ogimoto M, Mizuno K, Tate G, Takahashi H, Katagiri M, Hasegawa K et al. Regulation of lipopolysaccharide- and IL- 4-induced immunoglobulin heavy chain gene activation: differential roles for CD45 and Lyb-2. Int. Immunol. 1992; **4**: 651–659.

164. Mittler RS, Greenfield RS, Schacter BZ, Richard NF and Hoffamann MK. Antibodies to the common leukocyte antigen (T200) inhibit an early phase in the activation of resting human B cells. J. Immunol. 1987; **138**: 3159–3166.

165. Gruber MF, Bjorndahl JM, Nakamura S and Fu SM. Anti-CD45 inhibition of human B cell proliferation depends on the nature of activation signals and the state of B cell activation. J. Immunol. 1989; **142**: 4144–4152.

166. Ledbetter JA, Tonks NK, Fischer EH and Clark EA. CD45 regulates signal transduction and lymphocyte activation by specific association with receptor molecules on T or B cells. Proc. Natl. Acad. Sci. USA 1988; **85**: 8628–8632.

167. Justement LB, Campbell KS, Chien NC and Cambier JC. Regulation of B cell antigen receptor signal transduction and phosphorylation by CD45. Science 1991; **252** (5014): 1839–1842.

168. Kawauchi K, Lazarus AH, Rapoport MJ, Harwood A and Cambier JC. Tyrosine kinase and CD45 tyrosine phosphatase activity mediate p21^{ras} activation in B cell stimulated through the antigen receptor. J. Immunol. 1994; **152**: 3306–3316.

169. Kishihara K, Penninger J, Wallace VA, Kündiz TM, Kazuhiro K, Wakeham A et al. Normal B lymphocyte development but impaired T cell maturation in CD45-exon6 protein tyrosine phosphatase-deficient mice. Cell 1993; **74**: 143–156.

170. Ogimoto M, Katagiri T, Mashima K, Hasegawa K, Mizuno K, Yakura H. Antigen receptor-initiated growth inhibition is blocked in CD45-loss variants of a mature B lymphoma, with limited effects on apoptosis. Eur. J. Immunol. 1995; **25**: 2265–2271.

171. Kim K-M, Alber G, Weiser P, Reth M. Signalling function of the B-cell antigen receptors. Immum Rev. 1993; **132**: 125–146.

172a. Yakura H. The role of protein tyrosine phosphatases in lymphocyte activation and differentiation. Crit. Rev. Immunol. 1994; **14**: 311–336.

172b. Pao LI, Bedzyk W, Persin C, Reth M, Cambier JC. Molecular targets of CD45 regulation of B cell antigen receptor signal transduction. J. Immunol. (in press).

173. Ogimoto M, Katagiri T, Mashima K, Hasegawa K, Mizuno K and Yakura H. Negative regulation of apoptotic death in immature B cells by CD45. Int. Immunol. 1994; **6**: 647–654.

174. Mustelin T and Altman A. Dephosphorylation and activation of the T cell tyrosine kinase pp56^{lck} by the leukocyte common antigen (CD45). Oncogene 1990; **5**: 809–813.

175. Mustelin T, Coggeshall KM and Altman A. Rapid activation of the T-cell tyrosine protein kinase pp56^{lck} by the CD45 phosphotyrosine phosphatase. Proc. Natl. Acad. Sci. USA 1989; **86**: 6302–6306.

176. Mustelin T, Pessa-Morikawa T, Autero M, Gassmann M, Andersson LC, Gahmberg CG et al. Regulation of the p59^{fyn} protein tyrosine kinase by the CD45 phosphotyrosine phosphatase. Eur. J. Immunol. 1992; **22**: 1173–1178.

177. Ostergaard HL, Shackelford DA, Hurley TR, Johnson P, Hyman R, Sefton BM et al. Expression of CD45 alters phosphorylation of the lck-encoded tyrosine protein kinase in murine lymphoma T-cell lines. Proc. Natl. Acad. Sci. USA 1989; **86**: 8959–8963.

178. Hurley TR, Hyman R and Sefton BM. Differential effects of expression of the CD45 tyrosine protein phosphatase on the tyrosine phosphorylation of the lck, fyn, and c-src tyrosine protein kinases. Mol. Cell Biol. 1993; **13**: 1651–1656.

179. Cahir McFarland ED, Hurley TR, Pingel JT, Sefton BM, Shaw A and Thomas ML. Correlation between Src family member regulation by the protein-tyrosine-phosphatase CD45 and transmembrane signaling through the T-cell receptor. Proc. Natl. Acad. Sci. USA 1993; **90**: 1402–1406.

180. Schraven B, Kirchgessner H, Gaber B, Samstag Y and Meuer S. A functional complex is formed in human T lymphocytes between the protein tyrosine phosphatase CD45, the protein tyrosine kinase p56^{lck} and pp32, a

possible common substrate. *Eur. J Immunol.* 1991; **21**: 2469–2477.

181. Guttinger M, Gassmann M, Amrein KE, Burn P. CD45 phosphotyrosine phosphatase and p56^{lck} protein tyrosine kinase: a functional complex crucial in T cell signal transduction. *Int. Immunol.* 1992; **4**: 1325–1330.

182. Shiroo M, Goff L, Biffen M, Shivnan E, Alexander D. CD45 tyrosine phosphatase-activated p59^{fyn} couples the T cell antigen receptor to pathways of diacylglycerol production, protein kinase C activation and calcium influx. *EMBO J.* 1992; **11**: 4887–4897.

183. Biffen M, McMichael-Phillips D, Larson T, Venkitaraman A and Alexander D. The CD45 tyrosine phosphatase regulates specific pools of antigen receptor-associated p59^{fyn} and CD4-associated p56^{lck} tyrosine kinases in human T- cells. *EMBO J.* 1994; **13**: 1920–1929.

184. Burns CM, Sakaguchi K, Appella E, Ashwell JD. CD45 regulation of tyrosine phosphorylation and enzyme activity of *src* family kinases. *J. Immunol.* 1994; **269**: 13594–13600.

185. Brown VK, Ogle EW, Burkhardt AL, Rowley RB, Bolen JB and Justement LB. Multiple components of the B cell antigen receptor complex associate with the protein tyrosine phosphatase, CD45. *J. Biol. Chem.* 1994; **269**(25): 17238–17244.

186a. Lin J-J, Brown VK, Justement LB. Regulation of basal tyrosine phosphorylation of the B cell antigen receptor complex by the protein tyrosine phosphatase, CD45. *J. Immunol.* 1992; **149**: 3182–3190.

186b. Pao LI, Furlonger C, Paige CJ, Cambier JC. CD45 regulated Lyn activation and recruitment to BCR following antigen stimulation. *J. Immunol.* 1996 (in press).

187. Benatar T, Cassetti R, Furlonger C, Kamalia N, Mak T and Paige CJ. Immunoglobulin mediated signal transduction in CD45 deficient mice. *J. Exp. Med.* 1996; **183**: 329-334.

188. Boue DR and Lebien TW. Structural characterization of the human B lymphocyte-restricted differentiation antigen CD22. *J. Immunol.* 1988; **140**: 192–199.

189. Schwartz-Albiez R, Dörken B, Monner DA and Moldenhauer G. CD22 antigen: biosynthesis, glycosylation and surface expression of a B lymphocyte protein involved in B cell activation and adhesion. *Int. Immunol.* 1991; **3**: 623–633.

190. Clark EA. CD22, a B cell-specific receptor, mediates adhesion and signal transduction. *J. Immunol.* 1993; **150**: 4715–4718.

191. Leprince C, Draves KE, Geahlen RL, Ledbetter JA and Clark EA. CD22 associates with the human surface IgM-B-cell antigen receptor complex. *Proc Natl. Acad. Sci. USA* 1993; **90**: 3236–3240.

192. Pezzutto A, Dörken G, Moldenhauer G and Clark EA. Amplification of human B cell activation by a monoclonal antibody to the B cell-specific antigen CD22, Bp 130/140. *J. Immunol.* 1987; **138**: 2793–2799.

193. Pezzutto A, Rabinovitch PS, Dörken B, Moldenhauer G and Clark EA. Role of the CD22 human B cell antigen in B cell triggering by anti-immunoglobulin. *J. Immunol.* 1988; **140**: 1791–1795.

194. Schulte RJ, Campbell M-A, Fischer WH and Sefton BM. Tyrosine phosphorylation of CD22 during B cell activation. *Science* 1992; **258**: 1001–1004.

195. Law C-L, Sidorenko SP and Clark EA. Regulation of lymphocyte activation by the cell-surface molecule CD22. *Immunol. Today* 1994; **15**: 442–449.

196. Campbell M-A and Klinman NR. Phosphotyrosine-dependent association between CD22 and protein tyrosine phosphatase 1C. *Eur. J. Immunol* 1995; **25**: 1573–1579.

197. Doody GM, Justement LB, Delibrias CC, Matthews RJ, Lin J, Thomas ML *et al.* A role in B cell activation for CD22 and the protein tyrosine phosphatase SHP. *Science* 1995; **269**: 242–244.

198. D'Ambrosio D, Hippen KL, Minskoff SA, Mellman I, Pani G, Siminovitch KA *et al.* Recruitment and activation of PTP1C in negative regulation of antigen receptor signaling by FcγRIIB1. *Science* 1995; **268**: 293–297.

199. Chan PL and Sinclair NRSC. Regulation of the immune response: V. An analysis of the function of the Fc portion of antibody in suppression of an immune response with respect to interaction with components of the lymphoid system. *Immunology* 1971; **21**: 967–981.

200. Phillips NE and Parker DC. Fc-dependent inhibition of mouse B cell activation by whole anti-μ antibodies. *J Immunol* 1983; **130**: 602–606.

201. Bijsterbosch MK and Klaus GGB. Crosslinking of surface immunoglobulin and Fc receptors on B lymphocytes inhibits stimulation of inositol phosphate breakdown via the antigen receptors. *J Exp Med* 1985; **162**: 1825–1827.

202. Choquet D, Partisetti M, Amigorena S, Bonnerot C, Fridman WH and Korn H. Crosslinking of IgG receptors inhibits membrane immunoglobulin-stimulated calcium influx in B lymphocytes. *J. Cell Biol.* 1993; **121**: 355–363.

203. Wilson HA, Greenblatt D, Taylor CW, Putney JW, Tsien RY, Finkelman FD *et al.* The B lymphocyte calcium response is diminished by membrane immunoglobulin cross-linkage to the Fcγ receptor. *J. Immunol.* 1987; **138**: 1712–1718.

204. Amigorena S, Bonnerot C, Drake JR, Choquet D, Hunziker W, Guillet J-G *et al.* Cytoplasmic domain heterogeneity and functions of IgG Fc receptors in B lymphocytes. *Science* 1992; **256**: 1808–1812.

205. Muta T, Kurosaki T, Misulovin Z, Sanchez M, Nussenzweig MC and Ravetch JV. A 13-amino-acid motif in the cytoplasmic domain of FcγRIIB modulates B-cell receptor signalling. *Nature* 1994; **368**: 70–74.

206. Cooper NR, Moore MD and Nemerow GR. Immunobiology of CR2, the B lymphocyte receptor for Epstein–Barr virus and the C3d complement fragment. *Annl. Rev. Immunol.* 1988; **6**: 85–113.

207. Ahearn JM and Fearon DT. Structure and function of the complement receptors, CR1 (CD35) and CR2 (CD21). *Ad. Immunol.* 1989; **46**: 183–218.

208. Tedder TF and Isaacs CM. Isolation of cDNAs encoding the CD19 antigen of human and mouse B lymphocytes: a new member of the immunoglobulin superfamily. *J. Immunol.* 1989; **143**: 712–719.

209. Bradbury L, Kansas G, Levy S, Evans RL and Tedder TF. The CD19/CD21 signal transducing complex of B

lymphocytes includes the target of antiproliferative antibody-1 and Leu-13 molecules. *J. Immunol.* 1992; **149**: 2841–2850.

210. Matsumoto AK, Martin DR, Carter RH, Klickstein LB, Ahearn JM and Fearon DT. Functional dissection of the CD21/CD19/TAPA-1/Leu-13 complex of B lymphocytes. *J. Exp. Med.* 1993; **178**: 1407–1417.

211. Takahashi S, Doss C, Levy S and Levy R. TAPA-1, the target of an antiproliferative antibody, is associated on the cell curface with the Leu-13 antigen. *J. Immunol.* 1990; **145**: 2207–2213.

212. Carter RH, Spycher MO, Ng YC, Hoffman R and Fearon DT. Synergistic interaction between complement receptor type 2 and membrane IgM on B lymphocytes. *J. Immunol.* 1988; **141**: 457–463.

213. Carter RH and Fearon DT. Polymeric C3dg primes human B lymphocytes for proliferation induced by anti-IgM. *J. Immunol.* 1989; **143**: 1755–1760.

214. Pezzutto A, Dorken B, Rabinovitch PS, Ledbetter JA, Moldenhauer G and Clark EA. CD19 monoclonal antibody HD37 inhibits anti-immunoglobulin-induced B cell activation and proliferation. *J. Immunol.* 1987; **138**: 2793–2802.

215. Callard RE, Rigley KP, Smith SH, Thurstan S and Shields JG. CD19 regulation of human B cell responses. *J. Immunol.* 1992; **148**: 2983–2987.

216. Carter RH and Fearon DT. CD19: lowering the threshold for antigen receptor stimulation of B lymphocytes. *Science* 1992; **256**: 105–107.

217. Hebell T, Ahearn JM and Fearon DT. Suppression of the immunc response by a soluble complement receptor of B lymphocytes. *Science* 1991; **254**: 102–105.

218. Rickert RC, Rajewsky K and Roes J. Impairment of T-cell-dependent B-cell responses and B-1 cell development in CD19⁻ deficient mice. *Nature* 1995; **376**: 352–355.

219. Engel P, Zhou L-J, Ord DC, Sato S, Koller B and Tedder TF. Abnormal B lymphocyte development, activation, and differentiation in mice that lack or overexpress the CD19 signal transduction molecule. *Immunity* 1995; **3**: 39–50.

220. Uckun FM, Burkhardt AL, Jarvis L, Jun X, Stealey B, Dibirdik I *et al.* Signal transduction through the CD19 receptor during discrete developmental stages of human B-cell ontogeny. *J. Biol. Chem.* 1993; **268**: 21172–21184.

221. Chalupny NJ, Kanner SB, Schieven GL, Wee S, Gilliland LK, Aruffo A *et al.* Tyrosine phosphorylation of CD19 in pre-B and mature B cells. *EMBO J* 1993; **12**: 2691–2696.

222. van Noesel CJM, Lankester AC, van Schijndel GMW and van Lier RAW. The CR2/CD19 complex on human B cells contains the *src*-family kinase *Lyn. Int. Immunol.* 1993; **5**: 699–705.

223. Tuveson DA, Carter RH, Soltoff SP and Fearon DT. CD19 of B cells as a surrogate kinase insert region to bind phosphatidylinositol 3-kinase. *Science* 1993; **260**: 986–989.

224. Weng W-K, Jarvis L and LeBien TW. Signaling through CD19 activates Vav/mitogen-activated protein kinase pathway and induces formation of a CD19/Vav/phosphatidylinositol 3-kinase complex in human B cell precursors. *J. Biol. Chem.* 1994; **269**: 32514–32521.

225. Rogers J and Wall R. Immunoglobulin RNA rearrangements in B lymphocyte differentiation. *Ad Immunol.* 1984; **35**: 39–59.

226. Neuberger MS, Patel KJ, Dariavach P, Nelms K, Peaker CJG and Williams GT. The mouse B-cell antigen receptor: definition and assembly of the core receptor of the five immunoglobulin isotypes. *Immun. Rev.* 1993; **132**: 147–161.

2

Antigen Receptor Structure and Signaling in T Cells

David A. Taylor-Fishwick, Carl H. June and Jeffrey N. Siegel

Immune Cell Biology Program, Naval Medical Research Institute, Bethesda, Maryland, USA

INTRODUCTION

The T cell antigen receptor (TCR) is a complex multicomponent structure which, when stimulated by antigen, initiates a series of biochemical changes within the T cell that alters the activation state and function of the cell. As with other functional receptors located on the cell plasma membrane, the TCR complex has the ability both to recognize ligand (processed antigen) and to initiate signal transducing events. Intracellular changes initiated by the TCR are fundamental for T lymphocyte development decisions and the regulation of immune responses. For example, the diverse T cell biology of activation, anergy and apoptosis can all be a functional consequence of TCR stimulation, these diverse functions are defined as follows.

1. *T cell activation*—the induction of T cell effector function, proliferation and cytokine production.
2. *T cell anergy*—the inability of a T cell to respond to a subsequent antigen stimulation.
3. *T cell apoptosis*—the induction of T cell death through an intrinsic cell death program (Chapter 12).

In this chapter we first review the complex structure of the T cell antigen receptor and describe the transmembrane molecules associated with the receptor. We then discuss the current model of how stimulation of this structure initiates the signal transduction pathways required to activate the nuclear events associated with T cell activation. Finally, we consider some of the receptor-related phenomena which may regulate the functional consequences of T cell receptor stimulation, that is, activation, anergy and apoptosis.

T CELL RECEPTOR STRUCTURE

The T cell antigen receptor (TCR) is a multichain complex of at least six distinct polypeptides (Figure 2.1). Overall the structure can be considered as two functional units: a region that is variable allowing the receptor to interact with a diverse spectrum of antigens; and a region that is constant which initiates a signal transduction cascade. Through this regional specialization the structure of the TCR incorporates the distinct features required for antigen recognition, namely a plasticity for antigen recognition along with invariant signal transduction.

Antigen recognition is achieved by highly polymorphic alpha and beta chains (Ti) which associate in a non-covalent manner with the CD3 complex. Consistent with the concept that Ti is dedicated to antigen recognition, the chains of the alpha/beta heterodimer have short cytoplasmic domains, only five amino acids in length. The Ti is thus unlikely to initiate directly the intracellular phosphorylation events which are characteristic of TCR signal transduction. The alpha and beta chains are highly specialized for antigen recognition. Both the alpha and beta chains are products of genes that have multiple segments. In the developing thymocyte the gene segments are uniquely arranged for each T cell creating a distinct TCR. This combinational capacity of the multiple gene segments termed variable (V), diversity (D) and joining (J) (only V/J for the alpha chain) together

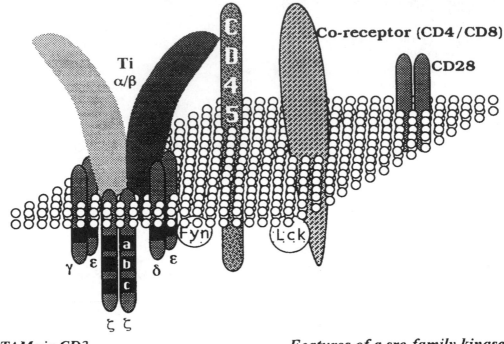

ITAMs in CD3

Consensus	γ	δ	ε	ζ–a	b	c
D/E	D	D	N	N	E	D
X	Q	Q	P	Q	G	G
X	L	V	D	L	L	L
Y	**Y**	**Y**	**Y**	**Y**	**Y**	**Y**
X	Q	Q	E	N	N	Q
X	P	P	P	E	E	G
L/I	**L**	**L**	**I**	**L**	**L**	**L**
X	K	R	R	N	Q	S
X	D	D	K	L	K	T
X	R	R	G	G	D	A
X	E	D	Q	R	K	T
X	D	D	R	R	M	K
X	D	A		E	Λ	D
X	Q	Q	L	E	E	T
					A	
Y	**Y**	**Y**	**Y**	**Y**	**Y**	**Y**
X	S	S	S	D	S	D
X	H	H	G	V	E	A
L/I	**L**	**L**	**L**	**L**	**I**	**L**

Features of a src-family kinase

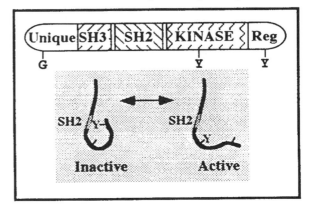

Figure 2.1 Structure of the T cell receptor and associated transmembrane proteins. The locations of the ITAMs are shown (black) and their amino acid sequences are given in the left panel (single letter code). The right panel illustrates the features of the Src-family kinases. The inset depicts the mode of activation for this kinase family

with the addition or deletion of bases at sites of joining (N-region diversity) generates the potential for at least 10^{10} binding pockets that can be formed by the disulfide-linked alpha/beta chains of the TCR. This colossal spectrum of TCRs within a T cell population enables the immune system to respond effectively to the myriad of antigens it may encounter. An additional subset of T cells is defined by gamma/delta T cells. The TCRs of these T cells also have polymorphic Ti-chains but they arise from different genes, the gamma and delta genes. Gamma/delta T cells may be specialized to recognize a distinct class of antigens compared to alpha/beta T cells [1]. Both lipid and non-lipid antigens can be presented to the alpha/beta TCR, either as processed peptides by class I or class II major histocompatibility complex (MHC) molecules expressed on the surface of antigen presenting cells (APC), or as lipids by the MHC-related molecule CB1b [2]. Peptides in complex with MHC bind to the polymorphic Ti and induce crosslinking of the TCR and initiation of an intracellular signal.

The signal transducing function of the TCR is mediated by the invariant chains of the CD3 complex: the gamma, delta, epsilon and zeta chains [3,4]. All of these chains are single transmembrane proteins. In contrast to the chains of the Ti, the CD3 chains have larger cytoplasmic domains, indeed more than 90% of the zeta chain is intracellular. The interchain arrangement of these proteins is as non-covalent heterodimers of gamma/epsilon and delta/epsilon and either a disulfide-linked zeta homodimer or a less common zeta/eta heterodimer. The eta chain is an alternately spliced product of the zeta chain that is present in up to 10% of mouse TCR but less frequently in human TCR [5]. Amino acid sequence analysis of these chains has revealed that they all share an homologous motif that consists of two YXXL(I) paired sequences [6] separated by six to eight amino acids (Figure 2.1). This arrangement is critical for function [7]. Historically, the motif has been termed variously, TAM (tyrosine-based activation motif), ARH-1 (antigen receptor homology) domain or ARAM (antigen recognition activation motif), and is present once in each of the gamma, delta and epsilon chains with three copies in the zeta chain (two in the eta chain). A consensus term for this motif, ITAM (Immuno-receptor Tyrosine-based Activation Motif) has recently been proposed [8]. The importance of the ITAM sequence for TCR signaling has been most clearly shown in studies using a chimeric surface protein whose cytoplasmic tail consisted solely of an isolated ITAM. Crosslinking the extracellular domain of this chimera was sufficient to induce PTK activity and TCR-mediated effector functions [9, 10]. Interestingly, the ITAM is also found in the cytoplasmic tail of a number of other receptors including the surface immunoglobulin associated chains Ig-alpha and Ig-beta (Chapter 1) and the gamma and beta subunits of the Fc receptor, FcRI [11]. This suggests that common mechanisms exist between receptors to couple them to cytoplasmic signal transducing molecules. The ITAM may be an important motif in viral transformation of lymphocytes since it is required for the *in vitro* infection and propagation of the Bovine Leukemia Virus [12].

Transmembrane Molecules Associated with the TCR

In addition to the polypeptide chains that make up the T cell receptor/CD3 complex, a number of other transmembrane proteins are also necessary for the physiologic activation of the T cell. Many of these proteins form the subject of other chapters in this book and will not be considered in depth in this chapter. These associated chains include the TCR coreceptors, CD4 and CD8 the cytoplasmic domains of which bind the Src-family protein kinase p56^{lck} (see below). Both CD4 and CD8, expressed on helper and cytotoxic T cells respectively, bind to the non- polymorphic regions of the MHC complex (class I for CD8 and class II for CD4). By binding to MHC the coreceptors help colocalize Lck into the microenvironment of the TCR. Additionally, the coreceptors along with cell surface adhesion molecules [13] help the TCR to overcome its low affinity for peptide-loaded MHC. These additional cell surface interactions increase avidity of the TCR/MHC interaction. The phosphatase CD45 (Chapter 8) has been shown to be required for T cell activation and it may be involved in activating the Src-family kinases, Lck and Fyn (see below). Finally, the successful activation of the unprimed T cell requires both stimulation through the TCR and activation of a second or costimulatory pathway. This second stimulation is provided principally by the engagement of the costimulatory molecule CD28 (Chapter 7) by ligands of the B7 family (reviewed in [14]).

T CELL RECEPTOR SIGNAL TRANSDUCTION

Crosslinking of the TCR is an essential step in the initiation of T cell activation and TCR signal transduction [15]. TCR crosslinking generates a signal which is then transduced from the plasma membrane to the nucleus of the T cell. In broad outline, stimulation of the TCR induces the activity of protein tyrosine kinases which in turn activate the membrane signaling molecules, PLCγ1 and Ras. Once activated, Ras and PLCγ1 transduce the signal from the plasma membrane to cytoplasmic proteins. Within the cytoplasm, signaling cascades involving proteins kinases and phosphatases are initiated which ultimately converge at the nucleus. Ras acts to switch on a number of important cytoplasmic effectors while PLCγ1 leads to an elevation in the intracellular calcium concentration. These cytoplasmic changes activate a number of nuclear factors to regulate the transcription of the genes involved in T cell activation. Below we consider in detail the key proteins that have been identified as components of the TCR signal transduction cascade.

Tyrosine Kinases are Essential for Early T Cell Signal Transduction

Stimulation of the TCR either through the physiological binding of antigen/MHC complexes, super-

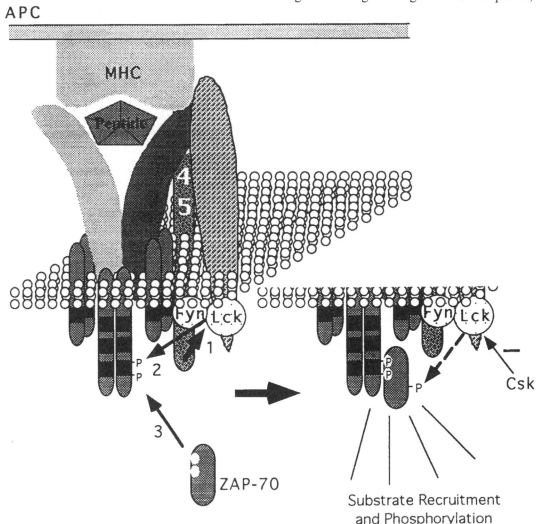

Figure 2.2 TCR induction of PTK activity. Crosslinking of the TCR induces receptor clustering. On the left, CD45-mediated activation of Src-kinase (1) induces phosphorylation of Src-kinases (2) and the subsequent recruitment of ZAP-70 (3). On the right, the binding of ZAP-70 to a phosphorylated ITAM results in its phosphorylation and activation. This subsequently results in the recruitment and activation of other signaling molecules (represented by the lines emanating from ZAP-70). Csk is a kinase that down-regulates these events

antigen/MHC complexes or by crosslinking the CD3 complex with monoclonal antibodies induces the phosphorylation on tyrosine residues of a number of intracellular substrates (Figure 2.2). These substrates include the chains of the CD3 complex. The induction of tyrosine phosphorylation occurs within 5 seconds of TCR triggering [16, 17] and is believed to be a critical proximal event. The importance of tyrosine phosphorylation for TCR activation has been shown through the use of inhibitors of protein tyrosine kinase activity, for example herbimycin A. Treatment of T cells with herbimycin A markedly diminishes the ability of the cells to respond to TCR stimulation [18]. Protein tyrosine kinase activity is therefore one of the first important signaling events to occur after TCR crosslinking.

The Significance of Tyrosine Phosphorylation

An important insight into the mechanism by which tyrosine phosphorylation promotes T cell activation was provided by the discovery that a number of cytoplasmic signaling proteins contain a sequence motif which enables them to bind to phosphotyrosine. This sequence motif, termed the Src-homology 2 (SH2) domain is characterized by homology to an approximately 100 amino acid region present in $p60^{src}$, the first Src-family PTK to be identified. SH2 domains are able to interact with specific phosphotyrosines through binding pockets which are created by the amino acids surrounding the phosphotyrosine. Through the use of SH2 affinity columns, and a degenerate phosphopeptide library, the specificity of many SH2-containing proteins that are important for TCR signal transduction has been determined [19, 20]. Signaling proteins with SH2 domains are thus able to interact with phosphorylated moieties in the TCR by binding specific phosphotyrosine residues.

Protein Tyrosine Kinases are Involved in TCR Signaling

The initial event in T cell activation, that is membrane aggregation of the receptor molecules, results in the association of the TCR with CD4/CD8 coreceptors. By bringing together the associated Src-family kinases, receptor aggregation leads to tyrosine phosphorylation. The Src-family tyrosine kinases $p56^{lck}$ and $p59^{fyn}$ are among the earliest kinases to be activated following TCR stimulation [21]. Generically, the Src-family members have a number of characteristic domains that effect their cellular localization and activation. Structurally, all Src-family kinases have a unique N-terminal domain that defines the individual kinase and several conserved regions (Figure 2.1). One SH2 domain and one SH3 domain are present in the conserved region of the kinase along with a kinase domain and a C-terminal regulatory domain that contains a major in vivo phosphorylation site. The phosphorylation state of the tyrosine in the regulatory domain in believed to be important in controlling the kinase activity (see below). In addition, Src-family kinases are anchored to the inner surface of the plasma membrane by an N-terminal myristic acid. Like SH2 domains, SH3 domains are also found in other signaling proteins. SH3 domains facilitate protein–protein interactions by binding to proline-rich protein domains. Lck and Fyn are thus versatile mediators of TCR triggering because of their potential for subsequent protein interaction with other signaling molecules through SH2 and SH3 domains.

A significant body of data exists to suggest that both Lck and Fyn are involved in T cell activation. Stimulation of the T cell receptor induces the protein tyrosine kinase activity of Fyn by two- to four-fold [22]. Moreover, Fyn co-immunoprecipitates with CD3 [23], albeit with low stoichiometry, and has been shown to associate with the zeta and eta chains and the epsilon and gamma chains of the CD3 complex [24]. In activated T cells, Fyn associates with two tyrosine phosphorylated proteins of 82 and 116 kDa [25]. Overexpression of an activated form of Fyn in a T cell hybridoma increased the sensitivity of the cells to TCR stimulation [26]. Additionally cotransfection of the genes for Fyn, PLCγ1 and the zeta chain of the TCR into the fibroblast cell line, COS, has identified Fyn as being important for coupling TCR-zeta to the PLCγ1 activation pathway [27] (discussed below). Clearly, Fyn has a role in T cell receptor signaling, however its function does not appear to be essential. Surprisingly, signal transduction in the mature peripheral T cells of mice lacking Fyn as a result of homologous recombination was near normal. In contrast, mature single positive thymocytes (CD4$^+$CD8$^-$ and CD4$^-$CD8$^+$) from these mice had abnormal responses to TCR stimulation [28, 29]. The expression of Fyn increased in mature single positive thymocytes when compared to immature double positive thymocytes. This increase in Fyn expression correlates with an increase in the response of thymocytes to TCR stimulation [30]. These results indicate that Fyn is involved in TCR signal transduction, however Fyn is not an essential kinase for the development of T cells for signaling of mature peripheral T cells. It is likely that other Src-family kinases replace the function of Fyn in Fyn knockout mice.

Unlike Fyn, Lck is crucial for thymocyte development since mice that lack Lck, or are transgenic for a kinase-deficient Lck, have an early arrest in their thymocyte development. The few T cells that do mature, and emigrate to the periphery, exhibit diminished responses to TCR stimulation [31, 32]. Lck is a lymphoid-specific PTK that associates with the coreceptors CD4 and CD8 through its unique N-terminus domain. This domain has two cysteine-rich regions which allow it to interact with a cysteine-containing region in the cytoplasmic tails of CD4 and CD8 [33]. In addition to its kinase activity, the non-catalytic domains of Lck are also important in TCR signaling [34]. Lck has a high stoichiometry for the TCR coreceptors and crosslinking of CD4 with antibodies results in autophosphorylation of Lck and an increased kinase activity for other cellular substrates [35]. However, Lck may also have an intrinsic ability to interact with the zeta chain and CD3 since overexpression of an active mutant of Lck in a CD4/CD8-deficient T cell hybridoma was still able to potentiate TCR signal transduction [36]. Moreover, Lck has been shown to associate with other surface receptors, for example IL-2 beta chain and CD28, where it is also implicated in the activation of protein tyrosine phosphorylation.

The Protein Tyrosine Phosphatase CD45 Regulates the Activity of the SRC-Family Kinases

Src-family kinases are therefore important for the intracellular PTK activity which follows TCR stimulation. Moreover, Fyn or Lck is required for the phosphorylation of the zeta chain and subsequent recruitment of another PTK, ZAP-70 (described below) suggesting that activation of Src-family PTK activity is the event which initiates the TCR

signal transduction cascade. Fyn is constitutively associated with the TCR in resting cells and Lck, by its association with the TCR coreceptor, is brought into the vicinity of the TCR either by the binding of CD4 or CD8 to the antigen presenting MHC molecules or by crosslinking of CD3 by antibody [37]. This then raises the question of what stimulates the tyrosine kinase activity of the Src-kinase once a T cell is activated? Critical to understanding the mechanism of Src-family-PTK activation are two tyrosine residues which are conserved in all Src-family kinases: one in the carboxy terminal, the other in the kinase domain. The kinase inactive form of Src-PTK is phosphorylated on the carboxy terminal tyrosine which is in the regulatory domain. This phosphotyrosine appears to interact with the autologous SH2 domain in the N-terminal region of the protein inducing a conformation change that inhibits kinase activity (Figure 2.1). Dephosphorylation of this tyrosine relaxes the protein resulting in kinase activation and autophosphorylation of the second conserved tyrosine. For Fyn and Lck, the conserved tyrosines in the autophosphorylation site are at positions 417 and 394 respectively and, for the C-terminal regulatory site, at positions 528 and 505 respectively.

For the Src-family kinase p60^{src} dephosphorylation of the tyrosine residues in the regulatory site results in a constitutively active kinase [38, 39]. Upon T cell activation the tyrosine in the regulatory site is dephosphorylated and the tyrosine at the autophosphorylated site becomes phosphorylated. Thus, the regulation of Fyn and Lck PTK activity requires a phosphatase for activation and a kinase for deactivation. The transmembrane protein phosphatase, CD45, has been implicated in the dephosphorylation event. In lymphocytes, more than 90% of the total membrane protein tyrosine phosphatase activity has been attributed to CD45. Expression of CD45 is required for T cell activation and there are several lines of evidence to indicate that it can activate Fyn and Lck. In the absence of CD45, Src-family kinases are hyperphosphorylated [40] preventing the response of T cells to CD3 crosslinking. Moreover, CD45 codistributes with Fyn and Lck on T cell activation suggesting a structural association between the two [41, 42]. In an *in vivo* system, direct tyrosine phosphorylation of CD45 induced its activity and association with Src-family PTKs [43]. Evidence suggests that the PTK, Csk (*C*-terminal *Src* k*inase*) is a regulating kinase for the Src-family PTK. Mutant mice that are Csk-deficient have increased phosphorylation and activation of Src-family kinases [44]. Moreover, overexpression of Csk inhibits Lck and Fyn activity and TCR signaling [45]. Csk is related to the Src-family PTK but lacks the conserved tyrosine residues and does not have a site for membrane anchorage. The identification of Lsk, a Csk-like molecule that is preferentially expressed in leukocytes, suggests there is a family of kinases with *C*-terminal Src kinase activity [46]. Csk, by phosphorylating the tyrosine in the *C*-terminal regulatory domain of the Src-family kinase deactivates the kinase domain.

Src-Family Kinases Activate a Second PTK Family

Several molecules have been identified as key substrates of the Src-family protein tyrosine kinases. One of these targets is the ITAM of the TCR, which for the zeta chain can involve the phosphorylation of up to six tyrosine residues (three motifs, each with two tyrosines). Identification of the two phosphotyrosines in the ITAM has led to the suggestion that the ITAM could function as a binding target for proteins that possess a tandem SH2 domain. Such a tandem SH2 domain is found in ZAP-70 (*z*eta-chain *a*ssociated *p*rotein), a 70 k Da kinase that co-immunoprecipitates with the zeta chain of the TCR in activated T cells. ZAP-70 is expressed selectively in T and NK cells [47]. Evidence for SH2 domains mediating the interaction between ZAP-70 and the zeta chain is provided by binding studies using tyrosine phosphorylated ITAMs and fusion proteins containing the ZAP-70 SH2 domains. High affinity binding was observed, only when the SH2 domains were present in tandem and not when the amino or carboxyl SH2 domain was present singly [48]. The importance of ZAP-70 in cell signal transduction is highlighted by patients with a form of severe combined immunodeficiency disease (SCID) [49–51]. These patients carry a mutation in the ZAP-70 gene which correlates with a marked reduction in T cell signal transduction [51]. ZAP-70 is a protein tyrosine kinase distinct from the Src-family kinases.

Another kinase that is closely related to ZAP-70 is p72^{syk}. However, Syk is expressed at low levels in

mature T cells and is predominantly expressed in B cells, thymocytes and myeloid cells. Thus unlike ZAP-70, Syk has a more diverse tissue distribution [52] and is unlikely to be the relevant kinase for TCR signaling in the majority of mature T cells. Binding of ZAP-70 to a phosphorylated ITAM is a kinase-independent event, but once bound, ZAP-70 becomes phosphorylated. The phosphorylation and activation of ZAP-70 kinase activity is required for TCR function [53]. Phosphorylation of ZAP-70 has been reported to require the expression of Lck, suggesting that the activating functions of the Src-family kinases following T cell stimulation occurs at two levels, that is, the recruitment of ZAP-70 to the TCR complex and the subsequent phosphorylation and activation of ZAP-70 kinase.

One of the curiosities of the TCR structure is why it contains multiple ITAMs, particularly since the expression of one motif in a chimeric transmembrane receptor appeared sufficient to initiate TCR-mediated effector functions in leukemia cells [9]. Moreover the binding of ZAP-70 either to the zeta chain or to the CD3 epsilon chain requires the tandem phosphorylation of the tyrosines in just one ITAM [54]. The simplest explanation of these observations is that the repeated ITAMs bind to multiple ZAP-70 proteins thus providing a mechanism for signal amplification. Autophosphorylation of ZAP-70 is enhanced when it binds to the multiple ITAM-containing zeta chains but not when it is bound to the gamma or epsilon chain [55]. These data suggest that the function of the tandem ITAM is to facilitate ZAP-70 autophosphorylation. However, the observation that ZAP-70 can bind to separate ITAMs with differential affinities [56] suggests the additional possibility that different ITAMs may recruit distinct signal transduction molecules. The finding that the lipid kinase PI-3K, preferentially binds to the membrane proximal ITAM in the zeta chain a (zeta-a) over the other zeta-ITAMs (zeta-b or zeta-c) (Figure 2.1) [57] also supports this model.

In summary, antigenic stimulation of T cells results in aggregation of the TCR and activation of protein tyrosine kinases of at least two distinct families: the Src-family PTKs Fyn and/or Lck and ZAP-70. Activation of ZAP-70 occurs in two distinct steps: first it is recruited to the tyrosine-phosphorylated TCR and then it is activated in an Lck-dependent manner. This cumulative PTK activity results in further membrane events that propagate the signal. The activation of two key membrane-bound signaling molecules, PLCγ1 and Ras, is discussed in the next section.

Events Induced at the Plasma Membrane by TCR-stimulated PTK Activity

Phosphorylation of the ITAMs and ZAP-70 leads to the activation of the protooncogene product Ras and activation of PLCγ1. The activation of these two signaling molecules is an essential step for T cell activation and results in a divergence of the signal generated at the TCR into two discrete signaling pathways: (i) a cascading activation of serine/threonine kinases; and (ii) an elevation in intracellular free calcium concentration (Figure 2.3). A number of proteins have been implicated as candidates to shepherd the correct signaling molecules into the vicinity of the TCR in order for the activation or RAS and PLCγ1 to occur. Col-

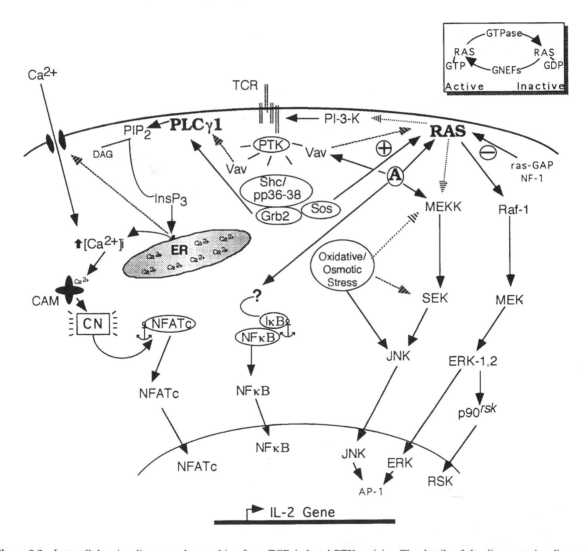

Figure 2.3   Intracellular signaling cascades resulting from TCR-induced PTK activity. The details of the divergent signaling pathways from the plasma membrane to the nucleus are given in the text. The hashed arrows are pathways that are implicated but not confirmed. **A** represents a CD28-mediated signal which coregulates SAPK, induces Vav activity and the activation of Ras in T cells [137]. The panel inset shows the Ras-regulation cycle. The anchor is representative of ankyrin sites

lectively, these protein candidates have been termed adaptor proteins.

Function of Adaptor Proteins

The mechanism by which the increased PTK activity induces the activation of Ras and PLCγ1 involves a number of adaptor molecules which undergo protein–protein interactions mediated by SH2 and SH3 domains. Several adaptor proteins have been identified and include Shc, pp36–38, Grb2, Nck, Crk and possibly Vav. While the exact sequence of events involving these proteins is still under investigation, a number of potential interactions have been identified. Grb2 is a protein comprising one SH2 domain and two SH3 domains. The SH3 domain of Grb2 binds to the proline-rich C-terminal domain of a guanine nucleotide exchange protein, Sos [58], named due to its homology to the drosophila protein, son of sevenless. The activity of Sos results in the activation of the membrane-bound guanine nucleotide binding protein, Ras (see below). In T cells the SH2 domain of Grb2 predominantly associates with a tyrosine phosphorylated protein of 36–38 kDa (pp36–38) [59, 60] and also to phosphorylated Shc proteins. Shc can associate with the zeta chain of the TCR after T cell activation [61] and has thus been thought to recruit Grb2 : Sos to the plasma membrane facilitating the activation of Ras. However, although Shc can activate Ras, it does not appear to be involved in the activation of Ras mediated by the TCR [62]. Thus while it has yet to be proved, pp36–38 may play a similar role to Shc for the activation of Ras stimulated through the TCR. Additional substrates of TCR-induced PTK activity have also been shown to bind the SH3 domain of Grb2. The function of these proteins of 75 and 116 kDa is unclear but their association underlines the potential importance of the adaptor proteins as a coupling step in the TCR signal transduction cascade. Crk and Nck are SH2-and SH3-containing proteins that are related to Grb2. Both Crk and Nck are able to bind Sos and are involved in growth factor mediated signaling. The site of activity of the 95 kDa protooncogene Vav has also been linked to the adaptor protein coupling step. Vav, which is selectively expressed in hematopoietic cells, contains both SH2 and SH3 domains, both of which appear to be involved in functional interactions. Co- immunoprecipitation studies show that Vav can associate with ZAP-70 through its SH2 domain in activated T cells [63] and with Grb2 through its SH3 domain [64]. Moreover, Vav has been shown to have guanine nucleotide exchange activity for Ras [65] though this may not be the essential property of Vav for its in vivo effects. Structurally Vav has greater homology to the guanine nucleotide exchange proteins for the related G-proteins, Rho and Rac [66]. Interestingly, in a non-T cell system, Rac has been identified to regulate signaling pathways associated with T cell activation [67]. Studies in mice that fail to express the Vav protein, as a result of homologous recombination, have emphasized the importance of Vav in T cell activation [68–70]. For example, Vav was found to be necessary for normal thymocyte development and the T cells that were able to mature in Vav-deficient mice responded poorly to antigen stimulation. These studies have confirmed Vav to function at the level of the adaptor proteins. TCR-induced PTK activity in cells was normal and the response to pharmacologic agents that stimulate T cell signaling elements which are distal to the activity of the adaptor proteins was intact. Vav and pp36–38 have also been implicated in the activation of PLCγ1 mediated pathways. Thus, a major consequence of the adaptor protein function is to couple PTK activity induced by TCR crosslinking to the activation of downstream pathways.

Regulation of Ras

The regulation of Ras activity is controlled by a number of opposing processes. Ras is a membrane-bound member of a family of small molecular weight proteins that bind guanine nucleotides. Ras is active when bound to guanine triphosphate (GTP) and inactive when bound to guanosine diphosphate (GDP). Whether Ras is active in a cell, is determined by the ratio of Ras-GTP to RAS-GDP. This ratio is set by the balance between the rate of hydrolysis of bound GTP and the rate of exchange of bound GDP for cytosolic GTP. This balance is under control of a complex array of proteins [71]. The Ras protein has an intrinsic GTPase activity, that is, an ability to hydrolyze bound GTP. This activity is further controlled by a group of proteins termed GTPase-activating proteins (GAPs). In mammalian cells two proteins with GAP activity have been identified: p120–GAP and neurofibromin. As already mentioned, mSos and possibly Vav are proteins that can regulate guanine nucleotide exchange on Ras. The importance of Ras activation in T cell signal transduction has been demonstrated most clearly in studies using mutant forms of the Ras protein that are either dominantly negative or constitutively active. Transfection of a dominant negative Ras inhibited the induction of transcriptional nuclear events including the induction of the IL-2 gene and induction of the nuclear factor NFAT [72]. In direct contrast, transfection of a constitutively active Ras could, in conjunction with a calcium signal, induce NFAT and IL-2 gene transcription. Upon T cell activation, the increased PTK activity induces the recruitment and activation of the guanine nucleotide exchange proteins, Sos and potentially Vav, leading to the activation of Ras. In addition, T cell triggering inhibits Ras-GAP activity [73]. Thus, activation of Ras in T cells is complex and occurs by a dual regulation: an increased activation signal and a decreased inhibitory signal.

Activation of PLCγ1

Which proteins link the activity of phospholipase Cγ1 (PLCγ1) to tyrosine phosphorylation is less established. However, Src-family PTKs and adaptor proteins have been implicated. Lck associates with PLCγ1 following TCR stimulation [74]. In addition, the adaptor protein pp36–38 associates with PLCγ1 [75] and Vav is required for the TCR-induced increase in intracellular calcium [70] suggesting that it may regulate the activity of PLCγ1. Active PLCγ1 cleaves the lipid, phosphatidylinositol 4,5-bis-phosphate (PIP$_2$) to yield two second messengers, lipid soluble diacylglycerol (DAG) and a water-soluble sugar, inositol 1,4,5- triphosphate (InsP$_3$). DAG activates a family of protein kinases, protein kinase C (PKC), and InsP$_3$ binds to a specific receptor on the endoplasmic reticulum. Binding of InsP$_3$ to its receptor induces a conformational change and results in the release of calcium into the cytoplasm from these intracellular stores. The activation of these two second messengers can be achieved using pharmacologic agents: phorbol esters which bind and activate

PKC and calcium inophores that artificially induce an elevation in intracellular calcium. The combination of these two agents is sufficient to induce many of the events of T cell activation. PKC may also play a role in Ras activation. However, the existence of other pathways, for example, those involving Grb2 : Sos, suggests that PKC activation may not be necessary to activate Ras in T cells. Thus the function of PKC following TCR stimulation is not clear. In contrast the elevation of intracellular calcium induced by PLCγ1 is essential for T cell activation [76].

Regulation of PI-3 Kinase

Phosphatidylinositol-3-OH kinase (PI-3K) is another key regulatory protein that is phosphorylated as a result of TCR mediated PTK activity [77]. PI-3K is the first reported dual-specificity kinase that possesses both lipid and serine kinase activity. PI-3K is rapidly activated following TCR stimulation and this activity is dependent upon the presence of Lck [78]. In addition, Ras has been shown to interact directly with PI-3K and has been implicated in the regulation of its lipid kinase activity [79]. Presently, there are at least three distinct PI-3 kinases described in mammalian cells: (i) p110/p85 consists of two subunits, an 85 kDa regulatory subunit of which there are two isoforms, p85 alpha and p85 beta and a 110 kDa catalytic subunit (p110); (ii) a PI-3K activated by G-protein βγ- subunits; and (iii) a phosphatidylinositol-specific PI-3K equivalent to Vps34p in yeast. SH2 domains are present in the p85 subunit facilitating protein–protein interaction. In distinction to the classical phosphorylation of phosphoinositides which results in phosphorylation at D1, D4 and D5 positions to form phosphoinositol (1,4,5) trisphosphate, PI-3K activity is defined by the production of D3 phosphoinositides, that is, the inositol ring is phosphorylated at the D3 position. While much is still unknown about the signal transduction pathway initiated by PI-3K, D3-lipids do appear to activate essential second messenger systems. One potential downstream target of PI-3K activation is the S6 ribosomal kinase p70^{s6k} [80] which is believed to be required for entry into cell cycle [81].

The plasma membrane events involved in TCR signal transduction can thus be summarized as follows. PTKs link the TCR to the GTP-binding protein Ras and phospholipase PLCγ1 via adaptor proteins, and other substrates, resulting in the activation of Ras and the hydrolysis of PIP_2 respectively. The hydrolysis of PIP_2 generates at least one essential arm of the TCR signal transduction cascade, that is an elevation of intracellular calcium. While other, as yet uncharacterized, substrates are also phosphorylated and shown to be associated with the adaptor molecules, the importance of both an increase in intracellular calcium and an activation of Ras for TCR activation has been established. T cells provided with a constitutively active form of Ras, by transfection, and a pharmacologically induced elevation of intracellular calcium undergo transcriptional events similar to those resulting from an activation of the TCR. The cytoplasmic events that result from the activation of these two pathways will now be considered.

Transduction of the TCR Signal through the Cytoplasm

Elevation in Intracellular Free Calcium

Two components to an increase in cellular calcium can be delineated: an intracellular and an extracellular component. Following $InsP_3$ receptor ligation, a rapid elevation in intracellular free calcium concentration occurs, consisting of a transient calcium peak lasting approximately one minute. This transient peak is followed by a lower but sustained elevation in intracellular calcium termed the plateau phase. The plateau phase is likely to result from the entry of extracellular calcium into the T cell, a process initiated by the release of intracellular calcium and possibly mediated by CIF (calcium influx factor) [82, 83]. Another mediator implicated in calcium influx is a protein termed calcium-signal modulating cyclophilin ligand (CAML) which acts downstream of the TCR and upstream of calcineurin (see below) [84]. Both intracellular and extracellular components of the cytosolic calcium influx are necessary for T cell activation. In the absence of extracellular calcium, T cells will not proliferate via TCR/CD3 stimulation. Moreover, the intracellular calcium component derived from intracellular stores is insufficient for T cell activation [85] since for IL-

2 production an elevation in intracellular calcium must be maintained for 1–2 hours [86].

Potential effectors of an increase in intracellular calcium are numerous [87]. An understanding of an essential role for calcineurin in TCR signal transduction arose from an exceptional series of studies aimed at elucidating the mechanism of action of the immunosuppressive drugs cyclosporin A (CsA) and FK506 (Chapter 23). Affinity studies of CsA and FK506 in complex with their intracellular binding proteins, showed that they both bound and inhibited the same intracellular protein, the calcium–calmodulin-regulated serine/threonine phosphatase, calcineurin (PP2B) [88, 89]. Calcineurin (CN) comprises two subunits, a calmodulin-binding 59 kDa catalytic subunit (CN-A) and an intrinsic calcium binding 19 kDa regulatory subunit (CN-B), which appears to be required for enzymatic activity. The likely importance of CN in the TCR signal transduction cascade was suggested by the observation that the concentrations of CsA and FK506 required to inhibit CN were similar to the doses required to inhibit lymphokine gene expression [90]. Moreover, the immunosuppressive potential of a panel of CsA/FK506 analogs correlated with their ability to inhibit CN [91]. Conclusive proof of the importance of calcineurin in TCR signaling came following the transfection of a CN mutant that is constitutively active. Of particular importance was the observation that co-expression of the constitutively active CN with constitutively active Ras could replace pharmacological stimuli to induce TCR-mediated nuclear events [92]. These observations showed that CN replaced the need for an increase in intracellular calcium. The phosphatase activity of CN is believed to regulate the nuclear factor, NFAT (nuclear factor of activated T cells) (Figure 2.1). NFAT is expressed predominantly in activated T cells and consists of two components: a constitutive cytoplasmic component NFATc and an induced nuclear component. In unstimulated T cells NFATc is anchored in the cytoplasm. Upon TCR stimulation the phosphatase activity of CN induces the release of NFATc from its cytoplasmic anchor allowing it to translocate to the nucleus. In the nucleus NFATc combines with the nuclear component to form an active NFAT complex that binds to the promotor of the IL-2 gene. The nuclear component of NFAT is believed to be an

AP-1 complex of Fos and Jun [93], the expression of which is regulated by Ras-mediated pathways [71]. Thus, PLCγ1 through its calcium-mediated activation of calcineurin and NFATc induces a pathway which combines at the nucleus with Ras-mediated pathways to regulate transcriptional events.

Effectors of Ras

Ras is arguably one of the most studied proteins involved in the T cell signal transduction cascade and it is associated with the activation of several cytoplasmic molecules involved in signal transduction [71, 94]. Possibly the best defined signaling cascade initiated by Ras in T cells is the activation of the extracellular signal regulated kinases (ERKs) which are members of a family of mitogen activated protein kinases (MAPKs). T cells express at least two ERKs, ERK-1 and ERK-2, both of which are activated in response to TCR stimulation [95]. Activation of the ERKs is achieved through a kinase cascade which is regulated by a balance of specific kinase and phosphatase activity. A dual function kinase, MAPK kinase (MEK or MKK-1), activates ERK by phosphorylating it on both tyrosine and threonine residues. Evidence also exists for dual function phosphatases, PAC-1 and VHR, that may function to deactivate ERK [96, 97]. MEK itself is regulated through phosphorylation by Raf-1 which is therefore a MAPK kinase kinase. Raf-1 is a 72 kDa serine/threonine kinase the amino terminal regulatory domain of which can interact directly with active (GTP-bound) Ras [98]. Raf-1 is both phosphorylated and activated following TCR stimulation [99]. Expression of a chimeric form of Raf-1 which had a membrane localization motif resulted in its activation in a Ras-independent manner [100, 101]. Thus Ras itself does not stimulate Raf-1 kinase activity but functions to localize the cytoplasmic Raf-1 to the plasma membrane for it to be activated through an undefined event [102]. Expression of dominant negative or constitutively active mutants of Raf-1 has demonstrated its importance for IL-2 production. Expression of the dominant negative mutant showed Raf-1 to be necessary for IL-2 production, and the constitutively active mutant of Raf-1 enhanced the production of IL-2 in response to TCR crosslinking [103]. Similar

modulations in IL-2 gene regulation were seen with constitutively active and dominant negative mutants of Ras [72]. These and other data show that Ras associates with and facilitates the activation of Raf-1, thus linking TCR receptor occupancy to activation of ERK1 and ERK2. Once activated, ERK2 induces the activity of p90^{rsk} (an S6 kinase) and both translocate into the nucleus. ERK2 is reported to regulate the Elk1 transcription factor and AP-1 activity [104, 105]. Phosphorylation of Elk1 by ERK appears necessary for the induction of the c-fos gene. In addition, Raf-1 increases the transcription of jun family members [106]. Thus Ras, by activating Raf-1 and ERK2, is able to affect the levels of Fos and Jun to regulate the components of the AP-1 element.

The regulation of transcription factors by Ras is complex and has been shown to involve several pathways emanating from the TCR. In addition to its activation of Raf-1, Ras has also been implicated in the stimulation of a distinct signal transduction pathway leading to the activation of Jun N-terminal kinases (JNK1 and JNK2). The JNK isoforms, of 46 kDa and 55 kDa for JNK1 and JNK2 respectively, are also activated by dual phosphorylation on tyrosine and threonine by SEK, a kinase homologous to MEK in the MAPK pathway [107]. This kinase is itself regulated by a serine/threonine kinase termed MEKK [108, 109] forming a cascade analogous to that linking Raf-1 and MAPK. Despite their similarities, the pathways to the activation of ERK and JNK are distinct. Unlike ERK, JNK is activated by UV irradiation, oxidative, and osmotic stress [110, 111]. Moreover, activation of JNK requires the costimulation of T cells [112] whereas activation of ERK1 and ERK2 do not. Interestingly, JNK activity appears to be required for IL-2 production. JNK phosphorylates Jun protein in its N-terminus on serine residues at positions 63 and 73. Phosphorylation at these sites promotes Jun-mediated transcription [111]. Furthermore, evidence exists in non-lymphoid cells for a third pathway for Ras regulation of transcription factors that is distinct from both ERKs and JNKs. Protein phosphorylation of Fos via a Fos-regulating kinase (FRK) can be stimulated by Ras [113]. Clearly, the activation of Ras by the TCR induces multiple pathways to influence discrete nuclear events.

As a result of the activation of Ras and PLCγ1, there is a split in the signal induced by stimulation of the TCR. The distinct effectors of the elevation in intracellular calcium and Ras each induce discrete sets of nuclear proteins. These nuclear proteins bind to the regulatory regions of the genes required for T cell activation and induce their transcription. Given that expression of IL-2 is essential for T cell activation, the transcriptional activation factors that regulate the IL-2 gene are now described.

Activation of Nuclear Events: Regulation of the IL-2 Gene

Nuclear targets of the TCR-mediated signal transduction pathways include the early genes asso-

transcription factors [114] (Figure 2.4). There are two NFAT binding sites, a site termed variously IL-2A, A site or ARRE-1 (antigen receptor recognition element); a site that binds a factor produced in response to CD28 costimulation, termed CD28RE (CD28 response element), and binding sites for the transcription factors, NFκB and AP-1. Occupancy of all these sites is required for full activation of the IL-2 gene. As described above, NFAT is a factor consisting of a nuclear and a cytoplasmic component. The cytoplasmic component consists of members of the NFAT gene family of Rel-related proteins. One member of the family, NFATc, is expressed primarily in T cells, but the other members of the gene family are expressed more widely [115, 116]. All can bind the NFAT site in the IL-2 gene and mediate transcription. A dominant negative NFATc will block acti-

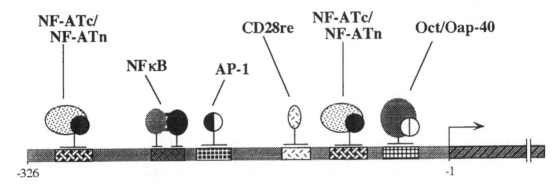

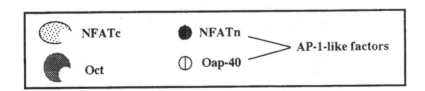

Figure 2.4 Transaction domains in the IL-2 enhancer. The transcription factors that bind to these domains in activated T cells are represented by the ovals and circles

ciated with T cell activation, for example, the genes for fos, jun and IL-2. Due to its critical role in T cell activation, much work has focused on the regulation of the IL-2 gene. The enhancer region for the IL-2 gene has been extensively characterized and contains several defined regions that bind

vation of the IL-2 promoter [117]. The IL-2A site binds the transcription factors, Oct-1 and Oct-2 [118]. Upon induction of the cell with PMA and calcium ionophore this site is co-occupied with a 40 kDa protein termed OAP-40 (octamer associated protein) [119]. The tissue distribution of Oct-1 is ubiquitous whereas Oct-2 is more restricted, being expressed in B cells and some T cells. Both transacting factors are regulated by stimuli that activate the IL-2 gene [120, 121]. OAP-40 is a critical factor for activation of the A-site and OAP-40 increases the strength of binding of Oct-1 to the IL-2 promoter site by up to 10-fold [119]. Moreover, the presence of OAP-40 correlates with the ability of Oct-1 to induce transcription from its DNA binding site. Purification of OAP-40 has identified JunD and c-Jun as its components [122]. Similar to the regulation of NFATc, the regulation of the Oct/OAP-40 is dependent on calcineurin [114], however, the details of the mechanism are not understood. The TCR activation signals that lead to an activation of NFκB are also unclear, but they may involve the CD28 costimulatory pathway [123, 124]. NFκB is a dimeric complex. Its components belong to a family of proteins with homology to the dorsal/rel-family of DNA-binding proteins [121]. NFκB proteins include p50 and p65 which can combine to form dimers that mediate different functions [125]. The p50 : p65 heterodimer is enhancing whereas the p50 : p50 homodimer exerts negative regulation. Like the cytoplasmic components of NFAT, the rel-protein dimers of NFκB exist in the cytoplasm as anchored inactive complexes. Two anchoring or inhibitory proteins have been identified: IκB-α and IκB-β, The phosphorylation and proteolytic degradation of IκB releases NFκB from the cytoplasmic inhibitory complex [126]. This process is coupled to the translocation of active NFκB to the nucleus where it binds to its cognate DNA-binding site in the IL-2 enhancer to induce gene transcription. An unconventional AP-1 site is also present in the IL-2 promoter. Unlike other AP-1 sites, activation of this AP-1 site is dependent on calcium and Ras-mediated pathways [112]. Studies investigating the full IL-2 promoter, or multiple elements from the IL-2 promoter, have indicated that occupancy of all of the described binding sites is required for optimal IL-2 production [127]. Indeed, the loss of just one of the binding sites drastically reduces the level of IL-2 gene transcription. Moreover, the discordant occupancy of transcription sites in the IL-2 enhancer has been proposed to account for the inhibition of IL-2 secretion seen in T cell anergy [128]. The regulation of the IL-2 gene is clearly tightly controlled

and dependent upon the activation of several discrete signal transduction pathways (discussed above) emanating from the TCR. Clearly the regulation of Fos and Jun family members, which are components of many of the IL-2 transcription factors, is a critical step in T cell activation. The induction and synthesis of IL-2 initiates the second phase of T cell activation. This phase is TCR-independent and involves the response of the cell to cytokines.

WHAT REGULATES THE FUNCTIONAL OUTCOME OF TCR STIMULATION?

We have described, above, current models of the "nuts and bolts" of T cell signal transduction. However, TCR engagement can lead to strikingly different outcomes, namely, activation, anergy or apoptosis. What parameters control the outcome of T cell stimulation?

Altered Peptide Ligands

The dogma has been that the TCR acts in a binary manner: it is either on or off. Control of the outcome following TCR activation has thus been presumed to be mediated by the additional signals the T cell is subjected to at the time of TCR stimulation, for example those signals provided by the microenvironment in which the T cell becomes stimulated. Costimulating signals which accompany TCR ligation, generated through either CD28 and/or adhesion molecules, are known to be key mediators in controlling the different responses of a T cell to TCR stimulation. Indeed crosslinking of the TCR in the absence of CD28 costimulation results in those cells being unresponsive (anergic) to additional, complete stimulation. In recent years several investigators [129, 130] have observed phenomena to suggest that the TCR is actually more versatile than has originally been proposed. Using antigenic peptides with sequential alterations in their amino acids it was observed that peptide sequences could be identified that initiated either a partial or total loss of T cell response despite binding to MHC molecules with affinities similar to wildtype peptides [131]. These peptides, termed altered peptide ligands (APL), were defined in functional studies as being partial agonists or antagonists respectively. Response of a T cell to receptor ligation can therefore be qualitatively different depending on the peptide. Costimulatory signals unquestionably contribute to regulating whether TCR stimulation results in activation, anergy or apoptosis, however for activation and anergy the type of peptide stimulus the T cells receives is also important.

APL-induced changes in the typical events characterizing early TCR signal transduction (see above) have been described. The patterns of phosphorylation for the CD3 epsilon and zeta chains stimulated with a partial agonist or an antagonistic APL are significantly different from the patterns detected with an agonistic peptide. Stimulation of T cells with an agonist peptide was shown to induce equivalent levels of a phosphorylated epsilon chain (27 kDA) and of two phosphorylated species of the zeta chain which migrate at apparent molecular weights of 21 and 23 kDa. However, a partial agonist or antagonist peptide, while inducing a strong increase in the levels of the p21 zeta chain species, failed to produce equivalent levels of either the p27 epsilon chain or the p23 zeta chain. Furthermore, this pattern of phosphorylation induced by the APL correlated with a lack of phosphorylation and kinase activation of ZAP-70. The biochemical mechanism that accounts for distinct signal transduction by different ligands, through the same receptor, remains to be determined. At the moment, the simplest explanation is that differences in off-rate between agonistic peptides and APL lead to the functional differences that have been observed. Unquestionably, these exciting observations have the potential to lead to a mechanistic understanding of many complex immune phenomena facilitated by the TCR, for example thymocyte maturation.

Strength of Signal and Determination of T Cell Differentiation

In thymocyte development immature T cells have to progress through two distinct stages of selection in the thymus in order to mature and be released into the periphery. Positive selection picks those thymocytes that are able to bind to peptide/MHC complexes and is believed to involve a weak interaction/stimulation of the TCR [132]. Thymocytes whose TCRs do not receive a stimulus die. In contrast, negative selection involves the removal of potentially autoreactive thymocytes and is thought to be in response to a strong TCR stimulation. Endogenously produced peptides in the thymus may act as APLs thus providing a potential mechanism of positive selection [133, 134]. Recent evidence indicates that the differentiation of $CD4^+$ T cells into T helper type 1 (Th1) cells or T helper type 2 (Th2) cells can also be determined by altered peptide ligands [135]. Peptides that had strong interactions with the TCR generated Th1-like cells while those that interacted weakly with the TCR led to the Th2-cell development. Moreover APL hold the prospect of a new family of therapeutic agents for immune-mediated pathologies such as cell mediated autoimmunity.

TCR Signal Transduction in Series or In Parallel

Many ligand receptor systems involve high affinity interactions. For example, the binding of IL-2 to the heterotrimeric IL-2 receptor has an equilibrium binding affinity in the picomolar range. It binds very rapidly ($t\frac{1}{2} = 5$ to 30s) but has a very slow off-rate ($t\frac{1}{2} = 5$h) and thus can be considered to be a nearly irreversible reaction. In striking contrast, preliminary data indicate that the interaction of the TCR with peptide and MHC is a low affinity interaction with a high off-rate. Recent studies indicate that a single peptide–MHC complex can serially engage and trigger up to 200 TCRs [136]. These results may account for previous observations that an antigen presenting cell, with only a few hundred peptide–MHC complexes, can fully activate a T cell. If these results can be generalized, then the peptide–MHC complex may emerge as a promiscuous ligand capable of specifically engaging many TCRs. Therefore, the signal transduction rules for the polygamous TCR may be very different from those of the relatively monogamous cytokine receptor signal transduction cascades.

FUTURE PROSPECTS

Through the intense research that has focused over the last decade on the TCR/CD3 complex, a molecule-by-molecule description of the intracellular signal transduction pathways that are activated following stimulation of the T cell antigen receptor can now be considered a realistic target. Transgenic studies and the cellular transfection of constitutively active and dominant negative signaling molecules have already aided the identification of many of the key players and the definition of critical signaling pathways. Several intracellular substrates that are phosphorylated in response to TCR stimulation are not yet characterized and alterations and additions to this model cannot be excluded. An example of such an alteration is the recent discovery that stimulation of the TCR can result in qualitatively different signals depending on the type of peptide stimulus it received. This detailed understanding will undoubtedly continue to contribute to our understanding of physiologic and pathologic processes and identify potential targets for therapeutic interventions to modulate immune responses.

ACKNOWLEDGEMENTS

This work was supported in part by Naval Medical Research and Development Command grant #90.0101.KHX.1437 and 1402. The views expressed in this chapter are those of the authors and do not reflect the official policy or position of the Department of the Navy, Department of Defense, or the United States Government.

REFERENCES

1. Havran WL and Boismenu R. Activation and function of gamma delta T cells. *Curr. Opin. Immunol.* 1994; **6**(3): 442–446.
2. Bendelac A. CD1: Presenting unusual antigens to unusual T lymphocytes. *Science*, 1995; **269**: 185–186.
3. Weiss A. Molecular and genetic insights into T cell antigen receptor structure and function. *Ann. Rec. Genet.* 1991; **25**(487): 487–510.
4. Ashwell JD and Klausner RD Genetic and mutational analysis of the T-cell antigen receptor. *Ann. Rev. Immunol.*, 1990; **8**(139): 139–167.
5. Jensen JP Cenciarelli C, Hou D, Rellahan, BL, Dean M and Weissman AM. T cell antigen receptor-eta subunit.

Low levels of expression and limited cross-species conservation. *J. Immunol.* 1993; **150**(1): 122–130.
6. Reth M. Antigen receptor rail clue. *Nature* 1989; **338**(6214): 383–384.
7. Letourneur F and Klausner RD. Activation of T cells by a tyrosine kinase activation domain in the cytoplasmic tail of CD3. *Science* 1992; **255**: 79–82.
8. Cambier JC. New nomenclature for the Reth motif (or ARH1/TAM/ARAM/YXXL). *Immunol. Today* 1995; **16**(2): 110.
9. Irving BA, Chan AC and Weiss A. Functional characterization of a signal transducing motif present in the T cell antigen receptor zeta chain. *J. Exp. Med.* 1993; **177**(4): 1093–1103.
10. Romeo C, Amiot M and Seed B. Sequence requirements for induction of cytolysis by the T cell antigen/Fc receptor zeta chain. *Cell* 1992; **68**(5): 889–897.
11. Reth M. Antigen receptors on B lymphocytes. *Ann. Rev. Immunol.* 1992; **10**(97): 97–121.
12. Willems L, Gatot JS, Mammerickx M, Portetelle D, Burny A, Kerkhofs P and Kettmann R. The YXXL signalling motifs of the bovine leukemia virus transmembrane protein are required for in vivo infection and maintenance of high viral loads. *J. Virol.* 1995; **69**(7): 4137–4141.
13. Collins TL, Kassner PD, Bierer BE and Burakoff SJ Adhesion receptors in lymphocyte activation. *Curr. Opin. Immunol.* 1994; **6**(3): 385–393.
14. June CH Bluestone JA, Nadler LM and Thompson CB. The B7 and CD28 receptor families. *Immunol. Today*, 1994; **15**(7): 321–331.
15. Heldin CH. Dimerization of cell surface receptors in signal transduction. *Cell* 1995; **80**(2): 213–223.
16. June CH Fletcher MC, Ledbetter JA and Samelson LE. Increases in tyrosine phosphorylation are detectable before phospholipasae C activation after T cell receptor stimulation. *J. Immunol.* 1990; **144**(5): 1591–1599.
17. Hsi ED, Siegel JN, Minami Y, Luong ET, Klausner RD and Samelson LE. T cell activation induces rapid tyrosine phosphorylation of a limited number of cellular substrates. *J. Biol. Chem.* 1989; **264**(18): 10836–10842.
18. June CH, Fletcher MC, Ledbetter JA, Schieven GL, Siegel JN, Phillips AF and Samelson LE. Inhibition of tyrosine phosphorylation prevents T-cell receptor-mediated signal transduction. *Proc. Natl. Acad. Sci. USA*, 1990; **87**(19): 7722–7726.
19. Songyang Z, Shoelson SE, Chaudhuri M, Gish G, Pawson T, Haser WG, King F, Roberts T et al. SH2 domains recognize specific phosphopeptide sequences. *Cell*, 1993; **72**(5): 767–778.
20. Songyang Z, Shoelson SE, McGlade J, Olivier P, Pawson T, Bustelo XR, Barcacid M, Sabe H. et al. Specific motifs recognized by the SH2 domains of Csk, 3BP2, fps/fes. GRB-2, HCP, SHC, Syk and Vav. *Mol. Cell. Biol.* 1994; **14**(4): 2777–2785.
21. Burkhardt AL, Stealey B, Rowley RB, Mahajan S, Prendergast M, Fargnoli J and Bolen JB. Temporal regulation of non-transmembrane protein tyrosine kinase enzyme activity following T cell antigen receptor engagement. *J. Biol. Chem.* 1994; **269**(38): 23642–23647.

22. Tsygankov AY, Broker BM, Fargnoli J, Ledbetter JA and Bolen JB. Activation of tyrosine kinase p60fyn following T cell antigen receptor cross-linking. *J. Biol. Chem.* 1992; **267**(26): 18259–18262.

23. Samelson LE, Phillips AF, Luong ET, and Klausner RD. Association of the fyn protein-tyrosine kinase with the T-cell antigen receptor. *Proc. Natl. Acad. Sci. USA*, 1990; **87**(11): 4358–4362.

24. Timson GL, Kong AN, Samelson LE and Shaw AS. p59fyn tyrosine kinase associates with multiple T-cell receptor subunits through its unique amino-terminal domain. *Mol. Cell. Biol.*, 1992; **12**: 5438–5446.

25. Tsygankov AY, Spana C, Rowley RB, Penhallow RC, Burkhardt AL and Bolen JB. Activation-dependent tyrosine phosphorylation of Fyn-associated proteins in T lymphocytes. *J. Biol. Chem.* 1994. **269**(10): 7792–7800.

26. Davidson D, Chow LM, Fournel M and Veillette A. Differential regulation of T cell antigen responsiveness by isoforms of the src-related tyrosine protein kinase p59fyn. *J. Exp. Med.* 1992; **175**(6): 1483–1492.

27. Hall CG, Sancho J and Terhorst C. Reconstruction of T cell receptor zeta-mediated calcium mobilization in non-lymphoid cells. *Science*, 1993; **261**(5123): 915–918.

28. Appleby MW, Gross JA, Cooke MP, Levin SD, Qian X and Perlmutter RM. Defective T cell receptor signalling in mice lacking the thymic isoform of p59fyn. *Cell*, 1992; **70**(5): 751–763.

29. Stein PL, Lee HM, Rich S and Soriano P. pp59fyn mutant mice display differential signaling in thymocytes and peripheral T cells. *Cell*, 1992; **70**(5): 741–750.

30. Cooke MP, Abraham KM, Forbush KA and Perlmutter RM. Regulation of T cell receptor signaling by a src family protein-tyrosine kinase (p59fyn). *Cell*, 1991; **65**(2): 281–291.

31. Molina TJ, Kishihara K, Siderovski DP, van Ewijk W, Narenndran A, Timms E, Wakeham A, Paige CJ *et al.* Profound block in thymocyte development in mice lacking p56lck. *Nature*, 1992; **357**(6374): 161–164.

32. Levin SD, Anderson SJ, Forbush KA and Perlmutter RM. A dominant-negative transgene defines a role for p56lck in thymopoiesis. *EMBO J.* 1993; **12**(4): 1671–1680.

33. Shaw AS, Chalupny J, Whitney JA, Hammond C, Amrein KE, Kavathas P, Sefton BM and Rose JK. Short related sequences in the cytoplasmic domains of CD4 and CD8 mediate binding to the amino-terminal domain of the p56lck tyrosine protein kinase. *Mol. Cell. Biol.* 1990; **10**(5): 1853–1862.

34. Xu H and Littman DR. A kinase-independent function of Lck in potentiating antigen-specific T cell activation. *Cell*, 1993; **74**(4): 633–643.

35. Veillette A, Bookman MA, Horak EM, Samelson LE and Bolen JB. Signal transduction through the CD4 receptor involves the activation of the internal membrane tyrosine-protein kinase p56lck. *Nature*, 1989; **338**(6212): 257–259.

36. Abraham N, Miceli MC, Parnes JR and Veillette A. Enhancement of T-cell responsiveness by the lymphocyte-specific tyrosine protein kinase p56lck. *Nature*, 1991; **350**(6313): 62–66.

37. Mittler RS, Goldman SJ, Spitalny GL and Burakoff SJ. T-cell receptor-CD4 physical association in a murine T-cell hybridoma: induction by antigen receptor ligation. *Proc. Natl. Acad. Sci. USA* 1989; **86**(21): 8531–8535.

38. Kmiecik TE and Shalloway D. Activation and suppression of pp60c-src transforming ability by mutation of its primary sites of tyrosine phosphorylation. *Cell*, 1987; **47**(1): 65–73.

39. Piwnica Worms H, Saunders KB, Roberts TM, Smith AE and Cheng SH. Tyrosine phosphorylation regulates the biochemical and biological properies of pp60c-src. *Cell*, 1987; **49**(1): 75–82.

40. Deans JP, Kanner SB, Torres RM and Ledbetter JA. Interaction of CD4:Lck with the T cell receptor/CD3 complex induces early signaling events in the absence of CD45 tyrosine phosphatase. *Eur. J. Immunol.* 1992; **22**(3): 661–668.

41. Mustelin T, Coggeshall KM and Altman A. Rapid activation of the T-cell tyrosine protein kinase pp56lck by the CD45 phosphotyrosine phosphatase. *Proc. Natl. Acad. Sci. USA* 1989; **86**(16): 6302–6306.

42. Mustelin T, Pessa MT, Autero M, Gassman M, Andersson LC, Gahmberg CG and Burn P. Regulation of the p59fyn protein tyrosine kinase by the CD45 phosphotyrosine phosphatase. *Eur. J. Immunol.* 1992; **22**(5): 1173–1178.

43. Autero M, Saharinen J, Pessa MT, Soula RM, Oetken C, Gassman M, Bergman M, Alitalo K *et al.* Tyrosine phosphorylation of CD45 phosphotyrosine phosphatase by p50csk kinase creates a binding site for p56lck tyrosine kinase and activates the phosphatase. *Mol. Cell. Biol.* 1994; **14**(2): 1308–1321.

44. Nada S, Yagi T, Takeda H, Tokunaga T, Nakagawa H, Ikawa Y, Okada M and Aizawa S. Constitutive activation of Src family kinases in mouse embryos that lack Csk. *Cell* 1993; **73**(6): 1125–1135.

45. Chow LM, Fournel M, Davidson D and Veillette A. Negative regulation of T-cell receptor signalling by tyrosine protein kinase p50csk. *Nature* 1993; **365**(6442): 156–160.

46. McVicar DW, Lal BK, Lloyd A, Kawamura M, Chen YQ, Zhang X, Staples JE, Ortaldo JR *et al.* Molecular cloning of lsk, a carboxyl-terminal src kinase (csk) related gene, expressed in leukocytes. *Oncogene* 1994; **9**(7): 2037–2044.

47. Chan AC, Iwashima M, Turck CW and Weiss A. ZAP-70: a 70 kd protein-tyrosine kinase that associated with the TCR zeta chain. *Cell* 1992; **71**(4): 649–662.

48. Wange RL, Lalek SN, Desiderio S and Samelson LE. Tandem SH2 domains of ZAP-70 bind to T cell antigen receptor zeta and CD3 epsilon from activated Jurkat T cells. *J. Biol. Chem.* 1993; **268**(26): 19797–19801.

49. Chan AC, Kadlecek TA, Elder ME, Filipovich AH, Kuo WL, Iwashima M, Parslow TG and Weiss A. ZAP-70 deficiency in an autosomal recessive form of severe combined immunodeficiency. *Science* 1994; **264**(5165): 1599–1601.

50. Elder ME, Lin D, Clever J, Chan AC, Hope TJ, Weiss A and Parslow TG. Human severe combined immunodeficiency due to a defect in ZAP-70, a T cell tyrosine kinase. *Science* 1994; **264**(5165): 1596–1599.

51. Arpaia E, Shahur M, Dadi H, Cohen A and Roifman CM. Defective T cell receptor signalling and thymic selection in humans lacking zap-70 kinase. *Cell* 1994; **76**(5): 947–958.

52. Chan AC, van Oers NS, Tran A, Turka L, Law CL, Ryan JC, Clark EA and Weiss A. Differential expression of ZAP-

70 and Syk protein tyrosine kinases, and the role of this family of protein tyrosine kinases in TCR signalling. *J. Immunol.* 1994; **152**(10): 4758–4766.

53. Chan AC, Dalton M, Johnson R, Kong GH, Wang T, Thoma R and Kurosaki T. Activation of ZAP-70 kinase activity by phosphorylation of tyrosine 493 is required for lymphocyte antigen receptor function. *EMBO* J. 1995; **14**(11): 2499–2508.

54. Gauen LK, Zhu Y, Letourneur F, Hu Q, Bolen JB, Matis LA, Klausner RD and Shaw AS. Interactions of p59fyn and ZAP-70 with T-cell receptor activation motifs: defining the nature of a signalling motif. *Mol. Cell. Biol.* 1994; **14**(6): 3729–3741.

55. Neumeister EN, Zhu Y, Richard S, Terhorst C, Chan AC and Shaw AS. Binding of ZAP-70 to phorphorylated T-cell receptor zeta and eta enhances its autophosphorylation and generates specific binding sites for SH2 domain-containing proteins. *Mol. Cell. Biol.* 1995; **15**(6): 3171–3178.

56. Isakov N, Wange RL, Burgess WH, Watts JD, Aebersold R and Samelson LE. ZAP-70 binding specificity to T cell receptor tyrosine-based activation motifs: the tandem SH2 domains of ZAP- 70 bind distinct tyrosine-based activation motifs with varying affinity. *J. Exp. Med.* 1995; **181**(1): 375–380.

57. Exley M, Varticovski L, Peter M, Sancho J and Terhorst C. Association of phosphatidylinositol 3-kinase with a specific sequence of the T cell receptor zeta chain is dependent on T cell activation. *J. Biol. Chem.* 1994; **269**(21): 15140–15146.

58. Ravichandran KS, Lorenz U, Shoelson SE and Burakoff SJ. Interaction of Shc with Grb2 regulates association of Grb2 with mSOS. *Mol. Cell. Biol.* 1995; **15**(2): 593–600.

59. Reif K, Buday L, Downward J and Cantrell DA. SH3 domains of the adapter molecule Grb2 complex with two proteins in T cells: the guanine nucleotide exchange protein Sos and a 75-kDa protein that is a substrate for T cell antigen receptor-activated tyrosine kinases. *J. Biol. Chem.* 1994; **269**(19): 14081–14087.

60. Buday L, Egan SE, Rodriguez VP, Cantrell DA and Downward J. A complex of Grb2 adaptor protein, Sos exchange factor, and a 36-kDa membrane-bound tyrosine phosphoprotein is implicated in ras activation in T cells. *J. Biol. Chem.* 1994; **269**(12): 9010–9023.

61. Ravichandran KS, Lee KK, Songyang Z, Cantley LC, Burn P and Burakoff SJ. Interaction of Shc with the zeta chain of the T cell receptor upon T cell activation. *Science* 1993; **262**(5135): 902–905.

62. Baldari CT, Pelicci G, Di SM, Milia E, Giuli S, Pelicci PG and Telford JL. Inhibition of CD4/p56lck signalling by a dominant negative mutant of the Shc adaptor protein. *Oncogene* 1995; **10**(6): 1141–1147.

63. Katzav S, Sutherland M, Packham G, Yi T and Weiss A. The protein tyrosine kinase ZAP-70 can associate with the SH2 domain of proto-Vav. *J. Biol. Chem.*, 1994; **269**(51): 32579–32585.

64. Ramos MF, Druker BJ and Fischer S. Vav binds to several SH2/SH3 containing proteins in activated lymphocytes. *Oncogene* 1994; **9**(7): 19917–19923.

65. Gulbins E, Coggeshall KM, Baier G, Katzav S, Burn P and Altman A. Tyrosine kinase-stimulated guanine nucleotide

exchange activity of Vav in T cell activation. *Science* 1993; **260**(5109): 822–825.

66. Adams JM, Houston H, Allen J, Lints T and Harvey R. The hematopoietically expressed vav proto-oncogene shares homology with the dbl GDP-GTP exchange factor, the bcr gene and a yeast gene (CDC24) involved in cytoskeletal organization. *Oncogene* 1992; **7**(4): 611–618.

67. Minden A, Lin A, Claret F-X, Abo A and Karin M. Selective activation of the JNK signalling cascade and c-jun transcriptional activity by the small GTPases Rac and Cdc42Hs. *Cell* 1995; **81**: 1147–1157.

68. Zhang R, Alt FW, Davidson L, Orkin SH and Swat W. Defective signalling through the T-and B-cell antigen receptors in lymphoid cells lacking the vav protooncogene. *Nature* 1995; **374**(6521): 470–473.

69. Tarakhovsky AM, Turner M, Schaal S, Mee PJ, Duddy LP, Rajewsky K and Tybulewicz VL. Defective antigen receptor-mediated proliferation of B and T cells in the absence of Vav. *Nature* 1995; **374**(6521): 467–470.

70. Fischer KD, Zmuldzinas A, Gardner S, Barbacid M, Bernstein A and Guidos C. Defective T-cell receptor signalling and positive selection of Vav-deficient CD4+ CD8+ thymocytes. *Nature* 1995; **374**(6521): 474–477.

71. Izquierdo PM Reif K, and Cantrell D. The regulation and function of p21ras during T-cell activation and growth. *Immunol. Today* 1995; **16**(3): 159–164.

72. Rayter SI, Woodrow M, Lucas SC, Cantrell DA and Downward J. p21ras mediates control of IL-2 gene promoter function in T cell activation. *EMBO* J. 1992; **11**(12): 4549–4556.

73. Graves JD, Downward J, Rayter S, Warne P, Tutt AL, Glennie M and Cantrell DA. CD2 antigen mediated activation of the guanine nucleotide binding proteins p21ras in human T lymphocytes. *J. Immunol* 1991; **146**(11): 3709–3712.

74. Weber JR, Bell GM, Han MY, Pawson T and Imboden JB. Association of the tyrosine kinase LCK with phospholipase C-gamma 1 after stimulation of the T cell antigen receptor. *J. Exp. Med.* 1992; **176**(2): 373–379.

75. Sieh M, Batzer A, Schlessinger J and Weiss A. GRB2 and phospholipase C-gamma 1 associate with a 36-to 38-kilodalton phosphotyrosine protein after T-cell receptor stimulation. *Mol. Cell. Biol.* 1994; **14**(7): 4435–4442.

76. Jayaraman T, Ondriasova E, Ondrias K, Harnick DJ and Marks AR. The inositol 1,4,5-trisphosphate receptor is essential for T-cell receptor signalling. *Proc. Natl. Acad. Sci. USA* 1995; **92**: 6007–6011.

77. Cantrell DA, Izquierdo M, Reif K and Woodrow M. Regulation of Ptdlns-3-kinase and the guanine nucleotide binding proteins p21ras during signal transduction by the T cell antigen receptor and the interleukin-2 receptor. *Semin. Immunol.* 1993; **5**(5): 319–326.

78. von Willebrand M, Baier G, Couture C, Burn P and Mustelin T. Activation of phosphatidylinositol-3 kinase in Jurkat T cells depends on the presence of the p56lck tyrosine kinase. *Eur. J. Immunol.* 1994; **24**(1): 234–238.

79. Rodriguez-Viciana P, Warne PH, Dhand R, Vanhaesebroeck B, Gout T, Fry MJ, Waterfield MD and Downward J. Phosphatidylinositol-3-OH kinase as a direct target of Ras. *Nature* 1994; **370**: 527–532.

80. Monfar M, Lemon KP, Grammer TC, Cheatham L, Chung J, Vlahos CJ and Blenis J. Activation of pp 70/85 S6 kinases in interleukin-2-responsive lymphoid cells is mediated by phosphatidylinositol 3-kinase and inhibited by cyclic AMP. *Mol. Cell. Bio.* 1995; **15**(1): 326–337.

81. Lane HA, Fernandez A, Lamb NJC and Thomas G. p70s6k function is essential for G1 progression. *Nature* 1993; **363**: 170–172.

82. Randriamampita C and Tsien RY. Emptying of intracellular Ca2+ stores releases a novel small messenger that stimulated Ca^{2+} influx. *Nature* 1993; **364**(6440): 809–814.

83. Randriamampita C and Tsien RY. Degradation of a calcium influx factor (CIF) can be blocked by phosphatase inhibitors or chelation of Ca^{2+}. *J. Biol. Chem.* 1995; **270**(1): 29–32.

84. Bram RJ and Crabtree GR. Calcium signalling in T cells stimulated by a cyclophilin B-binding protein. *Nature* 1994; **371**(6495): 355–358.

85. Gelfand EW, Cheung RK, Mills GB and Grinstein S. Uptake of extracellular Ca^{2+} and not recruitment from internal stores is essential for T lymphocyte proliferation. *Eur. J. Immunol.* 1988; **18**(6): 917–922.

86. Goldsmith MA and Weiss A. Early signal transduction by the antigen receptor without commitment to T cell activation. *Science* 1988; **240**(4855): 1029.

87. Clapham DE. Calcium signalling. *Cell* 1995; 80(2): 259–268.

88. Friedman J and Weissman I. Two cytoplasmic candidates for immunophilin action are revealed by affinity for a new cyclophilin: one in the presence and one in the absence of CsA. *Cell* 1991; **66**(4): 799–806.

89. Liu J, Farmer JJ, Lane WS, Friedman J, Weissman I and Schreiber SL. Calcineurin is a common target of cyclophilin–cyclosporin A and FKBP-FK506 complexes. *Cell* 1991; **66**(4): 807–815.

90. Fruman DA, Klee, CB, Bierer, BE and Burakoff, SJ. Calcineurin phosphatase activity in T lymphocytes is inhibited by FK 506 and cyclosporin A. *Proc. Natl. Acad. Sci. USA* 1992; **89**(9): 3686–3690.

91. Liu J, Albers MW, Wandless TJ, Luan S, Alberg DG, Belshaw PJ, Cohen P, MacKintosh C et al. Inhibition of T cell signalling by immunophilin ligand complexes correlates with loss of calcineurin phosphatase activity. *Biochemistry* 1992; **31**(16): 3896–3901.

92. Woodrow M, Clipstone NA and Cantrell D. p21ras and calcineurin synergize to regulate the nuclear factor of activated T cells. *J. Exp. Med.* 1993; **1678**(5): 1517–1522.

93. Jain J, McCaffrey PG, Valge AV and Rao A. Nuclear factor of activated T cells contains Fos and Jun. *Nature* 1992; **356**(6372): 801–804.

94. Cantrell D, G proteins in lymphocyte signalling. *Curr. Opin. Immunol.* 1994; **6**(3): 380–384.

95. Whitehurst CE, Boulton TG, Cobb MH and Geppert TD. Extracellular signal-regulated kinases in T cells. Anti-CD3 and 4 beta-phorbol 12-myristate 13-acetate-induced phosphorylation and activation. *J. Immunol.* 1992; **148**(10): 3230–3237.

96. Ward Y, Gupta S, Jensen P, Wartmann M, Davis RJ and Kelly K. Control of MAP kinase activation by the mitogen-induced threonine/tyrosine phosphatase PAC1. *Nature* 1994; **367**(6464): 651–654.

97. Denu JM, Zhou G, Wu L, Zhao R, Yuvaniyama J, Saper MA and Dixon JE. The purification and characterization of a human dual-specific protein tyrosine phosphatase. *J. Biol. Chem.* 1995; **270**(8): 3796–3803.

98. Zhang XF, Settleman J, Kyriakis JM, Takeuchi SE, Elledge SJ, Marshall MS, Bruder JT, Rapp UR et al. Normal and oncogenic p21ras proteins bind to the amino-terminal regulatory domain of c-Raf-1. *Nature* 1993; **364**(6435): 308–313.

99. Siegel JN, Klausner RD, Rapp UR and Samelson LE. T cell antigen receptor engagement stimulates c-raf phosphorylation and induces c-raf-associated kinase activity via a protein kinase C- dependent pathway. *J. Biol. Chem.* 1990; **265**(30): 18472–18480.

100. Stokoe D, Macdonald SG, Cadwallader K, Symons M and Hancock JF. Activation of Raf as a result of recruitment to the plasma membrane. *Science* 1994; **264**(5164): 1463–1467.

101. Leevers SJ, Paterson HF and Marshall CJ. Requirement for Ras in Raf activation is overcome by targeting Raf to the plasma membrane. *Nature* 1994; **369**(6479): 411–414.

102. Whitehurst CE, Owaki H, Bruder JT, Rapp UR and Geppert TD. The MEK kinase activity of the catalytic domain of RAF-1 is regulated independently of Ras binding in T cells. *J. Biol. Chem.* 1995; **270**(10): 5594–5599.

103. Owaki H, Varma R, Gillis B, Bruder JT, Rapp UR, Davis LS and Geppert TD. Raf-1 is required for T cell IL2 production. *EMBO J.* 1993; **12**(11): 4367–4373.

104. Hunter T and Karin M. The regulation of transcription by phosphorylation. *Cell* 1992; **70**(3): 375–387.

105. Marais R, Wynne J and Treisman R. The SRF accessory protein Elk-1 contains a growth factor-regulated transcriptional activation domain. *Cell* 1993; **73**(2): 381–393.

106. Wasylyk C, Wasylyk B, Heidecker G, Huleihel M and Rapp UR. Expression of raf oncogenes activates the PEA1 transcription factor motif. *Mol. Cell. Biol.* 1989; **9**(5): 2247–2250.

107. Kyriakis JM, Banerjee P, Nikolakaki E, Dai T, Rubie EA, Ahmad MF, Avruch J and Woodgett JR. The stress-activated protein kinase subfamily of c-Jun kinases. *Nature* 1994; **369**(6476): 156–160.

108. Minden A, Lin A, McMahon M, Lange CC, Derijard B, Davis RJ, Johnson GL and Karin M. Differential activation of ERK and JNK mitogen-activated protein kinases by Raf-1 and MEKK. *Science* 1994; **266**(5191): 1719–1723.

109. Yan M, Dai T, Deak JC, Kyriakis JM, Zon LI, Woodgett JR and Templeton DJ. Activation of stress-activated protein kinase by MEKK1 phosphorylation of its activator SEK1. *Nature* 1994; **372**(6508): 798–800.

110. Galcheva GZ, Derijard B, Wu IH and Davis RJ. An osmosensing signal transduction pathway in mammalian cells. *Science* 1994; **265**(5173): 806–808.

111. Derijard B, Hibi M, Wu IH, Barrett T, Su B, Deng T, Karin M and Davis RJ. JNKI: a protein kinase stimulated by UV light and Ha-Ras that binds and phosphorylates the c-Jun activation domain. *Cell* 1994; **76**(6): 1025–1037.

112. Su B, Jacinto E, Hibi M, Kallunki T, Karin M and Ben NY. JNK is involved in signal integration during costimulation of T lymphocytes. *Cell* 1994; **77**(5): 727–736.

113. Deng T and Karin M. c-Fos transcriptional activity stimulated by H-Ras-activated protein kinase distinct from JNK and ERK. *Nature* 1994; **371**: 171–175.

114. Crabtree GR and Clipstone NA. Signal transmission between the plasma membrane and nucleus of T lymphocytes. *Ann. Rev. Biochem.* 1994; **63**(1045): 1045–1083.

115. Hoey T, Sun YL, Williamson K and Xu X. Isolation of two new members of the NF-AT gene family and functional characterization of the NF-AT proteins. *Immunity* 1995; **2**(5): 461–472.

116. Masuda ES, Naito Y, Tokumitsu H, Campbell D, Saito F, Hannum C, Arai K and Arai N. NFATx, a novel member of the nuclear factor of activated T cells family that is expressed predominantly in the thymus. *Mol. Cell. Biol.* 1995; **15**(5): 2697–2706.

117. Northrop JP, Ho SN, Chen L, Thomas DJ, Timmerman LA, Nolan GP, Admon A and Crabtree GR. NF-AT components define a family of transcription factors targeted in T-cell activation. *Nature* 1994; **369**(6480): 497–502.

118. de Grazia U, Felli MP, Vacca A, Farina AR, Maroder M, Cappabianca L, Meco D, Farina M *et al.* Positive and negative regulation of the composite octamer motif of the interleukin 2 enhancer by AP-1, Oct-2, and retinoic acid receptor. *J. Exp. Med.* 1994; **180**(4): 1485–1497.

119. Ullman KS, Flanagan WM, Edwards CA and Crabtree GR. Activation of early gene expression in T lymphocytes by Oct-1 and an inducible protein, OAP40. *Science* 1991; **254**(5031): 558–562.

120. Kang SM, Tsang W, Doll S, Scherle P, Ko HS, Tran AC, Lenardo MJ, and Staudt LM. Induction of the POU domain transcription factor Oct-2 during T-cell activation by cognate antigen. *Mol. Cell. Biol.* 1992; **12**(7): 3149–3154.

121. Gilmore TD. NF-kappa B, KBF1, dorsal, and related matters. *Cell* 1990; **62**(5): 841–843.

122. Ullman KS, Northrop JP, Admon A and Crabtree GR. Jun family members are controlled by a calcium-regulated, cyclosporin A-sensitive signaling pathway in activated T lymphocytes. *Genes Dev.* 1993; **7**(2): 188–196.

123. Costello R, Lipcey C, Algarte M, Cerdan C, Baeuerle PA, Olive D and Imbert J. Activation of primary human T-lymphocytes through CD2 plus CD28 adhesion molecules induces long-term nuclear expression of NF-kappa B. *Cell Growth Differ.* 1993; **4**(4): 329–339.

124. Lai JH and Tan TH. CD28 signaling causes a sustained down-regulation of I kappa B alpha which can be prevented by the immunosuppressant rapamycin. *J. Biol. Chem.* 1994; **269**(48): 30077–30080.

125. Bours V, Franzoso G, Brown K, Park S, Azarenko V, Tomita YM, Kelly K and Siebenlist U. Lymphocyte activation and the family of NF-kappa B transcription factor complexes. *Curr Top Microbiol Immunol* 1992; **182**(411): 411–420.

126. Henkel T, Machleidt T, Alkalay I, Kronke M, Ben NY and Baeuerle PA. Rapid proteolysis of I kappa B-alpha is necessary for activation of transcription factor NF-kappa B. *Nature* 1993; **365**(6442): 182–185.

127. Chen D and Rothenberg EV. Interleukin 2 transcription factors as molecular targets of cAMP inhibition: delayed inhibition kinetics and combinatorial transcription roles. *J. Exp. Med.* 1994; **179**(3): 931–942.

128. Kang SM, Beverly B, Tran AC, Brorson K, Schwartz RH and Lenardo MJ. Transactivation by AP-1 is a molecular target of T cell clonal anergy. *Science* 1992; **257**(5073): 1134–1138.

129. Sloan LJ, Shaw AS, Rothbard JB and Allen PM. Partial T cell signaling: altered phospho-zeta and lack of zap 70 recruitment in APL-induced T cell energy. *Cell* 1994; **79**(5): 913–922.

130. Madrenas J, Wange RL, Wang JL, Isakov N, Samelson LE and Germain RN. Zeta phosphorylation without ZAP-70 activation induced by TCR antagonists or partial agonists. *Science* 1995; **267**(5197): 515–518.

131. Evavold BD, Sloan LJ and Allen PM. Tickling the TCR: selective T-cell functions stimulated by altered peptide ligands. *Immunol. Today* 1993; **14**(12): 602–609.

132. von Boehmer H. Thymic selection: a matter of life and death. *Immunol. Today* 1992; **13**(11): 454–458.

133. Ashton RP, Van KL, Schumacher TN, Ploegh HL and Tonegawa S. Peptide contributes to the specificity of positive selection of CD8+ T cells in the thymus. *Cell* 1993; **73**(5): 1041–1049.

134. Hogquist KA, Gavin MA and Bevan MJ. Positive selection of CD8+ T cells induced by major histocompatibility complex binding peptides in fetal thymic organ culture. *J. Exp. Med.* 1993; **177**(5): 1469–1473.

135. Pfeiffer C, Stein J, Southwood S, Ketelaar H, Sette A and Bottomly K. Altered peptide ligands can control CD4 T lymphocyte differentiation *in vivo*. *J. Exp. Med.* 1995; **181**(4): 1569–1574.

136. Valitutti S, Muller S, Cella M, Padovan E and Lanzavecchia A. Serial triggering of many T-cell receptors by a few peptide–MHC complexes. *Nature* 1995; **375**(6527): 148–151.

137. Nunes JA, Collette Y, Truneh A, Olive D and Cantrell DA. The role of p21ras in CD28 signal transduction: triggering of CD28 with antibodies, but not the ligand B7-1, activates p21ras. *J. Exp. Med.* 1994; **180**(3): 1067–1076.

Part II
Accessory Molecule Signalling in B and T Cells

3

CD19 and CR2 on B Lymphocytes: Structure, Biochemistry and Function

Robert H. Carter[1] and Douglas T. Fearon[2]

1 *Departments of Medicine and Microbiology, University of Alabama at Birmingham, Alabama, USA*
2 *Wellcome Trust Immunology Unit, University of Cambridge School of Clinical Medicine, Cambridge, UK*

INTRODUCTION

A central theme in lymphocyte biology is the need to modulate the response to antigen according to the context in which it is recognized. This could occur either through qualitative differences in the manner of ligation of the antigen receptor, of coligation of accessory surface receptors, or in the ambient cytokine environment. Studies of T cells have assumed the role for accessory receptor mechanisms for some time, perhaps because T cells can only recognize processed antigen in the context of an antigen presenting cell. Ligation of the T cell receptor involves not only the binding of the receptor itself by the peptide associated with an MHC molecule, but also the binding of CD4 or CD8. These interactions modulate both the development of the immature T cell and the response of the mature T cell to antigen [1–3] by the intracellular molecules which CD4 and CD8 recruit [4–7].

In B cells, such concepts were slower to evolve, perhaps because the capacity for recognition of soluble antigen did not suggest a context. However, several facts suggest that a similar mechanism must operate in the B cell. First, peripheral B cells that recognize self-antigens do become suppressed, and thus must be responsive to context [8]. Secondly, when B cells leave the bone marrow and enter the periphery, the antigen receptors, formed by DNA rearrangement independently of antigen, have low affinity for most antigens, yet, if only mIg is ligated, a high percentage of receptors must be ligated for full activation to ensure [9]. These B cells must have some means of enhancing their responsiveness to physiological levels of anti-gen. Finally, recent data demonstrate that accessory receptors have a function in determining the pathway of B cell development. These observations have led to the idea that B cell activation by antigen is regulated by accessory receptors, and that CD19 functions in such a capacity.

Before the discovery of the association between CR2 and CD19, published research into both molecules had accumulated two independent bodies of literature, but there was a common theme. In the case of CR2, although *in vivo* work had demonstrated a requirement for complement in enhancing an antibody response at limited concentrations of antigen [10], existing studies of molecular mechanisms were contradictory. For CD19, early work had discovered that it also played a role in modulating antigen receptor signaling, but the published observations were mostly on inhibitory effects [11, 12]. The discovery of the CR2/CD19/TAPA-1 complex suggested a new model for both receptors [13, 14]. CR2 could function as a ligand binding subunit for the complex. Binding of C3dg and antigen in immune complexes by CR2 and the antigen receptor [15], respectively, would serve to bridge the CR2/CD19 complex to mIg, providing a mechanism for signaling by CR2. Studies to determine the role of co-aggregation of CD19 and mIgM revealed a powerful enhancing effect of CD19 under these conditions [16, 17]. Mutational analysis revealed that the different effects known to occur after ligation of CD19 (enhancement, inhibition, aggregation) could be ascribed to different associations [18, 19], which perhaps are differentially recruited after different types of binding. Biochemical studies revealed that CD19 became phosphorylated on tyrosine following ligation of either CD19 or mIgM [20–23]. However, until recently little was known of the role of CD19 *in vivo*, except as could be inferred from studies of CD19 *in vitro* and the work on CR2.

These earlier studies have been extensively reviewed previously [24, 25], and will only be summarized here. This review will focus on recent evidence which has significantly shaped our model for how CR2 and CD19 function. New data have confirmed that CD19 acts as an accessory protein, directly interacting with the B cell antigen receptor complex. New biochemical studies have demonstrated interaction of CD19 with Vav in human cells, and work with murine CD19 has shown biochemical functions similar to those of the human molecule, as well as highlighting a potential role in enhancing signals through early μ chain complexes. Loss of functions in CR2 and CD19 deficient mice and the effects of systemic administration of anti-murine CD19 monoclonal antibody during fetal life have revealed biologic functions for CD19 both in B cell development and in antibody responses to limited antigen. These new findings underscore the role of CR2/CD19 in modulating antigen-driven B cell responses.

CD19 STRUCTURE

Immunoprecipitates prepared with anti-CD19 from ^{125}I-surface labeled B cells lysed in digitonin reveal not only CD19, but also other labeled proteins, the most abundant of which are TAPA-1 and CR2 [13, 14]. To determine which regions of CD19 mediated these interactions, chimeric molecules were constructed in which the extracellular domain of CD19 was replaced by that of CD4, and/or the transmembrane or the cytoplasmic regions of CD19 were replaced by the corresponding domains of the HLA-A2 class I molecule (Figure 3.1) [18]. The chimera were expressed in cell lines, and specifically immunoprecipitated with the appropriate antibody. The immunoprecipitates were examined for the presence of the associated molecules. Such mapping studies demonstrated that the association of CD19 with TAPA-1 requires the extracellular domain of CD19. The association with CR2 required the transmembrane

and extracellular regions of CR2. Loss of association in similar constructs using a small spacer suggests that a region just outside the membrane is involved [19].

Searching for additional associated proteins revealed an association between CD19 and the BCR complex (RH Carter, JB Bolen, DT Fearon, unpublished observations). Immunoprecipitates prepared with CD19 from large numbers (2×10^8 per assay) of B lymphoblastoid cells lysed in Brij 96 were analyzed by Western blotting with anti-phosphotyrosine antibodies. Proteins of approximately 120, 70, 50 and 40 kDa, as well as CD19 itself, were present with specific antibody and not with isotype-matched control antibody. This pattern was consistent with the M_r of proteins (CD22, syk, Igα and Igβ) known to associate with the BCR [26]. The presence of the first two in association with CD19 was confirmed by Western blotting, and that of Igα and Igβ by two-dimensional gel electrophoresis. Bands of the appropriate M_r for lgM heavy and light chains could also be identified in the 2-D gels, and the presence of lgM heavy chain in CD19 precipitates was confirmed by Western blotting. The region of CD19 mediating the association with the BCR complex was determined using the chimeric molecules shown in Figure 3.1. Coprecipitation of the BCR-associated molecules was observed with the chimera containing the extracellular domain of CD4 and the transmembrane and cytoplasmic regions of CD19, but not with a trimeric molecule containing the further substitution of the cytoplasmic region of HLA class I for that of CD19, demonstrating a requirement for the CD19 cytoplasmic domain.

A series of truncation mutants, containing varying lengths of the CD19 cytoplasmic region, were expressed and examined for their capacity to associate with the BCR complex. A construct containing the CD4 extracellular region, the CD19 transmembrane region, and only the 17 amino acids encoded by the first exon of the CD19 cytoplasmic domain was sufficient. Although these studies do not exclude a role for the CD19 transmembrane domain, they suggest that of the cytoplasmic domain only this juxtamembrane region is required for the association. This allows the construction of a model in which CD19 associates with TAPA-1 and CR2 by a region which includes the transmembrane and juxtamembrane

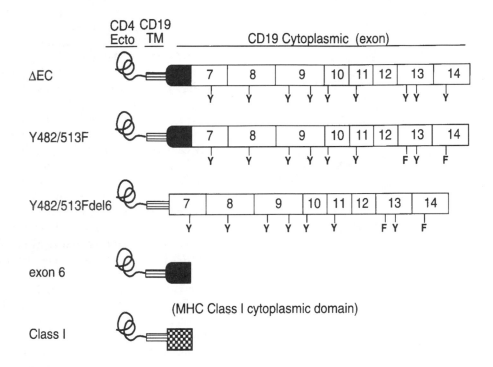

Figure 3.1 Models of chimeric molecules used to explore CD19 associations that are discussed in the text. These contain the extracellular domain of CD4, the transmembrane domain of CD19, and all or portions of the CD19 cytoplasmic domain, or the MHC class II cytoplasmic domain

extracellular portions of CD19, while the association with the BCR is by the juxtamembrane cytoplasmic region. The model makes sense topographically, as it leaves the *N*-terminal extracellular portions of CR2 and perhaps CD19 free to interact with ligands, and the *C*-terminal region of CD19 available for interactions with downstream signaling molecules. The component of the BCR with which CD19 interacts is unknown. CD22 appears less prominent relative to Igα and Igβ and syk in experiments in which the CD4/19 chimera are precipitated from B lymphoblastoid cells lysed in NP40. The retention of the Igα and Igβ association in NP40 suggests that the interaction of CD19 with these molecules is not via IgM, as the interaction of IgM with Igα and Igβ may be disrupted by NP40 [27]. The association is constitutive, without significant increase after ligation of mIg, and, as the exon-6-encoded region of CD19 does not contain tyrosines, is not mediated by tyrosine (RH Carter, JB Bolen, DT Fearon, unpublished observations).

The CD19/BCR association strengthens the conclusion that CD19 functions as an accessory protein for regulation of BCR-induced B cell activation. In addition, the association is consistent with an emerging paradigm for the tyrosine phosphorylation of B cell surface molecules; those molecules which physically associate with the BCR, such as Igα, Igβ, CD19 and CD22, become phosphorylated on tyrosine after ligation of mIg alone [20, 28, 29], while others such as FcγRIIB require co-crosslinkage [30, 31]. This concept is supported by the finding that crosslinking of either IgM or IgD on B cells which express both receptors results in tyrosine phosphorylation of only the Igα associated with the bound type of membrane Ig [32]. Similar findings have been demonstrated in studies using chimeric molecules containing the cytoplasmic domains of Igα and Igβ. Specific ligation of these chimeras results in phosphorylation on tyrosines of the Igα and Igβ portions of these molecules, but not of the native Igα and Igβ also present in the cells [33]. To test the hypothesis that

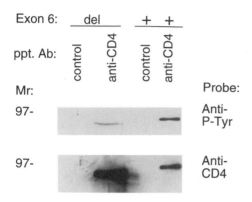

Exon 6:   del   + +

ppt. Ab: control anti-CD4 control anti-CD4

Mr: Probe:

97-   Anti-
                                           P-Tyr

97-   Anti-
                                           CD4

Figure 3.2 Daudi B lymphoblastoid cells (4×10^7 per sample) expressing either CD4/19 Y484/515F del6 or CD4/19 Y484/515F chimeras were stimulated with 20 μg/ml F(ab')^2 anti-human IgM for 60 s and lysed in 1% NP-40. Immunoprecipitates were prepared from each with either UPC-10 (control antibody) or with anti-CD4 (to precipitate the chimera). Eluates were resolved by SDS–PAGE, transferred to nitrocellulose, and probed with anti-phosphotyrosine (top) or anti-CD4 (bottom)

association of CD19 with the BCR accounts for tyrosine phosphorylation of CD19 after ligation of only mIg, a deletion mutant was constructed in which the exon-6-encoded region is removed from the CD4/19 chimera. In preliminary studies, the tyrosine phosphorylation of this deletion mutant appears significantly diminished relative to the full length chimera (Figure 3.2; RH Carter and DT Fearon, unpublished observations).

Another consequence of the low affinity interactions is the grouping, at least transiently, of the downstream effector molecules recruited by the 'accessory' proteins such as CD19 and CD22 with the activated BCR complex. Thus, PI3K and Vav, bound to CD19, would colocalize with the BCR. Indeed, in the coprecipitation studies discussed above, SHP, presumably bound to CD22 [62], could be detected in the CD19 precipitates.

BIOCHEMISTRY

With the development of a monoclonal antibody, studies of the biochemical functions of murine CD19 (mCD19) have became possible, and have confirmed and extended the findings in human cells. In an Abelson virus-transformed cell line, derived from a RAG-2 deficient mouse, reconsti-

tuted with murine μ chain, stimulation with either anti-mCD19 or anti-μ induced tyrosine phosphorylation of mCD19. The p85 subunit of PI3K coprecipitated with mCD19 from activated cells. In studies of PLC activation, mCD19 could induce an increase in cytosolic calcium by itself in murine spleen cells, and coligation of subthreshold amounts of mCD19 and membrane IgM resulted in a synergistically enhanced response. In progenitor B cells with μ chain but lacking light chain, coligation of mCD19 and μ chain also produced an enhanced calcium release (I. Krop, AL Shaffer, AR deFougerolles, RR Hardy, MS Schlissel and DT Fearon, submitted for publication). These studies, combined with those of human B cell precursors, raise the possibility that CD19 has a role in B cell development (see below), perhaps by modification of signals derived from μ itself or in combination with the surrogate light chain complex.

The binding of PI3K demonstrated the potential functional importance of phosphorylation of the cytoplasmic tyrosines of CD19 [20]. This association confirmed that these residues can act as sites for interactions with downstream effector molecules containing SH2 domains. The binding of PI3K was consistent with the presence at two tyrosines of the Y–X–X–M sequence reported as the consensus motif for association of the SH2 domains in the p85 subunit of PI3K with phosphorylated tyrosines in other proteins. However, CD19 has seven other cytoplasmic tyrosines. Studies of the CD4/19 truncation mutants discussed above suggest that at least several of the others are phosphorylated (RH Carter and DT Fearon, unpublished observations; see [25]). Songyang and others studied the affinities of peptides derived from a degenerate phosphopeptide library for the individual SH2 domains from a variety of proteins [34, 35]. This work identified motifs of amino acids, surrounding a tyrosine, which have a high affinity interaction with a given SH2 domain. The GenBank database was searched for proteins containing these sequences. Interestingly, the search predicted interaction of CD19 cytoplasmic tyrosines with several SH2 domains, including those of c-abl, Vav and 3BP2. Weng *et al.* [36] have demonstrated the association of Vav with both the p85 subunit of PI3K and CD19 in immature B cell lines. The former interaction did not seem to be tyrosine-phosphorylation dependent. The association between CD19 and Vav increased in parallel with the time course of tyrosine phosphorylation induced by crosslinking CD19. Although crosslinking CD19 under these conditions in mature cells would induce tyrosine phosphorylation of CD19, the phosphorylation status of CD19 in the immature cell lines was not addressed directly, and the mechanism of interaction of Vav and CD19 remains to be demonstrated.

All three signaling enzymes that are known to be regulated by or interact with CD19 (PLC, PI3K, Vav) might act on the Ras/MAP kinase pathway. PLC might activate PKC by the release of diacylglycerol and the increased concentration of intracellular free calcium concentration, leading to the serine/threonine phosphorylation of Raf [37]. PI3K associates physically with Ras [38, 39]. Vav is reported to increase Ras guanine nucleotide exchange [40, 41], although this has been disputed [42]. Ligation of CD19 in B cell precursors results in increased MEK activity [36], and co-crosslinking of CD19 and mIgM on B lymphoblastoid cells induces an enhanced retardation in Raf mobility in SDS–PAGE gels (A Sheehey, RH Carter, DT Fearon, unpublished observation). However, PLC, PI3K and Vav each have effects that do not seem to relate to the MAP kinase pathway, and the mechanism of CD19 enhancement of later stages of B cell activation remains unknown.

In Vav deficient B cells found in RAG knockout mice reconstituted with Vav$^{-/-}$ES cells, the increase in cytoplasmic free calcium concentration following ligation of the antigen receptor was blunted relative to Vav$^{+/-}$ B cells, suggesting a requirement for Vav for activation of PLC [43]. As Vav is known to associate with CD19, Vav may mediate the effects of CD19 in regulating calcium. The increase in inositol phosphate induced by CD19, either by itself or after coligation with mIgM, requires tyrosine kinase activity but occurs without increased tyrosine phosphorylation of PLC-γ [16], and Vav provides a potential mechanism. Perhaps the activation of Vav by the BCR complex (either directly or indirectly via association with molecules such as CD19) is blocked by treatment of B cells with PMA, while the CD19–Vav interaction is not, accounting for the failure of PMA to inhibit the rise in calcium induced by CD19, either by itself or following coligation with mIgM [16]. If the principal interaction between the BCR complex and Vav is indirect, a corollary would be that serine/threonine phosphorylation regulates intermolecular associations.

The biochemistry of the synergy between CR2 and mIg has been further explored by activating human B cells with latex beads coated with antibodies or natural ligands [44]. Coligation of CR2 and mIgM resulted in enhanced increases in c-fos mRNA levels. Ligands for CR2 had to bind at or near the C3dg site to be effective, and ligation of CR2 alone had no effect. The enhancement was blocked by the protein tyrosine kinase inhibitors herbimycin A or tyrphostin 25. CD19 ligation had no effect in this system, implying a CR2 signal transduction pathway that is independent of CD19. However, an effect of CD19 on fos DNA binding activity has been reported [45], and would be expected from the observations of the effects of CD19 on the Raf pathway ([36], and A Sheehey, RH Carter and DT Fearon, unpublished observation). The differences may relate to the mechanism of crosslinking of receptors. Coligation of CD19 and mIgM by antibodies bound to Sepharose was less effective than co-aggregation of the antibodies by culturing B cells with fibroblasts expressing transfected FcγRII (RH Carter and DT Fearon, unpublished observations). Restricted lateral diffusion of membrane proteins may affect complex formation and hence signal transduction.

CD19 *IN VIVO*

The development of the anti-mCD19 monoclonal antibody and the mCD19 knockout mice have at last established a role for CD19 *in vivo*. As one would expect from the pattern of expression of CD19, one function of CD19 relates to B cell development. In addition, as predicted from the studies of CD19 as an accessory protein for activation of B cells through the BCR, CD19 functions in the antibody response to challenge with antigen.

The studies of biochemical responses induced in B cell progeniters, described above, suggest that CD19 could deliver a signal that might regulate development. To test this hypothesis, anti-mCD19 was injected into pregnant mice. This treatment resulted in down-regulated expression of CD19 on B cells in the newborn pups. Total numbers of B-2 cells were no different from controls born of pregnant mice treated with isotype-matched irrelevant antibodies, but anti-mCD19 reduced splenic and peritoneal B-1a to 15% of normal. Prolonged treatment of mature mice with anti-mCD19 induced a gradual decline in peritoneal B-1a cells that reached 40% of normal after 5 weeks. Studies analyzing BrdU incorporation were consistent with a disruption in the process of self-renewal of these B cells rather than with acute elimination [63].

Two groups of researchers have recently constructed CD19-knockout mice by targeted disruption of the CD19 gene. Both groups found that the number of peritoneal B-1a cells in the CD19$^{-/-}$ mice was reduced to 10–15% of normal [64, 65]. Thus, maintenance of B-1a cells may be dependent on a low affinity interaction of the mIg on these cells with self-antigens, with CD19 being required to enhance the signal generated through this interaction, similar to the effects of CD19 *in vitro* on mIgM-induced DNA synthesis [17]. Interestingly, this type of enhancement appears not to be necessary for development of B-2 cells, in which ligation of mIg, without T cell help, leads to apoptosis.

The association of CD19 deficiency with lack of B-1a cell development is similar to the failure of development of B-1 cells in RAG-deficient mice reconstituted with Vav ES cells [43, 46, 47]. This might relate to the blunted receptor signaling in the Vav-deficient lymphocytes discussed above, supporting the hypothesis that a positive signal requiring CD19 and Vav is required for differentiation into B-1a cells. Whether the absence of Vav specifically associated with CD19 is the cause of the defect in either case cannot be determined at present, but this would make a unifying hypothesis.

A third molecule associated with defective development of B-1a cells is the tyrosine kinase btk. The *xid* mice, which carry a mutation in the gene for btk [48, 49], have a decreased total number of B cells, but the defect is more pronounced in the B-1 than B-2 subpopulation [50]. There is evidence for a defect in CD19 function in XLA patients (M. Cooper, personal communication). This suggests that CD19 is either involved in the activation of, or is a substrate for, btk. However, unlike the CD19-deficient mice (see below), the *xid* mice fail to respond to type II T-independent antigens and have reduced serum levels of

IgM and IgG [51], so CD19 appears not to be required for all functions in which btk is involved.

In addition to the findings related to development, the CD19-deficient mice also showed diminished responses to immunization. As one would predict in mice deficient in B-1 cells, the antibody response to the hapten NP coupled to dextran was reduced. The response to NP-ficoll, which is mediated by B-2 cells with little T cell help, was normal. In contrast, hapten-specific antibody production was reduced after immunization with a T-dependent antigen, particularly after secondary challenge. In addition, serum IgG1 and IgE and germinal center formation were reduced [64]. These deficiencies suggest lack of cytokines such as IL-4 and/or lack of a CD40 signal, consistent with the effect of the CD19 deficiency on T cell help. Thus, at the current level of knowledge, an effect of CD19 on activation of B-1 cells by antigen is suspected but unproven, and CD19 is not required for activation of B-2 cells by antigens which deliver a sufficient signal primarily through mIg, such as NP-ficoll. The effect of CD19 deficiency is most pronounced in responses requiring T–B collaboration. The most obvious ways this could happen are: (i) that CD19 enhances the effects of mIg ligation; or (ii) that CD19 has a direct or indirect interaction with T cells. The former possibility could occur through increased internalization and processing of antigen, or up-regulation of cytokine receptors or molecules required for T–B collaboration, such as MHC class II or B7. The role of PIK regulation of intracellular trafficking in other systems [52, 53] supports the concept of an effect of CD19 on antigen processing. The enhancing effects of CD19 on mIgM-induced proliferation required the presence of IL4 [17]. The coligation of CD19 with mIgM rendered the cells more responsive to IL4 without changing the IL4 dose–response curve. Optimal DNA synthesis by cells stimulated with either saturating or limited concentrations of anti-IgM, or the combination of limited anti-IgM together with anti-CD19 required at least 200 units IL4 (Figure 3). This observation is consistent with an up-regulation of either the cytokine receptors themselves and/or enhanced downstream signaling by these receptors (as opposed to increased receptor affinity).

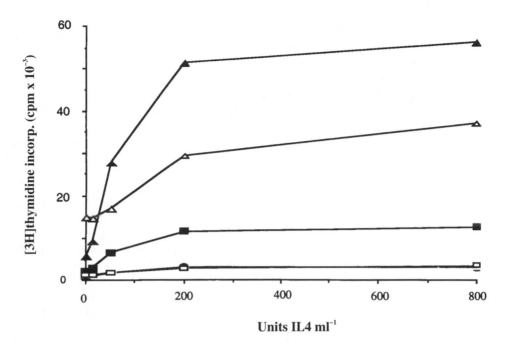

Figure 3.3 Normal peripheral blood B cells were cultured with L cells stably transfected with FcγRII and 0–800 units IL4ml^{-1}. The cells were stimulated with medium only (●), 1 mg ml^{-1}HD37 anti-CD19(□),1 mg ml^{-1} DA4.4 anti-IgM(△),10 ng ml^{-1} DA4.4 anti-IgM (■) or the combination of 1 mg ml^{-1}HD37 anti-CD19 and 10 ng ml^{-1} DA4.4 anti-IgM (▲). Cultures were pulsed with [3H] thymidine for the last 16 h of a 2.5 day culture

The hypothesis of an interaction of CD19 with T cells would go against the grain of current evidence, as ligation of CD19 independently from mIg shuts off B cell activation in most models. However, the effects of CD19 could depend on the activation stage of the B cell, similar to the role of MHC class II on B cells and CD4 on T cells, perhaps by regulated association of CD19 (and CD4 and MHC class II) with TAPA-1. The biochemical function of MHC class II molecules changes from cAMP generation and protein kinase C translocation, in resting cells, to PTK activation in B cells which have undergone previous activation [54]. The CD4 molecule changes from a costimulatory role in unprimed T cells to an inhibitory function in memory T cells [3]. Given the duration of expression of CD19 through several stages of B cell development, similar alterations in its functions would be consistent with this pattern.

A putative interaction of CD19 with T cells could be indirect. As a conjecture, CD19 might modulate MHC class II molecule function. The CD19 : TAPA-1 ratio might alter the MHC class

II : TAPA-1 complex, with an impact on expression, antigen presentation or signaling. A second possibility is that the CD19 deficiency interrupts the interaction between CR2 and CD23, which may be necessary for T–B collaboration [55].

CR2 *IN VIVO*

The enhancing effects of complement on the antibody response have been known since the demonstration by Pepys that treatment of mice with cobra venom factor to deplete C3 results in diminished specific antibody titers after primary and, more dramatically, secondary challenge with limited amounts of sheep red blood cells, a T-dependent antigen [10]. It is controversial whether this enhancement: (i) is mediated by complement receptors; (ii) involves trafficking of antigen or cell signaling; and (iii) is a function of B cells or dendritic cells. The initial report that treatment of mice with antibody that recognized murine CR1/ 2 blocked antibody responses after antigen challenge supported the concept that the effect is mediated by complement receptors [56, 57]. However, interpretation of this result is complicated by the possibility that binding of the antibody to CR1/2 might in itself down-regulate a response, independent of complement. Therefore, the finding that a soluble chimeric construct containing the C3dg-binding site of CR2 coupled to an IgG backbone could also inhibit antibody responses to the T dependent antigen KLH [58] established that the enhancing effect of complement was a function of CR1/2.

The role of C3 in the immune-enhancing effect of complement has now been shown to be related to its capacity to attach to the immunogen. Using recombinant DNA technology, a fusion protein was created by linking up to three copies of the cDNA for the C3d peptide in frame to the cDNA for hen egg lysozyme (HEL). Challenge of mice with the rcombinant protein in the absence of adjuvant demonstrated that the presence of two C3ds attached to the HEL lowered the dose for priming by 1000-fold, and that three C3ds attached to HEL lowered the dose by 10 000 fold. Treatment of the mice with anti-CD21 blocked the enhanced response to HEL-C3d, demonstrating that it was mediated by the complement receptor [66]. These findings suggest a means for enhancing the potency of vaccines.

To address the question of whether the enhancing effects of complement on antibody responses is a function of CR1/2 on B cells or on follicular dendritic cells, Ahearn, Carroll and coworkers reconstituted RAG 2 knockout mice with ES cells with targeted disruption of both alleles of CR1/2. As all mature B cells in these mice would be derived from the ES cells, all B cells are lacking CR1/2, while the follicular dendritic cells, which would not be effected by the RAG deficiency, appear to have normal levels of CR1/2. Such mice had greatly diminished antibody responses to the T dependent antigen KLH, compared to the normal response seen in mice reconstituted with CR1/2$^{+/+}$ ES cells [59, 67–69], demonstrating that the CR1/2 on B cells is required for the enhancing effects of complement. However, the data as yet do not permit distinction between a role for the CR1/2 on the B cells in transporting antigen to lymphoid centers and signaling by CR1/2 to enhance B cell activation. For example, the CR1/2 on follicular dendritic cells may be important in maintaining antigen in germinal centers, but in the B cell CR1/2$^{-/-}$ mice the antigen may never have arrived there.

In the studies of blocking the effects of complement (and CR2) on antibody responses, and in the CD19 deficient mice, the most dramatic effect was on T-dependent antigens, and on secondary responses associated with such functions of T cell help as class switching and memory. This correlation suggests that both operate through the same mechanism; signaling by the CR2/CD19 complex. The similarities between CR2 and CD19 deficiencies suggest that the effects of CR2 are through signaling the B cell by CD19. The evidence to date that this hypothesis is correct is the short cytoplasmic tail of CR2, the existence of the CR2/CD19 complex, the shared biochemical features such as synergistic enhancement of calcium responses after coligation with mIgM, and the shared ability to enhance mIg-induced proliferation of B cells *in vitro*. However, formal proof that the functions of CR2 are mediated by the complex (i.e. including the effects mediated by either CD19 or TAPA-1) is lacking. Certainly, CR2 on the B cell may have a dual role: CR2 on all B cells is capable of binding immune complexes containing antigens and multiple C3dg fragments and transporting them to a lymphoid center, while, on those B cells which have mIg specific for a particular antigen in the immune complex, CR2 functions to synergistically activate the cell (via CD19) when CR2 and mIg are both bound by the components of the immune complex. The CD19 deficient mice may provide insight to this question.

In humans, CR2 can associate with either CR1 or with CD19 [60]. Perhaps immune complexes are initially recognized by CR1, but the C3 fragments are rapidly cleaved to C3dg. Upon dissociation from CR1, the new C3dg is bound by that CR2 associated with CR1, but now the CR2 with bound C3dg releases from the CR1 and becomes associated with CD19. If the proper antigen is present, the CR2/CD19 complex becomes coligated with mIg, and enhanced activation ensues. Such an hypothesis would explain why antibodies which recognize the C3dg binding site on CR2 (e.g. OKB7) appear more potent than antibodies to other domains of CR2 (e.g. HB5) in certain assays [44, 61]. At the current time, however, there is little evidence on what regulates the association of CR2 with either CR1 (in humans) or CD19, and nothing to indicate "translocation" as hypothesized here. Furthermore, given the low stochiometry of the association between CD19 and the BCR complex, whether the CD19 associated with CR2 can constitutively interact with the BCR remains to be determined, although immune complexes containing C3dg and antigen could serve as a direct link.

CONCLUSION

Until recently, clues to the function of CD19 were drawn from studies of its expression, sequence, and *in vitro* reactions. The pattern of expression, from the time of IgM heavy chain rearrangement until final differentiation into a plasma cell, has suggested a role in differentiation. The cDNA revealed a large cytoplasmic domain with nine tyrosines, but without homologies to suggest a particular enzymatic or biochemical role. Initial functional studies revealed the potential for CD19 to inhibit B cell activation. More recent evidence demonstrated that CD19 is involved in activation triggered by the B cell antigen receptor (BCR) complex, and can act as a powerful enhancer. CD19 becomes rapidly phosphorylated on tyrosines after crosslinking of membrane immuno-globulin (mIg), and thereby associates with the cytoplasmic effector molecule phosphatidylinositol 3-kinase (PI3K). Coligation of CD19 and mIg could reduce by two orders of magnitude the amount of anti-Ig required to induce proliferation of B cells *in vitro*. Finally, CD19 associates with complement receptor type 2 (CR2) and might mediate the enhancing effects of the complement system on antibody responses. New studies of structural, biochemical and biologic functions of CD19 over the past year have confirmed this concept of CD19 as an "accessory" protein, and have highlighted the role of CD19 in B cell development. CD19 is physically as well as functionally associated with the mIg complex. CD19 also binds Vav, whose importance in B cell development and regulation has been demonstrated in RAG complementation studies. The role of CD19 *in vivo* has been investigated in mice using both anti-murine CD19 antibodies and in CD19 knockout mice. The effects of complement on antibody responses have been elegantly confirmed.

FUTURE PROSPECTS

Current understanding of CD19 function *in vivo* is summarized in the model shown in Figure 3.4. CD19 plays a role in B cell development, with an

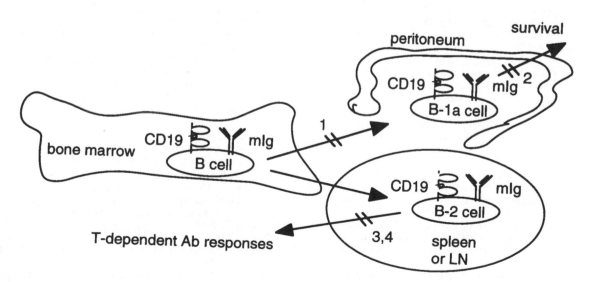

Figure 3.4 Model of role of CR2 and CD19 *in vivo*. Double slashes represent functions disrupted by blocking or disrupting CR2 or CD19. Numbers indicate functions described in text

essential function in B-1 cell (1) maturation and (2) maintenance, and that (3) CD19 and (4) CR2 are required for optimal antibody responses to limited amounts of T-dependent antigens. These observations are consistent with the concept of CD19 as an accessory protein for modulation of antigen-receptor-driven B cell responses, and the new biochemical data also support this hypothesis. However, alternative hypothesis exist for each of the *in vivo* observations, and the development of data that correlate biochemical effects with specific biologic functions will be exciting.

ACKNOWLEDGEMENTS

The authors thank Robert Rickert, Tom Tedder, Max Cooper and Mike Carroll for communicating recent findings. Douglas T. Fearon is a Principal Research Fellow of the Wellcome Trust. Robert H. Carter is supported in part by the Office of Research and Development, Medical Research Service, Dept. of Veteran Affairs, and the Arthritis Foundation.

REFERENCES

1. Bank I and Chess L. Perturbation of the T4 molecule transmits a negative signal to T cells. *J. Exp. Med.* 1985; **162**: 1294–1303.
2. Anderson P, Blue M-L, Morimoto C and Schlossman SF. Cross-linking of T3 (CD3) with T4 (CD4) enhances the proliferation of resting T lymphocytes. *J. Immunol.* 1987; **139**: 678–682.
3. Farber DL, Luqman M, Acuto O and Bottomly K. Control of memory CD4 T cell activation: MHC Class II molecules on APCs and CD4 ligation inhibit memory but not naive CD4 T cells. *Immunity* 1995; **2**: 249–259.
4. Rudd CE, Trevillyan JM, Dasgupta JD, Wong LL and Schlossman SF. The CD4 receptor is complexed in detergent lysates to a protein-tyrosine kinase (pp58) from human T lymphocytes. *Proc. Natl. Acad. Sci. USA* 1988; **85**: 5190–5194.
5. Veillette A, Bookman MA, Horak EM and Bolen JB. The CD4 and CD8 T cell surface antigens are associated with the internal membrane tyrosine-protein kinase p56lck. *Cell* 1988; **55**: 301–308.
6. Glaichenhaus N, Shastri N, Littman DR and Turner JM. Requirement for association of p56lck with CD4 in antigen-specific signal transduction in T cells. *Cell* 1991; **64**: 511–520.
7. Molina TJ, Kishihara K, Siderovski DP *et al.* Profound block in thymocyte development in mice lacking p56lck. *Nature* 1992; **357**: 161–164.
8. Goodnow CC, Crosbie J, Jorgensen H, Brink RA and Basten A. Induction of self-tolerance in mature peripheral B lymphocytes. *Nature* 1989; **342**: 385–391.
9. Maruyama S, Kubagawa H and Cooper MD. Activation of human B cells and inhibition of their terminal differentiation by monoclonal anti-μ antibodies. *J. Immunol.* 1985; **135**: 192–199.
10. Pepys MB. Role of complement in induction of antibody production *in vivo. J. Exp. Med.* 1974; **140**: 126.
11. Pezzutto A, Dorken B, Rabinovitch PS, Ledbetter JA, Moldenhauer G and Clark EA. CD19 monoclonal antibody HD37 inhibits anti-immunoglobulin-induced B cell activation and proliferation. *J. Immunol.* 1987; **138**: 2793–2799.
12. Pezzutto A, Barret TB, Ellingsworth L, Dorken B and Clark EA. Down-regulation of B-cell activation by CD19 mAb. In: Knapp W (ed.) *Leucocyte Typing IV.* Oxford: Oxford University Press, 1989: pp. 39–40.
13. Matsumoto AK, Kopicky-Burd J, Carter RH, Tuveson DA, Tedder TF and Fearon DT. Intersection of the complement and immune systems: a signal transduction complex of the B lymphocyte-containing complement receptor type 2 and CD19. *J. Exp. Med.* 1991; **173**: 55–64.
14. Bradbury LE, Kansas GS, Levy S, Evans RL and Tedder TF. The CD19/CD21 signal transducing complex of human B lymphocytes includes the target of antiproliferative antibody-1 and Leu-13 molecules. *J. Immunol.* 1992; **149**: 2841–2850.
15. Carter RH, Spycher MO, Ng YC, Hoffman R and Fearon DT. Synergistic interaction between complement receptor type 2 and membrane IgM on B lymphocytes. *J. Immunol.* 1988; **141**: 457–463.
16. Carter RH, Tuveson DA, Park DJ, Rhee SG and Fearon DT. The CD19 complex of B lymphocytes. Activation of phospholipase C by a protein tyrosine kinase-dependent pathway that can be enhanced by the membrane IgM complex. *J. Immunol.* 1991; **147**: 3663–3671.
17. Carter RH and Fearon DT. CD19: lowering the threshold for antigen receptor stimulation of B lymphocytes. *Science* 1992; **256**: 105–107.
18. Matsumoto AK, Martin DR, Carter RH, Klickstein LB, Ahearn J and Fearon DT. Functional dissection of the CD21/CD19/TAPA-1/Leu-13 complex of B lymphocytes. *J. Exp. Med.* 1993; **178**: 1407–1417.
19. Bradbury LE, Goldmacher VS and Tedder TF. The CD19 signal transduction complex of B lymphocytes. *J. Immunol.* 1993; **151**: 2915–2927.
20. Tuveson DA, Carter RH, Soltoff SP and Fearon DT. CD19 of B cells as a surrogate kinase insert region to bind phosphatidylinositol 3-kinase. *Science* 1993; **260**: 986–989.
21. Roifman CM and Ke S. CD19 is a substrate of the antigen receptor-associated protein tyrosine kinase in human B cells. *Biochem. Biophys. Res. Commun.* 1993; **194**: 222–225.
22. Uckun F, Burkhardt AL, Jarvis L *et al.* Signal transduction through the CD19 receptor during discrete developmental stages of human B-cell ontogeny. *J. Biol. Chem.* 1993; **268**: 21172–21184.
23. Chalupny NJ, Kanner SB, Schieven GL *et al.* Tyrosine phosphorylation of CD19 in pre-B and mature B cells. *EMBO J.* 1993; **12**: 2691–2696.

24. Tedder TF. The CD19/CD21 signal transduction complex of B lymphocytes. *Immunol. Today* 1994; **15**: 437–442.

25. Fearon DT and Carter RH. The CD19/CR2/TAPA-1 complex of B lymphocytes. *Ann. Rev. Immunol.* 1995; **13**: 127–149.

26. Law C-L, Sidorenko SP and Clark EA. Regulation of lymphocyte activation by the cell-surface molecule CD22. *Immunol. Today* 1994; **15**: 442–449.

27. Hutchcroft JE, Harrison ML and Geahlen RL. Association of the 72-kDa protein-tyrosine kinase PTK72 with the B cell antigen receptor. *J. Biol. Chem.* 1992; **267**: 8613–8619.

28. Peaker CJ and Neuberger MS. Association of CD22 with the B cell antigen receptor. *Eur. J. Immunol.* 1993; **23**: 1358–1363.

29. Leprince C, Draves KE, Geahlen RL, Ledbetter JA and Clark EA. CD22 associates with the human surface IgM–B-cell antigen receptor complex. *Proc. Natl. Acad. Sci. USA* 1993; **90**: 3236–3240.

30. Muta T, Kurosaki T, Misulovin Z, Sanchez M, Nussenzweig MC and Ravtech JV. A 13-amino-acid motif in the cytoplasmic domain of the FcγRIIB modulates B-cell receptor signalling. *Nature* 1994; **368**: 70–73.

31. D'Ambrosio D, Hippen KL, Minskoff SA *et al*. Recruitment and activation of PTP1C in negative regulation of antigen receptor signaling by FcγRIIb1. *Science* 1995; **268**: 293–297.

32. Gold MR, Matsuuchi L, Kelly RB and DeFranco AL. Tyrosine phosphorylation of components of the B-cell antigen receptors following receptor crosslinking. *Proc. Natl. Acad. Sci. USA* 1991; **88**: 3436–3440.

33. Kim KM, Alber G, Weiser P and Reth M. Differential signaling though the Ig-α and Ig-β components of the B cell antigen receptor. *Eur. J. Immunol.* 1993; **23**: 911–916.

34. Songyang Z, Shoelson SE, Chaudhuri M *et al*. SH2 domains recognize specific phosphopeptide sequences. *Cell* 1993; **72**: 767–778.

35. Songyang Z, Shoelson SE, McGlade J *et al*. Specific motifs recognized by the SH2 domains of Csk, 3BP2, fps/fes, GRB-2, HCP, SHC, Syk, and Vav. *Mol. Cell. Biol.* 1994; **14**: 2777–2785.

36. Weng W-K, Jarvis L and LeBien TW. Signaling through CD19 activates Vav/mitogen-activated protein kinase pathway and induces formation of a CD19/Vav/phosphatidylinositol 3-kinase complex in human B cell precursors. *J. Biol. Chem.* 1994; **269**: 32514–32521.

37. Kolch W, Heidecker G, Kochs G *et al*. Protein kinase Cα activates RAF-1 by direct phosphorylation. *Nature* 1993; **364**: 249–252.

38. Sjolander A, Yamamoto K, Huber BE and Lapetina EG. Association of p21 ras with phosphatidylinositol 3-kinase. *Proc. Natl. Acad. Sci. USA* 1991; **88**: 7908–7912.

39. Rodriguez-Viciana P, Warne PH, Dhand R *et al*. Phosphatidylinositol-3-OH kinase as a direct target of Ras. *Nature* 1994; **370**: 527–532.

40. Ye ZS and Baltimore D. Binding of Vav to Grb2 through dimerization of Src homology 3 domains. *Proc. Natl. Acad. Sci. USA* 1991; **26**: 12629–12633.

41. Gulbins E, Coggeshall KM, Baier G, Katzav S, Burn P and Altman A. Tyrosine kinase-stimulated guanine nucleotide exchange activity of Vav in T cell activation. *Science* 1993; **260**: 822–825.

42. Pastor MI, Reif K and Cantrell D. The regulation and function of p21ras during T-cell activation and growth. *Immunol. Today* 1995; **16**: 159–164.

43. Fischer K-D, Zmuidzinas A, Gardner S, Barbacid M, Bernstein A and Guidos C. Defective T-cell receptor signalling and positive selection of Vav-deficient CD4+ CD8+ thymocytes. *Nature* 1995; **374**: 474–477.

44. Luxembourg AT and Cooper NR. Modulation of signaling via the B cell antigen receptor by CD21, the receptor for C3dg and EBV. *J. Immunol.* 1994; **153**: 4448–4456.

45. Weng W-K, O'Brien D, VanNess B and LeBien TW. Signalling through CD19 activates AP1 transcription factors in human B lineage cells. *FASEB J.* 1995; **9**: A214.

46. Tarakhovsky A, Turner M, Schaal S *et al*. Defective antigen-receptor mediated proliferation of B and T cells in the absence of Vav. *Nature* 1995; **374**: 467–470.

47. Zhang R, Alt FW, Davidson L, Orkin SH and Swat W. Defective signalling through the T-and B-cell antigen receptors in lymphoid cells lacking the vav proto-oncogene. *Nature* 1995; 374: 470–473.

48. Thomas JD, Sideras P, Smith CI, Vorechovsky I, Chapman V and Paul WE. Colocalization of X-linked agammaglobulinemia and X-linked immunodeficiency genes. *Science* 1993; **261**: 355–358.

49. Rawlings DJ, Saffran DC, Tsukada S *et al*. Mutation of unique region of Bruton's tyrosine kinase in immunodeficient XID mice. *Science* 1993; **261**: 358–361.

50. Hayakawa K, Hardy RR and Herzenberg LA. Peritoneal Ly-1 B cells: genetic control, autoantibody production, and increased lambda light chain expression. *Eur. J. Immunol.* 1986; **16**: 450–456.

51. Scher I. The CBA/N mouse strain: an experimental model illustrating the influence of the X chromosome on immunity. *Adv. Immunol.* 1982; **33**: 1–71.

52. Okada T, Kawano Y, Sakakibara T, Hazeki O and Ui M. Essential role of phosphatidylinositol 3-kinase in insulin-induced glucose transport and antilipolysis in rat adipocytes. Studies with a selective inhibitor wortmannin. *J. Biol. Chem.* 1994; **269**: 3568–3573.

53. Yeh JI, Gulve EA, Rameh L and Birnbaum MJ. The effects of wortmannin on rat skeletal muscle. Dissociation of signaling pathways for insulin-and contraction-activated hexose transport. *J. Biol. Chem.* 1995; **270**: 2107–2111.

54. Andre P, Cambier JC, Wade TK, Raetz T and Wade WF. Distinct structural compartmentalization of the signal transducing functions of Major Histocompatibility Complex Class II (Ia) molecules. *J. Exp. Med.* 1994; **179**: 763–768.

55. Grosjean I, Lachaux A, Bella C, Aubry JP, Bonnefoy JY and Kaiserlian D. CD23/CD21 interaction is required for presentation of soluble protein antigen by lymphoblastoid B cell lines to specific CD4+ T cell clones. *Eur. J. Immunol.* 1994; **24**: 2982–2986.

56. Heyman B, Wiersma EJ and Kinoshita T. *In vivo* inhibition of the antibody response by a complement receptor-specific monoclonal antibody. *J. Exp. Med.* 1990; **172**: 665–668.

57. Thyphronitis G, Kinoshita T, Inoue K *et al*. Modulation of mouse complement receptors 1 and 2 suppresses antibody responses *in vivo*. *J. Immunol.* 1991; **147**: 224–230.

58. Hebell T, Ahearn JM and Fearon DT. Suppression of the immune response by a soluble complement receptor of B lymphocytes. *Science* 1991; **254**: 102–105.

59. Carroll MC, Ma M, Kelsoe GK, Han SW, Croix D and Ahearn J. Examination of the role of CR1/2 in acquired immunity. *FASEB J.* 1995; **9**: A777.

60. Tuveson DA, Ahearn JM, Matsumoto AK and Fearon DT. Molecular interactions of complement receptors on B lymphocytes: a CR1/CR2 complex distinct from the CR2/CD19 complex. *J. Exp. Med.* 1991; **173**: 1083–1089.

61. Bohnsack JF and Cooper NR. CR2 ligands modulate human B cell activation. *J. Immunol.* 1988; **141**: 2569–2576.

62. Doody GM, Justement LB, Delibrias CC, *et al.* A role in B cell activation for CD22 and the protein tyrosine phosphatase SHP. *Science* 1995; **269**: 242–244.

63. Krop I, de Fougerolles AR, Hardy RR, *et al.* Self-renewal of B-1 lymphocytes is dependent on CD19. *Eur.J. Immunol.* 1996; **26**(1): 238–42.

64. Rickert RC, Rajewsky K. Roes J. Impairment of T-cell-dependent B-cell responses and B-1 cell development in CD19-deficient mice. *Nature* 1995; **376**: 352–355.

65. Engel P, Zhou L-J, Ord DC, et al Abnormal B lymphocyte development, activation, and differentiation in mice that lack or overexpress the CD19 signal transduction molecule. *Immunity* 1995; **3**: 39–50.

66. Dempsey PW, Allison ME, Akkaraju S, et al C3d of complements as a molecular adjuvant: bridging innate and acquired immunity. *Science* 1996; **271**(5247) : 348–50.

67. Ahearn J, Fischer M, Croix D, *et al.* Disruption of the CR2 locus results in a reduction in B-1a cells and in an impaired B cell response to T-dependent antigen. *Immunity* 1996; **4** 251–261.

68. Croix D, Ahearn J, Rosengard A, *et al.* Antibody response to a T-dependent antigen requires B cell expression of complement receptors. *J. Exp. Med.* 1996; **183**: 1857–1864.

69. Molina H. Holers V, Li B, *et al.* Markedly impaired humoral immune response in mice deficient in complement receptors 1 and 2. *Proc. Natl. Acad. Sci., USA* 1996; **93**: 3357–3361.

4

CD23 and T–B Cell Interactions

Jean-Yves Bonnefoy, Jean-Pierre Aubry, Jean Francois Gauchat, Pierre Graber and Sybille Lecoanet-Henchoz

Glaxo Institute for Molecular Biology, Plan-les-Ouates, Geneva, Switzerland

INTRODUCTION

This chapter summarizes recent data on CD23 and its role in T cell/B cell interaction. CD23 is the only FcR which does not belong to the immunoglobulin gene superfamily. The CD23 molecule was discovered independently as an IgE receptor on human lymphoblastoid B cells [1], as a cell surface marker expressed on Epstein–Barr Virus transformed B cells (EBVCS) [2] and as a B-cell activation antigen (Blast 2) [3]. CD23 and the low affinity receptor for IgE (FcRII) were shown to be the same molecule [4, 5]. Similar to most FcR, soluble forms of CD23 (sCD23) are released into extracellular fluids. The soluble fragments formed by proteolytic cleavage of surface CD23 are not only capable of binding IgE (IgE binding factors) but also exhibit multiple functions *in vitro* that are not IgE related. These observations together with the finding that CD23 displays significant homology with Ca^{2+} dependent (C-type) animal lectins, suggested the existence of natural ligands other than IgE. The recent finding that CD23 interacts with membrane CD21 [6] indicates that CD23 should be viewed not only as a low affinity IgE receptor but also as an adhesion molecule involved in cell-cell interaction.

CD23 AND IgE PRODUCTION

The cytokines Il-4 and IL-13 which induce CD23 expression, were also found to induce IgE production by normal B cells by causing IgE isotype switching [7, 8]. IL-4-induced IgE synthesis was blocked by interferon-γ and α, and prostaglandin E_2 [9], factors which also inhibit the induction of CD23 by IL-4 on B cells [10]. sCD23 has been shown to augment ongoing spontaneous IgE production by B cells derived from atopic individuals [11] and to act synergistically with a suboptimal concentration of IL-4 to induce IgE production by normal human B cells [9]. The importance of CD23 or its soluble fragments in the synthesis of human IgE *in vitro* was highlighted by the demonstration that a specific subset of anti-CD23 monoclonal antibodies blocked IL-4-induced IgE production by normal B cells [9, 12, 13] as well as the spontaneous production of IgE by B cells from atopic patients [12]. *In vivo*, the findings of high cell-surface expression as well as of elevated serum CD23 levels in Hyper-IgE syndrome and various allergic and parasitic diseases [14–16], provided circumstantial evidence for CD23 involvement in IgE production. This was later substantiated by another correlative finding *in vivo*, that peak levels of serum CD23 in patients from autologous bone marrow transplantation anticipated the rise in IgE levels observed in these patients [17].

Physical interaction between T cells and B cells is a process known to be required for IgE production [18]. Therefore, one possible role for CD23 in IgE regulation would be in enhancing the T cell–B cell interaction [19]. In support for such a role, we have found that certain anti-CD23 monoclonal antibodies inhibit T/B cell conjugate formation *in vitro* [20] and that this inhibition is restricted mainly to CD4-positive T cells which form conjugates with B cells [21]. Interestingly, the CD23 antibodies inhibiting IgE production [9, 12, 22], and the homotypic [23] and heterotypic [20] cell–cell interactions, map to the lectin domain of CD23. This region shares homology with certain well-known adhesion molecules such as ELAM-1, MEL-14 and GMP-140 [24], and also embodies the IgE binding activity [25]. It was found only recently that this domain encompasses the binding to CD21 [6], which (unlike the IgE binding) is carbohydrate-dependent [26].

CD23 INTERACTIONS WITH CD21 TO MEDIATE T–B CELL COOPERATION

We have recently found that CD23 specifically interacts with another cell surface molecule in addition to IgE, that most probably contributes to T cell/B cell interaction. Recombinant full-length CD23 reconstituted into fluorescent liposomes was found to bind to a subset of T cells and B cells [26]. Studies using inhibitory anti-CD21 antibodies as well as the binding of the CD23 liposomes to recombinant CD21 transfected cells revealed that CD23 binds to a subtype of CD21 [6], a molecule that had been previously identified as the receptor for EBV and as the complement receptor-2.

Structurally, CD21 consists of 15 to 16 short consensus repeats of 60 to 75 amino acids (SCRs) followed by a transmembrane domain and an intracytoplasmic region. The sites of interaction of CD23 on CD21 were determined using purified recombinant CD23 incorporated into fluorescent liposomes and CD21 mutants bearing various deletions of extracytoplasmic SCRs. The sites of interaction of CD23 on CD21 are on SCRs 5–8 and 1–2. The first site is involved in a lectin–sugar type of interaction and the second site in a protein–protein interaction. By mutating, together or individually, the three asparagines present in SCRs 5–8, Asn 370 and 295 (but not Asn 492), were shown to be critically involved in the binding of CD23 [27]. SCRs 5–8 therefore represent a novel functional domain on the CD21 molecule and is the first demonstration of an activity in an extracytoplasmic region of CD21 outside of SCRs 1–4. Among the other ligands for CD21 (EBV, C3d,g and IFNα), only EBV virus which binds to SCR 2 can inhibit the binding of CD23 to CD21. Furthermore, even a peptide from gp350/220 of EBV that is known to bind to CD21 is able to decrease CD23 binding to CD21 [28]. Monoclonal antibodies which map in SCRs 5–8 inhibit the most CD23 binding to CD21 and mimic the most CD23 activities [27].

CD23/CD21 INTERACTIONS REGULATE IgE PRODUCTION

CD21 is expressed on B cells, follicular dendritic cells and on T cells. Anti-CD21 antibodies, like anti-CD23 antibodies, were found to decrease T/B conjugate formation [20], as would be expected if CD23/CD21 pairing has a role in the T cell/B cell interaction. Thus CD23 appears to be another molecule involved in the T cell / B cell interaction, along with LFA-1/ICAM-1, CD2/LFA-3, CD28/B7 and CD5/CD72. But the particular characteristic of the CD21/CD23 pair, when compared to the others, is that this pairing controls IgE synthesis in an isotype-specific manner.

Engagement of CD21 on B cells by some anti-CD21 mAbs increased IL-4-induced IgE production, as did treatment with recombinant sCD23 [6]. Molecular analysis revealed that triggering of CD21 increased the IL-4-induced germline ε transcription levels and had a synergistic effect on the expression of the ε transcript induced by T cells [29]. Increased IgE production by triggering of CD21 was observed in both T-cell-dependent and independent systems [29], suggesting that heterotypic T–B [20] and homotypic B–B [23] interactions involving CD23–CD21 can take place. In order to respond to CD23–CD21 stimulation and IL-4, purified B cells require either the presence of T cells or costimulation by anti-CD40 monoclonal antibody [29]. The CD40/CD40-ligand pair is likely to be a major cosignal required for CD21/CD23 pairing effect on IgE synthesis. Triggering of CD40 by CD40-ligand expressed on T cells [30] or basophils and mast cells [31] has been shown to induce IgE switching in purified B cells. Moreover, the recent findings that CD40-ligand mutations are responsible for the hyper-IgM syndrome, emphasize the central role of the CD40/CD40-ligand interaction in immunoglobulin class switch [32–36]. Based on the observations that triggering of CD21 by either anti-CD21 antibodies or by recombinant sCD23 enhances IgE production [6, 29], and based on the reported induction of CD23 on T cells by IL-4 and allergen [37], one can speculate that in allergic individuals, T-cell-associated

CD23 could interact with B-cell-associated CD21 leading to an increase in IgE production. Conversely, the T cell/B cell interaction leading to IgE production would not occur in normal individuals due to the absence of CD23 expression on T cells.

Clinical data support the role of CD23–CD21 in human IgE-mediated diseases. Allergen induces CD23 expression on CD4+T cells and CD21 expression on B cells in patients with allergic asthma [38]. Moreover, CD23 expression on lymphocytes is decreased in allergic children undergoing hyposensitization [39]. Interestingly, the EBV-peptide which inhibited CD23 binding to CD21, also inhibited IgE and IgG4 production induced by IL-4. Another CD21 ligand, C3, did not affect binding of CD23 to CD21 nor the production of IgE and IgG4 [28]. These results are in keeping with previous reports that EBV can substitute the costimulatory signal required in the IL-4-permissive system to induce IgE secretion by human B lymphocytes [40].

CELL-ASSOCIATED CD23 IS LIKELY TO PLAY A MAJOR ROLE IN IgE REGULATION

Polyclonal antibodies raised to the human CD23 lectin domain, highly conserved between murine and human CD23, were shown to recognize a surface molecule on rat B lymphocytes that have been stimulated *in vivo* with homologous IgE. This process has been previously shown to up-regulate the rat lymphocyte IgE receptor and widely used to study the rat FcERII [41]. Intact anti-CD23 antibodies or their Fab fragments were found highly efficient to diminish the IgE response induced by pertussis toxin (PT) + ovalbumin (OVA) in this animal model [42]. Moreover, expression of rat B cell FcERII showed a clear correlation to serum IgE levels, as it has been described for human CD23 in subjects with high IgE levels [14, 15, 16]. Cloning of rat CD23 revealed a high degree of protein homology to murine and human CD23. The sequence also revealed that rat CD23, in a similar fashion to its murine homolog but unlike the human counterpart, lacks the inverted RGD sequence close to the extracellular carboxy terminus [42]. This *in vivo* model does not exclude the possibility of CD23 also acting through soluble forms. However, the fact that CD23 has indeed been shown to participate in several cell–cell interactions including T–B cell conjugate formation [20], antigen presentation [43–46], B cell homotypic adhesion [23], and given its physical association to MHC class II [43, 47], indicates that cell-associated CD23 is likely to play a major role in IgE regulation.

CD23 KNOCKOUT AND TRANSGENIC MICE

Recently CD23-deficient and overexpressing transgenic mice have been generated. Overexpression of either membrane CD23 or soluble CD23 (38 kDa) did not alter lymphoid cell maturation [48]. The CD23 deficient mice do not display IgE-dependent augmentation of an immune response [49]. Inactivation of the CD23 gene does not modify the capacity of mice to develop an IgE response to *Nippostrongylus brasiliensis* [49–51]. This result is in line with previous findings using a rat anti-mouse CD23 antibody. Moreover, membrane CD23 transgenic mice showed an impairment of IgE responses. In contrast, soluble CD23 transgenic mice behaved like non-transgenic mice regarding IgE production [48], suggesting that in the mouse system at least, the soluble form of CD23 does not exhibit the same activity as the membrane form. Immunization of CD23 knockout mice with thymus-dependent antigen resulted in an increase of IgE production, which was interpreted as evidence for CD23 providing a negative feedback regulation for IgE synthesis [51].

CD23–CD21 INTERACTIONS IN GERMINAL CENTERS

The CD23–CD21 lymphocyte interaction may have other important effects independent of IgE regulation. Another CD23-positive cell type, crucially involved in the generation and the maintenance of immune responses is the Follicular Dendritic Cell (FDC). These cells, located exclusively in the germinal centers of secondary lymphoid tissues, are strong expressers of CD23 both in human [52] and mouse [53]. The function, however, of these high levels of CD23 on FDC is still unknown. It is probably involved in B cell rescue, activation and growth during germinal center reactions [54]. CD23 is one of the signals that can rescue germinal center B cells (GCC) from apoptosis [54], and this is mirrored by anti-CD21 antibodies [55], thus suggesting that CD23 on FDC may affect B cell development in germinal centers by acting through CD21. In mice, FDC-CD23 is up-regulated upon exposure to parasites, and clearly involved in the capture and retention of IgE-immune complexes [53]. Interestingly, both CD23 and IgE have been found strongly up-regulated on FDC from patients bearing high levels of serum IgE [56]. These observations open the possibility that immune complexes consisting of IgE and allergen, and retained (probably for long periods) on FDC through CD23, might be relevant in the maintenance of long-term IgE responses and in the generation of IgE-memory B cells.

NOVEL ROLES FOR CD23–CD21 INTERACTIONS

We also demonstrated that CD21 is expressed on basophilic cells and that CD21 controls histamine production upon ligand-induced stimulation (CD23 or anti-CD21 mAb) [57]. The CD23/CD21 interaction is therefore a novel mechanism for basophil activation and increases histamine release. The CD23/CD21 pair of adhesion molecules is able to regulate IgE production and is also able to act at another stage of the allergic reaction by regulating effector cells such as basophils. Interestingly, EBV-B cell lines and nasopharyngeal carcinoma cells express both CD23 and CD21, and therefore the CD23/CD21 interaction may well contribute to auto-stimulatory growth of these transformed cells.

FUTURE PROSPECTS

As a conclusion, CD23 becomes a more and more interesting molecule. It is an Fc receptor for IgE and as such is involved in IgE-mediated inflammatory responses and in protective immunity against parasites. It is also an adhesion molecule which interacts with the CD21 antigen expressed on T and B cells and by this means is involved in differentiation of B cells into IgE secreting plasma cells, in survival of germinal center B cells and in the release of histamine from basophils. Furthermore, CD23 is cleaved into soluble fragments acting as cytokines *in vitro* and could therefore be viewed as a membrane-bound cytokine, such as TNFα or CD40-ligand. Finally, because of its multiple roles on various target cells, it is quite likely that CD23 may well interact with yet other unidentified ligand(s). In this context, we recently demonstrated that CD23 interacts functionally with CD11b/CD18 and CD11c/CD18 on monocytes [58].

ACKNOWLEDGEMENTS

The authors wish to thank Drs K Hardy and J Knowles for reviewing the manuscript and support.

REFERENCES

1. Gonzalez-Molina A and Spiegelberg HL. Binding of IgE myeloma proteins to human cultured lymphoblastoid cells. *J. Immunol.* 1976; **117**: 1838–1845.
2. Kintner C and Sugden B. Identification of antigenic determinants unique to the surfaces of cells transformed by Epstein–Barr virus. *Nature* 1981; **294**: 458–460.
3. Thorley-Lawson TA, Nadler LM, Bahn AK and Scholley RT. BLAST-2 [EBVCS], an early cell surface marker of human B cell activation, is superinduced by Epstein–Barr virus, *J. Immunol.* 1985; **143**: 3007–3012.
4. Yukawa K, Kikutani H, Howaki H, Yamasaki K, Yokota A, Makamura H, Barsumian EL, Hardy RR, Suemura M and Kishimoto T. A B cell-specific differentiation antigen CD23, is a receptor for IgE (FcR) on lymphocytes, *J. Immunol.* 1987; **138**: 2576–2580.
5. Bonnefoy JY, Aubry JP, Peronne C, Wijdeness J and Banchereau J. Production and characterization of a monoclonal antibody specific for the human lymphocyte low affinity receptor for IgE: CD23 is a low affinity receptor for IgE. *J. Immunol.* 1987; **138**: 2970–2978.
6. Aubry JP, Pochon S, Graber P, Jansen KU and Bonnefoy JY. CD21 is a ligand for CD23 and regulates IgE production. *Nature* 1992; **358**: 505–507.
7. Lebman DA and Cofman RL. Interleukin 4 causes isotype switching to IgE in T cell stimulated clonal B cell cultures. *J. Exp. Med.* 1988; **168**: 853–862.
8. Punnonen J, Aversa G, Cocks BG *et al.* Interleukin 13 induces interleukin 4-independent IgG4 and IgE synthesis and CD23 expression by human B cells. *Proc. Natl. Acad. Sci. USA* 1993; **90**: 3730–3734.

9. Pene J, Rousset F, Briere F *et al.* IgE production by normal human lymphocytes is induced by interleukin 4 and suppressed by interferons γ and α and prostaglandin E2. *Proc. Natl. Acad. Sci. USA* 1988; **85**: 6880–6884.

10. Defrance T, Aubry JP, Rousset F, Vanbervliet B, Bonnefoy JY, Arai N, Takebe Y, Yokota T, Lee F, Arai K, de Vries J and Banchereau J. Human recombinant interleukin 4 induces Fc receptors (CD23) on normal B lymphocytes. *J. Exp. Med.* 1987, **165**: 1459–1467.

11. Sarfati M, Rector E, Rubio-Trujillo M, Wong K, Sehon AH and Delespesse G. *In vitro* synthesis of IgE by human lymphocytes. II. Enhancement of the spontaneous IgE synthesis by IgE-binding factors secreted by RPMI 8866 lymphoblastoid B cells. *Immunology* 1984; **53**: 197–205.

12. Sarfati M and Delespese G. Possible role of human lymphocyte receptor for IgE (CD23) or its soluble fragments in the in vitro synthesis of human IgE. *J. Immunol.* 1988; **141**: 2195–2219.

13. Bonnefoy JY, Shields JG and Mermod JJ. Inhibition of human interleukin-4-induced IgE synthesis by a subset of anti-CD23/FcRII monoclonal antibodies, *Eur. J. Immunol.* 1990; **20**: 139–144.

14. Thompson LF, Spiegelberg HL and Buckley RH. IgE Fc receptor positive T and B lymphocytes in patients with the Hyper IgE syndrome. *Clin. Exp. Immunol.* 1985; **59**: 77–84.

15. Nagai T, Adachi M, Noro N *et al.* T and B lymphocytes with immunoglobulin E Fc receptors (FcER) in patients with nonallergic hyperimmunoglobulinemia E: Demonstration using a monoclonal antibody against FcER-associated antigen. *Clin. Immunol. Immunopathol.* 1985; **35**: 261–275.

16. Yanagihara Y, Sarfati M, Marsh D *et al.* Serum levels of IgE-binding factor (soluble CD23) in diseases associated with elevated IgE. *Clin. Exp. Allergy* 1990; **20**: 395–401.

17. Bengtsson M, Gordon J, Flores-Romo L *et al.* B cell reconstitution after autologous bone marrow transplantation: Increase in serum CD23 precedes IgE and B cell regeneration. *Blood* 1989; **73**: 2139–2144.

18. Vercelli DH, Jabara H, Arai K and Geha RS. Induction of human IgE synthesis requires interleukin 4 and T/B cell interactions involving the T cell receptor/CD23 complex and MHC class II antigen. *J. Exp. Med.* 1989; **169**: 1295–307.

19. Bonnefoy JY, Pochon S, Aubry JP, Graber P, Gauchat JF, Jansen K and Flores-Romo L. A new pair of surface molecules involved in human IgE regulation. *Immunol. Today* 1993; **14**: 1–2.

20. Aubry JP, Shields G, Jansen KU and Bonnefoy JY. A multiparameter flow cytometric method to study surface molecules in interactions between subpopulations of cells. *J. Immunol. Methods* 1993; **159**: 161–71.

21. Bonnefoy JY. Immunoglobulin E FcRII: Molecular biology and role in immunoglobulin-mediated inflammation. In: Levinson AI and Paterson Y (eds) *Molecular and Cellular Biology of the Allergic Response*, 1st edn. New York: Marcel Dekker, Inc., 1994: pp. 245–259.

22. Bonnefoy JY, Mermod JJ and Shields J. Inhibition of human IL-4-induced IgE synthesis by a subset of anti-CD23/FcERII monoclonal antibodies. *Eur. J. Immunol.* 1990; **20**: 139–144.

23. Bjorck P, Elenstrom-Magnusson C, Rosen A *et al.* CD23 and CD21 function as adhesion molecules in homotypic aggregation of human B lymphocytes. *Eur. J. Immunol.* 1993; **23**: 1771–1775.

24. Bevilacqua MP, Stengelin S, Gimbrone MA *et al.* Endothelial leukocyte adhesion molecule 1: an inducible receptor for neutrophils related to complement regulatory proteins and lectins. *Science* 1989; **243**: 1160–1165.

25. Bettler B, Maier R, Ruegg D *et al.* Binding site for IgE of the human low affinity receptor (FcERII/CD23) is confined to the domain homologous with animal lectins. *Proc. Natl. Acad. Sci. USA* 1989; **86**: 7118–7122.

26. Pochon S, Graber P, Yaeger M *et al.* Demonstration of a second ligand for the low affinity receptor for IgE (CD23) using recombinant CD23 reconstituted into fluorescent liposomes. *J. Exp. Med.* 1992; **176**: 389–397.

27. Aubry JP, Pochon S, Gauchat JF, Nueda-Marin A, Holers VM, Graber P, Siegfried C and Bonnefoy JY. CD23 interacts with a new functional extracytoplasmic domain involving *N*-linked oligosaccharides on CD21. *J. Immunol.* 1994; **152**: 5806–5813.

28. Henchoz-Lecoanet S, Jeannin P, Aubry JP, Graber P, Bradshaw CG, Pochon S and Bonnefoy JY. The Epstein–Barr virus-binding site on CD21 is involved in CD23 binding and IL-4-induced IgE and IgG4 production by human B cells. Submitted for publication.

29. Henchoz S, Gauchat JF, Aubry JP, Graber P, Pochon S and Bonnefoy JY. Stimulation of human IgE production by a subset of anti-CD21 monoclonal antibodies: requirement of a co-signal to modulate epsilon transcripts. *Immunology* 1994; **81**: 285–290.

30. Armitage RJ, Fanslow WC, Strockbine L *et al.* Molecular and biological characterization of a murine ligand for CD40. *Nature* 1992; **357**: 80–82.

31. Gauchat, JF, Henchoz S, Mazzei G, Aubry JP, Brunner T, Blasey H, Life P, Talabot D, Flores-Romo L, Thompson J, Kishi K, Butterfield J, Dahinden C and Bonnefoy JY. Induction of human IgE synthesis in B cells by mast cells and basophils. *Nature* 1993; **365**: 340–343.

32. DiSanto JP, Bonnefoy JY, Gauchat JF *et al.* CD40 ligand mutations in X-linked immunodeficiency with hyper-IgM. *Nature* 1993; **361**: 541–343.

33. Korthauer U, Graf D, Mages HW *et al.* Defective expression of T-cell CD40 ligand causes X-linked immunodeficiency with hyper-IgM. *Nature* 1993; **361**: 539–541.

34. Aruffo A, Farrington M, Hollenbaugh D *et al.* The CD40 ligand, gp39, is defective in activated T cells from patients with X-linked hyper-IgM syndrome. *Cell* 1993; **72**: 291–300.

35. Allen RC, Armitage RJ, Conley ME *et al.* CD40 ligand deffects responsible for X-linked hyper-IgM syndrome. *Science* 1993; **259**: 990–993.

36. Fuleihan R, Ramesh N, Loh R *et al.* Defective expression of the CD40 ligand in X chromosome-linked immunoglobulin deficiency with normal or elevated IgM. *Proc. Natl. Acad. Sci. USA* 1993; **90**: 2170–2173.

37. Prinz JC, Bauer X, Mazur G, Rieber EP. Allergen-directed expression of Fc receptors for IgE (CD23) on human T lymphocytes is modulated by interleukin-4 and interferon-α. *Eur. J. Immunol.* 1990; **20**: 1259–1264.

38. Gagro A and Rabatic S. Allergen-induced CD23 on CD4+ T lymphocytes and CD21 on B lymphocytes in patients with allergic asthma: evidence and regulation. *Eur. J. Immunol.* 1994; **24**: 1109–1114.

39. Gagro A, Rabatic S, Trescec A, Dekaris D and Medar-Lasic M. Expression of lymphocytes Fc epsilon RII/CD23 in allergic children undergoing hyposensitization. *Int. Arch. Allergy Immunol.* 1993, **101**: 203–208.

40. Thyphronitis G, Tsokos GC, June C *et al*. IgE secretion by Epstein–Barr virus-infected purified human B lymphocytes is stimulated by interleukin 4 and suppressed by interferon γ. *Proc. Natl. Acad. Sci. USA* 1989; **86**: 5580–5584.

41. Yodoi J, Ishizaka T and Ishizaka K. Lymphocytes bearing receptors for IgE. II. Induction of Fcϵ-receptor bearing rat lymphocytes by IgE. *J. Immunol.* 1979; **123**: 455–462.

42. Flores-Romo L, Shields J, Humbert Y, Graber P, Aubry JP, Gauchat JF, Ayala G, Allet B, Chavez M and Bazin H. Inhibition of an *in vivo* antigen-specific IgE response by antibodies to CD23. *Science* 1993; **261**: 1038–1041.

43. Flores-Romo L, Johnson GD, Veronesi A *et al*. Functional implication for the topographical association between MHC Class II and the low affinity receptor for IgE: occupancy of CD23 blocks the ability of B cells to present alloantigen to T cells. *Eur. J. Immunol.* 1990; **20**: 2465–2469.

44. Kehry MR and Yamashita LC. FcERII (CD23) function on mouse B cells: role in IgE dependent antigen focusing. *Proc. Natl. Acad. Sci. USA* 1989; **86**: 7556–7560.

45. Pirron U, Schlunch T, Prinz JC *et al*. IgE-dependent antigen focusing by human B lymphocytes is mediated by the low affinity receptor for IgE. *Eur. J. Immunol.* 1990; **20**: 1547–1551.

46. Grosjean I, Lachaux A, Bella C *et al*. CD23/CD21 interaction is required for presentation of soluble protein antigen by lymphoblastoid B cell lines to specific CD4+ T cell clones. *Eur. J. Immunol.* 1994; **24**: 2982–2986.

47. Bonnefoy JY, Guillot O, Spits H *et al*. The low affinity receptor for IgE (CD23) on B lymphocytes is spatially associated with HLA-DR antigens. *J. Exp. Med.* 1988; **167**: 57–72.

48. Texido G, Eibel H, Le Gros G and van der Putten H. Transgene CD23 expression on lymphoid cells modulates IgE and IgG1 responses. *J. Immunol.* 1994; **153**: 3028–3042.

49. Fujiwara H, Kikutani H, Suematsu S, Naka T, Yoshida K, Tanaka T, Suemura M, Matsumoto N and Kojima S. The absence of IgE antibody-mediated augmentation of immune responses in CD23-deficient mice. *Proc. Natl. Acad. Sci. USA* 1994; **91**: 6835–6839.

50. Stief A, Texido G, Sansig G, Eibel H, Le Gros G and van der Putten H. Mice deficient in CD23 reveal its modulatory role in IgE production but no role in T and B cell development. *J. Immunol.* 1994; **152**: 3378–3390.

51. Yu P, Kosco-Vilbois M, Richards M, Kohler G and Lamers M C. Negative feedback regulation of IgE synthesis by murine CD23. *Nature* 1994; **369**: 753–756.

52. Gordon J, Flores-Romo L, Cairns J *et al*. CD23: A multifunctional receptor/lymphokine? *Immunol. Today* 1989; **10**: 153–157.

53. Maeda K, Burton GF, Padgett DA *et al*. Murine follicular dendritic cells and low affinity Fc receptors for IgE (FcERII). *J. Immunol.* 1992; **148**: 2340–2347.

54. Liu YJ, Cairns JA, Holder M *et al*. Recombinant 25 KDa CD23 and interleukin-1 promote the survival of germinal center B cells: evidence for bifurcation in the development of centrocytes rescued from apoptosis. *Eur. J. Immunol.* 1991; **21**: 1107–1114.

55. Bonnefoy JY, Henchoz S, Hardie D *et al*. A subset of anti-CD21 antibodies promote the rescue of germinal center B cells from apoptosis. *Eur. J. Immunol.* 1993; **23**: 969–972.

56. Mitani S, Takagi K, Oka T *et al*. Increased immunoglobulin E Fc receptor bearing cells in germinal centers of hyper-immunoglobulinemia E patients. *Int. Arch. Allergy Appl. Immunol.* 1988; **87**: 63–69.

57. Bacon K, Gauchat JF, Aubry JP, Pochon S, Graber P, Henchoz S and Bonnefoy JY. CD21 expressed on basophilic cells is involved in histamine release triggered by CD23 and anti-CD21 antibodies. *Eur. J. Immunol.* 1993; **23**: 2721–2724.

58. Lecoanet-Henchoz S, Gauchat JF, Aubry JP, Graber P, Life P, Paul-Eugene N, Ferrua B, Corbi A, Dugas B, Plater-Zyberk C and Bonnefoy JY. CD23 regulates monocyte activation through a novel interaction with the adhesion molecules CD11b/CD18 and CD11c/CD18. *Immunity* 1995 3:119–125.

5

The Enigma of CD38, a Multifunctional Ectoenzyme

R M E Parkhouse[1] and Maureen Howard[2]

[1]BBSRC Institute for Animal Health, Pirbright, Woking, Surrey, UK and [2]DNAX Research Institute, Palo Alto, California, USA

INTRODUCTION

Over ten years have passed since the first papers appeared defining CD38 [1, 2]. Originally identified in man by a monoclonal antibody, human CD38, or T10 as it was then named, is 45 kDa type II membrane protein, characteristically expressed by activated lymphocytes, monocytes and granulocyte precursors [3]. Recently, the murine homologue has been identified and at the same time, CD38 has been demonstrated to possess exciting functional [4], structural [5] and enzymatic [6] activities.

At first sight it appeared that CD38 offered a novel signal transduction pathway capable of controlling and or modulating lymphocyte activation and survival. Specifically, this possibility was provided by the clear-cut identification of two related enzyme activities, ADP-ribosyl cyclase and cyclic ADP-ribosyl glycohydrase, in murine recombinant CD38 [6] and in the immunologically identified CD38 of human erythrocytes [7]. Today, after a further three years, it has become clear that CD38 presents us with many more questions than answers; its participation in the control of lymphocyte physiology, however, is not in doubt.

STRUCTURE

Sequencing of both human [8] and mouse [5] cDNA revealed that CD38 was a typical type II membrane protein with a short intracellular domain, a typical transmembrane domain and an approximately 260 amino acid extracellular domain. The mouse protein sequence was 70% identical to the human and, as shown by chromosomal mapping, was also syntenic to the human counterpart. Upon these criteria, therefore, assignment of CD38 to the murine protein seemed justified, in spite of the different cellular distributions of human and murine CD38 molecules (see below).

As the intracellular domain is short and lacks consensus sequences typical for tyrosine kinases, nucleotide binding proteins and ion channels, signal transduction via CD38 would logically appear to proceed as a result of association between CD38 and another, yet to be defined, transducing molecule, perhaps interacting with the CD38 extracellular domain. However, the most interesting detail to emerge from analysis of the sequences was the striking conservation of cysteine residues between CD38 and a marine mollusc ADP-ribosyl cyclase [9]. This clue, suggesting that CD38 might indeed possess the same enzyme activity, was directly tested and affirmed using recombinant murine CD38 (6).

CELLULAR DISTRIBUTION OF CELL SURFACE CD38

The most obvious immediate difference between human and murine CD38 is that, typically, the former is expressed by thymocytes and activated B and T lymphocytes, whereas the latter is most easily detected on both resting and activated B lymphocytes [3, 4, 10]. In the mouse, CD38 is also a selective marker for the minor, but undoubtedly biologically significant, subpopulation of TCR +ve double negative thymocytes [11]. Recent work also suggests that murine CD38 may be detected on GM-CSF activated macrophages and *in vivo* activated T cells [10]. Human CD38 is perhaps more widely distributed, being observed on resting and activated cells of various lineages, and on erythrocytes and platelets [2, 3, 12, 13]. These differences in tissue distribution between the human and murine molecules have raised the possibility that they may not be true homologues; rather that they may represent two related members of a larger gene family, with the human homologue of the mouse, and vice versa, yet to be discovered. Whether this may, or may not be so, awaits inevitable further sequencing information. In the meantime, the important point to stress is that both the human and mouse CD38 molecules, as presently assigned, do possess the enzyme activities mentioned above and discussed below.

FUNCTIONAL CORRELATES

Perhaps the most dramatic findings with CD38 have been the functional consequences of its interaction with agonistic antibodies in both man [3] and mouse [4]. In the mouse, soluble anti-CD38 induces low, but significant, levels of activation of resting B cells. Such stimulation, as measured by conventional thymidine incorporation proliferation assays, is considerably enhanced by submitogenic amounts of IL-4 and lipopolysaccharide [4, 6]. The proliferation of B cells induced by anti-CD38 is preceded by an up-regulation of surface MHC class II and CD23 [4]. Similar to anti-Ig activation, the anti-CD38 induced activation of resting murine B cells is accompanied by prolonged influx of calcium from the exterior milieu and by an increase in tyrosine phosphorylation [4, 14]. Indeed, recent unpublished work in the lab of Kontani has shown that CD38 stimulation of human HL60 cells also induces increases in tyrosine phosphorylation of proteins with similar molecular weights to those seen in murine B cells by Kirkham *et al.* [14]. One of these phosphotyrosine proteins has been identified as the protooncogene p 120-CBL. However, unlike anti-Ig-mediated B cell stimulation, there is no hydrolysis of phosphoinositides and, thus, no release of calcium from intracellular stores sensitive to IP_3 when murine B cells are activated with soluble anti-CD38 [4, 14]. Interestingly, stimulation of murine B cells with anti-CD38 does not prejudice further stimulation with anti-Ig [4] and therefore CD38 constitutes an activation signal independent of receptor Ig. Anti-CD38 is also able to induce a vigorous proliferative response in germinal centre B cells (M Kosco-Vilbois and RME Parkhouse, unpublished work). Finally, and of particular interest, anti-CD 38 induces proliferation and rescue from apoptosis of B cells activated *in vitro* [4]. In all of these studies relatively high levels of anti-CD38 (20–50 μgml^{-1}) have been required for maximum stimulation, raising the possibility that the stimulation depends on univalent, rather than a bivalent (i.e. crosslinking), interaction between CD38 and the agonistic antibody. The demonstration that $F(ab)_2$, but not Fab, anti-CD38 stimulated murine B cells (RME Parkhouse, E Reid and M Howard, unpublished work), however, rules out this hypothesis and affirms the importance of receptor crosslinking in B cell activation.

As noted above, stimulation with anti-CD38 or anti-Ig is not entirely identical at the level of second messengers. There are indications, however, that these two modes of B cell stimulation do share part of a common intracellular signalling pathway. Thus in several models of murine B cell function, the B cells express CD38 at normal levels but are refractory to stimulation with both agonistic anti-CD38 and anti-Ig antibodies [10, 15, 16]. These include neonatal B cells, $CD5^+B - 1$ (peritoneal) cells, B cells from X-linked, immunodeficient *Xid* mice, and B cells from *in vivo* tolerized antigen-receptor-unresponsive "Goodnow" mice. Significantly, all of these deficient B cell population cells fail to respond to antibody while at the same time maintaining responsiveness to a variety of other stimuli, e.g. CD40 ligand, IL-4 or lipopolysaccharide. Moreover, recent experiments do indicate that signalling through CD38 requires the simultaneous presence of the Ig receptor complex [17]. It is therefore possible that signalling pathways via CD38 and receptor Ig converge at some, as yet, unidentified point in the signal transduction pathway downstream of the membrane proximal events. The information from the experiment with the *Xid* mouse, where the molecular lesion is known to be located in the unique region of a B cell tyrosine kinase (BTK), may point to direct, or indirect, regulation of the CD38 signalling pathway by BTK. To date, coprecipitation experiments designed to test this hypothesis have proved negative [10].

Whether CD38 and receptor Ig do, or do not, share part of their intracellular signalling pathways in mature B cells, this is certainly not the case in immature B cells, where anti-Ig, but not anti-CD38, will induce apoptosis [10]. This, then, provides another example of different functional activities of the same B cell receptor at different stages of B cell development.

In contrast to experiments in the mouse, agonistic antibody activation via human CD38 has focused more on the T cell, although anti-human CD38 will also activate NK cells and B cells [18, 19] and rescue germinal centre B cells from apoptosis [20] while suppressing B cell lymphopoiesis [21]. In the case of the human, activation of T cells via CD38 requires accessory cells and IL-2, and is thought to involve a signalling pathway distinct from CD3 [3, 18, 19], although there is no direct evidence for this assertion.

An additional difference between man and mouse is that human CD38 has been reported to associate/modulate with CD3 on T cells and receptor Ig, CD19 and CD21 on B cells [19], whereas in the mouse similar experiments have failed to detect any defined molecular associations. It is worth noting here, that in man CD38 is also expressed on erythrocytes [7] and platelets [13], where association with activation receptors such as the Ig and TCR receptor complex are not apparent. Finally, total human peripheral blood mononuclear cells were rapidly induced to secrete multiple lymphokines when treated with anti-CD38 *in vitro* [22]. Further work will establish responses of defined cell types in this system.

ENZYME ACTIVITIES OF CD38

As noted above, searches through sequence banks revealed an unexpected homology between both human and mouse CD38 and the ADP-ribosyl cyclase of a sea mollusc, *Aplysia californica* [6, 9]. Although the overall molecular homology was insignificant ($\sim 30°$), there was a remarkable conservation of the cysteine residues. As their participation in protein structure, through the formation of intramolecular disulphide bonds, is likely to be decisive in determining protein secondary and tertiary structure, it was clearly of immediate interest to investigate whether CD38 possessed ADP-ribosyl cyclase activity. The possibility that CD38 might be an ectoenzyme with ADP-ribosyl cyclase activity was immediately appealing as the enzyme catalyses the formation of cADP-ribose from NADH+. Specifically, cADP-ribose is thought to be the natural ligand of an IP$_3$-independent channel, the ryanodine receptor [23] and, as reported above, the anti-CD38-stimulated calcium influx from the exterior, does not generate IP$_3$ via phosphoinositide hydrolysis [4, 14].

In order to test this hypothesis, the extracellular domain of murine CD38 was cloned, expressed, purified and then employed in the standard assay for ADP-ribosyl cyclase [6]. Biochemical analysis after incubation of the recombinant murine CD38 extracellular domain provided conclusive evidence of two distinct enzyme activities, an ADP-ribosyl cyclase and a cyclic ADP-ribosyl glycohydrolase. Thus, there was detectable early conversion of NADH+ to cyclic ADP ribose, mediated by the cyclase. This was followed by hydrolysis of cyclic ADP ribose, yielding ADP-ribose as a result of the glycohydrolase activity.

Thus it is clear that in the mouse, the extracellular domain is entirely responsible for the two enzyme activities. In addition, the possibility that the extracellular domain may also function as an ADP ribosyl transferase, ribosylating substrates such a G proteins, as do cholera and pertussis toxins [10], adds yet another dimension to the range of functions addressed by CD38. In man, there are now also several reports that isolated or coprecipitated CD38 molecules have ADP-ribosyl cyclase and NADH+ glycohydrolase activities [7, 24–26]. Finally, CD38 cDNA from a human acute myelocytic leukaemia, yielded NADH+ glycohydrolase activity when expressed in *Escherichia coli* [27].

CD38 AS AN ADHESION MOLECULE

In addition to being a multifunctional ectoenzyme which also transduces activation signals upon interaction with agonistic antibodies, CD38 has recently been claimed, in man at least, to weakly mediate selectin-like adhesion of lymphoid cell lines to endothelial cells [28]. The interesting question of the endothelial cell counter-ligand for CD38 was resolved by the demonstrated inhibition of CD38-mediated adhesion by an anti-endothelial cell monoclonal antibody recognizing a 120 kDa molecule. This molecule was later shown to be frequently co-expressed with CD38 on vascular endothelium, lymphocytes, platelets, NK cells and, to a lesser extent T, B and myeloid cells [29]. It will be interesting to see if similar findings are established in the mouse.

THE CD38 FAMILY

Although it is still early days, the impressive structure/function conservation observed between the mammalian CD38 and the invertebrate (sea mollusc) ADP-ribosyl cyclase would argue that such a gene might be expected to have evolved, duplicated and diverged during evolution, leading to the establishment of a CD38 family of molecules. With respect to this, two further candidates for such a family have recently emerged, both having the characteristic cysteine-related homology exhibited by CD38. One (BST-1) is present in human bone marrow stromal cells and facilitates pre-B-cell growth [30, 31] and the other (BP-3) is a mouse protein (32) expressed by early B and T cells, a subpopulation of reticular cells in peripheral lymphoid organs, epithelial cells and kidney tubules. It may be significant that both BST-1 and BP-3 are GPI linked surface receptors. Further exploration of these and future members of the CD38 family are awaited with considerable interest.

FUTURE PROSPECTS

It is now clear that the structural motif originally placing CD38 and the *Aplysia californica* ADP-ribosyl cyclase in the same evolutionary, structural and functional context is even more widely spread and may be expected to grow. Quite probably, not all members of the group will have exactly the same functions as CD38 which, indeed, is almost embarrassingly rich in functional properties: an antigen mediating a variety of physiological responses upon interaction with agonistic antibodies, depending not only on the tissue, but also the stage of differentiation of a given cell type; a multifunctional enzyme with ADP-ribosyl cyclase, cyclic ADP-ribose glycohydrolase and, perhaps, ADP ribosylase activities; and a weak, selectin-type adhesion property mediating interactions between human lymphoid cells and endothelial cells.

Perhaps the four most fundamental questions to resolve are:

1. Does the enzyme activity of CD38 function extracellularly and / or intracellularly, and if intracellularly, how is it translocated to a subcellular compartment for interaction with its substrate?
2. What role, if any, do the enzyme activities and their reaction products play in the transduction process observed to follow cellular activation via CD38? Or, to restate the point, is the ADP-ribosyl cyclase activity, and resultant cyclic ADP-ribose generated, the signal responsible for anti-CD38-mediated calcium flux?
3. Is there a natural ligand for CD38 which is mimicked by agonistic antibodies?
4. Is the mechanism of activation via CD38 inevitably the same irrespective of the cell lineage or stage of differentiation?

An understanding of the mechanism of signalling is clearly urgent, although it is becoming apparent that both calcium influx and protein tyrosine phosphorylation may have major roles to play in CD38 signalling. Given that there are so many different haemotolymphoid cell surface receptors and correspondingly varied physiological response patterns, it has always been intriguing to note that there is a relatively restricted number of second messenger systems. Thus the possibility that CD38 may invoke another type of signal(s), for example through cADP-ribose, with attendant "cross-talk" to the existing already recognized second messenger systems, is of enormous interest and pharmacological potential. A major conundrum is the site of interaction between cell-bound CD38 and its substrate, $NADH^+$; as the metabolite $NADH^+$ is not characteristically present in body fluids.

It has been suggested that in areas of high apoptosis, such as the germinal centre, NADH+ might perhaps become available to surface CD38 via release through the leaky membranes of moribund apoptotic cells. On the other hand, however, it may be argued that such apoptotic cells are doomed to a very short half-life, being rapidly engulfed by macrophages. A second possibility is that extracellular cell-bound CD38 enzyme activity merely serves a scavenger function, degrading $NADH^+$ and cADP-ribose for metabolic re-use. The fact that human erythrocytes are CD38 positive would support the latter notion, as indeed does the recent finding of two enzymatically active, phosphatidylinositol-linked (as opposed to transmembrane) members of the CD38 family [30–32].

The logical counter-proposal, that CD38 is interiorized in order to encounter its substrate and thus deliver the calcium mobilizing cADP-ribose, is attractive, taking into account that mouse B cell stimulation via CD38 results in a prolonged calcium flux, in the absence of phosphoinositide hydrolysis to yield IP_3. This observation is perhaps dangerously titillating, as cyclic ADP-ribose is known to mobilize calcium independently of IP_3. Furthermore, cyclic ADP-ribose, together with submitogenic amounts of lipopolysaccharide and IL-4 is mitogenic for murine B cells [6]. A major problem, however, arises when considering how CD38 may be translocated from the surface to the cell interior. Apart from the current lack of agreement over whether this may or may not be achieved through ligation of associating molecules such as Ig [3, 10], any proposal based on internalization through "cocapping" specifically fails to achieve transport of the enzymatically active ectodomain, to the cytoplasm, where the NADH+ substrate resides. Thus, internalization via cocapping would result in localization of the ectodomain within the internalized vesicle and hence be inaccessible to the cytoplasmic source of NADH+. Orientation of internalized CD38 through cocapping to place the ectodomain in the cytoplasm would demand molecular "flip flop" within the membrane, a hitherto totally unknown and unlikely biophysical mechanism. Clearly, careful studies tracking the destiny, route and molecular associations of CD38 following interactions with agonist antibodies and potential associated proteins will greatly help in resolving some of these problems and, in addition, provide essential information for the pharmacological exploitation of the CD38 system, perhaps most profitably at first

through the testing and intervention of cADP-ribose analogues.

The problem of providing an intellectually convincing functional link between agonistic anti-CD38 stimulation and CD38 enzyme activity has raised sufficient doubts for serious contemplation of a CD38 counter-receptor. The latter would serve to fulfil the cell stimulation roles described above, leaving the enzyme activities to perform, as yet undefined, functions. To date, searches for such a counter-receptor in the mouse have proved disappointingly negative. In man, on the other hand, there is a very recent report of a possible CD38 ligand [29]. This 120 kDa molecule, expressed by endothelial cells, is postulated to mediate adherence of endothelial cells to lymphoid cells via CD38. As yet, no studies are available on the possible stimulatory consequences of its interaction with lymphocytes.

In conclusion, CD38 represents one of a family of proteins, with ancient origins reflected in a conserved structure, and an enzyme activity generating the known, IP_3 independent, calcium mobilizing reagent, cyclic ADP-ribose. On the surface of lymphoid cells it may act as an adhesion molecule but, more dramatically, it may serve as a signal transducer resulting in a variety of biological responses participating in the control of activation, proliferation and apoptosis. How all of these activities are integrated and controlled remains to be elucidated, but the system appears to offer a novel pathway of cell signalling with possible potential for pharmacological manipulation.

REFERENCES

1. Reinherz EL, Kung PC, Goldstein, G, Levey RH and Schlossman SF. Discrete stages of human intrathymic differentiation analysis of normal thymocytes and leukemic lymphoblasts of T-cell lineage. *Proc. Natl. Acad. Sci. USA* 1980; **77**: 1588–1592
2. Terhorst C, Van Agthoven, A, LeClair K, Snow P, Reinherz EL and Schlossman SF. Biochemical studies of the human thymocyte cell-surface antigens T6, T9 and T10. *Cell* 1981; **23**; 771–780.
3. Malavasi F, Funaro A, Roggero S, Horenstein AL, Calosso L and Mehta K. Human CD38: one molecule in search of a function. *Immunol. Today* 1994; **15**: 95–97.
4. Santos-Argumedo L, Teixeira C, Preece G, Kirkham PA and Parkhouse RME. A molecule on the surface of murine B lymphocytes mediating activation and protection from

apoptosis via calcium channels. *J. Immunol.* 1993; 151: 3119–3130.

5. Harada N, Santos-Argumedo L, Chang R, Grimaldi JC, Lund FE, Brannan CI, Copeland NG, Jenkins NA, Heath AW, Parkhouse RME and Howard M. Expression cloning of a cDNA encoding a novel murine B cell activation marker: homology to human CD38. *J. Immunol.* 1993: 151: 3111–3118.

6. Howard M, Grimaldi JC, Bazan JF, Lund FE, Santos-Argumedo L, Parkhouse RME, Waseth TF and Lee HC. Formation and hydrolysis of cyclic ADP-ribose is catalyzed by lymphocyte antigen CD38. *Science* 1993; 262: 1056–1059.

7. Zocchi E, Franco L, Guida L, Benatti U, Bargellesi A, Malavasi F, Lee HC, DeFlora A. A single protein immunologically identified as CD38 displays NAD+-glycohydrolase, ADP-ribosyl cyclase and cyclic ADP-ribose hydeolase activities in human red cells. *Biochem. Biophys. Res. Commun.* 1993; 196: 1459–1465.

8. Jackson DG and Bell JI. Isolation of a cDNA encoding the human CD38 (T10) molecule, a cell surface glycoprotein with an unusual discontinuous pattern of expression during lymphocyte differentiation. *J. Immunol.* 1990; 144: 2811–2815.

9. States DJ, Walseth TF and Lee HC. Similarities in amino acid sequences of *Aplysis* ADP-ribosyl cyclase and human lymphoctye antigen CD38. *Trends Biochem. Sci.* 1992; 17: 495–497.

10. Lund F, Solvason N, Grimaldi JC, Parkhouse RME and Howard M. Murine CD38: an immunoregulatory ectoenzyme. *Immunol. Today.* 1995; 16: 469–473.

11. Bean AGD, Godfrey DI, Ferlin WG et al. CD38 expression on mouse T-cells: CD38 defines functionally distinct subsets of $\alpha\beta$TCR$^+$CD4$^-$CD8$^-$ thymocytes. *Int. Immunol.* 1994; 7: 213–221.

12. Malavasi F, Caligaris-Cappio F, Milanesi C, Dellabona P, Richiardi P and Carbonara AO. Characterisation of a murine monoclonal antibody specific for human early lymphohemopoietic cells. *Hum. Immunol.* 1984; 9: 9–20.

13. Ramaschi G, Torti M, Tolnai Festetics E, Sinigaglia F, Malavasi F and Balduini C. Expression of cyclic ADP-ribose-synthesizing CD38 molecule on human platelet membrane, *Blood* 1995 (in press).

14. Kirkham PA, Santos-Argumedo L, Harnett MM and Parkhouse RME. Murine B cell activation via CD38 and protein tyrosine phosphorylation. *Immunology* 1994; 83: 513–516.

15. Santos-Argumedo L, Lund FE, Heath AW, Solvason N, Wu, WW, Grimaldi JC, Parkhouse RME and Howard M. CD38 unresponsiveness of xid B cells implicates Bruton's tyrosine kinase (btk) in CD38 induced signal transduction. *Int. Immunol.* 1995; 7: 163–170.

16. Lund FE, Solvason NW, Cooke MP, Heath AW, Grimaldi JC, Parkhouse RME, Goodnow CC and Howard MC. Signalling through murine CD38 is impaired in antigen receptor unresponsive B cells. *Eur. J. Immunol.* 1995; 25: 1338–1345.

17. Lund FE, Yu N and Howard M. Signalling through the ectoenzyme, murine CD38, requires the Ig receptor complex. The 9th International Congress of Immunology, San Francisco, California, USA 1995. Abstract No. 4030.

18. Funaro A, Spagnoli GC, Ausiello CM, Alessio M, Roggero S, Delia D, Zaccolo M and Malavasi F. Involvement of the multilineage CD38 molecule in a unique pathway of cell activation and proliferation. *J. Immunol.* 1990; 145: 2390–2396.

19. Funaro A, DeMonte LB, Dianzani U, Forni M and Malavasi F. Human CD38 is associated to distinct molecules which mediate transmembrane signalling in different lineages. *Eur. J. Immunol.* 1993; 23: 2407–2411.

20. Zupo S, Rugarli E, Dono M, Tamborelli G, Malavasi F and Ferrarini M. CD38 signalling by agonistic monoclonal antibodies prevents apoptosis of human germinal center B cells. *Eur. J. Immunol.* 1994; 24: 1218–1222.

21. Kumagai M, Coustan-Smith E, Murry DJ, Silvennoinen O, Murti KG, Evans WE, Malavasi F and Campana D. Ligation of CD38 suppresses human B lymphopoiesis. *J. Exp. Med.* 1995; 181: 1101–1110.

22. Ausiello CM, Urbani F, la Sala A, Funaro A and Malavasi F. CD38 ligation induces discrete cytokine mRNA expression in human cultured lymphocytes. *Eur. J. Immunol.* 1994; 25: 1477–1480.

23. Meszaros L, Bak JB and Chu A. Cyclic ADP-ribose as an endogenous regulator of the non-skeletal type ryanodine receptor CA^{2+} channel. *Nature* 1993; 364: 76–78.

24. Summerhill RJ, Kaclspm DG and Galione A. Human lymphocyte antigen CD38 catalyzes the production of cyclic ADP-ribose. *FEBS Lett.* 1993; 335: 231–233.

25. Takasawa S, Tohgo A, Noguchi N et al. Synthesis and hydrolysis of cyclic ADP-ribose by human leukocyte antigen CD38 and inhibition of the hydrolysis by ATP. *J. Biol. Chem.* 1993; 268: 26052–2654.

26. Gelman L, Deterre P, Gouy H, Boumsell L, Debré P and Bismuth G. The lymphocyte surface antigen CD38 acts as a nicotinamide adenine dinucleotide glycohydrolase in human T lymphocytes. *Eur. J. Immunol.* 1993; 23: 3361–3364.

27. Kontani K, Nishina H, Ohoka Y, Takahashi K and Katada T. NAD glycohydrolase specifically induced by retinoic acid in human leukemic HL-60 cells. *J. Biol. Chem.* 1993; 268: 16895–16898.

28. Dianzani U, Funaro A, DiFranco D, Garbarino G, Bragardo M, Redoglia V, Buonfiglio D, DeMonte L, Pileri A and Malavasi F. Interaction between endothelium and CD$^+$CD45RA$^+$ lymphocytes: role of the human CD38 molecule. *J. Immunol.* 1994; 153: 952–959.

29. Deaglio S, Dianzani U, Horenstein AL, Fernàndex JE, van Kooten C, Bragardo M, Garbarino G, Funaro A, Di Virgilio F, Banchereau J and Malavasi F. A human CD38 ligand: a 120 kDa protein predominantly expressed by endothelial cells. *J. Immunol.* 1996 (in press).

30. Itoh M, Ishihara K, Tomizawa H, Tanaka H, Kobune Y, Iskikawa J, Kaisho T and Hirano T. Molecular cloning of murine BST-1 having homology with CD38 and *Aplysia* ADP-ribosyl cyclase. *Biochem. Biophys. Res. Commun.* 1994; 203: 1309–1317.

31. Kaisho T, Ishikawa J, Oritani K, Inazawa J, Tomizawa H, Muraoka O, Ochi T and Hirano T. BST-1, a surface molecule of bone marrow stromal cell lines that facilitates pre-B-cell growth. *Proc. Natl. Acad. Sci. USA.* 1994; 91: 5325–5329.

32. Dong C, Wang J, Neame P and Cooper M. The murine BP-3 gene encodes a relative of CD38/NAD glycohydrolase family. *Int. Immunol.* 1994; 6: 1353–1360.

6

CD40 Signalling in T-dependent B Cell Responses

David Gray

Department of Immunology, Royal Postgraduate Medical School, Hammersmith Hospital, London, UK

INTRODUCTION

The CD40 antigen has evoked intense interest from immunologists since the time it was first recognized [1,2]. They have largely ignored that it was originally found on bladder carcinoma cells, because on B lymphocytes it mediated powerful costimulatory signals for proliferation and antibody production [1–5]. In this review we will perpetuate this narrow outlook by focusing on the role of CD40 in B cell responses that require T cell help. The cloning of the CD40 molecule and demonstration of its sequence similarity to low affinity nerve growth factor receptor, LNGFR [6], led to the notion that CD40 might be a receptor for a cytokine. At the same time as the costimulatory effects of anti-CD40 antibodies were being investigated in man, immunologists working in mouse models had recognized that not all of the help delivered to B cells by T cells was carried by soluble mediators [7–11]; fixed, activated T cells, or membrane preparations from these cells, were able to induce activation and proliferation of resting B cells [7–11]. It was soon noticed that soluble CD40 fusion proteins largely blocked the activity of membranes from activated T cells in the induction of B cell proliferation [12,13]. The gene for CD40 ligand was subsequently isolated from cDNA libraries derived from activated T cells [14,15]. Thus, CD40 was recognized as the major acceptor molecule for contact-mediated T cell help. The crucial importance of the CD40 interaction with its ligand, in T-dependent antibody responses, was underscored by the elucidation of the underlying lesion in X-linked hyper-IgM immunodeficiency syndrome (HIGM) [16–19].

Mutations or deletions in the CD40L gene cause a failure of isotype switching and germinal centre formation.

The function of CD40 in B cell responses is now yielding to investigation but during this process immunologists have noted that its function on other cell lineages may impinge upon their experiments. This recognition has already been a stimulus to investigate these functions; some will be alluded to throughout the text. This review represents the state of knowledge in the field as of July 1995.

THE STRUCTURE OF CD40

The human CD40 molecule is a 48 kDa protein that consists of a 193 amino acid (aa) extracellular domain, a 22 aa transmembrane region and a 62 aa cytoplasmic tail. Mouse CD40 is similar in size and bears a greater than 60% homology to the human molecule. Although the cytoplasmic domain of mouse CD40 is longer (74 aa) the greatest homology (78%) resides in this portion [20]. Both mouse and human CD40 carry *N*-linked sugars. The protein is encoded by a gene, spread over 9 exons (spanning 16.3 kb of genomic DNA), that is transcribed to give a message of 1.4 kb (humans) or 1.7 kb (mice). Interestingly, in mice, there is a minor mRNA transcript of 1.4 kb (from an alternative polyadenylation signal in the 3' untranslated region) and this is increased preferentially following B cell activation [20].

The extracellular portion of the CD40 molecule is composed of four similar, cysteine-rich domains. These domains are "imperfect repeats" of about 40 aa that each contain 6 cysteine residues that form 3 disulphide links. This distinctive domain

Lymphocyte Signalling: Mechanisms, Subversion and Manipulation. Edited by M. M. Harnett and K. P. Rigley © 1997 John Wiley & Sons Ltd.

structure places CD40 in the tumour necrosis factor receptor (TNFR) superfamily, that now consists of at least 10 members. These include LNGFR (the founder member), TNFR I, TNFR II, lymphotoxin-β receptor, OX-40 (a molecule expressed on rat activated T cells), 4-1BB (found on mouse T cells), CD95 (also known as Fas or Apo1), CD27 and CD30. Recently viral and even fungal proteins have been identified with this characteristic, cysteine-rich domain structure. The members of the family carry from one to four of these cysteine-rich extracellular domains that are linked to the transmembrane region by serine–threonine–proline-rich segments. No crystal structure is yet available for CD40, or any other member of the TNFR family and so the relationship of this domain structure to ligand binding is unclear. The evolutionary relationships of the molecules within the TNFR superfamily, as well as other structural considerations, are discussed in detail [21].

Within the cytoplasmic tail the family members show relatively little homology although some motifs are shared by certain members; for instance the cytoplasmic regions of LNGFR, TNFR I and CD95 show significant sequence homology that seems to be related to the transduction of signals inducing cell death. The cytoplasmic tail of CD40, on the other hand, bears similarity to OX-40 CD27 and CD30 and transduces survival or activation signals. The sequence of the cytoplasmic region gives little clue to the means by which CD40 transduces signals. There is no sequence within the tail that is recognizable as a protein tyrosine kinase and although the molecule is constitutively phosphorylated there are no tyrosines in the tail of the human molecule (in the mouse molecule there is one). However, there are abundant serine and threonine residues and a deletion analysis of the cytoplasmic tail, for regions important in signalling, has revealed the threonine at position 234 as crucial for the signal transduction [22].

Although normally a membrane protein, CD40, like other members of the TNFR family, is known to be released as a soluble molecule. Both EBV-transformed lines and activated normal B cells release soluble CD40 into culture supernatants [23]. The lack of an alternatively spliced mRNA suggests that the molecule may be solubilized by proteolytic cleavage during biosynthesis, as demonstrated for other transmembrane proteins [24].

EXPRESSION OF CD40

B lymphocytes express CD40 from the time they are born until the point at which they terminally differentiate into plasma cells. Most pre-B cells in the bone marrow express CD40 and it has been reported to be present on CD19+ pro-B cells in the fetal liver, that have yet to rearrange their IgH locus [25]. All populations of B cells in the peripheral lymphoid organs are reported to express CD40 equally; thus recirculating, lgD$^+$ lgM$^+$ B cells of the follicular mantle, centroblasts and centrocytes from the germinal centre and CD38$^-$ lgD$^-$ lgG$^+$ "memory cells" all bear similar levels of CD40 [21]. This is a curious finding, as, *in vitro*, activation of B cells, with a number of stimuli (anti-Ig, IL-4, LPS, anti-CD20 or phorbol esters), leads to an increase in CD40 expression. There are some unpublished observations (M. Casamayor-Palleja and ICM MacLennan, personal communication) that centrocytes (CD39$^+$, lgD$^-$, CD77$^+$ and CD44low) have two to three-fold more CD40 on their surface. It is only when B cells fully differentiate into plasma cells that CD40 is lost; a significant proportion of plasmablasts still express CD40.

Precursors of other cell lineages also express CD40 in the bone marrow; the majority of CD34+ cells are also CD40+. These precursors can give rise to cells of the myeloid lineage. Unlike the B cell lineage, myeloid cells lose CD40 expression during differentiation only to up-regulate it again during the later stages of maturation. For instance, monocytes have low levels of CD40 that increase after incubation with GM-CSF, IL-3 or IFNγ. Analysis of the dendritic cell lineage indicates that CD40 is expressed late in their life; Langerhans cells of the skin have barely detectable levels of CD40 on their surface that increase by the time the cells reach draining lymph nodes and settle in T zones as interdigitating dendritic cells. Langerhans cells within a short time in culture also increase surface CD40. The use of GM-CSF to grow dendritic cells from precursors in blood or bone marrow [26,27] has enabled an analysis of the control of expression; CD40 induction seems to parallel that of MHC class II in maturing dendritic cells, coming up late in the cultures and augmented by IL-4. Coculture with T cells is a powerful inducer of CD40 on dendritic cells. *In vivo*, dendritic cells from RAG-1-deficient mice, that have no lymphocytes, express very little CD40 until reconstituted with T cells (A. Volkmann and B. Stockinger, personal communication) indicating that the interaction with T cells causes differentiation of the dendritic cells as well as the potential activation of the T cells. The role of CD40 on dendritic cells is still open to speculation but it has been suggested that the signals enhance survival and/or alter the antigen-processing/presentation capacity of the cell [28,29]. Interestingly, T cells, after activation, may also express CD40, as they have been reported to respond to CD40 ligand [30]. However, there is no direct protein or RNA expression data.

The follicular dendritic cells (FDC) that are the focus of intense B cell proliferation during the germinal centre (GC) reaction (see later) also express CD40 [31]. Its function here is unknown. CD40 on thymic epithelial cells is clearly detectable in both the cortex and the medulla. It is upregulated on these cells by IL-1α, TNFα, and IFNγ [32]. Endothelial cells express CD40 constitutively, at low levels that are augmented by inflammatory cytokines (TNFα, IL-1, IFNβ and IFNγ). Stimulation of these cells with CD40 ligand brings about an increase in expression of the adhesion molecules, E-selectin, VCAM1 and ICAM1 [33], leading to the speculation that an inflammatory response may cause changes in endothelium that amplifies the influx of T cells to the lesion. The CD40 signals delivered to these non-lymphoid cells induce a form of activation and phenotypic change that indicates differentiation; by analogy with lymphocytes it seems likely that under some circumstances the CD40 stimulus will be important for maintenance of viability.

THE LIGAND FOR CD40

The ligand for CD40 (CD40L) is a 39 kDa molecule, consisting of a 215 aa (in humans; 214 aa in mice) extracellular domain (containing five cysteine residues) and a 22 aa cytoplasmic tail (in

mouse and man). It is a type II membrane protein, with an extracellular carboxy terminus. The molecule is now classified in the TNF family, that includes TNFα, lymphotoxin-α, lymphotoxin-$\alpha2$-β and the ligands for CD27, CD30, CD95, OX40 and 4-1BB [34]. The sequence homology between CD40L and other members of the family is relatively low, but an alignment that takes into account the structure of the putative binding site reveals clear similarities [21]. By analogy with the crystal structure of TNF, CD40L was predicted to form a trimer in the membrane, a proposal borne out by the trimerization of soluble CD40L fusion proteins [35]. A soluble form of CD40L is also generated physiologically [36], although the exact conditions in which this occurs and the reason are not known.

The ligand for CD40 (CD40L) is expressed on T cells soon after activation; it is not detectable on resting T cells. It appears within 1–2 hours of activation of T cells with phorbol esters and calcium ionophore, and peaks at around 4–6 hours (12–15). The expression is transient and is lost from the surface within 48–72 hours. The information on the dynamics of CD40L expression derives from *in vitro* stimulation studies, using either phorbol esters and calcium ionophore or anti-CD3 antibodies. It is difficult to know how closely this reflects physiological activation by antigen presenting cells. Some data suggest that T cells only express CD40L if both TcR-mediated signals and CD28 costimulation are delivered [37]. However, other workers find that the TcR signal is sufficient [38]. It is possible that naive T cells require costimulation while previously activated, memory T cells do not. The transience of CD40L expression depends upon the presence of CD40 in the environment; for instance CD40L is down-regulated during coculture with B cells [39]. B cells also release soluble CD40 which can bind to CD40L on T cells. Interestingly, the binding of soluble CD40 fusion protein to CD40L seems to stabilize its expression on the membrane [40]. These factors make the detection of CD40L potentially difficult. *In vivo* CD40L is expressed during the first 4 days after immunization, as administration of soluble CD40 or antibodies to CD40L during this time blocks the primary antibody response [41].

CD40L is expressed by all subsets of CD4+ T cells (Th0, Th1 and Th2) and a subset of CD8+ T cells. The sites of T cell expression of CD40L *in vivo* reflect the dual function of CD40 signals in B cell biology. Early in the response it is expressed in the T zones (e.g. in the periarteriolar lymphocytic sheath of the spleen) and in association with plasma cells close to terminal arterioles [42]. This reflects the role of CD40L in the initiation of the primary antibody response. Later expression can be detected around the edge of germinal centres by a specialized subset of activated CD4+ T cells [43]. At this site the role of CD40L seems to be distinct from that in the T zones; here it provides rescue signals to GC B cells that facilitate their entry into the memory B cell pool (see later). Although the physiological relevance is unclear, an accumulation of data indicates that CD40L is expressed on activated B cells [44] and on a number of non-lymphoid cell types, including monocytes, eosinophils, basophils and mast cells [45,46].

CD40L was identified as crucial for the delivery of contact-mediated T cell help by the ability of soluble CD40 fusion protein to block the B cell activation and antibody production that results from the coculture of T and B cells. This conclusion was amply confirmed by the discovery that the molecular lesion underlying the rare immunodeficiency, X-linked hyper-IgM syndrome, was a number of mutations/deletions in the CD40L gene [16–19]. As we will discuss in more detail later, these patients fail to make mature antibody responses, secreting IgM but not IgG, IgA or IgE. It has been assumed that the defect in these patients and in CD40L knockout mice [47,48] is solely due to the absence of the CD40 signal to B cells. This may not be the complete story as there are now at least two reports that the CD40L molecule can impart costimulatory signals to T cells [49,50].

CD40 SIGNAL TRANSDUCTION

One of the earliest observations in this field was that the threonine at position 234 in the cytoplasmic tail was necessary for signal transduction [22]. It is still not clear if this residue is important for association of CD40 with other molecules or if it is a substrate for phosphorylation; CD40 seems to be constitutively phosphorylated. The cytoplasmic tail of the CD40 molecule seems to carry no sequence motif that would suggest kinase activity

and the molecule possesses no such activity. Before we delve further into the available data on CD40 signal transduction, there are a number of considerations to discuss.

To draw any sort of cohesive picture of the signalling pathways utilized by the CD40 molecule is extremely difficult given the fragmentary nature of the published (and unpublished) data. The single most important message from the literature is that the CD40 signals differently in different B cell populations. Thus the results of the analysis of high density, resting B cells differ considerably from those in low density, activated B cells, including germinal centre (GC) B cells. Given the quite different consequences of CD40 ligation in these two populations (see later) this may not be entirely unexpected. We have already alluded to this dichotomy of function of CD40 that occurs during B cell differentiation. Initiation of the immune response involves a CD40-mediated proliferative signal, while later the development of memory B cells requires the delivery of an apoptotic rescue signal. If we approach the available data with this in mind, some sort of picture emerges.

Much of the information in this field has derived from Burkitt lymphoma cell lines (Raji, Daudi and Ramos) and where these have been studied in parallel with B cells, *ex vivo* [51,52], it is clear that the signalling pathways highlighted resemble those detected in activated B cells, but not in resting B cells. I suspect it is crucial when reading any study of signalling in lymphocytes to ascertain the type of cells under analysis. The use of cell lines clearly simplifies, experimentally, the study of signalling pathways. Another simplification, which is difficult to avoid, is the study of signals in isolation: this is a particular problem for CD40 as the first step on the pathway of T-dependent B cell activation is not CD40 ligation but a signal via sIg from antigen. Both of these technical problems lead to difficulties of interpretation, however, let us try. It is only possible to make sense of the CD40 signal transduction data, if we deal with resting B cells and "activated"/GC B cells separately.

Resting B Cells

Only two studies have addressed the CD40 signalling pathways utilized in resting B cells [51,52].

Uckun *et al.* [51] found no evidence of tyrosine phosphorylation of any cellular substrates following stimulation of dense tonsillar B cells with anti-CD40 antibodies. Faris *et al.* [52] found a weak tyrosine phosphorylation of four substrates [28, 32, 38, 42 and kDa) that was sustained for 5–10 minutes following stimulation. The 28 and 32 kDa molecules were not detectably phosphorylated prior to stimulation. The pattern and strength of tyrosine phosphorylation differed markedly from that observed in low density tonsillar B cells (see below). These authors also noted that the rapid dephosphorylation of a variety of proteins, observed in low density B cells, was not seen in resting cells. If it is possible to summarize these two pieces of work, it might be fair to say that tyrosine phosphorylation/dephosphorylation is probably not the major immediate event following CD40 ligation on resting B cells. Secondly, the phosphorylation of tyrosines that does occur targets quite different substrates from that in activated B cells.

An indication that protein tyrosine kinases (PTKs) do play some role in primary CD40 activation is that inhibitors of these enzymes preclude some of the normal consequences of CD40 stimulation of resting cells, such as homotypic adhesion [53]. A recent paper reports the adhesion to be dependent on protein kinase C (PKC) activation and this is, in turn, dependent on tyrosine phosphorylation events [54], however, the role of PKC in CD40 activation is controversial (see below). No detailed studies of mouse CD40 signalling have been published; tyrosine phosphorylation has been detected in resting B cells following incubation with plasma membranes from activated T cells and this is to some extent blocked by addition of antibodies to CD40 ligand. Our own analysis of phosphorylation following CD40 engagement on murine resting B cells tends to support the data from humans, that it is at a low level.

Uckun *et al.* [51] detected serine–threonine-specific kinase activity in "mature tonsillar" B cells, but it is unclear if these represent resting or GC B cells. The activation of a phospholipase C in resting cells seems to be ruled out, as no phosphoinositide turnover and no production of Ins-1,4,5-P3 could be detected [51]. Likewise there is no mobilization of intracellular calcium stores. In mice this may be different, as there is a report of a calcium-dependent CD40 activation pathway [55]. The role of cAMP-dependent second messenger pathways may be important, as levels of cAMP rise following CD40 stimulation [56]. This may implicate the cAMP-dependent protein kinase A that phosphorylates the CREB transcription factor [57]. However, in B cells CREB seems to be phosphorylated by PKC, not PKA, and neither PKC activation nor CREB phosphorylation occurs after CD40 signalling [58].

If the direct and immediate consequence of CD40 engagement on resting B cells is proliferation, then the signalling pathway that facilitates this has not yet been clearly delineated.

Activated B Cells, GC B Cells, and B Cell Lymphomas

Low buoyant density tonsillar B cells (including GC B cells) exhibit increases in tyrosine phosphorylation of a number of molecular species (67, 72, 96 and 113 kDa) following CD40 stimulation [51]. These increases are much stronger than those seen in resting B cells. The study of Faris *et al.* [52] concurs with this, but also identifies a 28 kDa molecule that is strongly, but transiently, phosphorylated in these cells (0.5–1 minute). This may be the same molecule that is tyrosine phosphorylated for a much longer period (10 minutes) in resting cells. These authors also note strong dephosphorylation of several substrates. In the Raji cell line, the PTKs *lyn, fyn* and *syk* were found to be transiently dephosphorylated (0.5–1 minute). A more thorough study of the *lyn src*-type kinase, in the Daudi cell line, showed that it was phosphorylated within 1 minute of CD40 stimulation and this was sustained for 20 minutes [59]; in addition the activity of *lyn* was enhanced. Other PTKs (*fyn, fgr, lck*) showed no change in phosphorylation or activity. Serine/threonine-specific kinase activity is particularly strong in activated B cells [51].

In the Daudi cell line two other substrates for CD40-mediated PTK activity are the 85 kDa subunit of phosphoinositide-3-kinase (PI-3-kinase) and phospholipase Cγ (PLCγ2). The phosphorylation of these enzymes increases their activity and is consistent with the increased phosphoinositide turnover seen in this cell line and in low buoyant density tonsillar B cells following anti-CD40 stimulation [59]. Not all workers agree on the production of Ins-1,4,5-P3 and the subsequent increase in cytosolic free Ca^{2+}, as purified GC B cells fail to do this [60]. This study differs from other studies in that the pattern of tyrosine phosphorylation in GC B cells following CD40 stimulation seems to be very similar to that induced by sIg crosslinking, although the consequences are quite different. Relating their signalling data to biological function, Knox and Gordon [60] showed that inhibitors of phosphorylation prevent the CD40-mediated rescue of GC B cells from apoptosis but that Ca^{2+} chelators do not. While these apparent contradictions in the published data may arise from experimental differences, they may represent subtle changes in CD40 signalling characteristics throughout the continuum of B cell differentiation. Analysis of fetal liver pro-B cells and pre-B-cell leukaemias indicates that CD40 signal transduction pathways in B cell precursors is similar those found in activated B cells.

In summary, the role of PTKs is much more important in activated B cells than in resting B cells. The action of PTKs seems to be crucially important to the function of CD40 in the rescue of GC B cells from apoptosis.

Downstream Events

The B-cell-restricted tyrosine kinase btk, may be involved in the CD40 signalling pathway as B cells from CBA/N mice, that carry a mutant btk gene (the *xid* mutation), are defective in CD40 signalling [61]. This is a surprising finding, given that CBA/N mice have a defect in T-independent responses but normal T-dependent responses where CD40 is clearly crucial. This presumably indicates that the *btk* molecule is involved in one of the downstream events in the CD40 signalling cascade.

A number of transcription factors are reported to be induced soon after CD40 stimulation: the nuclear factor of activated T cells (NF-AT) is induced by CD40 ligand + IL-4 [62]. This observation presumably explains the sensitivity to cyclosporin A (CsA) or FK506 of CD40-induced proliferation, reported by the same group [55]; the translocation of the NF-ATp complex (NF-AT + Fos and Jun proteins) to the nucleus is inhibited by the action of CsA on calcineurin. In sharp contrast to the up-regulation of transcription factors, AP-1 (Jun–Fos heterodimer) and $NF - \kappa B$, following sIg crosslinking, their appearance after CD40 stimulation is independent of PKC [63–65] and emphasizes again the lack of a role for PKC in CD40 signal transduction.

A novel molecule in the CD40 signal transduction pathway has recently appeared, following its cloning, using the yeast two-hybrid system, to pull out proteins that associate with the cytoplasmic tail of CD40 [66–68]. The molecule has been termed CRAF1 (CD40 receptor-associated factor 1) because of its similarity to the TNF receptor-associated factors (TRAF-1 and TRAF-2). The fact that it binds to the cytoplasmic tail of CD40 suggests that it is involved at the earliest stages of signal transduction, but its exact role has not been defined. The molecule has multiple domains: an *N*-terminus that has homology to a RING finger motif that is found in several DNA binding proteins, several zinc finger motifs, a isoleucine-rich central segment that would form a coiled coil and potentially mediate oligomerization and the CD40-binding *C*-terminus. It is intriguing that CRAF-1, having the potential to interact at the membrane with CD40 and also to bind to DNA response elements, may shuttle between the membrane and the nucleus. CRAF-1 does bind to other membrane proteins, such as p80 TNFR II, $LT\beta$ receptor and also the Epstein–Barr virus (EBV) latent infection membrane protein 1 (LMP1), a protein essential for the growth transformation of B cells by EBV. Indeed, the CRAF-1 molecule was independently cloned and called LMP1-associated protein 1 (LAP1), using the yeast two hybrid system to identify the products of cellular genes that bound to viral LMP1 [68]. This clearly implicates the CRAF-1 molecule in the proliferative effects of CD40 signal transduction. It will be interesting to see if its expression or association with CD40 is regulated during differentiation.

There is no published information concerning CD40 signal transduction in cells other than B cells.

CD40 IN B CELL DEVELOPMENT

A role for CD40 during the development of B cells in the bone marrow seems likely given its expression on the earliest identifiable B cell precursors and the demonstrable signal transduction in these cells [51]. Pro-and pre-B cells will proliferate on stromal cell layers (fibroblasts) presenting an array of anti-CD40 antibodies, the so-called "CD40 system" of Banchereau and coworkers [69], if IL-3 is also present. This proliferation is enhanced by addition of IL-7 and IL-10 [70, 71]. Even though pre-B cells will proliferate in response to CD40L expressed on activated T cells [72], this is unlikely to be the source of the CD40L signal in the bone marrow. It has been argued that the CD40 on B cell precursors might be engaged by a distinct ligand, but despite searches none has been identified. Whatever the source, the signal does not initiate differentiation as few surface lgM$^+$ cells emerge from the CD40-dependent cultures.

CD40 ACTIVATION OF B CELLS *IN VITRO*

The consequences of CD40 ligation of B cells *in vitro* depend upon a variety of factors including the form of the stimulus (e.g. soluble antibody or ligand versus extensive crosslinking), the costimulus provided and the species. In man CD40 ligation is rarely directly mitogenic, whether the stimulus is soluble antibody/ligand, or presented as a crosslinking array of antibodies on FcγRII (CD32) transfected L cells, or as CD40L-transfected L cells. Ligation causes increase in cell size and a low level of cell division. The proliferation of human B cells is synergistically enhanced by addition of IL-4, IL-10 (to a lesser extent IL-2), anti-Ig or phorbol esters as costimuli [1–5]. Most antibodies raised so far to the mouse CD40 molecule are directly mitogenic even in soluble form [73, 74]; it is also the case that transfectants bearing mouse CD40L induce proliferation in the absence of any costimuli [75]. This proliferation is augmented by the addition of the costimuli mentioned above [73, 74]. Most B cells proliferate in response to anti-CD40 stimulation, including lgM$^+$ lgD$^+$ resting B cells of the follicular mantle,

lgM$^+$ lgD$^-$ cells of the marginal zone and CD5$^+$ B cells. GC B cells proliferate to a much lesser extent (DG, unpublished observation). In mouse and man one of the hallmarks of CD40 stimulation is the formation of homotypic aggregates within the first 24 hours of culture. This adhesion is mediated by LFA1–ICAM1 pairing and also the interaction of CD23 and CD21 [76, 77]. The activation of resting B cells via CD40 brings about a number of phenotypic changes including up-regulation of MHC class II, CD23, B7 molecules, IL-4 receptor (IL-4R), IL-5R [74, 75].

One of the breakthroughs for the study of B cells *in vitro* was the development of long-term culture of non-transformed B cells, involving the use of CD32 (FγRII)-transfected fibroblasts to present an array of anti-CD40 antibodies to B cells [69]. To some extent this has been superseded by the use of CD40L-transfected fibroblasts. Both systems allow the growth/survival of B cells for periods in excess of 10 to 12 weeks. The human system is particularly efficient when used in conjunction with IL-4 (and even more so with addition of IL-10). However, this seems counterproductive in the cultures of mouse B cells with CD40L transfectants, as cytokines cause many of the cells to differentiate. At the start of the cultures, proliferation is intense, but falls away with time: [3H]-thymidine incorporation in the second week is at least half what it was in the first, the decay continuing until, by week 4–5, only a small proportion of the cells undergo division. In spite of the cessation of cell division, the cell viability remains high for several more weeks. The "maturation" of these cultures may relate to changes in the signalling pathways utilized by CD40.

CD40 INDUCED DIFFERENTIATION OF B CELLS *IN VITRO*

B cells stimulated solely via CD40 do not differentiate to plasma cells and secrete very little immunoglobulin. Stimulation of human B cells with both anti-Ig and anti-CD40 can give rise to secretion of IgM, IgG and IgA. Addition of IL-4 induces large amounts of IgE, with only a small increase in IgM and IgG [78, 79]. It is to a large extent the use of anti-CD40 antibodies that has allowed the elucidation of lymphokine control of isotype-directed B cell differentiation. Thus IL-13 elicits IgE and IgG4 secretion, TGFβ + IL-10 is a strong inducer of IgA production and IL-10 alone can elicit IgG1 and IgG3. The original references for these observations can be found in recent detailed reviews on the role of cytokines in isotype switching and B cell differentiation by Banchereau *et al.* [21, 80] and by Coffman *et al.* [81]. The role of CD40 in isotype switching will be covered in a later section. The necessity for CD40 ligation for *in vitro* antigen-specific antibody production has also been demonstrated [82]. It has to be borne in mind that the potency of cytokines in driving switching differs considerably, depending on the type of cells upon which they act. Thus highly purified, resting, IgD$^+$ B cells secrete different isotypes to pre-activated IgD$^-$ B cells (e.g. GC B cells) following anti-CD40 + cytokine stimulation.

In the mouse, CD40 signals seem to be strongly anti-differentiative; even in conjunction with IL-4 (and IL-5) the continued presence of anti-CD40 antibodies in the culture precludes significant antibody secretion. If the anti-CD40 antibody is washed away and IL-4 added back then secretion of IgG1 is vastly enhanced. This may relate to a difference in affinity of antibodies and the physiological ligand. However, a similar phenomenon may be happening in the human "CD40 system" of Banchereau *et al.* [69] where, even in the presence of IL-4, a significant proportion of the cultured cells maintain IgD on their surface [83]. The transience of the CD40 ligand expression may be an important factor in allowing B cells to differentiate. The major action of CD40 in B cell differentiation is to prime the cells for the subsequent reception of cytokine signals.

T-DEPENDENT ANTIBODY RESPONSES AND ISOTYPE SWITCHING *IN VIVO*

Resting B cells can be induced to proliferate by coculture with fixed, activated T cells or membrane preparations from such cells [7–11]. The proliferation is largely inhibited by addition of soluble CD40 fusion proteins [12, 13]. In the presence of cytokines the cultures will proceed to antibody secretion (see above). Thus, CD40 ligand on activated T cells was identified as the mediator

of contact-dependent T cell help for antibody responses. With the discovery that the lesion in X-linked hyper IgM syndrome is caused by point mutations or deletions in the CD40 ligand gene came confirmation that, *in vivo* also, CD40 ligand delivered the major costimulus to B cells during T-dependent responses. Patients suffering from this syndrome, among other serious symptoms, fail to make the isotype switch to produce IgG, IgA or IgE [15–19]. This is reflected in a lack of these isotypes in both the serum and antigen-specific antibody responses following immunization [84]. These people, however, make super-normal amounts of IgM. The syndrome is mimicked, in its immunological features, in mice, by targeted disruption of either the CD40 [85, 86] or the CD40L genes [47, 48] and also by the blockade of the CD40–CD40L interaction by *in vivo* administration of soluble CD40 fusion protein [41]. It is clear from these experiments that the IgM response, although T cell dependent, is independent of CD40 ligation. At present, we do not know if this T cell dependent IgM response requires any sort of contact-mediated costimulus or if cytokines are sufficient.

The full explanation of the role of CD40 in isotype switching is not available. It may be required to drive the level of B cell proliferation that is necessary to achieve efficient switching [87]. It does induce transcription from downstream C_H genes, as germline transcripts [88], and so prepares the Ig locus for the switch recombination that proceeds with the arrival of cytokines. It is possibly the only signal delivered to the B cell during the early stages of activation that up-regulates the expression of cytokine receptors (for instance IL-4R and IL-5R) [74, 89]. This not only changes the B cell's response to IL-4 in quantitative terms (sensitivity to IL-4 increased by 100–1000-fold) but also in qualitative terms (IL-4 signalling pathways are altered) [89]. In addition, costimulation of T cells via CD40L might be a means of regulating cytokine production [49, 50].

Given the potent and unrestricted manner in which CD40L will activate B cells, the problem of bystander activation, with all its deleterious consequences, becomes a serious worry. It has been argued that the transience of expression (possibly accelerated by ligation induced down-regulation/internalization) would limit bystander activation. Also, because the release of cytokines during B–T cell interaction is directional and localized, it was argued that bystander activation via CD40 is unlikely to lead to significant antibody secretion. In addition, an elegant potential failsafe mechanism has been highlighted recently in studies of Fas (CD95)-mediated killing of B cells by Th1 cells [90]: B cells activated via CD40 alone are very susceptible to Fas-mediated killing, while cells stimulated with both anti-IgM and CD40 are resistant. Thus, only B cells activated in an antigen-specific manner will survive the interaction with a T cell expressing Fas-ligand.

CD40 IN THE DEVELOPMENT OF MEMORY B CELLS AND GERMINAL CENTRES

A characteristic feature of T-dependent antibody responses is the generation of memory B cells. While this capacity has not been tested directly in HIGM patients or in CD40/40L knockout mice, it was assumed to be defective given the absence of germinal centres in these individuals [15–19, 47, 48, 85, 86]. This gave rise to the notion that signalling of B cells via CD40 is necessary for GC formation. The question of CD40 involvement in memory generation has been addressed directly by in vivo blockade of the CD40–CD40L interaction either using soluble CD40 fusion protein [41] or anti-CD40L antibody [91]. The two studies concur by showing that blockade during the induction phase of the response (the first 4–5 days) leads to an impairment in the development of memory populations when measured by adoptive transfer 2 months later. One of these studies, however (Gray et al) [41], found that GC formation proceeded normally if soluble CD40 was injected. In contrast to the knockout models, it could be argued that the injection of the soluble CD40 might potentiate costimuli to T cells via CD40L. This has since been shown to be so, by immunizing CD40KO mice and then injecting them with soluble CD40 [50]: the mice regenerate GC, in contrast to those injected with a control fusion protein. The conclusion from these experiments is that signals to B cells via CD40 are not required to initiate the GC reaction, but that costimulation of T cells via CD40L is crucial for the subsequent delivery of the

soluble mediators that drive the GC reaction [50]. Clearly, CD40 has an important role to play in GC, but it is at a later point when the B cells are undergoing selection (see next section). The recent observation that CD40L can be expressed on activated B cells [44] might explain one of the mysteries of GC concerning the driving force behind the intense proliferation occurring here; given the paucity of T cells in the GC, one could speculate that the B cell proliferation is regulated in an autocrine manner by expression of CD40L on centroblasts.

SURVIVAL SIGNALS AND THE ROLE OF CD40 IN GERMINAL CENTRES

A few simple arithmetical calculations based on the number of B cells produced during the proliferative phase of the GC reaction and the known turnover rate of the peripheral, recirculating pool makes it obvious that somewhat less than 10% of the B cells produced in the GC will ever leave. There is ample evidence of cell death occurring in situ (e.g. tingible body macrophages carrying engulfed lymphocytes). The biochemical basis of this phenomenon, although not completely explained, is related to the down-regulation in GC B cells of genes, such as bcl-2, that control apoptosis [92,93]. GC B cells are susceptible to apoptosis in vivo and in vitro. Thus, tonsillar GC B cells die within 24 hours of being placed into culture; mouse GC B cells die much more slowly (and seem to express more bcl-2) but are vastly inferior to resting B cells in their survival capacity. It was noted by MacLennan and coworkers [92–94] that tonsil B cells could be prevented from entering apoptosis by crosslinking sIg (a weak signal) or by engaging CD40 (a strong signal). While the former leads to only a 24 hour stay of execution for the GC B cell and no up-regulation of bcl-2, the CD40 signal promotes a much longer survival of the cells and the induction of bcl-2.

These two "survival signals" represent two quite distinct phases of the selection process, both of which are essential for memory B cell development [94,95]. The rapidly dividing sIg centroblasts are mutating their Ig V genes, as part of the process of affinity maturation. As they exit cell cycle they start to re-express sIg and become susceptible to apoptosis. The first stage of selection is based on the capacity of their mutated receptors to bind to antigen within the antibody–antigen complex displayed on the surface of follicular dendritic cells (FDC); crosslinking sIg mimics this. The sIg rescue signal is relatively short-lived and the cell must receive other longer-lasting survival signals. It has since been shown that the process of antigenic selection within GC is independent of bcl-2, as it is not overridden by the overexpression of this proto-oncogene [96]. The second stage of selection involves the facilitation of long-term survival, re-entry into the recirculating pool and the programming of cellular differentiation. Thus, while some signals delivered in the GC result in plasma cell development, the CD40 signal allows cells to enter the recirculating pool as memory B cells. Ligation of CD40 is the only signal so far tested *in vitro* that gives rise to cells that resemble memory B cells in morphology and surface phenotype [92,97]. It is also by far the most potent rescue signal.

Histological examination of CD40L on T cells reveals a pattern of expression that fits very well into this scenario for the selection of memory B cells for exit from the GC. The highest concentration of T cells in tonsil GC is around the edge (the so-called outer zone) and it is these cells that express CD40L, in contrast to the T cells scattered throughout the body of the GC. *In vitro* analysis of these T cells shows that they do not express CD40L on the membrane, but in cytoplasmic granules, and it is brought to the surface within minutes of activation via the TcR [43]. This provides an elegant means by which the process of selection for long-term survival is made antigen specific: only B cells that have taken up antigen from FDC and present the correct peptide in association with MHC class II will receive a CD40L rescue signal.

As discussed earlier, the CD40 signalling pathways that lead to this GC B cell rescue from apoptosis are quite different to those that initiate proliferation of resting B cells. In other words the CD40 molecule is "re-wired". The re-wiring of the molecule seems to be a consequence of the initial CD40 engagement on resting cells: in mice treated with soluble CD40 from days 0–4 of the immune response, to prevent the ligation of CD40 on resting B cells, the GC contain a three to four to fold

excess of apoptotic cells. This increased apoptosis seems to be due to an inability to receive the CD40 rescue signal: the GC B cells from these mice die very rapidly *in vitro* and, unlike cells from control mice, cannot be rescued by the addition of anti-CD40 to the cultures. The switch to the CD40-rescue pathway does not occur if the primary CD40 engagement is prevented [98]. The basis for the re-wiring is not known, but is likely to involve the differential association of the CD40 tail with different cytoplasmic proteins.

These experiments form part of the recurring theme of this review: the CD40 molecule couples to different signalling pathways in different B cell subpopulations. This multiplicity is likely to be the key to the large range of biological actions that CD40 exhibits.

REFERENCES

1. Clark EA and Ledbetter JA. Activation of human B cells mediated by two distinct cell surface differentiation antigens, Bp35 and Bp50. *Proc. Natl. Acad. Sci. USA* 1986; **83**: 4494–4498.
2. Paulie S, Ehlin-Henriksson B, Mellstedt H, Koho H, Ben-Aissa H and Perlmann P. A p50 surface antigen restricted to human urinary bladder carcinomas and B lymphocytes. *Cancer Immunol. Immunother.* 1985; **20**: 23–28.
3. Gordon J, Millsum MJ, Guy GR and Ledbetter JA. Synergistic interaction between interleukin-4 and anti-Bp50 (CDw40) revealed in a novel B cell restimulation assay. *Eur. J. Immunol.* 1987; **17**: 1535–1538.
4. Gordon J, Millsum MJ, Flores RL and Gillis S. Regulation of resting and cycling B cells via surface IgM and accessory molecules interleukin 4, CD23 and CD40. *Immunology* 1989; **68**: 526–531.
5. Vallé A, Zuber CE, Defrance T, Djossou O, De Rie M and Banchereau J. Activation of human B lymphocytes through CD40 and interleukin 4. *Eur. J. Immunol.* 1989; **19**: 1463–1467.
6. Stamenkovic I, Clark EA and Seed B. A B lymphocyte activation molecule related to the nerve growth factor receptor and induced by cytokines in carcinomas. *EMBO. J.* 1989; **8**: 1403–1410.
7. Brian AA. Stimulation of B cell proliferation by membrane-associated molecules from activated T cells. *Proc. Natl. Acad. Sci. USA.* 1988; **85**: 564–568.
8. Noelle RJ, McCann J, Marshall L and Bartlett WC. Cognate interactions between helper T cells and B cells. III. Contact-dependent, lymphokine-independent induction of B cell cycle entry by activated helper T cells. *J. Immunol.* 1989; **143**: 1807–1814.
9. Hodgkin PD, Yamashita LC, Coffman RL and Kehry M. Separation of events mediating B cell proliferation and Ig

production by using T cell membranes and lymphokines. *J. Immunol.* 1990; **145**: 2025–2034.

10. Tohma S and Lipsky PE. Analysis of the mechanisms of T-dependent polyclonal activation of human B cells. Induction of human B cell responses by fixed activated T cells. *J. Immunol.* 1991; **146**: 2544–2552.

11. Noelle RJ, Daum J, Bartlett WC, McCann J and Shepherd DM. Cognate interactions between T cells and B cells. V. Reconstitution of T helper cell function using purified plasma membranes from activated Th1 and Th2 helper cells and lymphokines. *J. Immunol.* 1991; **146**: 1118–1124.

12. Noelle RJ, Roy M, Shepherd D, Stamenkovic I, Ledbetter JA and Aruffo AA. 39-kDa protein on activated T cells binds CD40 and transduces the signal for cognate activation of B cells. *Proc. Natl. Acad. Sci. USA.* 1992; **89**: 6550–6554.

13. Lane P, Traunecker A, Hubele S, Inui S, Lanzavecchia A and Gray D. Activated human T cells express a ligand for the human B cell-associated antigen CD40 which participates in the T cell-dependent activation of B lymphocytes. *Eur. J. Immunol.* 1992; **22**: 2573–2578.

14. Armitage RJ, Fanslow WC, Stockbrine L, Sato TA, Clifford KN, Macduff BM *et al.* Molecular and biological characterization of a murine ligand for CD40. *Nature.* 1992; **357**: 80–82.

15. Graf D, Korthäuer U, Mages H W, Senger G and Kroczek R A. Cloning of TRAP, a ligand for CD40 on human T cells. *Eur. J. Immunol.* 1992; **22**: 3191–3194.

16. Korthäuer U, Graf D, Mages HW, Briere F, Padayachee M, Malcolm S *et al.* Defective expression of T cell CD40 ligand causes X-linked immunodeficiency with hyper-IgM. *Nature.* 1993; **361**: 539–541.

17. DiSanto JP, Bonnefoy JY, Gauchat JF, Fischer A and de Saint Basile G. CD40 ligand mutations in X-linked immunodeficiency with hyper-IgM. *Nature.* 1993; **361**: 541–543.

18. Fuleihan R, Ramesh N, Loh R, Jabara H, Rosen FS, Chatila T *et al.* Defective expression of the CD40 ligand in X chromosome-linked immunoglobulin deficiency with normal or elevated IgM. *Proc. Natl. Acad. Sci. USA.* 1993; **90**: 2170–2173.

19. Aruffo A, Farrington M, Hollenbaugh D, Li X, Milatovich A, Nonoyama S *et al.* The CD40 ligand, gp39, is defective in activated T cells from patients with X-linked hyper-IgM syndrome. *Cell* 1993; **72**: 291–300.

20. Torres RM and Clark EA. Differential increase of an alternatively polyadenylated mRNA species of murine CD40 upon B lymphocyte activation. *J. Immunol.* 1992; **148**: 620–626.

21. Banchereau J, Bazan F, Blanchard D, Brière F, Galizzi JP, van Kooten C *et al.* The CD40 antigen and its ligand. *Ann. Rev. Immunol.* 1994; **12**: 881–922.

22. Inui S, Kaisho T, Kikutani H, Stamenkovic I, Seed B, Clark EA *et al.* Identification of the intracytoplasmic region essential for signal transduction through a B cell activation molecule, CD40. *Eur. J. Immunol.* 1990; **20**: 1747–1753.

23. van Kooten C, Gaillard C, Galizzi J-P, Hermann P, Banchereau J and Blanchard D. B cells regulate expression of CD40-ligand on activated T cells by lowering the mRNA level and through release of soluble CD40. *Eur. J. Immunol.* 1994; **24**: 787–792.

24. Loenen W, de Vries E, Gravestein L, Hintzen R, van Lier R and Borst J. The CD27 membrane receptor, a lymphocyte-specific member of the nerve growth factor receptor family, gives rise to a soluble form by protein processing that does not involve receptor endocytosis. *Eur. J. Immunol.* 1992; **22**: 447–455.

25. Law C, Wörmann B and LeBien T. Analysis of expression and function of CD40 on normal and leukemic human B cell precursors. *Leukemia* 1990; **4**: 732–738.

26. Inaba K, Steinman R M, Pack M W, Aya H, Inaba M, Sudo T *et al.* Identification of proliferating dendritic cell precursors in mouse blood. *J. Exp. Med.* 1992; **175**: 1157–1167.

27. Caux C, Dezutter-Dambuyant C, Schmitt D and Bancher-eau J. GM-CSF and TNFα cooperate in the generation of dendritic Langerhans cells. *Nature* 1992; **360**: 258–261.

28. Sallusto F and Lanzavecchia A. Efficient presentation of soluble antigen by cultured human dendritic cells is maintained by granulocyte/macrophage colony-stimulating factor plus interleukin 4 and down regulated by tumor necrosis factor. *J. Exp. Med.* 1994; **179**: 1109–1118.

29. Caux C, Massacrier C, Vanbervliet B *et al.* Activation of human dendritic cells through CD40 cross-linking. *J. Exp. Med.* 1994; **180**: 1263–1272.

30. Armitage R J, Tough T W, Macduff B M, Fanslow W C, Spriggs M K, Ramsdell F *et al.* CD40 ligand is a T cell growth factor. *Eur. J. Immunol.* 1993; **23**: 2326–2331.

31. Schriever F, Freedman A S, Freeman G, Messner E, Lee G, Daley J *et al.* Isolated follicular dendritic cells display a unique antigenic phenotype. *J. Exp. Med.* 1989; **169**: 2043–2048.

32. Galy AHM, Spits H. CD40 is functionally expressed on human thymic epithelial cells. *J. Immunol.* 1992; **149**: 775–782.

33. Karmann K, Hughes CC, Schechner J, Fanslow WC and Pober JS. CD40 on human endothelial cells—inducibility by cytokines and functional regulation of adhesion molecules. *Proc. Natl. Acad. Sci. USA* 1995; **92**: 4342–4346.

34. Beutler B and van Huffel C. Unravelling function in the TNF ligand and receptor families. *Science* 1994; **264**: 667–668.

35. Fanslow W C, Srinivasan S, Paxton R, Gibson M G, Spriggs M K and Armitage RJ. Structural characteristics of CD40 ligand that determine biological function. *Semin. Immunol.* 1994; **6**: 267–278.

36. Graf D, Müller S, Korthäuer U, van Kooten C and Weise CRK. A soluble form of TRAP (CD40 ligand is rapidly released after T cell activation). *Eur. J. Immunol.* 1995; **25**: 1749–1754.

37. Ranheim EA and Kipps TJ. Activated T cells induce expression of B7/BB1 on normal or leukemic B cells through a CD40-dependent signal. *J. Exp. Med.* 1993; **177**: 925–935.

38. Roy M, Aruffo A, Ledbetter J, Linsley P, Kehry M and Noelle R. Studies on the interdependence of gp39 and B7 expression and function during antigen-specific immune responses. *Eur. J. Immunol.* 1995; **25**: 596–603.

39. Yellin MJ, Sippel K, Inghirami G, Covey LR, Lee JJ, Sinning J *et al.* CD40 molecules induce down-modulation and endocytosis of T cell surface T cell–B cell activating

molecule/CD40-L. Potential role in regulating helper effector function. *J. Immunol.* 1994; **152**: 598–608.

40. Castle BE, Kishimoto K, Stearns C, Brown ML and Kehry MR. Regulation of expression of the ligand for CD40 on T helper lymphocytes. *J. Immunol.* 1993; **151**: 1777–1788.

41. Gray D, Dullforce P, Jainandunsing S. Memory B cell development but not germinal centre formation is impaired by *in vivo* blockade of CD40–CD40 ligand interaction. *J. Exp. Med.* 1994; **180**: 141–155.

42. van den Eertwegh A, Noelle RJ, Roy M, Shepherd DM, Aruffo A, Ledbetter JA et al. *In vivo* CD40–gp39 interactions are essential for thymus-dependent humoral immunity. I. *In vivo* expression of CD40 ligand, cytokines, and antibody production delineates sites of cognate T–B cell interactions. *J. Exp. Med.* 1993; **178**: 1555–1565.

43. Casamayor-Palleja M, Khan M and MacLennan ICM. A subset of CD4+ memory T cells contains pre-formed CD40-ligand that is rapidly but transiently expressed on their surface after activation through the T cell receptor complex. *J. Exp. Med.* 1995; **181**: 1293–1301.

44. Grammer AC, Bergmann MC, Miuri Y, Fujita K, Davis LS and Lipsky PE. The CD40 ligand expressed by human B cells co-stimulates B cell responses. *J. Immunol.* 1995; **154**: 4996–5010.

45. Gauchat JF, Henchoz S, Mazzei G, Aubry JP, Brunner T, Blasey H et al. Induction of human IgE synthesis in B cells by mast cells and basophils. *Nature* 1993; **365**: 340–343.

46. Gauchat J F, Henchoz S, Fattah D, Mazzei D, Aubry J P, Jomotte T et al. CD40 ligand is functionally expressed on human eosinophils. *Eur. J. Immunol.* 1995; **25**: 863–865.

47. Xu J, Foy TM, Laman JD, Elliot EA, Dunn JJ, Waldschmidt TJ, Elsemore J, Noelle RJ and Flavell RA. Mice deficient for CD40 ligand. *Immunity* 1994; **1**: 423–431.

48. Renshaw BR, Fanslow WC, Armitage RJ, Campbell KA, Liggitt D, Wright B, Davidson BL and Maliszewski CR. Humoral immune responses in CD40 ligand-deficient mice. *J. Exp. Med.* 1994; **180**: 1889–1900.

49. Cayabyab M, Phillips JH and Lanier L. CD40 preferentially costimulates activation of CD4+T lymphocytes. *J. Immunol.* 1994; **152**: 1523–1531.

50. Van Essen D, Kikutani H and Gray D. CD40 ligand-transduced costimulation of development of helper function. *Nature* 1995; **378**: 620–623.

51. Uckun FM, Schieven GL, Dibirdik I, Chandan LM, Tuel AL and Ledbetter JA. Stimulation of protein tyrosine phosphorylation, phosphoinositide turnover, and multiple previously unidentified serine/theronine-specific protein kinases by the pan-B-cell receptor CD40/Bp50 at discrete developmental stages of human B-cell ontogeny. *J. Biol. Chem.* 1991; **266**: 17478–17485.

52. Faris M, Gaskin F, Parsons JT and Fu SM. CD40 signaling pathway: anti-CD40 monoclonal antibody induces rapid dephosphorylation and phosphorylation of tyrosine-phosphorylated proteins including protein tyrosine kinase Lyn, Fyn, and Syk and the appearance of a 28-kD tyrosine phosphorylated protein. *J. Exp. Med.* 1994; **179**: 1923–1931.

53. Kansas GS and Tedder TF. Transmembrane signals generated through MHC class II, CD19, CD20, CD39, and CD40 antigens induce LFA-1-dependent and independent adhesion in human B cells through a tyrosine kinase-dependent pathway. *J. Immunol.* 1991; **147**: 4094–4102.

54. Ren C, Fu S and Geha R. Protein tyrosine kinase activation and protein kinase C translocation are functional components of CD40 signal transduction in resting human B cells. *Immunol. Invest.* 1994; **23**: 437–448.

55. Klaus G, Choi M and Holman M. Properties of mouse CD40. Ligation of CD40 activates B cells via a dependent, FK506-sensitive pathway. *Eur. J. Immunol.* 1994; **24**: 3229–3232.

56. Knox KA, Johnson GD and Gordon J. Distribution of cAMP in secondary follicles and its expression in B cell apoptosis and CD40-mediated survival. *Int. Immunol.* 1993; **5**: 1085–1091.

57. Lee KAW. Transcriptional regulation by cAMP. *Curr. Opin. Cell Biol.* 1991; **3**: 953–959.

58. Xie H and Rothstein TL. Protein kinase C mediates activation of nuclear cAMP response element-binding protein (CREB) in B lymphocytes stimulated through surface Ig. *J. Immunol.* 1995; **154**: 1717–1723.

59. Ren CL, Morio T, Fu SM and Geha RS. Signal transduction via CD40 involves activation of lyn kinase and phosphatidylinositol-3-kinase, and phosphorylation of phospholipase Cγ2. *J. Exp. Med.* 1994; **179**: 673–680.

60. Knox KA and Gordon J. Protein tyrosine phosphorylation is mandatory for CD40-mediated rescue of germinal center B cells from apoptosis. *Eur. J. Immunol.* 1993; **23**: 2578–2584.

61. Hasbold J and Klaus GGB. B cells from CBA/N mice do not proliferate following ligation of CD40. *Eur. J. Immunol.* 1994; **24**: 152–157.

62. Choi S K, Brines R D, Holman M J and Klaus GGB. Induction of NF-AT in normal B lymphocytes by anti-immunoglobulin or CD40 ligand in conjunction with IL-4. *Immunity* 1994; **1**: 179–187.

63. Huo L and Rothstein TL. Receptor-specific induction of individual AP-1 components in B lymphocytes. *J. Immunol.* 1994; **154**: 3300–3309.

64. Lalmanach GA, Chiles TC, Parker DC and Rothstein TL. T cell-dependent induction of NF-kappa B in B cells. *J. Exp. Med.* 1993; **177**: 1215–1219.

65. Berberich I, Shu G L and Clark EA. Cross-linking CD40 on B cells rapidly activates nuclear factor-kappa B. *J. Immunol.* 1994; **153**: 4357–4366.

66. Cheng G, Cleary AM, Ye ZS, Hong DI, Lederman S and Baltimore D. Involvement of CRAF1, a relative of TRAF, in CD40 signaling. *Science* 1995; **267**: 1494–1498.

67. Hu HM, O'Rourke K, Boguski MS and Dixit VM. A novel RING finger protein interacts with the cytoplasmic domain of CD40. *J. Biol. Chem.* 1994; **269**: 30069–30072.

68. Mosialos G, Birkenbach M, Yalamanchili R, VanArsdale T, Ware C and Kieff E. The Epstein–Barr virus transforming protein LMP1 engages signaling proteins for the tumor necrosis factor receptor family. *Cell* 1995; **80**: 389–399.

69. Bancbereau J, de Paoli P, Vallé A, Garcia E and Rousset F. Long-term human B cell lines dependent on interleukin-4 and antibody to CD40. *Science.* 1991; **251**: 70–72.

70. Saeland S, Duvert V, Moreau I and Banchereau J. Human B cell precursors proliferate and express CD23 after CD40 ligation. *J. Exp. Med.* 1993; **178**: 113–120.

71. Larson AW and LeBien TW. Cross-linking CD40 on human B cell precursors inhibits or enhances growth depending on the stage of development and the IL costimulus. *J. Immunol.* 1994; **153**: 584–594.

72. Renard N, Duvert V, Blanchard D, Banchereau J and Saeland S. Activated cells induce CD40-dependent proliferation of human B cell precursors. *J. Immunol.* 1994; **152**: 1693–1701.

73. Heath AW, Wu WW and Howard M. Monoclonal antibodies to murine CD40 define two distinct functional epitopes. *Eur. J. Immunol.* 1994; **24**: 1828–1834.

74. Hasbold J, Johnson-Léger C, Atkins C J, Clark E A and Klaus GGB. Properties of mouse CD40: cellular distribution of CD40 and B cell activation by monoclonal anti-mouse CD40 antibodies. *Eur. J. Immunol.* 1994; **24**: 1835–1842.

75. Gray D, Siepmann K and Wohlleben G. CD40 ligation in B cell activation, isotype switching and memory development. *Semin. Immunol.* 1994; **6**: 303–310.

76. Björck P and Paulie S. Inhibition of LFA-1-dependent human B-cell aggregation induced by CD40 antibodies and interleukin-4 leads to decreased IgE synthesis. *Immunology* 1993; **78**: 218–225.

77. Björck P, Elenstrom MC, Rosen A, Severinson E and Paulie S. CD23 and CD21 function as adhesion molecules in homotypic aggregation of human B lymphocytes. *Eur. J. Immunol.* 1993; **23**: 1771–1775.

78. Rousset F, Garcia E and Banchereau J. Cytokine-induced proliferation and immunoglobulin production of human B lymphocytes triggered through their CD40 antigen. *J. Exp. Med.* 1991; **173**: 705–710.

79. Spriggs M, Armitage RJ, Stockbrine L, Clifford KN, Macduff BM, Sato TA *et al.* Recombinant human CD40 ligand stimulates B cell proliferation and immunoglobulin E secretion. *J. Exp. Med.* 1992; **176**: 1543–1550.

80. Banchereau J and Rousset F. Human B lymphocytes: phenotype, proliferation, and differentiation. *Adv. Immunol.* 1992; **52**: 125–262.

81. Coffman RL, Lebman DA and Rothman P. Mechanism and regulation of immunoglobulin isotype switching. *Adv. Immunol.* 1993; **54**: 229–270.

82. Nonoyama S, Hollenbaugh D, Aruffo A, Ledbetter JA and Ochs H. B cell activation via CD40 is required for the specific antibody production by antigen-stimulated human B cells. *J. Exp. Med.* 1993; **178**: 1097–1102.

83. Galibert L, Durand I, Banchereau J and Rousset F. CD40-activated surface IgD-positive lymphocytes constitute the long term IL-4-dependent proliferating B cell pool. *J. Immunol.* 1994; **152**: 22–29.

84. Ochs HD, Davis SD and Wedgwood RJ. Immunologic responses to bacteriophage ϕX174 in immunodeficiency disease. *J. Clin. Invest.* 1971; **50**: 2559–2565.

85. Kawabe T, Naka T, Yoshida K, Tanaka T, Fujiwara H, Suematsu S, Yoshida N, Kishimoto T and Kikutani H. The immune responses in CD40-deficient mice: Impaired immunoglobulin calls switching and germinal center formation. *Immunity* 1994; **1**: 167–178.

86. Castigli E, Alt F, Davidson L, Bottaro A, Mizoguchi E, Bhan AK and Geha RS CD40-deficient mice generated by recombination-activating gene deficient blastocyst complementation. *Proc. Natl. Acad. Sci. USA.* 1994; **91**: 12135–12139.

87. Severinson-Gronowicz E, Doss C and Schröder J. Activation to IgG secretion by lipopolysaccharide requires several proliferative cycles. *J. Immunol.* 1979; **123**: 2057–2062.

88. Jumper MD, Splawski JB, Lipsky PE and Meek K. Ligation of CD40 induces sterile transcripts of multiple Ig H chain isotypes in human B cells. *J. Immunol.* 1994; **152**: 438–445.

89. Siepmann K, Wohlleben G and Gray D. CD40-mediated regulation of IL-4 signalling pathways in B lymphocytes. *Eur. J. Immunol.* 1996; **26**: 1544-1552.

90. Rothstein T, Wang J, Panka D, Foote L, Wang Z, Stanger B *et al.* Protection against Fas-dependent Th1-mediated apoptosis by antigen receptor engagement in B cells. *Nature* 1995; **374**: 163–165.

91. Foy TM, Laman JD, Ledbetter JA, Aruffo A, Claassen E and Noelle RJ. gp39–CD40 interactions are essential for germinal center formation and the development of B cell memory. *J Exp Med.* 1994; **180**: 157–163.

92. Liu Y-J, Joshua DE, Williams GT, Smith CA, Gordon J and MacLennan ICM. Mechanisms of antigen-driven selection in germinal centers. *Nature.* 1989; **342**: 929–931.

93. Liu Y-J, Mason DY, Johnson GD, Abbot S, Gregory CD, Hardie DL *et al.* Germinal center cells express bcl-2 after activation by signals which prevent their entry into apoptosis. *Eur. J. Immunol.* 1991; **21**: 1905–1910.

94. MacLennan ICM. Germinal centers. *Ann. Rev. Immunol.* 1994; **12**: 117–139.

95. Gray D. Immunological memory. *Ann. Rev. Immunol.* 1993; **11**: 49–77.

96. Smith K, Weiss U, Rajewsky K, Nossal G and Tarlington D. Bcl-2 increases memory B cell recruitment but does not perturb selection in germinal centers. *Immunity* 1994; **1**: 803–813.

97. Arpin C, Déchanet J, van Kooten C, Merville P, Grouard G, Brière F *et al.* Generation of memory B cells and plasma cells *in vitro*. *Science* 1995; **268**: 720–722.

98. Siepmann K, van Essen D, Harnett MM and Gray D. CD40 signalling in resting and germinal centre B cells is functionally and biochemically distinct: Evidence for rewiring of CD40 for delivery of rescue signals. Manuscript submitted for publication.

7

The T Cell Costimulatory Molecule CD28 Couples to Multiple Signalling Pathways

David M Sansom[1], Christine Edmead[1], Richard Parry [2] and Stephen G Ward [2]

Bridge, Bath, UK and [2] Department of Pharmacology, University of Bath, Bath, UK

INTRODUCTION

Activation of resting T lymphocytes requires at least two signals: one provided by engagement of the T cell antigen receptor (TcR)/CD3 complex by foreign antigen associated with self-MHC and the second by a costimulatory molecule present on "professional" antigen presenting cells such as dendritic cells, monocytes and activated B cells [1,2]. These signals trigger transition of the cell cycle as well as induction of interleukin 2 (IL-2) secretion and IL-2 receptor expression [1,2]. TcR/CD3 signals alone are insufficient to allow T cell proliferation [1] and thus an understanding of the molecular interactions and signals generated by costimulation is of considerable importance. The identification of B7.1 (CD80), as a potent costimulatory molecule found on antigen presenting cells and a ligand for the T cell surface molecule CD28 focused attention on this pathway as a potential "second signal" required for T cell activation [3–7]. Subsequently, the identification of additional family members has given the potential for at least four interactions in T cell costimulation which involve the proteins B7.1 [3–7], B7.2 (CD86) [8–10], CD28 [2–4] and CTLA-4 [9–11].

The functional consequences of engagement of the CD28 family of receptors is still the subject of investigation. However, it is established that ligation of CD28 alone has little, if any, effect on resting T cell proliferation, but this interaction has been shown to control proliferation, and IL-2 production from TcR-stimulated CD28$^+$ T cells [6,7]. In addition, costimulation via CD28 also mediates strong up-regulation of other cytokines including IL-4, IL-8, IL-13, γ-interferon and GM-CSF [12–16] as well as up-regulation of the IL-2 receptor α, β and γ chains ([17,18] and C Cerdan, personal communication). Moreover, studies using soluble antagonists such as CTLA-4Ig fusion proteins or antibodies to block CD28/B7.1/B7.2 interactions have shown that T cells which are deprived of costimulatory signals enter into a state of proliferative hyporesponsiveness (anergy) or may undergo apoptosis [1,19–22]. These effects can largely be prevented by the delivery of a costimulatory signal via CD28 [22,23]. These data along with *in vivo* transgenic experiments provides ample evidence of a critical role for CD28 signals in controlling T cell responsiveness [24]. Manipulation of CD28 interactions has already provided exciting results in both transplantation and tumour therapy settings and is likely to be equally valuable in autoimmune diseases [2]. It is therefore of considerable interest to determine and understand the biochemical signals by which CD28 and CTLA-4 specify these functional outcomes. This review aims to summarize the current state of knowledge regarding the signalling pathways utilized by CD28 in T cells.

STRUCTURE AND EXPRESSION OF CD28 AND B7 RECEPTOR FAMILIES

CD28 and CTLA-4

The known members of this "family" of receptors have now been cloned and analysed in some detail

Lymphocyte Signalling: *Mechanisms, Subversion and Manipulation*. Edited by M. M. Harnett and K. P. Rigley © 1997 John Wiley & Sons Ltd.

at both the DNA and protein levels [3]. The genes for CD28 and CTLA-4 are closely linked on human chromosome 2 (2q33–34) and share the same genomic organization, suggesting a common evolutionary origin [25]. However, the overall homology between the genes is as little as 30%. CD28 is a homodimeric cell surface glycoprotein expressed on 95% of CD4+ T cells and approximately 50% of CD8$^+$ T cells [3]. The protein is composed of two glycosylated 44 kDa chains, which are members of the immunoglobulin superfamily, each containing a single disulphide-linked extracellular Ig V-like domain. The mature CD28 polypeptide contains 202 amino acids giving a molecular mass of 23 kDa which is then glycosylated to the molecular mass of the mature protein. The extracellular domain is linked via a single-pass transmembrance region to a 41 amino acid cytoplasmic domain which is presumed to be responsible for initiating costimulatory signals [3, 26].

Like CD28, CTLA-4 is a disulphide-linked homodimer of predicted polypeptide molecular mass of 20 kDa with only a single glycosylation site. This protein also contains a single disulphide-linked extracellular Ig V-like domain linked via a short amino acid stretch to a transmembrane region and a cytoplasmic domain of 36 amino acids. Interestingly there is 100% conservation between the cytoplasmic domains of human and mouse CTLA-4, suggesting strongly conserved function [3]. In contrast, there is only approximately 30% amino acid conservation between CD28 and CTLA-4 in the cytoplasmic domain, perhaps suggesting an ability to generate different signals. The functional role of CTLA-4 on human T cells has not been fully clarified but unlike CD28, it is not expressed constitutively and is only expressed after T cell activation by TcR/CD3 and CD28 ligation [2–4]. Interestingly, both CD28 and CTLA-4 can bind to the same ligands, namely the 60 kDa B7.1 and the 70 kDa B7.2 [2,4]. The biological consequences of CTLA-4–B7.1/B7.2 interactions are not clearly defined, although both amplification [27] and suppression [27–29] of the T cell immune response have been postulated. If indeed the main function of CTLA-4 is to act as a mediator of immune response suppression, it is interesting to note that CD28 is transiently downregulated upon binding to its ligand [30] and may become hyporesponsive to subsequent stimulation

at approximately the same time that CTLA-4 is normally expressed.

The cytoplasmic domains of both CD28 and CTLA-4 lack any direct enzymatic activity and they are therefore presumed to signal via the recruitment of cellular enzymes. Most notably, there are consensus sequence motifs within the cytoplasmic domains of both CD28 ([173]$YMNM$) and CTLA-4 ([164]YVKM), which form potential binding sites for interaction with signalling proteins [3,31]. Moreover, CD28 (but interestingly not CTLA-4) also contains two proline rich motifs ([178]PRRP and [190]PYAP) which conform to the PXXP SH3 binding consensus sequence [32] and these regions of the CD28 tail may mediate interactions with signalling proteins. While the extracellular domains have limited sequence homology, mapping of the ligand binding sites of CD28 and CTLA-4 has localized the interaction to a conserved sequence MYPPY which is found in both CD28 and CTLA-4 [2,3].

B7.1 and B7.2

B7.1 and B7.2 are also members of the immunoglobulin superfamily with two Ig-like domains (Ig V and Ig C) but they share only 25% sequence homology [2,3]. There is also evidence that an additional B7 molecule may exist which has been provisionally termed B7.3 [33]. B7.1 is a 60 kDa glycoprotein (30 kDa polypeptide) which consists of the two extracellular disulphide-linked Ig-like domains, a transmembrane region and a short 19 amino acid cytoplasmic domain. In contrast, B7.2 is a 70 kDa glycoprotein (34 kDa polypeptide) whose major feature is an extended cytoplasmic domain which contains phosphorylation sites for protein kinase C possibly indicating a signalling function for it in APCs. The published sequences of B7.2 [9] and the molecule termed B70 [10] are identical except for six additional N-terminal amino acids in the B70 (CD86) sequence.

Generally, both B7.1 and B7.2 are expressed on activated monocytes, activated B and T cells and activated natural killer cells. However, one major difference is that only B7.2 is expressed on resting monocytes [2,3,9,10]. Furthermore, B7.2 expression on B cells activated by immunoglobulin crosslinking peaks at 72 hours whereas maximum B7.1

expression occurs more than 3 days later [9,10], suggesting that B7.2 may play a critical role in initiating the immune response. B7.1 and B7.2 have overall similar receptor binding characteristics in that both bind CTLA-4 with 20–100-fold higher affinity than they bind CD28 [2–4], with B7.2 being at most two to three-fold less avid than B7.1 [34]. However, the B7.2–CTLA-4–Ig complex dissociates much faster than the B7.1–CTLA-4–Ig complex [34]. The fast on/off nature of B7.2/CTLA-4 interactions may serve to limit an immune response by permitting more efficient disengagement of B7.2 from the high avidity CTLA-4 receptor. Generally the subtle differences between B7.1 and B7.2, as well as CD28 and CTLA-4, allow a number of different interactions to occur. This may suggest differential functions for each molecule and, indeed, recent reports have shown that B7.1 preferentially acts as a costimulator of Th1 cells whereas B7.2 costimulates and induces Th2 cells [35,36].

SIGNALLING PATHWAYS COUPLED TO CD28

Several biochemical signalling pathways have been implicated as pivotal links in the control of cell activation, namely the lipid kinase phosphatidylinositol (PI) 3-kinase [37], the GTP-binding protein p21^{ras} [38] and the phosphatidylinositol specific phospholipase C (PLC) [39]. Although it is probable that all three pathways interact and complement each other at one or more levels, in this chapter we will consider the coupling of CD28 to each pathway separately. Until recently, studies addressing CD28-mediated signal transduction events were based on the use of specific CD28 mAbs as stimulators but the advent of specific ligands (B7.1 and B7.2) has provided potentially more interesting results and highlighted important discrepancies in the signalling pathways utilized by CD28 depending on its mode of activation (see Table 7.1).

Triggering of the TcR/CD3 complex initiates a cascade of biochemical events, the earliest of which is the activation of intracellular protein tyrosine kinases (PTK) [40]. This activation of PTK appears to be absolutely required for all subsequent T cell responses [40]. One substrate for TcR/CD3 complex-stimulated PTK is a phospha-

tidylinositol lipid-specific phospholipase C (PLCγ) [41]. Hence, PTK activation enables the TcR/CD3 complex to regulate the hydrolysis of membrane phosphatidylinositol lipids releasing the second messengers, diacylglycerol (DAG), which activate the serine/threonine kinase protein kinase C (PKC), and inositol 1,4,5-trisphosphate [Ins (1,4,5)P$_3$], which regulates intracellular calcium release [42]. While the CD28 cosignal appears to be insensitive to inhibition by cyclosporine A (CsA) [43] and thus independent of Ca^{2+} and calcineurin [44], CD28 signal transduction does appear to be regulated by PTKs [45, 46]. The tyrosine phosphorylation pattern induced by CD28 ligation is similar to that induced by TcR/CD3 ligation, but there is also unique CD28-induced tyrosine phosphorylation of 62 kDa (J Nunes and D Olive, unpublished observation), 75 kDa and 100 kDa proteins [45, 46], suggesting that unique pathways may be coupled to CD28. The use of distinct signal transduction pathways would lead to a complementation between TcR/CD3 and CD28 to allow full activation of T cells which would fit well with the idea of costimulation.

CD28 Activates the Signalling Pathway Regulated by Phosphatidylinositol 3-Kinase

Analysis of the cytoplasmic domain of CD28 reveals a motif (YMNM) around the tyrosine at position 173 [3]. This is similar to the core phosphotyrosine-containing consensus motif, namely YXXM [31] which has specificity for the SH2 domains of the p85 subunit of PI 3-kinase and hence suggested a direct interaction between PI 3-kinase and CD28 [3]. PI 3-kinase is a heterodimer which has a unique dual specificity as a lipid and protein serine kinase [47] and consists of an 85 kDa regulatory subunit containing two src-homology (SH2) domains and an SH3 domain [48], tightly associated with a catalytic 110 kDa subunit [49]. PI 3-kinase phosphorylates the membrane phosphatidylinositol lipids at the 3 position on the inositol ring, resulting in the generation of phosphatidylinositol-(3)-monophosphate [PtdIns(3)P], phosphatidylinositol-(3,4)-bisphosphate [PtdIns(3, 4)P$_3$] and phosphaidylinositol-(3,4,5)-bisphosphate [PtdIns (3,4,5) P$_3$] [50]. These D-3 phosphatidylinositol lipid isoforms represent only a minor

Table 7.1 Comparison of biochemical signals elicited by CD28 mAbs, B7.1 and B7.2 in resting or activated T cells and T cell lines

Response	Stimulus	Remarks	References
Tyrosine phosphorylation	B7.1, CD28 mAbs	Activated T cells, Jurkat, murine hybridomas	(D Olive) 3, 4, 45, 46
PLC-γ1 phosphorylation	B7.1, CD28 mAbs	Activated T cells, Jurkat and murine hybridomas	(D Olive) 2–4
$[Ca^{2+}]_i$ elevation	CD28 mAb (cross-linked), B7.1	Occurs in Jurkat cells without crosslinking; crosslinking is a prerequisite in resting T cells, activated T cells and murine hybridomas	3, 4, 73, 77, 78
Ins(1,4,5)P₃ elevation	CD28 mAb (cross-linked)	Jurkat or activated T cells; B7.1 elicits no detectable $PtdIns(4,5)P_2$ degradation	3, 4, 51, 77
CD28/PI 3-kinase association	B7.1, B7.2, CD28 mAb	CD28 associates with p85α subunit of PI 3-kinase. Occurs in resting and activated T cells and T cell lines. mAb cross-linking is not required	52–56
PI 3-kinase activation	B7.1, B7.2, CD28 mAb	$PtdIns(3,4,5)P_3$ accumulates in Jurkat. *In vitro* assays show activation in resting and activated T cells	51, 52, 56
PI 3-kinase inhibition by wortmannin or LY294002	B7.1, B7.2, CD28 mAb	Inhibition of $PtdIns(3,4,5)P_3$ accumulation; no effect on CD28/PI 3 kinase association	52, 56
IL-2 production	CD28 mAb, B7.1/CD3 mAb	Inhibition of IL-2 production on resting and activated T cells by wortmannin and LY294002	52, 56
	CD28 mAb, B7.1/CD3 mAb	Potentiation of IL-2 production from Jurkat by wortmannin	52
Tyrosine phosphorylation of:			
(i) vav	CD28 mAb, B7.1, B7.2	Jurkat	(J Nunes and D Cantrell) 82
(ii) p36 associated with Sos/Grb2	CD28 mAb	Jurkat	82
(iii) p62 (p120 GAP associated)	CD28 mAb, B7.1, B7.2	Jurkat; binds to GST-Grb-2 fusion proteins	(J Nunes and D Cantrell) 82
p21ras activation	CD28 mAb	Jurkat	82
Raf-1 activation	CD28 mAb	Jurkat	82
MAP kinase activation	CD28 mAb	Jurkat	82
Costimulation of IL-2 secretion	CD28 mAb + PMA/ CD3 B7.1, B7.2 + CD3	Resting T cells and Jurkat	52, 56, 82
Association with Grb2/Sos complex	CD28 mAb	Resting T cells, Jurkat and HPB-ALL	59
Sensitivity to herbimycin A	CD28 mAb	Inhibits elevation of and PI 3-kinase activation	(C June) 73
	B7.1, B7.2	No effect on PI 3-kinase association and kinase activation	(C June)
	PMA/CD3 + CD28/ ligands	Inhibits IL-2 production and substrate phosphorylation	45
p70S6 kinase	CD28 mAb (cross-linked)	Resting T cells; rapamycin-sensitive, CsA-insensitive	96
IκBα down-regulation	CD28 mAb	Jurkat T cells and resting T cells; requires presence of PMA; rapamycin sensitive, CsA-insensitive	101
Sphingomyelinase	CD28 mAb (cross-linked)	Resting T cells, activated T cells and Jurkat; acidic but not neutral sphingomyelinase is activated	95
JNK	CD28 mAb	Synergizes with CD3 mAb in Jurkat cells; CsA-sensitive	114
AP-1 regulation	CD28 mAb, B7.1, B7.2	Transcriptional activity requires TcR signal in murine T cells; down-regulated during anergy	107, 108

fraction of cellular phosphatidylinositiol lipids and are not substrates for PLC [37]. However, considerable importance has been attached to this pathway, particularly concerning their putative role as regulatory molecules [37,50]. Kinetic evidence has suggested that PtdIns(3,4,5)P$_3$, formed by a phosphatidylinositol-(4,5)-bisphosphate-specific 3-kinase, is the probable physiological intracellular mediator and this is subsequently degraded and inactivated by sequential dephosphorylation to PtdIns(3,4)P$_2$ and PtdIns(3)P [50].

The first indication that CD28 could indeed couple to PI 3-kinase was the demonstration that ligation of CD28 by B7.1 induced the accumulation of D-3 phosphatidylinositol lipids, the products of PI 3-kinase activation, in the leukaemic T cell line Jurkat [51]. Later studies revealed that B7.2 or mAb ligation of CD28 also induced formation of D-3 phosphatidylinositol lipids in Jurkat cells [52]; see also Table 7.1). Moreover, the p85 subunit of PI 3-kinase can associate with the cytosolic domain of CD28 via the YMNM motif [53, 54]. The interaction of the YMNM motif with the SH2 domains depends on phosphorylaton of the tyrosine residues within the motif and CD28 is indeed phosphorylated on tyrosine following ligation [53,55] and site directed mutagenesis within the YMNM motif showed that mutation of [173](p)Y results in the elimination of PI 3-kinase binding [53]. Interestingly, CD28 binds to the *C*-terminal SH2 domain of p85 with a 10-fold greater affinity than the *N*-terminal domain [54], but the relevence of this observation is not presently known.

The functional significance of the CD28/PI 3-kinase association was highlighted by reports showing that murine T cell hybridomas expressing point mutations of the [173](p)Y residue within the CD28 cytoplasmic tail are no longer able to produce IL-2 following CD28 ligation [53]. The relevence of PI 3-kinase association and activation, with respect to CD28-dependent IL-2 production from normal T cells, has also been demonstrated in studies using specific inhibitors of PI 3-kinase, namely the fungal metabolite wortmannin and the structurally unrelated PI 3-kinase inhibitor LY294002 [52,56]. These experiments demonstrated that CD28-mediated increases in D-3 phosphatidylinositol lipids, but not the association between CD28 and PI 3-kinase following ligation

by B7, can be specifically inhibited by the PI 3-kinase inhibitor wortmannin [52,56]. Furthermore, wortmannin abrogates B7.1/CD28-dependent T cell proliferation, but has no effect on T cell proliferation elicited by phorbol 12-myristate 13-acetate (PMA) and ionomycin [56]. These data demonstrate that recruitment and activation of PI 3-kinase by CD28 is of major importance in CD28-mediated T cell functions.

Certain discrepancies in the actions of PI 3-kinase inhibitors on CD28-dependent IL-2 production have arisen however. Firstly, it has been reported that while wortmannin inhibits CD28-associated PI 3-kinase activity in Jurkat cells, this inhibition appears to have no effect on CD28-mediated IL-2 production in Jurkat cells [57]. Indeed, subsequent studies revealed that wortmannin actually potentiates CD28-mediated IL-2 production from Jurkat cells [52]. Hence, PI 3-kinase appears to have inhibitory or stimulatory effects on IL-2 production depending on the model system used (Table 7.1). Heterogeneity in the regulation of biochemical signals between normal and Jurkat T cells has also been observed with respect to the regulation of PLC by phorbol esters [58]. The basis for the difference between Jurkat cells and primary T cells could be a reflection of the state of cellular differentiation. The ability of PI 3-kinase to regulate either a negative or positive signal for IL-2 production may be critical to understanding the proposed inhibition of T cell activation by CTLA-4 [27–29] which has also been reported to couple to PI 3-kinase [59]. Certain CD28 mAbs in the absence of other stimuli can elicit IL-2 production, albeit at much reduced levels compared to that induced by combining CD3 or PMA with CD28 stimuli (D Olive, personal communication). This production of IL-2 by CD28 mAbs has also been shown to be insensitive to wortmannin (D Olive, personal communication). Generally, these contrasting results with PI 3-kinase inhibitors highlight the differences in CD28 signalling in different model systems. One explanation for the discrepancies observed with wortmannin, however, may be due to the p85 subunit having dual functions. For instance, under certain circumstances (and with respect to certain receptors), the major function of the p85 subunit may actually be to act as an adaptor molecule, by virtue of interactions with other signalling molecules which are mediated by the SH2 and SH3 domains of the p85 subunit as has been demonstrated to occur in insulin receptor signalling [60]. Hence, even though the lipid kinase activity of the p110 subunit can be inhibited by wortmannin, the p85 subunit is still able to mediate other key interactions. In addition, multiple T cell isoforms of PI 3-kinase are known to exist in T cells [61] and these isoforms may have different sensitivities to wortmannin. These wortmannin insensitive isoforms may be preferentially recruited by CD28 following ligation by mAbs.

In addition to CD28 [51,52,56], other T cell molecules such as the IL-2 receptor [62], the TcR/CD3 complex and CD4 are known to associate with PI 3-kinase via interactions with the *src* family kinases p59[fyn] or p56[lck] [63,64]. Moreover, both the TcR [65] and CD7 [66] also induce D-3 phosphatidylinositol lipid production following ligation, albeit at much reduced levels compared to CD28-mediated increases which are generally four to five-fold greater than either CD3-or CD7-induced increases [66]. It is unclear therefore, why activation of this pathway, which is clearly coupled to and activated by multiple T cell surface receptors, should be a critical event in T cell costimulation when specifically activated by CD28. In this respect, heterogeneity of: (i) PI 3-kinase isoforms [67–69]; (ii) receptor coupling to PI 3-kinase (either tyrosine kinase regulated or G protein regulated mechanisms [50, 68]; (iii) PI 3-kinase substrate specificity [50, 67]; or (iv) magnitude and/or compartmentalization of receptor-induced D-3 phosphatidylinositol lipid formation, may determine distinct functions for different T cell surface receptors. While the CD28 activated pool of PI 3-kinase appears to be critical in determining the costimulation of IL-2 production, the functional relevence of the other pools of PI 3-kinase remains to be established. Another possibility is that while many interactions can be observed with "high-density ligands" such as crosslinked antibodies (particularly on transformed cells), these signals may not be efficiently generated by the small number of receptor engagements provided by TcR/MHC peptide interactions. Thus, the differences observed between CD3 and CD28 may simply be the efficiency of activation of each pathway under limiting conditions.

CD28 Interactions with Protein Tyrosine Kinases

The primary structure of CD28 indicates that the cytoplasmic tail has no obvious enzymatic activity but CD28 clearly couples to PTK, as evidenced by the tyrosine phosphorylation of a number of substrates [2–4, 45, 46]. In addition, CD28 must also act as a substrate for PTK since it is tyrosine phosphorylated [53,55] and this tyrosine phosphorylation appears critical for its interaction with PI 3-kinase [53]. From analogy with the TcR, CD4 and CD8 T cell surface molecules, which couple to the *src*-family non-receptor tyrosine kinases such as p59[fyn] and p56[lck][64], it was postulated that CD28 also coupled to similar non-receptor PTK. Indeed, p56[lck] has been demonstrated to be activated following crosslinking of CD28 mAb but not by CD28 mAb alone [70]. While, the association of CD28 with PI 3-kinase has been shown to occur in p56[lck] deficient cell lines [71,72], CD28 ligation by mAb is unable to elicit Ca^{2+} mobilization and tyrosine phosphorylation of certain substrates in these p56[lck] deficient cell lines [71]. Hence, it has been suggested that there are at least two PTK pathways activated in response to CD28 ligation only one of which is dependent on p56[lck] activity [71]. Studies with the *src* family tyrosine kinase inhibitor herbimycin A tend to support the hypothesis that several PTK pathways are activated by CD28. For instance, herbimycin A prevents tyrosine phosphorylation of substrates and IL-2 production following CD28 ligation [45]. However, while herbimycin A also prevents the elevation of intracellular calcium ($[Ca^{2+}]_i$) [73] and PI 3-kinase activation induced by crosslinked CD28 mAb, herbimycin A has no effect on $[Ca^{2+}]_i$ elevation or PI 3-kinase activation induced by B7.1 (Carl June, personal communication). Hence, CD28 is able to couple to PTKs which are herbimycin-A-sensitive (following ligation by mAb) as well as herbimycin-insensitive PTKs (following ligation by B7.1 or B7.2). The identity of the CD28-coupled PTKs remains unclear although studies have suggested an *in vitro* association of CD28 with p56[lck] and p59[fyn] following mAb ligation of CD28 [74]. CD28 activation by mAb in the absence or presence of crosslinking has also been shown to result in the association with, and immediate tyrosine phosphorylation and activation of the Tec family kinase p72$^{ITK/EMT}$ in the Jurkat T cell line [75]. Accordingly, it was postulated that p72$^{ITK/EMT}$ activation by CD28 may result in tyrosine phosphorylation of CD28 and the recruitment of PI 3-kinase. In addition, a novel member of the Tec/itk/Btk family of PTKs has been reported to exist in resting T cells and has appropriately been termed resting lymphocyte kinase (*rlk*) [76]. This newly identified protein kinase may prove to play an important role in T cell costimulation, although this hypothesis has yet to be verified.

CD28 Activates the Phospholipase C Pathway

It has been described that CD28 regulates tyrosine phosphorylation of PLCγ-1, PtdIns hydrolysis and [Ca^{2+}]$_i$ [2–4]. There appears to be differences however, depending on the type of T cell model used, in the relative ability of CD28 to elicit these responses (summarized in Table 7.1). For instance, in Jurkat cells, PtdIns hydrolysis, [Ca^{2+}]$_i$ and DAG generation can be elicited in the absence of crosslinking [77]. However, CD28 crosslinking is a prerequisite for eliciting PLC activity in purified resting T cells, activated T cells and human CD28 transfected murine T cell hybridoma [78]. Interestingly, ligation of CD28 by B7.1, but not B7.2, has been shown to result in elevation of [Ca^{2+}]$_i$ in activated T cells [73]. In marked contrast to these results, B7.1 failed to elicit any detectable activation of PLC as assessed by PtdIns(4,5)P$_2$ and phosphatidic acid generation [51] (Table 7.1) although B7.1 has been reported to induce PLCγ phosphorylation (D Olive, personal communication) and the final events of the activation process namely IL-2 production [52,56]. This suggests that PLC activation following mAb activation of CD28 may be functionally redundant. There is evidence to suggest, however, that there are both CsA- sensitive and insensitive components of the CD28 signalling responses depending on the activation state of the T cell such that the CsA-sensitive pathway is dominant in activated T cells, while the CsA-insensitive pathway is dominant in resting T cells [3]. Recent studies have shown that the CD28-induced calcium signal was mainly limited to T cells of the CD4+ subset [79] indicating that the costimulatory signal delivered by CD28 may have different biochemical properties in CD4+ and CD8+ T cell subsets and therefore the role of CD28 may differ in these cells.

CD28 Activates the p21^{ras} Pathway

In addition to PI 3-kinase, the GTP binding protein p21^{ras} has been established as a pivotal link in the control of cell growth [38,80]. The intracellular proteins linking receptor-associated PTKs to the regulation of p21^{ras} have been defined: Grb2, an adaptor protein, is composed of two SH3 domains that bind to Sos (a guanine nucleotide exchange protein that activates p21^{ras}) and an SH2 domain that interacts with tyrosine phosphoproteins [79]. The TcR/CD3 complex stimulates the guanine nucleotide binding proteins p21^{ras} and this is essential for the coupling of the TcR/CD3 complex to the mitogen activated protein (MAP) or extracellular signal regulated (ERK) kinase, ERK2 and hence to the regulation of transcription factors such as AP-1, NFAT and the IL-2 gene [80,81].

TcR/CD3 complex activation of p21^{ras} requires cellular PTK function and recent studies have identified two PTK substrates, Vav and a 36 kDa membrane protein, as potential mediators of the TcR/p21^{ras} link [80]. Interestingly, at least one 100 kDa substrate for CD28-induced tyrosine phosphorylation has been identified as Vav, a guanine nucleotide exchange factor [75,82]. Hence, the question was raised as to whether p21^{ras} can also be activated by CD28? Indeed, the CD28 mAb, CD28.2 can activate the p21^{ras} pathway, since it induced an increase of *ras*–GTP complexes and the phosphorylation of two substrates: namely, Vav and the Grb2/Sos-associated protein p36 [82] which is believed to be involved in p21^{ras} guanine nucleotide exchange. Moreover, the cascade of MAP kinases is also activated resulting in hyperphosphorylation of Raf-1 and activation of ERK2 [82]. In marked contrast, B7.1 stimulation of CD28 is unable to induce tyrosine phosphorylation of Grb2 or p36 [82] although it can tyrosine phosphorylate Vav to a higher extent than by CD28.2 mAb or anti-CD3 ligation [82]. Another CD28 mAb, CD28.5 which binds CD28 but is unable to induce mobilization or IL-2 production, is devoid of effect on p21^{ras} [82]. Thus, CD28 does have the capacity to couple to the p21^{ras} pathway as demonstrated by the effect of the CD28.2 mAb although CD28 may not activate it following ligation by B7.

Interestingly, the PI 3-kinase binding motif,^{173}(p) YMNM motif, present in the CD28 cytoplasmic tail also corresponds to the consensus binding motif (pYXNX) for Grb2 [83,84]. Indeed, a recent report demonstrated that CD28 binds to the Grb2/ Sos complex by means of the cytoplasmic ^{173}(p)YMNM motif although the Grb2 binding has much lower affinity (up to 100-fold lower) than CD28/PI 3-kinase binding [84]. The relevence of this association and the possible regulation of CD28 coupling to the p21^{ras} pathway and ultimately CD28-dependent IL-2 production, has yet to determined. One putative role for p21^{ras} activation by CD28 (and perhaps even CD3) may be to enhance and/or facilitate the PI 3-kinase activation induced by CD28 in light of recent reports which show that p21^{ras} can stimulate PI 3-kinase [85,86]. It should also be pointed out that the CD28/Grb2 association was observed using 1.5×10^8 cells (resting T cells, Jurkat or HPB-ALL) in the presence of extensively crosslinked CD28 mAb [84] and has not been observed following ligation of CD28 by either of the natural ligands (D Olive and J Nunes, personal communication). The p36 Grb2 SH2 domain binding protein and SLP, the 75 kDa Grb2 SH3 domain binding protein are selective substrates for TcR/CD3 complex but not CD28 signalling pathways [80]. However, a 62 kDa p120 GAP-associated molecule has been described as a multifunctional adaptor molecule in many different cells and receptor systems [87]. This p62 protein is a substrate for CD28 but not TcR regulated PTKs and a p62 tyrosine phosphoprotein binds to GST–Grb2 fusion proteins following CD28 activation, which implies that p62 may have a selective function in CD28 but not TcR signal transduction (J Nunes and D Cantrell, personal communication).

REGULATION OF TRANSCRIPTIONAL EVENTS BY CD28

A major outcome of CD28 signalling is the production of cytokines. This effect is mediated by an effect on transcriptional up-regulation as well as on mRNA stability of cytokine genes. Since CD28 signals are not generally mediated in isolation, the effects are only seen in combination with signals from the TcR pathway. In studies of transcriptional regulation, potent signals such as those induced by crosslinked mAbs, PMA with ionomycin or mitogens are commonly used to mimic activation signals. To date there are few data using natural ligands to investigate transcriptional regulation. Of the cytokines affected by CD28, the IL-2 gene is best characterized and will be used to exemplify the effects of CD28 costimulation. The 5' upstream region of the IL-2 promoter between −325 and the transcription start site contains a number of sites for the binding of transcription factor complexes including AP-1, NFκB, NFAT and OCT for transcriptional activation [88]. The binding of these proteins to their recognition sites within the IL-2 promoter results in the up-regulation of IL-2 gene transcription and removal of any of these binding sites generally has a profound negative effect on gene transcription. We will consider the transcription factors thought to be important for CD28 functions, as well as key protein kinases, namely p70S6 kinase [89] and c- Jun N-terminal kinase [90], which have been implicated as key enzymes in the regulation of CD28-mediated transcriptional events.

The CD28 Response Complex and NFκB

The existence of a CD28-regulated transcription factor was demonstrated following observations that mAb stimulation of CD28 increased the activity of IL-2-chloramphenicol acetyltransferase (CAT) reporter constructs above the levels seen with PMA and iononycin alone [87]. This binding site has been termed the CD28 response element (CD28RE) which is located between positions– 160 and – 152 relative to the transcription start site and the resulting protein complex is termed the CD28-responsive complex (CD28RC). Mutation of the CD28RE results in the loss of CD28-induced activity without affecting TcR-induced IL-2 promoter activity [88].

The CD28RE is distinct from, but related to, that of the NFκB element [88]. The NFκB/Rel family of transcription factors consists of five family members NFκB1 (p50), NFκB2 (p52), c-rel, p65 RelA, and RelB [81]. NFκB is a dimeric complex which is constitutively found in an inactive form in the cytoplasm [81]. The p50/p65 heterodimer is recognized as the main active form although c-rel-containing and homodimeric complexes have also been described [81]. Interestingly, three Rel family proteins, p50, p65 and c-Rel, are among the components of the CD28 RC [91] and the recruitment of these proteins was shown to be CsA-insensitive [91]. Evidence for the importance of the p50 in the immune response can be seen from knockout mice which show multiple B cell and T cell defects including defective CD28 responses [92]. Activation of NFκB can be brought about by a variety of agents including PMA and ionomycin, which result in the rapid and transient degradation of IκBα and dissociation of the active NFκB complex from its cytoplasmic inhibitor IκBα [81,93]. The regulation of NFκB may involve the phosphorylation and inducible decay of IκB by chymostrypsin-like proteases [93] and since the promoter region of the IκB gene contains an NFκB binding site, it appears that a negative feedback is in operation whereby NFκB switches on its own inhibitor [93]. Interestingly, CD28 can induce a sustained down-regulation of IκBα which is associated with an increase in the nuclear translocation of c-Rel [94]. The precise mechanisms by which CD28 modulates IκBα are unclear. However, CD28 has been reported to activate an acidic sphingomyelinase [95] which catalyses the breakdown of sphingomyelin to phosphorylcholine and ceramide, which have been shown to trigger the degradation of IκBα required for the subsequent nuclear translocation of NFκB [96].

Possible Regulation of NFκB by CD28-activated p70S6 Kinase

CD28-dependent proliferation is inhibited by rapamycin—a macrocyclic compound that binds to the family of intracellular receptors termed FKBP [44,97]. Rapamycin inhibits activation of p70/85 kDaS6 kinase (αI and αII isoforms, referred to collectively as p70S6 kinase), a member of a family of related serine/threonine protein kinases that catalyse the phosphorylation of the 40S ribosomal subunit S6 [89] and perhaps other as yet unknown substrates. Blocking of p70S6

kinase by either injection of antibodies or by the use of rapamycin has suggested a role for p70S6 kinase in $G_1 - S$ phase transition [89,98,99]. Given the significance of CD28-derived signals in regulating proliferation of resting T cells, a role for CD28-induced signals in p70S6 kinase activation therefore seems appropriate.

Crosslinked CD28 mAbs (but not soluble CD28 mAbs) have indeed been found to induce activation of p70S6 kinase in resting purified T lymphocytes [97]. This signal occurred with slower kinetics than those observed for CD3 ligation by mAbs and both pathways were inhibited by rapamycin but not by FK506 or CsA [97]. This observation is particularly interesting given the evidence that PI 3-kinase and its D-3 phosphatidylinositol lipid products may mediate p70S6 kinase activation, since the PI 3-kinase inhibitors wortmannin and LY294002 block PDGF and insulin-mediated p70S6 kinase activation [100]. While rapamycin does not prevent CD28-induced PI 3-kinase activation, at least implying that PI 3-kinase lies upstream of p70S6 kinase (S Ward, unpublished observations), the effects of CD28 (in the presence of PMA) on IκBα down-regulation and c-Rel translocation are blocked by rapamycin [101]. This implies that there may be an upstream regulation of IκBα and/or c-Rel involving p70S6 kinase and possibly the metabolic products of PI 3-kinase. Experiments that assess whether PI 3-kinase inhibitors prevent CD28-induced activation of p70S6 kinase and c-Rel translocation are in progress. However, the activation of p70S6 kinase by CD28 mAbs has only been observed under crosslinking conditions [97] and to date, there are no reports concerning the outcome of ligand activation of CD28 on p70S6 kinase activity. Since the activation of p70S6 kinase by CD3 and CD28 appears to be additive [97], this dual stimulation of a common pathway may be sufficient to cross critical thresholds which are not crossed upon ligation of individual molecules and, hence, this pathway may be a necessary component of growth regulation via either cell surface receptor. While not sufficient to drive proliferation, the induction of p70S6 kinase activity may be necessary for the initiation of cell cycle progression in resting T cells. Indeed, a requirement for p70S6 kinase activity has already been demonstrated for the G_1 to S progression triggered by IL-2 [102].

There is, however, some doubt as to whether CD28-activated p70S6 kinase actually mediates IκBα down-regulation, since the signals generated by PMA alone (which include p70S6 kinase activation) [100] are insufficient to induce down-regulation of IκB [101]. This discrepancy could have several interpretations. Firstly, CD28 may activate a phorbol-ester-insensitive PKC isoform. The role of PKC activation in CD28 signalling is unclear since CD28 activation of Jurkat cells is not associated with PKC translocation [103], although levels of DAG are elevated following CD28 ligation by mAb [77]. Moreover, PKC inhibitors have been reported to either inhibit [77] or have no effect [104] on CD28-dependent IL-2 production. However, certain PKC isoforms may not be sensitive to the inhibitors used in these studies. Secondly, the signal that mediates the effect of CD28 on IκBα may not involve p70S6 kinase but may instead involve the mammalian equivalent of the target of rapamycin (TOR) protein which is the site of action of rapamycin in yeast [89]. The activation of a "TOR-like" molecule by CD28 may be an alternative mechanism for the formation of D-3 phosphatidylinositols since it has been postulated to have PI 3-kinase activity [89]. One interesting possibility is that the D-3 phosphatidylinositol lipids formed by the activation of PI 3-kinase (or TOR-like proteins) following CD28 ligation, may activate the δ, ϵ, η and ζ isoforms of PKC [105,106] since these PKC isoforms are activated by D-3 phosphatidylinositols, at least *in vitro*. Thus, it is possible that one or more of these PKC isoforms mediate the activation of p70S6 kinase either by direct phosphorylation or by the activation of intermediate kinase cascades. This hypothesis may explain the apparent wortmannin insensitivity of T cell proliferation induced by PMA and ionomycin [56] since phorbol esters, which are known to activate p70S6 kinase, would bypass the requirement for D-3 phosphatidylinositol lipids. It should be noted, however, that the ζ isozyme of PKC at least, is phorbol-ester-insensitive [106].

Importance of AP-1 Regulation and c-Jun N-terminal Kinase to CD28 Function

In addition to binding a functionally important AP-1 site in the IL-2 promoter, AP-1 proteins participate in the formation of NF-AT and NF-IL-2 [81]. Interestingly, AP-1 transcriptional activity requires both TcR and CD28-mediated signals [107] and a lack of AP-1 induction has been suggested as one of the abnormalities found in anergic murine T cells [108]. Given that CD28 is the proposed costimulus required to prevent anergy, this observation appears to be somewhat at odds with a recent observation that CD28 stimulation also causes a down-regulation of AP-1 [109].

AP-1 activity is regulated both at the level of *jun* and *fos* gene transcription and by post-translational modification of their products. A key event in c-fos induction is phosphorylation of the transcription factor Elk-1 by the MAP kinases ERK 1 and ERK 2 [110]. Accordingly, the role of p21^{ras} in coupling the TcR/CD3 complex to ERK2 and the ability of ERK2 to translocate to the nucleus where it can directly regulate transcriptional factors like Elk-1 could explain the role of p21^{ras} in TcR/CD3 signal transduction [80]. However, induction and regulation of c-Jun appears to be mediated by a different protein kinase, namely c-Jun N-terminal kinase (JNK-1) which is a member of the stress-activated protein (SAP) kinase subfamily of MAP kinases [90,111,112]. JNKs display around 40% homology with ERKS and share with them the S/T-P minimal consensus site for substrate phosphorylation [90]. Similarly to ERKS they are regulated through phosphorylation of upstream kinases but differ from ERKS at their regulatory site displaying TPY rather than TEY [90]. These subtle differences subject JNKs to distinct upstream regulatory kinases, SEKs or JNK kinase(s) rather than MEK (also known as MAP kinase kinase 1) [90,113]. JNK is activated by costimulation with PMA and calcium ionophores or CD3 and CD28, whereas each stimulus alone resulted in little or no activation [114]. Similar to its effect on IL-2 producton, CsA also inhibited this synergistic activation of JNK [114]. This is a particularly interesting point of signal integration by these surface molecules, since one postulated function of D-3 phosphatidylinositols is the regulation of the complex kinase cascades involved in activation of MAP kinases [115]. Hence, D-3 phosphatidylinositols may be involved in the regulation of upstream JNKKs. The effects of PI 3-kinase inhibitors on the observed synergistic activation of

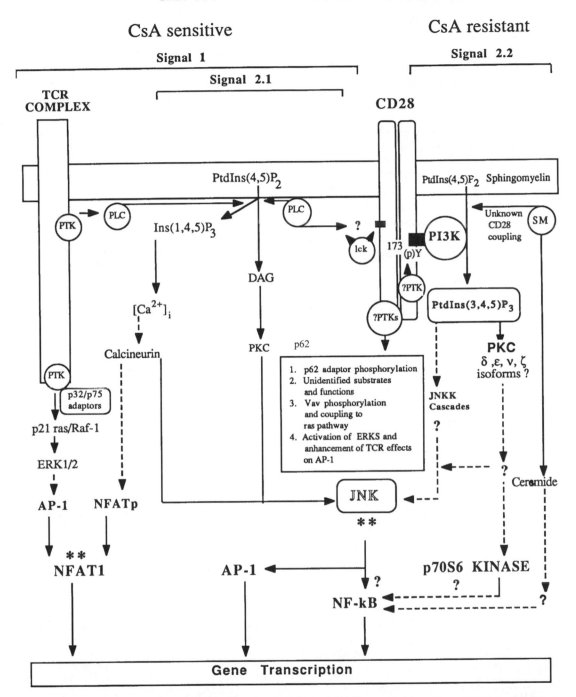

Figure 7.1 Proposed model of signalling events elicited by ligation of CD28. Cytokine genes such as IL-2 require the coordinate action of multiple transcription factors for transcriptional activation. Accordingly both the TcR/CD3 complex and CD28 can each activate several known signal transduction pathways as indicated. These pathways are thought to converge on key integration sites such as JNK and NF-AT1 (indicated by asterisks). The activation of PI 3-kinase by CD28 appears to be a particularly critical signal for CD28-dependent IL-2 production and is postulated to be the major component of the CsA- insensitive signal delivered by CD28. Solid lines indicate known pathways and interactions. Dashed lines indicate uncharacterized pathways. Abbreviations: NFATp, cytosolic component of nuclear factor of activated T cells; SM, acidic sphingomyelinase. For other abbreviations and further details see text

JNK by TcR/CD3 complex and CD28 have, however, yet to be determined.

FUTURE PROSPECTS

A proposed model of CD28 signal transduction events is shown in Figure 7.1. It appears that the state of activation of the T cells determines the ability of CD28 and natural ligands to elicit certain biochemical signals as evidenced by the use of several T cell models (Table 7.1). Clearly, CD28 has the capacity to couple to multiple signalling pathways depending on the overall state of T cell activation, which in turn appears to determine whether or not mAb crosslinking is required for certain signalling responses such as PLC activation and elevation of $[Ca^{2+}]_i$. It has previously been proposed that two primary signal transduction pathways are coupled to CD28, one that is dominant in activated T cells (signal 2.1) and is CsA-sensitive and one that occurs in naive T cells (signal 2.2) and is CsA-resistant [3]. We propose that the CsA-resistant pathway is most likely to be the signalling pathway which is mediated by PI 3-kinase, given the effects of pharmacological inhibitors of PI 3-kinase on IL-2 production in resting T cells [56]. The effects of CsA on PI 3-kinase activity have not been determined but, even if this experiment showed no effect of CsA on PtdIns $(3, 4, 5)P_3$ production, there could still be a possible CsA site of inhibition at the level of the PtdIns $(3, 4, 5)P_3$ target molecule(s), which is so far unidentified. To date, JNK appears to be the most convincing candidate for a molecule which integrates the signals delivered by the TcR/CD3 complex (CsA-sensitive) and CD28 (CsA-resistant) [114].

There are several ways in which CD28 may mediate costimulation: (i) CD28 may provide distinct signal transduction pathways which would lead to a complementation between TcR/CD3 and CD28 to allow full activation of T cells which would fit well with the idea of costimulation; (ii) CD28 may enhance the amplitude or duration of a TcR/CD3 complex-triggered signal, thereby crossing a threshold to activate downstream signalling cascades; or (iii) CD28 may provide similar signals to those provided by the TcR/CD3 complex but at different time points. Moreover, the uniqueness of the CD28 signal may be further determined by way of compartmentalization of CD28-activated signals or activation of distinct isoforms of downstream target molecules. These may have subtle differences from those molecules activated by the TcR/CD3 complex, both in activation requirements and substrate recognition. Finally, differential regulation of adaptor proteins by TcR (p36 and p75) and CD28 (p62) may prove to be a critical difference between these receptors and explain their ability to initiate divergent signal transduction responses during T cell activation.

ACKNOWLEDGEMENTS

We thank Carl June, Doreen Cantrell, Jacques Nunes and Daniel Olive for sharing unpublished data.

REFERENCES

1. Mueller DL, Jenkins MK and Schwartz RH. Clonal expression versus functional clonal inactivation: a costimulatory signalling pathway determines the outcome of T cell antigen receptor occupancy. *Ann. Rev. Immunol.* 1989; **7**: 445–475.
2. Guinan EC, Gribben JG, Boussiotis VA, Freeman GJ, and Nadler LM. Pivotal role of the B7:CD28 pathway in transplantation tolerance and tumor immunity. *Blood* 1994; **84**: 3261–3282.
3. June CH, Bluestone JA, Nadler LM and Thompson CB. The B7 and CD28 receptor families. *Immunol. Today* 1994; **15**: 321–331.
4. Linsley PS and Ledbetter JA. The role of CD28 during T cell responses to antigen. *Ann. Rev. Immunol.* 1993; **11**: 191–212.
5. Schlossman SF, Boumsell L, Gilks W, Harlan JM, Kishimoto T, Morimoto C, Ritz J, Shaw S, Silverstein RL, Springer TA, Tedder TF and Todd RF. CD antigens 1993. *J. Immunol.* 1994; **152**: 1–2.
6. Gimmi CD, Freeman GJ, Gribben JG, Sugita K, Freedman AS, Morimoto C *et al.* B-cell surface antigen B7 provides a costimulatory signal that induces T cells to proliferate and secrete interleukin 2. *Proc Natl. Acad. Sci. USA* 1991; **88**: 6575–6579.
7. Linsley PS, Bradey W, Grosmaire L, Aruffo A, Damle NK and Ledbetter JA. Binding of the B cell activation antigen B7 to CD28 costimulates T cell proliferation and interleukin 2 mRNA accumulation. *J. Exp. Med.* 1991; **173**: 721–730.
8. Engel P, Gribben JG, Freeman GJ, Zhou LJ, Nozawa Y, Abe M *et al.* The B7.2 (B70) costimulatory molecule expressed by monocytes and activated B lymphocytes is the CD86 differentiation antigen. *Blood* 1994; **84**: 1402–1407.
9. Freeman GJ, Gribben JG, Boussitois VA, Ng JW, Restivo VA, Lombard, LA *et al.* A CTLA-4 counter-receptor that

costimulates human T cell proliferation. *Science* 1993; **262**: 909.

10. Azuma M, Philips JH, Lanier L and Somoza C. B70 is a second ligand for CTLA-4 and CD28. *Nature* 1993; 366: 76–78.

11. Linsley PS, Brady W, Urnes M, Grosmaire L, Damle NK and Ledbetter JA. CTLA-4 is a second receptor for the B cell activation antigen B7. *J Exp Med* 1991; **174**: 561–569.

12. Thompson CB, Lindsten T, Ledbetter JA, Kunkel SL, Young HA, Emerson SG *et al.* CD28 activation pathway regulates the production of multiple T cell-derived lymphokines/cytokines. *Proc. Natl. Acad. Sci USA* 1989; **86**: 1333–1337.

13. Minty A, Chalon P, Derocq JM *et al.* Interleukin-13 is a new human lymphokine regulating inflammatory and immune responses. *Nature* 1993; **362**: 248–250.

14. Weschler AS, Gordon MC, Dendorfer U and LeClair KP. Induction of IL-8 expression in T cells uses the CD28 costimulatory pathway. *J. Immunol.* 1994; **153**: 2515–2523.

15. De-Boer M, Kasran A, Kwekkeboom J, Walter H, Vandenberghe P and Cueppens JL. Ligation of B7 with CD28 / CTLA-4 on T cells results in CD40 ligand expression, IL-4 secretion and efficient help for antibody production by B cells. *Eur. J. Immunol.* 1993; **23**: 3120–3125.

16. Seder RA, Germain RN, Linsley PS and Paul WE. CD28-mediated costimulation of IL-2 production plays a critical role in T cell priming for IL-4 and γ-interferon production. *J. Exp. Med.* 1994; **179**: 299–304.

17. Cerdan C, Martin Y, Courcoul M, Brailly H, Mawas C, Birg F and Olive D. Prolonged IL-2 receptor α/CD25 expression after T cell activation via the adhesion molecules CD2 and CD28. *J. Immunol.* 1992; **149**: 2255–2261.

18. Cerdan C, Martin Y, Courcoul M, Mawas C, Birg F and Olive D. CD28 costimulation up-regulates long-term IL-2Rβ expression in human T cells through combined transcriptional and post-transcriptional regulation. *J. Immunol.* 1995; **154**: 1007–1013.

19. Jenkins MK and Schwartz RH. Antigen presentation by chemically modified splenocytes induces antigen-specific T cell unresponsiveness *in vitro* and *in vivo*. *J. Exp. Med.* 1987; **165**: 302–319.

20. Tan P, Anasetti C, Hansen J, Melrose J, Brunvand M, Bradshaw J *et al.* Induction of alloantigen specific hyporesponsiveness in human T lymphocytes by blocking interaction of CD28 with its natural ligand B7/BB1. *J. Exp. Med.* 1993; **177**: 165–173.

21. Russell JH, White CL, Loh DY and Meleedy-Rey P. Receptor-stimulated death pathway is opened by antigen in mature T cells. *Proc. Natl. Acad. Sci. USA* 1991; **88**: 2151–2155.

22. Groux H, Torpier G, Monte D, Mouton Y, Capron A and Ameisen J-C. Activation-induced death by apoptosis in CD4+ T cells from Human Immunodeficiency Virus-infected individuals. *J. Exp. Med.* 1992; **175**: 331–340.

23. Harding F, McArthur JG, Gross JA, Raulet DH, Allison JP. CD28-mediated signalling co-stimulates murine T cells and prevents the induction of anergy in T cell clones. *Nature* 1992; 356: 607–609.

24. Shahinian A, Pfeffer K, Lee KP, Kundig TM, Kishihara K, Wakeham A, Kawai K, Ohashi PS, Thompson CB and Mak T. Differential T cell costimulatory requirements in CD28-deficient mice. *Science* 1993; **261**: 609–611.

25. Harper K, Balzano C, Rouvier E, Mattei M, Luciani M and Golstein P. CTLA-4 and CD28 activated lymphocyte molecules are closely related in both mouse and human as to sequence, message expression, gene structure and chromosomal location. *J. Immunol.* 1991; **147**: 1037–1044.

26. Aruffo A and Seed B. Molecular cloning of a CD28 cDNA by a high efficiency COS cell expression system. *Proc. Natl. Acad. Sci. USA* 1987; **84**: 8573–8577.

27. Jenkins M. The ups and downs of T cell regulation. *Immunity* 1994; **1**: 443–446.

28. Gribben J, Freeman G, Boussiotis V, Gray G and Nadler L. CTLA-4 mediates antigen-specific apoptosis of human T cells. *Proc. Natl. Acad. Sci. USA.* 1995; **92**: 811–815.

29. Walunas TL, Lenschow DJ, Bakker CY, Linsley PS, Freeman GJ, Green JM *et al.* CTLA-4 can function as a negative regulator of T cell activation. *Immunity* 1994; **1**: 405–413.

30. Linsley P, Bradshaw J, Urnes M, Grosmaire L and Ledbetter J. CD28 engagement by B7/BB1 induces transient down-regulation of CD28 synthesis and prolonged unresponsiveness to CD28 signalling. *J. Immunol.* 1993; **150**: 3161–3169.

31. Songyang Z, Shoelson SE, Chaudhuri M, Gish G, Pawson T, Haser WG *et al.* SH2 domains recognize specific phosphopeptide sequences. *Cell* 1993; **72**: 767.

32. Ren R, Mayer B and Baltimore D. A proline rich SH3 binding site. *Science* 1993; **259**; 1157–1159

33. Boussiotis VA, Freeman GJ, Gribben JG, Daley J, Gray GS and Nadler LM. Activated human B lymphocytes express CTLA-4 counter-receptors that costimulate T cell activation. *Proc. Natl. Acad. Sci. USA* 1993; **90**: 11054–11057.

34. Linsley P, Greene JL, Brady W, Bajorath J, Ledbetter JA and Peach R. Human B7.1 (CD80) and B7.2 (CD86) bind with similar avidities but distinct kinetics to CD28 and CTLA-4 receptors. *Immunity* 1994; **1**: 793–801.

35. Freeman GJ, Boussiotis VA, Anamanthan A, Bernstein GM, Ke X, Rennert PD *et al.* B71 and B7-2 do not deliver identical costimulatory signals since B7.2 but not B7.1 preferentially costimulates the initial production of IL-4. *Immunity* 1995; **2**: 523–532.

36. Kuchroo VK, Das MP, Brown JA, Ranger AM, Zamvil SS, Sobel RA *et al.* B7.1 and B7.2 costimulatory molecules activate differentially the Th1/Th2 developmental pathways: application to autoimmune disease therapy. *Cell* 1995; **80**: 707–718.

37. Kapeller R and Cantley, L.C. Phosphatidylinositol 3-kinase. *Bioessays* 1994; **16**: 565–576

38. McCormick F. Activators and effectors of ras p21 proteins. *Curr. Opin. Gen. and Dev.* 1994; **4**: 71–76.

39. Valius M and Kazlauskas A. Phospholipase Cγ1 and phosphatidylinositol 3-kinase are the mediators of the PDGF receptor''s mitogenic signal. *Cell* 1993; **73**: 321–334.

40. Samelson LE and Klausner RD. Tyrosine kinases and tyrosine-based activation motifs. *J. Biol. Chem.* 1992; **267**: 24913–24916.

41. Park D, Rho H and Rhee S. CD3 stimulation causes phosphorylation of PLCγ in a human T cell line. *Proc. Natl. Acad. Sci USA* 1991; **99**: 5453–5456

42. Berridge MJ. Inositol trisphosphate and calcium signalling. *Nature* 1993; **361**: 315–325

43. June CH, Ledbetter JA, Gillespie MM, Lindsten T and Thompson CB. T cell proliferation involving the CD28 pathway is associated with cyclosporine-resistant IL-2 gene expression. *Mol. Cell. Biol.* 1987; **7**: 4472–4481.

44. Sigal N and Dumont FJ. Cyclosporin A, FK506 and rapamycin: pharmacologic probes for lymphocyte signal transduction. *Ann. Rev. Immunol.* 1992; **10**: 519–560

45. Vandenberghe P, Freeman GJ, Nadler LM, Fletcher MC, Kamoun M, Turka LA *et al.* Antibody and B7/BB1-mediated ligation of the CD28 receptor induces tyrosine phosphorylation in human T cells. *J. Exp. Med.* 1992; **175**: 951–960

46. Lu Y, Granelli-Piperno A, Bjorndahl JM, Phillips CA and Trevillyan JM. CD28-induced T cell activation: evidence for a protein tyrosine kinase signal transduction pathway. *J. Immunol.* 1992; **149**: 24–29

47. Dhand R, Hara K, Hiles I, Bax B, Gout I, Panayatou G, Fry M, Yonezawa K, Kasuga M and Waterfield MD. PI 3-kinase is a dual specificity enzyme: autoregulation by an intrinsic protein serine kinase. *EMBO J.* 1994; **13**: 522–533.

48. Otsu M, Hiles I, Gout I, Fry MJ, Ruiz-Larrea F, Panayatou G *et al.* Two 85 kDa proteins that associate with receptor tyrosine kinases, middle T/pp60 c-src complexes and PI 3-kinase. *Cell* 1991; **65**: 91–104.

49. Hiles, ID, Otsu M, Volinia S, Fry MJ, Gout I, Dhand R, Panayatou G, Ruiz-Larrea F, Thompson A, Totty N, Hsuan J, Courtneidge SA, Parker PJ and Waterfield MD. Structure and expression of the PI 3-kinase 110kD catalytic subunit. *Cell* 1992; **70**: 419–429.

50. Stephens, L, Jackson T and Hawkins P. Agonist-stimulated synthesis of phosphatidylinositol 3,4,5-trisphosphate: a new intracellular signalling system. *Biochem. Biophys. Acta* 1993; **1179**: 27–75.

51. Ward SG, Westwick J, Hall N and Sansom DM. Ligation of CD28 receptor by B7 induces formation of D-3 phosphoinositides independently of TcR/CD3 activation. *Eur. J. Immunol.* 1993; **23**: 2572–2577

52. Ueda Y, Levine B, Freeman GJ, Nadler LM, June CH and Ward SG. Both the CD28 ligands CD80 (B7-1) and CD86 (B7-2) activate phosphatidylinositol 3-kinase and wortmannin reveals heterogeneity in the regulation of T cell IL-2 secretion. *Int. Immunol.* 1995; **7**: 957–966.

53. Pages F, Ragueneau M, Rottapel R, Truneh A, Nunes J, Imbert J and Olive D. Binding of PI 3-kinase to CD28 is required for T cell signalling. *Nature* 1994; **369**: 327–329.

54. Prasad KV, Cai Y, Raab M, Duckworth B, Cantley L, Shoelson S and Rudd CR. T cell antigen CD28 interacts with the lipid kinase PI 3-kinase by a cytoplasmic (p)Tyr-Met-X-Met motif. *Proc. Natl. Acad. Sci. USA* 1994; **91**: 2834.

55. August A and Dupont B. CD28 of T lymphocytes associates with PI 3-kinase. *Int. Immunol.* 1994; **6**: 769–774.

56. Ward SG, Wilson A, Turner L, Westwick J and Sansom D. Inhibition of CD28-mediated T cell costimulation by the phosphoinositide 3-kinase inhibitor wortmannin. *Eur. J. Immunol.* 1995; **25**: 526–532.

57. Lu Y, Phillips CA and Trevillyan JM. Phosphatidylinositol 3-kinase activity is not essential for CD28 costimulatory activity in Jurkat cells: studies with a selective inhibitor. *Eur. J. Immunol.* 1995; **25**: 533–537.

58. Ward SG and Cantrell D. Heterogeneity of the regulation of phospholipase C in T lymphocytes. *J. Immunol.* 1990; **144**: 3651–3658.

59. Schneider H, Prasad K, Shoelson S and Rudd C. CTLA-4 binding to the lipid kinase PI 3-kinase in T cells. *J. Exp. Med.* 1994; **181**: 351–355.

60. Sung CK, Sanchez-Margalet V and Goldfine ID. Role of p85 subunit of phosphatidylinositol 3-kinase as an adaptor molecule linking the insulin receptor, p62 and GTPase activating protein. *J. Biol. Chem.* 1994; **269**: 12503–12507

61. Reif K, Gout I, Waterfield MD and Cantrell DA. Divergent regulation of phosphatidylinositol 3-kinase p85α and p85β isoforms upon T cell activation. *J. Biol. Chem.* 1993; **268**: 10780–10788.

62. Remillard B, Petrillo R, Maslinski W, Tsudo M, Strom TB, Cantley LC and Varticovski L. IL-2 regulates activation of phosphatidylinositol 3-kinase. *J. Biol. Chem.* 1991; **266**: 14167–14170.

63. Karnitz LM, Sutor SL and Abraham RT. The *src*-family kinase *fyn* regulates the activation of phosphatidylinositol 3-kinase in an IL-2-responsive cell line. *J. Exp. Med.* 1994; **179**: 1799–1808.

64. Prasad KV, Kapeller R, Janssen O, Repke H, Duke-Cohan J, Cantley LC. and Rudd C. Two step TCRζ/CD3-CD4 and CD28 signalling in T cells: SH2/SH3 domains, protein tyrosine and lipid kinases. *Immunol. Today* 1994; **15**: 226–233.

65. Ward SG, Ley SC, MacPhee C and Cantrell DA. Regulation of D-3 phosphoinositides during T cell activation via the T cell antigen receptor/CD3 complex and CD2 antigens. *Eur. J. Immunol.* 1992; **22**: 45–49.

66. Ward SG, Parry R, Lefeuvre C, Sansom D, Westwick J and Lazarovits A. Antibody ligation of CD7 leads to association with phosphoinositide 3-kinase and formation of PtdIns(3,4,5)P3. *Eur. J. Immunol.* 1995; **25**: 502–507.

67. Stephens L, Cooke FT, Walters R, Jackson T, Volinia S, Gout I *et al.* Characterization of a phosphatidylinositol specific phosphoinositide 3-kinase from mammalian cells. *Curr. Biol.* 1994; **4**: 203–214.

68. Stephens L, Smrcka A, Cooke FT, Jackson TR, Sternweis PC and Hawkins PT. A novel phosphoinositide 3-kinase activity in myeloid-derived cells is activated by G protein $\beta\gamma$ subunits. *Cell* 1994; **77**: 83–93.

69. Hu P, Mondino A, Skolnik EY and Schlessinger J. Cloning of a novel, ubiquitously expressed human phosphatidylinositol 3-kinase and identification of novel polyphosphoinositides in intact cells. *Mol. Cell. Biol.* 1993; **13**: 7677–7688.

70. August A. and Dupont B. Activation of the *src* family kinase *lck* following CD28 cross-linking in the Jurkat leukaemic cell line. *Biochem. Biophys. Res. Commun.* 1994; **199**: 1466–1473.

71. Lu Y, Phillips CA, Bjorndahl JM and Trevillyan JM. CD28 signal transduction: tyrosine phosphorylation and receptor association of phosphoinositide 3-kinase correlate with Ca^{2+} independent costimulatory activity. *Eur. J. Immunol.* 1994; **24**: 2732–2739.

72. Stein PH, Fraser JD and Weiss A. The cytoplasmic domain of CD28 is both necessary and sufficient for costimulation of IL-2 secretion and association with PI 3-kinase. *Mol. Cell. Biol.* 1994; **14**: 3392.

73. Ueda Y, Freeman G, Levine B, Ward SG, Huang ML, Abe R, Nadler LM and June CH. Distinct mechanisms of T cell signal transduction by CD28 and CTLA-4 and by ligands B7-1 and B7-2. *Clin. Res.* 1994; **42**: 309.

74. Hutchcroft J and Bierer B. Activation-dependent phosphorylation of the T lymphocyte surface receptor CD28 and associated proteins. *Proc. Natl. Acad. Sci. USA* 1994; **91**: 3260–3264.

75. August A, Gibson S, Kawakami Y, Kawakami T, Mills G and Dupont B. CD28 is associated with and induces the immediate tyrosine phosphorylation and activation of the Tec family kinase ITK/EMT in Jurkat cells. *Proc. Natl. Acad. Sci. USA* 1994; **91**: 9347–9351.

76. Hu Q, Davidson D, Schwartzberg PL, Macchiarini F, Lenardo M, Bluestone G and Matis L. Identification of *Rlk*, a novel protein tyrosine kinase with predominant expression in the T cell lineage. *J. Biol. Chem.* 1995; **270**: 1928–1934.

77. Nunes J, Klasen S, Franco MD, Lipcey C, Mawas C, Bagnasco M and Olive D. Signalling through CD28 T cell activation pathway involves a InsPL-specific PLC activity. *Biochem. J.* 1993; **293**: 835–842.

78. Couez D, Pages F, Ragueneau M, Nunes J, Klasen S, Mawas C, Truneh A and Olive D. Functional expression of human CD28 in murine T cell hybridomas. *Mol. Immunol.* 1994; **31**: 47–56.

79. Abe R, Vandenberge P, Craighead N, Smoot D, Lee KP and June CH. Distinct signal transduction in mouse CD4+ and CD8+ splenic T cells after CD28 receptor ligation. *J. Immunol.* 1995; **154**: 985–997

80. Izquierdo M, Reif K and Cantrell DA. The regulation and function of $p21^{ras}$ during T cell activation and growth. *Immunol. Today* 1995; **16**: 159–164.

81. Liou H and Baltimore D. Regulation of the NFκB/rel transcription factor and IκB inhibitor system. *Curr. Opin. Cell. Biol.* 1993; **5**: 477–487.

82. Nunes J, Collette Y, Truneh A, Olive D and Cantrell DA. The role of $p21^{ras}$ in CD28 signal transduction: triggering of CD28 with antibodies, but not the ligand B7-1, activates $p21^{ras}$. *J. Exp. Med.* 1994; **180**: 1067–1076.

83. Songyang Z, Shoelson SE, McGlade J, Olivier P, Pawson T, Bustelo XR *et al.* Specific motifs recognized by the SH2 domains of Csk, 3BP2, fps/fes, Grb-2, HCP, SHC, Syk and Vav. *Mol. Cell. Biol.* 1994; **14**: 2777–2785.

84. Schneider H, Cai YC, Prasad KVS, Shoelson SE and Rudd CE. T cell antigen CD28 binds to the Grb-2/Sos complex, regulators of $p21^{ras}$. *Eur. J. Immunol.* 1995; **25**; 1044–1050.

85. Rodriguez-Viciana P, Warne PH, Dhand R, van Haesebroeck B, Gout I, Fry MJ *et al.* Phosphatidylinositol 3-kinase acts as a direct target for $p21^{ras}$. *Nature* 1994; **370**: 494–499.

86. Kodaki T, Woscholski R, Hallberg B, Rodriguez-Viciana P, Downward J and Parker P. The activation of phosphatidylinositol 3-kinase by *Ras. Curr. Biol.* 1994; **4**: 798–806.

87. Richard S, Yu D, Blumer KJ, Hausladen D, Olszowy MW, Connelly PA *et al.* Association of p62, a multifunctional SH2-and SH3-domain binding protein with src family tyrosine kinases Grb2 and phospholipase Cγ-1. *Mol. Cell. Biol.* 1995; **15**: 186–187.

88. Fraser JD, Straus D and Weiss A. Signal transduction events leading to T cell lymphokine gene expression. *Immunol. Today* 1993; **14**: 357–362.

89. Downward J. Regulating S6 kinase. *Nature* 1994; **371**: 378–379.

90. Cano E and Mahadevan LC. Parallel signal processing among mammalian MAPKs. *Trends Biochem.* 1995; **20**: 117–122.

91. Ghosh P, Tan T, Rice NR, Sica A and Young HA. The interleukin 2 CD28-responsive complex contains at least three members of the NFκB family: c-Rel, p50, and p-65. *Proc. Natl. Acad. Sci. USA* 1993; **90**: 1696–1700.

92. Sha WC, Liou H, Tuomanen EI and Baltimore D. Targeted disruption of the p50 subunit of NFκB leads to multifocal defects in immune responses. *Cell* 1995; **80**: 321–330.

93. Henkel T, Machleidt T, Alkalay I, Kronke M, Ben-Neriah Y and Baeuerle PA. Rapid proteolysis of IκBα is necessary for activation of transcription factor NF-kB. *Nature* 1993; **365**: 182–185.

94. Bryan RG, Li Y, Lai JH, Van M, Rice N, Rich R and Tan TH. Effect of CD28 signal transduction on c-Rel in human peripheral blood T cells. *Mol. Cell. Biol.* 1994; **14**: 7933–7942.

95. Boucher LM, Wiegman K, Futterer A, Pfeffer K, Machleidt, T, Schutze S *et al.* CD28 signals through acidic sphingomyelinase. *J. Exp. Med.* 1995; **181**: 2059–2068.

96. Schutze S, Potthoff K, Machleidt T, Berkovic D, Wiegman K and Kronke M. TNF activates NFκB by phosphatidylcholine-specific phospholipase C-induced acidic sphingomyelin breakdown. *Cell* 1992; **71**: 765–776.

97. Pai S, Calvo V, Wood M and Bierer B. Cross-linking CD28 leads to activation of 70kDa S6 kinase. *Eur. J. Immunol.* 1994; **24**: 2364–2368

98. Chung J, Kuo CJ, Crabtree GR and Blenis J. Rapamycin-FKBP specifically blocks growth dependent activation of and signalling by the 70kd S6 protein kinases. *Cell* 1992; **69**: 1227–1236.

99. Maurice WG, Brunn GJ, Wiederrecht G, Siekierka JJ and Abraham RT. Rapamycin-induced inhibition of p34cdc2 kinase activation is associated with G1-S phase growth arrest in T lymphocytes. *J. Biol. Chem.* 1993; **268**: 3734–3738.

100. Chung J, Grammer T, Lemon C, Kazlauskas A and Blenis J. PDGF and insulin-dependent pp70S6K activation mediated by PI 3-kinase. *Nature* 1994; **370**: 71–73.

101. Lai JH and Tan TH. CD28 signalling causes a sustained down-regulation of IkBα which can be prevented by the immunosuppressant rapamycin. *J. Biol. Chem.* 1994; **269**: 30077–30080.

102. Kuo CJ, Chung J, Fiorentino DF, Flanagan WM, Blenis J and Crabtree GR. Rapamycin inhibits IL-2 activation of p70S6 kinase. *Nature* 1992; **358**: 70–73.

103. Weiss A, Manger B and Imboden J. Synergy between the T3/antigen receptor complex and Tp44 in the activation of human T cells. *J. Immunol.* 1986; **137**: 819–825

104. Van Lier RAW, Brouer M, De Groot E, Kramer I, Aarden LA and Verhoeven AJ. T cell receptor/CD3 and

CD28 use distinct intracellular signalling pathways. *Eur. J. Immunol.* 1991; **21**: 1775–1778

105. Toker A, Meyer M, Reddy K, Falck J, Aneja R, Aneja S, Parra A, Burns D, Ballas M and Cantley L. Activation of PKC family members by the novel polyphosphoinositides PtdIns(3,4)P$_2$ and PtdIns(3,4,5)P$_3$. *J. Biol. Chem.* 1994; **269**: 32358–32367.

106. Nakanishi H. and Exton J.H. Activation of the zeta iso-zyme of PKC by phosphatidylinositol 3,4,5-trisphosphate. *J. Biol. Chem.* 1992; **267**: 16347–16354.

107. Rincon M and Flavell RA. AP-1 transcriptional activity requires both T cell receptormediated and costimulatory signals in primary T lymphocytes. *EMBO J.* 1994; **13**: 4370–4381.

108. Kang S, Beverly B, Tran A, Brorson K, Schwartz RH and Lenardo MJ. Transactivation by AP-1 is a molecular target of T cell clonal anergy. *Science* 1992; **257**: 1134–1138.

109. Los M, Droge W and Schulze-Osthoff K. Inhibition of activation of transcription factor AP-1 by CD28 signalling in human T cells. *Biochem. J.* 1994; **302**: 119–123.

110. Angel P and Karin M. The role of Jun, Fos and the AP-1 complex in cell proliferation and transformation. *Biochim. Biophys. Acta* 1991; **1072**: 129–157.

111. Hibi M, Lin A, Smeal T, Minden A and Karin M. Identi-fication of an oncoprotein and UV-responsive protein kinase that binds and potentiates the c-Jun activation domain. *Genes Dev.* 1993; **7**: 2135–2148.

112. Kyriakis J, Banerjee P, Nikolakaki E, Dal T, Rubie E, Ahmad A, Avruch J and Woodgett JR. The stress-acti-vated sub-family of c-Jun kinases. *Nature* 1994; **369**: 156–159.

113. Lin A, Minden A, Martinetto H, Claret FX, Lange-Carter C, Mercurio F, Johnson GL and Karin M. Identification of a dual specificity kinase that activates the Jun kinases and p38-Mpk2. *Science* 1995; **268**: 286–288.

114. Su B. Jacinto E, Hibi M, Kailunki T, Karin M and Ben-Neriah Y. JNK is involved in signal integration during costimulation of T lymphocytes. *Cell* 1994; **77**: 727–736.

115. Downes P and Carter N. Phosphoinositide 3-kinase: a new effector in signal transduction. *Cell Signalling* 1991; **3**: 501–521.

8

The Role of the CD45 Phosphotyrosine Phosphatase in Lymphocyte Signalling

Denis R. Alexander

The T Cell Laboratory, Department of Immunology, The Babraham Institute, Cambridge, UK

INTRODUCTION

The CD45 phosphotyrosine phosphatase (PTPase) is an abundant transmembrane glycoprotein expressed on all nucleated haematopoietic cells [1–6]. So abundant is CD45 on the surface of lymphocytes, where it comprises about 5–10% of the cell surface membrane, that it was one of the earliest lymphocyte antigens to be characterized. In the earlier literature CD45 was also called T200, B220, Ly-5 or leukocyte common antigen. As illustrated in Figure 8.1, CD45 has an overall structure similar to the receptor protein tyrosine kinases, with a large heavily glycosylated ectodomain (391–552 amino acids) and an extensive cytoplasmic tail (700 amino acids) containing tandem repeat domains with PTPase homology (termed "Domain 1" and "Domain 2"). Alternative splicing of a single gene, located on human chromosome 1 q31–q32 [7], generates up to eight CD45 isoforms which vary in the *N*-terminal portions of their ectodomains.

The CD45 antigen was known to immunologists for about 15 years before it was unequivocally shown to be a signalling molecule. The key steps leading to the elucidation of this role are summarized in Table 8.1. The signalling role of CD45 which has been most clearly established so far is the regulation of the *src* family of tyrosine kinases utilized by receptors such as the T cell antigen receptor (TCR) and B cell antigen receptor (BCR). It is now also clear that the signal transduction coupling mechanisms of certain other receptors are also regulated by CD45. This chapter will focus: (i) on the molecular mechanisms whereby CD45 regulates receptor-mediated signalling pathways; (ii)

the possibility that CD45 isoforms differentially affect these pathways; and (iii) on the question of how the CD45 PTPase itself may be regulated.

CD45 AS A RECEPTOR

The definition of a transmembrane glycoprotein as a signalling molecule implies that its ectodomain functions as a receptor and that binding of one or more ligands to this receptor mediates signals to the cell interior. However, physiologically significant CD45 ligands have not yet been unequivocally identified, and therefore CD45 does not yet completely fulfil these classical criteria although, as discussed further on page 131, CD45 mAbs can bind to the ectodomain and trigger various intracellular events. Three proposals have been made about the possible status of CD45 as a receptor. The first is that the CD45 ectodomain interacts with external ligand(s), either soluble or cell-bound, in a *trans* configuration, the second is that the CD45 ectodomain interacts with one or more molecules on the same cell surface in a *cis* configuration, and the third is that CD45 is not a receptor at all. The first two of these proposals are clearly not mutually exclusive and all three possibilities are considered below.

Structure of the CD45 Ectodomain

Three main structural considerations suggest that the CD45 ectodomain may function as a receptor (Figure 8.1). First, by electron microscopy it has been estimated that the ectodomain has a rod-like

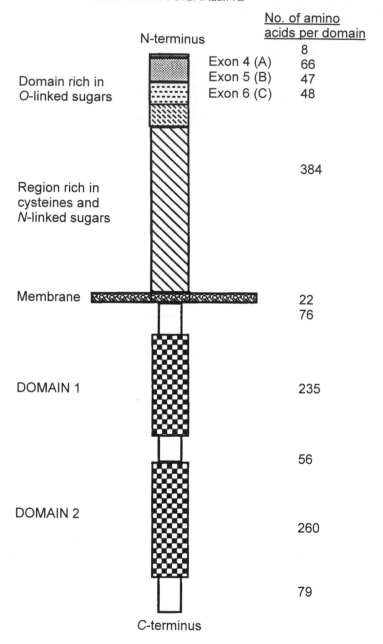

Figure 8.1   Domain structure of human CD45

structure extending 51 nm from the cell surface, or 28 nm for the smallest CD45 isoform [25], and the abundance of CD45, together with its elongated structure, could therefore provide ample possibilities for interactions with either soluble ligands or ligands expressed on interacting cells.

Second, although the ectodomain is only 35% conserved between human, rat and mouse CD45, several features of the overall structure are highly conserved and are characteristic of other glycoproteins known to be receptors. Thus the ectodomain contains a variable N-terminal domain rich in O-

Table 8.1 Key steps in the characterization of CD45 as a signalling molecule

Date	Finding	Reference
1975–1978	Identification as differentiation antigens (termed T200 or leukocyte common antigen) of abundant surface glycoproteins on T and B cells in range 170–220 kDa using T200 mAbs	[8–10]
1983–1986	Development of isoform-specific mAbs, e.g. OX-22 against rat CD45RC, 2H4 against human CD45RA and UCHL1 against human CD45RO	[11–13]
1985–1986	Cloning and sequencing of CD45 gene; discovery of alternative splicing	[14–16]
1988	Identification of CD45 as a PTPase	[17, 18]
1989–1990	Demonstration that the antigen receptor complex is uncoupled from intracellular mitogenic signalling pathways in CD45-negative mutant T cells	[19, 20]
1989–1992	First results suggesting that CD45 may regulate the $p56^{lck}$ and $p59^{fyn}$ tyrosine kinases *in situ*	[21, 22]
1993	Knockout of CD45 exon-6 in transgenic mice generates a phenotype largely but not completely null for CD45, characterized by few mature peripheral T cells but a normal B cell repertoire, so demonstrating an important role for CD45 in T cell development	[23]
1996	Knockout of CD45 exon-9 in transgenic mice generates a completely CD45-null phenotype revealing a role for CD45 at two distinct stages of T cell development and signalling	[24]

linked sugars, containing no cysteine residues, in which (in human CD45) 63 out of the first 169 residues (37%) are serine/threonine and 17 residues (10%) are proline. This is followed by an extensive cysteine-rich region containing 16 conserved cysteine residues (between human, rat and mouse) in which the numbers of amino acid spaces between each cysteine have also been highly conserved, suggesting an important contribution to tertiary structure [1]. Out of the 15 spaces between the 16 conserved cysteine residues, the numbers of amino acids in 8 spaces are identical, and the variation in numbers in the other spaces is small. The CD45 ectodomain also contains up to 16 NXS/T motifs that are signals for N-linked glycosylation, distributed throughout much of the ectodomain, and in the rat, 13 of these sites have been shown to be actually glycosylated [15].

Although N-linked glycosylation sites are present in variably spliced exons 4 and 6, no N-linked sites are present in the domain encoded by exon 5. The heavy glycosylation of the CD45 ectodomain presumably contributes to its extended structure.

The third structural consideration which supports the idea that CD45 may be a receptor is the variable splicing of exons 4–6 to generate up to eight distinct isoforms varying at their N-termini (Figure 8.2). Exons 4, 5 and 6 are also termed A, B and C, giving rise to the most widely used terminology for the CD45 isoforms currently in use. The abbreviations CD45RABC and CD45RA therefore refer to isoforms containing either all three variable exon products or only the exon-4 encoded product, respectively. The CD45RO isoform refers to the "null" isoform in which exons 4–6 have all been spliced out. It should also be

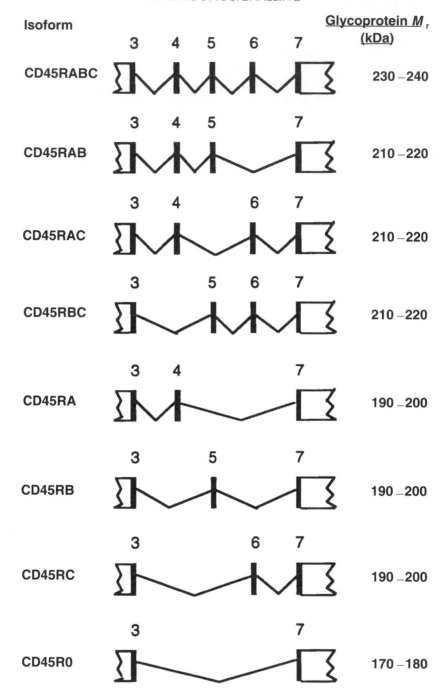

Figure 8.2 The CD45 isoforms generated by alternative splicing of exons 4–6 with the approximate   values for the mature glycoproteins upon separation by SDS–PAGE. The figure is based on reference [1]

noted that exon 7 may be alternatively spliced in mouse CD45, although this event has been observed by the polymerase chain reaction and the putative new isoform has yet to be detected as a protein [26]. Exons 4–6 encode only about 50–60 amino acids each, but the effects of splicing out these variable regions is likely to be greatly amplified by the loss of their multiple O-linked glycosylation sites. Therefore a major difference between the CD45RO and CD45RABC isoforms is the markedly lower abundance of O-linked sugars on CD45RO. This loss of sugars upon splicing out all the variable exons likely explains the observed 45% reduction in length in the ectodomain as measured by electron microscopy [25], a greater reduction than might be expected by splicing out of the variable polypeptide backbone alone which only comprises about 33% of the total amino acids in the ectodomain.

As discussed in greater detail on page 129, CD45 splicing events occur during thymocyte differentiation and the activation of mature T cells. In addition, the differential expression of CD45 isoforms characterizes T cell subsets with distinctive physiological properties. It is therefore possible that each CD45 isoform interacts with distinct ligands in *trans* and/or *cis* configurations. Considering that similar patterns of alternative splicing are conserved, for example, between shark and human [27], it is difficult to believe that the generation of CD45 isoforms differing only in their ectodomains does not play a critical role in T cell function, even though the distinctive roles of each isoform remain to be elucidated.

Our current lack of understanding about the precise role of the CD45 ectodomain stems partly from our ignorance about the detailed topography of the T cell surface during the presentation of MHC-restricted antigenic peptides to the TCR by antigen presenting cells (APC). It is possible that the extended rod structure and heavy glycosylation of the CD45RABC isoform, for example, hinders antigenic peptide binding to the TCR due to steric and/or charge effects, since the TCR-$\alpha\beta$ heterodimer only extends from the cell surface a relatively short distance compared to CD45. Such effects would presumably be less significant for the CD45RO isoform with its truncated ectodomain. According to this scenario the removal of the higher molecular weight CD45 isoforms from the region of apposition between the APC and the interacting T cell would be of importance in increasing the efficiency of antigen presentation,

and such a view makes unnecessary the idea that CD45 necessarily binds to specific ligands. At present such ideas remain speculative and indeed are not consistent with the considerable body of data, discussed further below, that indicates a role for the CD45 cytoplasmic tail in regulating the TCR- and CD4/CD8-associated tyrosine kinases. Such a role requires that CD45 be in close association with the TCR and/or its CD4/CD8 coreceptors during engagement of these receptors with MHC-bound peptides. It is possible that the CD45 ectodomain, normally extended away from the cell, is bent over towards the T cell surface by means of a membrane-proximal hinge-region during cell–cell interactions. A resolution of such issues will likely come with increasing understanding of the architecture of the T cell surface which surrounds the TCR-CD4/CD8 receptors as they interact with MHC-peptides.

Elucidation of the functions of the CD45 ectodomain may also be facilitated by comparison with molecules bearing structural similarities. CD45 is the prototype member of a growing family of transmembrane PTPases which are highly conserved in their cytoplasmic tails, but remarkably diverse in their ectodomains. Most members bear variable numbers of immunoglobulin (Ig)-like motifs and fibronectin type-III repeats [28–30], structures very distinct from the CD45 ectodomain in which Ig motifs have not been detected. Several members of the transmembrane PTPase family, such as mRPTPμ and R-PTPκ, are involved in homophilic interactions [31,32]. Of perhaps greater relevance to the CD45 ectodomain is the chicken transmembrane PTPase, PTPλ, highly expressed in spleen, which is 70% homologous to CD45 in its cytosolic tail and 20% homologous in its ectodomain [33], a modest level of homology which understates some striking similarities. Thus the PTPλ ectodomain contains a ser/thr/pro-rich domain (47% of the first 135 N-terminal amino acids) followed by a cys-rich region in which seven of the conserved cys residues found in CD45 are also conserved. Furthermore, alternative splicing generates several protein species in the M_r range from 170 000 to 210 000, and the presence of spectrin-like repeats in both PTPλ and CD45 has also been suggested [33]. PTPλ therefore qualifies as a member of the CD45 family and the identification of ligand(s) for this PTPase may facilitate the characterization of CD45 ligands, and vice versa.

Putative CD45 Ligands in *Trans*

The only CD45 ligand which has been proposed to date is the B cell activation and differentiation marker CD22. In 1991 it was suggested that CD22 specifically interacts with the CD45RO isoform [34], but more recent work has shown that CD22 is a lectin which interacts with a wide range of glycoproteins bearing N-linked α-2, 6-sialylated oligosaccharides, including all the CD45 isoforms which have been investigated so far [35–37]. The possible physiological significance of these interactions therefore remains to be determined.

In light of the abundance of CD45 on the cell surface it is possible that putative ligands will bind with low affinity but high avidity due to the potentially multimeric nature of the interactions. This may explain why the use of recombinant soluble monomeric forms of the CD45 ectodomain has not, as yet, proved successful as a way of identifying ligands. Multimeric probes might mimic more closely the high density of CD45 ectodomains on the cell surface. The possibility of homotypic interactions between CD45 molecules has also not been excluded.

The multiple effects arising from the binding of CD45 mAbs to CD45 provide strong support for its putative role as a receptor and for the existence of possible ligands in *trans*. These effects are discussed below on page 131.

Putative CD45 Ligands in *Cis*

The idea that significant interactions occur between CD45 and other transmembrane molecules on the same cell surface in a *cis* configuration is supported by an extensive literature. Using chemical crosslinking to identify near-neighbours, coprecipitation in detergent lysates, and fluorescence energy transfer (FRET) experiments, CD45 has been reported to associate with CD2 [38], the TCR [39], the CD4 and CD8 coreceptors [40], the BCR [41], Thy-1 [39], CD16 [42] and CD26 [43]. Furthermore, chemical crosslinking studies suggest that dimeric forms of CD45 may be present at the

T cell surface [44]. The reported associations between CD45 and CD4/CD8, as determined by FRET analysis, appeared to be late activation events, maximal association not occurring until several days after primary human T cells had been activated with CD3 mAb [40]. However, other investigators have reported significant interaction between CD45 and CD4 following detergent lysis of primary resting rat lymph node T cells [45]. A concern common to all these co-association studies arises from the abundance of CD45 on the cell surface and the possibility that not all the reported associations are physiologically significant. A resolution of this question will require detailed analysis of the putative binding domains between CD45 and its co-associating molecules and subsequent genetic deletion of these domains to determine the possible consequences *in situ*.

Of particular interest are reports suggesting that CD45 isoforms differentially interact with the TCR and its CD4 coreceptor [46–48]. It has been shown that the CD45RB^{lo} population of murine CD4$^+$ T cells purified from immunized mice proliferate more vigorously in response to antigenic challenge than the CD45RB^{hi} population. In human T cells the equivalent populations are the CD45RO^{hi} (memory) and CD45RA^{hi} (naive) subsets, respectively. Cocapping studies on the two T cell subsets have suggested that CD4, CD45 and the TCR are co-associated in the murine CD45RB^{lo} or human CD45RO^{hi} (memory) subsets to a much greater extent than in the murine CD45RB^{hi} or CD45RA^{hi} (naive) cell populations. However, capping studies were only achievable following prior polyclonal activation of both subsets with a mitogen [46,47]. Furthermore, in a murine cell line transfected with different CD45 isoforms it was found that CD45RO preferentially cocaps with CD4 when compared with CD45RABC [48]. Since cap formation is a time (10–30 minutes) and energy-dependent process, and requires an intact cytoskeleton, a possible caveat to such studies is the difficulty of determining whether molecules are recruited into a cap on the cell surface as a result of signalling events or whether these associations are already present prior to mAb-mediated receptor aggregation. This question is of particular importance in light of the current model regarding the intracellular actions of the CD45 PTPase activity, reviewed further below, which requires that CD45 be in close association with the TCR and its CD4/CD8 coreceptors.

THE SIGNALLING FUNCTIONS OF THE CD45 CYTOPLASMIC TAIL

CD45 Mutants

CD45 Mutants and Signalling in T Cells

Results obtained using CD45 mutants have clearly established that CD45 positively regulates TCR signal transduction coupling. In CD45$^-$ CD4$^+$ and CD8$^+$ murine T cell clones, antigen-induced proliferation, IL2 secretion and cytolysis of targets (for the CD8$^+$ cells) were markedly defective in the CD45$^-$ cells, whereas signals were restored in CD45$^+$ revertants [19,49]. Proliferation induced by a Thy-1 mAb was also defective in CD45$^-$ cells. In contrast the proliferative response of CD45$^-$ CD4$^+$ clones to IL-2 was 77% of normal, suggesting that CD45 expression is not essential for signalling through the IL-2 receptor [19]. Proliferation in CD45-clones was not defective when triggered by mitogenic lectins [50], suggesting that signalling through coreceptors and/or crosslinking of multiple receptors is sufficient to overcome the deficiencies caused by lack of CD45 expression.

Results obtained from several CD45$^-$ T leukaemia cell lines are consistent with these findings. In CD45$^-$ HPB-ALL, Jurkat and CB1 cells, TCR ligation triggered negligible protein tyrosine phosphorylation, inositol phosphate production, calcium signals, protein kinase C activation or, in the case of Jurkat cells, IL-2 production, whereas all these signals were restored upon transfection of CD45 cDNA [20,22,51–54]. CD45 expression was likewise shown to be essential for signalling mediated via the CD2 receptor [51], via MHC class I [55] and via the transfected Fc receptor [56] in Jurkat cells. The induction of tyrosine phosphorylation by UV irradiation also required the expression of both the TCR and CD45 [57]. Basal levels of protein tyrosine phosphorylation were not increased in these CD45$^-$ HPB-ALL, Jurkat and CB1 cell lines, suggesting that the loss of the CD45 PTPase did not lead to amplified phosphorylation levels of proteins already being constitutively phosphorylated,

but rather was involved in regulating the tyrosine kinase(s) utilized by the TCR and CD2 during signal transduction coupling. Interestingly, in CD45$^-$ CD4$^+$CD8$^+$ HPP-ALL cells, coligation of either the CD4 or CD8 coreceptors with the TCR was sufficient to restore TCR-induced protein tyrosine phosphorylation and calcium signalling, suggesting that in these cells, at least, CD45 expression was not essential for the actions of the CD4/CD8 coreceptors [22,58].

Two CD45$^-$ transformed cell lines have been investigated which have phenotypes quite distinct from those described above. In the first, a CD45 Jurkat T cell sub-clone, TCR-induced protein tyrosine phosphorylation was only partially reduced, whereas IL2 secretion was defective [59]. The reason for this discordance between tyrosine phosphorylation and IL2 secretion in this particular Jurkat sub-clone remains unknown. However, it is possible that another transmembrane PTPase in these cells, such as PTPλ which is known to regulate members of the src tyrosine kinase family [60], may be able to partially restore TCR signalling but to an extent insufficient to promote IL2 secretion. The second CD45$^-$ cell line which has revealed distinctive properties is the retrovirally transformed murine YAC-1 lymphoma cell line [61]. In YAC-1 cells expressing < 5% of wildtype levels of CD45, the basal tyrosine phosphorylation level of TCR-ζ was greatly amplified, although the TCR was uncoupled from inositol phosphate production and calcium signalling as for the other CD45$^-$ cell lines described above. In contrast, basal phosphorylation of TCR-ζ was barely detectable in CD45$^+$ wildtype or revertant cells [61]. It has been suggested that this unusual CD45$^-$ phenotype may be related to the retroviral transformation events which resulted in the immortalization of this cell line with the consequent activation of one or more tyrosine kinases which could hyperphosphorylate TCR-ζ [62]. In primary wildtype murine thymocytes, TCR-ζ is indeed tyrosine phosphorylated under basal conditions [63,64], but in CD45$^-$ transgenic murine thymocytes basal tyrosine phosphorylation of TCR-ζ is not detectable (see below), inconsistent with the idea that TCR-ζ phosphorylation is a direct consequence of CD45 loss.

The finding that transfection of CD45 cDNA into CD45$^-$ cells restores TCR and CD2 signal transduction coupling [52, 22] raises the question as to whether restoration requires the actions of the CD45 ectodomain, cytoplasmic tail, or both of these regions of the molecule. The use of chimeric molecules has clearly established that the CD45 cytoplasmic tail is both necessary and sufficient for the restoration of TCR-and CD2-mediated signalling events [65–68]. Whether expressed in CD45⁻ cells as a myristylated polypeptide lacking its transmembrane and ectodomains [66], or with the EGF-receptor [67] or class I [65] as ectodomains, the CD45 cytoplasmic tail restored receptor-induced signalling. However, it should be noted that in the case of the class I-CD45-tail chimera, restoration of both TCR and CD2-mediated signals was considerably less than in CD45⁺ wildtype cells, suggesting that the CD45 transmembrane and/or ectodomain plays an important role also in promoting signalling via these receptors [65, 68].

Intriguingly, addition of EGF to the cells expressing the EGF-receptor–CD45-tail chimera prior to TCR stimulation resulted in a striking inhibition of TCR-triggered signals, including a decline in the elevation of intracellular Ca⁺ and the dephosphorylation of several phosphoproteins. These inhibitory effects were specific to the actions of the chimeric receptor, since no inhibitory effects were observed when the chimera was expressed in cells expressing wildtype CD45 [67]. Although the mechanism whereby EGF causes inactivation of TCR-mediated signals has not yet been elucidated, it is possible that addition of EGF causes rapid dimerization of the chimeric molecules with a consequent sequestration of CD45 away from its relevant tyrosine kinase substrates. If this interpretation is correct, it has the important implication that CD45 regulates not only the earliest events of TCR signalling, but is also necessary for maintaining these events for some period of time [62]. Whether there is any soluble CD45 ligand which might interact with the CD45 ectodomain on the intact molecule and mediate similar inhibitory effects at present remains speculative.

CD45 Mutants and Signalling in B Cells

The investigation of mutant CD45⁻ B cell lines has clearly established a positive role for CD45 in regulating BCR-mediated signalling pathways [5].

Ligation of sIgM in a CD45⁻ sub-clone of J558Lμm3 plasmacytoma cells failed to trigger Ca²⁺ influx whereas this signal was readily detectable in CD45⁺ sub-clones [69]. Maximal Ca²⁺ mobilization in these cells was observed when CD45 expression reached about 15% of that observed on mature resting B cells or B cell lymphomas. sIgM-triggered *ras* activation [70], as well as phospholipase C and mitogen-activated protein (MAP) kinase activation [71], were also found to be defective in the CD45⁻ J558Lμm3 cells, although partial induction of protein tyrosine phosphorylation was observed, indicating that the uncoupling of the BCR in the absence of CD45 was incomplete in this cell line [71].

In contrast to these results, ligation of sIgM in a CD45⁻ sub-clone of the immature B cell line WEHI-231 resulted in larger and more prolonged calcium signals than those detected in CD45⁻ cells [72]. Furthermore, basal tyrosine phosphorylation was increased in CD45⁻ cells and was not enhanced upon sIgM stimulation, but sIgM-triggered apoptosis was amplified in these cells.

Although B cell CD45 mutant cell lines have not been as extensively investigated as their T cell equivalents, overall the results support a role for CD45 in regulating BCR-mediated signals, particularly since sIgM is uncoupled from proliferative signals in primary mature B cells from CD45⁻ transgenic mice ([23]; also see section below). However, it appears likely that the actions of CD45 depend on the stage of B cell maturation and that, in contrast to the situation pertaining to T cells, CD45 may not regulate all BCR-mediated signals to an equivalent extent.

CD45-deficient Transgenic Mice

The generation of homozygous CD45⁻/⁻ knockout transgenic mice has provided major insights into our understanding of the role of CD45 in lymphocyte development and signalling [23, 24]. Somewhat surprisingly, the excision of CD45 exon-6, which might have been expected to result in a transgenic mouse line deficient only in the CD45RC isoform, in fact resulted in a partial CD45-null phenotype in which CD45 expression ranged from 5 to 40% depending on the lymphocyte subset and the individual mouse being analysed [23]. T cell development in these mice was somewhat impaired so that the percentages of CD8$^+$ cells, in particular, were lower in peripheral blood, lymph nodes and spleen. However, it should be noted that in terms of absolute cell numbers, the peripheral blood and lymph node compartments were unchanged, whereas there was a two-fold increase in the numbers of spleen cells in the CD45 exon-6 deficient mice [23]. There was also impaired development of Vγ3 dendritic epidermal $\gamma\delta$ T cells, an impairment shared with mice deficient in p56^{lck} [73]. The major block in thymocyte development in CD45 exon-6 deficient mice was at the double positive (CD4$^+$ CD8$^+$) to single positive (CD$^+$ or CD8$^+$, SP) transition, whereas B cell development was normal. The proliferative response of CD45$^-$ mature peripheral T cells to mitogens in exon-6 deficient mice was much reduced compared to wildtype but, curiously, the proliferative response of the CD45$^+$ peripheral cells from the same mice was equally defective. The responses of both populations to addition of PMA plus ionomycin, which activate T cells independently of TCR-signalling, were also reduced, suggesting that the T cells which escape to the periphery in exon-6 deficient mice may have more than one abnormality. Although B cell development was normal, the proliferative response of B cells upon anti-immunoglobulin crosslinking was much reduced, whereas the response to LPS was normal. Immunoglobulin-E-mediated degranulation of mast cells from exon-6 deficient mice was defective [74], showing that CD45 is required for sIgE as well as sIgM signalling. Clonal deletion of the relevant thymocyte compartments by the Mls-1^a superantigen appeared to be normal [23]. A caveat to the interpretation of results obtained from the exon-6 deficient mice is the presence of significant CD45 expression remaining on various cell populations, levels which might be sufficient to promote normal developmental and/or signalling events.

More recently a CD45 exon-9 deficient transgenic mouse line has been generated which demonstrates similarities as well as some important differences when its phenotype is compared with the exon-6 deficient line [24]. These mice are completely null for CD45, which cannot be detected at either the DNA or protein levels. This complete absence of CD45 has revealed a double-block in thymocyte development, at the double-negative (DN) to DP transition, as well as at the DP to SP transition, allowing only about 5–10% of the normal level of mature T cells to escape to the periphery. Investigation of fetal thymic organ cultures (FTOC) from these mice has shown that the basal apoptotic rate in the CD45$^{-/-}$ CD4$^+$ CD8$^+$ compartment is elevated, leading to a loss of about 50% of DP cells during 7 days of organ culture, whereas no such losses were observed in wildtype FTOC. Addition of CD3 mAb to ligate the TCR in CD45$^{-/-}$ FTOC caused little further apoptosis of DP cells, in contrast to wildtype cells in which 50% or more of the cells were depleted by this protocol [75]. In contrast, ligation of the Fas receptor, or the addition of pharmacological reagents such as ionomycin or dexamethasone, induced normal levels of apoptosis in the CD45$^{-/-}$ DP cells, showing that CD45 is not required for signalling via Fas and that no generalized defect in pathways leading to apoptosis occurs when CD45 is absent. Overall these results show that CD45 is involved in thymocyte apoptosis in two distinct ways. First, CD45 expression is required in order to rescue DP cells from "spontaneous" depletion, an observation which may explain the marked block in DP to SP development in CD45$^{-/-}$ mice. It may be speculated that there is a receptor on the thymocyte surface which requires CD45 expression in order to mediate "rescue" signals from surrounding thymic stromal cells. Second, as with cell lines and mature peripheral T cells, the TCR is uncoupled from intracellular signalling pathways, including those that lead to apoptosis, in the absence of CD45.

The inability of cell-surface receptors to trigger intracellular signalling pathways has been demonstrated more directly in exon-9 deficient CD45$^{-/-}$ thymocytes by showing that neither TCR nor Thy-1 ligation induces proliferation, and that TCR stimulation results in very little induction of protein tyrosine phosphorylation in these cells. Furthermore, whereas in wildtype thymocytes there is a pronounced basal level of TCR-ζ tyrosine phosphorylation [63, 64], in CD45$^{-/-}$ thymocytes no basal TCR-ζ tyrosine phosphorylation is observed (J Stone, K Byth, N Holmes and D Alexander, unpublished observations), consistent with the idea that the tyrosine kinase(s) responsible for phosphorylating TCR-ζ are inactive in CD45$^{-/-}$ cells (see also page 116).

The development of B cells in exon-9 deficient transgenic mice is normal but, as with the exon-6 deficient mice, ligation of sIgM does not trigger B cell proliferation. Likewise NK cell development is normal and, in contrast to a study using an NK cell line [76], NK functions also appear normal in CD45$^{-/-}$ mice [77].

Overall the investigations carried out on CD45-deficient transgenic mice have provided striking confirmation of the earlier results using mutant cell lines showing that CD45 positively regulates signalling mediated by the TCR and BCR, and substantiate the use of cell lines as relevant model systems for the study of CD45 function. However, it should also be emphasized that results obtained using cell lines should be treated with some caution until confirmation is obtained by the use of primary non-transformed cells.

CD45 Substrates

Considerable evidence suggests that the physiologically significant substrates for CD45 *in vivo* include various members of the *src* family of tyrosine kinases and the probable role for CD45 in activating these kinases in T cells is illustrated in Figure 8.3. As shown in Figure 8.4, these kinases contain regulatory tyrosine residues at their *C*-terminus (tyrosine 505 for p56^{lck} and tyrosine 528 for p59^{fyn}) which, when phosphorylated, are thought to interact by an intramolecular association with the SH2 domain, thereby sterically hindering interaction of the kinase domain with its substrates and at the same time lessening the possibility of interaction between the SH2 domain and other tyrosine phosphorylated molecules. The regulation of the phosphorylation of these critical *C*-terminus tyrosines results from the dual actions of the c-*src* kinase (Csk) and the CD45 PTPase. An important role for Csk in this process is supported by the constitutive activation of the src, fyn and lyn tyrosine kinases, which has been noted in Csk-deficient murine embryos [78]. Several findings support such a model for the regulation of the *src* tyrosine kinases. For example, it has been shown that synthetic tyrosine phosphorylated *C*-terminal peptide homologues of p60^{c-src} or p56^{lck} can bind directly to these kinase SH2 domains *in vitro* [79, 80] and deletion of the SH2 domain reveals the

oncogenic actions of *src* kinases in a manner similar to that caused by mutation of their *C*-terminal regulatory tyrosine residues [81]. It should be noted that this model for the regulation of the *src* kinases predicts competition effects between binding of exogenous tyrosine phosphorylated proteins and the endogenous phosphorylated *C*-terminal to the same SH2 domain. Binding with high affinity of exogenous phosphorylated proteins to the SH2 domain might promote dephosphorylation of the *C*-terminus by decreasing its own interaction with this domain, so making it more available as a phosphatase substrate.

We will now review the evidence suggesting that CD45 is involved in dephosphorylating the *C*-terminal tyrosine residues of *src* family kinases and also consider whether there might be other physiologically relevant CD45 substrates.

Substrates in T Cells

The first data suggesting that the p56^{lck} tyrosine kinase is a substrate for CD45$^-$ *in situ* were obtained using CD45$^-$ murine lymphoma cell lines in which it was shown that phosphorylation of the *C*-terminal tyrosine 505 is increased [21]. Similar findings have also been obtained for the *C*-terminal regulatory tyrosines of both p56^{lck} and p59^{fyn} using sub-clones of murine T-lymphoma cell lines [82] and a cytotoxic T cell clone [83], and for p56^{lck} in HPB-ALL T cells [84], Jurkat cells [85] and YAC-1 lymphoma cells [86]. Significantly, no change in the phosphorylation level of the p60^{c-src} *C*-terminal peptide was noted in CD45$^-$ cells [82], and the transmembrane PTPase RPTα was unable to restore TCR signalling in Jurkat cells [62], suggesting that the actions of CD45 in T cells are selective and are directed specifically to p56^{lck} and p59^{fyn} [82]. Furthermore, a synthetic tyrosine phosphorylated *C*-terminal lck peptide was found to bind considerably more p56^{lck} in CD45$^+$ wild-type Jurkat cells compared to CD45$^-$ cells. This is consistent with the idea that a pool of p56^{lck} molecules is hyperphosphorylated at Tyr-505 in the absence of CD45, so causing tight intramolecular associations between the *C*-terminal and SH2 domains of the kinase, thereby making the SH2 domain inaccessible for binding by the synthetic peptide [85]. In this study it was demonstrated that the pool of p56^{lck} which bound to the phosphory-

117

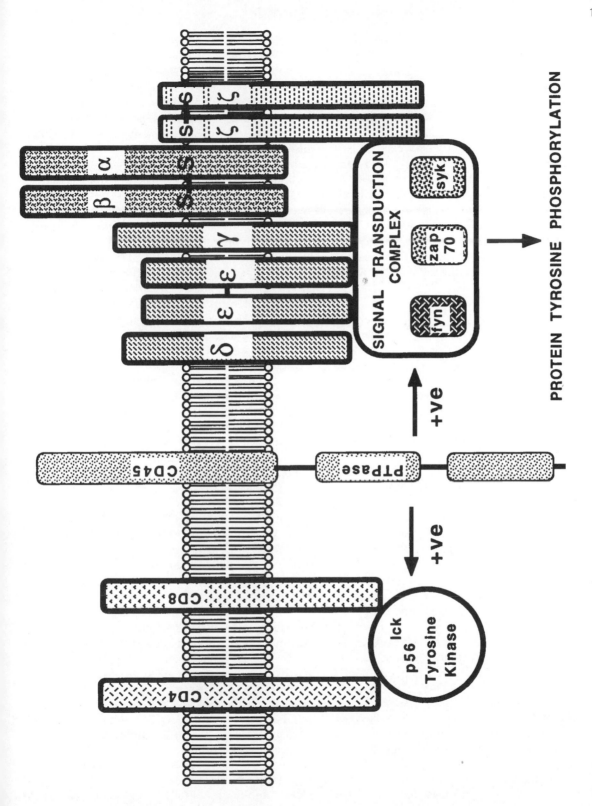

Figure 8.3 The positive effects of CD45 on signalling mediated by the T cell antigen receptor and its CD4/CD8 coreceptors

lated lck peptide was indeed dephosphorylated at Tyr-505. That such findings are not merely relevant to transformed cell lines is supported by the observation that a synthetic tyrosine phosphorylated C-terminal lck peptide was likewise found to bind p56[lck] in primary CD45$^+$ murine thymocytes but was unable to interact significantly with p56[lck] in CD45$^{-/-}$ thymocytes from CD45 exon-9 deficient transgenic mice, showing that CD45 also regulates p56[lck] in primary T cells (J Stone, K Byth, N Holmes and D Alexander, unpublished observations). These *in situ* findings are consistent with the *in vitro* observations that purified CD45 PTPase dephosphorylates the C-terminal regulatory tyrosines of both p56[lck] and p59[fyn] and causes a concomitant increase in their kinase activities [87–89].

Considering the highly consistent finding that the C-terminal regulatory tyrosine residues of p56[lck] and p59[fyn] are hyperphosphorylated in CD45$^-$ cells, it might at first appear surprising that direct assays of the activities of these enzymes, as measured by immunoprecipitating the kinases using specific antisera from detergent lysates, followed by *in vitro* kinase assays, have produced rather conflicting results. For example, using this approach: (i) no differences in p56[lck] or p59[fyn] activities were detected between CD45$^-$ and CD45$^+$ murine T-lymphoma cells [82, 90]; (ii) a three to four-fold increase in p59[fyn] activity was noted upon transfecting CD45 cDNA into CD45$^-$ HPB-ALL cells, whereas no change in p56[lck] activity was detected in these cells [22]; (iii) both p56[lck] and p59[fyn] were more active in CD45$^+$ compared to CD45$^-$ cytotoxic T cell clones [83]; (iv) the p59[fyn] activity was comparable, but p56[lck] activity unexpectedly *greater* in CD45$^+$ than in CD45$^+$ sub-clones of CB1 T-leukaemia cells [54]; (v) both p56[lck] and p59[fyn] activities were reported to be *greater* in CD45$^-$ YAC-1 and HPB-ALL subclones than in CD45$^+$ sub-clones [86]; and (vi) no consistent differences were noted between p56[lck] and p59[fyn] kinase activities in CD45$^{-/-}$ thymocytes from CD45 exon-9 deficient transgenic mice when compared to wildtype (J Stone, K Byth, N Holmes and D Alexander, unpublished observations).

Since a considerable body of data suggests that the activity of the p56[lck] and p59[fyn] kinases is critical for TCR-mediated signalling, in those cases cited here where *increases* in kinase activities were

reported in CD45$^-$ cells, it is not clear why this was associated with *uncoupling* of the TCR from intracellular events. Furthermore, if the model illustrated in Figure 8.3 is correct and phosphorylation of the C-terminal tyrosine residues of the p56[lck] and p59[fyn] kinases is associated with a reduction in their kinase activities, how may these discrepant findings be reconciled with the very consistent data showing that the regulatory C-terminal tyrosines are hyperphosphorylated in CD45$^-$ cells? One possible caveat is the fact that many of these comparisons of kinase activities between CD45$^-$ and CD45$^+$ sub-clones have been carried out on revertants or by comparing wildtype cells with mutagenized CD45$^-$ cells. It is well known that transformed cell lines can undergo spontaneous and striking changes in phenotype during prolonged culture, and it is therefore possible that some of the differences in kinase activities noted above are due to factors other than loss of CD45 and/or to differential CD45 isoform expression in the CD45$^-$ cells. Comparisons made between CD45$^-$ cells and the same cells transfected with CD45-isoform cDNA may be more reliable for such studies. A further caveat is that p56[lck] and p59[fyn] tyrosine kinases are under multiple regulatory controls and other factors besides phosphorylation at their C-termini may be dominant in some contexts. For example, p56[lck] isolated from CD45$^-$ YAC-1 cells was found to be hyperphosphorylated at both Tyr^{394} and Tyr^{505} [91], unlike the situation in other CD45$^-$ cell lines in which only hyperphosphorylation at Tyr^{505} has been noted [84]. In cells in which phosphorylation at Tyr^{394} is a dominant feature, possibly as a result of phosphorylation by a further tyrosine kinase [92] activated as a result of retroviral transformation, it is possible that the increase in p56[lck] activity which occurs upon phosphorylation at this site becomes dominant over the negative effects resulting from Tyr^{505} phosphorylation [91].

However, a third complication has been revealed by the finding, illustrated in Figure 8.5, that large pools of both p56[lck] and p59[fyn] are found in intracellular locations in the CD45-mutant cells which have been used for such studies [54] and these intracellular pools are also found in non-transformed T cells [93]. Investigation of the differential regulation of these kinase pools in CD45$^-$ and CD45$^+$ sub-clones of CB1 T cells has revealed that

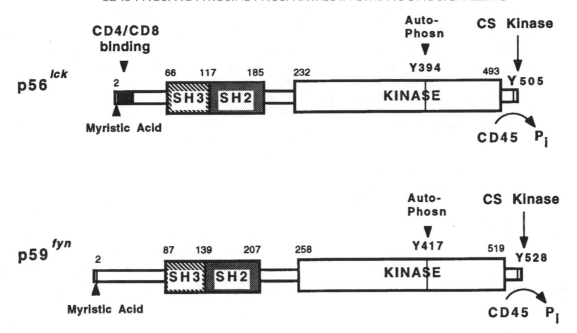

Figure 8.4   The regulation by CD45 and Csk of the p56[lck] and p59[fyn] tyrosine kinases

whereas *total* kinase activities were comparable or even greater in the CD45+ compared to CD45+ cells, the *specific* pools of TCR-associated-p59[fyn] and CD4-associated p56[lck], purified by means of a novel technique involving receptor antibodies bound to magnetic beads, were strikingly inactive in the CD45− when compared to the CD45+ cells [54]. These results suggest that the actions of CD45 are directed towards particular pools of kinases, most likely those which interact with receptors at the plasma membrane, whereas other PTPase(s) may be involved in regulating intracellular kinase pools, as illustrated in Figure 8.6. Such a scenario may also provide a simple explanation for the discrepancy between measurements of kinase activities and kinase *C*-terminal phosphorylation levels. In Figure 8.7 the total pool of cellular p56[lck] is represented by the complete box, whereas the box in one corner represents the specific receptor-associated kinase pool at the cell surface (given an arbitrary value of 10% of the total pool for the purposes of this illustration). If much of the intracellular p56[lck] pool in T cells is dephosphorylated at Tyr-505, as appears likely from site-mapping studies [83, 84], then assays of the phosphorylation level of Tyr-505 measured in ^{32}P-labelled cells

under equilibrium conditions will detect largely the 10% of p56[lck] molecules in the receptor-associated pool, since the intracellular kinase pool will be largely "invisible" under these assay conditions (viz. little ^{32}P at Tyr-505). In contrast, when *in vitro* p56[lck] kinase assays are carried out by immunoprecipitation of the kinase from detergent lysates, in which all p56[lck] pools are mixed, the activity measured will reflect largely the 90% of the intracellular p56[lck] molecules. This likely explains the discrepancies between kinase phosphorylation states and kinase activities described above. To confirm such a model it will be necessary to compare the phosphorylation status of the cell-surface TCR-associated-p59[fyn] and CD4-associated p56[lck] *C*-terminal residues with their equivalent phosphorylation status in intracellular kinase pools. The possible access of intracellular pools of CD45 [94] to intracellular tyrosine kinases also requires careful consideration. The reports that *total* p56[lck] kinase activities may be greater in the CD45− than in the CD45+ cells of certain cell types is intriguing [54, 86], and might even reflect some mechanism for amplifying intracellular kinase activities upon loss of CD45 from the cell surface. Further work will be needed to assess such a possibility.

The notion that CD45 regulates specific kinase pools finds further support in the observation that membrane association of the CD45 cytoplasmic tail is essential for restoration of TCR signalling in CD45$^-$ cells [95], indicating that the localization of the CD45 PTPase with reference to its substrates is critical. Furthermore, a pool of p56^{lck} has also been found to associate with CD45 in T cells (96–98). This association does not require expression of the TCR [98], nor CD4 or CD8 [99], suggesting that the association is not mediated via interactions of these coreceptors with CD45. The association also occurs independently of TCR signalling [99]. Whether the CD45-associated p56^{lck} pool is regulated differently from the CD4/CD8-associated p56^{lck} pools is an important question which requires further investigation.

Besides the well-established role of CD45 in regulating p56^{lck} and p59^{fyn}, several other possible CD45 substrates in T cells have been suggested. Interestingly, a PTPase inactive form of CD45, in which Cys^{828} had been mutated to a Ser, was found to bind with high affinity to the phosphorylated TCR-ζ chain, and wildtype CD45 was able to dephosphorylate the ζ chain relatively selec-

tively *in vitro* [100]. In CD45$^-$ cells the possible actions of CD45 in dephosphorylating the ζ chain would not normally be detectable, since active pools of p56^{lck} and p59^{fyn} are required in order for TCR-stimulated ζ chain phosphorylation to occur [54]. However, as noted on page 113, in CD45$^-$YAC-1 cells the ζ chain is constitutively phosphorylated, whereas little ζ phosphorylation is observed in CD45$^+$ cells [61]. If, as seems possible, the kinase(s) phosphorylating the ζ chain in these cells are dysregulated due to retroviral transformation, then these findings may support the idea that TCR-ζ is a physiological substrate for CD45 *in situ*. It may therefore be speculated that CD45 has both positive and negative roles in TCR signal transduction coupling: positive as a result of promoting ζ chain phosphorylation due to its activating effects on p56^{lck} and p59^{fyn}, and negative by switching off TCR signalling by the subsequent dephosphorylation of ζ. Such dual positive and negative effects could be involved in the incomplete states of TCR-ζ tyrosine phosphorylation which characterize various states of T cell nonresponsiveness [101].

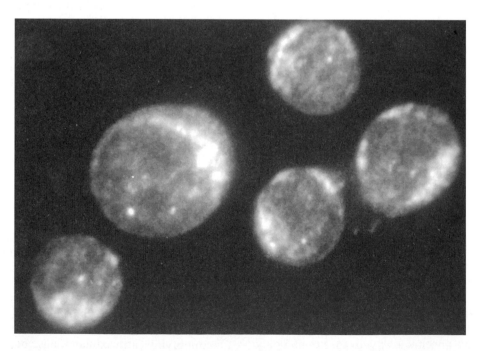

Figure 8.5   Intracellular pools of p56^{lck} revealed by confocal microscopy in CD45$^-$HPB-ALL cells. Whole cell projections are shown. Note the heavy staining for p56^{lck} in the perinuclear region

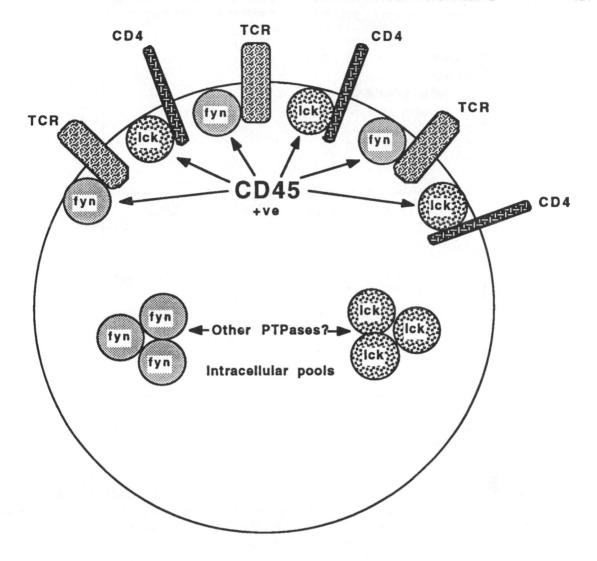

Figure 8.6   A model suggesting that CD45 may positively regulate specific cell-surface receptor-associated pools of tyrosine kinases, whereas intracellular kinase pools may be under different regulatory controls

Substrates in B Cells

Investigations to identify physiologically relevant CD45 substrates in B cells have been relatively sparse compared to those performed in T cells. In primary quiescent B cells the Ig-α/Ig-β (MB-1/B29) heterodimer couples the antigen recognition structure to intracellular signalling pathways by means of several *src* kinase members, including p53/p56^{lyn}, p55^{blk} and p59^{fyn} [5]. Multiple compo-

nents of the BCR, including the Lg-α/Lg-β heterodimer, associate with CD45 in these cells, and p53/p56^{lyn}, but not p55^{blk} or p59^{fyn}, is associated with this complex [41]. As noted on page 114, signalling via the BCR in CD45$^-$ B cell lines is not completely abrogated, suggesting that the actions of CD45 on the kinases associated with the BCR may be rather selective. The specific association of p53/p56^{lyn} with CD45 might indicate that it is a specific substrate for the PTPase, and data supporting such a possibility have been reported [30].

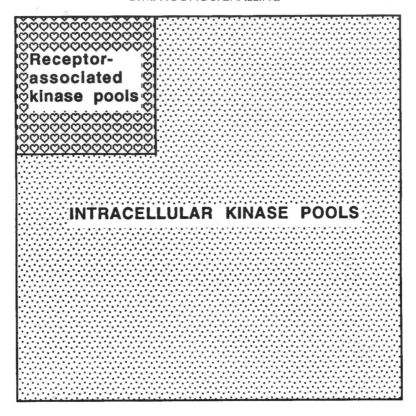

Figure 8.7 The distribution of intracellular tyrosine kinase pools may explain reported differences between kinase activities and phosphorylation states. See the text for further details

Overall, it appears likely that BCR signalling in B cells is regulated by CD45 in a manner analogous to that elucidated in T cells, namely, by activation of one or more receptor-associated tyrosine kinases. Although CD45 dephosphorylates Ig-α/ Ig-β *in vitro* [69], as with the dephosphorylation of TCR-ζ in T cells, it remains to be determined whether the Lg-α/Lg-β heterodimer represents a physiologically significant substrate for CD45 in B cells *in situ*.

The PTPase Activity of the CD45 Cytoplasmic Tail

Several studies have confirmed that the PTPase activity of the CD45 cytoplasmic tail is essential in order for CD45 to positively regulate TCR-mediated signalling [84, 95, 102] and without CD45 PTPase activity *in situ* dephosphorylation

of the *C*-terminal regulatory Tyr-505 in p56^{lck} does not occur [84]. An understanding of the regulation of this PTPase activity is therefore critical for the elucidation of CD45 function.

The Structure and Mutational Analysis of the CD45 Cytoplasmic Tail

As shown in Figure 8.8, the CD45 cytoplasmic tail comprises a short membrane proximal region followed by two tandem PTPase-homology repeats of about 240 residues, termed Domain 1 and Domain 2, separated by a spacer region of 56 residues, and a *C*-terminal tail of 79 residues. A highly acidic region of 24 amino acids within Domain 2 close to the spacer region is not found within other transmembrane PTPases and provides multiple potential sites of serine phosphorylation. Deletion of this acidic insert reduced the PTPase activity of a recombinant cytoplasmic tail protein by 75% [103]. Deletion of the 79-residue *C*-terminus made no difference to activity, whereas deletion of the membrane proximal region ablated activity [103].

A 157-residue segment of the canonical cytosolic PTP1B shows 40% and 33% identity with CD45 Domains 1 and 2, and if conservative substitutions are considered this homology rises to 54% and 48%, respectively [17, 104]. Comparison of human, mouse and rat CD45 sequences shows a very high degree of conservation between the cytoplasmic tails comprising 85% identical residues, or more than 90% if conservative substitutions are included [1]. In a 27 amino acid region adjacent to the Domain 1 PTPase active site only three differences were noted between the human and green iguana (*Iguana iguana*) sequences, of which two are conservative replacements, demonstrating a very high level of evolutionary conservation [27].

Domain 1 contains the PTPase active site signature motif [V/I]HCXAGXGR[S/T]G which, without exception, is found in all PTPases described to date [105]. During the initial purification of CD45 it was shown that its activity was dependent on sulphydryl reducing reagents [106] and it is now known that reduction of the conserved cysteine residue found at the active site of all PTPases is essential for activity. Mutation of this Cys^{828} in Domain 1 of CD45 followed by bacterial expression resulted in a recombinant protein without PTPase activity [107]. This was a rather surprising result since it was expected that the high degree of homology between Domain 2 and PTP1B was a predictor of Domain 2 PTPase activity, but the use of a range of PTPase substrates likewise did not reveal any Domain 2 activity in the Cys^{828} mutated protein, and neither did Domain 2 have measurable activity when expressed alone [103, 108]. This is in contrast to other transmembrane PTPases in which Domain 2 expressed alone does have PTPase activity [109, 110]. Although it is difficult to exclude the formal possibility that CD45 Domain 2 has PTPase activity towards *in vivo* substrates that have not been utilized in these *in vitro* studies, the fact that PTPases are normally active towards artificial substrates *in vitro* makes this possibility unlikely.

Further study of Domain 2 has shown that the sequence surrounding the conserved cysteine residue (Cys^{1144}) at the putative PTPase active site differs at 6 out of the 11 positions when compared to Domain 1. Substitution of 3 out of 5 of these altered amino acids from Domain 2 into the equivalent positions in Domain 1 was sufficient to completely abrogate the PTPase activity of Domain 1, again strongly suggesting that the differences in amino acid sequence in the putative PTPase active site in Domain 2 disallow the potential PTPase activity of this Domain. Mutation of Cys^{1144} was found either to make no difference to the *in vitro* PTPase activity of CD45 or to reduce its activity by up to 50% depending on the substrate used [84, 108]. This mutant CD45 also restored TCR signalling to CD45$^-$ cells as efficiently as wildtype CD45 [84], consistent with the idea that Domains 1 + 2 function adequately *in vivo* even when the Cys known to be essential for any putative PTPase activity in Domain 2 has been mutated.

One finding which has raised questions about these results is that proteolysis of purified CD45 generated a fragment containing a segment of Domain 1, lacking Cys^{828}, together with most of Domain 2, and this proteolytic fragment was shown to contain PTPase activity. Furthermore, transfection of CD45 cDNA lacking the catalytic domain of Domain 1 resulted in the transfected cells expressing a higher PTPase activity [111]. However, insufficient data were provided in this study to assess the purity of the truncated forms of CD45 being analysed, and further work will therefore be necessary to establish whether Domain 2 expresses PTPase activity under certain conditions. It is theoretically possible that removal of Domain 1 could reveal a PTPase activity in Domain 2 which would otherwise be suppressed, but at present this possibility remains speculative.

If Domain 2 does not express PTPase activity *in situ*, then what is its role? Interestingly, removal of the complete Domain 2, or an internal sequence within Domain 2 (924–1109), was found to severely reduce but not completely abrogate the PTPase activity of Domain 1 [108]. Mixing of recombinant Domains 1 and 2 *in vitro* did not restore PTPase activity [103], and removal of the spacer region separating the two domains ablated activity [112], showing that the tertiary structure relationship between the two domains is critical. These results are quite distinct from those described for other transmembrane PTPases. For example, a recombinant protein containing Domain 1 of LAR had similar PTPase activity to a protein expressing Domains 1 + 2 [110, 113], whereas the individual domains of HPTPα had considerably less activity than a recombinant protein incorporating both domains [114]. An important role for CD45 Domain 2 is also suggested by the observation that mutation of various conserved residues surrouding Cys^{1144} caused a 50% reduction in activity [103] and an even more striking result was obtained when a single amino acid change in Domain 2 (mouse Glu^{1180} to Gly) abrogated CD45 PTPase activity completely [112]. This contrasts with mutation of the neighbouring Tyr which had no effect [103]. Both these residues are highly conserved in a wide range of PTPases indicating critical roles in PTPase function.

Important insights into the mechanisms of PTPase actions have come from the analysis of three cytosolic PTPase crystal structures, each containing a single copy of the conserved PTPase signature sequence [115–117], and these findings have provided some clues about the possible relationship between CD45 Domains 1 and 2. All three PTPases have a common structural core comprising four parallel β-strands with surrounding α-helices. The signature sequence, termed the P-loop, lies at the carboxyl end of the parallel β-strands, and forms the base of the catalytic cleft. The P-loop provides a circular array of amino-nitrogen hydrogen bonds plus a critical arginine side-chain interaction which together activate the catalytic cysteine thiolate and stabilize binding to the phosphate anion. The surrounding loops provide a deep phosphate-binding site that prevents entry by the shorter phosphoserine and phosphothreonine side-chains, so ensuring the specificity of catalysis for phosphotyrosine. The mechanism of catalysis involves a nucleophilic attack by a cysteine residue to form a thiol-phosphate intermediate followed by phosphate displacement by a water molecule [118]. By using tungstate as a phosphate analogue, a second catalytically important loop has also been identified containing an aspartate residue which folds over the active site and positions the aspartate side-chain for proton transfer to the tyrosine leaving group. In *Yersinia* PTPase the binding of tungstate triggers a conformational change which involves the movement of this conserved Asp^{356} about 6 Å into the active site [116].

Comparison of the putative PTPase active site in CD45 Domain 2 with the sequences of the P-loops in these three low molecular weight PTPases shows that the critical arginine is missing from the CD45 sequence, inconsistent with a separate PTPase activity for this domain. However, the involvement in catalysis of residues far from the P-loop in the low molecular weight PTPases, at least as far as primary sequence is concerned, suggests that critical residues within CD45 Domain 2, such as murine Glu^{1180} (Glu^{1193} in the human sequence), could be directly involved in catalysis. Certain residues in Domain 2 might also be of particular importance in maintaining the correct conformation of Domain 1 for substrate binding and catalysis. Analysis of the crystal structure of the CD45 cytoplasmic tail will be necessary to assess such possibilities.

Regulation of CD45 PTPase Activity

Four major mechanisms have been described for the regulation of phosphatases: (i) regulation by targeting to substrates [119]; (ii) regulation by post-translational modifications such as phosphorylation; (iii) regulation by the binding of inhibitors; and (iv) regulation by changes in the phospholipid milieu. Although much further work remains to be carried out, the results so far suggest that the main mechanism for regulation of CD45 is by targeting its PTPase-expressing Domain 1 to relevant substrates. Several considerations support such a view. First, either when immunoprecipitated from primary resting T cells or following expression as a recombinant protein, CD45 expresses high levels of PTPase activity. Although such *in vitro* measurements may not accurately reflect the PTPase activity of CD45 *in vivo*, such results point to a significant level of basal CD45 activity. Second, transfection of CD45 cDNA into CD45− cell lines is sufficient to cause dephosphorylation and activation of $p56^{lck}$ and $p59^{fyn}$ without any necessity for ligation of the TCR [22, 84], strongly implying that expression of CD45 at the cell surface *per se* is sufficient to target its cytoplasmic domain to these kinase targets and that no further activation step is required to induce a functional CD45 PTPase. Third, CD45 demonstrates a broad specificity towards artificial PTPase substrates *in vitro* [114, 120], indicating that its regulation of action lies in targeting rather

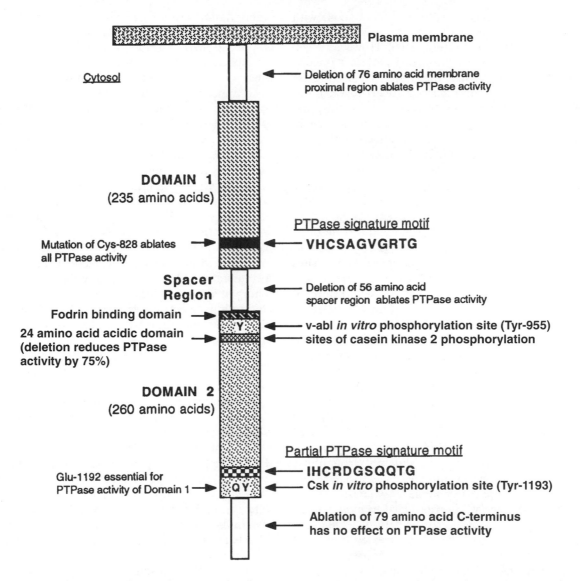

Figure 8.8 Domain structure of the CD45 cytoplasmic tail. Note that the human sequence numbering is shown and that mutational results have been converted into the human numbering even though the mutational analysis was carried out on murine cDNA

than precise substrate specificity. Fourth, a chimeric protein containing a class I ectodomain and a CD45 cytoplasmic tail was considerably less efficient than wildtype CD45 at restoring TCR signalling in CD45⁻ Jurkat cells [65], suggesting that the CD45 ectodomain is required for optimal targeting of the cytoplasmic tail PTPase to relevant substrates.

Should further work establish that CD45 targeting to substrates is indeed the main way in which

its actions are regulated, this is certainly not incompatible with a role for modification of its PTPase activity by phosphorylation, or even by other post-translational modifications such as myristylation [121]. Indeed, several investigations have indicated that phosphorylation plays such a role, although at present the conclusions remain rather dependent on pharmacological and *in vitro* data. Purification and sequencing of CD45 revealed several putative phosphorylation sites for ser/thr kinases, but phosphorylation of CD45 with casein kinase 2 or with glycogen synthase kinase failed to alter CD45 activity [106]. However, when sequential phosphorylation of CD45 *in vitro* was carried out first with the v-abl tyrosine kinase followed by casein kinase 2, then a seven-fold activation towards RCM-lysozyme as substrate was observed, although there was no change in activity towards another artificial substrate, myelin basic protein [122]. Treatment of human T cells *in vivo* with the mitogen phytohaemagglutinin caused phosphorylation of serine residues 976, 979, 980 and 984, and the same sites were found to be phosphorylated by casein kinase 2 *in vitro*, whereas the phosphorylation site of v-abl was determined as Tyr^{955} [122]. It should be noted that all these phosphorylation sites are located in Domain 2 close to the spacer (Figure 8.8). The possible *in vivo* significance of these phosphorylation events remains to be elucidated.

An increase in CD45 phosphorylation has also been observed following the activation of protein kinase C by phorbol esters *in vivo* [123, 124] and this was reported to inhibit the CD45 PTPase activity about two-fold [124], although phosphorylation of CD45 by protein kinase C *in vitro* did not change its PTPase activity [106]. IL-2 also caused phosphorylation of CD45 in serine residues in CTLL-2 cells, but at sites different from those phosphorylated upon protein kinase C activation, causing no apparent change in PTPase activity [125]. When ionomycin was added to mouse T cell lines or to thymocytes, the activity of CD45 subsequently recovered from the treated cells was found to be inhibited by 50–90%, and this correlated with a reduction in serine phosphorylation at specific sites [126]. However, it has not yet been reported that the serine phosphorylation of CD45 induced by TCR-dependent signalling, with a consequent increase in intracellular calcium, has any effects on its PTPase activity [122].

Phosphorylation of CD45 in tyrosine residues *in vivo* in response to CD3 mAb or mitogenic lectin has so far only been observed when the cells were treated with non-specific PTPase inhibitors such as phenylarsine oxide [122, 127]. It has been suggested that the addition of such a PTPase inhibitor is necessary to prevent potential CD45 autodepho-

sphorylation. However, when a PTPase inactive form of the EGF-receptor-CD45 tail chimera was expressed in a $CD45^+$ T cell line, no tyrosine phosphorylation of the chimeric protein was observed following TCR stimulation [84]. It therefore appears from this study that no TCR-triggered tyrosine phosphorylation of CD45 took place under conditions in which CD45 autodephosphorylation was impossible, raising questions about the mechanism of action of phenylarsine oxide on intact cells. It is possible that CD45 phosphorylation is under stringent regulation by a cytosolic PTPase and that this needs to be inhibited before CD45 tyrosine phosphorylation can be detected. In a separate study tyrosine phosphorylation of CD45 was observed following treatment of T cells with the cell-permeable PTPase inhibitor, pervanadate [128]. In this study CD45 was phosphorylated by the Csk tyrosine kinase both *in vitro* and following cotransfection into COS-1 cells, and the major phosphorylation site was identified in human CD45 as Tyr^{1193}. Phosphorylation of CD45 in this system caused a several-fold increase in its PTPase activity. Furthermore, phosphorylation increased the association of $p56^{lck}$ with CD45 in T cells treated with phenylarsine oxide, and also following phosphorylation of CD45 by Csk *in vitro*. It should be noted that mutation of the equivalent murine CD45 Tyr^{1181} does not alter its PTPase activity [103]. Since $p56^{lck}$ has been found by other workers to associate with CD45 in cells that had been neither stimulated nor treated with PTPase inhibitors [98, 99], it will be of interest to determine whether mutation of CD45 Tyr^{1193} affects its functions *in vivo*, in particular its regulation of $p56^{lck}$.

Overall the various CD45 phosphorylation events that have been reported have not yet supported the idea that phosphorylation plays an important role in regulating CD45 functions. Given that non-phosphorylated CD45 has high basal PTPase activity, and that this activity restores the signalling capacity of receptors such as the TCR and BCR merely by expressing CD45 at the cell surface, phosphorylation (or other) events which markedly *inhibit* this activity would appear to be the type of regulation with the greatest potential for *in vivo* relevance. Whether such regulatory mechanisms exist remains to be demonstrated.

Candidate Regulatory Proteins which Associate with CD45

Besides p56[lck], several other proteins of unknown function have been demonstrated to associate with CD45 by co-immunoprecipitation from detergent cell lystates. One of these proteins, a 32 kDa phosphoprotein named LPAP (lymphocyte phosphatase-associated phosphoprotein), or CD45-AP, has been cloned from both mouse [129] and human [130]. A striking feature of the sequence is a stretch of 9 leucine residues which starts 11 amino acids downstream from the N-terminus and is part of a 23-residue stretch of hydrophobic amino acids which could form a transmembrane domain. The sequence contains no conserved regions characteristic of GTP-binding proteins and the protein is therefore distinct from a 32 kDa protein that was earlier described to associate with CD4/CD8-p56[lck] [131]. LPAP was found associated with CD45 irrespective of whether CD45 was present at the cell surface in monomeric or dimeric forms [44]. Interestingly, LPAP expression appears to be restricted to T and B cells and it is absent from CD45$^-$ sub-clones of these cells [130]. In the absence of its CD45 binding partner the protein is synthesized but then rapidly degraded, strongly supporting the concept that the binding of LPAP to CD45 occurs in vivo and is of physiological significance. It is therefore formally possible that some of the effects of CD45 deficiency in CD45$^-$ cell lines and transgenic mice may be ascribed to a concomitant secondary LPAP deficiency. LPAP is expressed in at least two different forms (pp29 and pp32) in resting T cells. These forms shift to pp30 and pp31 upon phosphorylation stimulated by phorbol-ester-induced activation of protein kinase C [130, 132]. The association of LPAP with CD45 occurs independently of the expression of p56[lck] [98, 130]. Treatment of T cells with the PTPase inhibitor pervanadate caused tyrosine phosphorylation of LPAP but this was not observed in a cell line lacking p56[lck] expression, suggesting that LPAP may be a substrate for this kinase [130]. Elucidation of the possible role of LPAP in regulating CD45 function should prove to be of great interest.

A protein of 116 kDa has also been reported to associate with CD45, although it is less well char-acterized than LPAP [133]. Like LPAP, the protein may have a transmembrane domain since it is a glycoprotein and appears to be biotinylated in intact cells. Furthermore, the protein associates with a mutant form of CD45 in which the cytoplasmic tail has been truncated, although this association was not CD45-isoform specific [133]. The protein is also tyrosine phosphorylated and may therefore be the same 116 kDa protein that is tyrosine phosphorylated in response to TCR signalling and that associates with Grb-2 [134] and Crk [135] in T cells.

The association of the CD45 cytoplasmic tail with fodrin has been reported [136] and in vitro it has been shown that CD45 binds with high affinity to both fodrin and spectrin, causing a stimulation of its PTPase activity [137]. Binding of both fodrin and spectrin has been localized to the sequence ^{930}EENKKNRN^{939} S (murine sequence), which is located close to the spacer region at the beginning of Domain 1 [138]. A synthetic peptide containing this sequence blocked the three to four-fold activation of PTPase activity that resulted from addition of fodrin or spectrin to CD45 in vitro. Since this sequence likely mediates the interaction of CD45 with the cytoskeleton in situ, it will be of interest to determine whether the function of CD45 is changed upon excision of these residues.

DO CD45 ISOFORMS DIFFERENTIALLY REGULATE TCR SIGNALLING?

As outlined on page 195, alternative splicing generates up to eight different CD45 isoforms (Figure 8.2) and major questions remain as to the physiological significance of these splicing events [4, 139, 140]. In practice, in human and murine T cells, only the CD45RB, CD45RAB, CD45RBC, CD45RABC and CD45RO isoforms are expressed at significant levels. However, minor levels of the transcripts of the CD45RC and CD45RA isoforms have been observed, whereas no transcript for CD45RAC was detected [141]. All of these isoforms have identical cytoplasmic tails and appear to have similar PTPase activities [21]. Nevertheless there is some experimental support for the idea that CD45 isoforms differentially regulate TCR-mediated signalling.

Signalling in Cell Lines Expressing Single CD45 Isoforms

Definitive answers concerning the role of CD45 isoforms are likely to arise from experimental systems in which the role of each isoform is studied individually. Transfection of either the CD45RO isoform into CD45$^-$ Jurkat T cells [52] or of the CD45RABC, CD45RBC or CD45RO isoforms into CD45$^-$ HPB-ALL cells [22] was sufficient to restore TCR signal transduction coupling, although in the case of HPB-ALL cells the efficiency of restoration was not the same for all isoforms (S Doe, N Holmes and D Alexander, unpublished observations).

The clearest data so far pointing to a differential role for individual isoforms come from a study in which cDNAs encoding the CD45RABC, CD45RBC, CD45RC and CD45RO isoforms were transfected into a TCR$^-$CD4$^-$ CD45$^-$ sub-clone of the BW5147 murine thymoma cell line cotransfected with CD4 and a clonotypic TCR [142]. Comparison of the four CD45 isoform transfected subclones showed that presentation of specific antigen induced IL2 secretion in the order CD45RO > CD45RC > CD45RABC = CD45RBC, whereas IL2 secretion in response to an immobilized TCR mAb was approximately equivalent. These results do not reveal whether the positive effects of the CD45RO and CD45RC isoforms were mediated at the T cell surface or by increasing the efficiency of antigen peptide presentation, or at both these levels. However, as described in on page 112, CD45RO preferentially cocaps with CD4 when compared with CD45RABC in these BW5147 cells, and if this reflects a preferential CD45RO–CD4 interaction already present in the non-stimulated cells, this might provide a potential mechanism for differential control of TCR signalling [48]. According to this model (Figure 8.9), CD45RO in association with CD4 would generate a pool of CD4-associated p56^{lck} which should be more active in the CD45RO$^+$ cells. Upon binding of the TCR by antigenic peptide and simultaneous engagement of CD4 by class II, the CD45-activated CD4-p56^{lck} would be brought into close association with the TCR, so promoting phosphorylation of the TCR polypeptide ITAM motifs and thereby amplifying TCR-mediated signals. The lack of differential effects in terms of IL2

secretion upon challenge with the TCR mAb would be explained by the dominant effect of this mAb in binding with high affinity and crosslinking the TCR, which would tend to render CD4 coreceptor-mediated events less significant.

Rather different results from the above were obtained upon investigating TCR signalling in transgenic mouse lines made by overexpressing either CD45RABC or CD45RO under the control of the *lck* promoter [143]. Expression of the transgenes did not change the expression levels of endogenous CD45 isoforms which consisted mainly of CD45RO (about 70%) and CD45RB (about 20%). Although the expression of CD45RABC at the protein level was only 10–20% above endogenous levels, compared to 15–30% overexpression for CD45RO, TCR-stimulated proliferation was increased up to seven-fold in the CD45RABC transgenic CD4$^+$ CD8$^+$ thymocytes. The TCR-triggered calcium and tyrosine phosphorylation signals were also elevated in these cells, whereas signals in the equivalent CD45RO transgenic thymocyte subset were similar to wildtype (proliferation, calcium) or even less than wildtype (phosphorylation of certain proteins) [143]. A more detailed investigation of the CD45RO overexpressing mice revealed that there was a 40–60% reduction in thymocytes in these mice compared to controls, a reduction which was particularly pronounced in the CD4$^+$ CD8$^+$ (DP) compartment in which negative selection events occur. TCR-stimulated apoptosis was significantly increased in these DP cells overexpressing CD45RO. This correlated with an increase in p56^{lck} tyrosine kinase activity, a slightly increased calcium signal but, surprisingly, no greater induction of tyrosine phosphorylation with the exception of an increase in a 32 kDa protein [144]. Evidence for enhancement of negative selection was also obtained by breeding the CD45RO overexpressing transgenic mice with HY TCR transgenic mice [144]. Overall, therefore, these results suggest that overexpression of the CD45RABC isoform up-regulates TCR-stimulated proliferative signals in mature thymocytes, whereas CD45RO overexpression up-regulates TCR coupling to apoptotic signals in DP thymocytes. In this respect it is of interest that DP thymocytes have been reported to down-regulate their CD45RA expression and up-regulate CD45RO just prior to TCR-stimulated apoptosis [145].

Clearly the results obtained using the CD45 isoform overexpressing transgenic mice are distinct from those described above using the BW5147 cell line. No doubt these differences are at least partly due to the expression of a mixture of endogenous CD45 isoforms in the transgenic mice in contrast to the single isoforms expressed in the BW 5147 cells. Reconstitution of CD45 knockout transgenic mice with specific CD45 isoforms will no doubt help to resolve some of these discrepancies. However, at present there are already sufficient data to suggest that CD45 isoforms differentially regulate TCR signalling and that the outcome of this regulation may depend on the differentiation stage of the T cell.

Signalling in T Cell Subsets Expressing Different CD45 Isoforms

As already noted on page 112, human T cells may be separated into "memory/effector" CD45RO[hi] and "naive/quiescent" CD45RA[hi] subsets, the equivalent murine subsets being defined as CD45RB[lo] and CD45RB[hi], respectively [146]. At birth >90% of T cells are CD45RA[hi], but with time and increasing exposure to antigens there is a gradual conversion to the CD45RO[hi] phenotype until in adulthood about 40–60% of T cells are CD45RO[hi][147]. Upon activation of mature peripheral CD4[+] CD45RA[hi] cells *in vitro* there is a partial down-regulation of CD45RA and an up-regulation of CD45RO over a period of several days [148], although it should be emphasized that at no time during these isoform transitions do T cell subsets express a single CD45 isoform. Indeed, CD45RB expression is high in both CD45RO[hi] and CD45RA[hi] human T cell subsets. While the terminology used to describe these subsets remains somewhat controversial due to uncertainties about the extent of interconversion between the two cell populations, many studies demonstrate that the CD45RO[hi] subset contains a population of cells which respond more vigorously to recall antigens in proliferation assays [149]. This subset also shows a greater proliferative response to suboptimal concentrations of CD3 and CD2 mAbs [150–152] and to superantigen [153] when compared with CD45RA[hi] cells. CD45RO[hi] but not CD45RA[hi] T cells could also be efficiently costimulated with the

CD28 ligand B7 [154]. Is it possible that the differential CD45 isoform expression on these subsets has direct molecular effects on the efficiency of TCR and CD2 signal transduction coupling which could explain the greater proliferative responses of the CD45RO[hi] cells? Such a scenario might be suggested by the greater IL2 secretion noted in response to antigen in the CD45RO-transfected BW 5147 cells described above [142]. However, this increased response was not observed upon stimulation with a TCR mAb, which contrasts with the greater proliferative response of primary CD45RO[hi] cells upon TCR ligation.

Direct comparisons of TCR-mediated signalling in primary human CD45RO[hi] and CD45RA[hi] T cell subsets have suggested that whereas certain signalling pathways are up-regulated in CD45RO[hi] cells, this up-regulation does not appear to be due to increased efficiency of the very early signals mediated by the TCR. For example, TCR-triggered calcium signals and protein kinase C activation were significantly higher on average in CD45RO[hi] than in the CD45RA[hi] subset [155], but TCR-induced protein tyrosine phosphorylation was comparable between the two subsets ([156]; also J Stone, A Robinson and D Alexander, unpublished observations). Furthermore, the protein tyrosine phosphorylation triggered by the HIV gp120 protein via the CD4 coreceptor was also similar in the two subsets [157]. These results are inconsistent with the idea that there is a generalized increase in the earliest TCR-mediated signals in the CD45RO[hi] subset, at least as investigated using triggering CD3 mAbs. It is possible that more subtle differences might emerge in an analysis employing physiological TCR ligands. But, at present, it appears likely that the greater proliferative response of CD45RO[hi] cells in response to mitogenic mAbs is due to the up-regulation of possibly more than one pathway downstream of the TCR. For example, the increased basal level of diacylglycerol in the CD45RO[hi] subset may contribute to the observed increase in CD3-induced protein kinase C activation and may reflect the fact that a population of cells within this subset has been recently activated. This is consistent with data showing that pharmacological activation of protein kinase C by phorbol esters in the presence of submitogenic doses of CD3 mAbs can restore proliferation of CD45RA[hi] cells to the

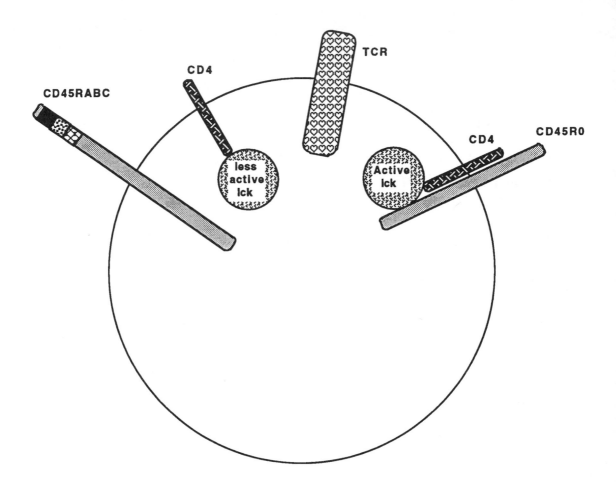

Figure 8.9 A model illustrating the possible way in which CD45 isoforms could differentially associate with cell- surface receptors. The model suggests that if CD45RO is expressed and preferentially associates with CD4, then the CD4-associated pool of p56[lck] may be more active, whereas in cells expressing the CD45RABC isoform which could associate less with CD4, this CD4-associated p56[lck] pool would be less active

levels observed in CD45RO[hi] cells. Similarly the increased CD3-triggered calcium signals could be due to up-regulation and/or pre-activation of any one of the many proteins involved in the cascade of events leading to calcium signals in T cells, such as phospholipase Cγ1 or IP$_3$ receptors [155].

Overall it appears that the CD45RO[hi] subset represents a primed population of cells which have

a lower threshold for activation than CD45RA[hi] cells and which therefore presumably respond to fewer numbers and/or to fewer cycles of presentation of MHC–peptide ligands [158]. Further work will be necessary to determine whether the CD45RO isoform *per se* contributes to this process at the molecular level, or whether its expression is more related to other aspects of T cell physiology, such as the targeting of CD45RO^{hi} cells to sites of inflammation.

SIGNALLING BY CD45 MONOCLONAL ANTIBODIES

Considering the lack of clearly defined physiological ligand(s) for CD45 it is not surprising that there has been a large number of investigations on the role of CD45 mAbs in modulating CD45 function. This has given rise to a large and somewhat confusing literature (earlier papers were reviewed in reference [1]). However, a few generalizations may be made about the effects of CD45 mAbs which may act as useful guidelines for assessing this literature.

Coligation of CD45 mAbs with Cell-surface Receptors Produces Non-specific Effects

A consistent finding in the use of CD45 mAbs has been that when a CD45 mAb of IgG isotype is coligated with the TCR using a secondary cross-linking antibody, then TCR-mediated activation signals are dramatically inhibited [159–163]. Prior to the investigation of mutant CD45$^-$ T cells, these results were interpreted by suggesting that when CD45 was brought into association with the TCR it had a negative effect on TCR signal transduction. However, as discussed on page 113, when CD45$^-$ T cells were examined it became apparent that CD45 exerted a positive effect on TCR signalling and this effect was later shown to involve tyrosine kinase activation. A paradoxical situation therefore arose in which two experimental protocols suggested two quite different functions for CD45. A possible resolution of this paradox was uncovered by the finding that when CD45 and the TCR were crosslinked using chemically constructed heteroconjugate bivalent or trivalent mAbs containing fixed numbers of CD45 and CD3 mAb F(ab') fragments, then no inhibition of CD3-induced signals was observed. Indeed, crosslinking by these reagents tended to increase CD3-triggered proliferation of primary T cells when compared with relevant controls. Furthermore, when an IgM CD45 mAb was coligated with

an IgM TCR mAb, then once again no inhibition of TCR-mediated signals occurred [163]. These results may be rationalized by the model illustrated in Figure 8.10 [2]. Since T cell activation with CD3 or TCR mAbs involves TCR aggregation, any experimental protocol which lessens aggregation will also tend to inhibit the TCR-triggered signals. Since CD45 is about 10-fold more abundant on the T cell surface compared to the TCR, if both CD45 and the TCR are bound with saturating amounts of mAbs, then addition of a crosslinking antibody will be much more likely to coligate a TCR with a CD45 molecule than with another TCR. Therefore TCR aggregation will be reduced by coligation with CD45 and activation signals will decrease. In contrast, when TCR aggregation is maintained by use of heteroconjugate mAbs or by crosslinking multivalent IgM mAbs, then no inhibitions are observed [2].

It is likely that similar explanations apply to similar experimental protocols in which inhibitory effects of CD45 mAbs or its putative ligand have been observed. For example, the inhibition of Fcγ receptor signalling by coligation with CD45 [164] could result from the prevention of Fcγ receptor oligomerization, and the inhibitory effects observed upon coligation of CD45 with the TCR by crosslinking a CD22–Ig fusion protein with a TCR mAb would likewise be expected to prevent TCR aggregation non-specifically [165].

These considerations do not exclude the possibility that CD45 may have specific inhibitory effects on TCR signalling *in situ* under some circumstances. For example, if further work shows that the TCR-ζ chain is an *in vivo* substrate for CD45 (see page 120) then TCR-ζ dephosphorylation might be expected to have negative effects on TCR signalling and could lead to T cell non-responsiveness [101].

The Effects of CD45 mAbs Are Often Epitope Specific

Antigenic determinants encoded by alternatively spliced exons of CD45 have been shown to be largely determined by the polypeptide backbone, but are also influenced by glycosylation, thereby generating an enormous potential for CD45 mAbs selective not only for individual CD45 isoforms

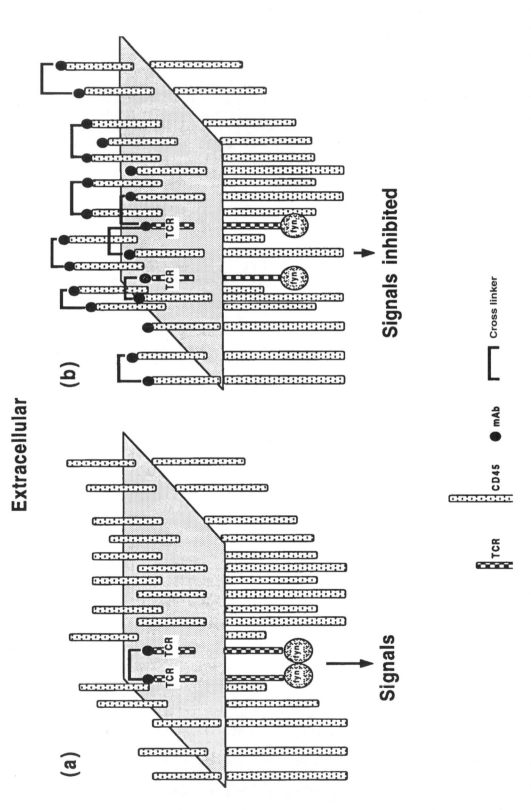

Figure 8.10 A model illustrating the possible non-specific inhibitory effects that may arise when the TCR is aggregated with CD45 using crosslinking antibodies: (a) coligation of the TCR by crosslinking of a TCR mAb triggers intracellular signals; (b) both CD3 and CD45 mAbs have been added to the same cells and then crosslinked. Due to the greater abundance of CD45 relative to the TCR it is expected that the TCR will be more likely to coligate with CD45 molecules than with other TCRs, so preventing effective TCR aggregation and inhibiting TCR-triggered signals. Further details are in the text

but also for isoforms which differ slightly in their oligosaccharide profiles [166]. In contrast to the generalized inhibitory effect on TCR signalling which results from crosslinking virtually any pan-CD45 mAb with the TCR, certain CD45 mAbs, but not others, have been reported to exert striking effects on TCR signalling in the *absence* of cross-linking. For example, CD3-mediated activation signals in a murine T cell hybridoma were strikingly inhibited by pretreatment with one particular pan-CD45 mAb while another pan-CD45 mAb had little effect [167] and a CD45RB selective mAb (MT3) that bound to a sialic acid dependent epitope was found to inhibit T cell alloreactivity and prevented IL-2 receptor expression, whereas other CD45 mAbs were ineffective in this respect [168]. Similar epitope-specific effects of CD45 mAbs have been noted on B cells [169]. One intriguing possibility is that certain CD45 mAbs cause efficient dimerization/oligomerization of CD45 molecules thereby sequestering them away from the TCR and/or its CD4/CD8 coreceptors, so causing inhibition of $p56^{lck}$ and/or $p59^{fyn}$ kinase activities and consequently inhibiting TCR signalling. Such inhibitory effects could be similar to those resulting from EGF binding to the EGF-receptor–CD45-tail chimeric receptor [67], as described on page 114, except that EGF might achieve very rapidly what certain CD45 mAbs achieve only slowly.

Certain epitope-specific CD45 mAbs activate rather than inhibit T cell activation. For example, the 4H2D mAb was found to react with what appeared to be a particular glycosylation epitope of the CD45RB isoform and treatment of primary human CD4+ T cells with this mAb up-regulated calcium signals and proliferation induced by CD3 and CD2 mAbs [170].

The Effects of CD45 mAbs Are Subset Specific

It has frequently been observed that the effects of CD45 mAbs are selective for particular cellular subsets. Thus, CD45 mAbs were found to increase the proliferative response of CD45RA^{hi} but not CD45RO^{hi} cells to a mitogenic pair of CD2 mAbs [171] and similar differences were noted upon sti-

mulation of T cells via the TCR [172]. If the model illustrated in Figure 8.9 is correct, and there is a preferential association of CD45RO rather than CD45RABC with CD4, then such findings could be rationalized by postulating that in these experiments the deficit in CD45RABC–CD4 association is being overcome by induced receptor crosslinking.

Some CD45 mAb effects are also unique to the CD4$^+$ rather than the CD8$^+$ T cell subset. For example, pretreatment of CD4$^+$ but not CD8$^+$ primary human T cells with some, but not all, pan-CD45 mAbs, caused inhibition of subsequent TCR mAb induced activation signals [173].

CD45 mAbs Have Selective Effects on Cell Physiology

There are numerous examples in the literature of CD45 mAbs exerting selective effects on particular aspects of cellular physiology, and in several cases these effects were not due to the modulation by CD45 of activation signals being mediated by other receptors, but were triggered via CD45 in apparent independence of other signalling pathways. These results support the idea that CD45 can act as a receptor "in its own right" to induce intracellular signals and that its role is not only to regulate the signal transduction coupling of other receptors. For example, resting NK cells, like T cells, express predominantly CD45RA-containing isoforms, whereas IL-2 activated NK cells express CD45RO. CD45 mAbs specific for CD45RO, but not for CD45RA, were found to stimulate IFN-γ production in NK cells [174]. CD45 mAbs have also been found to trigger the release of several cytokines from monocytes, including TNF-α and IL-1β [175] as well as IL-6 [176]. In other quite different contexts, ligation of CD45 on B cells was found to facilitate production of secondary Lg isotypes [177], whereas the injection of CD45 mAb *in vivo* was reported to inhibit the differentiation of immature DP thymocytes into mature T cells [178]. In no case yet has the CD45-triggered signalling pathway leading to such changes in cell differentiation or function been delineated, and this remains a challenge for future research.

CD45 mAbs Cause Changes in Cell Adhesion

Numerous papers have described the effects of CD45 mAbs on changes in cell adhesion and many of these effects are highly selective for CD45 mAbs. For example, intercellular adhesion molecule (ICAM)-3, a counter-receptor for the lymphocyte function-associated antigen (LFA)-1 integrin, appears to play an important role in the initial phase of the immune response. ICAM-3 was shown to mediate the aggregation of a T cell line with a B cell line and certain CD45 mAbs were found to block this aggregation [179]. CD45 mAbs were also found to induce homotypic adhesion of human thymocytes through an LFA-1/ICAM-3 dependent interaction, again pointing to an important role for CD45 in regulating adhesion via this pair of receptors [180]. A role for CD45 mAbs in regulating ICAM-1/LFA-1 interactions has also been suggested [181]. Only 1 CD45 mAb (HAB-1) out of a total panel of 12 CD45 mAbs tested was able to inhibit homotypic adhesion in B cell lines induced through various cell-surface receptors [182].

Further work will be necessary to determine to what extent CD45 mAbs act on cell–cell adhesion events by means of their signalling properties and to what extent CD45 mAbs may be mediating changes in cell adhesion by blocking cell–cell interactions. Identification of the physiological ligand(s) for CD45 would help to explain the mechanisms underlying such phenomena.

FUTURE PROSPECTS

Although CD45 is one of the most abundant molecules on the lymphocyte surface and has been used now as a marker for identifying leukocyte populations for more than two decades, its secrets as far as functions are concerned are only gradually being revealed. However, it is now clear that CD45 expression is essential for normal T cell development and for normal signalling through several of the receptors, such as the TCR and BCR, which play critical roles in both normal and pathological immune responses. The PTPase activity of CD45 is also essential for its actions on TCR signalling, although the regulation of this activity

remains poorly understood. Precise splicing events generate an array of CD45 isoforms which characterize various stages of lymphocyte differentiation and activation, yet the physiological significance of these events in terms of the potentially different signalling properties of the different isoforms remains largely unknown. Physiological ligand(s) for CD45 have not as yet not been identified, but interactions between CD45 and other molecules on the lymphocyte surface provide great scope for potential regulatory mechanisms. A bewildering array of selective effects may be triggered using CD45 mAbs, many of them specific to particular CD45 epitopes or cellular subsets, suggesting the potential of CD45 isoforms as possible targets for therapeutic intervention. It is expected that the powerful tools of molecular genetics at both the transgenic animal and cell-line levels, coupled to detailed biochemical analysis, will gradually elucidate the remaining secrets of this fascinating molecule.

ACKNOWLEDGEMENTS

I would like to thank the following colleagues for their valued contributions to the work described in this chapter: Louise Alldridge, Mark Biffen, John Coadwell, Louise Conroy, Senam Doe, Julie Frearson, Rosie Hederer, Karen Keating, Danielle McMichael-Phillips, Anne Robinson, Masahiro Shiroo, Emer Shivnan, John Stone and Margaret Woods. I would also like to thank Keith Burridge for his helpful comments on the manuscript. I am indebted to the Arthritis and Rheumatism Research Council, the Biotechnology and Biological Sciences Research Council, the Cancer Research Campaign, Cantab Pharmaceuticals, the Leukaemia Research Fund, the Medical Research Council and NATO for funding various aspects of our cited work.

REFERENCES

1. Thomas ML. The leukocyte common antigen family. *Ann. Rev. Immunol.* 1989; 7: 339–369.
2. Alexander D, Shiroo M, Robinson A, Biffen M and Shivnan E. The role Of CD45 In T-cell activation—resolving the paradoxes. *Immunol. Today* 1992; **13**: 477–481.
3. Koretzky GA. Role of the CD45 tyrosine phosphatase in signal transduction in the immune-system. *FASEB J.* 1993; 7: 420–426.
4. Trowbridge IS and Thomas ML. CD45—an emerging role as a protein-tyrosine-phosphatase required for lymphocyte-

activation and development. *Ann. Rev. Immunol.* 1994; **12**: 85–116.

5. Justement LB, Brown VK and Lin JJ. Regulation of B-cell activation by CD45—a question of mechanism. *Immunol. Today* 1994; **15**: 399–406.

6. Thomas ML. The regulation of B-lymphocyte and T-lymphocyte activation by the transmembrane protein-tyrosine-phosphatase CD45. *Curr. Opin. Cell. Biol.* 1994; **6**: 247–252.

7. Fernandezluna JL, Matthews RJ, Brownstein BH, Schreiber RD and Thomas ML. Characterization and expression of the human leukocyte-common antigen (CD45) gene contained in yeast artificial chromosomes. *Genomics* 1991; **10**: 756–764.

8. Trowbridge IS, Ralph P and Bevan MJ. Differences in the surface proteins of mouse B and T cells. *Proc. Natl. Acad. Sci. USA* 1975; **72**: 157–161.

9. Trowbridge IS. Interspecies spleen-myeloma hybrid producing monoclonal antibodies against mouse lymphocyte surface glycoprotein, T200. *J. Exp. Med.* 1978; **148**: 313–323.

10. Standring R, McMaster WR, Sunderland CA and Williams AF. The predominant heavily glycosylated glycoproteins at the surface of rat lymphoid cells are differentiation antigens. *Eur. J. Immunol.* 1978; **8**: 832–839.

11. Spickett GP, Brandon MR, Mason DW, Williams AF and Wollett GR. MRC OX-22, a monoclonal antibody that labels a new subset of T lymphocytes and reacts with the high molecular weight form of the leucocyte-common antigen. *J. Exp. Med.* 1983; **158**: 795–810.

12. Morimoto C, Letvin NL, Distaso JA, Aldrich WR and Schlossman SF. The isolation and characterization of the human suppressor inducer T cell subset. *J. Immunol.* 1985; **134**: 1508–1515.

13. Smith SH, Brown MH, Rowe D, Callard RE and Beverley PCL. Functional subsets of human helper-inducer cells defined by a new monoclonal antibody, UCHL1. *Immunology* 1986; **58**: 63–70.

14. Saga Y, Tung J-S, Shen F-W and Boyse EA. Sequences of Ly-5 cDNA: isoform-related diversity of Ly-5 mRNA. *Proc. Natl. Acad. Sci. USA* 1986; **83**: 6940–6944, and correction (1987) **84**: 1991.

15. Barclay AN, Jackson DI, Willis AC and Williams AF. Lymphocyte specific heterogeneity in the rat leucocyte common antigen (T200) is due to differences in polypeptide sequences near the NH2-terminus. *EMBO J.* 1987; **6**: 1259–1264.

16. Streuli M, Hall LR, Saga Y, Schlossman SF and Saito H. Differential usage of three exons generates at least five different mRNAs encoding human leukocyte common antigens. *J. Exp. Med.* 1987; **166**: 1548–1566.

17. Charbonneau H, Tonks NK, Walsh KA and Fischer EH. The leukocyte common antigen (CD45)—a putative receptor-linked protein tyrosine phosphatase. *Proc. Natl. Acad. Sci. USA* 1988; **85**: 7182–7186.

18. Tonks NK, Charbonneau H, Diltz CD, Fischer EH and Walsh KA. Demonstration that the leukocyte common antigen CD45 is a protein tyrosine phosphatase. *Biochemistry* 1988; **27**: 8696–8701.

19. Pingel JT and Thomas ML. Evidence that the leukocyte-common antigen is required for antigen-induced lymphocyte-T proliferation. *Cell* 1989; **58**: 1055–1065.

20. Koretzky GA, Picus J, Thomas ML and Weiss A. Tyrosine phosphatase CD45 is essential for coupling T-cell antigen receptor to the phosphatidyl inositol pathway. *Nature* 1990; **346**: 66–68.

21. Ostergaard HL, Shackelford DA, Hurley TR *et al.* Expression of CD45 alters phosphorylation of the Lck-encoded tyrosine protein-kinase in murine lymphoma T-cell lines. *Proc. Natl. Acad. Sci. USA* 1989; **86**: 8959–8963.

22. Shiroo M, Goff L, Biffen M, Shivnan E and Alexander D. CD45-tyrosine phosphatase-activated P59fyn couples the T-cell antigen receptor to pathways of diacylglycerol production, protein-kinase-C activation and calcium influx. *EMBO J.* 1992; **11**: 4887–4897.

23. Kishihara K, Penninger J, Wallace VA *et al.* Normal B-lymphocyte development but impaired T-cell maturation in CD45-Exon6 protein-tyrosine-phosphatase deficient mice. *Cell* 1993; **74**: 143–156.

24. Byth K F, Conroy LA, Howlett S *et al.* CD45-null transgenic mice reveal a positive regulatory role for CD45 in early thymocyte development in the selection of CD4$^+$ CD8$^+$ thymocytes and in B cell maturation. *J. Exp. Med.* 1996; **183**: 1707–1718.

25. McCall MN, Shotton DM and Barclay AN. Expression of soluble isoforms of rat CD45—analysis by electron-microscopy and use In epitope mapping of anti-CD45r monoclonal-antibodies. *Immunology* 1992; **76**: 310–317.

26. Chang HL, Lefrancois L, Zaroukian MH and Esselman WJ. Developmental expression of CD45 alternate exons in murine T-cells—evidence of additional alternate exon use. *J. Immunol.* 1991; **147**: 1687–1693.

27. Matthews RJ, Pingel JT, Meyer CM and Thomas ML. Studies on the leukocyte-common antigen—structure, function, and evolutionary conservation. *Cold Spring Harbor Symposia on Quantitative Biology* 1989; **54**: 675–682.

28. Alexander DR. The role of phosphatases in signal transduction. *New Biologist* 1990; **12**: 1–16.

29. Walton KM and Dixon JE. Protein-tyrosine phosphatases. *Ann. Rev. Biochem.* 1993; **62**: 101–120.

30. Yakura H. The role of protein-tyrosine-phosphatases in lymphocyte-activation and differentiation. *Crit. Revs. Immunol.* 1994; **14**: 311–336.

31. Gebbink M, Zondag GCM, Wubbolts RW, Beijersbergen RL, Vanetten I and Moolenaar WH. Cell–cell adhesion mediated by a receptor-like protein-tyrosine-phosphatase. *J. Biol. Chem.* 1993; **268**: 16101–16104.

32. Bradykalnay SM, Flint AJ and Tonks NK. Homophilic binding of Ptp-Mu, a receptor-type protein-tyrosine-phosphatase, can mediate cell–cell aggregation. *J. Cell. Biol.* 1993; **122**: 961–972.

33. Fang KS, Barker K, Sudol M and Hanafusa H. A transmembrane protein-tyrosine-phosphatase contains spectrin-like repeats in its extracellular domain. *J. Biol. Chem.* 1994; **269**: 14056–14063.

34. Stamenkovic I, Sgroi D, Aruffo A, Sy MS and Anderson T. The lymphocyte-B adhesion molecule CD22 interacts with leukocyte common antigen CD45RO on T-cells and alpha-2-6 sialyltransferase, Cd75, on B-cells. *Cell* 1991; **66**: 1133–1144.

35. Sgroi D, Varki A, Braeschandersen S and Stamenkovic I. Cd22, a B-cell-specific immunoglobulin superfamily

member, is a sialic acid-binding lectin. *J. Biol. Chem.* 1993; **268**: 7011–7018.

36. Powell LD, Sgroi D, Sjoberg ER, Stamenkovic I and Varki A. Natural ligands of the B-cell adhesion molecule Cd22-beta carry *N*-linked oligosaccharides with alpha-2, 6-linked sialic acids that are required for recognition. *J. Biol. Chem.* 1993; **268**: 7019–7027.

37. Powell LD, Jain RK, Matta KL, Sabesan S and Varki A. Characterization of sialyloligosaccharide binding by recombinant soluble and native cell-associated CD22. *J. Biol. Chem.* 1995; **270**: 7523–7532.

38. Schraven B, Samstag Y, Altevogt P and Meuer S. Association of CD2 and CD45 on human T lymphocytes. *Nature* 1990; **345**: 71–74.

39. Volarevic S, Burns CM, Sussman JJ and Ashwell JD. Intimate association of Thy-1 and the T-cell antigen receptor with the CD45 tyrosine phosphatase. *Proc. Natl. Acad. Sci. USA* 1990; **87**: 7085–7089.

40. Mittler RS, Rankin BM and Kiener PA. Physical associations between CD45 and CD4 or CD8 occur as late activation events in antigen receptor-stimulated human T-cells. *J. Immunol.* 1991; **147**: 3434–3440.

41. Brown VK, Ogle EW, Burkhardt AL, Rowley RB, Bolen JB and Justement LB. Multiple components of the B-cell antigen receptor complex associate with the protein-tyrosine-phosphatase, CD45. *J. Biol. Chem.* 1994; **269**: 17238–17244.

42. Altin JG, Pagler EB, Kinnear BF and Warren HS. Molecular associations involving CD16, CD45 and zeta-chain and gamma-chain on human natural-killer-cells. *Immunol. Cell. Biol.* 1994; **72**: 87–96.

43. Torimoto Y, Dang NH, Vivier E, Tanaka T, Schlossman SF and Morimoto C. Coassociation of CD26 (dipeptidyl peptidase-lv) with CD45 on the surface of human lymphocytes-T. *J. Immunol* 1991; **147**: 2514–2517.

44. Takeda A, Wu JJ, Maizel AL. Evidence for monomeric and dimeric forms of CD45 associated with a 30-kDa phosphorylated protein. *J. Biol. Chem.* 1992; **267**: 16651–16659.

45. Carmo AVXM. Molecular associations of proteins involved in signal transduction in T lymphocytes [PhD]: University of Oxford, 1995.

46. Dianzani U, Luqman M, Rojo J *et al.* Molecular associations on the T cell surface correlate with immunological memory. *Eur. J. Immunol.* 1990; **20**: 2249–2257.

47. Dianzani U, Redoglia V, Malavasi F *et al.* Isoform-specific associations of CD45 with accessory molecules in human lymphocytes-T. *Eur. J. Immunol.* 1992; **22**: 365–371.

48. Leitenberg D, Novak T, Grant CF, Smith BR and Bottomly K. Molecular basis for the co-association between CD4 and specific isoforms of the CD45 tyrosine phosphatase. *J. Cell. Biochem.* 1995; **21A**: 76.

49. Weaver CT, Pingel JT, Nelson JO and Thomas ML. CD8[+] T-cell clones deficient In the expression of the CD45 protein tyrosine phosphatase have impaired responses to T-cell receptor stimuli. *Mol. Cell. Biol.* 1991; **11**: 4415–4422.

50. Pingel JT, McFarland EDC and Thomas ML. Activation of CD45-deficient T-cell clones by lectin mitogens but not anti-Thy-1. *Int. Immunol.* 1994; **6**: 169–178.

51. Koretzky GA, Picus J, Schultz T and Weiss A. Tyrosine phosphatase CD45 is required for T-cell antigen receptor and CD2-mediated activation of a protein tyrosine kinase and interleukin-2 production. *Proc. Natl. Acad. Sci. USA* 1991; **88**: 2037–2041.

52. Koretzky GA, Kohmetscher MA, Kadleck T and Weiss A. Restoration of T-cell receptor-mediated signal transduction by transfection of CD45 Cdna into a CD45-deficient variant of the Jurkat T-cell line. *J. Immunol.* 1992; **149**: 1138–1142.

53. Biffen M, Shiroo M and Alexander DR. Selective coupling of the T-cell antigen receptor to phosphoinositide-derived diacylglycerol production in HPB-ALL T-cells correlates with CD45-regulated P59(Fyn) activity. *Eur. J. Immunol.* 1993; **23**: 2980–2987.

54. Biffen M, McMichaelphillips D, Larson T, Venkitaraman A and Alexander D. The CD45 tyrosine phosphatase regulates specific pools of antigen receptor-associated P59(Fyn) and Cd4-associated P56(Lck) tyrosine kinases in human T-cells. *EMBO J.* 1994; **13**: 1920–1929.

55. Skov S, Odum N and Claesson MH. MHC class-I signaling in T-cells leads to tyrosine kinase-activity and PLC-gamma-1 phosphorylation. *J. Immunol.* 1995; **154**: 1167–1176.

56. Adamczewski M, Numerof RP, Koretzky GA and Kinet JP. Regulation by CD45 of the tyrosine phosphorylation of high-affinity lg receptor beta-chains and gamma-chains. *J. Immunol.* 1995; **154**: 3047–3055.

57. Schieven GL, Mittler RS, Nadler SG *et al.* Zap-70 tyrosine kinase, CD45, and T-cell receptor involvement in UV-induced and H_2O_2-induced T-cell signal-transduction. *J. Biol. Chem.* 1994; **269**: 20718–20726.

58. Deans JP, Kanner SB, Torres RM and Ledbetter JA. Interaction of CD4-Lck with the T-cell receptor CD3 complex induces early signaling events in the absence of CD45 tyrosine phosphatase. *Eur. J. Immunol.* 1992; **22**: 661–668.

59. Peyron JF, Verma S, Malefyt RD, Sancho J, Terhorst C and Spits H. The CD45 protein tyrosine phosphatase is required for the completion of the activation program leading to lymphokine production in the Jurkat human T-cell line. *Int. Immunol.* 1991; **3**: 1357–1366.

60. Fang KS, Sabe H, Saito H and Hanafusa H. Comparative-study of 3 protein-tyrosine phosphatase—chicken protein-tyrosine-phosphatase-lambda dephosphorylates C-Src tyrosine-527. *J. Biol. Chem.* 1994; 269: 20194–20200.

61. Volarevic S, Niklinska BB, Burns CM *et al.* The CD45 tyrosine phosphatase regulates phosphotyrosine homeostasis and its loss reveals a novel pattern of late T-cell receptor induced Ca2+ oscillations. *J. Exp. Med.* 1992; **176**: 835–844.

62. Chan AC, Desai DM and Weiss A. The role of protein-tyrosine kinases and protein-tyrosine phosphatases in T-cell antigen receptor signal-transduction. *Ann. Rev. Immunol.* 1994; **12**: 555–592.

63. Nakayama T, Singer A, Hsi ED and Samelson LE. Intrathymic signaling in immature CD4[+]CD8[+] thymocytes results in tyrosine phosphorylation of the T-cell receptor zeta-chain. *Nature* 1989; **341**: 651–654.

64. van Oers NSC, Killeen N and Weiss A. ZAP-70 is constitutively associated with tyrosine-phosphorylated TCR ζ in murine thymocytes and lymph node T cells. *Immunity* 1994; **1**: 675–685.

65. Hovis RR, Donovan JA, Musci MA et al. Rescue of signaling by a chimeric protein containing the cytoplasmic domain of CD45. Science 1993; 260: 544–546.

66. Volarevic S, Niklinska BB, Burns CM, June CH, Weissman AM and Ashwell JD. Regulation of TCR signaling by CD45 lacking transmembrane and extracellular domains. Science 1993; 260: 541–544.

67. Desai DM, Sap J, Schlessinger J and Weiss A. Ligand-mediated negative regulation of a chimeric transmembrane receptor tyrosine phosphatase. Cell 1993; 73: 541–554.

68. Donovan JA, Goldman FD and Koretzky GA. Restoration of CD2-mediated signaling by a chimeric membrane-protein including the cytoplasmic sequence of CD45. Hum. Immunol. 1994; 40: 123–130.

69. Justement LB, Campbell KS, Chien NC and Cambier JC. Regulation of B-cell antigen receptor signal transduction and phosphorylation by CD45. Science 1991; 252: 1839–1842.

70. Kawauchi K, Lazarus AH, Rapoport MJ, Harwood A, Cambier JC and Delovitch TL. Tyrosine kinase and CD45 tyrosine phosphatase-activity mediate P21 (Ras) activation in B-cells stimulated through the antigen receptor. J. Immunol. 1994; 152: 3306–3316.

71. Kim KM, Alber G, Weiser P and Reth M. Signaling function of the B-cell antigen receptors. Immunol. Revs. 1993; 132: 125–146.

72. Ogimoto M, Katagiri T, Mashima K, Hasegawa K, Mizuno K and Yakura H. Negative regulation of apoptotic death in immature B-cells by CD45. In Immunol. 1994; 6: 647–654.

73. Kawai K, Kishihara K, Molina TJ, Wallace VA, Mak TW and Ohashi PS. Impaired development of V-gamma-3 dendritic epidermal T-cells in P56(Lck) protein-tyrosine kinase-deficient and CD45 protein-tyrosine phosphatase-deficient mice. J. Exp. Med. 1995; 181: 345–349.

74. Berger SA, Mak TW and Paige CJ. Leukocyte common antigen (CD45) is required for immunoglobulin E-mediated degranulation of mast-cells. J. Exp. Med. 1994; 180: 471–476.

75. Conroy LA, Howlett S, Byth K, Holmes N and Alexander D. Apoptosis during thymic development in CD45-deficient transgenic mice. British Society of Immunology Spring Meeting, Birmingham, March 1995: Abstract 1.1.

76. Bell GM, Dethloff GM and Imboden JB. CD45-negative mutants of a rat natural-killer-cell line fail to lyse tumor target-cells. J. Immunol. 1993; 151: 3646–3653.

77. Byth K, Stone J, Howlett S et al. The role of CD45 in thymic development, B cell activation and NK cell triggering. European Network of Immunology Institutes Conference, Les Embiez, France. 1995:36.

78. Nada S, Yagi T, Takeda H et al. Constitutive activation of Src family kinases in mouse embryos that lack Csk. Cell 1993; 73: 1125–1135.

79. Peri KG, Gervais FG, Weil R, Davidson D, Gish GD and Veillette A. Interactions of the Sh2 domain of lymphocyte-specific tyrosine protein-kinase P56 (Lck) with phosphotyrosine-containing proteins. Oncogene 1993; 8: 2765–2772.

80. Payne G, Shoelson SE, Gish GD, Pawson T and Walsh CT. Kinetics of P56(Lck) and P60(Src) Src homology-2 domain binding to tyrosine-phosphorylated peptides determined by a competition assay or surface-plasmon resonance. Proc. Natl. Acad. Sci. USA 1993; 90: 4902–4906.

81. Veillette A, Caron L, Fournel M and Pawson T. Regulation of the enzymatic function of the lymphocyte-specific tyrosine protein-kinase P56lck by the noncatalytic Sh2 and Sh3 domains. Oncogene 1992; 7: 971–980.

82. Hurley TR, Hyman R and Sefton BM. Differential-effects of expression of the CD45 tyrosine protein phosphatase on the tyrosine phosphorylation of the Lck, Fyn, and C-Src tyrosine protein-kinases. Mol. Cell. Biol. 1993; 13: 1651–1656.

83. McFarland EDC, Hurley TR, Pingel JT, Sefton BM, Shaw A and Thomas ML. Correlation between Src family member regulation by the protein-tyrosine-phosphatase CD45 and transmembrane signaling through the T-cell receptor. Proc. Natl. Acad. Sci. USA 1993; 90: 1402–1406.

84. Desai DM, Sap J, Silvennoinen O, Schlessinger J and Weiss A. The catalytic activity of the CD45 membrane-proximal phosphatase domain is required for Tcr signaling and regulation. EMBO J. 1994; 13: 4002–4010.

85. Sieh M, Bolen JB and Weiss A. CD45 specifically modulates binding of Lck to a phosphopeptide encompassing the negative regulatory tyrosine of Lck. EMBO J. 1993; 12: 315–321.

86. Burns CM, Sakaguchi K, Appella E and Ashwell JD. CD45 regulation of tyrosine phosphorylation and enzyme-activity of Src family kinases. J. Biol. Chem. 1994; 269: 13594–13600.

87. Mustelin T, Coggeshall KM and Altman A. Rapid activation of the T-cell tyrosine kinase pp56^{lck} by the CD45 phosphotyrosine phosphatase. Proc. Natl. Acad. Sci. USA 1989; 86: 6302–6306.

88. Mustelin T and Altman A. Dephosphorylation and activation of the T-cell tyrosine kinase Pp56lck by the leukocyte common antigen (CD45). Oncogene 1990; 5: 809–813.

89. Mustelin T, Pessamorikawa T, Autero M et al. Regulation of the P59(Fyn) protein tyrosine kinase by the CD45 phosphotyrosine phosphatase. Eur. J. Immunol. 1992; 22: 1173–1178.

90. Trowbridge IS, Ostergaard H, Shackleford D, Hole N and Johnson P. A leukocyte-specific protein tyrosine phosphatase. Adv. Prot. Phosphatases 1991; 6: 227–250.

91. D'Oro U and Ashwell JD. CD45 tyrosine phosphatase decreases the enzymatic activity of lck by dephosphorylation of the autophosphorylation site. Cold Spring Harbor Laboratory Conference 'Tyrosine Phosphorylation and Cell Signaling" 3–7 May, 1995: 42.

92. Hardwick JS and Sefton BM. Activation of the Lck tyrosine protein kinase by hydrogen peroxide requires the phosphorylation of Tyr-394. Proc. Natl. Acad. Sci. USA 1995; 92: 4527–4531.

93. Ley SC, Marsh M, Bebbington CR, Proudfoot K and Jordan P. Distinct intracellular-localization of Lck and Fyn protein tyrosine kinases in human T-lymphocytes. J. Cell. Biol. 1994; 125: 639–649.

94. Minami Y, Stafford FJ, Lippincottschwartz J, Yuan LC and Klausner RD. Novel redistribution of an intracellular pool of CD45 accompanies T-cell activation. J. Biol. Chem. 1991; 266: 9222–9230.

95. Niklinska BB, Hou D, June C, Weissman AM and Ashwell JD. CD45 tyrosine phosphatase-activity and membrane anchoring are required for T-cell antigen receptor signaling. *Mol. Cell. Biol.* 1994; **14**: 8078–8084.

96. Schraven B, Kirchgessner H, Gaber B, Samstag Y and Meuer S. A functional complex is formed in human lymphocytes-T between the protein tyrosine phosphatase CD45, the protein tyrosine kinase P56lck and Pp32, a possible common substrate. *Eur. J. Immunol.* 1991; **21**: 2469–2477.

97. Guttinger M, Gassmann M, Amrein KE and Burn P. CD45 phosphotyrosine phosphatase and P56lck protein tyrosine kinase—a functional complex crucial in T-cell signal transduction. *Int. Immunol.* 1992; **4**: 1325–1330.

98. Koretzky GA, Kohmetscher M and Ross S. CD45-associated kinase-activity requires Lck but not T-cell receptor expression in the Jurkat T-cell line. *J. Biol. Chem.* 1993; **268**: 8958–8964.

99. Ross SE, Schraven B, Goldman FD, Crabtree J and Koretzky GA. The association between CD45 and Lck does not require CD4 or CD8 and is independent of T-cell receptor stimulation. *Biochem. Biophys. Res. Commun.* 1994; **198**: 88–96.

100. Furukawa T, Itoh M, Krueger NX, Streuli M and Saito H. Specific interaction of the CD45 protein-tyrosine-phosphatase with tyrosine-phosphorylated CD3 zeta-chain. *Proc. Natl. Acad. Sci. USA* 1994; **91**: 10928–10932.

101. Madrenas J, Wange RL, Wang JL, Isakov N, Samelson LE and Germain RN. ζ Phosphorylation without ZAP-70 activation induced by TCR antagonists or partial agonists. *Science* 1994; **267**: 515–518.

102. Motto DG, Musci MA and Koretzky GA. Surface expression of a heterologous phosphatase complements CD45 deficiency in a T-cell clone. *J. Exp. Med.* 1994; **180**: 1359–1366.

103. Johnson P, Ostergaard HL, Wasden C and Trowbridge IS. Mutational analysis of CD45—a leukocyte-specific protein tyrosine phosphatase. *J. Biol. Chem.* 1992; **267**: 8035–8041.

104. Charbonneau H, Tonks NK, Kumar S *et al.* Human-placenta protein-tyrosine-phosphatase—amino-acid sequence and relationship to a family of receptor-like proteins. *Proc. Natl. Acad. Sci. USA* 1989; **86**: 5252–5256.

105. Stone RL and Dixon JE. Protein-tyrosine phosphatases. *J. Biol. Chem.* 1994; **269**: 31323–31326.

106. Tonks NK, Diltz CD and Fischer EH. CD45, an integral membrane-protein tyrosine phosphatase—characterization of enzyme-activity. *J. Biol. Chem.* 1990; **265**: 10674–10680.

107. Streuli M, Krueger NX, Tsai AYM and Saito H. A family of receptor-linked protein tyrosine phosphatases in humans and *Drosophila*. *Proc. Natl. Acad. Sci. USA* 1989; **86**: 8698–8702.

108. Streuli M, Krueger NX, Thai T, Tang, M and Saito H. Distinct functional roles of the two intracellular phosphatase like domains of the receptor-linked protein tyrosine phosphatases LCA and LAR. *EMBO J.* 1990; **9**: 2399–2407.

109. Cho HJ, Krishnaraj R, Kitas E, Bannwarth W, Walsh CT and Anderson KS. Isolation and structural elucidation of a novel phosphocysteine intermediate in the LAR protein tyrosine phosphatase enzymatic pathway. *J. Amer. Chem. Soc.* 1992; **114**: 7296–7298.

110. Itoh M, Streuli M, Krueger NX and Saito H. Purification and characterization of the catalytic domains of the human receptor-linked protein tyrosine phosphatases HPTP-Beta, leukocyte common antigen (Lca), and leukocyte common antigen-related molecule (Lar). *J. Biol. Chem.* 1992; **267**: 12356–12363.

111. Tan XH, Stover DR and Walsh KA. Demonstration of protein tyrosine phosphatase-activity in the 2nd of 2 homologous domains of CD45. *J. Biol. Chem.* 1993; **268**: 6835–6838.

112. Ng DHW, Maiti A and Johnson P. Point mutation In the 2nd phosphatase domain of CD45 abrogates tyrosine phosphatase-activity. *Biochem. Biophys. Res. Commun.* 1995; **206**: 302–309.

113. Cho HJ, Ramer SE, Itoh M *et al.* Catalytic domains of the Lar and CD45 protein tyrosine phosphatases from *Escherichia-coli* expression systems—purification and characterization for specificity and mechanism. *Biochemistry* 1992; **31**: 133–138.

114. Wang Y and Pallen CJ. The receptor-like protein tyrosine phosphatase Hptp-alpha has 2 active catalytic domains with distinct substrate specificities. *EMBO J.* 1991; **10**: 3231–3237.

115. Barford D, Flint AJ and Tonks NK. Crystal-structure of human protein-tyrosine-phosphatase 1b. *Science* 1994; **263**: 1397–1404.

116. Stuckey JA, Schubert HL, Fauman EB, Zhang ZY, Dixon JE and Saper MA. Crystal-structure of *Yersinia* protein-tyrosine-phosphatase at 2.5-angstrom and the complex with tungstate. *Nature* 1994; **370**: 571–575.

117. Su X-D, Taddei N, Stefani M, Ramponi G and Nordlund P. The crystal structure of a low-molecular-weight phosphotyrosine protein phosphatase. *Nature* 1994; **370**: 575–578.

118. Guan KL and Dixon JE. Evidence for protein-tyrosine-phosphatase catalysis proceeding via a cysteine-phosphate intermediate. *J. Biol. Chem.* 1991; **266**: 17026–17030.

119. Mauro LJ and Dixon JE. Zip codes direct intracellular protein-tyrosine phosphatases to the correct cellular address. *Trends Biochem. Sci.* 1994; **19**: 151–155.

120. Pot DA, Woodford TA, Remboutsika E, Haun RS and Dixon JE. Cloning, bacterial expression, purification, and characterization of the cytoplasmic domain of rat LAR, a receptor-like protein tyrosine phosphatase. *J. Biol. Chem.* 1991; **266**: 19688–19696.

121. Takeda A and Maizel AL. An unusual form of lipid linkage to the CD45 peptide. *Science* 1990; **250**: 676–679.

122. Stover DR and Walsh KA. Protein-tyrosine-phosphatase activity of CD45 is activated by sequential phosphorylation by 2 kinases. *Mol. Cell. Biol.* 1994; **14**: 5523–5532.

123. Autero M and Gahmberg CG. Phorbol diesters increase the phosphorylation of the leukocyte common antigen CD45 in human T-cells. *Eur. J. Immunol.* 1987; **17**: 1503–1506.

124. Yamada A, Streuli M, Saito H, Rothstein DM, Schlossman SF and Morimoto C. Effect of activation of protein-

kinase-C on CD45 isoform expression and CD45 protein tyrosine phosphatase-activity in T-cells. *Eur. J. Immunol.* 1990; **20**: 1655–1660.

125. Valentine MA, Widmer MB, Ledbetter JA *et al*. Interleukin-2 stimulates serine phosphorylation of CD45 in CTLL-2.4 cells. *Eur. J. Immunol.* 1991; **21**: 913–919.

126. Ostergaard HL and Trowbridge IS. Negative regulation of CD45 protein tyrosine phosphatase-activity by ionomycin in T-cells. *Science* 1991; **253**: 1423–1425.

127. Stover DR, Charbonneau H, Tonks NK and Walsh KA. Protein-tyrosine-phosphatase-CD45 is phosphorylated transiently on tyrosine upon activation of Jurkat T-cells. *Proc. Natl. Acad. Sci. USA* 1991; **88**: 7704–7707.

128. Autero M, Saharinen J, Pessamorikawa T *et al*. Tyrosine phosphorylation if CD45 phosphotyrosine phosphatase by P50(Csk) kinase creates a binding-site for P56(Lck) tyrosine kinase and activates the phosphatase. *Mol. Cell. Biol.* 1994; **14**: 1308–1321.

129. Takeda A, Maizel AL, Kitamura K, Ohta T and Kimura S. Molecular-cloning of the CD45-associated 30-kDa protein. *J. Biol. Chem.* 1994; **269**: 2357–2360.

130. Schraven B, Schoenhaut D, Bruyns E *et al*. Lpap, a novel 32-kDa phosphoprotein that interacts with CD45 in human lymphocytes. *J. Biol. Chem.* 1994; **269**: 29102–29111.

131. Telfer JC and Rudd CE. A 32-kDa GTP-binding protein associated with the CD4-P56lck and CD8-P56lck T-cell receptor complexes. *Science* 1991; **254**: 439–441.

132. Schraven B, Schirren A, Kirchgessner H, Siebert B and Meuer SC. 4 CD45/P56lck-associated phosphoproteins (Pp29–Pp32) undergo alterations in human T-cell activation. *Eur. J. Immunol.* 1992; **22**: 1857–1863.

133. Arendt CW and Ostergaard HL. CD45 protein-tyrosine-phosphatase is specifically associated with a 116-kDa tyrosine-phosphorylated glycoprotein. *J. Biol. Chem.* 1995; **270**: 2313–2319.

134. Motto DG, Ross SE, Jackman JK *et al*. In-vivo association of Grb2 with Pp116, a substrate of the T-cell antigen receptor-activated protein-tyrosine kinase. *J. Biol. Chem.* 1994; **269**: 21608–21613.

135. Sawasdikosol S, Ravichandran KS, Lee KK, Chang J-H and Burak SJ. Crk interacts with tyrosine-phosphorylated p116 upon T cell activation. *J. Biol. Chem.* 1995; **270**: 2893–2896.

136. Bourguignon LYW, Suchard SJ, Nagpal ML and Glenney JR. A T-lymphoma transmembrane glycoprotein (gp180) is linked to the cytoskeletal protein, fodrin. *J. Cell. Biol.* 1985; **101**: 477–487.

137. Lokeshwar VB and Bourguignon LYW. Tyrosine phosphatase-activity of lymphoma CD45 (Gp180) is regulated by a direct interaction with the cytoskeleton. *J. Biol. Chem.* 1992; **267**: 21551–21557.

138. Iida N, Lokeshwar VB and Bourguignon LYW. Mapping the fodrin binding domain in CD45, a leukocyte membrane-associated tyrosine phosphatase. *J. Biol. Chem.* 1994; **269**: 28576–28583.

139. Beverley PCL, Daser A, Michie CA and Wallace DL. Functional subsets of T-cells defined by isoforms of CD45. *Biochem. Soc. Trans.* 1992; **20**: 184–187.

140. Mason D. Subsets of CD4[+] T-cells defined by their expression of different isoforms of the leukocyte-common antigen, CD45. *Biochem. Soc. Trans.* 1992; **20**: 188–190.

141. Rogers PR, Pilapil S, Hayakawa K, Romain PL and Parker DC. CD45 alternative exon expression in murine and human CD4[+] T-cell subsets. *J. Immunol.* 1992; **148**: 4054–4065.

142. Novak TJ, Farber D, Leitenberg D, Hong SC, Johnson P and Bottomly K. Isoforms of the transmembrane tyrosine phosphatase CD45 differentially affect T-cell recognition. *Immunity* 1994; **1**: 109–119.

143. Chui D, Ong CJ, Johnson P, Teh HS and Marth JD. Specific CD45 isoforms differentially regulate T-cell receptor signaling. *EMBO J.* 1994; **13**: 798–807.

144. Ong CJ, Chui D, Teh HS and Marth JD. Thymic CD45 tyrosine phosphatase regulates apoptosis and MHC-restricted negative selection. *J. Immunol.* 1994; **152**: 3793–3805.

145. Merkenschlager M and Fisher AG. CD45 isoform switching precedes the activation-driven death of human thymocytes by apoptosis. *Int. Immunol.* 1991; **3**: 1–7.

146. Beverley P. Immunological memory In T-cells. *Curr. Opin. Immunol.* 1991; **3**: 355–360.

147. Hayward AR, Lee J and Beverley PCL. Ontogeny of expression of UCH11 antigen on TCR-1+ (CD4/8) and TCR-delta+ T-cells. *Eur. J. Immunol.* 1989; **19**: 771–773.

148. Akbar AN, Terry L, Timms A, Beverley PCL and Janossy G. Loss of CD45R and gain of UCH11 reactivity is a feature of primed T-cells. *J. Immunol.* 1988; **140**: 2171–2178.

149. Merkenschlager M, Terry L, Edwards R and Beverley PCL. Limiting dilution analysis of proliferative responses in human lymphocyte populations defined by the monoclonal antibody UCHL1, implications for differentiation. *Eur. J. Immunol.* 1988; **18**: 1653–1661.

150. Byrne JA, Butler JL and Cooper MD. Differential activation requirements for virgin and memory T cells. *J. Immunol.* 1988; **141**: 3249–3257.

151. Sanders ME, Makgoba MW, June CH, Young HA and Shaw S. Enhanced responsiveness of human memory T cells to CD2 and CD3 receptor-mediated activation. *Eur. J. Immunol.* 1989; **19**: 803–808.

152. Dejong R, Brouwer M, Miedema F and Vanlier RAW. Human CD8+ lymphocytes-T can be divided into CD45ra+ and CD45RO+ cells with different requirements for activation and differentiation. *J. Immunol.* 1991; **146**: 2088–2094.

153. Fischer H, Gjorloff A, Hedlund G *et al*. Stimulation of human naive and memory T helper cells with bacterial superantigen. *J. Immunol.* 1992; 148: 1993–1998.

154. Vandevelde H, Lorre K, Bakkus M, Thielemans K, Ceuppens JL and Deboer M. CD45RO(+) memory T-cells but not CD45RA(+) naive T-cells can be efficiently activated by remote co-stimulation with B7. *Int. Immunol.* 1993; **5**: 1483–1487.

155. Robinson AT, Miller N and Alexander DR. CD3 antigen-mediated calcium signals and protein-kinase-C activation are higher in CD45RO+ than In CD45RA+ human lymphocyte-T subsets. *Eur. J. Immunol.* 1993; **23**: 61–68.

156. Schwinzer R, Siefken R, Franklin RA, Saloga J, Wonigeit K and Gelfand EW. Human CD45RA+ and CD45RO+ T

cells exhibit similar CD3/T cell receptor-mediated transmembrane signaling capacities but differ in response to co-stimulatory signals. *Eur. J. Immunol.* 1994; 24: 1391–1395.

157. Lecomte O, Hivroz C, Mazerolles F, Fischer A. Differential CD4-dependent regulation of naive and memory CD4(+) T-cell adhesion is not related to differences in expression and function of CD4 and P56(Lck). *Int. Immunol.* 1994; 6: 551–559.

158. Valitutti S, Muller S, Cella M, Padovan E and Lanzavecchia A. Serial triggering of many T-cell receptors by a few peptide-MHC complexes. *Nature* 1995; 375: 148–151.

159. Ledbetter JA, Tonks NK, Fischer EH, Clark EA. CD45 regulates signal transduction and lymphocyte-activation by specific association with receptor molecules on T-cells or B cells. *Proc. Natl. Acad. Sci. USA* 1988; 85: 8628–8632.

160. Ledbetter JA, Schieven GL, Uckun FM and Imboden JB. CD45 cross-linking regulates phospholipase-C activation and tyrosine phosphorylation of specific substrates in CD3 Ti stimulated T-cells. *J. Immunol.* 1991; 146: 1577–1583.

161. Pollack S, Ledbetter JA, Katz R *et al.* Evidence for involvement of glycoprotein-CD45 phosphatase in reversing glycoprotein-CD3-induced microtubule-associated protein-2 kinase-activity in Jurkat T-cells. *Biochem. J.* 1991; 276: 481–485.

162. Kanner SB, Deans JP and Ledbetter JA. Regulation of CD3-induced phospholipase-C-gamma1 (PLC-gamma-1) tyrosine phosphorylation by CD4 and CD45 receptors. *Immunology* 1992; 75: 441–447.

163. Shivnan E, Biffen M, Shiroo M, Pratt E, Glennie M and Alexander D. Does coaggregation of the CD45 and CD3 antigens inhibit T-cell antigen receptor complex-mediated activation of phospholipase-C and protein-kinase-C. *Eur. J. Immunol.* 1992; 22: 1055–1062.

164. Rankin BM, Yocum SA, Mittler RS and Kiener PA. Stimulation of tyrosine phosphorylation and calcium mobilization by Fc-gamma receptor cross-linking—regulation by the phosphotyrosine phosphatase CD45. *J. Immunol.* 1993; 150: 605–616.

165. Aruffo A, Kanner SB, Sgroi D, Ledbetter JA and Stamenkovic I. CD22-mediated stimulation of T-cells regulates T-cell receptor CD3-induced signaling. *Proc. Natl. Acad. Sci. USA* 1992; 89: 10242–10246.

166. Cyster JG, Fowell D and Barclay AN. Antigenic determinants encoded by alternatively spliced exons of CD45 are determined by the polypeptide but influenced by glycosylation. *Int. Immunol.* 1994; 6: 1875–1881.

167. Goldman SJ, Uniyal S, Ferguson LM, Golan DE, Burakoff SJ and Kiener PA. Differential activation of phosphotyrosine protein phosphatase-activity in a murine T-cell hybridoma by monoclonal-antibodies to CD45. *J. Biol. Chem.* 1992; 267: 6197–6204.

168. Lazarovits Al, Poppema S, White MJ and Karsh J. Inhibition of alloreactivity *in vitro* by monoclonal-antibodies directed against restricted isoforms of the leukocyte-common-antigen-(CD45). *Transplantation* 1992; 54: 724–729.

169. Alsinet E, Inglesesteve J, Vilella R *et al.* Differential-effects of anti-CD45 monoclonal-antibody on human B-cell proliferation—a monoclonal-antibody recognizing a neuraminidase-sensitive epitope of the T200 molecule enhances anti-immunoglobulin-induced proliferation. *Eur. J. Immunol.* 1990; 20: 2801–2804.

170. Torimoto Y, Dang NH, Streuli M *et al.* Activation of T-cells through a T-cell-specific epitope of CD45. *Cell. Immunol.* 1992; 145: 111–129.

171. Schraven B, Roux M, Hutmacher B and Meuer SC. Triggering of the alternative pathway of human T cell activation involves members of the T200 family of glycoproteins. *Eur. J. Immunol.* 1989; 19: 397–403.

172. Welge T, Wolf M, Jaggle C and Luckenbach GA. Human naive T-cells are preferentially stimulated by cross-linking of CD3 and CD45RA with monoclonal-antibodies. *Cell. Immunol.* 1993; 148: 218–225.

173. Maroun CR and Julius M. Distinct involvement of CD45 in antigen receptor signaling in CD4(+) and CD8(+) primary T-cells. *Eur. J. Immunol.* 1994; 24: 967–973.

174. Shen F, Xu XL, Graf LH and Chong ASF. CD45-cross-linking stimulates IFN-gamma production in Nk cells. *J. Immunol.* 1995; 154: 644–652.

175. Webb DSA, Shimizu Y, Van Seventer GA, Shaw S, Gerrard TL. LFA-3, CD44, and CD45: physiologic triggers of human monocyte TNF and IL-1 release. *Science* 1990; 249: 1295–1297.

176. Corvaia N, Reischl IG, Kroemer E and Mudde GC. Modulation of receptor-mediated early events by the tyrosine phosphatase CD45 in primary human monocytes. Consequences for interleukin-6 production. *Eur. J. Immunol.* 1995; 25: 738–744.

177. George A, Rath S, Shroff KE, Wang M and Durdik JM. Ligation of CD45 on B-cells can facilitate production of secondary Ig isotypes. *J. Immunol.* 1994; 152: 1014–1021.

178. Benveniste P, Takahama Y, Wiest DL, Nakayama T, Sharrow SO and Singer A. Engagement of the external domains of CD45 tyrosine phosphatase can regulate the differentiation of immature CD4+CD8+ thymocytes into mature T-cells. *Proc. Natl. Acad. Sci. USA* 1994; 91: 6933–6937.

179. Arroyo AG, Campanero MR, Sanchezmateos P *et al.* Induction of tyrosine phosphorylation during ICAM-3 and LFA-1-mediated intercellular-adhesion, and its regulation by the CD45 tyrosine phosphatase. *J. Cell. Biol.* 1994; 126: 1277–1286.

180. Bernard G, Zoccola D, Ticchioni M, Breittmayer JP, Aussel C and Bernard A. Engagement of the CD45 molecule induces homotypic adhesion of human thymocytes through a LFA-1 ICAM-3-dependent pathway. *J. Immunol.* 1994; 152: 5161–5170.

181. Lorenz HM, Harrer T, Lagoo AS, Baur A, Eger G and Kalden JR. CD45-Mab induces cell-adhesion in peripheral-blood mononuclear-cells via lymphocyte function-associated antigen-1 (LFA-1) and intercellular cell-adhesion molecule-1 (ICAM-1). *Cell. Immunol.* 1993; 147: 110–128.

182. Wagner N, Engel P and Tedder TF. Regulation of the tyrosine kinase-dependent adhesion pathway in human-lymphocytes through CD45. *J. Immunol.* 1993; 150: 4887–4899.

9

Signalling via MHC Molecules

Tania H. Watts

Department of Immunology, University of Toronto, Toronto, Ontario Canada

INTRODUCTION

MHC classes I and II molecules are polymorphic cell surface glycoproteins whose primary function is to display peptide fragments on the surface of antigen presenting cells (APC) for recognition by T cells. MHC class I proteins bind peptides that are synthesized within the cell and bring them to the APC surface for recognition by $CD8^+$ T cells. MHC class II proteins bind peptides derived from antigens taken up by endocytosis and present these peptides to $CD4^+$ T cells [1].

MHC class I proteins (human HLA A, B, C and mouse H-2 K, D and L) consist of a polymorphic 45 kDa heavy chain encoded in the MHC locus non-covalently linked to the non-polymorphic 12 kDa soluble protein, β2-microglobulin. The class I heavy chain consists of three extracellular domains (α1, α2, α3), a transmembrane region and a cytoplasmic domain of 35–40 amino acids. The X-ray structures of the extracellular domains of several class I MHC molecules have been solved [2–8]. The α1 and α2 domains combine to form a peptide binding groove consisting of two α-helical segments sitting on top of an eight-stranded β-sheet. Most of the polymorphic amino acids in MHC class I molecules line this groove. Both the α3 domain and β2-microglobulin have an immunoglobulin-like fold (Figure 9.1a).

MHC class II proteins (HLA-DP, DQ, DR; H-2 A, E) are composed of two polymorphic chains, α and β, of 34 and 29 kDa both encoded in the MHC. Each chain consists of two extracellular domains, a transmembrane region and a short cytoplasmic domain of 12–18 amino acids. The cytoplasmic domains do not possess known signalling motifs. X-ray crystallography of the extracellular domain of HLA-DR1 reveals that the

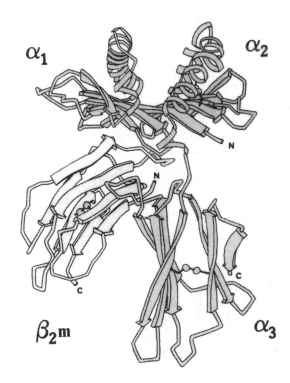

Figure 9.1a Structure of the extracellular domains of HLA class I proteins. (a) Structure of HLA-A2. Reproduced with permission from Bjorkman *et al. Nature*, **329**: 50 6–12. Copyright 1987 MacMillan Magazines Ltd.

membrane proximal α2 and β2 domains have an Ig-like fold, while the *N*-terminal α1 and β1 domains combine to form a peptide binding domain, similar to that of the class I structure, except that the ends of the MHC II binding groove are more open to accommodate longer peptides [9, 10].

It is well established that binding of the TCR to its ligand (Ag/MHC) on the APC results in a signal transduction cascade in the T cell [11] leading to T cell activation or inactivation, depending on the circumstances. There is now a large body of data suggesting that engagement of MHC molecules by reagents which mimic the binding of TCR can also lead to signals within the antigen presenting cell. Crosslinking of MHC molecules expressed on B cells and T cells has been shown to result in signal transduction events which lead to a variety of functional consequences including modulation of homotypic adhesion and proliferation. In this chapter, I review the functional consequences that have been documented in response to MHC crosslinking as well the early biochemical events associated with MHC signalling. Mutagenesis studies of MHC molecules to assess the role of transmembrane and cytoplasmic domains in signalling as well as the search for MHC-associated proteins will also be reviewed.

IS MHC CROSSLINKING REQUIRED FOR MHC SIGNALLING?

Most experiments on MHC signalling have used anti-MHC antibodies to initiate signal transduction. Other experiments have involved the use of specific T cells or the CD4 or CD8 coreceptors to crosslink the MHC molecules. In addition, MHC class II signalling can be initiated by the binding of superantigens to the class II molecules. Superantigens (SAg) [12] are proteins that can activate T cells bearing particular Vβ segments. SAg bind with high affinity to a relatively conserved region of the MHC class II protein in the absence of T cells. In addition, SAg have a lower affinity binding site for a segment of particular Vβ chains. Therefore, SAg can link the MHC to the TCR in the absence of specific antigenic peptide. This results in a much higher frequency of T cells responding to SAg as compared to conventional Ag. A major class of SAg are the bacterial exotoxins, such as the staphylococcal enterotoxins (SEs) [12]. As will be discussed below, SAg binding to class II molecules in the absence of T cells can transduce signals in APC leading to induction of inflammatory mediators which may have implications for the pathogenesis of staphylococcal infections.

In many experiments on MHC signalling, antibodies are crosslinked either by a second step (sec-

ondary antibody or biotin–avidin) or by immobilizing the antibody on plastic culture dishes. In other experiments, simply using bivalent antibody is sufficient to induce MHC signalling. In some studies this might be due to Fc receptors on B cells in the culture providing a means of crosslinking [13]. However, in several studies the role of Fc receptors has been ruled out by use of (Fab')2 fragments [14, 15]. In most studies MHC signalling has not been observed with Fab fragments [14]. However, Mourad et al. found that Fab fragments of the anti-class II antibody L243, but not Fab fragments of another class II specific antibody, ST2-59, could induce signal transduction via HLA-DR in human B cells [16]. It may be that antibodies specific for particular class II epitopes promote aggregation of class II while others do not. However, until aggregation of antibodies or Fabs used in such studies has been rigorously ruled out, it is premature to conclude that a monovalent interaction with MHC II leads to signal transduction.

There are limited data available on MHC aggregation on cells. Aggregates of MHC I have been observed, however these appear to consist of free heavy chains lacking β2-microglobulin, so their significance for antigen presentation is unclear [17]. Edidin's lab has shown that the mobility of both MHC class I and MHC class II on the cell surface is constrained by their cytoplasmic domains presumably due to their interaction with the cytoskeleton [18, 19]. The importance of cytoskeletal attachment of MHC molecules is not well established. However, cytochalasin treatment of antigen presenting cells inhibits antigen presentation to low avidity T cells, suggesting that the cytoskeleton may play a role in antigen presentation, perhaps by facilitating MHC clustering [20].

The X-ray structure of MHC class II proteins revealed that the MHC class II protein is arranged in the crystal as a dimer of α/β heterodimers $(\alpha\beta)_2$ (Figure 9.1b) [9]. Although these dimers may simply reflect crystal packing forces, the observation that the dimer occurs in several crystal forms and that the two MHC II heterodimers are orientated with their peptide binding sites facing the same way makes the structure appear physiologically plausible.

Schafer and Pierce [21] have suggested the existence of the $(\alpha\beta)_2$ dimer of MHC class II on the

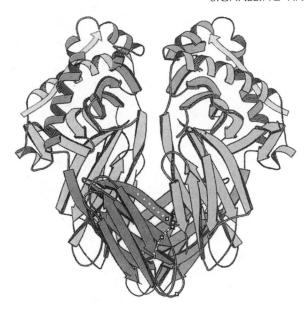

Figure 9.1b Structure of the extracellular domains of HLA class II proteins. (b) Structure of HLA-DR1 showing two MHC complete HLA-DR1 molecules as a dimer of heterodimers. Reproduced with permission from Brown *et al. Nature*, **364**: 33–39. Copyright 1993 MacMillan Magazines Ltd.

surface of B cells. After surface iodination of B cells a species of 120 kDa could be detected in immunoprecipitates provided the samples had not been exposed to temperatures above 37°C during immunoprecipitation or preparation for electrophoresis. Schafer and Pierce suggested that these were $(\alpha\beta)_2$ dimers based on Western blotting of unlabelled precipitates with anti-class-II antibodies and the absence of invariant chain in the 120 kDa complex. Furthermore, in pulse chase experiments, they did not observe other labelled species. However, insufficient evidence was provided to show that other proteins that might not label well did not contribute to this class-II-containing complex or that the dimer was not formed in the lysate during the precipitation with bivalent antibody. Other labs have failed to show the existence of class II multimers even after crosslinking [22]. MHC II molecules on the surface of APC display a variety of peptides. If class II dimers pre-existed on APC, it is not clear how one would ensure that each heterodimer in the dimer of dimers displayed the same peptide. If, on the other hand, MHC clustering was promoted by TCR engagement, then presumably only those MHC–peptide com-

plexes which contain the correct peptide would enter the aggregate. The tendency of MHC II molecules to form dimers such as occurs at very high concentrations in a crystal, may facilitate this aggregation process: i.e. when the TCR and its coreceptors (CD4 or CD8) bind to a specific peptide/MHC complex, this might serve to form a high local concentration of class II molecules at the T cell–APC interface, thereby promoting dimerization or oligomerization.

Similarly, SAg induced MHC II signalling often requires crosslinking of the SAg to detect a signal [23]. However, in other studies crosslinking was not needed, although signals were improved by crosslinking [24]. Therefore, the question of how one gets aggregation of MHC II in response to monovalent toxin binding must also be considered. A difficulty with this area is that commercially available SEs may be aggregated (M Davis, personal communication) so this may explain the discrepancy as to whether crosslinking is required for MHC signalling by SE. X-ray crystallographic analysis of the SAg SEB has shown that two regions implicated in interaction with class II molecules by mutagenesis segregate to different regions of the molecule. Thus one SEB might interact with two different MHC II molecules [25]. The crystal structure of a complex between SEB and DR1 shows interaction of SEB with only the α chain of DR1 [26a]. However, recent data from mutagenesis experiments suggests that a related SAg, SEA can bridge two different MHC class II molecules [26b].

MHC CLASS II SIGNALLING.

Crosslinking of MHC class II molecules on B and T lymphocytes has been shown to induce a variety of outcomes depending on the antibody used, the activation state of the cell and whether other stimuli are given simultaneously. MHC signalling can influence these events both positively and negatively, again depending on the antibody and the activation state of the cell. Many of these studies have been carried out on B lymphocytes which are highly responsive to LPS. Therefore, it is critical for these studies that antibodies and reagents used to stimulate the B cells are free from contaminating endotoxin. We have found that

Protein A or Protein G purified antibodies can have significant endotoxin contamination. This can be removed using commercially available detoxification columns. In recent studies investigators have analysed reagents for endotoxin contamination. However, some of the earlier studies on MHC signalling, where LPS levels were not reported, must be viewed with caution. Although these studies often use isotype matched control antibodies, it is conceivable that the antibodies used could differ in their levels of endotoxin contamination.

Effects of MHC II Signalling on B Lymphocyte Growth and Differentiation

Some of the earliest studies of MHC signalling suggested that anti-MHC II antibodies antagonize the effects of the B cell mitogen LPS on murine B cell proliferation [27, 28]. However, since T cell contact was thought to promote B cell growth and differentiation, this was somewhat surprising. In contrast, Baluyat and Subbarao [13] showed that anti-MHC class II could augment proliferation induced by anti-IgM on murine B cells by 2.6-fold. Cambier and Lehmann analysed the effects of anti-Ia antibodies after B cell priming with IL-4 and anti-immunoglobulin [29]: B cells primed in this way proliferated in response to anti-MHC class II, but not to anti-MHC class I antibodies immobilized on plastic dishes. In the presence of IL-4 and IL-5, anti-MHC II antibodies promoted Ig secretion by the cultured B cells. Similarly, Fuleihan et al. [30] have shown that engagement of MHC class II molecules on human B cells by the SAg TSST-1 synergizes with anti-μ or PMA treatment to enhance B cell proliferation.

In another example, Bishop and colleagues have characterized a B lymphoma, CHB3, which undergoes growth inhibition in response to sIgM signalling [31]. In this model, antibodies to class II molecules appear to counteract anti-IgM mediated growth inhibition. Thus on pre-activated or co-activated B cells, anti-MHC class II induced signalling appears to promote B cell growth and differentiation.

Bishop and collaborators have also demonstrated a role for MHC II signalling in differentiation of the B lymphoma, CH12.LX. This lymphoma can be induced to differentiate in vitro into an Ig secreting cell in response to antigen (SRBC) binding to its surface IgM together with a signal delivered through $I - E^k$ but not $I - A^k$ molecules on its surface [32–34]. CH12.LX cells transfected with A^b molecules also respond to anti-MHC II signalling [33]. It is not clear why A^k molecules fail to signal in this system, since the sequence of A^k in these cells is wildtype [34]. Moreover, in the presence of anti-IgM, CD40 signalling on CH12 cells as well as normal B cells is also enhanced in the presence of anti-class II antibodies [35].

In contrast to the effects of MHC II crosslinking on antigen and IL-4 primed B cells, resting murine B cells undergo apoptosis in response to MHC class II crosslinking (15). Freed and collaborators have suggested that this serves as a mechanism of B cell tolerance: Antigen-specific B cells take up antigen via their sIg receptors and thereby efficiently present the processed antigen to T lymphocytes [36–38]. This in turn activates the T lymphocyte to express CD40 ligand and release cytokines which together stimulate the B cell to differentiate into an Ig secreting cell [39, 40]. However, at much higher antigen concentrations (about 1000-fold higher), non-specific B cells can take up antigen by fluid phase pinocytosis and also present them to T cells. If, under conditions of high antigen concentration, the activation of T cells by non-specific B cells led to activation of bystander B cells this could lead them to secrete an immunoglobulin with a specificity unrelated to the antigen which activated the T cell. The induction of apoptosis in response to MHC II engagement on B cells which have not taken up antigen via their sIg receptor may represent a means of preventing this bystander activation of B cells. One must ask, however, whether such a tolerance mechanism is needed or does death in response to anti-MHC II ligation simply represent a non-physiological response to a signal delivered out of context. After all, MHC II ligation would normally take place in the presence of other receptor–ligand interactions that take place between the T cell and the APC. Furthermore, since naive T cells are tolerized by resting B cells, which lack costimulatory molecules until they receive a signal through their surface Ig, it is unlikely that the interaction between a resting B cell and a resting

T cell would be productive [39]. However, a T cell that has been first activated on a competent APC could interact with a resting B cell and the above tolerance mechanism might be relevant in this case. Engagement of CD40 on B cells in the presence of IL-4 leads to limited Ig secretion, whereas treatment of B cells with anti-CD40 and SAC particles results in much higher Ig secretion [40]. These data suggest that in the absence of signals through the BCR, differentiation of bystander B cells due to pinocytic uptake of antigen and recognition by activated T cells will result in rather limited Ig secretion. Nevertheless, the potential is there for such antibodies to be autoreactive. The induction of apoptosis on resting B cells by MHC II crosslinking in the absence of Ig signalling may represent a failsafe mechanism.

In human B cells, anti-DR induced cell death by apoptosis has also been observed [41]. However, in this case resting B cells were resistant to anti-DR induced cell death but a subpopulation of more buoyant human B cells were susceptible to anti-DR induced death as was a B lymphoblastoid line. These data were interpreted as representing a means of eliminating activated B cells once they had played their role in the immune response.

Although the above studies clearly show that under appropriate circumstances *in vitro*, one can show enhancement of B cell growth and differentiation in response to MHC II signalling, recent studies of B cell development and function in MHC class-II-deficient mice, raise the question as to whether this signalling plays an essential role in B cell function *in vivo*. Laurie Glimcher's laboratory created MHC class-II-deficient mice by targeted disruption of the class II Aβ^b gene [42]. When these mice were assessed for B cell development [43] it was found that B cell numbers and surface phenotype were normal. Furthermore, *in vitro* responses of class-II-deficient B cells to anti-CD3 activated T cells were indistinguishable from wildtype. These studies do not rule out a critical role for MHC II signalling within the APC in enhancing activation of the T cell. However, once the T cell has been activated by other means, B cell responses can be MHC-II-independent. On the other hand, anti-CD3 is a very strong T cell activator and likely induces higher expression of CD40 ligand by the T cells than one would obtain in an antigen-specific response. It is possible that

under conditions of suboptimal engagement of TCR and B cell surface receptors that MHC II signalling might well contribute to the amplification of signals involved in T and B cell activation.

Effects of MHC II Signalling on T Lymphocyte Growth

In contrast to murine T lymphocytes which are class II negative, activated T cells from human, rat and several other species express MHC class II molecules. These MHC class II molecules might play a role in regulatory interactions between T cells. Crosslinking of MHC class II on activated human T cells has been shown to synergize with signals delivered via the TCR/CD3 complex to induce proliferation, cytokine gene expression and increased intracellular Ca^{2+} levels [44,45]. Odum *et al.* further analysed this effect of anti-MHC II on T cell proliferation and found that class II signalling enhanced the responsiveness of T cells to IL-2 [46]. No effect was observed on IL-4-driven proliferation. This enhancement required immobilized class II antibody, whereas soluble anti-class II antibodies had inhibitory effects on T cell proliferation. This may explain earlier reports that class II signalling antagonizes T cell proliferation [47, 48]. The enhancement of T cell proliferation to IL-2 by class II signalling appeared to be due to increased expression of the IL-2 receptor β chain and also led to enhanced tyrosine phosphorylation and increased intracellular Ca^{2+} fluxes [46].

Induction of Homotypic Adhesion of T and B Lymphocytes by MHC II Signalling

Activation through the T cell receptor has been shown to augment T cell–APC cell adhesion by increasing the avidity of LFA-1 on the T cell for its ligand ICAM-1 on the APC [49]. The reciprocal event, MHC-II-mediated activation of adhesion molecules on APC has also been demonstrated in several labs [50–55]. In general these adhesion events are not measured as T–B adhesion, but are studied as a homotypic adhesion event with a single type of cell. This homotypic adhesion is visualized as a tight clumping of the cells, similar to what is seen when T or B lymphocytes are treated with phorbol esters. It is thought that these homotypic adhesion events actually mimic heterotypic adhesion that might occur during antigen presentation. Enhancement of adhesion after appropriate signal transduction through the TCR or through MHC molecules, might serve to amplify and reinforce APC–T cell interactions once it has been established that an antigen-specific interaction has taken place.

It should be noted that the fact that many anti-class II antibodies promote homotypic adhesion means that when one uses class II crosslinking to study late events of MHC II signalling, it may be difficult to distinguish those consequences emanating directly from the MHC from those induced via engagement of adhesion molecules, such as LFA-1.

Mourad et al. [50] showed that incubation of human B lymphoblastoid cell lines, monocytes and activated but not resting T cells with either the SAg TSST-1 or with an Fab of the anti-DR antibody L243 led to homotypic adhesion that was partly blocked by antibodies to LFA-1 or ICAM-1. This homotypic aggregation was blocked by the protein kinase C inhibitor sphingosine. In addition to the LFA-1 dependent component of anti-class II induced homotypic adhesion, others have found that there is an LFA-1 independent component of this adhesion as well, and this adhesion persists in B cells isolated from patients with leukocyte adhesion deficiency, who lack surface LFA-1 [51–53]. Odum et al. [51] provided evidence that this LFA-1 independent aggregation was an active process: it was blocked by low temperature, azide, cytochalasins B and E, and partly by the inhibitors of serine/threonine kinases, but not by EDTA or the tyrosine kinase inhibitors genestein or herbimycin A. Furthermore, use of $F(ab')_2$ ruled out a role for FcR in this process. Odum et al. (51) did not observe the LFA-1 dependent component of the adhesion in their study. This may be due to the different incubation times used. The studies showing LFA-1 dependent adhesion utilized incubation times of one to a few hours, whereas the LFA-1 independent adhesion was observed after overnight incubation of cells with antibodies.

Kansas and Tedder [53] showed that mAb binding to several B cell surface proteins including HLA-D, CD19, CD20, CD40 and CD43 led to homotypic adhesion. The induction of this adhesion appeared to be dependent on tyrosine kinases since it was inhibited by herbimycin. The discrepancy as to whether anti-class II induced homotypic adhesion is dependent on tyrosine kinases [53] or not [51] may be reflective of the two types of adhesion event (LFA-1-dependent vs. independent) which may have different kinetics or be differentially invoked by different antibodies. Kansas et al. (55) have also used a soluble recombinant form of CD4 to induce homotypic adhesion in B lymphoblastoid lines. This homotypic adhesion was observed in B lymphoblastoid lines but not in class II$^+$ pre-B cells.

Tedder's lab has recently generated a series of antibodies that block B cell homotypic adhesion [56, 57]. Several of these antibodies block homotypic adhesion induced via anti-class II as well as MHC class I, CD19, CD20, CD21, CD40 and leu-13 induced adhesion. One of these adhesion blocking antibodies, HAB-1, reacts with CD45, suggesting regulation of homotypic adhesion by CD45 [56]. Since induction of homotypic adhesion can involve tyrosine kinases [53], this is likely mediated by the effects of the CD45 tyrosine phosphatase on tyrosine kinases. Interestingly, three other homotypic adhesion blocking antibodies turned out to bind to MHC class II (HAB-2, -3) or MHC class I (HAB-4) [57]. The binding of these antibodies to their ligands resulted in inhibition of homotypic adhesion induced by other B cell surface molecules, including class I, class II, CD19, CD23, CD39, CD40 and leu 13. Furthermore, PMA-mediated adhesion was inhibited. The significance of this heterologous desensitization is not clear, although the authors suggested that it might be part of a tolerance mechanism for B cells. In the same study, Wagner et al. [57] showed that engagement of the same MHC class II molecule by different antibodies could either induce, inhibit or have no effect on adhesion. This led the authors to suggest that binding of antibodies to different epitopes of the class II molecule might induce different signal transduction events, perhaps by influencing the way in which two class II molecules are juxtaposed when aggregated by antibodies.

All of the above studies on MHC class-II-mediated homotypic adhesion have been done on human B cells including EBV lines, B lymphoblastoid lines, and resting as well as activated splenic B cells. To date homotypic adhesion in response to class II signalling has not been reported in mouse B cells. However, plastic immobilized anti-class II antibody has been shown to induce spreading of mouse B cells and formation of pseudopods. This morphological transformation was not observed with anti-class I mAb [29].

Induction of Cytokines, Chemokines and Costimulatory Molecules by MHC II Crosslinking

Signalling through MHC class II can induce a number of molecules that may be involved in amplifying the immune response. B7-1(CD80) and B7-2 (CD86) are Ig family members expressed on dendritic cells, activated B cells and activated macrophages. B7 molecules bind to CD28 and CTLA-4 molecules on T cells and thereby deliver costimulatory signals [58]. Kouvlova *et al.* demonstrated that crosslinking of HLA-DR on human tonsillar B cells using biotinylated anti-class II followed by avidin, results in induction of the costimulatory molecule B7/BB1 (now called B7-1) [59]. In this experiment B7-1 was induced rapidly, with optimal induction after 6 h and decreased expression by 18 h. However, other studies have shown markedly slower kinetics of B7-1 induction after various stimuli so this aspect of the Kouvlova *et al.* study remains puzzling [60]. As will be discussed below, M12 B lymphoma cells expressing murine class II molecules with truncated cytoplasmic domains have a defect in antigen presentation to some T hybrids. These T cells appear to be dependent on the expression of costimulatory molecules B7-1 and B7-2. Incubation of M12 cells expressing wildtype MHC II with T cells leads to induction of B7 molecules on M12 cells. However, incubation of truncated MHC expressing M12 cells with T cells does not lead to B7 induction on the B cells, suggesting that signalling through the MHC class II cytoplasmic domains leads to B7 induction [61, 62]. However, direct induction of B7 by MHC II crosslinking in M12 lymphomas could not be demonstrated [62]. In contrast, Smiley *et al.* [63] recently reported the induction of B7-1 by MHC II crosslinking on B cells from mice expressing wildtype MHC II or MHC II with a truncated β chain. No induction was seen in MHC II knock-

out mice. Thus in normal B cells, truncation of the MHC II cytoplasmic domain does not prevent B7 induction.

Poudrier and Owens [64] have examined the effects of crosslinking MHC II together with CD54 (ICAM-1) on resting murine B cells. They find that in the presence of IL-5, simultaneous crosslinking of CD54 and MHC II leads to induction of a functional IL-2 receptor. This study also reports that the combination of anti-CD54 and anti-class II induces B7 in resting murine B cells. This signal was enhanced by the presence of IL-5. Thus signals through MHC II may synergize with other signals induced in the B cell to up-regulate costimulatory molecules.

In monocytes, the SAg SEB and TSST-1 as well as anti-class II antibodies, have been shown to induce the expression of the monokines IL-1 [65–67] and TNFα [67]. Trede *et al.* [67] provided evidence that this was not due to contaminating LPS. Polymyxin B, which binds and inactivates LPS, blocked LPS but not TSST-1 induction of IL-1β. SE appear to induce TNFα transcription via activation of the transcription factor NF-κBB [68]; see also N Trede, AV Tsytskyova, T Chatila, AE Goldfeld and RS Geha, manuscript submitted). Likewise, in fibroblast-like synoviocytes, ligation of MHC II with the SAg SEA induces the inflammatory chemokines Rantes, MCP-1 and IL-8 [69]. The SAg was found to be free of LPS, using the Limulus amoebocyte lysate test.

Early Biochemical Events Associated with MHC Class II Crosslinking

A number of early biochemical events have been documented in response to MHC II ligation. As is the case with late events, early biochemical events associated with MHC class II ligation appear to depend on the activation state of cell.

One of the earliest reports on second messengers involved in class II signalling came from Cambier *et al.* [70] who showed that engagement of MHC II molecules by anti-class II F(ab')$_2$ on resting mouse B cells caused an increase in intracellular cAMP and translocation of PKC from the cytosol to a Triton X-100 insoluble fraction. This detergent-insoluble fraction contains the cytoskeleton and the nucleus. Analogues of cAMP could also mediate this PKC translocation suggesting that MHC II crosslinking leads to cAMP production followed by PKC translocation. As discussed above, Newell *et al.* [15] have shown that anti-class II treatment of resting mouse B cells leads to apopotosis. Interestingly, treatment of the resting B cells with cAMP analogues had the same effect [15].

Lane *et al.* [71] studied the effect of MHC II crosslinking on high density B cells isolated from human tonsils. The high density population is considered to represent the resting B cell population. These authors observed an increase in tyrosine phosphorylation, phosphatidyl inositol metabolism and a rise in intracellular Ca^{2+} in response to MHC II signalling. Detection of these signals required secondary crosslinking of the primary anti-class II antibody. Mooney *et al.* [72] obtained similar results using high density B cells from human spleen. Cambier *et al.* [70] had reported that they failed to detect this signalling pathway in resting mouse B cells. Later, however, Cambier *et al.* [73] showed that if high density resting mouse B cells were primed with anti-μ and IL-4, then they responded to biotinylated anti-class II plus avidin by mobilizing Ca^{2+}. Thus the signals observed in high density human tonsillar or human splenic B cells in response to MHC II crosslinking appear to differ from those in high density murine splenic B cells, but resemble signals observed in antigen plus IL-4 pre-activated murine B cells. Similarly, high density human tonsillar B cells and anti-CD40-treated mouse B cells are resistant to killing by dibutyryl-cAMP treatment, whereas resting mouse B cells are killed by cAMP treatment (M Goldstein and TH Watts, unpublished observation). Thus it seems that high density human tonsillar B cells are at a different stage of differentiation/activation than high density mouse splenic B cells. Whether this reflects a fundamental difference in human and mouse B cell differentiation/activation or the existence of a previously primed B cell subset in the human high density B cell population remains to be determined.

Brick-Ghannam *et al.* have examined PKC activation in B cells after HLA-DR crosslinking. They found maximal PKC activity after 30 minutes of anti-DR treatment. At least some of this increased activity was found to be due to increased levels of PKCα, β and δ mRNA as measured in a RT-pcr assay [74,75].

In addition to studies with normal B cells, several groups have looked at signal transduction in B lymphomas. M12 and K46 are spontaneously derived BALB/c lymphomas [76]. Wade and collaborators have shown that M12 cells transfected with wildtype A^k molecules and crosslinked with anti-A^k antibody translocate protein kinase C (PKC) to the detergent-insoluble fraction of the cell [77,78]. In contrast, crosslinking of A^d or of transfected A^k on K46J B lymphomas results in Ca^{2+} mobilization and protein tyrosine phosphorylation [79]. Thus the two kinds of MHC-II-associated signal transduction pathway found in normal B cells can be found in different B lymphomas.

Using the CH12.LX lymphoma model, Bishop and collaborators have provided further evidence for the role of the cAMP pathway in murine MHC II signalling. As discussed above, CH12.LX cells secrete IgM in response to Ag (SRBC) and a signal delivered through their I-E molecules [32–34]. This secretion of IgM is abrogated by the adenylate cyclase inhibitor 2' 5'-dideoxyadenosine [14]. However, cAMP analogues do not replace anti-Ia mediated B cell differentiation in this model, suggesting that cAMP plus additional signals induced via MHC II signalling are required for induction of IgM secretion. A puzzling observation is that while signalling through A^k in M12 cells seems to involve the cAMP pathway, as shown indirectly using the PKC translocation assay, in CH12 the endogenous A^k molecules cannot signal via this pathway even though they have a wildtype sequence. However, after transfection of $A\beta^b$ which allows the formation of a hybrid $A\alpha^k A\beta^b$ molecule, crosslinking with $A\beta^b$-specific antibody leads to increases in intracellular cAMP in CH12 [14]. Perhaps this failure of A^k to signal in CH12 reflects a difference in a post-translational modification of A^k in the two different lymphomas or a polymorphism in an MHC-associated molecule.

A number of groups have implicated tyrosine kinases in signalling via MHC II molecules, either by measuring the appearance of new tyrosine phosphorylated bands in anti-ptyr immunoblots or by inhibiting late events such as homotypic adhesion or cytokine secretion by tyrosine kinase inhibitors [24,53,54,79–81]. Morio *et al.* [80] have looked at src family kinase autophosphorylation in response to MHC II interaction with the SAg

TSST-1 or SEA in monocytes or the B cell lymphoma line Raji. In monocytes, the autophosphorylation of the src kinases hck and fgr is increased five-and four-fold respectively by TSST-1 binding. In Raji cells, SEA induces increased autophosphorylation of fgr and caused modest increases in autophosphorylation of the src family member, lyn. This effect was not seen on a class II^- variant of Raji. In T lymphocytes signalling via HLA-DR has also been shown to be dependent on tyrosine kinases and is modulated by CD45. Odum *et al.* [81], showed that the tyrosine kinase inhibitor, herbimycin could abrogate Ca^{2+} signalling induced by anti-DR antibodies. Co-crosslinking of HLA-DR with the tyrosine phosphatase, CD45 also inhibited the Ca^{2+} response. In T cells, HLA-DR ligation also induces increased tyrosine phosphorylation of phospholipase-Cγ1 [82].

The Role of the Cytoplasmic Domains of Class II in Antigen Presentation and Signalling

The cytoplasmic domains of MHC class II molecules are fairly conserved within isotypes and across species. However there are significant differences between the different isotypes (Figure 9.2).

The β chain of HLA-DQ lacks the fifth exon due to a point mutation in the splice acceptor site. This results in a DQβ molecule with a cytoplasmic domain that is eight amino acids shorter than that of other class II β chains. However, this deletion does not appear to prevent the DQ molecule from presenting antigen to T cells [83].

The effect of truncation of the cytoplasmic domains of murine class II molecules has been analysed in several studies. Laurie Glimcher and collaborators isolated a B hybridoma by immunoselection for mutations in Eβ and fortuitously obtained a mutation in the Aα^k gene resulting in an Aα^k molecule that was truncated by 12 amino acids in the cytoplasmic domain [84]. This mutation resulted in decreased A^k expression and a limited defect in antigen presentation: 2 out of 29 T hybrids tested failed to recognize the mutant molecule. The same group went on to engineer A^k molecules lacking the C-terminal 10 amino

acids of the 18 amino acid β chain cytoplasmic domain [78]. The truncated Aβ was expressed together with wildtype Aα^k (αWT/β-10) or with Aα^k molecules with a 12 amino acid truncation in the cytoplasmic domain (α-12/β-10). Both truncated forms of A^k were found to be expressed at the cell surface after transfection into Ia negative M12.C3 cells. Although the truncated A^k molecules presented antigen efficiently to several HEL-specific T hybrids, there was a defect in antigen presentation to four out of five autoreactive T cell hybrids as well as to two KLH specific hybrids. Truncation of both the β and α chains gave the more profound defect. In addition to the defect in antigen presentation, Nabavi *et al.* [78] found that M12 B cells expressing the α-12/β-10 form of A^k had a quantitative defect in PKC translocation to the insoluble fraction after MHC II crosslinking, as compared to B cells expressing wildtype A^k. The αWT/β-10 form showed normal levels of PKC translocation but with delayed kinetics.

A similar analysis on the effect of cytoplasmic domain truncation on MHC II signalling in M12 cells was carried out by Wade *et al.* [77]. They engineered A^k molecules truncated by 6 or 12 amino acids in the α chain and truncated by 12 or 18 amino acids in the β chain. Various combinations of the mutants were transfected into the Ia negative cell line M12.C3 and tested for T cell activation as well as for MHC II signalling using the PKC translocation assay. Truncation of the α chain by 12 amino acids resulted in a modest decrease in antigen presentation to some T hybrids. The α-12/β-18 form of A^k was not very well expressed and therefore relatively poor at antigen presentation. In the PKC translocation assay, it was found that truncation of 6 or 12 amino acids from the α chain of A^κ did not impair PKC translocation. The αWT/β-12 form of A^κ showed a delay in the kinetics of PKC translocation, while the α-12/β-12 form showed markedly impaired translocation of PKC to the insoluble compartment after a crosslinking of the A^κ molecules.

André *et al.* [79] have also transfected these truncated A^κ constructs into another B lymphoma, K46J. This lymphoma does not respond to MHC II crosslinking by PKC translocation to the insoluble fraction. Instead, increases in intracellular

MHC class II cytoplasmic domains

α chains

	222			225					230			
Aα b,d,f,k,q,r,s,u	R	S	G	G	T	S	R	H	P	G	P	L
DQα 1,2,3,4,5	R	S	**V**	G	**A**	S	R	H	**Q**	G	P	L
DQα 6	-	-	-	-	-	-	-	-	-	-	P	

	219						225					230			
Eα d,k	K	G	I	K	K	R	N	V	V	E	R	R	Q	G	A	L
DRα 1,2	K	G	**L**	**R**	K	**S**	N	A	A	E	R	R	**G**	**P**	**L**	
DRα 3,4,5	K	G	**V**	**R**	K	**N**	A	A	A	E	R	R	**G**	**P**	**L**	

β chains

	221				225					230					235			
Aβ b,d,k	R	H	R	S	Q	K	G	P	R	G	P	P	P	A	G	L	L	Q
DQβ 1,3	**H**	H	R	S	Q	K	-	-	-	-	-	-	-	-	G	L	L	**H**
DQβ 2	**R**	**Q**	R	S	**R**	K	-	-	-	-	-	-	-	-	G	L	L	**H**
DQβ 4	**H**	H	R	S	Q	K	-	-	-	-	-	-	-	-	-	-	-	-

	221				225					230					235			
Eβ b,d,q,k,s	Y	F	**R**	N	Q	K	G	Q	S	G	L	Q	P	T	G	L	L	S
Eβ u	Y	F	**A**	N	Q	K	G	Q	S	G	L	Q	P	T	G	L	L	S
DRβ 1	Y	F	R	N	Q	K	G	**S**	**H**	G	L	Q	P	**T**	G	F	L	S
DRβ 2	Y	F	R	N	Q	K	G	**H**	**S**	G	L	Q	P	**R**	G	F	L	S
DRβ 3,4	Y	F	R	N	Q	K	G	**H**	**S**	G	L	Q	P	**T**	G	F	L	S
DRβ 5	-	-	-	-	-	-	-	-	-	-	-	-	-	-	-	-	-	-
DRβ 6	Y	F	R	N	Q	K	G	**H**	S	G	L	Q	P	**T**	G	**L**	L	S

Figure 9.2 Cytoplasmic domains of selected MHC class II proteins. Sequences are arranged to compare human and mouse α and β chains for each of the two murine isotypes. Amino acids which differ within each group are in bold face. These sequences were taken from a compilation [136] made by Figueroa and Klein (1986) *Immunology Today*, **7**.

Ca^{2+} are observed. In these cells it was found that truncation of the A$^\kappa$ cytoplasmic domain had no effect on Ca^{2+} fluxes. Thus the two types of signalling pathway associated with murine MHC II molecules appear to segregate to different regions of the class II molecule.

Harton and Bishop [85a] have carried out mutational analysis of the cytoplasmic domain of Aβ^b expressed in CH12 cells. As discussed above, in CH12 lymphomas, transfected A^b molecules signal via a cAMP-dependent pathway. It was found that truncation of 10 amino acids from the Aβ chain had no effect, whereas truncation of 12 amino acids reduced signalling (as measured by Ig secretion) by more than 90%. A more recent study by this group [85b] has shown that the position and sequence of gly 227 and pro 228 are critical to this signalling pathway.

The studies of Nabavi *et al.* [78] from Laurie Glimcher's laboratory suggested that truncation of the cytoplasmic domain of class II molecules resulted in a defect in antigen presentation to some but not all T hybrids. In general, the T cells which were sensitive to truncation of the cytoplasmic domain were autoreactive T hybrids. St-Pierre *et al.* [86] found that treatment with dibutyryl-cAMP of M12 lymphomas expressing the α-12/β-10 form of A^κ corrected this antigen presentation defect. However, in the presence of cycloheximide the effects of db-cAMP were eliminated [20]. These data suggested, indirectly, that truncation of the cytoplasmic domain leads to a defect in antigen presentation due to loss of the cAMP-signalling pathway and that MHC signalling via this pathway led to induction of a protein or proteins involved in antigen presentation. Subsequently, it was found that treatment of M12 lymphomas with cAMP led to the induction of the costimulatory molecules B7-1 and B7-2 [61,62,87]. Incubation of A^κ-restricted T cells with M12 cells expressing wildtype A^κ also led to B7 induction, whereas incubation of T cells with M12 cells expressing truncated A^κ did not [61,62]. This led to a model suggesting that signalling via the MHC II cytoplasmic domain led to induction of cAMP and then B7. However, in M12 lymphomas it has not been possible to show B7 induction directly by MHC II engagement [62]. It may be that other molecules, such as ICAM-1, synergize with MHC II signalling during T cell–APC contact and are required to get detectable B7 induction [64]. Alternatively, the cytoplasmic domains of class II may not be involved directly in B7 induction, but in favouring a conformation or allowing aggregation of MHC II so as to induce initial signals through the TCR. Once activated via the TCR, T cells can induce B7 via an MHC-independent pathway [62]. Signalling through the TCR induces the expression of CD40 ligand on the T cell, which in turn binds to CD40 on the APC and thereby induces B7 molecules [88,89]. Watts *et al.* [62] found that T cells that did not require the MHC II cytoplasmic domains for activation could induce B7 in truncated MHC-II-bearing B cells. This induction of B7 was dependent on CD40–CD40L interaction (MA Debenedette and TH Watts, unpublished observation). The observation that the CD40 pathway of signalling may also involve cAMP [90,91] makes it difficult to resolve these two models of B7 induction with the currently available data.

Another series of studies which illustrate the importance of the cytoplasmic domain of MHC II molecules in antigen presentation come from Ostrand-Rosenberg and collaborators [92,93]. They have studied the ability of mice to reject a syngeneic tumour SaI. This tumour is lethal unless transfected with wildtype A^κ molecules, which allow the tumour to be rejected. However, transfection of the tumour cells with truncated (α-12,β-10) A^κ molecules does not result in tumour rejection [92], unless the SaI cells are also transfected with B7 [93]. Additional unpublished studies from this group (Suzanne Ostrand-Rosenberg, personal communication), show that when SaI cells, expressing wildtype MHC II, are removed from the mice during the process of tumour rejection they are now found to express B7-1 and B7-2.

Wade *et al.* [94] have shown that M12 cells, transfected with a site-directed mutant of A^κ truncated in the α chain, have a limited defect in antigen presentation to some peptide-specific T hybrids. In contrast to the B7-dependent hybrids that require cAMP treatment of M12 cells to respond [61,62], the T hybrids used in these studies respond to fixed M12 cells and do not require costimulatory molecules. In this case, fixation of the cells abrogated the effect of cytoplasmic domain truncation, suggesting that immobilization of MHC II molecules, perhaps via their cytoplasmic domains, might be important for T cell activation.

In contrast to the results in M12 cells, K46J B lymphoma cells transfected with truncated MHC II molecules do not show a defect in antigen presentation to the same autoreactive T cells that require the cytoplasmic domain in M12 cells [95]. However, it has recently been determined that K46J cells constitutively express an alternative costimulatory molecule, 4-1BB ligand, which appears to obviate the need for B7 molecules in this system [95].

The above studies illustrate the difficulty of interpreting the mechanism of effects of mutations when a read-out such as IL-2 production by T cells is used. The T cells used in the studies may differ in their avidity and in their requirements for costimulatory molecules. Furthermore, MHC class II molecules undergo a complex intracellular assembly process during which peptides are loaded into the MHC binding groove. A number of proteins are involved in MHC folding and peptide loading, including calnexin, invariant chain and the newly discovered HLA-DM molecules [96]. The lack of cytoplasmic domains may influence interactions at any of these steps and therefore affect the final conformation of the class II molecules. The ability of MHC class II to cluster in the membrane may also be influenced by changes in cytoplasmic domains. Finally, there are the effects on signalling pathways. Therefore, it is clear that more detailed analysis of the effects of cytoplasmic domain mutations on different aspects of MHC function will be required before a clear consensus emerges as to how these mutations influence antigen presentation and intracellular signalling.

MHC Class II Associated Proteins

As discussed above there are two major signalling pathways that have been implicated in class II signalling, a pathway involving cAMP and PKC translocation to the insoluble fraction (cytoskeleton or nucleus) and a pathway that involves activation of src-family tyrosine kinases, phosphatidyl inositol metabolism and release of intracellular Ca^{2+}. To date, little is known about the nature of the linkage of MHC II with either of its associated signalling pathways. The latter pathway does not appear to require the cytoplasmic domain of class II, so it seems likely that class II must associate in the membrane with another transmembrane protein that could serve as a link to the cytoplasmic tyrosine kinases.

MHC signalling via the cAMP pathway appears to require eight amino acids from the cytoplasmic domain of the Aβ chain, however, there is nothing in this sequence that suggests a means of linkage to a cAMP-generating pathway: the classical cAMP-generating pathway involves a multispanning membrane receptor interacting with trimeric G proteins, to release a Gα subunit which in turn activates adenylate cyclase. In addition, it has been found that a 13 aa segment of the cytoplasmic domain of EGF receptor can activate Gs [97]. This segment contains a motif found in other receptors that activate G proteins: BBXB or BBXXB, where B is a basic residue and X is any residue. However, the class II cytoplasmic domains do not appear to possess this motif. On the other hand, interaction with Gα protein is not the only way in which adenylate cyclases are regulated: they can be activated or inhibited by PKC, Ca^{2+} or the $\beta\gamma$ subunits of G proteins [98].

Newell et al. (99) found proteins in the range of 41–42 kDa and 56–58 kDa copurifying with A$^{\kappa}$ molecules isolated from an AKR B lymphoma. The intensity of the copurifying bands was increased by treating cells with a bifunctional crosslinker. One of the proteins in the 41–42 kDa range was identified as actin. A protein in the 56–58 kDa range could be photoaffinity labelled with 8-azido-cAMP, suggesting that it might be a regulatory subunit of a cyclic-AMP-dependent protein kinase. Except for actin, these bands were not found in isolated MHC I preparations. In contrast, Schafer and Pierce [21] did not find any other proteins copurifying with class II in their study looking for dimers of class II heterodimers. Both of these studies used NP40 for their lysis buffer.

Bonnefoy et al. [100] have identified an association between HLA-DR and low affinity FcR receptors (CD23) on human B cells. In screening hybridomas for blocking antibody binding to the FcR on RPMI 8866B cells, they isolated an anti-HLA-DR antibody, 135, that interfered with IgE binding to its receptor. F(ab')$_2$ fragments of MAb 135 gave a similar level of inhibition (about 40% inhibition at 50 $\mu g\ ml^{-1}$). Other anti-class II antibodies did not have this effect. Immunoprecipitation of the 42 kDa CD23 (FcR) from digitonin lysates, or from NP40 lysates crosslinked with DSP, showed coprecipitation with proteins comigrating with DRα and β chains. Similarly precipitation with anti-class II brought down a band comigrating with CD23. This study is consistent with earlier reports of murine MHC II molecules associating with the Fc receptor [101–103]. Interestingly, CD23 has been shown to associate physically with the src tyrosine kinase fyn [104]. Therefore, an association between class II and CD23 may be one means by which anti-HLA-DR could activate a tyrosine kinase pathway.

More recently, an association of HLA-DR molecules with the TAPA-1 molecule on the surface of B cells has been reported [105, 106]. TAPA-1 is a 26 kDa member of the transmembrane 4 superfamily, a group of proteins characterized by four highly conserved hydrophobic domains [107]. There are at least 15 members including CD9, CD37, CD53, CD63, CD81 (TAPA-1) and

CD82. The association between TAPA-1 and HLA-DR has been demonstrated by co-immuno-precipitation in mild detergents, such as 1% CHAPS [105,106]. Antibodies to TAPA-1 have been shown to induce homotypic adhesion and to induce an antiproliferative response in some cells [108]. TAPA-1 is also a component of the CD19/CD21 signalling complex on B cells [109]. Although one must be cautious in interpreting associations detected in mild detergents where incomplete solubilization can be a problem, the above studies do suggest that an association between TAPA-1 and a subset of HLA-DR molecules might explain some of signalling properties of class II molecules.

MHC CLASS I SIGNALLING

Functional Consequences of MHC I Signalling

The literature on MHC class I signalling is less extensive than that on MHC II signalling. However, a number of studies have shown effects of anti-MHC I on homotypic adhesion and cell growth. Induction of homotypic adhesion by anti-class I antibodies has been demonstrated on human monocytes, B cells, resting and activated T cells, with B cells and activated T cells showing the most striking effect [51,52,57,110]. The induction of homotypic adhesion depends on the antibody used. Odum et al. (51) showed that an anti-HLA-G specific antibody as well as the broadly reactive anti-class I antibody IOT2 induced homotypic adhesion, whereas the anti-class I antibody W6/32 did not. In contrast, Spertini et al. (110) found that W6/32 did induce homotypic adhesion in normal cells but not in LFA-1-deficient cells. In contrast, antibodies to β2 microglobulin (6B7 and BBM1) induced homotypic adhesion that was independent of LFA-1 and Ca^{2+}. In the Odum et al. study [51] the adhesion was Ca^{2+} independent and insensitive to anti-LFA-1. This may explain the differences in the results with W6/32. The differences in induction of LFA-1-dependent versus independent adhesion appear to be related to the timing of the adhesion assays: Odum et al. used a 16 h time point to measure adhesion, whereas

Spertini et al. observed homotypic adhesion between 90 minutes and 4 h. It seems that the LFA-1-dependent component of the adhesion is observed early but disappears by the 16 h time point, whereupon the LFA-1-independent component predominates.

On T cells, crosslinking of MHC class I in the presence of PMA or anti-CD3 can augment T cell proliferation [111–115]. Crosslinking of MHC class I on T cells induced a rise in intracellular free Ca^{2+}, increased IL-2 production and increased proliferation [112,113]. The rise in intracellular Ca^{2+} is dependent on tyrosine kinase activation of PLC-γ1 [114]. On mutant Jurkat cells which lack CD3 molecules, this MHC I signalling function is lost, suggesting that CD3 is essential for transduction of activation signals via MHC class I molecules on T cells [113]. Bluestone's lab [115] identified a role for MHC I in murine T cell activation, while screening mAbs for their ability to activate T cells in the presence of PMA. One antibody, UC3-10H3 could induce IFN-γ production and proliferation of T cells in the presence of PMA. This antibody was found to react with the majority of murine class I molecules and a number of human class I molecules. In another study, Geppert et al. [116] used CD8-transfected CHO cells to deliver a costimulatory signal to anti-CD3 or PMA treated CD4+ T cells. This costimulation was blocked by anti-CD8 or anti-class I.

Other studies have suggested a negative regulatory role for CD8–class I interaction. Precursors of cytotoxic T cells (CTLp) are sensitive to deletion if at the same time that they receive a signal through their TCR they also receive a signal through MHC I, a concept known as the Veto hypothesis [117]. Tykocinski and collaborators used a CD8+T cell clone and generated a CD8− copy of this clone with antisense RNA. They showed that the CD8+ clone could down-regulate the generation of cytotoxicity in a MLR, suggesting that CD8–MHC I interaction plays a role in inactivation of CTPp [118]. Sambhara and Miller [119] showed that treatment of CTLp with antibody to the α3 domain of D^d induced apoptosis. This study was carried out using a culture supernatant. However, attempts to repeat this work with purified antibody have been unsuccessful (RG Miller, personal communication), suggesting that some other component of the supernatant also contributed.

MHC class I cytoplasmic domains

		Exon 6		Exon 7	Exon 8
			320	335	
H–2K^{b,k}	KM--RRRNT	GGKGGDYALAP		GS---QTSDLSLPDCK	VMVHDPHSLA
H–2K^d	KM---RRNT	GGKG**VN**YALAP		GS---QTSDLSLPD**GK**	VMVH**P**PHSLA
H–2D^{b,}L^d	---KRRRNT	GGKGGDYALAP		GS---Q**SSEM**SLR**D**CK	**A**
H–2D^d	---KRRRNT	GGKGGDYALAP		GS---Q**SSDM**SLPDCK	**A**
HLA–A2	**WRRKSS**---	**DR**KGG**S**Y**S**QAA		**SSDSAQGSDVSLTA**CK	V
HLA–A3	**WRRKSS**---	**DR**KGG**S**Y**T**QAA		**SSDSAQGSDVSLTA**CK	V
HLA–B7	**CRRKSS**---	GGKGG**S**Y**S**QAA		**C**SDSAQGSDVSL**TA**	
HLA–Cw3	**CRRKSS**---	**DR**KGG**SCS**QAA		**S**S_N_SAQGSD**E**SL_IT_CK	**A**

Figure 9.3 Comparison of the cytoplasmic domains, exons 6, 7 and 8, of selected MHC class I proteins. Amino acids which differ from the H-2K^b sequence are indicated in bold face. "–" indicates deletions from the sequence. Amino acids which differ between the human sequences are underlined. These sequences were taken from a compilation [137] made by Figueroa and Klein (1986) *Immunology Today*, **7**.

Role of the Cytoplasmic Domain in MHC Class I Signalling

The cytoplasmic domains of MHC class I molecules consist of 30–40 amino acids encoded in three separate exons. While exon 6 and to a lesser extent exon 7 are well conserved among the different mouse alleles, exon 8 in mouse is quite variable and can consist of a single amino acid or up to 10 amino acids (Figure 9.3) [120]. In addition, there are extensive differences between mouse and human MHC class I cytoplasmic domains. Both human and murine MHC class I can be phosphorylated on serine residues in the cytoplasmic domain, although the significance of this phosphorylation remains to be determined. The major *in vivo* site for phosphorylation of MHC class I molecules is ser 335 [121]. In addition, a conserved tyrosine at position 320 of the MHC class I cytoplasmic domain can serve as an *in vitro* substrate for tyrosine phosphorylation by Rous sarcoma virus [122]. Exon 7 of the class I cytoplasmic domain is important for endocytosis of MHC class I molecules [123].

Little is known about the tertiary structure of the cytoplasmic domains of class I molecules, however there is some evidence to suggest that the conformation or accessibility of the cytoplasmic domain is influenced by the conformation of the extracellular domains of MHC I. Smith and Bar-

ber [124] prepared anti-peptide antisera to exon 6, 7 and 8 sequences of H − 2K^b and showed that antisera to the exon 8 peptide, recognized both β2-microglobulin associated class I as well as free heavy chain whereas the exon 6 and 7 determinants were only recognized by antisera in the free heavy chain. This led the authors to suggest that the conformation of the class I extracellular domains can influence the conformation of the intracytoplasmic tail, an idea that has implications for signal transduction. Similar results have been obtained by Capps and Zuniga using antisera raised against a 13 amino acid segment of the L^d cytoplasmic domain [125].

The main function of MHC class I molecules is to present peptides to CD8$^+$, cytotoxic T lymphocytes. Upon recognition of the MHC/peptide complex, the activated CTL then delivers a "lethal hit" to the target cell, and the target subsequently dies by apoptosis. In the past there has been speculation that signal transduction via the MHC I molecules themselves might contribute to this death signal. However, there is little evidence to date to support this hypothesis [126].

A number of groups have tested the effects of cytoplasmic domain deletion on MHC I function. Zuniga *et al.* [127] deleted 24 amino acids from the carboxy terminus of L^d gene and transfected the truncated cDNA into L cells. They found no defect in recognition of the truncated L^d by allogeneic, LCMV or VSV-specific cytotoxic T cells. In contrast, Murré et al. [128] deleted 20 amino acids from the carboxy terminus of L^d and found that while recognition of the truncated molecule by allo-and influenza-specific CTL was normal, killing by VSV-specific CTL was significantly impaired. These authors also showed that truncation of the L^d cytoplasmic domain had no effect on antibody mediated capping of the class I molecules.

Lipsky's group [129] has looked at the effect of truncation of the cytoplasmic domain of human class I molecules on signal transduction in T lymphocytes. They truncated all but four of the amino acids of the cytoplasmic domain of HLA-A2 and HLA-B27 and transfected them into Jurkat cells. The truncated molecules were expressed normally on the cell surface and were indistinguishable from wildtype in terms of Ca^{2+} responses. In addition, IL-2 production induced by PMA plus anti-class I crosslinking was not affected by truncation of the class I cytoplasmic domain.

Gao et al. [130] engineered an HLA-B27 molecule so that it is anchored on the cell surface by a phosphatidyl inositol linkage. They found that cells expressing this PI-linked form of B27 could present antigenic peptide as well as endogenously synthesized viral antigen to antigen-specific CTL. Furthermore, the PI-HLA-B27 expressing cells underwent apoptosis in response to CTL interaction. In another study, Huang et al. [131] created two GPI-anchored forms of HLA-A2, by fusing the gene of HLA-A2 to the 3' end of the human DAF gene, which contains the GPI modification sequence. Two forms were engineered, differing by 53 amino acids of additional DAF sequence. Transfectants expressing the longer variant could not serve as an allotarget for CTL killing, but transfectants expressing the shorter version could. Thus if signalling via the MHC I is involved in killing of target cells by CTL, that signalling must be independent of the transmembrane and cytoplasmic domains of the class I molecule.

MHC I Association with other Signalling Molecules

As is the case for MHC II, little is known about the coupling of MHC class I molecules to signal trans-

duction pathways. MHC class I molecules have been shown to associate with insulin receptor and with the epidermal growth factor receptor, both receptors with tyrosine kinase activity [132–134]. Antibodies against MHC I can interfere with binding of insulin to its receptor and peptides derived from the α1 domain of class I molecules can influence the function of insulin receptor as well as the EGF receptor [135]. The significance of these observations to MHC I signalling is not clear.

FUTURE PROSPECTS

There is now a large amount of data showing that crosslinking of MHC proteins in lymphocytes has consequences within the cell, both in terms of early biochemical events and late events such as up-regulation of adhesion, gene expression and regulation of proliferation or cell death. In the case of MHC class II molecules, at least two types of signalling pathway have been documented: a cAMP-dependent pathway found in resting mouse B cells, as well as in some murine B lymphomas; and a tyrosine kinase pathway found in human B cells and activated mouse B cells as well as in the B lymphoma K46J (Figure 9.4). The cAMP pathway of MHC II signalling appears to require class II cytoplasmic domains, whereas the tyrosine kinase pathway appears to be independent of the cytoplasmic domains of class II.

Although numerous studies have implicated the MHC II cytoplasmic domains in T cell activation in vitro, studies with transgenic mice suggest that truncation of the cytoplasmic domain of the class II β chain does not affect immune responses in vivo. Whether further studies with weak antigens or with low doses of antigen will reveal a role for the cytoplasmic domains of MHC II in vivo remains to be determined. For MHC I, the signalling events that have been documented to date also appear to be independent of the cytoplasmic domain. Again, the significance of these signal transduction events in vivo is unknown.

Both MHC I and MHC II proteins have been shown to associate with other cell surface receptors which might explain some of their signal transduction properties. However, much remains to be learned about the way in which MHC I and II molecules couple to their respective signal trans-

Resting mouse B cells, M12 and CH12B lymphomas

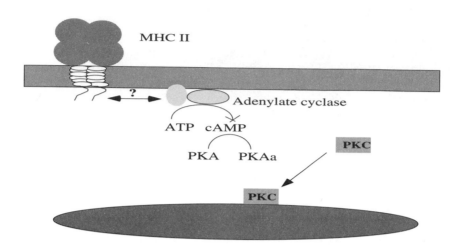

Human: B cells, activated T cells. Mouse: activated B cells, K46J lymphomas

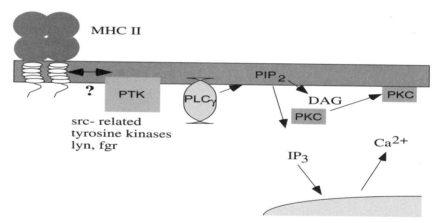

Figure 9.4 Signal transduction pathways associated with MHC class II proteins. Two different signalling pathways have been identified for MHC II signalling. One involving cAMP production and translocation of PKC to the Triton X-100 insoluble fraction, has been documented in resting mouse B cells and in M12 and CH12 murine B lymphomas. The other pathway involving tyrosine kinase activation and phosphatidyl inositol metabolism has been documented in human B and T cells, antigen and IL-4 primed murine B cells as well as the murine B lymphoma K46J.

duction pathways. Will they turn out to be associated with dedicated signalling molecules, such as the CD3 complex of the TCR or Igα and β of the BCR, or does signal transduction via MHC molecules occur via transient association with other signal transducing complexes on the T or B cell? So far co-immunoprecipitation experiments and crosslinking have failed to reveal an MHC associated signalling complex, so if one exists, it may

only be weakly associated with unaggregated MHC molecules.

Another area where much remains to be learned is in understanding how the interaction of MHC molecules with their natural ligands (SAg, TCR, CD4 or CD8) affects MHC aggregation and the initiation of the signal transducing pathways.

In conclusion, the field of signalling via MHC molecules is at a relatively early stage, where phenomenology has been established but much of the molecular detail and significance to the overall immune response remains to be elucidated.

ACKNOWLEDGEMENTS

I thank Peter Cresswell, Mark Davis, John Cambier, David Williams and Mark DeBenedette for helpful discussions and Mark DeBenedette and Marni Goldstein for critical reading of the manuscript. My research in this area is funded by the Medical Research Council of Canada and the National Cancer Institute of Canada with funds from the Canadian Cancer Society.

REFERENCES

1. Germain RN. MHC-dependent antigen processing and peptide presentation: providing ligands for T lymphocyte activation. *Cell* 1994; **76**: 287–299.

2. Bjorkman, PJ, Saper MA, Samraoui B, Bennet WS, Strominger JL and Wiley DC. Structure of the human class I histocompatibility antigen, HLA-A2. *Nature* 1987; **329**: 506–512.

3. Bjorkman, PJ, Saper MA, Samraoui B, Bennet WS, Strominger JL and Wiley DC. The foreign antigen binding site and T cell recognition of class I histocompatibility antigens. *Nature* 1987; **329**: 512–518.

4. Garret TPJ, Saper MA, Bjorkman PJ, Strominger JL and Wiley DC. Specificity pockets for the side chains of peptide antigens in HLA-Aw68. *Nature* 1989; **342**: 692–696.

5. Madden, DR, Gorga JC, Strominger JC and Wiley DC. The structure of HLA-B27 reveals nonamer self-peptides bound in an extended conformation. *Nature* 1991; **353**: 321–325.

6. Fremont DH, Matsumara M, Stura EA, Peterson PA and Wilson IA. Crystal structures of two viral peptides in complex with murine MHC class I H-2K^b. *Science* 1992; **257**: 919–927.

7. Zhang W, Young ACM, Imarai M, Nathenson SG and Sacchettini JC. Crystal structure of the major histocompatibility complex class I H-2K^b molecule containing a single viral peptide: Implications for peptide binding and T-cell receptor recognition. *Proc. Natl. Acad. Sci. USA* 1992; **89**: 8403–8407.

8. Madden DR, Garoboczi DN and Wiley DC. The antigenic identity of peptide–MHC complexes: a comparison of the conformations of five viral peptides presented by HLA-Ad. *Cell* 1993; **75**: 693–708.

9. Brown JH, Jardetsky TS, Gorga JC, Stern LJ, Urban RG, Strominger JL and Wiley DC. Three dimensional structure of the human class II histocompatibility antigen HLA-DR1. *Nature* 1993; **364**: 33–39.

10. Stern LJ, Brown JH, Jardetsky TS, Gorga JC, Urban RG, Strominger JL and Wiley DC. Crystal structure of the human class II MHC protein HLA-DR1 complexed with influenza virus peptide. *Nature* 1994; **368**: 215–221.

11. Chan AC, Desai DM and Weiss A. The role of protein tyrosine kinases and protein tyrosine phosphatases in T cell antigen receptor signal transduction. *Ann. Rev. Immunol.* 1994; **12**: 555–592.

12. Herman A, Kappler JW, Marrack P and Pullen A. Superantigens: mechanisms of T-cell stimulation and role in Immune responses. *Ann. Rev. Immunol.* 1991; **9**: 745–772.

13. Baluyut A and Subbarao B. The synergistic effects of anti-IgM and monoclonal anti-Ia antibodies in induction of murine B lymphocyte activation. *J. Mol. Cell. Immunol.* 1988; 4: 45–57.

14. Bishop GA. Requirements of class II-mediated B cell differentiation for class II cross-linking and cyclic AMP. *J. Immunol.* 1991; **147**: 1107–1114.

15. Newell MK, Vanderwall J, Beard KS and Freed JH. Ligation of major histocompatibility complex class II molecules mediates apoptotic death in resting B lymphocytes. *Proc. Natl. Acad. Sci. USA* 1993; **90**: 10459–10463.

16. Mourad W, Geha RS and Chatila T. Engagement of major histocompatibility complex class II molecules induces sustained lymphocyte function associated molecule 1-dependent cell adhesion. *J. Exp. Med.* 1990; **172**: 1513–1516.

17. Matko J, Bushkin Y, Wei T and Edidin M. Clustering of class I HLA molecules on the surfaces of activated and transformed human T cells. *J. Immunol.* 1994; **152**: 3353–3360.

18. Wade WF, Freed JH and Edidin M. Translation diffusion of class II major histocompatibility complex molecules is constrained by their cytoplasmic domains. *J. Cell. Biol.* 1989; **109**: 3325–3331.

19. Edidin M, Zuniga MC and Sheetz MP. Truncation mutants define and locate cytoplasmic barriers to lateral mobility of membrane glycoproteins. *Proc. Natl. Acad. Sci. USA* 1994; **91**: 3378–3382.

20. St-Pierre Y and Watts TH. Characterization of the signalling function of MHC class II molecules during antigen presentation by B cells. *J. Immunol.* 1991; **147**: 2875–2882.

21. Schafer PH and Pierce SK. Evidence for dimers of MHC class II molecules in B lymphocytes and their role in low affinity T cell responses. *Immunity* 1994; **1**: 699–707.

22. Lamb CA and Cresswell P. Assembly and transport properties of invariant chain trimers and HLA-DR-invariant chain complexes. *J. Immunol.* 1992; **148**: 3478–3482.

23. Damaj B, Mourad W and Naccache PH. Superantigen-mediated human monocyte-T lymphocyte interactions are associated with an MHC class II-, TCR/CD3-and CD4-dependent mobilization of calcium in monocytes. *J. Immunol.* 1992; **149**: 1497–1503.

24. Scholl PR, Trede N, Chatila TA and Geha RS. Role of protein tyrosine phosphorylation in monokine induction by the staphylococcal superantigen toxic shock syndrome toxin 1. *J. Immunol.* 1992; **148**: 2237–2241.

25. Swaminathan S, Furey W, Pletcher J and Sax M. Crystal structure of staphylococcal enterotoxin B, a superantigen. *Nature* 1992; **359**: 801–805.

26a. Jardetsky TS, Brown JH, Gorga JC, Stern LJ, Urban RG, Chi Y-i, Stauffacher C, Strominger JL and Wiley DC. Three dimensional structure of a human class II histocompatibility molecule complexed with superantigen. *Nature* 1994; **368**: 711–718.

26b. Mehindate K, Thibodeau J, Dohlsten M, Kalland T, Sékaly R-P, and Mourad W. Cross-linking of major histocompatibility complex class II molecules by Staphylococcal Enterotoxin A superantigen is a requirement for inflammatory cytokine gene expression. *J. Exp. Med.* 1995; 1573–1577.

27. Niederhuber JE, Frelinger JA, Dugan E, Coutinho A, Shreffler DC. Effects of anti-Ia serum on mitogenic responses. I. Inhibition of proliferative response to B cell mitogen, LPS, by specific anti-Ia sera. *J. Immunol.* 1975; **115**: 1672–1676.

28. Forsgren S, Poboer G, Coutinho A and Pierres M. The role of I-A/E molecules in B lymphocyte activation I. Inhibition of lipopolysaccharide-induced responses by monoclonal antibodies. *J. Immunol.* 1984; **133**: 2104–2110.

29. Cambier JC and Lehmann KR. Ia-mediated signal transduction leads to proliferation of primed B lymphocytes. *J. Exp. Med.* 1989; **170**: 877–886.

30. Fuleihan R, Mourad W, Geha R and Chatila T. Engagement of MHC-class II molecules by staphylococcal exotoxins delivers a comitogenic signal to human B cells. *J. Immunol.* 1991; **146**: 1661–1666.

31. Bishop GA, Ramirez LM and Koretzky GA. Growth inhibition of a B cell clone mediated by ligation of IL-4 receptors or membrane IgM. *J. Immunol.* 1993; **150**: 2565–2574.

32. Bishop GA and Haughton G. Induced differentiation of a transformed clone of Ly-1+B cells by clonal T cells and antigen. *Proc. Natl. Acad. Sci. USA* 1986; **83**: 7410–7414.

33. Bishop GA and Frelinger JA. Haplotype-specific differences in signalling by transfected class II molecules to a Ly-1+B-cell clone. *Proc. Natl. Acad. Sci. USA* 1989; **86**: 5933–5937.

34. Bishop, GA, McMillan MS, Haugton G and Frelinger JA. Signalling to a B-cell clone by E^k, but not A^k, does not reflect alteration of A^k genes. *Immunogenetics* 1988; **28**: 184–192.

35. Bishop GA, Berton MT and Warren WT. Signalling via MHC class II molecules and antigen receptors enhances the B cell response to gp39/CD40 ligand. *Eur. J. Immunol.* 1995 (in press).

36. Rock KL, Benacerraf B and Abbas AK. Antigen presentation by hapten-specific B lymphocytes. I. Role of surface immunoglobulin receptors. *J. Exp. Med.* 1984; **160**: 1102–1113.

37. Tony H-P and Parker DC. Major histocompatibility complex-restricted polyclonal B cell responses resulting from helper T cell recognition of antiimmunoglobulin presented by small B lymphocytes. *J. Exp. Med.* 1985; **161**: 223–241.

38. Lanzavecchia A. Antigen-specific interaction between T and B cells. *Nature* 1985; **314**: 537–559.

39. Parker DC. T cell-dependent B cell activation. *Ann. Rev. Immunol.* 1993; **11**: 331–360.

40. Banchereau J, Bazan F, Blanchard D, Briére F, Galizzi JP, van Kooten C, Liu YJ, Rousset F and Saeland S. The CD40 antigen and its ligand. *Ann. Rev. Immunol.* 1994; **12**: 881–922.

41. Truman JP, Ericson ML, Choqueax-Séébold CJM, Charron DJ and Mooney NA. Lymphocyte programmed cell death is mediated via HLA class II DR. *Int. Immunol.* 1994; **6**: 887–896.

42. Grusby MJ, Hohnson RS, Papaioannou VE and Glimcher LH. Depletion of CD4+ cells in major histocompatibility complex class II-deficient mice. *Science* 1991; **253**: 1417–1420.

43. Markowitz JS, Rogers PR, Grusby MJ, Parker DC and Glimcher LH. B lymphocyte development and activation independent of MHC class II expression. *J. Immunol.* 1993; **150**: 1223–1233.

44. Odum N, Martin PJ, Schieven GL, Maewicz S, Hansen S, Hansen JA and Ledbetter, JA. HLA-DR molecules enhance signal transduction through the CD3/Ti complex in activated T cells. *Tissue antigens* 1991; **38**: 72–77.

45. Spertini F, Chatila T and Geha RS. Signals delivered via MHC class II molecules synergize with signals delivered via TCR/CD3 to cause proliferation and cytokine gene expression in T cells. *J. Immunol.* 1992; **149**: 65–70.

46. Odum N, Kanner SB, Ledbetter JA and Svejgaard A. MHC class II molecules deliver costimulatory signals in human T cells through a functional linkage with IL-2 receptors. *J. Immunol.* 1993; **150**: 5289–5298.

47. Moretta A, Acolla RS and Cerottini JC. IL-2 mediated T cell proliferation in humans is blocked by monoclonal antibody directed against monomorphic HLA-DR antigens. *J. Exp. Med.* 1982; **155**: 599–604.

48. Racioppi L, Moscarella A, Ruggiero G, Manzo C, Ferrone S, Fontana S and Zappacosta S. Inhibition by anti-HLA class II monoclonal antibodies of monoclonal antibody OKT3 induced T cell proliferation: studies at the mRNA level. *J. Immunol.* 1990; **145**: 3635–3640.

49. Dustin ML and Springer TA. T cell receptor cross linking transiently stimulates adhesiveness through LFA-1. *Nature* 1989; **341**: 619–624.

50. Mourad W, Geha RS and Chatila T. Engagement of major histocompatibility complex class II antigens induces sustained, lymphocyte function-associated molecule 1-dependent cell adhesion. *J. Exp. Med.* 1990; **172**: 1513–1516.

51. Odum N, Ledbetter JA, Martin P, Geraghty D, Tsu T, Hansen JA and Gladstone P. Homotypic aggregation of human cell lines by class II-, class Ia-and HLA-G specific monoclonal antibodies. *Eur. J. Immunol.* 1991; **21**: 2121–2131.

52. Alcover A, Juillard V and Acuto O. Engagement of major histocompatibility complex class I and class II molecules up-regulates intercellular adhesion of human B cells via CD11/CD18-independent mechanism. *Eur. J. Immunol.* 1992; **22**: 405–412.

53. Kansas GS and Tedder TF. Transmembrane signals generated through MHC class II, CD19, CD20, CD39, and CD40 antigens induce LFA-1-dependent and independent adhesion in human B cells through a tyrosine kinase-dependent pathway. *J. Immunol.* 1992; **147**: 4094–4102.

54. Ramirez R, Carracedo J, Mooney N and Charron D. HLA class II-mediated homotypic aggregation: involvement of protein tyrosine kinase and protein kinase C. *Hum. Immunol.* 1992; **34**: 115–125.

55. Kansas GS, Cambier JC and Tedder TF. CD4 binding to major histocompatibility complex class II antigens induces LFA-1 dependent and independent homotypic adhesion of B lymphocytes. *Eur. J. Immunol.* 1992; **22**: 147–152.

56. Wagner N, Engel P and Tedder TF. Regulation of the tyrosine kinase-dependent adhesion pathway in human lymphocytes through CD45. *J. Immunol.* 1993; **150**: 4887–4899.

57. Wagner N, Engel P, Vega M and Tedder TF. Ligation of MHC class I and class II molecules can lead to heterologous desensitization of signal transduction pathways that regulate homotypic adhesion in human lymphocytes. *J. Imunol.* 1994; **152**: 5275–5287.

58. June CH, Bluestone JA, Nadler LM and Thompson CB. The B7 and CD28 receptor families. *Immunol. Today* 1994; **15**: 321–331.

59. Kouvlova L, Clark EA, Shu G and Dupont B. The CD28 ligand B7/BB1 provides costimulatory signal for alloactivation of CD4+ T cells. *J. Exp. Med.* 1991; **173**: 759–762.

60. Hathcock KS, Laszlo G, Pucillo C, Linsley P and Hodes RJ. Comparative analysis of B7-1 and B7-2 costimulatory ligands: expression and function. *J. Exp. Med.* 1994; **180**: 631–640.

61. Nabavi N, Freeman GH, Gault A, Godfrey D, Nadler LM and Glimcher LH. Induction of costimulatory molecule B7 in M12 B lymphomas by cAMP or MHC-restricted T cell interaction. *Nature* 1993; **360**: 266–268.

62. Watts TH, Alaverdi N, Wade WF and Linsley PS. Induction of costimulatory molecule B7 in M12 B lymphomas by cAMP or MHC-restricted T cell interaction. *J. Immunol.* 1993; **150**: 2192–2202.

63. Smiley ST, Laufer TM, Lo D, Glimcher LH and Grusby MJ. Transgenic mice expressing MHC class II molecules with truncated Ab cytoplasmic domains reveal signalling-independent defects in antigen presentation. *Int. Immunol.* 1995; **7**: 665–677.

64. Poudrier J and Owens T. CD54/intercellular adhesion molecule 1 and major histocompatibility complex II signalling induces B cells to express interleukin 2 receptors and complements help provided through CD40 ligation. *J. Exp. Med.* 1994; **179**: 1417–1427.

65. Palacios R. Monoclonal antibodies against human Ia antigens stimulate monocytes to secrete interleukin 1. *Proc. Natl. Acad. Sci. USA* 1985; **82**: 6652–6656.

66. Palkama T, Sihvola M and Hurme M. Induction of interleukin 1α (IL-1α) and IL-1β expression and cellular IL-1 production by anti-HLA-DR antibodies in human monocytes. *Scand. J. Immunol.* 1989; **29**: 609–615.

67. Trede NS, Geha RS and Chatila T. Transcriptional activation of IL-1β and tumour necrosis factor-α genes by MHC class II ligands. *J. Immunol.* 1991; **146**: 2310–2315.

68. Trede NS, Castigli E, Geha RS and Chatila TA. Microbial superantigens induce NF-kB in human monocytic cell line THP-1. *J. Immunol.* 1993; **150**: 5604–5613.

69. Mehindate K, Al-Daccak R, Schall TJ and Mourad W. Induction of chemokine gene expression by major histocompatibility complex class II ligands in human fibroblast- like sinoviocytes. *J. Biol. Chem.* 1994; **269**: 32063–32069.

70. Cambier JC, Newell MK, Justement LB, McGuire JC, Leach KL and Chen ZZ. Ia binding ligands and cAMP stimulate nucelar translocation of PKC in B lymphocytes. *Nature* 1987; **327**: 629–632.

71. Lane PJL, McConnell FM, Schieven GL, Clark EA and Ledbetter JA. The role of class II molecules in human B cell activation. Association with phosphatidyl inositol turnover, protein tyrosine phosphorylation and proliferation. *J. Immunol.* 1990; **144**: 3684–3692.

72. Mooney NA, Grillot-Courvalin C, Hivroz C, Ju L-Y and Charron D. Early biochemical events after MHC class II-mediated signalling on human B lymphocytes. *J. Immunol.* 1990; **145**: 2070–2076.

73. Cambier JC, Morrison DC, Chien MM and Lehman KR. Modelling of T cell contact-dependent B cell activation. IL-4 and antigen receptor ligation primes quiescent B cells to mobilize calcium in response to Ia cross-linking. *J. Immunol.* 1991; **146**: 2075–2082.

74. Brick-Ghannam C, Huang FL, Temime N and Charron D. Protein kinase C (PKC) activation via human leukocyte antigen class II molecules. *J. Biol. Chem.* 1991; **266**: 24169–24175.

75. Brick-Ghannam C, Ericson EL, Schelle I and Charron D. Differential regulation of mRNAs encoding prtein kinase C isoenzumes in activated human B cells. *Hum. Immunol.* 1994; **41**: 216–224.

76. Kim KJ, Kanellopoulos-Langevin C, Merwin RM, Sachs DM and Asofsky R. Establishment and characterization of Balb/c lymphoma lines with B cell properties. *J. Immunol.* 1979; **122**: 549–554.

77. Wade WF, Chen ZZ, Maki R, McKercher S, Palmer E, Cambier JC and Freed JH. Altered I-A protein-mediated transmembrane signalling in B cells that express truncated I-A^k protein. *Proc. Natl. Acad. Sci. USA* 1989; **86**: 6297–6301.

78. Nabavi N, Ghogawala Z, Meyer A, Griffith IJ, Wade WF, Chen ZZ, McKean DJ and Glimcher LH. Antigen presentation abrogated in cells expressing truncated Ia molecules. *J. Immunol.* 1989; **142**: 1444–1447.

79. André P, Cambier JC, Wade TK, Raetz T and Wade WF. Distinct structural compartmentalization of the signal transduction functions of major histocompatibility complex class II (Ia) molecules. *J. Exp. Med.* 1994; **179**: 763–768.

80. Morio T, Geha RS and Chatila TA. Engagement of MHC class II molecules by staphylococcal superantigens activates src-type protein tyrosine kinases. *Eur. J. Immunol.* 1994; **24**: 651–658.

81. Odum N, Martin PJ, Schieven GL, Norris NA, Grosmaire LS, Hansen JA, Ledbetter JA. Signal transduction by HLA-DR is mediated by tyrosine kinase(s) and regulated by CD45 in activated T cells. *Hum. Immunol.* 1991; **32**: 85–94.

82. Kanner SB, Odum N, Grosmaire L, Masewicz S, Svejgaard A and Ledbetter JA. Superantigen and HLA-DR ligation induce phospholipase-cγ1 activation in class II+ T cells. *J. Immunol.* 1992; **149**: 3482–3488.

83. Senju S, Kimura A, Yasunami M, Kamikawaji N, Yoshizumi H, Nishimura Y and Sasazuki T. Allele-specific expression of the cytoplasmic exon of the HLA-DQβ1 gene. *Immunogenetics* 1992; **36**: 319–325.

84. Griffith IJ, Ghogawala Z, Nabavi N, Golan DE, Myer A, McKean DJ and Glimcher LA. Cytoplasmic domain affects membrane expression and function of an Ia molecule. *Proc. Natl. Acad. Sci. USA* 1988; **85**: 4847–4851.

85a. Harton JA and Bishop GA. Length and sequence requirements of the cytoplasmic domain of the Ab molecule for class II-mediated B cell signalling. *J. Immunol.* 1993; **151**: 5282–5289.

85b. Harton JA, Van Hagen AE, and Bishop GA. The cytoplasmic and transmembrane domains of MHC class IIβ chains deliver distinct signals required for MHC class II-mediated B cell activation. *Immunity.* 1995; **3**: 349–358.

86. St-Pierre Y, Nabavi N, Ghogawala Z, Glimcher LH and Watts TH. A functional role for signal transduction via the cytoplasmic domains of MHC class II proteins. *J. Immunol.* 1989; **143**: 808–812.

87. Freeman GJ, Borriello F, Hodes RJ, Reiser H, Gribben JG, Ng JW, Kim J, Goldberg JM, Hathcock K, Laszlo G, Lombard LA, Wang S, Gray GS, Nadler LM and Sharpre AH. Murine B7-2, an alternative CTLA4 counter-receptor that costimulates T cell proliferation and interleukin 2 production. *J. Exp. Med.* 1993; **178**: 2185–2192.

88. Durie RH, Foy TM, Masters SR, Laman JD and Noelle RJ. The role of CD40 in regulation of humoral and cell mediated immunity. *Immunol. Today* 1994; **13**: 431–436.

89. Ranheim EA and Kipps TJ. Activated T cells induce expression of B7/BB1 on normal or leukemic B cells through a CD40-dependent signal. *J. Exp. Med.* 1993; **177**: 925–935.

90. Knox KA, Johnson GD and Gordon J. Distribution of cAMP in secondary follicles and its expression in B cell apoptosis and CD40-mediated survival. *Int. Immunol.* 1993; **5**: 1085–1091.

91. Kato T, Kokuho T, Tamura T and Nariuchi H. Mechanisms of T cell contact-dependent B cell activation. *J Immunol* 1994; **152**: 2130–2138.

92. Ostrand-Rosenberg S, Roby CA, Clements VK. Abrogation of tumorigenicity by MHC class II antigen expression requires the cytoplasmic domain of the class II molecule. *J. Immunol.* 1991; **147**: 2419–2422.

93. Baskar S, Ostrand-Rosenberg S, Nabavi N, Nadler LM, Freeman GJ and Glimcher LH. Constitutive expression of B7 restores immunogenicity of tumor cells expressing truncated major histocompatibility complex class II molecules. *Proc. Natl. Acad. Sci. USA* 1993; **90**: 5687–5690.

94. Wade WF, De Pirro Ward E, Rosloniec EF, Barisas BG and Freed JH. Truncation of the Aα chain of MHC class II molecules results in inefficient antigen presentation to antigen-specific T cells. *Int. Immunol.* 1994; **6**: 1457–1465.

95. DeBenedette MA, Chu NR, Pollok KE, Hutado J, Wade WF, Kwon BS and Watts TH. Role of 4-1BB ligand in costimulation of T lymphocyte growth and its upregulation on M12 B lymphomas by cAMP. *J. Exp. Med.* 1995; **181**: 985–992.

96. Busch R and Mellins ED. Developing and shedding inhibitions: How MHC class II molecules reach maturity. *Curr. Op. in Immunol.* 1996; **8**: 51–58.

97. Sun H, Seyer JM and Patel TB. A region in the cytosolic domain of the epidermal growth factor receptor antithetically regulates the stimulatory and inhibitory guanine nucleotide-binding regulatory proteins of adenylate cyclase. *Proc. Natl. Acad. Sci. USA* 1995; **92**: 2229–2233.

98. Cooper DMF, Mons N and Karpen JW. Adenylyl cyclases and the interaction between calcium and cAMP signalling. *Nature* 1995; **374**: 421–424.

99. Newell MK, Justement LB, Miles CR and Freed JR. Biochemical characterization of proteins that co-purify

100. Bonnefoy JY, Guillot O, Spits H, Blanchard D, Ishizaka K and Banchereau J. The low affinity receptor for IgE (CD23) on B lymphocytes is spatially associated with HLA-DR antigens. *J. Exp. Med.* 1988; **167**: 57–72.

101. Dickler HB and Sachs DH. Evidence for identity or closs association of the Fc receptor of B lymphocytes and alloantigens determined by the Ir region of the H-2 complex. *J. Exp. Med.* 1974; **140**: 779–796.

102. Schirrrmacher V, Halloran P and David CS. Interactions of Fc receptors with antigodies against Ia antigens and other cell surface components *J. Exp. Med.* 1975; **141**: 1201–1209.

103. Basten A, Miller JFAP and Abraham R. Relationship between Fc receptors, antigen-binding sites on T and B cells, and H-2 complex-associated determinants. *J. Exp. Med.* 1975; **141**: 547–560.

104. Sugie K, Kawakami T, Maeda Y, Kawabe T, Uchida A and Yodoi J. Fyn tyrosine kinase associated with FceRII/CD23: possible multiple roles in lymphocyte activation. *Proc. Natl. Acad. Sci. USA* 1991; **88**: 9132–9135.

105. Schick MR and Levy S. The TAPA-1 molecule is associated on the surface of B cells with HLA-DR molecules. *J. Immunol.* 1993; **151**: 4090–4097.

106. Angelisova P, Hilgert I and Horejsi V. Association of four antigens of the tetraspans family (CD37, CD53, TAPA-1, and R2/C33) with MHC class II glycoproteins. *Immunogenetics* 1994; **39**: 249–256.

107. Wright MD and Tomlinson, MG. The ins and outs of the transmembrane 4 superfamily. *Immunol. Today* 1994; **15**: 588–594.

108. Takahashi S, Doss C, Levy S and Levy R. Tapa-1 the target of an antiproliferative antibody is associated on the cell surface with the leu-13 antigen. *J. Immunol.* 1990; **145**: 2207–2213.

109. Bradbury LE, Kansas GS, Levy S, Evans RL and Tedder TF. The CD19/CD21 signal transducing complex of human B lymphocytes includes the target of antiproliferative antibody-1 and leu-13 molecules. *J. Immunol.* 1992; **149**: 2841–2850.

110. Spertini F, Chatila T and Geha RS. Engagement of MHC class I molecules induces cell adhesion via both LFA-1-dependent and LFA-1-independent pathways. *J. Immunol.* 1992; **148**: 2045–2049.

111. Geppert TD, Wacholtz MC, Davis CS and Lipsky PE. Activation of human T4 cells by cross-linking class I MHC molecules. *J. Immunol.* 1988; **140**: 2155–2164.

112. Geppert TD, Wacholtz MC, Patel SS, Lightfoot E and Lipsky PE. Activation of human T cell clones and Jurkat cells by cross-linking class I MHC molecules. *J. Immunol.* 1989; **142**: 3763–3772.

113. Wacholtz MC, Patel SS and Lipsky PE. Patterns of costimulation by cross-linking CD3, CD4/8 and class I MHC molecules. *J. Immunol.* 1989; **142**: 4201–4212.

114. Skov S, Odum N and Claesson MH. MHC class I signalling in T cells leads to tyrosine kinase activity and PLC-gl phosphorylation. *J. Immunol.* 1995; **154**: 1167–1176.

115. Houlden BA, Widacki SM and Bluestone JA. Signal transduction through class I MHC by a monoclonal anti-

body that detect multiple murine and human class I molecules. *J. Immunol.* 1991; **146**: 425–430.

116. Geppert TD, Nguyen H and Lipsky PE. Engagement of class I major histocompatibility complex molecules by cell surface CD8 delivers an activation signal. *Eur. J. Immunol.* 1992; **22**: 1379–1383.

117. Fink PJ, Shimonkevitz RP and Bevan MJ. Veto cells. *Ann. Rev. Immunol.* 1988; **6**: 115–137.

118. Kaplan DR, Hambor JE and Tykocinski ML. An immunoregulatory function for the CD8 molecule. *Proc. Natl. Acad. Sci. USA* 1989; **86**: 8512–8515.

119. Sambhara SR and Miller RG. Programmed cell death of T cells signaled by the T cell receptor and the α3 domain of class I MHC. *Science* 1991; **252**: 1424–1427.

120. Kimball ES and Coligan JE. Structure of class I major histocompatibility antigens. *Contemp. Topics Mol. Immunol.* 1983; **9**: 1–63.

121. Guild BC and Strominger JL. Human and murine class I MHC antigens share conserved serine 335, the site of phosphorylation *in vivo. J. Biol. Chem.* 1984; **259**: 9235–9240.

122. Guild BC, Erikson RL and Strominger JL. HLA-A2 and HLA-B7 antigens are phosphorylated *in vitro* by rous sarcoma virus (pp60v-src) at a tyrosine residues encoded in a highly conserved exon of the intracellular domain. *Proc. Natl. Acad. Sci. USA* 1983; **80**: 2894–2898.

123. Vega MA and Strominger JL. Constitutive endocytosis of HLA class I antigens requires a specific portion of the intracytoplasmic tail that shares structural features with other endocytosed molecules. *Proc. Natl. Acad. Sci. USA* 1989; **86**: 2688–2692.

124. Smith MH and Barber BH. The conformational flexibility of class I H-2 molecules as revealed by anti-peptide antibodies specific for intracytoplasmic determinants: differential reactivity of β2-microglobulin "bound" and "free" H-2K^b heavy chains. *Mol. Immunol* 1990; **27**: 169–180.

125. Capps CG and Zuniga MC. The cytoplasmic domain of the H-2L^d class I major histocompatibility complex molecule is differentially accessible to immunological and biochemical probes during transport to the cell surface. *J. Biol. Chem.* 1993; **268**: 21263–21270.

126. Berle G. The CTL's kiss of death. *Cell* 1995; **81**: 9–12.

127. Zuniga MC, Malissen B, McMillan M., Brayton PR, Clark SS, Forman J and Hood L. Expression and function of transplantation antigens with altered or deleted cytoplasmic domains. *Cell* 1983; **34**: 535–544.

128. Murré C, Reiss CS, Bernabeu C, Chen LB, Burakoff SJ and Seidman JG. Construction, expression and recognition of an H-2 molecule lacking its carboxyl terminus. *Nature* 1984; **307**: 432–436.

129. Gur H, El-Zaatari F, Geppert TD, Wacholtz MC, Taurog JD and Lipsky PE. Analysis of T cell signalling by class I MHC molecules: The cytoplasmic domain is not required for signal transduction. *J Exp Med* 1990; **172**: 1267–1270.

130. Gao X-M, Quinn CL, Bell JI and McMichael AJ. Expression and function of HLA-B27 in lipid-linked form: implications for cytotoxic T lymphocyte induced apoptosis signal transduction. *Eur. J. Immunol.* 1993; **23**: 653–658.

131. Huang JH, Greenspan NS and Tycocinski ML. 1994. Alloantigenic recognition of artificial glycosyl phosphatidylinositol-anchored HLA-A2.1. *Mol. Immunol.* 1994; **31**: 1017–1028.

132. Schreiber AB, Schlessinger J and Edidin M. Interaction between major histocompatibility complex antigens and epidermal growth factor receptors on human cells. *J. Cell. Biol.* 1984; **98**: 725–31.

133. Due C, Simonsen M and Olsson L. The major histocompatibility complex class I heavy chain as a structural subunit of the human cell membrane insulin receptor: implications for the range of biological functions of histocompatibility antigens. *Proc. Natl. Acad. Sci. USA* 1986; **83**: 6007–6011.

134. Phillips ML, Moule ML, Delovitch TL and Yip CC. Class I histocompatibility antigens and insulin receptors: evidence for interactions. *Proc. Natl. Acad. Sci.* 1986; **83**: 3474–3478.

135. Stagsted J, Ziebe S, Satoh S, Holman GD, Cushman SW and Olsson L. Insulinomimetic effect on glucose transport by epidermal growth factor when combined with a major histocompatibility complex class I-derived peptide. *J. Biol. Chem.* 1993; **268**: 1770–1774.

136. Figueroa F and Klein K. The evolution of MHC class II genes. *Immunol. Today* 1986; **7**: 78–79.

137. Figueroa F and Klein K. The evolution of MHC class I genes. *Immunol. Today* 1986; **7**: 76–77.

Part III

Life or Death: Signalling through B and T Cell Development

10

Scenes from a Short Life: Checkpoints and Progression Signals for Immature B Cell Life Versus Apoptosis

David W. Scott, Sergei Ezhevsky, Bourke Maddox, Karen Washart, Xiao-rui Yao and Yufang Shi

Immunology Department, Holland Laboratory, American Red Cross, Rockville, Maryland, USA

INTRODUCTION

Historically, tolerance is induced more readily in neonatal animals. A prediction of Burnet's clonal selection theory [1], the ease of tolerance induction in immature animals has been a reproducible finding in both B cells and T cells. The first definitive studies in this area were performed in 1976 by Metcalf and Klinman [2]. Using a splenic focus system, they found a dose-dependent decrease in antibody forming cell precursors with neonatal murine B cells after treatment with haptenated tolerogens, whereas adult B cells appeared to be resistant. Our laboratory provided evidence that this unresponsiveness may be due to the elimination of neonatal B cells since antigen binding cells (ABC) disappeared from the spleens of neonatal rats given a tolerogenic injection of haptenated IgG, whereas ABC persisted and re-bound antigen in tolerized adult animals; these "tolerant" cells in the adult were considered to be anergic since they persisted but were unable to respond to challenge [3]. Pike and coworkers [4] showed that the decision between clonal death and life was dose related: high doses of antigen led to the loss of B cells, whereas lower doses merely arrested their development; although these authors termed the latter process "clonal abortion", it was more akin to anergy. Hindsight interpretation of the data of Cooper and colleagues [5] also suggests that mice, treated from birth with anti-IgM antibodies, had undergone a polyclonal deletion of all B cells; indeed, *in vitro* treatment of neonatal B cells with

anti-IgM leads to a loss of responsiveness to mitogens like LPS [6,7]. Now, we know that they die via apoptosis (see below).

One of the model systems developed by our group and others to study this process has been a series of B cell lymphomas engendered in a variety of mouse strains during the last two decades. Death of so-called "immature" lymphoma cells had been reported as early as 1979 by Peter Ralph [8] and has been repeatedly confirmed by numerous investigators. The initial observation in these studies is that crosslinking of surface IgM, but not IgD, leads to growth arrest in these cells, followed by the loss of viability of the vast majority of these cell [9,10]. Most investigators type these cells as immature based on the fact that this functional response to anti-IgM is similar to that of neonatal B cells, cited above, although in fact the phenotype of these lymphomas ranges from that which we would predict for virgin, IgM only B cells to much more mature IgM/IgD expressing B cells in the adult. Thus, one should be cautious about interpreting phenotypic characterization of transformed cells. Nonetheless, ligation of surface IgM leads to the death of a subset of these lymphomas via apoptosis.

As will be discussed below, our laboratory demonstrated that the induction of apoptosis in immature B lymphomas is a consequence of crosslinking of IgM only in early G_1 [11]. That this led to apoptosis was demonstrated by Benhamou *et al.* [12] and Hasbold and Klaus [13], although the appearance of condensed nuclei, cytoplasmic vacuolization and nuclear membrane dissolution were observed as early as 1987 [14]. Indeed, the appearance of apoptotic bodies even by light microscopy was apparent to all groups working with these cells. Since a number of reviews have appeared recently on this subject [7,15,16,17], our goal herein is to present an overview of this process. We will propose hypotheses which may explain the checkpoints in the cell cycle that are critical in controlling if a B cell goes on to live (whether it is anergic or not) or if it dies by apoptosis. A critical part of this hypothesis is that as cells progress through the cycle, certain checkpoints must be passed or else the cell is committed to die. That is, the only thing certain about life (other than taxes) is death; hence, cells go through life trying to prevent death.

One of the first observations relevant to this point is that when either B cells or T cells are explanted from the host, they usually die within hours. As cellular immunologists, we always had assumed that the loss of cell viability in culture was a normal process resulting from the lack of appropriate nutrients. In fact, the process may be occurring quite naturally *in vivo*, but the efficient macrophage uptake of cells in the early stages of apoptosis may eliminate our detection of this process. Dilution of the cells *in vitro* would permit a greater number of cells "to be caught in the act of dying". Nonetheless, there are a number of controls that involve cell cycle regulation that may prevent this from occurring and these will be discussed below.

were not able to detect an increase in apoptosis as measured by DNA laddering, due to the high background in the control spleen cultures. In contrast, direct flow cytometric analysis by Brown *et al.* [18] demonstrated that neonatal B cells indeed underwent apoptosis after treatment with anti-IgM *in vitro*, as they appear to do *in vivo*. Recent data from Monroe's laboratory [19] have provided a detailed confirmation of this process although, in their studies, they suggest that qualitative differences may exist between neonatal and adult B cell tolerance. The thesis of this chapter, however, is that both neonatal and adult B cells undergo apoptosis through qualitatively similar pathways, but that the quantitative thresholds for such signaling are significantly different.

The experiments of Parry *et al.* [20], as well as our own experience [21], indicate that adult B cells also undergo apoptosis when their receptors are sufficiently crosslinked by anti-IgM *in vitro*. Consistent with the exquisite sensitivity of neonatal B cells in tolerance induction [2,4], we found that the doses of anti-IgM required to induce apoptosis were 10–100 fold higher in adult versus neonatal B cells. This is equally true when one considers B-lymphoma apoptosis (below). Carsetti *et al.* [22], using Ig transgenic mice specific for the TNP hapten, demonstrated that this could occur *in vivo* or *in vitro* using specific antigen; recent data suggest that a subset of adult B cells undergoes apoptosis *in vivo* and that these appear to be transitional cells that are IgM only, though not necessarily the most immature B cells [23].

APOPTOSIS IN B CELL SUBSETS: A HYPOTHESIS

Do neonatal B cells undergo apoptosis more readily than adult B cells? The answer to this question is an unequivocal "yes". However, we propose that this is primarily a quantitative difference rather than a physiologically qualitative control process. What is the evidence for neonatal B cell apoptosis after treatment with anti-IgM? Several years ago, to extend previous descriptive results [5], we demonstrated that *in vivo* treatment with anti-IgM led to the disappearance of IgM positive and B220 positive cells within 24–36 hours [7]. However, we

EARLY SIGNALS FOR APOPTOSIS IN B-CELL LYMPHOMAS: ROLE OF TYROSINE PHOSPHORYLATION IN GROWTH ARREST AND APOPTOSIS

My laboratory has studied the WEHI-231 line as a model for unresponsiveness, as has been done by several other groups [9,12–14,24,25]. In addition, we have focused on two of the CH series isolated by the Haughton lab [26] (CH31 and CH33) that show the same sensitivity to growth arrest by anti-IgM crosslinking as WEHI-231; an IgD-transfected derivative of CH33, called ECH408, is also anti-IgM sensitive [27] as would be expected. As stated above, the inhibition of growth and subsequent apoptosis is typical of these cells, which have very little in common phenotypically, other than high expression of surface IgM. What is most typically seen with these cells, however, is that treatment with anti-IgM leads to a decrease in cell size, an increase in buoyant density, and a loss of cells in S phase, with an accumulation of cells in G_1 (see below and references [9–11] and [28]). Subsequently, they undergo apoptosis.

The initiating steps in activation-induced apoptosis in B-cell lymphomas require tyrosine phosphorylation, as shown with pharmacologic inhibitors and antisense oligonucleotides [29–31]. Indeed, recent studies indicate that growth arrest and apoptosis are dependent on the function of two of the *src*-family of protein tyrosine kinases (PTK). Yao *et al.* [32] correlated the ability to undergo anti-IgM-induced apoptosis with the activation of the blk kinase in clones of CH31 B-lymphoma cells, and extended these observations through the use of antisense oligonucleotides specific for *blk* [30]. In fact, we were able to block cell cycle arrest *and* apoptosis in the CH31 line with antisense against *blk*. In the BCL$_1$ lymphoma, antisense oligonucleotides were found to specifically implicate fyn in anti-IgM mediated growth arrest, but not apoptosis [31]. This suggests that growth arrest and apoptosis can be separated in BCL$_1$, but not in CH31. This is important since it may allow one to dissect these processes from each other.

Because the Ig receptors on B cells lack a cytoplasmic domain, there must be associated molecules responsible for signaling via the *src* kinases. These coreceptors, termed Igα and Igβ, are required both for surface expression and PTK signaling for growth and apoptosis. Importantly, the cytoplasmic domains of Igα and Igβ subunits each has one copy of a consensus sequence, now called the Immunoreceptor Tyrosine-based Activation Motif (ITAM), which is conserved in signaling subunits of a variety of other receptors, such as TCR/CD3-ζ, γ, δ, , FcγRIII-γ, and FcRI-β and γ, to name but a few [33–35]. This motif possesses a unique tandem duplication of a tyrosine-containing sequence YXXL (or YXXI), that purportedly serves as a structure and substrate for the docking of the full receptor-associated signaling mechanism [34, 36]. Based on the seminal observations of

Reth (33), we wished to determine the abilities of Igα and Igβ to signal for growth arrest and apoptosis, and the requirement for the YXXL motif in this process. To test this, we obtained CD8 fusion proteins from Heinrich Flaswinkel and Michael Reth (Freiburg) and inserted them in retroviral vectors for high level expression in CH31 cells [37]. These constructs encode the extracellular and transmembrane regions of CD8 in frame with the cytoplasmic domains of Igα and Igβ.

After verifying high level expression of CD8 : Igα and CD8 : Igβ, as well as the lack of heterodimer formation with the endogenous complementary coreceptor, we tested whether anti-CD8 ligation would lead to growth arrest and apoptosis. Crosslinking of CD8 : Igα or CD8: Igβ with anti-CD8, followed by an anti-rat Ig "piggyback", led to growth arrest and apoptosis in both cases. This indicates that CD8 : Igα, as well as CD8 : Igβ, can independently trigger life and death regulatory signaling cascades [37].

To establish the role of the ITAM, mutations of the conserved tyrosines in the YXXL/I sequences of the Igα were established and also used to transfect CH31 lymphoma cells. Changing either or both conserved tyrosines to phenylalanine completely ablated the induction of growth arrest and apoptosis by these mutant constructs (Table 10.1). Interestingly, mutation of both Ys led to a complete loss of the initial tyrosine phosphorylation of multiple substrates by anti-CD8 with this construct, whereas a single Y → F change led to "normal" PTK activity, indistinguishable from the wildtype chimera or a mutant with a non-conserved Y → F residue change. Therefore, these changes must affect downstream events, including docking with p72^{syk} and complete activation of intracellular signaling. It is curious that phosphorylation of the chimeric protein *per se* does not occur when Y34 is mutated alone or in combination with Y23. This result implicates this tyrosine at position 34 as critical for the induction of signals leading to apoptosis. Studies are in progress to establish the steps which are blocked by these changes. At the present time, our results indicate that activation of *src* kinases is a critical first step leading to growth arrest and apoptosis. Moreover, these are the first to define this phosphorylation pathway, which directly links the ITAM in Ig-α in forming the complexes that signal for cell death.

Table 10.1   Signal transduction mediated by ITAM: role of conserved tyrosines in IgM-associated coreceptors

Chimeric receptor	Tyrosine phosphorylation	Ca^{2+} ↑	Apoptosis
CD8:Igα			
wild-type	+	+	+
Y23/34F	-	-	-
Y17/34F	+	+	-
Y34F	+	+	-
Y23F	+	+	-
Y17F	+	+	+
δ6–57	-	-	-
CD8:Igβ	+	+	+
CD8:γ2a	-	-	-

Results are data from references [34] and [37].

ROLE OF MYC IN SIGNALING B CELL GROWTH VERSUS APOPTOSIS

Over a decade ago, McCormick *et al.* [38] demonstrated that treatment of WEHI-231 B-lymphoma cells with anti-IgM leads to a transient increase in *c-myc* transcription, followed within hours by a drop in *c-myc* message levels to below background. Indeed, transcription of *c-myc* is undetectable by 8–24 hours following stimulation with anti-IgM in WEHI-231. We found a similar pattern for the modulation of *c-myc* message and protein levels in CH31, although the kinetics were more rapid ([39]; see also S Kent and D Scott, in preparation). Thus, anti-IgM-induced apoptosis was accompanied by both a rise and then a loss of myc expression. Which result is more important in immature B cell apoptosis? Several recent studies have a bearing on this issue.

Although c-myc plays an important role in cell cycle progression and survival, recent data in T cells and fibroblasts implicate c-myc as a negative regulator of cell growth and an inducer of apoptosis. For example, Evan and colleagues [40] found overexpression of myc in rat fibroblasts leads to rapid apoptosis at cell cycle restriction points. Moreover, Shi *et al.* [41] showed that anti-CD3 activation-induced apoptosis in T cell hybridomas could be blocked by antisense oligonucleotides for *c-myc*; moreover, constitutive *bcl-2* expression interfered with myc-induced apoptosis [41,42]. Therefore, overexpression of *myc* could predispose a cell to apoptosis.

Interestingly, *c-myc* message levels can be transiently increased by both anti-IgM and anti-IgD in B lymphomas that express both types of receptors [25]. Transcription of *c-myc* returns to baseline with anti-IgD treatment, in contrast to the dramatic loss of myc after anti-IgM. Notably, anti-IgD treated cells continue to grow, whereas anti-IgM crosslinking leads to apoptosis, as shown by several groups [25,27,43] Since anti-IgD-stimulation of increased expression of *c-myc* did not lead to apoptosis, this suggests that overexpression of myc *per se* is insufficient for apoptosis.

To further test the role of myc in our B-lymphoma model of activation-induced apoptosis, we added antisense (AS) or nonsense (NS) oligonucleotides designed against the first translational start site in exon 2 for murine *c-myc* [39], and as a control constructed AS against *c-fos*, which transiently rises in WEHI-231 cells stimulated with anti-IgM [44]. We found that AS to *c-myc*, but not NS *myc* or AS *fos*, prevented both growth arrest and apoptosis induced by IgM crosslinking [39]. Surprisingly, rather than blocking myc-specific RNA and protein synthesis, AS *c-myc* actually led to stabilization of *c-myc* message and protein ([39]; see also G. Sonenshein, personal communication).

TGF-β, an important regulatory cytokine [45], is a powerful inducer of growth arrest and apoptosis in B cell lymphomas [46]. We found AS oligos against *c-myc* also protected against TGF-β-induced apoptosis [39]. It is noteworthy that this cytokine does not induce even an evanescent increase in *c-myc* transcription, but only causes its

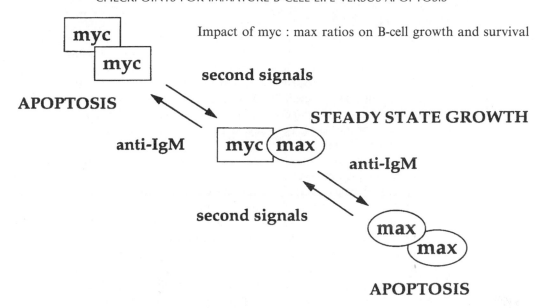

Figure 10.1 The stochiometry of myc : max controls both B cell growth and apoptosis. Anti-IgM drives B cells, both immature and mature, out of G_0 and into cycle in a dose-dependent fashion. In the absence of second (costimulatory) signals, these cells either produce excess myc or max, which shifts the cell's pathway toward apoptosis. While there is no evidence for myc : myc homodimers, as there is for max, the double box for myc is representative of overexpression of this oncoprotein. This hypothesis is reminiscent of the critical balance between bcl-2 and bax in apoptosis [48]

loss (S Kent and D Scott, unpublished; see also references [45,47]); As *c-myc* prevented the disappearance of myc from TGF-β treated B lymphomas. Therefore, the induction of apoptosis in these B-lymphoma cells does not appear to be due to the increased expression of myc protein, but is likely to be related to its loss. Stabilization of myc, therefore, is critical for cell cycle progression presumably via the myc-regulated synthesis of factors necessary for transit through a G_1 restriction checkpoint. Consistent with this view is the fact that AS *myc*-treated lymphomas shown normal phosphorylation of the retinoblastoma gene product (pRB; see below) even in the presence of normally inhibitory concentrations of anti-IgM and TGF-β.

It is ironic that we and others (D Scott and A Hayday, unpublished; N Hozumi, personal communication) had tried in vain to transfect *c-myc* under different promoter control into WEHI-231 in order to block the loss of myc protein and protect against growth arrest and apoptosis. No viable clones were ever obtained. These results could be explained, with 20/20 hindsight, by the

fact that overexpression of myc under drug-selective pressure for clones led to apoptosis. Thus, maintenance of stoichiometrically regulated levels of c-myc is critical for cell life; dysregulation leads to death via apoptosis. Indeed, transgenic mice expressing myc under the control of the immunoglobulin heavy chain promoter/enhancer (Eμmyc) show very high levels of spontaneous B cell apoptosis, compared to non- transgenic normal controls [21]. A simplified scheme for the importance of myc in apoptosis, modeled on data with bcl-2 and bax [48], is shown as Figure 10.1.

ANTI-IgM TREATED B LYMPHOMAS ACCUMULATE AT A G_1 CHECKPOINT BEFORE COMMITTING TO APOPTOSIS: ROLE OF RETINOBLASTOMA PHOSPHORYLATION

The accumulation of cells in G_1 and loss of cells in S is a gradual observation in unsynchronized cells, with a 10–12 hour lag phase, but is dramatically seen with elutriated cells. Therefore, examination of log phase B lymphomas treated with anti-IgM shows a delayed accumulation of cells in G_1, followed by an even slower decrease of cells in G_2/M; presumably this reflects the fact that cells which have passed a critical restriction point for growth arrest ([11]; also see below) continue to cycle through S and into G_2/M. Analysis of elutriated cells or ones synchronized with nocodazole [11], indicate that the critical signals for growth arrest (in apoptosis) occur in early G_1, prior to the onset of phosphorylation of the retinoblastoma gene product, pRB. As will be discussed below, phosphorylation of pRB is controlled by a series of cyclins and cyclin-dependent kinases, whose activity and complex composition is modulated by anti-IgM. In parallel with c-myc regulation, we consider the phosphorylation of pRB to be one of the most critical factors controlling cell cycle progression and apoptosis in (immature) B cells; underphosphorylated pRB is always present in growth-arrested cells [46] and cannot be separated from the process leading to apoptosis. It should be noted that this is different from apoptosis in T cells in that growth arrest and apoptosis can be clearly separated. Further differences between T and B cell apoptosis are discussed later.

The retinoblastoma gene product is a nuclear phosphoprotein, whose state of phosphorylation is modulated in the cell cycle-dependent manner. The hypophosphorylated form of pRB is growth suppressive and exists only in early to mid-G_1, when it is tightly associated with the nuclear membrane [49]. Phosphorylation of pRB begins in mid-to late G_1 and allows cells to enter S; this state of phosphorylation persists until mitosis, when pRB is dephosphorylated and converted to the original growth-suppressive form [50–53].

One of the well-established cellular targets of pRB is the DRTF1/E2F transcriptional factor complex: only the hypophosphorylated form of pRB can effectively bind E2F, thereby inhibiting transcription of cellular genes containing E2F-binding sites and thus preventing S phase entry [54]. We suggest that pRB participates in regulating the differentiation state of lymphocytes [55], for example, by driving such cells "out of cycle" and into quiescence. In fact, the lineage-restricted transcription factor, Elf-1, can modulate gene expression during T cell activation; like E2F, Elf-1 binds only to the hypophosphorylated form of pRB in resting human T lymphocytes, but is subsequently released upon T cell stimulation and pRB phosphorylation [56]. The opposite occurs in B cell lymphomas treated with anti-IgM: pRB becomes hypophosphorylated [46] and can now bind E2F to arrest these cells in G1 and drive them into quiescence.

If phosphorylation of pRB is required for cell cycle progression, which is blocked in anti-IgM-treated lymphomas, how is this accomplished? We know that the cell cycle and pRB phosphorylation are controlled by a series of cell-cycle-dependent kinases (cdks) and their cyclin partners. To explore the action of anti-IgM during early G_1, Sergei Ezhevsky and Luc Joseph first established that pRB phosphorylation was modulated in elutriated, early G_1 cells by anti-IgM treatment, but had no effect when added to fractions derived from mid-to late G_1, when pRB phosphorylation had already begun [57]. This was important since it established that growth arrest was the result of the activity of anti-IgM affecting pRB phosphorylation, not that growth arrest itself leads to the hypophosphorylation of pRB via an aberrantly activated phosphatase.

Phosphorylation of pRB occurs on serine and threonine residues, which correspond to the consensus sequence for Cdc2 or related kinases [58, 59]. Among these are cyclin-dependent kinase 2 (Cdk2), Cdk4 and Cdk6 [60–62]. At least three kinase complexes, including cyclin A, cyclin E and cyclin D, may phosphorylate pRB in vitro; indeed, all of the above, including cyclins D1, D2 and D3, have been shown to function to phosphorylate and inactivate pRB in an in vivo expression system [63]. Therefore, collaboration of all of these kinase complexes is necessary to maintain the state of pRB and other regulators in the cell cycle.

How does triggering surface IgM receptors affect the cyclin–Cdk complexes? And which of these kinases act on pRB in these lymphomas? Since anti-IgM antibodies prevent phosphorylation of pRB when added in early G_1, without affecting the duration of G_1, we wished to determine whether negative signaling affects the accumulation and activity of cyclin A or cyclin E kinase complexes, because these cyclins have been shown to participate in G_1 to S transition, in contrast to D-type cyclins, which have been implicated in the regulation of G_1 progression.

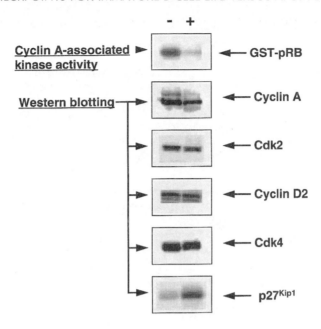

"-" Untreated WEHI-231 cells

"+" Treated with anti-IgM (24 hours)

Figure 10.2 Anti-IgM treatment results in inhibition of cyclin-A-associated kinase activity by up-regulating 27^{Kip1}. Exponentially growing WEHI-231 cells were treated with anti-IgM for 24 hourse, lysed and immunoprecipitated with anti-cyclin A antibodies. Immune complexes were assayed for GST-pRB kinase activity *in vitro* (top panel). In parallel, a second aliquot was lysed in SDS-buffer and the total cell lysates were probed with anti-cyclin A, anti-Cdk2, anti-cyclin D2, anti-Cdk4 and anti-p27 antibodies by Western blotting (lower five panels, respectively). This figure shows that anti-IgM abolished cyclin-A-associated kinase activity, but had no effect on the total amount of cyclin A (except for upper band), cyclins D and E (not shown), as well as Cdk4 and Cdk2 proteins. Since kinase activity associated with Cdk-2 [64] is diminished, we assume that inhibition of cyclin A/Cdk2 complex is due to the action of p27, whose synthesis and association with cyclin A is increased in growth inhibited WEHI-231 cells [65]. The p27 could be effectively sequestered by binding to a cyclin D2/Cdk4 complex and released upon diminishing expression of cyclin D2 or Cdk4 in the cell cycle

Using a GST–pRB fusion protein as a substrate, we next established that kinase activity on pRB was significantly impaired in extracts of anti-IgM-treated WEHI-231 lymphomas (Figure 10.2). This loss of kinase activity correlated with the appearance of the hypophosphorylated form of pRB and the latency period before G_1-arrested cells appeared. In addition, a slower migrating form of cyclin A disappeared from these cells with precisely the same kinetics [64]. It soon became clear from *in vitro* immunoprecipitation experiments that kinase activity associated with both cyclins A and E, as well as with cdk2 was modulated in anti-IgM-treated cells (see Figure 10.2).

Since this could be demonstrated with G_1 elutriated cells, we concluded that anti-IgM induces G_1 arrest (and apoptosis) by acting upon the cyclin A : cdk2 kinase complex that was capable of phosphorylating pRB *in vitro*, and presumably *in vivo*.

Recent data obtained by Sergei Ezhevsky [65], also shown in Figure 10.2, indicate that this may be due to an accumulation of the kinase inhibitor protein, $p27^{Kip}$, in these complexes. This protein is a member of the family of proteins that bind to and inhibit cyclin-dependent kinases required for the initiation of S phase in a variety of cells, including lymphocytes [66–68]. Interestingly, we found that p27 exists as a complex with cyclin A,

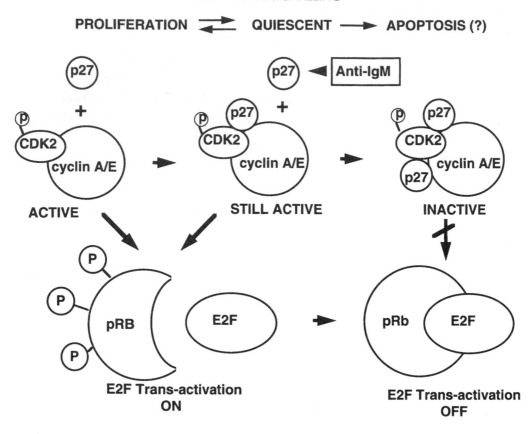

Figure 10.3 Schematic model of cyclin : cdk : Kip regulation of cell cycle progression and the points at which anti-IgM blocks this process. Cyclin-dependent kinases (cdk-2 is shown) are activated by phosphorylation, for example, via CAK in G_1 and form multimeric complexes with D-type cyclins, as well as with cyclins E and A. Kipl (p27) may be recruited to this complex, without affecting kinase activity, along with p21 and PCNA (not shown). In addition, the E2F transcription factor complex is synthesized and bound by underphosphorylated pRB. If pRB is phosphorylated by these kinase complexes, then E2F (for example) is released and cells progress into S. If not, the cells become quiescent and apoptosis is initiated. Treatment with anti-IgM blocks the activation of cyclin A (Figure 10.2) and the cyclin A-, E-and cdk2-associated kinase activity, but not cdk4; synthesis and/or release of sequestered p27 is also increased leading to a cdk/cyclin complex that is inactive on pRB. It is not clear whether myc acts to regulate p27 or cyclin synthesis.

even in actively growing WEHI-231 cells, but after anti-IgM treatment the ratio of p27/cyclin A increases dramatically. This results in an inhibition of kinase activity. The second messenger pathways by which anti-IgM drives the stoichiometric changes in cyclin : kinase complex composition in B cell lymphomas is unknown. Suffice it to say, the accumulation of the underphosphorylated form of pRB and loss of myc, in the absence of other cell cycle progression signals, leads to immature B cell arrest and apoptosis. The central role of pRB phosphorylation in this process is illustrated as Figure 10.3.

ARE P53 OR OTHER GENE PRODUCTS LINKED TO THE B CELL APOPTOTIC PATHWAY?

There is no doubt that p53 is an important gene product regulating the apoptotic response to, for example, DNA damage. In animals which are knockouts for p53, the induction of apoptosis in thymocytes by phorbol esters and ionomycin to mimic T cell receptor stimulation was not affected [69]. Moreover, glucocorticoid-induced apoptosis was normal in T cells from these mice, despite the fact that T cell radiosensitivity was blocked. Is p53 involved in B cell apoptosis? In our lab, Xiao-rui Yao examined the levels of p53 in CH31 and WEHI-231, and its modulation by anti-IgM or other agents. By Western blotting, we found that wildtype p53 was present in both cell lines, although it was expressed at much higher levels in CH31, which also produces mutant p53. Using WEHI-231, in which background levels of p53

of our goals has been to determine the role of these gene products in anti-IgM induced B-lymphoma apoptosis. Interestingly, Ishida *et al.* [71] found that RNA synthesis was not required for apoptosis in WEHI-231, although it was needed for T cell apoptosis. This suggests that the message and perhaps the proteins necessary for the apoptotic process are already present in cycling B-lymphoma cells. Nonetheless, Osborne and coworkers (B

Table 10.2 Modulation of p53 expression in B lymphomas

Stimulus	B-lymphoma cell line	
	CH31	WEHI-231
Endogenous		
wildtype	++	+/−
mutant	++	?
γ-Radiation	↑	↑↑↑
5-Fluorouracil	↑	↑↑↑
Anti-IgM	−	−
Cyclosporin A	↓↓	−

Unpublished data from Yao and Scott, 1995.

were low, we found that 5-fluorouracil and irradiation led to a transient increase in p53 expression, whereas anti-IgM effected no change; all of these reagents caused the appearance of apoptosis as assayed by DNA laddering (unpublished data from Yao and Scott, 1995; see also Table 10. 2). In CH31, it was difficult to observe changes in p53 due to the high background signal; however, cyclosporin A caused a drop in p53 protein expression, while inducing apoptosis. In general, there was no correlation with either p53 expression in a large series of B cell lines, nor with its modulation by agents known to induce apoptosis, other than 5FU and irradiation, which produced the predicted changes. Thus, activation-induced apoptosis in immature B cell lymphomas appears to be p53 independent.

Although a number of genes that may be functionally anti-apoptotic have been described, recently a novel group of apoptosis genes have been isolated. For example, several groups have used subtractive hybridization or differential display to isolate several candidate clones that were specific for the apoptosis process. These include *RP-8*, *PD-1*, *bad*, *bax*, *nur77* and *ICE*-related genes, in addition to p53 product [48, 70–74]. One

Osborne, personal communication) have found *nur-77* message in WEHI-231 cells treated with anti-IgM. Therefore, it would be timely to test the roles of these products in our B cell lymphoma models. Finally, we need to understand the interplay between putative apoptosis genes with anti-apoptotic products of *bcl-2* gene family members [75]. Note that overexpression of bcl-2 alone does not protect WEHI-231 lymphoma cells from growth inhibition by anti-IgM, although bcl-xL transfection does [76]. Moreover, we now realize that the context in which bcl-2 (and related genes) are expressed is important. That is, the stoichiometric amounts of inhibitory partners, such as bad, bax or bcl-xs, as well as newly described synergistic proteins like BAG-1, may determine the ultimate fate of each cell [48], just as the ratios of myc and max also regulate apoptosis.

HOW DO B CELLS EVADE APOPTOSIS?

Although immature B cells show a lower threshold for tolerance and apoptosis, they do not have an obligate commitment to unresponsiveness. In fact, it has been known for nearly twenty years that provision of T cell help protected neonatal B cells from tolerance in the splenic focus system and converted this stimulus into an immune response [2]. In addition, we previously reported that activated T_H2 clones or their products also could block growth arrest and apoptosis in anti-IgM-treated WEHI-231 and CH31 B cell lymphomas [14, 77, 78]. IL-4 was identified as the most active cytokine in these supernatants. Although IL-4 could partially prevent growth inhibition by anti-IgM, it has very little effect on the ability of TGF-β to block these cells. However, LPS protects B cell lymphomas against growth inhibition by both reagents [78]. These results also support the notion that TGF-β is not primarily responsible for anti-IgM-induced growth arrest. Interestingly, direct T cell contact plays an important role in protection because viable T cells were always more efficient than mixed cytokine supernatants or recombinant interleukins (unpublished data). Thus, we suspected that cell interaction molecules, like CD40, might play a role in blocking apoptosis; evidence for a role of CD40 in this process has recently been reported [79–82]. We have recently verified this pathway with anti-CD40 (from Dr Maureen Howard, DNAX) and will analyze the mechanism of anti-CD40, as well as IL-4, protection in terms of myc, p27 and cdk : cyclin activation as one of our current goals.

It is worth noting that extensive crosslinking of B cell receptors by anti-IgM also can lead to apoptosis in *mature B cells*, although the doses required are significantly higher than those needed for immature B cell apoptosis [21]; paradoxically, these doses are in the mitogenic range of anti-IgM with adult B cells. Recently, we found that anti-CD40 or soluble gp39 (CD40 ligand) both could prevent apoptosis in adult splenic B cells (Figure 10.4; B Maddox, K Washart, and D W Scott, unpublished data); interestingly, anti-gp39 actually augmented background apoptosis in whole spleen cultures. This suggests that T cell : B cell interactions capable of modulating apoptosis may be occurring even *in vitro*, and are likely to be operative *in vivo*.

Can we explain these results in terms of cell cycle regulation by p27 and c-myc? We know that p27 is highly expressed and active in resting B cells [65], and that it is down-regulated when these cells are stimulated with LPS alone or in combination with IL-4, but not by IL-4 alone (SE Ezhevsky,

DW Scott and AD Keegan, in preparation). This result is interesting because IL-4 alone can reduce spontaneous apoptosis in cultured splenic B cells. Moreover, anti-IgM treatment may also decrease p27 activity [65]. In contrast, resting B cells express very little myc protein, but transiently increase synthesis of *c-myc* message within 30–60 minutes after anti-IgM [83]. We suggest that in non-transformed B cells, the expression of c-myc in the absence of costimulation may drive these cells to commit to apoptosis. This notion is consistent with the fact the Eμ-myc transgenic mice have increased levels of spontaneous apoptosis. Moreover, antisense oligonucleotides against *c-myc* are able to block both spontaneous and anti-IgM-induced apoptosis in normal and Eμ-myc transgenic B cells [21].

ARE T CELLS AND B CELLS DIFFERENT IN THEIR APOPTOSIS PATHWAYS?

Like B cell lymphomas, thymomas and T cell hybridomas are often used as *in vitro* models for activation-induced apoptosis and tolerance. In addition, studies on thymocyte apoptosis have been useful to understand negative selection in a model system in which 95% are immature T cells at various stages, in contrast to B cells which develop in disparate locations that are quite heterogeneous (such as bone marrow). *In vitro* culture of thymocytes is a widely used model system for studying apoptosis. Thymocytes have been induced to undergo apoptosis by treatments with glucocorticoids [84], low-dose γ-irradiation [85], or calcium ionophores [86]. Antibodies to the T cell antigen receptor complex have been shown to induce apoptosis in thymus *in vivo* [87] or in thymic organ cultures [88]. Similarly, when thymocytes from H-Y specific T-cell antigen receptor (TCR) transgenic female mice were cultured in suspension with antigen presenting cells from syngeneic male mice, there was a significant antigen-specific induction of apoptosis in CD4$^+$, CD8$^+$ double-positive T cells [89]. Thus, these experiments clearly indicate that autoreactive T cells can be eliminated or negatively selected via apoptosis at immature stages after being activated by self-antigen presented by antigen presenting cells in the thymus.

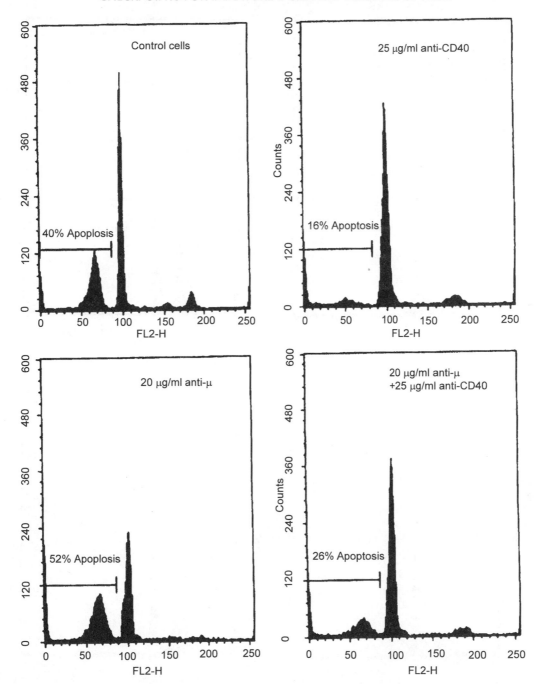

Figure 10.4 Induction of apoptosis by polyclonal anti-IgM and its prevention by ligation of CD40. Normal splenic B cells were incubated either with medium (control; upper panels) or with 20 μg ml^{-1} goat anti-IgM (anti-μ; lower panels), in the presence of anti-CD40 (second row), soluble gp39 (20% SF9 supernatant of CD5/gp39; third row, over page) or anti-gp39 (last row, over page). After 24 hours, these groups of cells were washed, fixed with ethanol and incubated with propidium iodide for flow cytometric analysis [21]. Apoptotic cells appear to the left of the G_0/G_1 peak. Apoptosis induced by anti-IgM was partially blocked by anti-CD40 and soluble gp39

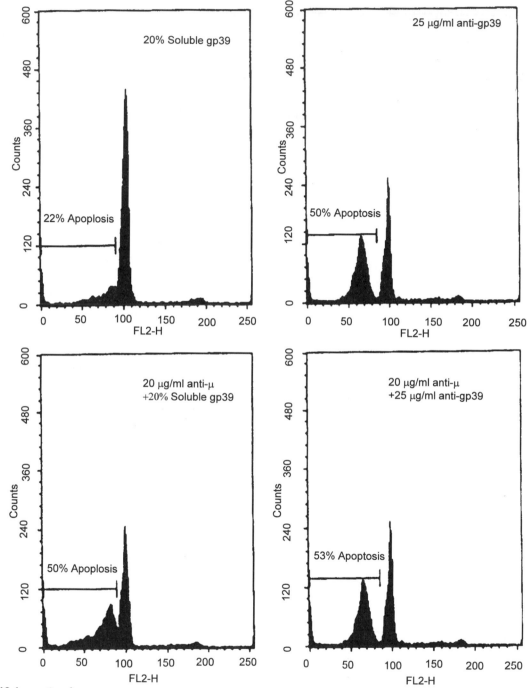

Figure 10.4 *continued*

Many T cell hybridomas were created by fusion of a mature T cell and a thymoma. The thymoma partner may represent a T cell developmental stage that contributes to the property of activation-induced cell death, presumably resembling apoptosis in immature thymocytes. Ashwell *et al.* [90] reported that some T cell hybridomas cease to incorporate tritiated thymidine after being activated, but it was not clear whether this was due to inhibition of cell division or death of these cells. Studies by Shi *et al.* [87] determined whether these cells undergo apoptosis upon activation and provide a model which the mechanism of negation selection in the thymus can be elucidated. Using a peptide-specific I-A^d restricted T-cell hybridoma, A1.1, they found that cells undergo apoptosis following activation with antigen or other activating signals [91], with the first signs of cell death, namely membrane blebbing and genomic DNA fragmentation, being detected at approximately 7 hours.

It now appears that mechanisms involved in the T cell and B cell apoptosis may not be identical. A summary of differences is presented in Table 10.3. In T cell hybridomas and thymocytes, glucocorticoids and antibodies to the TCR induce apoptosis. However, when the two signals are combined, they can actually act antagonistically [92]. We have demonstrated that glucocorticoids also induced apoptosis as measured by PI uptake in B cell lymphomas (T Grdina and D Scott, unpublished). However, steroids did not antagonize anti-IgM-induced apoptosis; instead, glucocorticoids actually synergized with the effects of anti-IgM! As

cyclosporin A has been shown to inhibit activation-induced apoptosis in T-hybridoma cells [94]. Under the same conditions, cyclosporin A induces apoptosis in B-lymphomas.

At the molecular level, Fas and Fas ligand are up-regulated in T cell hybridomas, and their interaction clearly initiates the apoptosis process upon the ligation of the TCR complex [95–97]. We have been unable to find induction of either Fas or its ligand in B cell lymphomas upon activation with anti-IgM [98]. Thus, B cell apoptosis, at least in lymphomas, appears to be Fas-independent and different from the pathway in T cells.

Differences between immature B cell lymphomas and T cell hybridomas and fibroblasts, and even resting B cells, in terms of the role of myc may be revealing. As mentioned earlier, antisense oligodeoxynucleotides to *c-myc* could block activation-cell death in both T and B cells. However, in T cells it appears to be due to the reduction of c-myc protein levels while in B cells it is due to the stabilization of the c-myc protein. Furthermore, actinomycin D and cycloheximide, inhibitors of transcription and translation, respectively, have been shown to block apoptosis in T cells but

Table 10.3 Comparison of activation-induced apoptosis in B and T cells

Stimuli/gene products	T cells	B cells
Glucocorticoid	+	+/-
Receptor crosslinking	+	+
Glucocorticoid + receptor crosslinking	-	++
TGF-β	-	+
TGF-β + receptor crosslinking	-	++
Cyclosporin A	-	+
Cyclosporin A + receptor crosslinking	-	?
Nur 77	+	-?
c-Myc	↑	-
Fas + Fas ligand	↑	-

*++ indicates synergistic effects;? indicates unknown

noted earlier, in B-lymphoma cells, TGF-β alone can induce apoptosis, whereas it is a powerful inhibitor of activation-induced apoptosis in T cell hybridomas. Furthermore, TGF-β can render apoptosis-resistant B-lymphomas sensitive to anti-IgM-mediated apoptosis [93]. In addition,

appear to induce it in B cell lymphomas. We propose that the level of c-myc transcription is not only higher, it is dysregulated in transformed B cells. Antisense shifts this regulatory pathway by an unknown mechanism to stabilize rather than knockout myc. In T cells, fibroblasts and resting

B cells with low levels of transcription, AS-myc prevents translation of new myc protein at cell cycle restriction points to prevent apoptosis. This agrees further with the differences between B cells and T cells in terms of the requirements for apoptosis. Indeed, results with actinomycin D/cyclohex-imide, cited above, suggest that apoptosis is tightly regulated by short-lived inhibitory proteins, with a half-life similar to that of myc. We would suggest that transcription factor complexes of the myc/max, fos/jun, nm2 [99] and NFκ-B [100] families or controlled by myc, for example, are intimately involved in this process leading to growth arrest and apoptosis. Moreover, we propose that apoptosis is a consequence of growth arrest (e.g. at $G_1 : S$) in B cells, but that cell-cycle-independent apoptosis is required only in T cells to prevent auto-immunity.

CONCLUSIONS

In B-lymphoma cells, the levels and activity of myc and myc-related proteins, as well as cyclin kinase components, are tightly regulated. Myc and pRB become central players not only for growth control, but also for apoptosis since abortive B cell activation into cycle can lead to G_1 arrest, quiescence and apoptosis. Apoptosis is required to purge the B cell repertoire and can only be prevented by appropriate T cell signals, including cytokines and direct T cell contact [14,77,78] and CD40 interactions [79–82]. While the initiating events for apoptosis involve tyrosine phosphoryla-tion and downstream sequelae affecting pRB phosphorylation complexes and c-myc transcrip-tion, the molecular signals to overcome growth arrest and apoptosis are still largely unknown. It is possible that T cell and cytokine interactions modulate the "negative" phosphorylation events via PKC activation, based on preliminary studies in synchronized WEHI-231 B-lymphoma cells [101]. However, further generalizations must await definitive data on the precise sequence of events leading to arrest and apoptosis of immature B cell lymphomas.

Finally, while apoptosis induced by the ligation of the antigen receptor in immature T and B cells plays an important role in the maintenance of self-tolerance, the mechanisms operating in the two cells appear different. Such pathways might be fundamental during the evolution of the immune system, as T and B cells develop quite differently, carry out distinct functions, and need to be regulated independently to avoid autoimmunity.

ACKNOWLEDGEMENTS

This is the second publication of the Department of Immunology, Holland Laboratory of the American Red Cross. The author's work reported herein has been supported by grants from the National Institutes of Health (AI29691, CA55644).

REFERENCES

1. Burnet FM. *The Clonal Selection Theory of Acquired Immunity*. Nashville: Vanderbilt University Press, 1959: 202pp.
2. Metcalf E and Klinman N. *In vitro* tolerance induction of neonatal murine spleen cells. *J. Exp. Med.* 1976; **143**: 1327–1340.
3. Scott DW, Venkataraman M and Jandinski J. Multiple pathways of B-cell tolerance. *Immunol. Rev.* 1979; 43: 241–280.
4. Pike B, Boyd A and Nossal GJV. Clonal anergy: The universally anergic B lymphocyte. *Proc. Nat. Acad. Sci. USA* 1982; **79**: 2013–2017.
5. Cooper MD, Kearney JF, Gathing WE and Lawton AR. Effects of anti-Ig antibodies on the development and differentiation of B cells. *Immunol. Rev.* 1980; **52**: 29–53.
6. Kearney JF, Lawton AR and Cooper MD. B lymphocyte differentiation induced by lipopolysaccharide. II. Response of fetal lymphocytes. *J. Immunol.* 1975; **115**: 677–683.
7. Scott DW, Borrello M, Liou L-B, Yao X-r and Warner GL. B- cell tolerance: Life or death? *Adv. Mol. Cell. Immunol.* 1993; **1**: 119–143.
8. Ralph P. Functional subsets of murine and human B lymphocyte cell lines. *Immunol. Rev.* 1979; **48**: 107–121.
9. DeFranco AL, Kung JT and Paul WE. Regulation of growth and proliferation in B cell subpopulations. *Immunol. Rev.* 1982; **64**: 161–182.
10. Pennell C and Scott DW. Models and mechanisms for signal transduction in B cells. *Surv. Immunol. Res.* 1986; **5**: 61–70.
11. Scott DW, Livnat D, Pennell CA and Keng P. Lymphoma models for B-cell activation and tolerance III. Cell cycle dependence for negative signaling of WEHI-231 B lymphoma cells by anti-μ. *J. Exp. Med.* 1986; **164**: 156–164.
12. Benhamou L, Casenave P and Sarthou P. Anti-immunoglo-bulins induce death by apoptosis in WEHI-231 B lymphoma cells. *Eur. J. Immunol.* 1990; **20**: 1405–1407.
13. Hasbold J and Klaus GGB. Anti-immunoglobulin antibodies induce apoptosis in immature B-cell lymphomas. *Eur. J. Immunol.* 1990; **20**: 1685–1690.

14. Scott DW, Chace J, Warner G, O'Garra A, Klaus GGB and Quill H. Role of T cell-derived lymphokines in two models of B-cell tolerance. *Immunol. Rev.* 1987; **99**: 153–171.

15. Scott, DW. B-lymphoma models for tolerance: The good, the bad and the apoptotic. *Immunol. Methods* 1993; **2**: 105–112.

16. Scott DW. Apoptosis in immature B-cell lymphomas. In: Gregory C (ed.) *Apoptosis and the Immune Response.* New York: John Wiley, 1994: 187–000.

17. Green D. R. and Scott, D.W. Activation-induced apoptosis in lymphocytes. *Curr. Opin. Immunol.* 1994; **6**: 476–487.

18. Brown D, Warner G, Alés-Martínez J-E, Scott DW and Phipps RP. Prostaglandin E2 induces apoptosis in normal and malignant B lymphocytes. *Clin. Immunol. Immunopathol.* 1992; **63**: 221–229.

19. Norvell A, Mandik and Monroe J. Engagement of the antigen-receptor on immature murine B lymphocytes results in death by apoptosis. *J. Immunol.* 1995; **154**: 4405.

20. Parry SL, Hasbold J, Holman M and Klaus GGB. Hyper-cross-linking surface IgM or IgD receptors on mature B cells induces apoptosis that is reversed by costimulation with IL-4 and anti-CD40. *J. Immunol.* 1994; **152**: 2821–2829.

21. Scott DW, Lamers M, Köhler G, Sidman C, Maddox B, Wang, R, and Carsetti R. Regulation of spontaneous and anti-receptor induced apoptosis in adult murine B-cells by *c-Myc. Inter. Immunology* 1996; **8**: 1375- 1385.

22. Carsetti R, Köhler G and Lamers M. A role for immuno-globulin D: Interference with tolerance induction. *Eur. J. Immunol.* 1993; **23**: 168–178.

23. Carsetti R, Köhler G and Lamers M. Transitional B cells are the target for tolerance in the B-cell compartment. *J. Exp. Med.* 1995; **181**: 2129- 40.

24. Monroe J. Molecular basis for unresponsiveness and tolerance induction in immature stage B lymphocytes. *Adv. Cell. Mol. Immunol.* 1993; **1b**: 1–32.

25. Tisch R, Roifman C and Hozumi N. Functional differences between immunoglobulins M and D expressed on the surface of an immature B-cell line. *Proc. Natl. Acad. Sci. USA* 1988; **85**: 6914–6918.

26. Haughton G, Arnold LW, Bishop GA and Mercolino TJ. The CH series of murine B lymphomas: neoplastic analogues of Ly-1+ normal B cells *Immunol. Rev.* 1986; **93**: 35–51.

27. Alés-Martinez J-E, Warner GL and Scott DW. Immuno-globulin D and M mediate signals that are qualitatively different in B cells with an immature phenotype. *Proc. Natl. Acad. Sci. USA* 1988; **85**: 6919–6923.

28. Scott DW, Tuttle J, Livnat D, Haynes W, Coggswell J and Keng P. Lymphoma models for B-cell activation and tolerance II. Growth inhibition by anti-m of WEHI-231 and the selection of properties of resistant mutants. *Cell. Immunol.* 1985; **93**: 124–128.

29. Beckwith M, Urba W, Ferris D, Kuhns D, Moratz C and Longo D. Anti-Ig-mediated growth inhibition of a human B lymphoma cell line is dependent of phosphatidylinositol turnover and protein kinase C activation and involves tyrosine phosphorylation. *J. Immunol.* 1991; **147**: 2411–2418.

30. Yao X-r and Scott DW. Antisense oligodeoxynucleotides to the blk tyrosine kinase prevent anti-μ-chain-mediated growth inhibition and apoptosis in a B-cell lymphoma. *Proc. Natl. Acad. Sci. USA* 1993; **90**: 7946–7950.

31. Scheuermann RH, Racila E, Tucker P, Yefenof E, Street N, Vitetta ES, Picker L and Uhr JW. Lyn tyrosine kinase signals cell cycle arrest but not apoptosis in B-lineage lymphoma cells. *Proc. Natl. Acad. Sci. USA* 1994; **91**: 4048–4052.

32. Yao X-r and Scott DW. Expression of protein tyrosine kinases in the Ig complex of anti-μ-sensitive and anti-μ-resistant B-cell lymphomas: role of the p55blk kinase in signaling growth arrest and apoptosis. *Immunol. Rev.* 1993; **132**: 163–186.

33. Reth M. Antigen receptors on B lymphocytes. *Ann. Rev. Immunol.* 1992; **10**: 97–121.

34. Flaswinkel H and Reth M. Dual role of the tyrosine activation motif of the Ig-α protein during signal transuction via the B cell antigen receptor. *EMBO J.* 1993; **13**: 83–89.

35. Alber G, Kim K-M, Weiser P, Riesterer C, Carsetti R and Reth M. Molecular mimicry of the antigen receptor signaling motif by transmembrane proteins of Epstein–Barr virus and the bovine leukemia virus. *Curr. Biol.* 1993; **3**: 333–339.

36. Pleiman CM, D'Ambrosio D and Cambier JC. The B-cell antigen receptor complex: structure and signal transduction. *Immunol Today* 1994; **15**: 393–398.

37. Yao X-R, Hanswinkel H, Reth M and Scott DW. Igα or Igβ cytoplasmic tail containing an immunoreceptor tyrosine-based activation motif (ITAM) is required to signal pathways of receptor-mediated growth arrest and apoptosis in murine B-lymphoma cells. *J. Immunol.* 1995; **155**: 652–661.

38. McCormick JE, Pepe VH, Kent BR, Dean M, Marshak-Rothstein A and Sonenshein G. Specific regulation of c-myc oncogene expression in a murine B-cell lymphoma line. *Proc. Natl. Acad. Sci. USA* 1984; **81**: 5546–5550.

39. Fischer G, Kent SC, Joseph L, Green DR and Scott DW. Lymphoma models for B-cell activation and tolerance. X. Anti-μ-mediated growth arrest and apoptosis of murine B-cell lymphomas is prevented by the stabilization of myc. *J. Exp. Med.* 1994; **179**: 221–228.

40. Evan GI, Wyllie AH, Gilbert CS, Littlewood TD, Land H *et al.* Induction of apoptosis in fibroblasts by c-myc protein. *Cell* 1992; 69: 119–128.

41. Shi Y, Glynn M, Guilbert LJ, Cotter TG, Bissonette RP and Green DR. Role for c-myc in activation induced apoptotic cell death in T-cell hybridoma. *Science* 1992; **257**: 212–214.

42. Bissonette RP, Echeverri F, Mahboubi A and Green DR. Apoptotic cell death induced by c-myc is inhibited by bcl-2. *Nature* 1992; **359**: 552–554.

43. Mongini P, Blessinger C, Posnett D and Rudich S. Membrane IgD and membrane IgM differs in capacity to transduce inhibitory signals with the same human B-cell clonal populations. *J. Immunol.* 1989; **143**: 1565–1574.

44. Seyfert V, Sukhatme V and Monroe J. Differential expression of a zinc finger-encoding gene in response to positive versus negative signaling through receptor immunoglobulin in murine B lymphocytes. *Mol. Cell. Biol.* 1989; **9**: 2083–2088.

45. Pietenpol J, Stein RW, Moran E, Yaciuk P, Schlegel R, Lyons RM, Pittelkow M, Munger K, Howley P and Moses

HL. TGF-b1 inhibition of c-myc transcription and growth in keratinocytes is abrogated by viral transforming proteins with pRB binding domains. *Cell* 1991; **61**: 777–785.

46. Warner GL, Nelson, D, Ludlow J and Scott DW. Anti-immunoglobulin treatment of murine B-cell lymphomas induces active TGF-β but pRB-hypophosphorylation is TGF-β-independent. *Cell Growth and Diff*. 1992; **3**: 175–181.

47. Pietenpol J, Holt JT, Stein RW and Moses HL. Transforming growth factor-b1 suppression of c-myc gene transcription: Role in inhibition of keratinocyte proliferation. *Proc. Natl. Acad. Sci. USA* 1991; **87**: 3758–3762.

48. Yang E, Zha J, Jockel J, Boise L, Thompson C and Korsmeyer S. Bad, a heterodimeric partner for Bcl-xL and Bcl-2, displaces Bax and promotes cell death. *Cell* 1995; **80**: 285–291.

49. Mittnacht S and Weinberg RA. G1/S phosphorylation of the retinoblastoma protein is associated with an altered affinity for the nuclear compartment. *Cell* 1991; **65**: 381–393.

50. Buchkovich K, Duffy LA and Harlow E. The retinoblastoma protein is phosphorylated during specific phases of the cell cycle. *Cell* 1989; **58**: 1097–1105.

51. DeCaprio J.A., Ludlow JW, Lynch D, Furukawa Y, Griffin J, Piwnica-Worms H, Huang CM and Livingston DM. The product of the retinoblastoma susceptibility gene has properties of a cell cycle regulatory element. *Cell* 1989; **58**: 1085–1095.

52. Durfee T, Becherer K, Chen PL, Yeh SH, Yang Y *et al.* The retinoblastoma protein associates with the protein phosphatase type 1 catalytic subunit. *Genes Dev* 1993; **7**: 555–569.

53. Ludlow JW, Glendening CL, Livingston DM and DeCaprio JA. Specific enzymatic dephosphorylation of the retinoblastoma protein. *Mol. Cell. Biol.* 1993; **13**: 367–372.

54. La Thangue NB. DRTF1/E2F: an expanding family of heterodimeric transcription factors implicated in cell-cycle control. *Trends Biochem. Sci.* 1994; **19**: 108–114.

55. Cooper, JA and Whyte P. Rb and the cell cycle: entrance or exit? *Cell* 1989; **58**: 1009–1011.

56. Wang CY, Petryniak B, Thompson CB, Kaelin WG and Leiden JM. Regulation of the Ets-related transcription factor Elf-1 by binding to the retinoblastoma protein. *Science* 1993; **260**: 1330–1335.

57. Joseph L, Ezhevsky S and Scott DW. Lymphoma models for B-cell activation and tolerance. XI. Anti-IgM treatment induces growth arrest by preventing the formation of an active kinase complex which phosphorylates pRB in G1. *Cell Growth Differ.* 1995; **6**: 51–57.

58. Hu, QJ, Lees JA, Buchkovich KJ and Harlow E. The retinoblastoma protein physically associates with the human cdc2 kinase. *Mol. Cell. Biol.* 1992; **12**: 971–980.

59. Lees JA, Buchkovich KJ, Marshak DR, Anderson C and Harlow, E. The retinoblastoma protein is phosphorylated on multiple sites by human cdc2. *EMBO J* 1991; **10**: 4279–4290.

60. Akiyama T, Ohuchi T, Sumida S, Matsumoto K and Toyoshima K. Phosphorylation of the retinoblastoma protein by cdk2. *Proc. Natl. Acad. Sci. USA* 1992; **89**: 7900–7904.

61. Matsushime H, Quelle DE, Shurtleff SA, Shibuya M, Sherr CJ and Kato JY. D-type cyclin-dependent kinase activity in mammalian cells. *Mol. Cell. Biol.* 1994; **14**: 2066–2076.

62. Meyerson M and Harlow E. Identification of G1 kinase activity for cdk6, a novel cyclin D partner. *Mol. Cell. Biol.* 1994; **14**: 2077–2086.

63. Horton L, Qian Y and Temptelton DJ. G1 restrictions control the retinoblastoma gene product growth regulation activity via upstream mechanisms. *Cell Growth Differ.* 1995; **6**: 395–407.

64. Ezhevsky S, Toyoshima H, Hunter T, and Scott DW. Role of cyclin A and p27 in anti-IgM-induced G_1 growth arrest of murine B-cell lymphomas. *Molec. Biol. of Cell* 1996; **7**: 553–564.

65. Ezhevsky SE and Scott DW. p27, a cyclin-cdk inhibitor, is a new regulator of transition into and out of quiescence. *FASEB J.* 1995; **9**: A505, p. 2922.

66. Firpo EJ, Koff A, Solomon MJ and Roberts JM. Inactivation of a Cdk2 inhibitor during interleukin 2-induced proliferation of human T lymphocytes. *Mol. Cell. Biol.* 1994; **14**: 4889–4901.

67. Polyak K, Kato JY, Solomon MJ, Sherr, CJ *et al.* p27Kip1, a cyclin-Cdk inhibitor, links transforming growth factor-beta and contact inhibition to cell cycle arrest. *Genes Dev.* 1994; **8**: 9–22.

68. Toyoshima H and Hunter T. p27, a novel inhibitor of G1 cyclin-Cdk protein kinase activity, is related to p21. *Cell* 1994; **78**: 67–74.

69. Osborne B, Smith S, Liu Z-G, McLaughlin KA, Grimm L and Schwartz L. Identification of genes induced during apoptosis in T lymphocytes. *Immunol. Rev.* 1994; **142**: 301–320.

70. Owens G, Hahn W and Cohen JJ. Identification of mRNAs associated with programmed cell death in immature thymocytes. *Mol. Cell. Biol.* 1991; **11**: 4177–4188.

71. Ishida Y, Agata Y, Shibahara K and Honjo T. Induced expression of PD-1, a novel member of the immunoglobulin gene superfamily, upon programmed cell death. *EMBO J.* 1992; **11**: 3887–3895.

72. Liu Z-G, Smith SW, McLaughlin KA, Schwartz LM and Osborne BA. Apoptotic signals delivered through the T-cell receptor of a T-cell hybrid require the immediate-early gene nur77. *Nature* 1994; **367**: 281–284.

73. Woronicz JD, Calnan B, Ngo V and Winto A. Requirement for the orphan steroid receptor Nur77 in apoptosis of T-cell hybridomas. *Nature* 1994; **367**: 277–281.

74. Wang L, Miura M, Bergeron L, Zhu H and Yuan J. Ich-1, an ice/ced-3-related gene, encodes both positive and negative regulators of programmed cell death. *Cell* 1994; **78**: 739–741.

75. Takayama S, Sato T, Krajewski S, Kochel K, Irie S, Millan J and Reed J. Cloning and functional analysis of BAG-1: A novel Bcl-2-binding protein with anti-cell death activity. *Cell* 1995; **80**: 279–284.

76. Gottschalk A, Boise L, Thompson C and Quintans J. Identification of immunosuppressant-induced apoptosis in a murine B-cell line and its prevention by bcl-x but not bcl-2. *Proc. Natl. Acad. Sci. USA* 1994; **91**: 7350–7354.

77. Alés-Martínez J, Silver L, LoCascio N and Scott DW. Lymphoma models for B-cell activation and tolerance. IX. Efficient reversal of anti-Ig-mediated growth inhibition by an activated TH2 clone. *Cell. Immunol.* 1990; **135**: 402–409.

78. Alés-Martínez J-E, Cuende E, Gaur A and Scott DW. Prevention of B cell clonal deletion and anergy by activated T cells and their lymphokines. Semin. Immunol. 1992; **4**: 195–202.

79. Valentine M and Licciardi K. Rescue from anti-IgM-induced programed cell death by the B cell surface proteins CD20 and CD40. Eur. J. Immunol. 1992; **22**: 3141–3148.

80. Heath AW, Wu W and Howard MC. Monoclonal antibodies to murine CD40 define two distinct functional epitopes. Eur. J. Immunol. 1994; **24**: 1828–1834.

81. Tsubata T, Wu J and Honjo T. B-cell apoptosis induced by antigen receptor crosslinking is blocked by a T-cell signal through CD40. Nature 1993; **364**: 645–648.

82. Parry SL, Hasbold J, Holman M and Klaus GGB. Hyper-cross-linking surface IgM or IgD receptors on mature B cells induces apoptosis that is reversed by costimulation with IL-4 and anti-CD40. J. Immunol. 1994; **152**: 2821–2829.

83. Phillips NE, Gravel KA, Tumas K and Parker DC. IL-4 overcomes Fcγ receptor-mediated inhibition of mouse B lymphocyte proliferation without affecting inhibition of c-myc mRNA induction. J. Immunol. 1984; **132**: 627–634.

84. Wyllie AH. Glucocorticoid induced thymocyte apoptosis is associated with endogenous endonulcease activation. Nature 1980; **284**: 555–556.

85. Sellins K and Cohen JJ. Gene induction by gama-irradiation leads to DNA fragmentation in lymphocytes. J. Immunol. 1987; **139**: 3199–3206.

86. McConkey DJ, Hartzell P, Nicotcra P and Orrenius S. Calcium activated DNA fragmentation kills immature thymocytes. FASEB J 1989; **3**: 1843–1849.

87. Shi Y, Bissonnette RP, Parfrey N, Szalay M, Kubo, RT and Green DR. In vivo administration of monoclonal antibodies to the CD3 T cell receptor complex induces cell death in immature thymocytes. J. Immunol. 1992; **146**: 3340–3346.

88. Smith CA, Williams GT, Kingston R, Jenkinson EJ and Owen JJT. Antibodies to CD3/T cell receptor complex induce death by apoptosis in immature T cells in thymic cultures. Nature 1989; **337**: 181–184.

89. Swat WS, Ignatowicz L and Kisielow, P. Detection of apoptosis of immature CD4+CD8+ thymocytes by flow cytometry. J. Immunol. Methods 1991; **137**: 79–84.

90. Ashwell JD, Cunningham, RE, Noguchi PO and Hernandez D. Cell growth cycle block of T cell hybridomas upon activation with antigen. J. Exp. Med. 1987; **165**: 173–194.

91. Fotedar A, Boyer M, Smart W, Widtman J, Fraga E and Singh B. Fine specificity of antigen recognition by T cell hybridoma clone specific for poly 18: a synthetic polypeptide of defined sequence and conformation. J. Immunol. 1985; **135**: 3028–3033.

92. Zacharchuk CM, Mercep M, Chakraborti PK, Simons SS and Ashwell JD. Programmed T lymphocyte death: Cell activation-and steroid-induced pathways are mutually antagonistic. J. Immunol. 1990; **12**: 4037–4045.

93. Warner GL, Nelson D and Scott DW. Synergy between TGF-β and anti-IgM in growth inhibition of CD5+ B-cell lymphomas. Ann. NY Acad. Sci. 1992; **651**: 274–277.

94. Shi Y, Sahai BM and Green DR. Cyclosporin A inhibits activation induced cell death in T-cell hybridomas and thymocytes. Nature. 1989; **339**: 625–626.

95. Ju S-T, Panka D, Cui H, Ettinger R, el-Khatib M, Sherr D and Marshak-Rothstein A. Fas (CD95)/fas-ligand interactions required for programmed cell death after T-cell activation. Nature 1995; **373**: 444–447.

96. Brunner T, Mogil R, LaFace D et al. and Green D. Cell-autonomous Fas (CD95)/fas-ligand interaction mediates activation induced apoptosis in T-cell hybridomas. Nature 1995; **373**: 441–444.

97. Dhein J, Walczak H, Baümler C, Debatin K-M and Krammer P. Autocrine suicide by APO-1/(Fas/CD95). Nature 1995; **373**: 438–441.

98. Scott, DW, Grdina T and Shi Y. T cells commit suicide but B cells are murdered. J. Immunol. 1996; **156**: 2352–2356.

99. Postel EH, Berberich SJ, Flint SJ and Ferrone CA. Human c-myc transcription factor PuF identified as nm23-H2 nucleoside diphosphate kinase, a candidate suppressor of tumor metastasis. Science 1993; **261**: 478–480.

100. Lee H, Arsura M, Wu M, Duyao M, Buckler A and Sonenshein G. Role of Rel-related factors on control of c-myc gene transcription in receptor mediated apoptosis of the murine B-cell WEHI-231 line. J. Exp. Med. 1995; **181**: 1169–1177.

101. Scott DW, Livnat D, Whitin J, Dillon SB, Snyderman R and Pennell CA. Lymphoma models for B-cell activation and tolerance V. Anti-Ig mediated growth inhibition is reversed by phorbol myristate acetate but does not involve changes in cytosolic free calcium. J. Mol. Cell. Immunol. 1987; **3**: 109–120.

11

Early Response Genes and B Cell Activation

John J. Murphy,[1] Joshua S. Newton[1] and John D. Norton[2]

[1]*Infection and Immunity Research Group, Division of Life Science, King's College London, London. UK and* [2]*CRC Department of Gene Regulation, Paterson Institute for Cancer Research, Christie Hospital NHS Trust, Manchester, UK*

INTRODUCTION

The biological responses of numerous cell types when induced to divide, differentiate or undergo apoptosis are preceded by the rapid, often transient expression of a set of cellular genes whose encoded proteins mediate programmed events in the cell in response to extracellular signals. Such genes, which are typically induced through transcriptional mechanisms within the first few hours following stimulation, have been collectively referred to as early response genes (ERGs) [1,2]. Since ERG expression occurs without a requirement for *de novo* protein synthesis, and therefore depends on the activities of pre- existing transcription factors, expression of this set of genes is directly coupled to third messenger signalling pathways in the cell's signal transduction machinery (see Figure 11.1). Interest in lymphocyte ERGs was initially prompted by the observation that nuclear protoncogenes, exemplified by *c-myc* and *c-fos*, display rapid, transient induction of expression following mitogenic stimulation of T and B cells [3–5] as previously shown to occur in other cell types [1,2,6]. Subsequently, a number of other genes encoding oncoproteins or proteins regulating cell growth/differentiation were found to share these expression characteristics [1,7] and, with the advent of differential cDNA cloning technology, several groups isolated panels of ERGs from lymphocytes activated with phorbol 12-myristate 13-acetate (PMA), phytohaemagglutinin (PHA) or IL-2 [8–13]. Perhaps the most remarkable initial finding to emerge from these studies was the large number of ERGs that are turned-on following activation of lymphocytes. As in fibroblasts [14,15], the complexity of lymphocyte ERG expression represents well in excess of 100 distinct genes. Inevitably, many of the ERGs expressed in lymphocytes are also common to other cell types such as fibroblasts and this is reflected in the frequent, repeated cloning of the same gene by several different groups. Most studies have employed resting (G_0) cells for activation and thus, many ERGs, like the prototype nuclear protoncogenes, are "ubiquitously" coupled to the $G_0 - G_1 - S$ phase of cell-cycle control. However, a substantial minority of ERGs are relatively unique to B and/or T cells [8,10,11]. Also, as might be expected, a number of ERGs cloned as "anonymous" cDNAs have turned out to be previously known genes, often with well-defined functions [1,2]. In other cases, ERGs encode novel proteins whose functions are only just beginning to be elucidated. Thus, cDNA cloning of ERGs provides an approach to identifying important regulatory molecules which may not be achievable by more traditional biochemical or immunological means. Furthermore, the study of ERG expression, both at the level of *cis–trans* mechanisms controlling individual genes and at the gross level of coordinate induction/repression of sets of genes, provides a rational strategy for analysing how different second messenger signal transduction pathways are integrated into the genetic programme coordinating cellular functions in response to the plethora of different ligand–receptor interactions in lymphopoiesis.

EARLY RESPONSE GENES EXPRESSED IN LYMPHOCYTES

The following section describes some of the better characterized ERGs that are known to be

Lymphocyte Signalling: Mechanisms, Subversion and Manipulation. Edited by M. M. Harnett and K. P. Rigley © 1997 John Wiley & Sons Ltd.

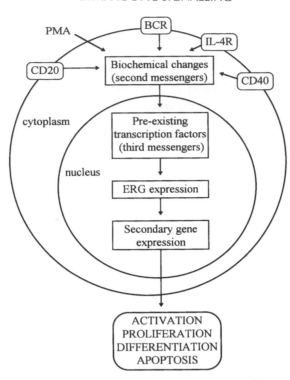

Figure 11.1 Signalling cascade leading to B cell activation. B cell surface receptors interact with either antigen (in the case of BCR), soluble cytokines or cell-bound ligands to generate intracellular signals which induce changes in early response gene expression. Surface receptors shown are those through which signalling has been shown to induce ERG expression. Many early response genes encode transcription factors which regulate expression of secondary genes. In this way rapid intracellular biochemical changes are converted into long-term phenotypic changes in the cell

expressed in lymphocytes and their functions. For a more detailed discussion, readers are referred to reference [16]. In the interest of clarity, data relating to genes with many homologues are discussed under one name in each subsection. In many instances, their functions have been inferred from studies in other cell types, particularly fibroblasts. As can be seen from Table 11.1, a majority of these ERGs encode transcription factors. However, others encode cytokines, growth factor receptors or second messenger proteins. One unifying theme is that most (but by no means all) ERG proteins function at various stages in the process of signal transduction (see Figure 11.1).

c-myc

The *c-myc* protoncogene encodes a 439 amino acid transcription factor protein, which contains trans-

activation and specific DNA-binding and dimerization domains (the last two are encompassed within the b-HLH-Zip region) [17] (see Table 11.1). c-Myc alone neither forms homodimers nor binds to DNA, except at high concentrations [18, 19]. The functional activity of c-Myc is dependent on heterodimerization with another b-HLH-Zip protein, Max [19,20]. Virtually all of the c-Myc protein within the cell is found in the form of Myc–Max dimers [19,20]. Max itself can form homodimers and they are capable of binding the same DNA sequences as their Myc–Max counterparts, but have only weak transactivating potential [18,21,22]. This provides a mechanism of transcriptional regulation whereby the activity of Myc is determined by the ratio of Myc–Max to Max–Max complexes [18,22].

c-Myc expression appears to be necessary for cell proliferation and down-regulation of c-Myc is associated with and indeed is essential for

Table 11.1 Early response genes* expressed in lymphocytes.

Name	Related genes	Function	References
Transcription factors			
c-fos	v-fos, fosB, fra-1, fra-2	basic leucine-zipper transcriptional activator/repressor—T/B cell development, lymphocyte activation and differentiation	31, 36–39
c-jun	junB, junD	basic leucine-zipper transcriptional activator—lymphocyte activation	31–35
c-myc	v-myc	basic-HL H-zip transcriptional activator—required for cell proliferation, induces apoptosis and blocks differentiation	17
egr-1 (Zif268, NGFI-A, Krox-24, TIS8, pAT225, d-2, GOS30, ETR103)	egr-2, egr-3, egr-4	zinc-finger transcriptional activator/repressor – BCR signalling – proliferation	1,2,57–65
NGFI-B (nur77, N10, NAK1, ST-59, TR3)	NOR-1, NURR-1, RNR-1, SF-1, TINUR	orphan steroid receptor zinc-finger transcriptional activator —induces apoptosis of thymocytes	81–86, 88, 90–93
TIS11	nup475, TTP, TIS11b, ERF-1, cMG1, Berg36, TIS11d, GOS24	zinc-finger transcriptional activator	1, 61, 133–136, 180
Blimp-1	PRDI-BfI	transcriptional repressor—drives plasma cell differentiation	181, 182
A20	–	putative zinc-finger transcription factor—inhibits lymphocyte apoptosis	165, 183
chx-1	ETR101	putative zinc-finger transcription factor	184, 185
HB24	HB9	DNA binding homeodomain protein— lymphocyte activation, enhances proliferation	140, 186
Intracellular signalling molecules			
PAC-1	3CH134, CL100, HVH2	dual specificity threonine/tyrosine nuclear phosphatase—inhibits MAP kinase activity—putative negative cell cycle regulator	8, 113–115, 187
Gem	ras superfamily	monomeric GTP-binding protein— putative negative cell cycle regulator	8, 117
BL11	–	unknown	139
Cytokines			
IL-2	IL-1,-3,-4,-5,-6,-8, others?	secreted growth factor—lymphocyte survival and proliferation	100–106, 108–110, 188, 189
GM-CSF	–	secreted growth factor—affects multiple haemopoietic cell types	52, 100, 108–110
TNF-α	Lymphotoxin	secreted growth factor—cytotoxic and inflammatory effects	190–192
Interferon-α	Interferon-β, Interferon-γ	secreted growth factor – antiproliferative agent	193, 194
MIP1-α, LD78, pAT464, JE, GOS19	MIP1-β, KC, TCA3, ACT-2/pAT744	secreted growth factors – pleiotropic effects on numerous cell types	8, 111, 120, 195–197
Cytokine receptors			
IL-2Rα (CD25)	Il-1R, IL-4R, others?	receptor for IL-2—lymphocyte survival and proliferation	7, 106, 107, 198

Name	Related genes	Function	References
Cell bound ligand CD40L	–	cell contact dependent signalling through CD40 interaction—B cell survival and proliferation	199
Others ornithine decarboxylase	–	first, rate-limiting enzyme in polyamine biosynthesis—proliferation and differentiation	200, 201

*In the interests of brevity, this table is not a comprehensive list of all ERGs, but is presented as an illustration of the range of ERGs and their functions particularly relevant to lymphocytes. Only representative references are given. Additional genes cloned in the authors' laboratory are shown in Table 11.2.

growth arrest and differentiation. Thus, c-Myc activity is normally associated with mitogenic stimuli in both normal and malignant cells [3,17]. Antisense c-Myc oligonucleotides prevent entry of activated lymphocytes from both normal and leukaemic individuals into the S phase of the cell cycle [23,24] and are capable of inducing differentiation of lymphoid and embryonic cells [25,26]. Conditional expression of a Myc–oestrogen receptor chimera is sufficient to drive quiescent, growth-factor-deprived fibroblasts into the cell cycle [27] and Rat-1 fibroblasts constitutively expressing c-Myc are unable to undergo growth arrest upon serum withdrawal, but continue to cycle and undergo apoptosis [20,28]. c-Myc expression is also necessary for apoptosis of T-cell hybridomas [29] and apoptosis in Burkitt lymphoma cells is similarly driven by c-Myc [30]. A model has been proposed to explain the apparently contradictory functions of c-Myc as a means of controlling cell fate and avoiding oncogenesis [28]. For successful proliferation of normal cells, apoptosis must be inhibited, perhaps by complementary signal transduction pathways, since c-Myc activates both processes. In the absence of such cosignals, the cells will undergo apoptosis, thereby providing an in-built safeguard against uncontrolled proliferation.

c-Fos and c-Jun Families

The c-Fos and c-Jun families of proteins, which together represent the transcription factor "AP-1", were isolated as the cellular homologues of viral oncogenes, *v-fos* and *v-jun* [31]. The Jun family comprises *c-jun* [32,33], *jun* B [34] and *jun*D [35]; the Fos family includes *c-fos* [36], *fos*B [37], *fra-1* [38] and *fra-2* [39] (see Table 11.1). All contain basic and leucine zipper domains associated with DNA binding and dimerization, respectively and transactivation domains [31]. All of the Jun proteins can form both homo-and heterodimers (with members of the Fos or Jun families) which can bind DNA [40,41], whereas the Fos proteins cannot homo-or heterodimerize with other Fos members and so are unable to bind DNA in the absence of a heterodimer partner [40,42], but they can heterodimerize with the Jun proteins. Fos–Jun heterodimers are more efficient DNA-binding proteins than are Jun dimers [40,42,43] and are more stable [44]. All these complexes bind to a similar DNA recognition sequence, referred to as the TPA response element (TRE) [41]. The AP-1 transcription factor is also capable of binding to the cyclic adenosine monophosphate (cAMP) response element (CRE) and can also regulate gene transcription through binding to CRE [45].

c-jun is subject to autostimulation by virtue of AP-1 sites in its promoter [46], whereas *c-fos* is the subject of auto-inhibition [47–49] and the Fos proteins and mRNA have shorter half-lives than the Jun proteins [50–52]. As a consequence, the composition of the "AP-1 complex" changes from Jun homo-and heterodimers prior to induction (there is basal AP- 1/TRE activity in most cells) [46] to Fos–Jun heterodimers immediately post-induction, followed by a return to Jun–Jun as Fos proteins decay [31]. Not all the Jun families have equivalent transactivating potential [35,53] and JunB in particular possesses weak transactivating function in Jun–Jun dimers [54], though it is stronger in combination with c-Fos [53]. Consequently, JunB–Jun will reduce transactivation by, for example, Jun–Jun by occupying TRE sites [54]. Thus Fos–Jun AP-1 is probably responsible for initial responses while Jun AP-1 complexes are responsible for maintaining them. In addition, the DNA-binding and transactivating functions of the Jun proteins appear to be subject to post-translational regulation through phosphorylation/dephosphorylation mechanisms [44] possibly so that active AP-1 can be up-regulated rapidly prior to the appearance of new proteins following mitogenic stimulation.

AP-1 activity is involved in the regulation of proliferation and differentiation. As with c-Myc, AP-1 is in general associated with the induction and maintenance of proliferation responses and repression of differentiation. Thus, AP-1 activity is rapidly elevated after stimulation with mitogens [34] and proliferating cells contain higher levels of AP-1 mRNA/proteins than their quiescent counterparts [46,50]. Crosslinking of surface immunoglobulin (Ig) on B lymphocytes induces AP-1 activity as does phorbol ester and ionizing radiation [55,56].

egr-1 [57] (Zif268 [58]/NGFI-A [59]/ Krox24 [60]/TIS8 [61]/pAT225 [62]/d-2 [63]/GOS30 [64]/ETR103 (65))

This gene encodes a protein containing three zinc-fingers of the C_2H_2 type through which it binds DNA, and the sequences of which have been used to clone at least three related zinc-finger proteins (Egr-2, Egr-3 and Egr-4 [1,2]) (see Table 11.1). The zinc-finger domain also contains a nuclear localization signal [66]. Egr-1 expression is induced in response to a wide variety of agents which are associated with increased intracellular calcium and protein kinase C (PKC) activation, cAMP and protein kinase A (PKA) activation, phosphatase inhibitors, heavy metals and ionizing radiation [56,67–72]. It appears to be expressed in most cells which can be induced to undergo proliferative or differentiative responses in vitro. The Egr proteins are described as members of the "GSG" binding family because of the DNA sequence to which they bind: GCGGGGGCG. It has been shown recently

that there may be some flexibility in the recognition sequence for different members of the family, and each distinct member may have distinct preferences [73].

Antisense oligonucleotide experiments have demonstrated that blockade of Egr-1 results in a dramatic reduction of the proliferative response in activation of mature murine B cells [74]. In human tonsillar B cells activated with a combination of anti-μ, IL-4 and CD40 antibody, antisense but not sense oligonucleotides to the related Egr-2 transcription factor, which binds to the same consensus sequence as Egr-1 (see above), inhibit ^3H-thymidine uptake (Figure 11.2), but have no effect on the induced expression of the B cell activation marker CD23 (data not shown). Taken together, these results are consistent with a critical role for induced expression of the Egr transcription factors in mature B cell proliferation.

Interestingly, anti-Ig does not induce egr-1 expression in the immature WEHI-231 murine B cell line which undergoes growth arrest following this stimulus [75] and lack of induction of this gene has been proposed as a mechanism to account for tolerance induction in immature B cells [76]. It has recently been reported that a ternary complex-factor-dependent mechanism comprising a homodimer of serum response factor and an unidentified member of the ETS family of transcription factors mediates induction of egr-1 following antigen receptor crosslinking on B lymphocytes [77]. Antisense oligonucleotides to egr-1 have also been reported to reduce proliferation of T cells [78] and Egr-1 is essential for and restricts differentiation along the macrophage lineage [79]. However, using homologous recombination to generate Egr-1 negative embryonic stem (ES) cells [80], it was found that growth rates, serum-induced gene expression profiles and differentiative potential were similar to those exhibited by wild-type cells. Indeed, the proliferative responses of T cells to mitogenic stimuli of normal and Egr-1 deficient mice were identical [80]. Growth and histological analyses of homozygous mutant mice, generated from the Egr-1 negative ES cells, confirmed that Egr-1 was not indispensable for many of the processes to which it had previously been ascribed and there thus appears to be functional redundancy between the different members of the EGR family in vivo [80].

NGFI-B [81] (nur77 [82]/N10 [83]/NAKI [84]/TR3 [85]/ST-59 [86])

This gene encodes an inducible zinc-finger transcription factor which is a member of the steroid hormone receptor family (see Table 11.1). It is expressed in a wide variety of cell types including lymphocytes, in response to various agents including epidermal growth factor, fibroblast growth factor, nerve growth factor, as well as heavy metals [70], PKC and PKA activating agents, serum [86] and phosphatase inhibition [87]. NGFI-B binds to DNA at a defined site similar to, but distinct from that of other steroid hormone receptor binding sites [88,89]. This consensus site is also shared with other members of the "superfamily" including NOR-1 [90], NURR-1 [91], RNR-1 [92], SF-1 [88] and TINUR [93]. As yet, no ligand for NGFI-B has been identified, but the protein appears to be functional in effecting transcriptional activation with or without its ligand-binding domain [88]. NGFI-B is unusual in that it can bind to DNA either as a monomer [89] or as a dimer with other members of the steroid hormone receptor superfamily [94]. Competition for binding sites or dimerization partners may be a means of regulating differential gene transcription. The transactivation domain resides in the amino terminus, but its activity is regulated by the carboxyl-terminal end of the protein, which is also the putative ligand binding site [95,96]. NGFI-B is post-translationally modified by phosphorylation near the amino terminus [95].

A role for NGFI-B in apoptosis has been demonstrated by antisense inhibition of expression

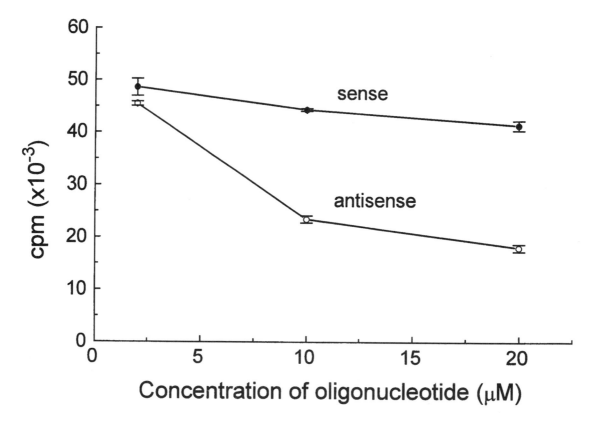

Figure 11.2 Dose responsive inhibition of B cell proliferation by antisense *egr-2* phosphorothioate oligonucleotides. B cells were stimulated with anti-μ, IL-4 and CD40 antibody (G28–5) in combination. Oligonucleotides were added to the cells 90 minutes before stimulation and DNA synthesis was measured by adding ^{3}H-thymidine to cells for the last 4 hours of a 72 hour culture period. Mean ±SEM are shown for triplicate cultures

of this protein which protected T cell antigen-receptor (TCR)-stimulated immature thymic T cells from apoptosis [97] and likewise, an NGFI-B dominant-negative mutant protein protected T cell hybridomas from activation-induced apoptosis [98]. Furthermore, cyclosporin protects T cells from TCR-mediated apoptosis and inhibits TCR-mediated activation of NGFI-B protein in T cell hybridomas by blocking the DNA binding activity of NGFI-B (and not its *de novo* synthesis) [99]. The effects of cyclosporin (thought to involve inhibition of the calcium-activated phosphatase calcineurin) are mediated through the amino-terminal region, which is the region subjected to post-translational phosphorylation [95].

Cytokines and Cytokine Receptors

The expression of several cytokine genes has been shown to conform to the ERG paradigm. IL-1, IL-2, IL-3, IL-4, IL-5, IL-6, and IL-8, IFN-α, IFN-β and GM-CSF have all been reported to have ERG characteristics, as have macrophage inhibitory factor (MIP)-1-α and MIP-1-β, TNF-α and lymphotoxin (TNF-β) [100] (see Table 11.1). For example, the expression of IL-2 in T lymphocytes is induced rapidly in response to a variety of agents evoking PKC activation, in particular TCR crosslinking, PHA, concanavalin A, PMA and calcium inophore, and anti-CD3 [100,101]. Most of these agents also activate the transcription factors NF-AT and AP-1, both of which appear to be necessary for IL-2 expression. High affinity IL-2 receptor (R) expression is also induced by similar stimuli [7]. Although IL-2 and IL-2R expression have been demonstrated in activated (not resting) B lymphocytes [102–106], relatively little is known about the molecular mechanisms regulating expression of this ligand–receptor combination in these cells. Dower and Urdahl (107) have suggested that IL-1R is coordinately regulated with the IL-2R, following a similar time course of induction. IL-1 and GM-CSF mRNAs are induced in lymphocytes in response to PMA and cytokines (including positive feedback loops) which activate PKC. The principal mechanism for this is post-transcriptional stabilization of mRNA. While the mRNAs may be easily induced (even in response to tissue culture medium) [108] or constitutively

expressed [109], they are also very rapidly degraded [110]. Thus, IL-1-α and IL-1-β, GM-CSF and IL-3 mRNAs are almost undetectable in unstimulated cells, but are rapidly induced if the cells are incubated in the presence of the protein synthesis inhibitor, cycloheximide [108–110]. PKC agonists, in the absence or presence of cycloheximide, evoke higher levels of expression of IL-1 and GM-CSF mRNA, consistent with increased transcription rate together with a substantial degree of mRNA stabilization [52,109,110].

NOVEL EARLY RESPONSE GENES ISOLATED FROM LYMPHOCYTES AND THEIR FUNCTIONS

cDNA libraries generated from activated B and T cells have been used by several groups to isolate panels of ERGs [8–13]. A variety of cloning strategies employing subtractive hybridization, differential screening and prior enrichment of newly synthesized transcripts have been used, often in conjunction with protein synthesis inhibition to "super-induce" transiently expressed genes. As in the preceding section, several examples illustrate how the functions of ERGs isolated from lymphocytes have been deduced from their properties and expression characteristics in other cell types. Although the main focus of the following discussion is on B cell ERGs, in several instances novel ERGs expressed in B cells were cloned initially from T cells (and *vice versa*). Therefore, in the following examples, which in the interests of brevity are rather selective, some of the more prominent ERGs isolated from T cells have also been included.

Early Response Genes Isolated from T Cells

Cytokines

A family of genes exemplified by *pAT464*, *pAT744* [8] and *TCA-3* [111] encode low molecular weight secreted proteins which are up-regulated following mitogenic stimulation of lymphocytes (see Table 11.1). The prototype of this family is the well-known (MIP-1-α)/LIF which elicits pleiotropic effects on multiple cell lineages. Functional data on the mouse TCA-3 encoded protein indicate it too possesses inflammatory properties in an *in vivo* assay of inflammation [112].

PAC-1

Originally designated *pAT120* [8], this gene encodes a 32 kDa dual specificity (threonine/tyrosine) nuclear phosphatase [113] (PAC = phosphatase of activated cells) (see Table 11.1). It appears to represent one member of a family of such inducible nuclear phosphatases, each with a distinctive pattern of tissue-specific expression. PAC-1 expression is largely restricted to haemopoietic cells [113]. A closely related phosphatase encoded by the mouse *3CH134* [114] and human *CL100* [115] genes was also isolated as an ERG from serum stimulated fibroblasts. Since phosphorylation/dephosphorylation, particularly at tyrosine residues, is known to be a universal regulatory mechanism in second and third messenger signalling, a role in cell cycle control was suggested for this family of ERG proteins. Indeed, recent studies have shown that overexpression of PAC-1 *in vivo* leads to inhibition of mitogen activated protein (MAP)-kinase activity, normally stimulated by mitogenic activation induced, for example, by TCR crosslinking [116]. This in turn antagonizes MAP-kinase-regulated reporter gene activity, implicating PAC-1 as a negative regulator or "cellular brake" in cell cycle progression [116].

GEM

The ras family of guanosine-triphosphate-binding proteins is well known for its functions in signal transduction and oncogenesis. *GEM* (*pAT270*) encoding a 35 kDa phosphoprotein [8,117] represents one member of a new family of these plasma membrane proteins which is notable for being phosphorylated on tyrosine residues (see Table 11.1). GEM is also expressed in some non-lymphoid cell types such as kidney, lung and testis. Overexpression of this gene in fibroblasts leads to inhibition of cell proliferation suggesting that the GEM-encoded protein facilitates negative regulation of cell cycle progression through participation in receptor-mediated signal transduction at the plasma membrane [117].

Early Response Genes Isolated from B Cells

Id3 (1R21/heir-1/HLH462)

The gene encoding the human Id3 helix–loop–helix protein (originally designated 1R21) [10,118] was isolated as one of a panel of cDNAs from phorbol-ester-stimulated B chronic lymphocytic leukaemia cells (see Table 11.2). This gene was also isolated independently by cloning human 1p36-specific CpG islands (*heir-1*) [119]. The murine homologue of *Id3*, *HLH462* was cloned from serum-stimulated fibroblasts [120]. The Id (inhibitor of differentiation) family of nuclear transcription factors, of which there are four members (Id1-4) (reviewed in reference [121]) share a similar overall gene organization, suggesting evolution from a common ancestral *Id*-like gene and conservation of function [122]. These proteins, which range in size from 15 to 20 kDa, are unusual among HLH transcription factors for the absence of a basic, DNA-binding domain and for their consequent inability to bind DNA [121]. Instead, Id-HLH proteins function as dominant negative antagonists of basic HLH transcription factors, exemplified by the "E"-protein family (E47, E12, E2-2 and HEB) by sequestering them into non-functional heterodimers which are unable to bind DNA [123,124]. The prototype Id1, has been most studied and was originally cloned as a negative regulator of muscle cell differentiation [123], a function it also serves in several other cell types including immature B cells [125]. More mature B cells isolated from human tonsil express only Id3 in their proliferating state and terminal plasma cell differentiation and cessation of cell division is accompanied by down-regulation of Id3 expression (J Norton and RW Deed, unpublished). Since Id3, like other Id proteins abrogates binding of E proteins to regulatory sequences of immunoglobulin genes ([118,120]; see also K Meyer, personal communication), it probably plays a specific role in regulating terminal plasma cell differentiation of B cells.

Recent data from antisense oligonucleotide blockade and antibody micro-injection experiments designed to ablate Id function in cell line models also suggests a role for this protein family in G_1 to S cell cycle progression [126], analogous

Table 11.2 Sequence homologies and possible functions of some B cell early response genes isolated in the authors' laboratory

Clone designation*	Related/identical genes	Possible function
1R21	HLH462/HEIR-1 (Id-3)	Id HLH transcription factor—negative regulator of cell differentiation
1R20	BL34,GOS8	Alpha helical, basic phosphoprotein transcription factor—putative cell cycle regulator
10A	Tis11b/cMG1/Berg36/ERF-1	Zinc-finger protein transcription factor—positive regulator of B cell apoptosis
1L3	PTD	Phosphatidylinositol cytosolic transfer protein (reference 202)
4B1	Nrf2	NF-E2-like basic leucine zipper transcriptional activator (reference 203)
2L12	CPSF	Cleavage and polyadenylation specificity factor (reference 204)
1R19	HLA-DQalpha	MHC class II antigen
3L3	hm01e11	"anonymous" cDNA clone from human liver
5L3	ye47e09.r1	"anonymous" cDNA clone from human liver
3L2	—	—
1L4	—	—
19A	—	—

*Unpublished observations of the authors.

to that documented for some cyclin genes. Indeed, forced overexpression of *Id* genes promotes cell growth in some, but not all cell types [118,127] and the human *Id3* functions as a weakly-cooperating oncogene in transformation assays of rodent fibroblasts ([118]; see also unpublished observations). These growth promoting functions of Id proteins also explain mechanistically how they negatively regulate differentiation. Interestingly, many tumour cell types overexpress or inappropriately express *Id* genes. The human *Id3* gene maps to chromosome 1p36.1 [122], a region often affected in diverse turnour types. This, together with the recent observation that the related Id2 protein binds the prototype turnour suppressor retinoblastoma pocket protein [127], hints at a causal role for Id3 and its relations in turnourigenesis.

1R20 (BL34)

The gene encoding this 196 amino acid basic, alpha helical, nuclear phosphoprotein, is one of several that are expressed selectively or exclusively in B lymphocytes following stimulation with mitogen or activation signals [10,11,128] (see Table 11.2). The protein is nuclear localized (JH Kehrl, personal communication) and bears distinct structural resemblance to the helix–loop–helix family of transcription factors [11,129]. Although convincing evidence for sequence-specific binding to DNA has not yet been obtained, it is very likely that the 1R20/BL34- encoded protein functions as a transcription factor. *In situ* hybridization analysis of human tonsil sections has revealed expression predominantly in a subset of germinal centre-region cells and in cells immediately adjacent to the mantle zone, suggestive of function at a specific stage of germinal cell maturation [11]. For reasons that are not entirely clear, enforced overexpression of the 1R20/BL34 protein in cell line models has proved problematical to achieve. However, preliminary data suggest that, like some of its T cell counterparts, the *1R20/BL34* B cell ERG may function as a negative regulator of cell cycle progression (J Norton, unpublished observations).

Several other ERGs very closely related to *1R20/BL34* such as *GOS8* [130] and *CR1* [13] have recently been cloned and there are estimated to be around six members of this family of putative transcription factors ([9,130,131]; see also D Forsdyke, personal communication). Interestingly, a rodent homologue of the human *GOS8* gene has been found to repress transcription of a MAP-kinase-dependent pheromone signalling pathway in yeast (K Blumer, personal communication), further suggesting a role in cell cycle control. One other feature shared by some members of the *1R20/BL34* family is their constitutive high level expression in some cases of primary leukaemia and lymphoma [11,128,132]. It may therefore be significant that both *1R20/BL34* [129] and *GOS8* [132] genes have been localized to human chromosome 1q31, a region known to be involved in haemopoietic malignancies.

Berg36 (Tis11b/cMG1/ERF-1)

Members of the *Tis11* gene family were originally cloned as ERGs from phorbol-ester-stimulated rat intestinal epithelial cells [1,61] (see Table 11.1). They all share a common zinc-finger motif typical of many transcription factor proteins. The human *Berg36* gene (originally designated 10A—see Table 11.2) was isolated from phorbol-ester-stimulated B-CLL cells [10,133] and, among its close relatives, the human *ERF-1* gene [134], rat *cMG1* [135] and murine *TIS11b* [136], there is a high degree of sequence conservation between rodents and humans. Significantly, forced overexpression of a murine homologue of *Berg36* in yeast suppresses growth [137] and recent data have also implicated this 36 kDa nuclear protein in B cell apoptosis. In response to calcium ionophore treatment, Ramos B lymphoblastoid cells rapidly die through apoptosis and this process is preceded by induced expression of several ERGs including *Berg36* [138]. Ablation of *Berg36* expression by antisense gene blockade suppresses apoptosis in these cells [133]. Interestingly, the cytokine IL-4 which also suppresses apoptosis coupled to calcium signalling, inhibits ionophore- induced *Berg36* expression. These data implicate *Berg36* as an effector of B cell apoptosis and its induced expression as a target for IL-4.

BL11

Expression of the 23 kDa protein encoded by *BL11* is, like *1R20/BL34*, almost exclusive to activated primary B cells [139] (see Table 11.1). The protein itself displays homology to several members of the Ig superfamily and contains a candidate signal peptide, a single membrane-spanning region and one "V"- like Ig domain with three predicted *n*-glycosylation sites [139]. Although functional data on this putative cell membrane protein is still lacking, it is quite likely to perform a specialized role in B cell activation.

HB24

Isolated as a new homeo-domain containing gene from activated B lymphocytes [140], the 52 kDa, *HB24*-encoded protein provides another example of a transcription factor ERG (see Table 11.1). In addition to the homeo-domain characteristic of many transcription factor families regulating developmental processes, HB24 also harbours a CAX repeat, a proline-rich region and an acidic domain, commonly found in other transcription factor proteins [140]. *In situ* hybridization analysis has revealed an expression pattern largely, but not exclusively, restricted to lymphopoietic organs [140]. Following mitogenic stimulation, the *HB24* gene is strongly up-regulated in both B and T lymphocytes suggestive of a role in lymphocyte activation. Consistent with this, enforced overexpression of HB24 in Jurkat T cells stimulates cell growth and confers a phenotype similar to that of activated T cells [141].

REGULATION OF EARLY RESPONSE GENE EXPRESSION IN B CELLS BY EXTRACELLULAR AGENTS

Early Response Gene Induction in Response to Non-physiological Agents

PMA-induced plasmacytoid differentiation of B-CLL cells [142] has provided a useful *in vitro* model system to study the gene regulatory pathways controlling activation and terminal differentiation of B lymphocytes. Most B-CLL populations do not divide in response to phorbol esters but do undergo G_0 to G_1 transition [143]. Early studies revealed that PMA-induced up-regulation of *c-myc*, *c-fos* and *c-jun* in this cell type preceded induced plasmacytoid differentiation [4,5,143–145]. Analysis of the complexity of the early gene regulatory pathways activated by PMA in B-CLL cells indicated that over 100 distinct ERGs were induced prior to the onset of differentiation [10]. This estimate was similar to those reported for studies in quiescent fibroblasts stimulated with serum [14,15] and in mitogen-stimulated T lymphocytes [8]. The regulation of expression of 13 novel ERGs isolated from PMA-induced B-CLL cells (see Table 11.2) has been analysed in greater detail. Intriguingly, PMA-induced expression of this panel of ERGs appeared to be highly cell-type specific [10]. In a series of human cell lines representing diverse stages of B cell differentiation, together with primary malignant cells from single cases of B-CLL, hairy cell leukaemia and centroblastic–centrocytic non-Hodgkin's lymphoma, PMA treatment revealed a spectrum of ERG expression profiles for the different cell types which appeared to be correlated to stage of differentiation relative to B-CLL cells [128]. Comparison of PMA-induced ERG expression in B-CLL cells with normal tonsillar B cells using a panel of 20 ERG probes including 12 B-CLL-derived ERGs (see Table 11.2), together with *c-jun*, *jun-B*, *Jun-D*, *c-fos*, *egr-1*, *egr-2*, *fos-B* and *fra-1*, revealed that 11 displayed a comparable pattern and magnitude of induction in both cell types, while 2 (*fra-1* and *fos-b*) were not induced in either cell type [146]. B-CLL-derived ERGs (*Id3*, *berg36*, *1R19* and *3L11*) were constitutively expressed in tonsillar B cells with a concomitant reduction in inducibility compared with leukaemic cells, one B-CLL-derived ERG (*5L3*) was not detectably expressed in normal B cells, and *jun-B* and *jun-D* which were PMA-inducible in tonsillar B cells were constitutively expressed and only marginally PMA-inducible in B-CLL cells [146]. Thus, as well as cell-type-specificity of induced expression there appears to be B cell differentiation stage-specificity of induced expression for many of these ERGs.

By comparing PMA-induced expression of B-CLL-derived ERGs (see Table 11.2) in a proliferating variant B-CLL population and a more representative B-CLL population that did not undergo DNA synthesis in response to PMA, it was shown that PMA-induced up-regulation of these genes was not correlated with cell proliferation and is most likely coupled to G_0 to G_1 transition [10]. Interestingly, the level of basal and PMA-induced levels of *c-myc* expression did correlate with proliferative response in this study [144]. It was subsequently shown that phosphorothioate antisense oligonucleotides which span the ATG start site of the *c-myc* gene can inhibit ^3H-thymidine uptake in B-CLL cells [24]. Intriguingly, PKC inhibitor studies suggested that PMA was able to induce ERG expression in B-CLL cells by PKC-dependent and independent mechanisms [147]. However, PMA, anti-μ and bryostatin [148] all induce very similar spectrums of ERG expression in B-CLL cells ([24]; see also Murphy *et al.*, unpublished) and since signalling with these three agents is associated with PKC activation, it appears that PKC activation is a primary route of induction. In B-CLL cells, the calcium ionophore A23187 was shown to induce a subset of PMA-inducible ERGs implying that different signalling pathways may activate distinct but overlapping sets of ERGs [149]. Calcium ionophore treatment of the human Burkitt's lymphoma Ramos cell line also induces a characteristic ERG response, which precedes ionophore-induced apoptosis in this cell type [138]. In human tonsillar B cells, agents which activate diverse signalling pathways (see below) are also capable of inducing overlapping, but largely distinct sets of ERGs [150]. Interestingly, ionizing radiation has been reported to induce a characteristic ERG response in tonsillar B cells with delayed kinetics, implying that the effects of radiation on this cell type may be mediated through a distinct cellular signalling cascade [56].

Early Response Gene Induction in Response to Physiological Agents

Many of the earlier studies utilized non-physiological agents such as lectins or phorbol esters to study ERG induction. A question of vital importance was whether ERGs are induced by more physiological agents which provide signals during the B cell immune response *in vivo*. B cell activation is initiated by the interaction of antigen with surface antibody and its associated proteins which form the B cell antigen receptor (BCR). In recent years, much effort has gone into the characterization of the biochemical changes which take place in B cells following this interaction. These changes include increased tyrosine kinase activity, hydrolysis of phosphatidylinositol 4,5-bisphosphate generating second messengers diacylglycerol and inositol trisphosphate, which activate PKC and increase intracellular calcium concentrations respectively, activation of the G protein $p21^{ras}$ and activation of phosphatidylinositol 3-kinase [151,152]. Following these rapid biochemical changes, DNA binding proteins including NF-κB [153,154], Ets-1 [155], AP-1 [55], CREB [156] and SRF [77] are activated and function as "third messengers" (as discussed above) to mediate changes in gene expression. In this way, short-term biochemical changes in the cell give rise to changes in ERG expression. As is becoming increasingly clear, many ERGs themselves encode transcription factors which regulate expression of secondary/effector genes (Figure 11.1).

Crosslinking surface Ig on B cells with antibodies has been reported to induce ERGs including c-fos, c-myc, egr-1, egr-2, jun-B, nur77, nup475, pip92, 3CH134, Id3, 1R20 and five ERGs previously cloned from PMA-stimulated B-CLL cells, 1L4, 1R19, 3L2, 3L3 and 19A [3,10,150,157–161]. For most of these genes, phorbol esters induce equally high levels as does anti-Ig stimulation, implying that PKC is probably the primary mediator of induction [146]. For maximal induction of c-fos, it has been reported that elevated intracellular calcium and activation of PKC is required [161,162]. Staurosporine, a broad spectrum protein kinase inhibitor, failed to inhibit anti-μ-induced c-fos and 1R20 expression in tonsil B cells implying that PKC- independent signalling through surface Ig can also induce expression of these genes [150].

Transient transfection of murine B lymphocyte blasts, revealed that expression of an activated form of $p21^{ras}$ resulted in egr-1 induction and that both antigen receptor crosslinking and $p21^{ras}$ used the same element in the egr-1 promoter to exert their effects [163]. In addition, the use of dominant-negative mutants of $p21^{ras}$ and raf-1, showed that induction of egr-1 after antigen receptor crosslinking is mediated by activation of the $p21^{ras}$/mitogen-activated protein kinase signalling pathway [163].

There is limited information on ERG induction in response to signalling through other B cell surface receptors. IFN-α was reported to induce c-fos and three B-CLL derived ERGs (1R19, 1R20 and 3L11) in B-CLL cells [149]. Induction of c-fos and c-myc has been reported following interleukin-4 (IL-4) stimulation of murine B cells [162]. Anti-CD20 antibody has been reported to induce c-myc RNA and protein in tonsillar B cells, while anti-CD40 antibody had only a minimal effect [160,164]. CD40 stimulation induces A20, a novel zinc-finger protein that inhibits apoptosis [165]. Important B cell activation signals are provided by IL-4 and the interaction of CD40 ligand (on activated T cells) with their specific receptors on B cells. Signalling through either of these pathways provides "primary" stimuli to B cells. Signal transduction through the IL-4 receptor involves receptor phosphorylation, activation of receptor associated kinases and phosphorylation of a 170 kDa 4PS/IRS-1 substrate which may activate members of the STAT family of transcription factors [166]. To date, fewer details are known about the signalling pathway(s) operating through CD40 although tyrosine phosphorylation of multiple substrates has been demonstrated [167]. More recently, activation of NF-κB, AP-1 and NF-AT has been reported following stimulation of CD40 on B cells [168–170]. To what extent do the signalling pathways activated by either IL-4 or CD40 ligand or antibody overlap with signalling through the BCR at the level of ERG induction? Of a panel of 11 anti-μ inducible ERGs (c-fos, egr-1, egr-2, c-myc, 1L4, 1R19, 1R20, 1R21, 3L2, 3L3, 19A) only two, c-fos and 1R20, were induced by IL-4 stimulation of tonsillar B cells [150] and CD40 antibody or ligand induced only c-myc and one anonymous ERG, 3L3 (Murphy et al., unpublished). These results suggest that early gene regulatory pathways activated by either IL-4 or CD40 stimulation are overlapping, but largely distinct from those activated through the BCR and this probably reflects the operation of distinct signal transduction cascades through each of these receptors. These results are consistent with ligand-specific ERG induction which may provide one possible mechanism for ligand-specific functional effects in cells, as has been discussed in an earlier review [1]. In an attempt to investigate whether the synergistic effect of IL-4 and anti-μ together on B cell proliferation could be mediated at the level of early response gene expression, the induction of a panel of ERGs in response to either anti-μ or IL-4 or the two agents together was investigated. No significant differences in the level of induction of ERGs, beyond that induced by anti-μ alone, was evident with the two agents together implying that the synergy between these two signalling pathways on B cell function is probably not mediated at the level of ERG expression [150].

FUTURE PROSPECTS

As in other cell types, the study of function and mechanisms of regulation of ERGs in lymphocytes is a rapidly expanding area of interest which impinges on both the areas of lymphocyte signalling and the molecular genetic control of cell growth and differentiation. The last few years have witnessed some notable advances in the understanding of signal transduction—the process through which information is relayed from ligand–receptor interactions at the cell surface to the transcriptional apparatus in the nucleus, responsible for transducing second messenger codes into changes in cellular protein expression, which ultimately underlie the pre-programmed biological response in the cell. The complexity of ERG signalling, representing the "third messenger pathway and beyond" is now beginning to rival that of its second messenger counterpart. A major challenge now exists to understand precisely how ERGs integrate lymphocyte signalling pathways into the cellular responses of activation, proliferation, differentiation and apoptosis. One emerging concept is that many ERG proteins appear to function as negative regulators of cell cycle progression/proliferation, perhaps as part of a feedback mechanism controlling signalling cascades in the cell, analogous to the family of CDK inhibitor ERG proteins described in other cell types [171].

A number of recent technical innovations should greatly enhance the pace of research in this field. For example, identifying ERGs coupled to the large number of distinct ligand–receptor interactions on lymphocytes by traditional cDNA cloning–screening methodology represents a daunting task. While sets of ERGs can, in many cases, be expected to be common to multiple signalling pathways, some individual ERGs which are likely to be functionally critical, may be rather specific to individual lymphocyte subsets/differentiation stage and extracellular stimulus. However, the use of PCR-based differential display analysis should provide a useful shortcut to generating comprehensive "libraries" of ERGs that are specific to the individual cellular context [172]. In this regard, comparisons between ERGs of normal and malignant lymphocytes promises to be an interesting approach for elucidating molecular mechanisms of lymphoid tumourigenesis. Several well-known ERGs such as, c-myc and cyclin D, are also distinguished for being oncogenes, playing a direct role in malignant transformation. An increasing number of examples now illustrate differences in ERG expression between normal and malignant cells either in cell-line models or in primary cell populations [146,173,174].

Finally, the biggest challenge and indeed the rate limiting step in this field of research is the elucidation of ERG function, particularly of novel genes whose expression is restricted to lymphocytes. A variety of cell-line models that can be manipulated *in vitro* have been used to probe the functions of several genes involved in B lymphopoiesis. More recently it has become possible to introduce genes into primary B cells thereby permitting the effects of forced overexpression/ablation of function to be studied in a more physiological context [175,176]. For *in vivo* studies, in addition to traditional transgenic mice engineered to overexpress the gene of interest in the lymphoid compartment, targeted ablation/modification of gene function either in reconstituted RAG-2-minus mice [177,178] or utilizing the Cre/lox P recombinase system [179] should all provide useful approaches to investigating ERG function in the future.

ACKNOWLEDGEMENTS

The authors would like to thank Drs Beadling, Blumer, Deed, Forsdyke, Kehrl, Meyer and Monroe for communication of unpublished or newly published material. In addition, we are very grateful to Mr Zhi-Qiang Ning for help in preparation of the illustrations and for critical reading of the manuscript. Work in the authors' laboratories is funded by the UK Medical Research Council and Cancer Research Campaign.

REFERENCES

1. Herschman, HR. Primary response genes induced by growth factors and tumour promoters. *Ann. Rev. Biochem.* 1991; **60**: 281–319.

2. McMahon SB and Monroe JG. (1992) Role of primary response genes in generating cellular responses to growth factors. *FASEB J.* 1992; **6**: 2707–2715.

3. Kelly K, Cochran BH, Stiles CD and Leder P. Cell-cycle regulation of the c-myc gene by lymphocytic mitogens and platelet-derived growth factor. *Cell* 1983; **35**: 603–610.

4. Larsson LG, Gray HE, Totterman T, Peterson U and Nillson K. Drastically increased expression of MYC and FOS protooncogenes during in vitro differentiation of chronic lymphocytic leukaemia cells. *Proc. Natl. Acad. Sci. USA* 1987; **84**: 223–227.

5. Norton JD, Leber B and Yaxley JC. Patterns of gene expression during plasmacytoid differentiation of chronic lymphocytic leukaemia cells. *FEBS Lett.* 1987; **215**: 127–131.

6. Greenberg ME and Ziff EB. Stimulation of NIH 3T3 cells induces transcription of the c-fos proto-oncogene, *Nature* 1984; **311**: 433–438.

7. Reed JC, Knowle PC and Hoover RG. (1986) Sequential expression of proto-oncogenes during lectin-stimulated mitogenesis of normal human lymphocytes. *Proc. Nat. Acad. Sci., USA* 1986; **83**: 3982–3987.

8. Zipfel PF, Irving SG, Kelly K and Siebenlist U. Complexity of the primary genetic response to mitogenic activation of human T cells. *Mol. Cell. Biol.* 1989; **9**: 1041–1048.

9. Siderovski DP, Blum S, Forsdyke RE and Forsdyke DR. (1990). A set of human putative lymphocyte G0/G1 switch genes includes genes homologous to rodent cytokine and zinc finger protein-encoding genes. *DNA Cell. Biol.* 1990; **9**: 579–587.

10. Murphy JJ and Norton JD. Cell-type-specific early response gene expression during plasmacytoid differentiation of human B lymphocytic leukemia cells. *Biochim. Biophys. Acta* 1990; **1049**: 261–271.

11. Hong JX, Wilson GL, Fox CH and Kehrl JH. Isolation and characterisation of a novel B cell activation gene. *J. Immunol.* 1993; **150**: 3895–3904.

12. Sabath DE, Podolin PL, Comber PG and Prystowsky MB. cDNA cloning and characterisation of an interleukin-2-induced genes in a cloned T helper lymphocyte. *J. Biol. Chem.* 1990; **265**: 12671–12678.

13. Beading C, Johnson KW and Smith KA. Isolation of interleukin-2-induced immediate early genes. *Proc. Nat. Acad. Sci., USA* 1993; **90**: 2719–2723.

14. Lau LF and Nathans D. Identification of a set of genes expressed during the G0/G1 transition of cultured mouse cells. *EMBO J.* 1985; **4**: 3145–3151.

15. Almendral JM, Sommer D, MacDonald-Bravo H, Burckhardt J, Perera J and Bravo R. Complexity of the early genetic response to growth factors in mouse fibroblasts. *Mol. Cell. Biol.* 1988; **8**: 2140–2148.

16. Hesketh R. *The Oncogene Handbook*. London: Academic Press, 1994.

17. Dang CV. c-Myc oncoprotein function. *Biochim. Biophys. Acta* 1991; **1072**: 103–113.

18. Amati B, Dalton S, Brokes MW, Littlewood TD, Evan GI and Land H. Transcriptional activation by the human c-Myc oncoprotein in yeast requires interaction with Max. *Nature* 1992; **359**: 423–426.

19. Amati B, Brooks MW, Levy N, Littlewood TD, Evan GI and Land H. Oncogenic activity of the c-Myc protein requires dimerisation with Max. *Cell* 1993; **72**: 233–245.

20. Amati B, Littlewood TD, Evan GI, Land H. The c-Myc protein induces cell cycle progression and apoptosis through dimerisation with Max. *EMBO J.* 1993; **12**: 5083–5087.

21. Kretzner L, Blackwood EM and Eisenman RN. Myc and Max proteins possess distinct transcriptional activities. *Nature* 1992; **359**: 426–429.

22. Fisher F, Crouch DH, Jayaraman PS, Clark W, Gillespie DAF and Goding CR. Transcription activation by myc and max—flanking sequences target activation to a subset of CACGTG motifs *in vivo*. *EMBO J.* 1993; **12**: 5075–5082.

23. Heikkila R, Schwab G, Wickstrom E, Loke SL, Pluznik D, Watt R and Neckers L. A c-Myc antisense oligonucleotide inhibits entry into S-phase, but not progress from $G_0 - G_1$. *Nature* 1987; **328**: 445–449.

24. Ning Z-Q, Hirose T, Deed R, Newton J, Murphy JJ and Norton JD. Early response gene signalling in bryostatin-stimulated B chronic lymphocytic leukemia cells in vitro. *Biochem J.* 1996; **319**: 59–65.

25. Griep AE and Westphal H. Antisense Myc sequences induce differentiation of F9 cells. *Proc. Natl. Acad. Sci. USA* 1988; **85**: 6806–6810.

26. Prochownik EV, Kukowska J and Rodgers C. c-Myc antisense transcripts accelerate differentiation and inhibits G_1 progression in murine erythroleukaemia cells. *Mol. Cell. Biol.* 1988; **8**: 3683–3695.

27. Eilers M, Schim S and Bishop JM. The Myc protein activates transcription of the alpha-prothymosin gene. *EMBO J.* 1991; **10**: 133–141.

28. Evan GI, Wyllie AH, Gilbert CS, Littlewood TD, Land H, Brooks MW, Waters CM, Penn LZ and Hancock DC. Induction of apoptosis in fibroblasts by c-Myc protein. *Cell* 1993; **69**: 119–128.

29. Shi Y, Glynn JM, Guilbert LJ, Cotter TG, Bissonette RP and Green DR. Role for c-Myc in activation-induced apoptotic cell death in T cell hybridomas. *Science* 1992; **257**: 212–214.

30. Milner AE, Grand RJA, Waters CM and Gregory CD. Apoptosis in Burkitt lymphoma cells is driven by c-myc. *Oncogene* 1993; **8**: 3385–3391.

31. Angel P and Karin M. The role of Jun, Fos and the AP-1 complex in cell-proliferation and transformation. *Biochim. Biophys. Acta* 1991; **1072**: 129–157.

32. Maki Y, Bos TJ, Davis C, Starbuck M and Vogt PK. Avian sarcoma virus 17 carries the *jun* oncogene. *Proc. Natl. Acad. Sci. USA* 1987; **84**: 2848–2852.

33. Vogt PK, Bos TJ and Doolittle RF. Homology between the DNA-binding domain of the GCN4 regulatory protein of yeast and the carboxyl-terminal region of a protein coded for by the oncogene jun. *Proc. Natl. Acad. Sci. USA* 1987; **84**: 3316–3319.

34. Ryder K, Lau LF and Nathans D. A gene activated by growth factor is related to the oncogene *v-jun*. *Proc. Natl. Acad. Sci. USA* 1988; **85**: 1487–1491.

35. Hirai S-T, Ryseck R-P, Mechta F, Bravo R and Yaniv M. Characterisation of *junD*: a new member of the *jun* proto-oncogene family. *EMBO J.* 1989; **8**: 1433–1439.

36. Curran T, MacConnell WP, Van Straaten F and Verma IM. Structure of the FBJ murine osteosarcoma virus genome: molecular cloning of its associated helper virus and the cellular homolog of the v-fos gene from mouse and human cells. *Mol. Cell. Biol.* 1983; **3**: 914–921.

37. Zerial M, Toschi Ryseck R-P, Schermann M, Muller R and Bravo R. The product of a novel growth factor acivated gene, *fosB*, interacts with JUN proteins enhancing their DNA binding activity. *EMBO J.* 1989; **8**: 805–813.

38. Cohen DR, Curran T. *fra-1*: a serum-inducible, cellular immediate-early gene that encodes a *fos*-related antigen. *Mol. Cell. Biol.* 1988; **8**: 2063–2069.

39. Nishina H, Sato H, Suzuki T, Sato M and Iba H. Isolation and characterisation of *Fra-2*, an additional member of the fos gene family. *Proc. Natl. Acad. Sci. USA* 1990; **87**: 3619–3623.

40. Halzonetis TD, Georgopoulos K, Greenberg ME and Leder P. c-Jun dimerizes with itself and with c-fos, forming complexes of different DNA binding affinities. *Cell* 1988; **55**: 917–924.

41. Nakabeppu H, Ryder K and Nathans D. DNA binding activities of three murine Jun proteins: stimulation by fos. *Cell* 1988; **55**: 907–915.

42. Sassone-Corsi P, Ransone LJ, Lamph WW and Verma IM. Direct interaction between fos and jun nuclear oncoproteins: Role of the "leucine zipper" domain. *Nature* 1988; **336**: 692–695

43. Kouzarides T, Ziff E. The role or the leucine zipper in the fos–jun interaction. *Nature* 1988; **336**: 646–651.

44. Boyle WJ, Smeal T, Defize LHK, Angel P, Woodgett JR, Karin M and Hunter T. Activation of protein kinase C decreases phosphorylation of c-Jun at sites that negatively regulate its DNA-binding activity. *Cell* 1991; **64**: 573–584.

45. Ryseck R and Bravo R. c-JUN, JUNB, and JUND differ in their binding affinities to AP-1 and CRE consensus sequences: effect of FOS proteins. *Oncogene* 1991; **6**: 533–542.

46. Angel P, Hattori K, Smeal T and Karin M. The *jun* proto-oncogene is positively autoregulated by its product Jun/AP1. *Cell* 1988; **55**: 875–885.

47. Sassone-Corsi P, Sisson JC and Verma IM. Transcriptional autoregulation of the proto-oncogene *fos*. *Nature* 1988; **334**: 314–319.

48. Schonthal A, Heurlich P, Rahmsdorf HJ and Ponta H. Requirement for *fos* gene expression in the transcriptional activation of collagenase by other oncogenes and phorbol esters. *Cell* 1988; **54**: 325–334.

49. Lamb NJC, Fernandez A, Tourkine N, Jeanteur P and Blanchard J-M. Demonstration in living cells of an intragenic negative regulatory element within the rodent *c-fos* gene. *Cell* 1990; **61**: 485–496.

50. Kruijer W, Cooper JA, Hunter J and Verma IM. PDGF induces rapid but transient expression of the c-fos gene and protein. *Nature* 1984; **312**: 711–716.

51. Muller R, Bravo R, Burckhardt J and Curran T. Induction of c-fos gene and protein by growth factors precedes activation of c-myc. *Nature* 1984; **312**: 716–720.

52. Shaw G and Kamen R. A conserved AU sequence from the 3' untranslated region of GM-CSF mRNA mediates selective mRNA degradation. *Cell* 1986; **46**: 659–667.

53. Schutte J, Viallel J, Nav M, Segal S, Fedorko J and Minna J. *junB* inhibits and *c-fos* stimulates the transforming and *Trans*-activating activities of *c-jun*. *Cell* 1989; **59**: 987–997.

54. Chiu R, Angel P and Karin M. Jun-B differs in its biological properties from, and is a negative regulator of, c- Jun. *Cell* 1989; **59**: 979–986.

55. Chiles TC and Rothstein TL. Surface Ig receptor-induced nuclear AP-1-dependent gene expression in B lymphocytes. *J. Immunol.* 1992; **149**: 825–831.

56. Wilson RE, Taylor SL, Atherton GT, Johnston D, Waters CM and Norton JD. Early response gene signalling cascades activated by ionising radiation in primary human B cells. *Oncogene* 1993; **8**: 3229–3237.

57. Sukatme VP, Kartha S, Toback FG, Taub R, Hoover RG and Tsai-Morris CH. A novel early growth response gene rapidly induced by fibroblast, epithelial cell and lymphocyte mitogens. *Oncogene Res.* 1987; **1**: 343–355.

58. Christy BA, Lau LF and Nathans D. A gene activated in mouse 3T3 cells by serum growth factors encodes a protein with "zinc finger" sequences. *Proc. Natl. Acad. Sci. USA* 1988; **85**: 7857–7861.

59. Milbrandt J. A nerve growth factor-induced gene encodes a possible transcriptional regulatory factor. *Science* 1987; **238**: 797–799.

60. Lemaire P, Revelant O, Bravo R and Chanay P. Two mouse genes encoding potential transcription factors with identical DNA-binding domains are activated by growth factors in cultured cells. *Proc. Natl. Acad. Sci. USA* 1988; **85**: 4691–4695.

61. Lim RW, Varnum BC and Herschman HR. Cloning of tetradecanoyl phorbol ester-induced "primary response" sequences and their expression in density-arrested Swiss 3T3 cells and a TPA non-proliferative variant. *Oncogene* 1987; **1**: 263–270.

62. Wright JJ, Gunter KC, Mitsuya H, Irving SG, Kelly K and Siebenlist U. Expression of a zinc finger gene in HTLV-1- and HTLV- II-transformed cells. *Science* 1990; **248**: 588–591.

63. Cho KO, Skarnes WC, Minsk B, Palmieri S, Jackson-Grusbe L and Wagner JA. Nerve growth factor regulates gene expression by several distinct mechanisms. *Mol. Cell. Biol.* 1989; **9**: 135–143.

64. Siderovski DP, Blum S, Forsdyke RE and Forsdyke DR. A set of human putative lymphocyte G_0/G_1 switch genes includes genes homologous to rodent cytokine and zinc finger-encoding genes. *DNA Cell. Biol.* 1990; **9**: 579–587.

65. Shimizu N, Ohta M, Fujiwara C, Sagara J, Mochizuki N, Oda T and Utiyama H. A gene coding for a zinc-finger protein is induced during 12-O-tetradecanoylphorbol-12-acetate- stimulated HL60 cell-differentiation. *J. Biochem.* 1992; **111**: 272–277.

66. Matheny C, Day ML and Milbrandt, J. Nuclear-localisation signal of NGFI-A is located within the zinc-finger DNA-binding domain. *J. Biol. Chem.* 1994; **11**: 8176–8181.

67. Gottschalk AR, Joseph LJ and Quintans J. Fc-gamma-RII cross-linking inhibits anti-Ig-induced egr-1 and egr-2 expression in BCL₁. *J. Immunol.* 1994; **152**: 2115–2122.

68. Enslen H and Soderling TR. Roles of calmodulin-dependent protein kinases and phosphatases in calcium-dependent transcription of immediate early genes. *J. Biol. Chem.* 1994; **269**: 20872–20877.

69. Tominaga T, De la Cruz J, Burrow GN and Meinkoth, JL. Divergent patterns of immediate early gene expression in response to thyroid-stimulating hormone and insulin-like growth factor I in Wistar rat thymocytes. *Endocrinology* 1994; **135**: 1212–1219.

70. Epner DE and Herschman HR. Heavy-metals induce expression of the TPA-inducible sequence (TIS) genes. *J. Cell. Physiol.* 1991; **148**: 68–74.

71. Kharbanda S, Saleem A, Shafman T, Emoto Y, Weischelbaum R and Kufe D. Activation of the *pp90^ras* and mitogen-activated serine threonine protein kinases by ionising radiation. *Proc. Natl. Acad. Sci. USA,* 1994; **91**: 5416–5420.

72. Crosby SD, Puertz JJ, Simburger KS, Fahrner TJ and Milbrandt, J. The early response gene NGFI-C encodes a zinc finger transcriptional activator and is a member of the GCGGGGGCG (GSG) element-binding protein family. *Mol. Cell Biol.* 1991; **11**: 3835–3841.

73. Swirnoff AH and Milbrandt, J. DNA-binding specificity of NGFI-A and related zinc-finger transcription factors. *Mol. Cell Biol.* 1995; **15**: 2275–2287.

74. Monroe J, Yellen-Shaw AJ and Seyfert VL. Molecular basis for unresponsiveness and tolerance induction in immature stage B-lymphocytes. *Adv. Mol. Cell. Immunol.,* 1993; **1B**: 1–32.

75. Seyfert VL, Sukhatme VP and Monroe JG. Differential expression of a zinc finger-encoding gene in response to positive versus negative signalling through receptor immunoglobulin in murine B lymphocytes. *Mol. Cell. Biol.* 1989; **9**: 2083–2088.

76. Seyfert VL, McMahon SB, Glenn WD, Yellen AJ, Sukhatme VP, Cao X and Monroe JG. Methylation of an immediate-early gene as a mechanism for B cell tolerance induction. *Science* 1990; **250**: 797–800.

77. McMahon SB and Monroe JG. A ternary complex factor-dependent mechanism mediates induction of egr-1 through selective serum response elements following antigen receptor cross-linking in B lymphocytes. *Mol. Cell. Biol.* 1995; **15**: 1086–1093.

78. Perezcastillo A, Pipaon C, Garcia I and Alemany S. NGFI-A gene expression is necessary for T-lymphocyte proliferation. *J. Biol. Chem.* 1993; **268**: 19445–19450.

79. Nquyen HQ, Hoffman-Liebermann B and Liebermann DA. The zinc-finger transcription factor egr-1 is essential for and restricts differentiation along the macrophage lineage. *Cell,* 1993; **72**: 197–209.

80. Lee SL, Tourtellotte LC, Wesselschmidt RL and Milbrandt J. (1995) Growth and differentiation proceeds normally in cells deficient in the immediate-early gene NGFI-A. *J. Biol. Chem.,* 1995; **270**: 9971–9977.

81. Milbrandt J. Nerve growth factor induces a gene homologous to the glucocorticoid gene. *Neuron* 1988; **1**: 183–188.

82. Hazel TG, Nathans D and Lau LF. A gene inducible by growth factors encodes a member of the steroid and thyroid receptor superfamily. *Proc. Natl. Acad. Sci. USA* 1988; **85**: 8444–8448.

83. Ryseck R-P, Macdonald-Bravo H, Mattei MG, Ruppert S and Bravo R. Structure, mapping and expression of a growth factor inducible gene encoding a nuclear hormonal binding receptor. *EMBO J.* 1989; **8**: 3327–3335.

84. Mapara MY, Weinmann P, Daniel P, Bargou R, Bommert K and Dorken B. Mechanisms of activation-induced cell-death in human B-cell lymphoma—possible involvement of NAK-1, the human nur77 homolog. *Blood* 1994; **84**: A291.

85. Uemura H, Mizokami A and Chang CS. Identification of a new enhancer in the promoter region of human TR3 orphan receptor gene—a member of steroid-receptor superfamily. *J. Biol. Chem.* 1995; **270**: 5427–5433.

86. Bondy GP. Phorbol ester, forskolin, and serum induction of a human colon nuclear hormone receptor gene related to the nur77/NGFI-B genes. *Cell Growth Differ.* 1991; **2**: 203–208.

87. Garcia I, Pipaon C, Alemany S and Perezcastillo A. Induction of NGFI-B gene-expression during T-cell activation—role of protein phosphatases. *J. Immunol.* 1994; **153**: 3417–3425.

88. Wilson TE, Fahrner TJ, Johnston M and Milbrandt J. Identification of the DNA-binding site for NGFI-B by genetic selection in yeast. *Science* 1991; **252**: 1296–1300.

89. Wilson TE, Fahrner TJ and Milbrandt J. The orphan receptors NGFI-B and Steroidogenic Factor 1 establish monomer binding as a third paradigm of nuclear receptor–DNA interaction. *Mol. Cell. Biol.* 1993; **13**: 5794–5804.

90. Ohkura N, Hijikuro M, Yamamoto A and Miki K. Molecular cloning of a novel thyroid/steroid receptor superfamily gene from cultured rat neuronal cells. *Biochem. Biophys. Res. Commun.* 1994; **205**: 1959–1965.

91. Law SW, Conneely OM, DeMayo FJ and O'Malley BW. Identification of a new brain-specific transcription factor, NURR1. *Mol. Endocrinol.* 1992; **6**: 2129–2135.

92. Scearce LM, Laz TM, Hazel TG, Lau LF and Taub R. RNR-1, a nuclear receptor in the NGFI-B/nur77 family that is rapidly induced in regenerating liver. *J. Biol. Chem.* 1993; **268**: 8855–8861.

93. Okabe T, Takayanagi R, Imasaki K, Haji M, Nawata H and Watanabe T. cDNA cloning of a NGFI-B/Nur77-related transcription factor from an apoptotic T-cell line. *J. Immunol.* 1995; **154**: 3871–3879.

94. Perlmann T and Jansson L. A novel pathway for vitamin-A signalling mediated by RXR heterodimerisation with NGFI-B and NURR1. *Gene Dev.* 1995; **9**: 769–782.

95. Davis FJ, Hazel TG, Chen RH, Blenis J and Lau LF. Functional domains and phosphorylation of the orphan receptor nur77. *Mol. Endocrinol.* 1993; **7**: 953–961.

96. Paulsen RE, Weaver CA, Fahrner TJ and Milbrandt J. Domains regulating transcriptional activity of the inducible orphan receptor NGFI-B. *J. Biol. Chem.* 1992; **267**: 16491–16496.

97. Liu ZG, Smith SW, McLaughlin KA, Schwarz LM and Osborne BA. Apoptotic signals delivered through the T-cell receptor of a T-cell hybrid require the immediate-early gene nur77. *Nature* 1994; **367**: 281–284.

98. Woronicz JD, Calnan B, Ngo V and Winoto A. Requirements for the orphan steroid-receptor nur77 in apoptosis of T-cell hybridomas. *Nature* 1994; **367**: 277–281.

99. Yazdanbakhsh K, Choi JW, Li YZ, Lau LF and Choi YW. Cyclosporine-A blocks apoptosis by inhibiting the DNA-binding activity of the transcription factor nur77. *Proc. Natl. Acad. Sci. USA* 1995; **92**: 437–441.

100. Ullman KS, Northrop JP, Vrweij CL and Crabtree GR. Transmission of signals from the T lymphocyte antigen receptor to the genes responsible for cell proliferation and immune function: the missing link. *Ann. Rev. Immunol.* 1990; **8**: 421–452.

101. Umlauf SW, Beverly B, Kang S-M, Brorson K, Tran A-C and Schwartz RH. Molecular regulation of the IL-2 gene: rheostatic control of the immune system. *Immunol. Rev.* 1993; **133**: 178–197.

102. Taira S, Matsui M, Hayakawa K, Yokohama T and Nariuchi H. Interleukin secretion by B cell lines and splenic B cells stimulated with calcium ionophore and phorbol ester. *J. Immunol.* 1987; **139**: 2957–2964.

103. Jung LKL, Hara T and Fu SM. Detection and functional studies of p60–65 (Tac antigen) on activated human B cells. *J. Exp. Med.* 1984; **160**: 1597–1602.

104. Waldmann TA, Goldman CK, Korsmeyer SJ and Greene WC. Expression of interleukin 2 receptors on activated human B cells. *J. Exp. Med.* 1984; **160**: 1450–1466.

105. Nakanishi K, Malek TR, Smith KA, Hamaoka T, Shevach EM and Paul WE. Both interleukin 2 and second Tcell-derived factor in EL-4 Supernatant have activity as differentiation factors in IgM synthesis *J. Exp. Med.* 1984; **160**: 1605–1619.

106. Waldmann TA. The multi-subunit interleukin-2 receptor. *Ann. Rev. Biochem.* 1989; **58**: 875–911.

107. Dower SK and Urdal DL. The interleukin-1 receptor. *Immunol. Today* 1987; **5**: 46–51.

108. Dinarello CA. IL-1 and its biologically related cytokines. *Adv. Immunol.* 1989; **44**: 153–205.

109. Wodnar-Filipowicz A and Moroni C. Regulation of interleukin 3 mRNA expression in mast cells occurs at the posttranscriptional level and is mediated by calcium ions. *Proc. Natl. Acad. Sci. USA* 1990; **87**: 777–781.

110. Gorospe M, Kumar S and Baglioni C. Tumour necrosis factor increases stability of interleukin-1 mRNA by activating protein kinase C. *J. Biol. Chem.* 1993; **268**: 6214–6220.

111. Burd PR, Freeman GJ, Wilson SD, Berman M, DeKruyff R, Billings PR and Dorf ME. Cloning and characterization of a novel T cell activation gene. *J. Immunol.* 1987; **139**: 3126–3131.

112. Wilson SD, Kuchroo VK, Israel DI and Dorf ME. Expression and characterisation of TCA3: A murine inflammatory protein. *J. Immunol*, 1990; **145**: 2745–2750.

113. Rohan PJ, Davis P, Moscaluk A, Kearns M, Krutzsch H, Siebenlist U and Kelly K. PAC-1: A mitogen-induced nuclear protein tyrosine phosphatase, *Science* 1993; **259**: 1763–1766.

114. Charles CH, Abler AS and Lau LF. cDNA sequence of a growth factor inducible immediate early gene and characterization of its encoded protein. *Oncogene* 1992; **7**: 187–190.

115. Keyse SM and Emslie EA. Oxidative stress and heat shock induce a human gene encoding a protein tyrosine phosphatase. *Nature* 1992; **359**: 644–647.

116. Ward Y, Gupta S, Jenson P, Wartmann M, Davis RJ and Kelly K. Control of MAP kinase activation by the mitogen-induced threornine/tyrosine phosphotase, PAC1. *Nature* 1994; **367**: 651–654.

117. Maguire J, Santoro T, Jenson P, Siebenlist U, Yewdell J and Kelly K. Gem: an induced immediate early protein belonging to the Ras family. *Science* 1994; **265**: 241–244.

118. Deed RW, Bianchi SM, Atherton GT, Johnston D, Santibanez-Koref M, Murphy JJ, and Norton JD. An immediate early human gene encodes an Id-like helix-loop-helix protein and is regulated by protein kinase C activation in diverse cell types. *Oncogene* 1993; **8**: 599–607.

119. Ellmeier W, Aguzzi A, Kleiner E, Kurzbauer R and Weith A. Mutually exclusive expression of a helix-loop-helix gene and N-myc in human neuroblastomas and in normal development. *EMBO J.* 1992; **11**: 2563–2571.

120. Christie BA, Saunders LK, Lau LF, Copeland NG, Jenkins NA and Nathans D. An Id-related immediate early protein encoded by a growth factor inducible gene. *Proc. Nat. Acad. Sci., USA* 1991; **88**: 1815–1819.

121. Littlewood T.D. and Evan G.I. Transcription factors 2: Helix-loop-helix; *Protein Profile* 1994; **1**: 639–707.

122. Deed RW, Hirose T, Mitchell ELD, Santibanez-Koref MF and Norton JD. Structural organization and chromosomal mapping of the human Id-3 gene, *Gene* 1994; **151**: 309–314.

123. Benezra R, Davis RL, Lockshon D, Turner DL and Weintraub H. The protein Id: a negative regulator of helix-loop- helix DNA binding proteins. *Cell* 1990; **61**: 49–59.

124. Sun X.H. Constitutive expression of the Id-1 gene impairs mouse B cell development, *Cell* 1994; **79**: 893–900.

125. Sun XH, Copeland NG, Jenkins NA and Baltimore D. (1991) Id proteins, Id-1 and Id2 selectively inhibit DNA binding by one class of helix-loop-helix proteins. *Mol. Cell. Biol.* 1991; **11**: 5603–5611.

126. Peverali FA, Rampvist Saffrich R, Peppercok R, Barone MV and Philipson L. Regulation of G1 progression by E2A and Id helix-loop-helix proteins *EMBO J.* 1994; **13**: 4291–4301.

127. Iavarone A, Garg P, Lasorella A, Hsu J and Israel MA. The helix-loop-helix protein Id-2 enhances cell proliferation and binds to the retinoblastoma protein. *Genes* Dev. 1994; **8**: 1270–1284.

128. Green RM, Murphy JJ and Norton JD. Use of cDNA probes for typing cells of B cell lineage: Application of early response genes to the analysis of mature B cell malignancies. Leuk. Lymph. 1991; **3**: 325–329.

129. Newton JS, Deed RW, Mitchell ELD, Murphy JJ and Norton JD. A B cell specific immediate early human gene is located on chromosome band lq31 and encodes an α helical basic phosphoprotein. *Biochim. Biophys. Acta* 1993; **1216**: 314–316.

130. Siderovski DP, Heximer SP and Forsdyke DR. A human gene encoding a putative basic helix-loop-helix phosphoprotein whose mRNA increases rapidly in cycloheximide-treated blood mononuclear cells. *DNA Cell. Biol.* 1994; **12**: 125–145.

131. Druey KM, Kang V, Harrison K and Kehrl KH. Isolation of a new member of the BL34 gene family. *J. Allerg. Clin. Immunol.* 1995; **95**: 176.

132. Wu HK, Heng HHQ, Forsdyke DR, Tsui LC, Mak TW, Mindon MD and Siderovski DP. Differential expression of a putative basic helix-loop-helix phosphoprotein gene, GOS8 in acute leukaemia and localisation to human chromosome 1q31. *Leukaemia* 1995 (in press).

133. Ning Z-Q, Norton JD, Li J and Murphy JJ. An early response gene encoding a zinc finger protein, Berg 36, is a target for interleukin 4 but not for CD40-mediated inhibition of apoptosis in human B cells. *Eur. J. Immunol.* 1996; **26**: 2356–63.

134. Barnard RC, Pascall JC, Brown KD, McKay IA, Williams NS and Bustin SA. Coding sequence of ERF-1, the human homologue of TIS11b/cMG1, members of the TIS11 family of early response genes. *Nucl. Acids Res.* 1993; **21**: 3580.

135. Gomperts MJ, Pascall JC and Brown KD. The nucleotide sequence of a cDNA encoding an EGF-inducible gene indicates the existence of a new family of mitogen-induced genes. *Oncogene* 1990; **5**: 1081–1083.

136. Varnum BC, Ma Q, Chi T, Fletcher B and Herschman H. The TIS11 primary response gene is a member of a gene family that encodes proteins with a highly conserved sequence containing an unusual Cys-His repeat. *Mol. Cell. Biol.* 1991; **11**: 1754–1758.

137. Ma Q and Herschman HR. The yeast homolog, YTIS 11, of the mammalian TIS gene family is a non-essential glucose repressible gene. *Oncogene* 1995; **10**: 487–494.

138. Ning Z-Q and Murphy JJ. Calcium ionophore-induced apoptosis of human B cells is preceded by the induced expression of early response genes. *Eur. J. Immunol.* 1993; **23**: 3369–3372.

139. Koslow EJ, Wilson GL, Fox CH and Kehrl JH. Subtracted cDNA cloning of a novel member of the Ig gene super family expressed at high levels in activated B lymphocytes. *Blood* 1993; **81**: 454–461.

140. Deguchi Y, Moroney JF, Wilson GL, Fox CH, Winter HS and Kehrl JH. Cloning of a human homeobox gene that resembles a diverged *Drosophila* homeobox gene and is expressed in activated lymphocytes. *New Biol.* 1991; **3**: 353–363.

141. Deguchi Y, Thevenia C and Kehrl JH. Stable expression of HB24, a diverged human homeobox gene in T lymphocytes induces genes involved in T cell activation and growth. *J. Biol. Chem.* 1992; **267**: 8222–8229.

142. Totterman TH, Nilsson K and Sundstrum C. Phorbol ester-induced differentiation of chronic lymphocytic leukaemia cells. *Nature* 1980; **288**: 176–178.

143. Drexler HG, Janssen JWG, Brenner MK, Hoffbrand AV and Bartram CR. Rapid induction of protooncogenes c-fos and c-myc in B-chronic lymphocytic leukemia cells during differentiation induced by phorbol ester and calcium ionophore. *Blood* 1989; **73**: 1656–1663.

144. Murphy JJ, Tracz M and Norton JD. Patterns of nuclear proto-oncogene expression during induced plasmacytoid differentiation and proliferation of human B chronic lymphocytic leukaemia cells. *Immunology* 1990; **69**: 490–493.

145. Gignac SM, Buschle M, Pettit GR, Hoffbrand AV and Drexler HG. Expression of proto-oncogene c-jun during differentiation of B chronic lymphocytic leukemia. *Leukemia* 1990; **4**: 441–444.

146. Murphy JJ and Norton JD. Phorbol ester induction of early response gene expression in lymphocytic leukemia and normal human B cells. *Leuk. Res.* 1993; **17**: 657–662.

147. Murphy JJ, Yaxley JC and Norton JD. Evidence for protein kinase C-independent pathways mediating phorbol ester induced plasmacytoid differentiation of B chronic lymphocytic leukemia cells. *BBA* 1991; **1092**: 110–118.

148. Pettit GR, Day JF, Hartwell JL and Wood HB. Antineoplastic components of marine animals. *Nature* 1970; **227**: 962–965.

149. Murphy JJ and Norton JD. Coupling of early response gene expression to distinct regulatory pathways during α-interferon and phorbol ester-induced plasmacytoid differentiation of B chronic lymphocytic leukaemia cells. *FEBS Letts.* 1990; **267**: 242–244.

150. Murphy JJ and Norton JD. Multiple signalling pathways mediate anti-Ig and IL-4-induced early response gene expression in human tonsillar B cells. *Eur. J. Immunol.* 1993; **23**: 2876–2881.

151. Cambier JC, Pleiman CM and Clark MR. Signal transduction by the B cell antigen receptor and its coreceptors. *Ann. Rev. Immunol.* 1994; **12**: 457–486.

152. Gold MR and DeFranco AL. Biochemistry of B lymphocyte activation. *Adv. Immunol.* 1994; **55**: 221–295.

153. Liu J, Chiles TC, Sen R and Rothstein TL. Inducible nuclear expression of NF-kappa B in primary B cells stimulated through the surface Ig receptor. *J. Immunol.* 1991; **146**: 1685–1691.

154. Rooney JW, Dubois PM and Sibley CH. Cross-linking of surface IgM activates NF-kappa B in B lymphocytes. *Eur. J. Immunol.* 1991; **21**: 2993–3008.

155. Fisher CL, Ghysdael J and Cambier JC. Ligation of membrane Ig leads to calcium-mediated phosphorylation of the proto-oncogene product, Ets-1. *J. Immunol.* 1991; **146**: 1743–1749.

156. Xie H, Chiles TC and Rothstein TL. Induction of CREB activity via the surface Ig receptor of B cells. *J. Immunol.* 1993; **151**: 880–889.

157. Monroe JG. Up-regulation of c-fos expression is a component of the mIg signal transduction mechanism but is not indicative of competence for proliferation. *J. Immunol.* 1988; **140**: 1454–1460.

158. Seyfert VL, McMahon S, Glenn W, Cao X, Sukhatme VP and Monroe JG. Egr-1 expression in surface Ig-mediated B cell activation. Kinetics and association with protein kinase C activation. *J. Immunol.* 1990; **145**: 3647–3653.

159. Tilzey JF, Chiles TC and Rothstein TL. Jun-B gene expression mediated by the surface immunoglobulin receptor of primary B lymphocytes. *Biochem. Biophys. Res. Commun.* 1991; **175**: 77–83.

160. White MW, McConnell F, Shu GL, Morris DR and Clark EA. Activation of dense human tonsillar B cells: Induction of c-myc gene expression via two distinct signal transduction pathways. *J. Immunol.* 1991; **146**: 846–853.

161. Mittelstadt PR and DeFranco AL. Induction of early response genes by cross-linking membrane Ig on B lymphocytes. *J. Immunol.* 1993; **150**: 4822–4832.

162. Klemsz MJ, Justement LB, Palmer E and Cambier JC. Induction of c-fos and c-myc expression during B cell

activation by IL-4 and immunoglobulin binding ligands. *J. Immunol.* 1989; **143**: 1032–1039.

163. McMahon SB and Monroe JG. Activation of the p21[ras] pathway couples antigen receptor stimulation to induction of the primary response gene egr-1 in B lymphocytes. *J. Exp. Med.* 1995; **181**: 417–422.

164. Clark EA, Shu GL, Luscher B, Draves KE, Banchereau J, Ledbetter JA and Valentine MA. Activation of human B cells: Comparison of the signal transduced by IL-4 to four different competence signals. *J. Immunol.* 1989; **143**: 3873–3880.

165. Sarma V, Lin Z, Clark L, Rust BM, Tewari M, Noelle RJ and Dixit VM. Activation of the B-cell surface receptor CD40 induces A20, a novel zinc finger protein that inhibits apoptosis. *J. Biol. Chem.* 1995; **270**: 12343–12346.

166. Keegan AD, Nelms, K, Wang L-M, Pierce J and Paul WE. Interleukin 4 receptor: signaling mechanisms. *Immun. Today* 1994; **15**: 423–431.

167. Durie FH, Foy FM, Masters SR, Laman JD and Noelle RJ. The role of CD40 in the regulation of humoral and cell-mediated immunity. *Immunol. Today* 1994; **15**: 406–410.

168. Berberich I, Shu GL and Clark EA. Cross-linking CD40 on B cells rapidly activates nuclear factor-kappa B. *J. Immunol.* 1994; **153**: 4357–4366.

169. Klaus GGB, Choi MSK and Holman M. Properties of mouse CD40. Ligation of CD40 activates B cells via a Ca^++-dependent FK506-sensitive pathway. *Eur. J. Immunol.* 1994; **24**: 3229–3232.

170. Francis DA, Karras JG, Ke XY, Sen R and Rothstein TL. Induction of the transcription factors NF-kappa B, AP-1 and NF-AT during B cell stimulation through the CD40 receptor. *Int. Immunol.* 1995; **7**: 151–161.

171. Scherr CJ and in Roberts JM. Inhibitors of mammalian G_1 cyclin-dependent kinases. *Genes Dev.* 1995; **9**: 1149–1163.

172. Liang P and Pardee AB. Differential display of eukaryotic messenger RNA by means of polymerase chain reaction. *Science* 1992; **257**: 967–971.

173. Sonobe MH, Bravo R and Armelin MS. Imbalanced expression of cellular nuclear oncogenes caused by v-sis/PDGF-2. *Oncogene* 1991; **6**: 1531–1537.

174. Kelly K, Davis P, Mitsuya H, Irving S, Wright J, Grassmann R, Fleckenstein V, Wano Y, Greene W and Siebenlist U. A high proportion of early response genes are constitutively activated in T cells by HTLV1. *Oncogene* 1992; **7**: 1463–1470.

175. Peng M and Lundgren E. Transient expression of the Epstein Barr virus LMP 1 gene in human primary B cells induces cellular activation and DNA synthesis. *Oncogene* 1992; **7**: 1775–1782.

176. McMahon SB, Norvell A, Levine KJ and Monroe JG. Transient transfection of murine B lymphoblasts as a method for examining gene regulation in primary B cells. *J. Immunol. Methods* 1995; **179**: 251–259.

177. Chen J, Lansford R, Stewart V, Young F and Alt FW. RAG-2-deficient blastocyst complementation: an assay of gene function in lymphocyte development. *Proc. Natl. Acad. Sci. USA* 1993; **90**: 4528–4532.

178. Kehrl JH. Haematopoietic lineage commitment: Role of transcription factors. *Stem Cells* 1995; **13**: 223–241.

179. Gu H, Marth JD, Orban PC, Mossman H and Rajewsky K. Deletion of a DNA polymerase beta gene segment in T cells using cell type specific gene targeting *Science* 1994; **265**: 103–106.

180. Heximer SP and Forsdyke DR. A human putative lymphocyte G0/G1 switch gene homologous to a rodent gene encoding a zinc-binding potential transcription factor. *DNA* and *Cell Biol.* 1993; **12**: 73–88.

181. Turner CA, Mack DH and Davis MM. Blimpl, a novel zinc- finger-containing protein that can drive the maturation of B-lymphocytes into immunoglobulin-secreting cells. *Cell* 1994; **77**: 297–306.

182. Huang S. Blimp-1 is the murine homolog of the human transcriptional repressor PRDI-BF1. *Cell* 1994; **78**: 9.

183. Tewari M, Wolf FW, Seldin MF, O'Shea KS, Dixit VM and Turka LA. Lymphoid expression and regulation of A20, an inhibitor of programmed cell death. *J. Immunol.* 1995; **154**: 1699–1706.

184. Coleclough C, Kuhn L and Lefkowits I. Regulation of mRNA abundance in activated T-lymphocytes: Identification of mRNA species affected by the inhibition of protein synthesis. *Proc. Natl. Acad. Sci. USA* 1990; **87**: 1753–1757.

185. Scott JL, Dunn SM, Zeng T, Baker E, Sutherland GR and Burns GF. Phorbol ester-induced transcription of an immediate-early response gene by human T-cells is inhibited by co-treatment with calcium ionophore. *J. Cell. Biochem.* 1994; **54**: 135–144.

186. Najfeld V, Menninger J, Ballard SG, Deguchi Y, Ward DC and Kehrl JH. 2 diverged human homebox genes involved in the differentiation of human hematopoietic progenitors map to chromosome-1, bands q41-42.1. *Genes Chrom. Cancer* 1992; **5**: 343–347.

187. Guan KL and Butch E. Isolation and characterisation of a novel dual specific phosphatase, HVH2, which selectively dephosphorylates the mitogen-activated protein-kinase. *J. Biol. Chem.* 1995; **270**: 7197–7203.

188. Walther Z, May LT and Sehgal PB. Transcriptional regulation of the interferon-β_2 cell differentiation factor BSF-2/hepatocyte-stimulating factor gene in human fibroblasts by other cytokines. *J. Immunol.* 1988; **140**: 974–977.

189. Mielke V, Bauman JGJ, Sticherling M, Ibs T, Zomershoe AG, Seligmann K, Henneike H-H, Schroder J-M, Sterry W and Christophers E. Detection of neutrophil activating peptide NAP/IL-8 and NAP/IL-8 mRNA in human recombinant IL-1alpha-and human recombinant TNF-alpha-stimulated human dermal fibroblasts. *J. Immunol.* 1989; **144**: 153–161.

190. Goldfeld AE, Flemington EK, Boussiotis VA, Theodos CM, Titus RG, Strominger JL and Speck SH. Transcription of the tumour necrosis factor alpha gene is rapidly induced by anti- immunoglobulin and blocked by cyclosporin A and FK506 in human B cells. *Proc. Natl. Acad. Sci. USA* 1993; **89**: 12198–12201.

191. Gray PW, Aggarwal BB, Benton CV, Bringman TS, Henzel WJ, Jarret JA, Leung DW, Moffat B, Ng P, Svedersky LP, Palladino MA and Nedwin GE. Cloning and expression of cDNA for human lymphotoxin, a lymphokine with tumour necrosis activity. *Nature* 1984; **312**: 721–724.

192. Pennica D, Nedwin GE, Hayflick JS, Seeburg PH, Derynck R, Palladino MA, Kohr WJ, Aggarwal BB and

Goeddel DV. Human tumour necrosis factor: precursor structure, expression and homology to lymphotoxin. *Nature* 1984; **312**: 724–729.

193. Sen GC and Lengyel P. The interferon system. *J. Biol. Chem.* 1992; **267**: 5017–5020.

194. Penix L, Weaver WM, Pang Y, Young HA and Wilson CB. Two essential regulatory elements in the human interferon-gamma promoter confer activation specific expression in T-cells. *J. Exp. Med.* 1993; **178**: 1483–1496.

195. Obaru K, Hattori T, Yamamura Y, Takatsuki K, Nomiyama H, Maeda S and Shimada K. A cDNA clone inducible in human tonsillar lymphocytes by a tumour promoter codes for a novel protein of the β-thromboglobulin superfamily. *Mol. Immunol.* 1989; **26**: 423–426.

196. Widmer U, Yang Z, Vandeventer S, Manogue KR, Sherry B and Cerami A. Genomic structure of murine macrophage inflammatory protein-1-alpha and conservation of potential regulatory sequences with a human homolog, LD78. *J. Immunol.* 1991; **146**: 4031–4040.

197. Russell L and Forsdyke DR. The 3rd human homolog of a murine gene encoding an inhibitor of stem-cell proliferation is truncated and linked to a CpG island-containing upstream sequence. DNA *Cell Biol.* 1993; **12**: 157–175.

198. Gelfand EW, Domenico J and Renz H. IL-4 regulates IL4 receptor (IL4R) expression in murine T and B cells.

FASEB J 1991; **5**: A970

199. Graf D, Korthauer U, Mages HW, Senger G and Kroczek RA. Cloning of TRAP, a ligand for CD40 on human T cells. *Eur. J. Immunol.* 1992; **22**: 3191–3194.

200. Pollok KE, O'Brien V, Marshall L, Olson JW, Noelle RJ and Snow EC. The development of competence in resting B cells. *J. Immunol.* 1991; **146**: 1633–1641.

201. Laitinen J and Holtta E. Methylation status of an early response gene (ornithine decarboxylase) in resting and stimulated NIH-3T3 fibroblasts. *J. Cell. Biochem.* 1994; **55**: 155–167.

202. Tanaka S and Hosaka K. Cloning of a cDNA encoding a second phosphatidylinositol transfer protein of rat brain by complementation of the yeast sec 14 mutation *J. Biochem.* 1994; **115**: 981–984.

203. Moi P, Chan K, Asunis I, Cao A and Wei Kan, Y. Isolation of NF-E2-related factor 2 (Nrf2), a NF-E2-like basic leucine zipper transcriptional activator that binds to the tandem NF-E2/AP1 repeat of the β-globin locus control region. *Proc. Natl. Acad. Sci. USA* 1994; **91**: 9926–9930.

204. Jenny A, Hauri H-P and Keller W. Characterisation of cleavage and polyadenylation specificity factor and cloning of its 100-kilodalton subunit. *Mol. Cell. Biol.* 1994; **14**: 8183–8190.

12

Signaling for Thymocyte and T Cell Apoptosis

Yili Yang and Jonathan D. Ashwell

Laboratory of Immune Cell Biology, National Cancer Institute, National Institutes of Health, Bethesda, Maryland, USA

INTRODUCTION

Extensive cell death during embryonic development has been observed by developmental biologists since the nineteenth century, and was clearly defined more than 40 years ago [1]. In the mid-1960s, experimental approaches by Saunders, Lockshin, and their colleagues led to the concept that some cells are fated to die during development, and the term programmed cell death (PCD) was coined [2, 3]. The term apoptosis was first used by pathologists Kerr, Wyllie, and Currie to describe the morphology of physiological cell death, which is characterized by condensation and margination of chromatin along the nuclear envelope, cell shrinkage, membrane blebbing, and breakdown of the nucleus into discrete membrane-bounded fragments (apoptotic bodies) [4, 5]. In most cases, initiation of the cell death process requires *de novo* gene expression and protein synthesis, and is subject to regulation by numerous intracellular and extracellular signals [2, 3, 6–9]. However, in many cells the components of the basic cell death machinery are constitutively present, and in these cases, active cell death can take place without synthesis of new proteins, as exemplified by the death mediated by Fas or induced by cytolytic T lymphocytes [10, 11]. It is becoming apparent that while apoptosis reflects programmed forms of cell death, not all examples of PCD result in an apoptotic appearance [12, 13]. However, for lymphocytes, most conditions that cause cell death result in an apoptotic morphology [12, 14], and thus the terms PCD and apoptosis are often used interchangeably to describe the process of lymphocyte cell death.

The phenomenon of regulated cell death is found widely in the immune system, nowhere more so than the thymus. During the course of their maturation and differentiation, thymocytes express CD8, CD4, and a stochastically-generated antigen-specific receptor (TCR). The vast majority (95%) of the so-called double-positive thymocytes ($CD4^+CD8^+$) are destined to die in the thymus [15]. In fact, the default outcome for these cells appears to be death: thymocytes die if their TCRs fail to engage self-ligands (the combination of antigenic peptides and major histocompatibility complex (MHC)-encoded proteins, or antigen/MHC), a process that has been called "death by neglect" [16, 17]. Some fraction of developing thymocytes recognize self-ligands with high avidity and undergo activation-induced apoptosis [18, 19], a process known as negative selection. The small number of remaining thymocytes that either recognize self-ligands with low-to-moderate avidity [20], or that perhaps encounter partial agonistic/antagonist peptides [21] are rescued from the default death pathway and differentiate into $CD4^+ CD8^+$ or $CD4^- CD8^+$ (single-positive) thymocytes. Thus, death of thymocytes is critical for the establishment of an antigen-specific and self-tolerant immune repertoire. Further, recent data have established that antigen-driven apoptosis of peripheral activated T cells is also important for maintenance of self-tolerance and termination of an ongoing immune response [22–24]. In addition to undergoing apoptotic cell death, certain T cells (such as $CD8^+$ cytolytic T lymphocytes) can themselves induce apoptosis in target cells, which may be important in the elimination of neoplastic or virally infected cells [25, 26]. Finally, alterations in apoptosis are accompanied by diseases of the immune system. For example, prevention of apoptosis by up-regulation of Bcl-2 activity results in autoimmunity, and in combination with a "second hit" such as *c-myc* rearrangement, lymphoid malignancies [27]. On the other side of the coin, lymphocytes from individuals infected with HIV are abnormally prone to undergo apoptosis when cultured or activated *in vitro*, leading to the speculation that inappropriate cell death may contribute to the immunodeficiency of AIDS [28].

In this chapter we will deal with some of the extracellular signals that induce T cell apoptosis and present evidence for the involvement of particular intracellular signals and genes in T cell apoptotic death. Regulation of antigen-specific negative thymocyte selection and the analogous but apparently mechanistically distinct deletion of peripheral T cells will be discussed based on our current understanding of apoptosis.

INDUCTION OF T CELL APOPTOSIS BY EXTRACELLULAR STIMULI

Activation-induced apoptosis

T cell recognition of antigen/MHC induces a series of intracellular events, collectively known as cellular activation, that in the presence of proper costimulatory signals typically leads to lymphokine production and cell proliferation [29, 30]. However, in certain circumstances T cell activation may give rise to apoptosis rather than proliferation, a phenomenon that has been appreciated only recently. For instance, restimulation of pre-activated mature T cells with antigen or anti-CD3 results in apoptotic cell death (31–36). Activation-induced apoptosis can also be observed under certain experimental conditions when T cell clones are activated *in vitro* with specific antigen, anti-TCR antibodies, or superantigens [22, 37–39], and *in vivo* when high-dose antigen is administrated [40, 41]. It has recently been shown that activation-induced apoptosis of mature T cells is mediated by the interaction of Fas with Fas ligand (FasL) [23, 42–44]. Activation of resting (non-pre-activated) T cells can also result in apoptosis under certain circumstances. For example, crosslinking of CD4 followed by crosslinking the TCR results in apoptosis of mouse CD4+ peripheral T cells [45], and antibody crosslinking of HIV gp120

bound to CD4 on human T cells predisposes these cells to die when they are stimulated via the TCR [46, 47].

In contrast to mature T cells, immature thymocytes readily undergo apoptosis when exposed to anti-TCR antibodies in fetal thymic organ culture [18] or *in vivo* [48] without a requirement for pre-activation (although in the latter experimental example it is possible that circulating glucocorticoids, to which thymocytes are very sensitive, and which can be elevated due to cytokine production by anti-TCR-activated peripheral T cells, contribute to this response [49, 50]). T cell Vβ-expressing subsets can be activated by bacterial or endogenous (retrovirally encoded) superantigens. Injection of superantigen leads to deletion of thymocytes with the corresponding Vβs, a finding that provided the first clear example of establishment of self-tolerance by the elimination of self-reactive T cells [51, 52]. The apoptotic nature of superantigen-induced cell death was indicated by the appearance of fragmented DNA and its susceptibility to blockade by RNA and protein synthesis inhibitors [53]. Furthermore, stimulation by the physiological ligand for the TCR, antigen/MHC, causes apoptosis of double-positive thymocytes [19, 54, 55]. These studies clearly indicate that, in contrast to the typical proliferative response of resting peripheral T cells, activation via the TCR largely promotes cell death of immature thymocytes.

TCR crosslinking alone does not seem to be enough to cause thymocyte death. When cultured in suspension, thymocytes are relatively insensitive to anti-TCR-induced death, although they die readily when cultured with PMA (phorbol ester) and/or ionomycin (Ca^{2+} ionophore) [56]. This rather unexpected result suggested that costimulatory signals might be required for activation-induced thymocyte death, a speculation that was supported by the finding that anti-TCR activation of thymocytes in suspension culture killed the thymocytes in the presence, but not in the absence, of fibroblasts [57]. Although no requirement for the prototypic costimulatory molecule, CD28, was found in these studies, another group demonstrated that anti-TCR antibodies were able to kill double-positive thymocytes in suspension culture when stimulatory anti-CD28 antibodies were added [58]. In fact, in addition to CD28, crosslinking of a number of cell surface molecules has been shown to provide "cosignals" for thymocyte death, including LFA-1, CD4, and CD8 [59, 60]. Which of these molecules, if any, is necessary to allow activation-mediated negative selection is not known.

Intracellular events such as phosphorylation of proteins on tyrosine residues, activation of phospholipase Cγ and protein kinase C (PKC), and an increase in free intracellular Ca^{2+} are part of the T cell activation cascade, and are required for the biological responses of the cell [61, 62]. These same events are also required for activation-induced T cell apoptosis. For example, protein tyrosine kinase inhibitors [63], PKC inhibitors or depletion of PKC [64], and inhibition of calcineurin [48, 65] block activation-induced apoptotic death of T cell hybridoma and/or thymocytes. It has recently been suggested that superantigens may induce T cell apoptosis by activating phospholipase A_2, resulting in the generation of hydroxyl radicals and the activation of guanylate cyclase, something that TCR occupancy with antigen/MHC (which did not lead to apoptosis) did not do [66]. Particular attention has been paid to the possible role of increased intracellular Ca^{2+} in T cell apoptosis, since not only is this observed in activated thymocytes and T cell hybridomas, but it has been reported in cells undergoing apoptotic death after exposure to cytotoxic T cells, glucocorticoids, and various chemical agents [67–70]. Chelation of Ca^{2+} prevents T cell hybridoma and thymocyte apoptosis [65, 71], and Ca^{2+} ionophores cause thymocyte death [7]. In addition, increasing intracellular Ca^{2+} activates a variety of intracellular proteases, phospholipases, protein kinases, and phosphatases, as well as Ca^{2+} and Mg^{2+} dependent endonucleases [72], enzymes that have been implicated in apoptosis [73–75]. Although Ca^{2+} is required for apoptosis, there is some controversy about whether Ca^{2+} elevations play a causative role in T cell death [76–78]. One problem in this area is that it is difficult to distinguish between the biological effects of transient Ca^{2+} elevations and the requirement for a minimal concentration of intracellular Ca^{2+} to allow certain biochemical processes to occur. At this time the issue of whether receptor-induced elevations of intracellular Ca^{2+} initiate apoptosis is still unresolved.

Great effort has been made to identify the so-called "death genes" that are induced by activation and presumably cause apoptosis in T cells. One of the better characterized candidates is the orphan steroid receptor Nur 77, also known as NGFI-B, N10, or TIS1, which is highly expressed in thymocytes and T hybridoma cells that are undergoing apoptosis [79, 80]. A dominant negative form of Nur77, or antisense oligonucleotides that are able to inhibit Nur77 expression, prevents activation-induced T hybridoma cell death [79, 80]. However, it is unlikely that the only function of Nur77 is as part of a cell death pathway, since it can be induced in normal T cells with mitogenic (non-apoptotic) stimuli such as PMA or A23187. In addition, up-regulation of several other genes in dying T cells was also reported [81, 82], but their roles in apoptotic death are even less well understood.

Glucocorticoids

Thymocytes exposed to glucocorticoids die by apoptosis [83, 84]. This phenomenon may be important under physiological conditions, since adrenalectomy of adult rats results in thymic hypertrophy [85]. Moreover, stress-induced elevation of glucocorticoid hormone can lead to atrophy of thymus [86, 87]. Largely on the basis of such data, it has been speculated that glucocorticoids contribute to the "death by neglect" of thymocytes bearing TCRs with subthreshold avidity for self-antigen/MHC [88]. Among thymocytes, the double-positive thymocytes are the most sensitive to killing by glucocorticoids. Single-positive cells are relatively insensitive to steroid-induced apoptosis, despite the fact that they have similar levels of glucocorticoid receptors [89, 90]. Peripheral T cells are also relatively resistant to glucocorticoid-induced apoptosis, although they are sensitive to glucocorticoid-mediated suppression of, among other things, IL-2 production and T cell proliferation [91]. Since peripheral T cells from Bcl-2 knockout mice are as sensitive to the lethal effects of glucocorticoids as are double-positive thymocytes from normal animals, it appears that the relative resistance of single-positive thymocytes and mature T cells to glucocorticoids is that they, but not double-positive thymocytes, constitutively express Bcl-2 [92].

Glucocorticoids bind to an intracytoplasmic receptor that then translocates to the nucleus and binds to specific DNA sequences and regulates gene expression [93]. Mutational analyses have shown that the glucocorticoid receptor is required for the lethal effects of glucocorticoids [94]. Several groups have used subtractive hybridization to isolate genes whose expression increases during steroid-induced death. Among the genes characterized are Tcl-30 (a T-cell-specific gene expressed in immature thymocytes), calmodulin, a mitochondrial phosphate carrier protein, and several other genes of unknown function [95, 96]. Whether any of these genes is involved in glucocorticoid-induced apoptosis, or whether they are regulated genes involved in other aspects of glucocorticoid action, is not known at this time. Interestingly, the glucocorticoid receptor can also exert negative effects on gene expression through interaction with other transcription factors, notably AP-1 [97–99]. It was originally reported that glucocorticoid receptor transactivation-defective (N-terminal truncation) mutants were unable to induce glucocorticoid-mediated apoptosis when expressed in the S49 lymphoma cell line [100]. More recently, however, other transactivation-defective glucocorticoid receptor mutants that still retain the ability to inhibit AP-1 activity were shown t oinduce glucocorticoid-mediated apoptosis in Jurkat T cells [101], raising the intriguing possibility that rather than directly regulating the expression of "death genes", glucocorticoids might induce apoptosis by interfering with other transcription factors that are essential for cell survival. These apparently contradictory results may reflect the different mutations and/or cell lines used.

Glucocorticoid-induced apoptosis can be prevented by occupancy or perturbation of a variety of cell surface receptors. One striking example is that while both glucocorticoids and activation with anti-TCR antibodies cause apoptotic death of T cell hybridomas, these cells survive when exposed to the two stimuli simultaneously. Activation also prevents glucocorticoid-induced death of thymocytes and Th1 and Th2 cell clones [102, 103]. This phenomenon, termed mutual antagonism, and its possible biological significance will be discussed later in this chapter. Certain cytokines can also prevent glucocorticoid-induced apoptosis. For thy-mocytes, it has been reported that IL-4 rescues double-negative and CD4$^+$ single-positive thymocytes from killing by glucocorticoids [104]. IL-2 was found to protect T cell clones and hybridomas from glucocorticoid-induced apoptosis [105, 106], and IL-2 but not IL-4 rescued Th1 cells from glucocorticoid-induced death, while IL-4 and not IL-2 rescued Th2 cells [107]. Since glucocorticoid levels increase during times of stress, such as an infection, these results raise the possibility that activated (i.e. antigen-specific) T cells are selectively protected from the effects of glucocorticoids by the lymphokines they produce in order to initiate or maintain an immune response.

Fas and FasL

In the course of screening for molecules that inhibit the growth of lymphoid tumor cells, two groups independently derived monoclonal antibodies that induce apoptosis in susceptible cells [10, 108]. The molecule these antibodies recognize, Fas (also known as APO-1 or CD95), is a type I transmembrane protein that belongs to the TNF/NGF receptor family [109, 110]. The ligand for Fas, FasL, is a type II transmembrane protein that belongs to TNF family [111], and cells that express FasL are able to induce apoptotic death of Fas-bearing cells [111]. The tissue distribution of Fas and FasL is quite different: Fas mRNA is constitutively expressed in the thymus, liver, lung, heart, and ovaries, but not unactivated spleen, whereas FasL mRNA is constitutively expressed in the testis, but not thymus or unactivated spleen [111, 112]. However, activation of spleen cells and, to a lesser extent, thymocytes induces FasL expression, which suggested that both Fas and FasL are expressed by T cells upon activation [111]. In fact, Fas is not expressed, or only expressed at low levels, on the majority of peripheral blood T cells, but is highly expressed on activated T cells [108]. It will be possible to define better the cellular distribution of FasL once antibodies to it become available. Interestingly, both Fas and FasL exist in soluble forms [111, 113–115]. The production of soluble Fas results from alternative splicing of its mRNA [113, 114]. Increased serum levels of soluble Fas have been reported in a subset of patients with systemic lupus erythematosis, leading to the speculation that this soluble apoptosis inhibitor might play a role in autoimmunity by preventing the death of autoreactive T cells [113]. However, it has not been shown that the relatively low level of soluble Fas measured in these patients ($\sim 0.2 \mu g \ ml^{-1}$) can block apoptosis *in vivo* or *in vitro*. In contrast to Fas but similarly to TNF [116], soluble FasL is most likely generated by proteolysis of the transmembrane molecule, since it occurs even with cells transfected with FasL cDNA [111, 115, 117]. Moreover, shed FasL can kill Fas$^+$ target cells. This finding is of interest because many normal tissues, as well as transformed and virus-infected cells, express high levels of Fas [109, 118–122]. It is possible that this mechanism accounts for some of the bystander killing and tissue damage caused by activated T cells in an immune response.

Induction of apoptosis by anti-Fas antibodies requires multivalent binding [123]. Based upon the similarity between Fas/FasL and TNFR/TNF [124], it seems likely that the functional signaling unit also consists of trimeric FasL bound to three Fas molecules. In fact, three molecular weight forms of FasL, with migrations consistent with monomers, dimers, and trimers, have been found in supernatants of COS cells transfected with FasL cDNA [115]. The intracellular domain of Fas is critical for its ability to transduce apoptotic signals. This was clearly illustrated by the lpr^{cg} mouse, which expresses a Fas molecule with a single amino acid substitution (Ile to Asn) at position 235, in the intracellular portion of the molecule. Cells transfected with Fas cDNA containing this mutation are resistant to killing by anti-Fas antibody [125]. Moreover, cells expressing chimeric molecules composed of the extracellular portion of CD40 and the transmembrane and intracellular domains of Fas or TNF receptor die when cultured with cells expressing CD40 ligand [126]. Mutagenic analysis of the 55 kDa TNF receptor found an intracellular domain of about 70 amino acids near the *C*-terminus that is required for the cytotoxic and the anti-viral activity of TNF [127]. A homologous domain exists in the cytoplasmic portion of Fas, and using a similar approach it has been found to be required for Fas-induced apoptosis [128]. Studies using the yeast two-hybrid system found that one Fas molecule can bind another via their intracellular portions,

and that Fas binds to the intracellular domain of the p55 TNF receptor [129]. Further, although expression of the Fas intracellular domain did not itself cause cell death in HeLa cells, it enhanced killing mediated by the p55 TNF receptor. Since Fas-mediated apoptosis is enhanced rather than inhibited by blocking protein synthesis [10], these data suggest that signaling for death by ligated p55 TNF receptor or Fas requires the association of their intracellular domains with pre-existing proteins, and that labile proteins may suppress signaling. Cloning and characterization of these associated proteins will certainly provide insight into the mechanisms of Fas-mediated cell death.

The cytotoxic activity of TNF has been linked to the metabolism of sphingomyelin, a membrane sphingolipid consisting of a polar head group esterified to ceramide, which is the amino alcohol sphingosine attached to a long-chain fatty acid by an amide bond [130–132]. Hydrolysis of sphingomyelin by sphingomyelinase generates free ceramides, which can activate protein kinases and phosphatases and regulate the activity of transcription factors such as NF-κB and c-Myc [133]. Moreover, synthetic cell-permeant ceramide, or treatment with sphingomyelinase, can induce apoptotic death of some cells [134, 135]. Interestingly, crosslinking of Fas also results in activation of an acid sphingomyelinase and generation of ceramides, suggesting the cytotoxicity of anti-Fas may also be mediated by ceramides [136]. In the B lymphoma WEHI 231, ceramide is also induced by apoptotic stimuli such as anti-IgM, irradiation, and glucocorticoids [137]. At this time it has not been proved that ceramides mediate Fas-induced cell death, and the mechanism by which ceramides might induce apoptosis is not well understood. In fact, the activities of ceramides are likely to be complex, since they also seem to be mediators for the actions of IL-1 and vitamin D$_3$ and TNF induction of differentiation in certain cells [138]. One intriguing hypothesis is that Fas-generated ceramides activate Ras which then couples to an apoptotic pathway. Consistent with this, ceramides drive Ras to the GTP-bound state and activate MAP kinase, and interfering with Ras signaling with a dominant negative mutant Ras inhibits Fas-mediated apoptosis of Jurkat T cells [139]. It seems likely that characterization of the newly appreciated role of ceramides as second messengers will be necessary to understand the mechanism of Fas-mediated apoptosis.

Simply expressing Fas does not necessarily confer susceptibility to Fas-mediated death. For example, human peripheral blood T lymphocytes express high levels of Fas within 24 hours of activation, yet are resistant to killing by anti-Fas antibodies. Only after 4–6 days of continuous culture does anti-Fas cause death [140, 141]. Restimulation of these cells with anti-TCR antibodies renders the surviving cells (a substantial subset) once again resistant to anti-Fas killing. In a similar vein, *in vivo* administration of LPS prevents superantigen-induced, $V\beta$-specific T cell deletion and confers resistance to anti-Fas killing, although the superantigen-activated cells express high level of Fas [142]. It has also been found that CD40 ligand-stimulated B cells were sensitive to Fas-mediated cytotoxic action of CD4+ Th1 cells, whereas anti-IgM-stimulated B cells are resistant [143]. How might Fas-induced death be prevented? One possible mechanism is the regulated expression of molecules with anti-apoptotic activity. Fap-1 is a protein tyrosine phosphatase that physically binds the last 15 amino acids of the intracytoplasmic domain of Fas [144], a region that has previously been shown to have negative regulatory activity [128]. Expression of Fap-1 in different cell lines correlated inversely with their susceptibility to Fas-induced killing, and enforced expression of Fap-1 made cells less sensitive to Fas ligation [144]. Another protein tyrosine phosphatase, hematopoietic cell phosphatase (HCP or PTP-1C), has also been implicated in Fas signaling, although in this case as a positive regulator of death [145]. Cell lines that are deficient in HCP and lymphoid cells from motheaten mice (which express mutated HCP) are relatively resistant to Fas-mediated killing. Therefore, regulated expression of the negative regulatory Fap-1, or the positive regulatory HCP, tyrosine phosphatases might account for differences in susceptibility to killing via Fas in different cells and among similar cells at varying states of activation. The effect of another class of anti-apoptotic molecules, of which Bcl-2 is the prototype, on Fas-mediated cell death has been studied by several groups. In a murine lymphoma and IL-3-dependent cell lines, transfection of human *fas* results in susceptibility of the cells to anti-human Fas antibody, which can be partially inhibited by enforced expression of Bcl-2 [146]. Overexpression of *bcl-2* in HeLa cells also confers resistance to TNFα as well as anti-Fas antibody [147]. Further, co-expressed Bag-1, a Bcl-2 binding protein, synergizes with Bcl-2 and completely blocks anti-Fas-induced apoptosis of Jurkat cells [148]. However, it has also been found that Fas-mediated apoptosis induced by cytotoxic T cells cannot be inhibited by overexpression of Bcl-2 [149]. This finding is consistent with the results with T cell hybridomas, in which overexpression of Bcl-2 blocks glucocorticoid-induced apoptosis but not activation-driven apoptosis, which is mediated by Fas and FasL interactions [150–153]. It is possible that the differences in these results are due to engagement of Fas by soluble antibody versus membrane-bound FasL, or that the cells express different Bcl-2 family members.

That Fas-mediated signaling might be regulated at multiple levels is not unexpected. Given the drastic and irreversible outcome of Fas signaling and the rapidity with which it occurs (hours), one might predict that its activity would be highly regulated. If expression of Fas was the only determinant of a cell's susceptibility to undergo FasL-induced death, any peripheral T cell responding to its specific antigen would be expected to survive only a short time, because Fas is up-regulated within one day after activation [141, 154]. Since it takes approximately a day for a resting T cell to be driven through the cell cycle, it would be difficult to expand antigen-specific T cells during an immune response. It appears, therefore, that multiple levels of regulation have developed: expression of Fas itself, expression of FasL, and susceptibility to Fas-mediated signaling. The last may reflect the antagonizing activities of soluble effector molecules such as TNF [142], the coupling of Fas to intracellular signaling pathways, and/or the protective effects of anti-apoptotic molecules such as Bcl-2 family members or Bag-1.

MOLECULES THAT REGULATE APOPTOSIS

c-Myc

The *c-myc* protooncogene is a short-lived nuclear protein that is induced by mitogenic stimuli and is required for cells to enter the cell cycle [155]. Deregulated expression of c-Myc can prevent cell differentiation and lead to transformation [156]. One function of Myc is as a transcription factor, for which it must dimerize with another protein, Max [157, 158]. Myc also functions by interacting with other proteins, such as the tumor suppressor protein Rb and the transcription factor TFII-I, to influence intracellular events [159, 160]. In non-lymphoid cells, constitutive expression of Myc suppresses the growth arrest and accelerates apoptosis caused by deprivation of growth factors [161, 162]. Myc is constitutively expressed in cycling T hybridoma cells, and decreasing its expression with antisense oligonucleotides prevents activation-induced apoptosis [163]. Therefore, in addition to promoting cell proliferation, Myc appears to be required (perhaps being permissive) for certain types of apoptotic death. Analysis of fibroblasts transfected with mutated *c-myc* found that Myc-dependent apoptosis induced by withdrawal of serum requires its dimerization with Max [162, 164]. Furthermore, transfection of a dominant-negative form of *max* into T hybridoma cells inhibits activation-induced apoptosis [165]. These results indicate that Myc/Max heterodimers are required for promoting cell death, probably because of their ability to regulate gene expression. Like cell death induced by many other stimuli, Myc-dependent apoptosis is inhibited by Bcl-2 [166, 167]. Interestingly, the apoptosis-promoting action of Myc requires p53, because expression of *c-myc* in p53-deficient serum-starved mouse embryo fibroblasts does not lead to apoptosis [168]. Whether p53 is required for activation-induced death of peripheral T cells is not known.

Which Myc-regulated genes might be involved in apoptosis? One candidate is ornithine decarboxylase (ODC), a rate-limiting enzyme in polyamine biosynthesis [169, 170]. Like Myc overexpression, enforced expression of ODC accelerated the factor withdrawal-induced death of IL-3-dependent cells [170]. Furthermore, cells transfected with *c-myc* expressed relatively high levels of ODC, and an ODC inhibitor blocked the death-promoting effect of Myc, raising the possibility that ODC is a mediator for the apoptotic activity of Myc. It was speculated that increased ODC results in accumulation of its enzymatic product, polyamines, whose oxidation would generate reactive oxygen species

and lead to cell death [170]. However, it seems that not all the apoptotic effects of Myc can be attributed to ODC, because *c-myc* transfected cells died faster than ODC transfectants, and the ODC inhibitor blocked ODC death more efficiently than Myc-dependent apoptosis [170]. In addition, it has been shown that the polyamine spermine prevents glucocorticoid-induced apoptosis of thymocytes, presumably due to its binding to DNA [171]. Depletion of intracellular polyamines also results in increased spontaneous DNA fragmentation of these cells [171], and prooxidant-induced apoptosis has been reported to be associated with inhibition of ODC activity and decreased polyamine levels [172]. It seems, therefore, that the relationship between ODC and apoptosis is complex, and that Myc likely acts through the regulation of other molecules in the apoptotic pathway.

Bcl-2 and Related Proteins

Bcl-2 (B cell lymphoma/leukemia-2) gene was first identified as a gene involved in the chromosome t(14;18) translocation observed in the majority of human follicular lymphomas [173–175]. It is a 26 kDa membrane-associated protein, and has been found in mitochondrial membranes, endoplasmic reticulum, and the nuclear envelope [176–178]. In immune tissues, Bcl-2 is normally expressed in the lymph node follicular mantle and the thymic medulla and is not expressed in germinal centers and the thymic cortex, areas in which apoptosis occurs at a high rate [179]. Initial studies demonstrating an effect of Bcl-2 on apoptosis were performed in hematopoietic cell lines, in which overexpression of *bcl-2* prolonged the survival of cells deprived of growth factors [176, 180, 181]. Subsequently, studies using a variety of cell lines demonstrated that enforced *bcl-2* expression makes cells more resistant to many apoptotic stimuli, including chemotherapeutic agents, anti-Fas antibody, overexpression of IL-1β converting enzyme, viral protein E1A, and cytolysis by cytotoxic T cells [146, 182–185]. The most striking examples of the potency of Bcl-2 come from studies in which it was transgenically expressed. B cells from transgenic mice expressing a Bcl-2-Ig fusion product survived for prolonged periods in a non-cycling state [186], and the lymphoid hyperplasia that resulted progressed to malignant lymphoma when a "second hit" such as *c-myc* rearrangement occurred, proving that Bcl-2 is indeed a protooncogene [27]. Thymocytes expressing *bcl-2* survived longer than their normal counterparts in culture, and were resistant to induction of apoptosis by glucocorticoids, anti-CD3, and γ-radiation, hydrogen peroxide, and even sodium azide [187–189]. Therefore, Bcl-2 suppresses apoptotic cell death induced by a wide range of stimuli.

A large (and ever-expanding) number of *bcl-2*-related genes has been found, and many of them share homology with *bcl-2* in the regions named the Bcl-2 homology 1 and 2 (BH1 and BH2) domains [190]. Bcl-x has extensive homology with Bcl-2, but unlike Bcl-2, expression of Bcl-x is not constitutive but is induced by mitogenic stimuli [191]. *bcl-x* undergoes alternative splicing to yield two mRNAs, $bcl-x_L$ and $bcl-xS$, the latter encoding a protein with a 63 amino acid deletion spanning the BH1 domain and most of the BH2 domain. Transfection with $bcl-x_L$ inhibits IL-3-withdrawal death of dependent cells, while when overexpressed, $bcl-x_S$ is a dominant negative that inhibits the ability of *bcl-2* to prevent apoptosis. In human T cells, expression of *bcl-xS* correlates with susceptibility to death, being expressed in immature thymocytes and activated peripheral T cells, but not in single positive thymocytes and resting peripheral T cells. In mice, however, cortical thymocytes express high levels of $bcl-x_L$ and low levels of $bcl-x_S$ [192]. The physiologic significance of this dichotomy, if any, is unknown. Another bcl-2 homolog protein Bax, like Bcl-xS, can promote apoptotic cell death and inhibit the anti-apoptotic action of Bcl-2 [193]. Bax can either homodimerize or form heterodimers with Bcl-2 or Bcl-x_L, and in fact the ability of the latter to prevent apoptosis requires their binding to Bax [194, 195]. It is therefore possible that Bax homodimers promote cell death, and heterodimerization with Bcl-2 or Bcl-x blocks this activity.

Three new members of the Bcl-2 family, Bad, Bag-1, and Bak, have been identified [148, 195–198]. Overexpression of Bad had no effect on IL-3-withdrawal cell death, but suppressed the anti-apoptotic action of Bcl-x_L but not Bcl-2 [195]. Since Bad can dimerize with Bcl-x_L and, to a lesser extent, with Bcl-2, but not with Bax, Bcl-x_S, or itself, one possible explanation for its pro-apoptotic action is that Bad displaces Bax from Bax/Bcl-x_L dimers, resulting in an increase of Bax homodimers. Bag-1 is another Bcl-2-binding protein but has little homology with Bcl-2 [148]. Enforced expression of *bag-1* makes cells more resistant to apoptotic stimuli, especially when co-expressed with *bcl-2*. Surprisingly, Bak appears to have diametrically opposite effects on apoptosis depending on the cell type. Bak can bind to Bcl-x_L but apparently not to Bcl-2, and promotes cell death when expressed in neurons, fibroblasts, and IL-3 dependent cells [196–198]. When expressed in an EBV-transformed human B cell line, however, Bak actually prevented apoptosis [198]. Although the basis for this is unknown, it is presumably due to interactions between Bak and other Bcl-2-related proteins expressed in a tissue-specific manner, and serves to underscore the complexity of the regulation of apoptosis. Finally, there are a host of molecules expressed by adenovirus (E1B 19K), Epstein–Barr virus (BHRF1), African swine fever virus (Lmw5-h1), *Caenorhabditis elegans* (Ced-9), and hematopoietic cells (A1 and MCL1) that are also related to Bcl-2 in terms of sequence similarity and anti-apoptotic activity [184, 199–203], emphasizing the importance of anti-apoptotic genes throughout evolution.

Despite intense interest in the Bcl-2 family, the biochemical activity of these related proteins has not been determined. Bcl-2 protects cells from death induced by γ-radiation, hydrogen peroxide, and menadione, all generators of reactive oxygen species (ROS) [187, 188, 204], and two reports have suggested that it is at this level that Bcl-2 may exert its effects. In a T cell hybridoma, Bcl-2 had no detectable effect on the generation of ROS but nonetheless prevented lipid peroxidation caused by dexamethasone [205], and it was suggested that Bcl-2 may be an antioxidant, acting after the generation of $0_2.^-$ and its conversion to peroxides [205]. Another study with a neural cell line found that glutathione depletion caused a rapid rise of intracellular ROS, lipid peroxidation, and cell death [206]. In this case, Bcl-2 inhibited the increase of ROS and lipid peroxides, and prevented cell death. This study implied that Bcl-2 acts by decreasing the generation of ROS. It is unlikely, however, that the function of Bcl-2 is simply to interfere with either the generation or activity of ROS. Staurosporine and anti-Fas antibodies administered in an oxygen-depleted atmosphere still induce fibroblast apoptosis, which is prevented by Bcl-2 [207]. In fact, hypoxia itself, which actually decreases the formation of oxygen free radicals, can induce apoptotic cell death, an outcome that is inhibited by expression of Bcl-2 or Bcl-x_L [208]. Therefore, an anti-oxidant effect alone cannot account for the protective effect of Bcl-2. Other activities attributed to Bcl-2 include inhibition of Ca^{2+} flux from the endoplasmic reticulum [209], and interactions with R-Ras and Raf-1, two important intracellular signaling molecules [210, 211]. The importance of these findings in the prevention of apoptosis, if any, remains to be determined. Interestingly, apoptotic changes can be induced in isolated *Xenopus* egg nuclei by adding soluble cell extracts, and such changes are inhibited by exogenous Bcl-2 [212]. This should prove to be a powerful system in which to explore the nuclear events of apoptosis and the mechanism of Bcl-2 action. What is clear is that characterization of the biochemical activity of Bcl-2 is a prerequisite to understanding its complex cellular effects.

p53

Mice deficient in the tumor suppressor p53 develop tumors at an early age [213]. Interestingly, among these tumors thymic lymphomas are predominant [213–215]. It has become clear that p53 can act as sequence-specific transcription factor to regulate the expression of a variety of genes, including Bax [216–218]. It may also function by directly interacting with various cellular and viral proteins [219–222]. In most cells, p53 is expressed at a low level and is induced by irradiation or DNA-damaging agents [223, 224]. Its relationship to apoptosis was first found in murine myeloid cells transfected with temperature-sensitive p53 (tsp53), a mutated molecule with wildtype activity at 32 °C but not at 37.5 °C [225]. These transfectants grew normally at 37.5 °C but underwent apoptotic death when cultured at 32 °C [226]. This was reproduced in a variety of cell types, including a T cell lymphoma [227–231]. Thymocytes from p53-deficient mice are dramatically resistant to radiation-and etoposide-induced apoptosis, although they are still susceptible to corticosteroids and phorbol ester [232, 233]. The thymus in

these mice appears to be normal, suggesting that p53 does not play an essential role in T cell development. DNA damage can also induce T cell apoptosis via a p53-independent pathway, since activated peripheral T cells from p53-deficient mice are sensitive to radiation and genotoxic drugs [234]. Despite many studies finding that p53 is required for the apoptotic effects of *c-myc*, E1A, E2F-1, and lack of Rb expression [168, 235–239], little is known about how p53 works in this regard. Some possibilities include its negative regulatory effect on *bcl-2*, and its positive regulatory effect on *bax* [217].

Abl

The Abelson murine leukemia virus (A-MuLV)-encoded transforming oncogene v-Abl and its normal cellular counterpart (c-Abl) are nuclear tyrosine kinases [240, 241]. Pre-B cells infected with a temperature-sensitive A-MuL V undergo apoptotic death at the non-permissive temperature due to loss of kinase activity of v-Abl [242], indicating that v-Abl normally prevents the death of infected cells. This was further supported by the finding that cells directly transfected with temperature sensitive *v-abl* were more resistant to apoptosis induced by anti-cancer drugs and growth factor deprivation at the permissive but not the restrictive temperature [243, 244]. Chronic myeloid leukemia cells often harbor a (9;22) chromosomal translocation that results in deregulated expression of a Bcr–Abl chimeric molecule [245]. Such cells have a prolonged survival in the absence of growth factors, and inhibition of Bcr–Abl expression by antisense oligonucleotides in these cells was able to induce apoptosis [246, 247]. Consistent with this, *c-abl*-deficient mice die 1–2 weeks after birth, and among the major changes observed are a hypocellular thymus and lymphopenia [248]. These results suggest that the ability of c-Abl to suppress apoptotic cell death is important in maintaining T cell homeostasis and may account for its oncogenic ability.

Proteases

Cloning of the *C. elegans ced-3* gene, which is required for developmentally programmed cell death, revealed that it shares some homology to the cysteine protease ICE (IL-1β converting enzyme) [249]. Transfection of wildtype ICE, but not a catalytically inactive mutated form of the molecule, into fibroblasts causes their apoptotic death [183]. Homologs of ced-3/ICE, Nedd-2 and Ich-1, have been isolated from mouse and human cDNA libraries, respectively [250, 251], and enforced expression of Nedd-2 in fibroblast and neuroblastoma cells results in apoptotic cell death [252]. The *ich*-1 gene encodes two products, Ich-1L and Ich-1S, due to alternative splicing of its mRNA. Overexpression of Ich-1L promotes cell death, while Ich-1S overexpression inhibits cell death induced by serum deprivation [251]. Consistent with the notion that proteases are important in apoptosis, studies on the mechanisms of CTL killing have demonstrated that granzymes, a group of serine proteases packaged in the granules of CTL, are required for induction of DNA fragmentation and apoptotic death of target cells [253, 254]. Induction of apoptosis in this case apparently involves the rapid activation of p34^{cdc2}, a cyclin-regulated serine–threonine kinase that controls entry into mitosis [255]. What are the potential targets of these proteases? Poly (ADP-ribose) polymerase, which is involved in DNA repair, is cleaved early in cells induced to undergo apoptosis, and inhibition of this event prevents etoposide-induced DNA fragmentation, although it does not block cell death as judged by trypan blue exclusion [256]. The enzyme responsible for the cleavage of poly (ADP-ribose) polymerase has ICE-like activity, cleaving at a tetrapeptide sequence identical to one of two sites in pro-IL-1β recognized by ICE, but is not ICE [257]. Cleavage of the 70 kDa U1 small nuclear ribonucleoprotein and lamin has also been observed in apoptotic cells, and the corresponding proteolytic enzymes are similar but distinct from ICE [256, 258].

The requirement for proteases in physiologic cell death has been demonstrated in a variety of ways. Several non-selective protease inhibitors have been shown to prevent thymocyte apoptosis induced by glucocorticoids or inhibitors of DNA topoisomerases [259]. Activation-induced apoptotic death of T cell hybridomas and peripheral T cells is blocked by inhibitors of cysteine and serine proteases [260]. Micro-injection of crmA, a cowpox virus gene product that inhibits ICE activity, into dorsal root

ganglion neurons prevents cell death induced by withdrawal of growth factor [261]. Furthermore, enforced expression of crmA or addition of inhibitors of ICE-activity block Fas and TNF-induced apoptosis in cell lines [262–264], and thymocytes from ICE-deficient mice are resistant to anti-Fas antibodies, although they are still susceptible to killing by radiation or glucocorticoids [265, 266]. These studies did not deal with sensitivity of peripheral T cells, for which Fas-mediated death has been shown to be physiologically important. Where proteases might function in signaling for death is not yet known, but it has been reported that protease inhibitors can prevent Ca^{2+} induced DNA fragmentation in isolated rat liver nuclei, supporting a direct role for proteases in the process of DNA degradation [267]. Together, the data generated by genetic and pharmacologic means strongly suggest the involvement of multiple types of proteases, at perhaps multiple levels in the signaling/effector pathways, in T cell apoptosis.

DNA FRAGMENTATION

In most cases, a change in nuclear morphology is an early and characteristic marker of apoptosis, and is coincident with the fragmentation of nuclear DNA. When separated by agarose gel electrophoresis, these low molecular weight DNA fragments yield the classic "DNA ladder", often regarded as a biochemical hallmark of apoptosis [74, 83]. More recently, with the use of field inversion gel electrophoresis, it has been shown that the initial DNA fragments generated in glucocorticoid- or etoposide-treated thymocytes are quite large (30–50 kb and 190–240 kb), and these large fragments are subsequently cleaved into oligonucleosomal fragments [268, 269]. The appearance of high molecular weight fragments is associated with the early nuclear morphologic changes, and is dependent on protein synthesis, while the generation of internucleosomal fragments does not require protein synthesis [269]. Since glucocorticoid-induced thymocyte apoptosis requires protein synthesis, these results are consistent with the notion that generation of large DNA fragments is a key step in the commitment of cells to die. Relatively low concentrations of Zn^{2+} (25–50 μM) that do not block glucocorticoid-induced thymocyte

DNA fragmentation [6] do not inhibit the formation of high molecular weight fragments, although they do prevent the progression to internucleosomal fragments [269]. Furthermore, similar concentrations of Zn^{2+} inhibit internucleosomal fragmentation but not chromatin condensation of isolated nuclei incubated with Ca^{2+} and Mg^{2+}; generation of high molecular weight DNA fragments was not assessed in this study [270]. Together, these data suggest that it is the large rather than the low molecular weight DNA fragments that indicate the commitment to apoptosis.

Early studies found the internucleosomal cleavage of nuclear DNA is dependent on Ca^{2+} and Mg^{2+}, and therefore extensive efforts have been made to isolate endonucleases dependent upon these cations [74]. One candidate enzyme is an 18 kDa nuclear protein called NUC-18, which was identified from glucocorticoid-treated thymocytes based on digestion of immobilized DNA in SDS gels (SDS–DNA-PAGE). NUC-18 activity requires Ca^{2+} and Mg^{2+} and is inhibited by Zn^{2+} [271]. NUC-18 is present in the nucleus of non-apoptotic cells, but in this case appears to be in a high molecular weight complex of unknown composition [271, 272]. Whether NUC-18 cleaves nuclear DNA to yield internucleosomal fragments has not been reported. Surprisingly, NUC-18 was found to be homologous to cyclophilins, the intracellular cyclosporin-A-binding proteins, and recombinant cyclophilins A, B, and C have all been shown to have nuclease activity in the DNA–SDS–PAGE assay [273]. Whether NUC-18 and cyclophilins play a role in DNA fragmentation during apoptosis of thymocytes is not known at this time. Another candidate enzyme is DNase I, a Ca^{2+} and Mg^{2+}/Mn^{2+} dependent endonuclease expressed in a wide variety of tissues, including thymus and lymph node [274, 275]. Nuclear extracts of thymocytes or lymph node cells can induce internucleosomal cleavage of nuclei, and this DNA ladder-forming activity was found to be reduced by anti-DNase I antibodies and prevented by G-actin, a DNase I inhibitor [275]. It is not obvious how this secreted endonuclease might gain entry to the nucleus to cleave chromosomal DNA, or how nucleosomal-sized fragments would be generated, since DNase I has no preference for intranucleosomal sites [276], although it has been suggested that access might be gained due to dis-

solution of the ER and the nuclear envelope, and that enzyme specificity might be modified by interactions with other proteins present in the nucleus [275, 277]. In contrast to NUC-18 and DNase I, DNase II is an ubiquitous acidic endonuclease whose activity does not require Ca^{2+} or Mg^{2+}. It can be activated by reducing intracellular pH with proton ionophores, and causes internucleosomal cleavage of nuclear DNA at low pH (below pH6.5) [278]. However, it is not known whether decrease of intracellular pH occurs in the processes such as glucocorticoids or activation-induced cell death. Furthermore, since DNase II is a 3'-endonuclease that produces 5'-OH and 3'-P DNA termini [279] rather than the 5'-P and 3'-OH termini found in apoptotic cells, it seems unlikely that it can be a predominant factor in DNA cleavage in apoptotic cells. Thus, at this time no endonuclease has been definitively implicated in the process of apoptosis. It is certainly possible, if not likely, that multiple endonucleases are utilized in different cells and by apoptotic pathways initiated by different stimuli.

The findings that nuclear changes and DNA fragmentation precede cell death, and that DNA fragmentation and cell death are both prevented by inhibitors of protein synthesis and endonuclease activity, are consistent with the hypothesis that apoptotic cell death requires DNA degradation [6, 11, 83, 280]. Further support comes from studies of somatic cell fusions made between thymocytes and NIH 3T3 fibroblasts or rat gliosarcoma cells [281]. Following exposure of these heterokaryons to glucocorticoids, the thymocyte nuclei, but not the nuclei from the cell line, developed apoptotic morphology, and the cells remained viable with some even undergoing mitosis, indicating that nuclear apoptotic changes can be triggered in the absence of cytoplasmic changes. Despite this evidence, there is reason to believe that DNA fragmentation does not play an essential role in apoptosis. For instance, a subset of *C. elegans* cells undergo PCD with an apoptotic morphology during development [282]. In *C. elegans* mutants that lack *nuc-1*, a Ca^{2+} and Mg^{2+} -independent endonuclease, the cells still die and are engulfed normally, although the DNA of dead cells remains undegraded [283]. Furthermore, mammalian cells such as fibroblasts, T cell hybridomas, and leukemia cells that have been enucleated with cytochalasin B still undergo apoptotic death (as judged by membrane blebbing, cytoplasmic condensation, loss of mitochondrial activity, and leakage of ^{52}Cr) after staurosporine treatment, engagement of Fas, and culture with cytotoxic cells [284–286]. Moreover, aurintricarboxylic acid can inhibit DNA fragmentation but not cell death induced by anti-Fas antibody [285]. Thus, apoptosis can take place in the absence of a nucleus. Given the fact that some cells died rapidly when treated with RNA or protein synthesis inhibitors [287], while erythrocytes survive without a nucleus, it is likely that whether the disintegration of the nucleus initiates, contributes to, or is irrelevant to apoptosis depends upon the cell type and, perhaps, the apoptotic stimulus.

APOPTOSIS IN T CELL DEVELOPMENT

In mice, lymphoid stem cells begin to appear in the thymus on day 11 of a normal 21 day gestation, corresponding to about the 7–8th week of gestation in humans [288]. Shortly after entering the thymus, these stem cells express high levels of the surface marker Thy-1 and proliferate extensively [16]. The most immature thymocytes do not express CD4, CD8, or the TCR until they have successfully rearranged TCR β and α [289, 290]. TCR β rearrangement is particularly important for this progression, since mice deficient in genes required for TCR gene rearrangement only have CD4$^-$ CD8$^-$ thymocytes, whereas introduction of a TCR β transgene into such mice restores the generation of CD4$^+$ CD8$^-$ cells [291, 292]. The TCR β chain can be expressed on the surface of immature T cells in association with a developmentally regulated protein, gp33 or pre-Tα, and CD3 to form a pre-TCR [293, 294]. The pre-TCR can function, since calcium is mobilized after its crosslinking with antibody [293, 295], and anti-CD3 treatment of cultured fetal thymus can promote the progression of CD4 CD8 cells to CD4$^+$ CD8$^+$ thymocytes [295]. Therefore, it appears that the pre-TCR binds a putative ligand in the thymus and delivers a signal that suppresses cell death and promotes further proliferation and differentiation of thymocytes.

CD4$^+$ CD8$^+$ thymocytes express low to intermediate levels of cell surface TCR, and are precursors of CD4$^+$ CD8$^-$ and CD4$^-$ CD8$^+$ cells. It is at the transitional stage (CD4$^+$ CD8$^+$TCR^{intermediate}) that positive or negative selection is thought to occur [20]. In the "avidity" model of thymocyte selection, thymocytes expressing TCRs with subthreshold avidity for self-antigen/MHC, presumably the majority of the cells that are generated, die in the thymus, a phenomenon that has been called "death by neglect" [17, 296–298]. Those thymocytes with high avidity for self-antigen/MHC, and which therefore are potentially autoreactive, undergo apoptotic cell death (negative selection). Finally, those thymocytes with low-to-moderate avidity for self-antigen/MHC are rescued from the default death pathway, and go on to become mature peripheral T cells (positive selection).

How thymocytes interpret TCR-mediated signals to yield such different outcomes, survival versus induction of death, is a matter of great speculation. One possibility is that engagement of TCR by low avidity versus high avidity ligands generates qualitatively different signals. Indeed, there is some evidence that positive and negative selection of thymocytes use different intracellular pathways. For example, activation of MAP kinase (MAPK) by MAP kinase kinase (MEK) is required for replication and differentiation of many cells [299, 300]. Expression of an inactive form of MEK in the thymus blocks MAPK activation and inhibits positive but not negative selection [301], as does expression of a dominant negative mutant Ras [302]. Thus, since MEKs are activated by Ras, it appears that positive selection may require the GTP–Ras activation pathway, while negative selection does not.

If qualitatively different signals indeed mediate positive and negative selection, how are they generated? One possibility is that the nature of the peptide that binds MHC affects TCR signaling. Analogs of antigenic peptides can be agonistic, partially agonistic, or antagonistic for mature T cell activation [303]. T cell clones respond to agonist peptides by proliferation and lymphokine production, whereas partial agonists stimulate lymphokine production without promoting proliferation [304]. Pure antagonist peptides block the proliferative response of co-administered agonist peptides. In contrast to agonist peptides, antagonist or partial agonist peptides have been reported to induce a different pattern of CD3ζ tyrosine phosphorylation and do not cause the activation of the ZAP-70 tyrosine kinase [305]. The possible relevance of this for thymocyte selection was suggested by fetal thymic organ culture (FTOC) experiments using thymi from β_2-microglobulin-deficient mice transgenic for MHC class I-restricted TCR recognizing ovalbumin or viral peptides. One group found that addition of agonist peptide plus β_2 resulted in negative selection, whereas antagonist peptides led to positive selection [306]. Strikingly, agonist peptides never caused positive selection, even at very low concentrations. In contrast, the other group found that a single agonist peptide caused positive selection at low concentrations and negative selection at high concentrations, compatible with the simple avidity model of selection [307]. The same phenomenon, peptide-concentration-dependent positive versus negative selection, was observed with fetal thymi from TAP-1-deficient mice bearing transgenic TCRs that recognize MHC class I plus viral peptide [308]. These different results are difficult to reconcile, but may indicate that factors other than simply the degree of receptor occupancy determine the outcome of thymocyte TCR recognition of self.

Another possibility is that the signals generated by the TCR/MHC/peptide interaction are directly or indirectly influenced by signals from other thymic cells. The thymic microenvironment clearly has a major influence on the life and death of thymocytes. For example, progression of thymocyte differentiation can only be observed in FTOC, since thymocytes die quickly in suspension culture. Furthermore, while double-positive thymocytes readily undergo apoptosis when their TCRs are crosslinked in FTOC, they are relatively resistant when cultured as single cells in suspension. Although the mechanisms of these phenomena are not yet understood, they cannot be fully attributed to the expression of MHC molecules on thymic epithelial cells. A host of accessory molecules are expressed by thymic epithelium, and as mentioned above with regard to inducing cell death, it is likely that one (or many) of these provides non-antigen-specific signals that nonetheless modulate the signals generated by the TCR. It is also possible that soluble molecules produced by the thymus itself play a role in the determination of the outcome

of TCR signaling. Based on mutual antagonism between glucocorticoids activation [102, 103], it has also been proposed that thymocytes expressing TCRs with low-to-moderate avidity for self-antigen/MHC survive because ambient glucocorticoids block activation-induced death. High avidity interactions would lead to cell death, because they would be too potent to be antagonized by glucocorticoids [309]. Another way of viewing this is that glucocorticoids might be a significant cause of "death by neglect", and that cells with low-to-moderate avidity TCRs would be rescued by interaction with ligand. In support of these possibilities, it has been found that thymic epithelium, especially that from fetal mice, can produce steroids [310]. In FTOC, inhibition of endogenous corticosterone production with the drug metyrapone, or blockade of the glucocorticoid receptor with RU486, potentiates the loss of CD4$^+$CD8$^+$ thymocytes caused by TCR signaling; the former is reversed by the addition of exogenous corticosterone. The role of glucocorticoids in thymocyte development was further studied in transgenic mice expressing antisense glucocorticoid receptor transcript driven by the thymocyte-specific *lck* proximal promoter. Expression of this antisense RNA resulted in a $\sim 50\%$ reduction in thymocyte glucocorticoid receptor levels. Mice that are homozygous for the transgene have greatly decreased numbers of immature CD4$^+$ CD8$^+$ cells, which correlates with an increased level of spontaneous apoptosis in this population (LB King and JDA, unpublished observation). Although the reasons for the apoptotic loss of thymocytes is still being explored, the results strongly suggest that glucocorticoids are required for normal thymocyte development.

Although the large majority of thymocytes die during development, dead cells are difficult to detect in the unmanipulated thymus with conventional histological methods. This problem was solved using terminal deoxynucleotide transferase (TdT)-mediated dUTP-biotin nick end-labeling (TUNEL) to detect breaks in the DNA of apoptotic cells. In thymi from normal untreated mice, a small number of apoptotic cells (0.5–1.0% of the mononuclear cells) were found mainly in the cortex, where the CD4$^+$CD8$^+$ cells reside [17]. In mice deficient in both MHC class I and class II molecules, in which neither positive nor negative selection occurs, the incidence and pattern of cell death did not change, suggesting that these cells died because of a lack of survival signals (death by neglect). To examine cell death in a negatively selecting situation, transgenic mice expressing a Vβ5-containing TCR that interacts with superantigens encoded by MTV 6 and 9, resulting in thymocyte deletion, were studied. The thymic cortex of both control and superantigen-expressing transgenic mice were similar. In contrast, the medulla of the superantigen-positive transgenic mice contained large numbers of apoptotic cells. Based on these results, it seems that cell death during negative selection takes place largely in the medulla, although the triggering event may occur in CD4$^+$CD8$^+$ cortical cells that are transiting to the medulla.

As detailed above, Bcl-2 (and its family members) and Fas are very important in the life-or-death decisions a peripheral T cell must make. Surprisingly, as yet there is no evidence that these molecules play a similarly important role in thymocyte development. Bcl-2 is expressed in early CD4$^-$CD8$^-$ cells, down-regulated as the cells become CD4$^+$CD8$^+$ and up-regulated again during positive selection, which correlates with their susceptibility to various apoptotic stimuli [179, 311–313]. Although this highly regulated pattern of expression suggested that Bcl-2 might rescue activated thymocytes from deletion, enforced expression of Bcl-2 in CD4$^+$CD8$^+$ cells did not block negative selection, although the thymocytes are resistant to many apoptotic stimuli, including TCR engagement and culture with glucocorticoids [187, 188, 312, 314]. Furthermore, T cells from Bcl-2-deficient mice die more rapidly in suspension culture, and are more sensitive to glucocorticoids and γ-radiation, than their normal counterparts, yet thymocyte selection appears to be normal [92, 315]. Interestingly, although fetal and neonatal lymphocyte development is normal in these mice, cells in the bone marrow and lymphoid organs undergo massive apoptosis and disappear several weeks after birth. The mechanism for this is unknown, but in light of the fact that Bcl-2 is a potent inhibitor of steroid-induced cell death, it may reflect the developmentally regulated increase in circulating steroids that occurs after birth. Since deficiency of bcl-x proved to be embryonically lethal, the effect of Bcl-x on T cell development

was assessed by injecting bcl-x$^{-/-}$ ES cells into blastocysts of normal or rearrangement-deficient mice [316]. The number of Bcl-x-deficient CD4$^+$CD8$^+$ and CD4$^+$CD8$^+$ thymocytes was reduced, while the number of CD4$^+$CD8$^+$ cells was unchanged. However, the CD4$^+$CD8$^+$ cells died more quickly than the more mature thymocytes when cultured in vitro, raising the possibility that the apparent block in thymocyte maturation resulted from an increased susceptibility of the bcl-x-deficient CD4$^+$8$^+$ cells to apoptotic stimuli. Moreover, the life-span of bcl-x$^{-/-}$ mature T cells in culture was longer than that of bcl-2$^{-/-}$ T cells, indicating that Bcl-2 and Bcl-x play different roles in regulating mature T cell viability.

In contrast to peripheral T cells, in which activation-induced apoptosis is mediated via the interaction of Fas and FasL, death of thymocytes does not seem to be mediated by these molecules. lpr and gld mice, which have very low levels of Fas expression and express a mutated FasL respectively, exhibit apparently normal thymocyte development, although the in vitro deletion of peripheral T cells is affected [42, 43, 317]. The relative importance of Fas in thymocyte versus T cell death was also analyzed in TCR-transgenic lpr mice, in which administration of a high-dose of specific peptide caused deletion of thymocytes bearing the transgenic TCR but not peripheral T cells [23]. The mechanisms of neglected thymocyte death and why TCR engagement can block this type of cell death are even less understood. Among the major questions that remain concerning T cell development are: which molecules regulate activation-induced thymocyte death and how are they different from those that regulate peripheral T cell apoptosis?

FUTURE PROSPECTS

While considerable information about the signals that induce and regulate apoptosis has been accumulated, the intracellular mechanisms of apoptotic death are still poorly understood. The fact that many molecules with widely diverse functions participate in the apoptotic response indicates the involvement of a variety of apparently independent cellular processes. Moreover, it appears that even in the same cell lineage multiple mechanisms may account for *in vivo* cell death. For example, Fas and Fas ligand are critical for the normal homeostasis and activation-induced cell death of mature peripheral T cells. Activation-induced deletion of thymocytes, however, which express high levels of Fas and are sensitive to Fas-mediated killing, does not seem to involve Fas. Likewise, although Bcl-2 potently prevents cell death (including that of thymocytes) caused by many distinct stimuli, its enforced expression in CD4+CD8+ thymocytes does not prevent negative selection. Why immature and mature T cells apparently utilize different pathways to lead to the elimination of unwanted cells is not obvious. Given the tremendous rate of progress in characterizing the molecules involved in signaling for cell death, however, it seems safe to predict that the answers will not be long in coming.

REFERENCES

1. Glücksmann A. Cell death in normal vertebrate ontogeny. *Biol Rev.* 1951; **26**: 59–86.
2. Saunders, Jr JW. Death in embryonic systems. *Science* 1966; **154**: 604–612.
3. Lockshin RA and Zakeri Z. *Programmed cell death and apoptosis.* New York: Cold Spring Harbor Laboratory Press, 1991: 47.
4. Kerr JFR, Wyllie AH and Currie AR. Apoptosis: a basic biological phenomenon with wide-ranging implications in tissue kinetics. *Br. J. Cancer* 1972; **26**: 239–257.
5. Wyllie AH, Kerr JFR and Currie AR. Cell death: the significance of apoptosis. *Int. Rev. Cytol.* 1980; **68**: 251–306.
6. Cohen JJ and Duke RC. Glucocorticoid activation of a calcium-dependent endonuclease in thymocyte nuclei leads to cell death. *J. Immunol.* 1984; **132**: 38–42.
7. Wyllie AH, Morris RG, Smith AL and Dunlop D. Chromatin cleavage in apoptosis: association with condensed chromatin morphology and dependence on macromolecular synthesis. *J. Pathol.* 1984; **142**: 67–77.
8. Sellins KS and Cohen JJ. Gene induction by gamma-irradiation leads to DNA fragmentation in lymphocytes. *J. Immunol.* 1987; **139**: 3199–3206.
9. Ucker DS, Ashwell JD and Nickas G. Activation driven T cell death. I. Requirements for *de novo* transcription and translation and association with genome fragmentation. *J. Immunol.* 1989; **143**: 3461–3469.
10. Yonehara S, Ishii A and Yonehara M. A cell-killing monoclonal antibody (anti-Fas) to a cell surface antigen co-down-regulated with the receptor of tumor necrosis factor. *J. Exp. Med.* 1989; **169**: 1747–1756.
11. Duke RC, Chervenak R and Cohen JJ. Endogenous endonuclease-induced DNA fragmentation: an early event in cell-mediated cytolysis. *Proc. Natl. Acad. Sci. USA.* 1983; **80**: 6361–6365.
12. Cohen JJ, Duke RC, Fadok VA and Sellins KS. Apoptosis and programmed cell death in immunity. *Ann. Rev. Immunol.* 1992; **10**: 267–293.
13. Schwartz LM, Smith SW, Jones MEE and Osborne BA. Do all programmed cell deaths occur via apoptosis? *Proc. Natl. Acad. Sci. USA* 1993; **90**: 980–984.
14. Ishigami T, Kim K-M, Horiguchi Y *et al.* Anti-IgM antibody-induced cell death in a human B lymphoma cell line, B104, represents a novel programmed cell death. *J. Immunol.* 1992; **148**: 360–368.
15. Egerton M, Scollay R and Shortman K. Kinetics of mature T-cell development in the thymus. *Proc. Natl. Acad. Sci. USA* 1990; **87**: 2579–2582.
16. Shortman K, Egerton M, Spangrude GJ and Scollay R. The generation and fate of thymocytes. *Semin. Immunol.* 1990; **2**: 3–12.
17. Surh CD and Sprent J. T-cell apoptosis detected in situ during positive and negative selection in the thymus. *Nature* 1994; **372**: 100–103.
18. Smith CA, Williams GT, Kingston R, Jenkins EJ and Owen JJT. Antibodies to CD3/T-cell receptor complex induce death by apoptosis in immature T cells in thymic cultures. *Nature* 1989; **337**: 181–184.
19. Murphy KM, Heimberger AB and Loh DY. Induction by antigen of intrathymic apoptosis of CD4+CD8+ thymocytes in vivo. *Science.* 1990; **250**: 1720–1723.
20. Robey E and Fowlkes BJ. Selective events in T cell development. *Ann. Rev. Immunol.* 1994; **12**: 675–705.
21. Allen PM. Peptides in positive and negative selection: a delicate balance. *Cell* 1994; **76**: 593–596.
22. Kabelitz D, Pohl T and Pcchhold K. Activation-induced cell death (apoptosis) of mature peripheral T lymphocytes. *Immunol. Today* 1993; **14**: 338–339.
23. Singer GG and Abbas AK. The Fas antigen is involved in peripheral but not thymic deletion of T lymphocytes in T cell receptor transgenic mice. *Immunity* 1994; **1**: 365–371.
24. Wu J, Zhou T, Zhang J, He J, Gause WC and Mountz JD. Correction of accelerated autoimmune disease by early replacement of the mutated lpr gene with the normal Fas apoptosis gene in the T cells of transgenic MRL-lpr/lpr mice. *Proc. Natl. Acad. Sci. USA* 1994; **91**: 2344–2348.
25. Russell JH and Dobos CB. Mechanisms of immune lysis. II. CTL-induced nuclear disintegration of the target begins within minutes of cell contact. *J. Immunol.* 1980; **125**: 1256–1261.
26. Henkart P. Lymphocyte-mediated cytotoxicity: two pathways and multiple effector molecules. *Immunity* 1994; **1**: 343–346.
27. McDonnell TJ and Korsmeyer SJ. Progression from lymphoid hyperplasia to high-grade malignant lymphoma in mice transgenic for the t(14;18). *Nature* 1991; **349**: 254–256.
28. Ameisen JC. Programmed cell death and AIDS: from hypothesis to experiment. *Immunol. Today* 1992; **13**: 388–391.
29. Ullman KS, Northrop JP, Verweij CL and Crabtree GR. Transmission of signals from the T lymphocyte antigen receptor to the genes responsible for cell proliferation and

immune function: the missing link. *Ann. Rev. Immunol.* 1990; **8**: 421–452.

30. Schwartz RH. Costimulation of T lymphocytes: the role of CD28, CTLA-4, and B7/BB1 in interleukin-2 production and immunotherapy. *Cell* 1992; **71**: 1065–1068.

31. Russell JH, White CL, Loh DY and Meleedy-Rey P. Receptor-stimulated death pathway is opened by antigen in mature T cells. *Proc. Natl. Acad. Sci. USA* 1991; **88**: 2151–2155.

32. Janssen O, Wesselborg S and Kabelitz D. The immunosuppressive action of OKT3. OKT3 induces programmed cell death (apoptosis) in activated human T cells. *Transplantation.* 1992; **53**: 233.

33. Wesselborg S, Janssen O and Kabelitz D. Induction of activation-driven death (apoptosis) in activated but not resting peripheral blood T cells. *J. Immunol.* 1993; **150**: 4338–4345.

34. Radvanyi LG, Mills GB and Miller RG. Religation of the T cell receptor after primary activation of mature T cells inhibits proliferation and induces apoptotic cell death. *J. Immunol.* 1993; **150**: 5704–5715.

35. Boehme SA and Lenardo MJ. Propriocidal apoptosis of mature T lymphocytes occurs at S phase of the cell cycle. *Eur. J. Immunol.* 1993; **23**: 1552–1560.

36. D'Adamio L, Awad KM and Reinherz EL. Thymic and peripheral apoptosis of antigen-specific T cells might cooperate in establishing self-tolerance. *Eur. J. Immunol.* 1993; **23**: 747.

37. Liu Y and Janeway, Jr CA. Interferon γ plays a critical role in induced cell death of effector T cell: a possible third mechanism of self-tolerance. *J. Exp. Med.* 1990; **172**: 1735–1739.

38. Lenardo MJ. Interleukin-2 programs mouse $\alpha\beta$ T lymphocytes for apoptosis. *Nature.* 1991; **353**: 858–861.

39. Kabelitz D and Wesselborg S. Life and death of a superantigen reactive human CD4$^+$ T cell clone: staphylococcal enterotoxins induce death by apoptosis but simultaneously trigger a proliferative response in the presence of HLA-DR$^+$ antigen-presenting cells. *Int. Immunol.* 1992; **4**: 1381–1388.

40. Critchfield JM, Racke MK, Zuniga-Pflucker JC *et al.* T cell deletion in high antigen dose therapy of autoimmune encephalomyelitis. *Science.* 1994; **263**: 1139–1143.

41. Liblau RS, Pearson CI, Shokat K, Tisch R, Yang X-D and McDevitt HO. High-dose soluble antigen: peripheral T-cell proliferation or apoptosis. *Immunol. Rev.* 1994; **142**: 193–208.

42. Russell JH, Rush B, Weaver C and Wang R. Mature T cells of autoimmune lpr/lpr mice have a defect in antigen-stimulated suicide. *Proc. Natl. Acad. Sci. USA* 1993; **90**: 4409–4413.

43. Bossu P, Singer GG, Andres P, Ettinger R, Marshak-Rothstein A and Abbas AK. Mature C48$^+$ T lymphocytes from MRL/lpr mice are resistant to receptor-mediated tolerance and apoptosis. *J. Immunol.* 1993; **151**: 7233–7239.

44. Alderson MR, Tough TW, Davis-Smith T *et al.* Fas ligand mediates activation-induced cell death in human T lymphocytes. *J. Exp. Med.* 1995; **181**: 71–77.

45. Newell MK, Haughn LJ, Maroun CR and Julius MH. Death of mature T cells by separate ligation of CD4 and the T-cell receptor for antigen. *Nature* 1990; **347**: 286–289.

46. Banda NK, Bernier J, Kurahara DK *et al.* Crosslinking CD4 by human immunodeficiency virus gp120 primes T cells for activation-induced apoptosis. *J. Exp. Med.* 1992; **176**: 1099–1106.

47. Oyaizu N, McCloskey TW, Coronesi M, Chirmule N, Kalyanaraman VS and Pahwa S. Accelerated apoptosis in peripheral blood mononuclear cells (PBMCs) from human immunodeficiency virus type-1 infected patients and in CD4 cross-linked PBMCs from normal individuals. *Blood* 1993; **82**: 3075–3080.

48. Shi Y, Sahai BM and Green DR. Cyclosporin A inhibits activation-induced cell death in T-cell hybridomas and thymocytes. *Nature* 1989; **339**: 625–626.

49. Urba WJ, Ewel C, Kopp W *et al.* Anti-CD3 monoclonal antibody treatment of patients with solid tumors: a phase IA/B study. *Cancer Res.* 1992; **52**: 2394–2401.

50. Jondal M, Okret S and McConkey D. Killing of immature CD4+CD8+ thymocytes *in vivo* by anti-CD3 or 5'-(N-ethyl)-carboxamido-adenosine is blocked by glucocorticoid receptor antagonist RU-486. *Eur. J. Immunol.* 1993; **23**: 1246–1250.

51. Kappler JW, Roehm N and Marrack P. T cell tolerance by clonal elimination in the thymus. *Cell* 1987; **49**: 273–280.

52. Herman A, Kappler JW, Marrack P and Pullen AM. Superantigens: mechanism of T-cell stimulation and role in immune responses. *Ann. Rev. Immunol.* 1991; **9**: 745–772.

53. MacDonald HR and Lees RK. Programmed death of auto-reactive thymocytes. *Nature* 1990; **343**: 642–644.

54. Swat W, Ignatowicz L, von Boehmer H, Kisielow P. Clonal deletion of immature CD4+CD8+ thymocytes in suspension culture by extrathymic antigen-presenting cells. *Nature* 1991; **351**: 150–153.

55. Vasquez NJ, Kaye J and Hedrick SM. In vivo and in vitro clonal deletion of double-positive thymocytes. *J. Exp. Med.* 1992; **175**: 1307–1316.

56. Tadakuma T, Kizaki H, Odaka C *et al.* CD4+CD8+ thymocytes are susceptible to DNA fragmentation induced by phorbol ester, calcium ionophore and anti-CD3 antibody. *Eur. J. Immunol.* 1990; **20**: 779–784.

57. Page DM, Kane LP, Allison JP and Hedrick SM. Two signals are required for negative selection of CD4+CD8+ thymocytes. *J. Immunol.* 1993; **151**: 1868–1880.

58. Punt JA, Osborne BA, Takahama Y, Sharrow SO and Singer A. Negative selection of CD4+CD8+ thymocytes by T cell receptor-induced apoptosis requires a costimulatory signal that can be provided by CD28. *J. Exp. Med.* 1994; **179**: 709–713.

59. Iwata M, Mukai M, Nakai Y and Iseki R. Retinoic acids inhibit activation-induced apoptosis in T cell hybridomas and thymocytes. *J. Immunol.* 1992; **149**: 3302–3308.

60. McConkey DJ, Fosdick L, D'Adamio L, Jondal M and Orrenius S. Co-receptor (CD4/CD8) engagement enhances CD3-induced apoptosis in thymocytes. Implications for negative selection. *J. Immunol.* 1994; **153**: 2436–2443.

61. Crabtree GR. Contingent genetic regulatory events in T lymphocyte activation. *Science* 1989; **243**: 355–361.

62. Weiss A and Littman DR. Signal transduction by lymphocyte antigen receptors. *Cell* 1994; **76**: 263–274.

63. Migita K, Eguchi K, Kawabe Y, Mizokami A, Tsukada T and Nagataki S. Prevention of anti-CD3 monoclonal anti-

body-induced thymic apoptosis by protein tyrosine kinase inhibitors. *J. Immunol.* 1994; **153**: 3457–3465.

64. Jin L-W, Inaba K and Saitoh T. The involvement of protein kinase C in activation-induced cell death in T-cell hybridoma. *Cell Immunol.* 1992; **144**: 217–227.

65. Mercep M, Noguchi PD and Ashwell JD. The cell cycle block and lysis of an activated T cell hybridoma are distinct processes with different Ca^{2+} requirements and sensitivity to cyclosporin A. *J. Immunol.* 1989; **142**: 4085–4092.

66. Weber GF, Abromson-Leeman S and Cantor H. A signaling pathway coupled to T cell receptor ligation by MMTV superantigen leading to transient activation and programmed cell death. *Immunity* 1995; **2**: 363–372.

67. Kaiser N and Edelman IS. Calcium dependence of glucocorticoid-induced lymphocytolysis. *Proc. Natl. Acad. Sci. USA* 1977; **74**: 638–642.

68. Allbritton NL, Verret CR, Wolley RC and Eisen HN. Calcium ion concentrations and DNA fragmentation in target cell destruction by murine cloned cytotoxic T lymphocytes. *J. Exp. Med.* 1988; **167**: 514–527.

69. Zheng LM, Zychlinsky A, Lui C-C, Ojcius DM and Young JD-E. Extracellular ATP as a trigger for apoptosis or programmed cell death. *J. Cell. Biol.* 1991; **112**: 279–288.

70. Bellomo G, Perotti M, Taddei F *et al.* Tumor necrosis factor a induces apoptosis in mammary adenocarcinama cells by an increase in intracellular free Ca2+ concentration and DNA fragmentation. *Cancer Res.* 1992; **52**: 1342.

71. McConkey DJ, Hartzell P, Amador-Pérez JF, Orrenius S and Jondal M. Calcium-dependent killing of immature thymocytes by stimulation via the CD3/T cell receptor complex. *J. Immunol.* 1989; **143**: 1801–1806.

72. McConkey DJ, Nicotera P and Orrenius S. Signalling and chromatin fragmentation in thymocyte apoptosis. *Immunol. Rev.* 1994; **142**: 343–363.

73. Liu J, Farmer JDJ, Lane WS, Friedman J, Weissman I and Schreiber SL. Calcineurin is a common target of cyclophilin-cyclosporin A and FKBP-FK506 complexes. *Cell* 1991; **66**: 807.

74. Wyllie AH, Arends MJ, Morris RG, Walker SW and Evan G. The apoptosis endonuclease and its regulation. *Semin. Immunol.* 1992; **4**: 389–397.

75. Squier MKT, Miller ACK, Malkinson AM and Cohen JJ. Calpain activation in apoptosis. *J. Cell. Physiol.* 1994; **159**: 181.

76. McConkey DJ, Hartzell P, Jondal M and Orrenius S. Inhibition of DNA fragmentation in thymocytes and isolated thymocyte nuclei by agents that stimulate protein kinase C. *J. Biol. Chem.* 1989; **264**: 13399–13402.

77. Iseki R, Kudo Y and Iwata M. Early mobilization of Ca2+ is not required for glucocorticoid-induced apoptosis in thymocytes. *J. Immunol.* 1993; **151**: 5198–5207.

78. Iwata M, Iseki R, Sato K, Tozawa Y and Ohoka Y. Involvement of protein kinase C-epsilon in glucocorticoid-induced apoptosis in thymocytes. *J. Immunol.* 1994; **6**: 431–438.

79. Woronicz JD, Calnan B, Ngo V and Winoto A. Requirement for the orphan steroid receptor Nur77 in apoptosis of T-cell hybridomas. *Nature* 1994; **367**: 277–281.

80. Liu ZG, Smith SW, McLaughlin KA, Schwartz LM and Osborne BA. Apoptotic signals delivered through the T-cell receptor of a T-cell hybrid require the immediate-early gene nur77. *Nature* 1994; **367**: 281–284.

81. Ishida Y, Agata Y, Shibahara K and Honjo T. Induced expression of PD-1, a novel member of the immunoglobulin gene superfamily, upon programmed cell death. *EMBO J.* 1992; **11**: 3887–3895.

82. Schwartz LM and Osborne BA. Programmed cell death, apoptosis and killer genes. *Immunol. Today* 1993; **14**: 582–590.

83. Wyllie AH. Glucocorticoid-induced thymocyte apoptosis is associated with endogenous endonuclease activation. *Nature* 1980; **284**: 555–556.

84. Compton MM and Cidlowski JA. Rapid in vivo effects of glucocorticoids on the integrity of rat lymphocyte genomic deoxyribonucleic acid. *Endocrinology* 1986; **118**: 38–45.

85. Jaffe HL. The influence of the suprarenal gland on the thymus. III. Stimulation of the growth of the thymus gland following double supradrenalectomy in young rats. *J. Exp. Med.* 1924; **40**: 753–759.

86. Selye H. Thymus and adrenals in the response of the organism to injuries and intoxications. *Br. J. Exp. Path.* 1936; **17**: 234–248.

87. Dougherty TF and White A. Functional alterations in lymphoid tissue induced by adrenal cortical secretion. *Am. J. Anat.* 1945; **77**: 81–116.

88. Sprent J, Lo D, Gao E-K and Ron Y. T cell selection in the thymus. *Immunol. Rev.* 1988; **101**: 172–190.

89. Duval D, Dardenne M, Dausse JP and Homo F. Glucocorticoid receptors in corticosensitive and corticoresistant thymocyte subpopulations. *Biochim. Biophys. Acta* 1977; **496**: 312–320.

90. Screpanti I, Morrone S, Meco D *et al.* Steroid sensitivity of thymocyte subpopulations during intrathymic differentiation. Effects of 17 beta-estradiol and dexamethasone on subsets expressing T cell antigen receptor or IL-2 receptor. *J. Immunol.* 1989; **142**: 3378–3383.

91. Cupps TR and Fauci AS. Corticosteroid-mediated immunoregulation in man. *Immunol. Rev.* 1982; **65**: 133.

92. Nakayama K, Nakayama K, Negishi I *et al.* Disappearance of the lymphoid system in Bcl-2 homozygous mutant chimeric mice. *Science* 1993; **261**: 1584–1588.

93. Beato M. Gene regulation by steroid hormones. *Cell.* 1989; **56**: 335–344.

94. Cohen JJ. Lymphocyte death induced by glucocorticoids. In: Schleimer RP, Clamdi HN, Oronsky AL (eds), *Antiinflammatory steroid action-basic and clinical aspects.* San Diego: Academic Press, 1989: 110–131.

95. Baughman G, Harrigan MT, Campbell NF, Nurrish SJ and Bourgeois S. Genes newly identified as regulated by glucocorticoids in murine thymocytes. *Mol. Endocrinol* 1991; **5**: 637–644.

96. Owens GP, Hahn WE and Cohen JJ. Identification of mRNAs associated with programmed cell death in immature thymocytes. *Mol. Cell. Biol.* 1991; **11**: 4177–4188.

97. Jonat C, Rahmsdorf HJ, Park K-K *et al.* Antitumor promotion and antiinflammation: down-modulation of AP-1 (Fos/Jun) activity by glucocorticoid hormone. *Cell* 1990; **62**: 1189–1204.

98. Schüle R, Rangarajan P, Kliewer S et al. Functional antagonism between oncoprotein c-Jun and the glucocorticoid receptor. Cell 1990; 62: 1217–1226.

99. Yang-Yen H-F, Chambard J-C, Sun Y-L et al. Transcriptional interference between c-Jun and the glucocorticoid receptor: mutual inhibition of DNA binding due to direct protein–protein interaction. Cell 1990; 62: 1205–1215.

100. Dieken ES and Miesfeld RL. Transcriptional transactivation functions localized to the glucocorticoid receptor N terminus are necessary for steroid induction of lymphocyte apoptosis. Mol. Cell. Biol. 1992; 12: 589–597.

101. Helmberg A, Auphan N, Caelles C and Karin M. Glucocorticoid-induced apoptosis of human leukemic cells is caused by the repressive function of glucocorticoid receptor. EMBO J. 1995; 14: 452–460.

102. Zacharchuk CM, Mercep M, Chakraborti P, Simons, Jr SS and Ashwell JD. Programmed T lymphocyte death: cell activation- and steroid-induced pathways are mutually antagonistic. J. Immunol. 1990; 145: 4037–4045.

103. Iwata M, Hanaoka S and Sato K. Rescue of thymocytes and T cell hybridomas from glucocorticoid-induced apoptosis by stimulation via the T cell receptor/CD3 complex: a possible in vitro model for positive selection of the T cell repertoire. Eur. J. Immunol. 1991; 21: 643–648.

104. Migliorati G, Nicoletti I, Pagliacci MC, D'adamio L and Riccardi C. Interleukin-4 protects double-negative and CD4 single-positive thymocytes from dexamethasone-induced apoptosis. Blood. 1993; 81: 1352–1358.

105. Fernandez-Ruiz E, Rebollo A, Nieto MA et al. IL-2 protects T cell hybrids from the cytolytic effect of glucocorticoids. Synergistic effect of IL-2 and dexamethasone in the induction of high-affinity IL-2 receptors. J. Immunol. 1989; 143: 4146–4151.

106. Nieto MA and López-Rivas A. IL-2 protects T lymphocytes from glucocorticoid-induced DNA fragmentation and cell death. J. Immunol. 1989; 143: 4166–4170.

107. Zubiaga AM, Munoz E and Huber BT. IL-4 and IL-2 selectively rescue Th cell subsets from glucocorticoid-induced apoptosis. J. Immunol. 1992; 149: 107–112.

108. Trauth BC, Klas C, Peters AMJ et al. Monoclonal antibody-mediated tumor regression by induction of apoptosis. Science 1989; 245: 301–304.

109. Itoh N, Yonehara S, Ishii A et al. The polypeptide encoded by the cDNA for human cell surface antigen Fas can mediate apoptosis. Cell 1991; 66: 233–243.

110. Oehm A, Behrmann I, Falk W et al. Purification and molecular cloning of the APO-1 cell surface antigen, a member of the tumor necrosis factor/nerve growth factor receptor superfamily-sequence identity with Fas antigen. J. Biol. Chem. 1992; 267: 10709–10715.

111. Suda T, Takahashi T, Golstein P and Nagata S. Molecular cloning and expression of the Fas ligand, a novel member of the tumor necrosis factor family. Cell 1993; 75: 1169–1178.

112. Watanabe-Fukunaga R, Brannan CI, Itoh N et al. The cDNA structure, expression, and chromosomal assignment of the mouse Fas antigen. J. Immunol. 1992; 148: 1274–1279.

113. Cheng J, Zhou T, Liu C et al. Protection from Fas-mediated apoptosis by a soluble form of the Fas molecule. Science 1994; 263: 1759–1762.

114. Cascino I, Fiucci G, Papoff G and Ruberti G. Three functional soluble forms of the human apoptosis-inducing Fas molecule are produced by alternative splicing. J. Immunol. 1995; 154: 2706–2713.

115. Tanaka M, Suda T, Takahashi T and Nagata S. Expression of the functional soluble form of human Fas ligand in activated lymphocytes. EMBO J. 1995; 14: 1129–1135.

116. Perez C, Albert I, DeFay K, Zachariades N, Gooding L and Kriegle M. A nonsecretable cell surface mutant of tumor necrosis factor (TNF) kills by cell-to-cell contact. Cell 1990; 63: 251–258.

117. Dhein J, Walczak H, Bäumler C, Debatin K-M and Krammer PH. Autocrine T-cell suicide mediated by APO-1/(Fas/CD95). Nature 1995; 373: 438–441.

118. Leithauser F, Dhein J, Mechtersheimer G et al. Constitutive and induced expression of APO-1, a new member of the nerve growth factor/tumor necrosis factor receptor superfamily, in normal and neoplastic cells. Lab. Invest. 1993; 69: 415–429.

119. Kobayashi N, Hamamoto Y, Yamamoto N, Ishii A, Yonehara M and Yonehara S. Anti-Fas monoclonal antibody is cytocidal to human immunodeficiency virus-infected cells without augmenting viral replication. Proc. Natl. Acad. Sci. USA 1990; 87: 9620–9624.

120. Debatin KM, Goldman CK, Waldmann TA and Krammer PH. APO-1-induced apoptosis of leukemia cells from patients with adult T-cell leukemia. Blood 1993; 81: 2972–2977.

121. Debatin KM, Fahrig-Faissner A, Enenkel-Stoodt S, Kreuz W, Benner A and Krammer PH. High expression of APO-1 (CD95) on T lymphocytes from human immuno deficiency virus-1-infected children [letter]. Blood 1994; 83: 3101–3103.

122. Hiramatsu K, Aoki Y, Makino M et al. Increased Fas antigen expression in murine retrovirus-induced immunodeficiency syndrome. MAIDS. Eur. J. Immunol. 1994; 24: 2446–2451.

123. Dhein J, Daniel PT, Trauth BC, Oehm A, Moller P and Krammer PH. Induction of apoptosis by monoclonal antibody anti-APO-1 class switch variants is dependent on cross-linking of APO-1 cell surface antigens. J. Immunol. 1992; 149: 3166–3173.

124. Banner DW, D'Arcy A, Janes W et al. Crystal structure of the soluble human 55 kd TNF receptor-human TNFβ complex: implications for TNF receptor activation. Cell 1993; 73: 431–445.

125. Watanabe-Fukunaga R, Brannan CI, Copeland NG, Jenkins NA and Nagata S. Lymphoproliferation disorder in mice explained by defects in Fas antigen that mediates apoptosis. Nature 1992; 356: 314–317.

126. Clement MV and Stamenkovic I. Fas and tumor necrosis factor receptor-mediated cell death: similarities and distinctions. J. Exp. Med. 1994; 180: 557–567.

127. Tartaglia LA, Ayres TM, Wong GHW and Goeddel DV. A novel domain withn the 55 KD TNF receptor signals cell death. Cell 1993; 74: 845–853.

128. Itoh N and Nagata S. A novel protein domain required for apoptosis. Mutational analysis of human Fas antigen. *J. Biol. Chem.* 1993; **268**: 10932–10937.

129. Boldin MP, Mett IL, Varfolomeev EE *et al.* Self-association of the death domains of the p55 tumor necrosis factor (TNF) receptor and Fas/APO1 prompts signaling for TNF and Fas/APO1 effects. *J. Biol. Chem.* 1995; **270**: 387–391.

130. Kim M-Y, Linardic C, Obeid L and Hannun YA. Identification of sphingomyelin turnover as an effector mechanism for the action of tumor necrosis factor-α and γ-interferon. Specific role in cell differentiation. *J. Biol. Chem.* 1991; **266**: 484–489.

131. Mathias S, Dressler KA and Kolesnick RN. Characterization of a ceramide-activated protein kinase: stimulation by tumor necrosis factor alpha. *Proc. Natl. Acad. Sci. USA* 1991; **88**: 10009–10013.

132. Dressler KA, Mathias S and Kolesnick RN. Tumor necrosis factor-α activates the sphingomyelin signal transduction pathway in a cell free system. *Science* 1992; **255**: 1715–1718.

133. Hannun YA. The sphingomyelin cycle and the second messenger function of ceramide. *J. Biol. Chem.* 1994; **269**: 3125–3128.

134. Obeid LM, Linardic CM, Karolak LA and Hannun YA. Programmed cell death induced by ceramide. *Science* 1993; **259**: 1769–1772.

135. Jarvis WD, Kolesnick RN, Fornari FA, Traylor RS, Gewirtz DA and Grant S. Induction of apoptotic DNA damage and cell death by activation of the sphingomyelin pathway. *Proc. Natl. Acad. Sci. USA* 1994; **91**: 73–77.

136. Cifone MG, De Maria R, Roncaioli P *et al.* Apoptotic signaling through CD95 (Fas/Apo-1) activates an acidic sphingomyelinase. *J. Exp. Med.* 1994; **180**: 1547–1552.

137. Quintans J, Kilkus J, McShan CL, Gottschalk AR and Dawson G. Ceramide mediates the apoptotic response of WEHI 231 cells to anti-immunoglobulin, corticosteroids and irradiation. *Biochem. Biophys. Res. Commun.* 1994; **202**: 710–714.

138. Hannun YA and Bell RM. The sphingomyelin cycle: a prototypic sphingolipid signaling pathway. *Adv. Lipid Res.* 1993; **25**: 27–41.

139. Gulbins E, Bissonnette R, Mahboubi A *et al.* Fas-induced apoptosis is mediated via a ceramide-initiated ras signaling pathway. *Immunity* 1995; **2**: 341–351.

140. Owen-Schaub LB, Yonehara S, Crump WL and Grimm EA. DNA fragmentation and cell death is selectively triggered in activated human lymphocytes by Fas antigen engagement. *Cell. Immunol.* 1992; **140**: 197–205.

141. Klas C, Debatin KM, Jonker RR and Krammer PH. Activation interferes with the APO-1 pathway in mature human T cells. *Int. Immunol.* 1993; **5**: 625–630.

142. Vella AT, McCormack JE, Linsley PS, Kappler JW and Marrack P. Lipopolysaccharide interferes with the induction of peripheral T cell death. *Immunity* 1995; **2**: 261–270.

143. Rothstein TL, Wang JLM, Panka DJ *et al.* Protection against Fas-dependent Th1-mediated apoptosis by antigen receptor engagement in B cells. *Nature* 1995; **374**: 163–165.

144. Sato T, Irie S, Kitada S and Reed JC. FAP-1: a protein tyrosine phosphatase that associated with Fas. *Science.* 1995; **268**: 411–415.

145. Su X, Zhou T, Wang Z, Yang PA, Jope RS and Mountz JD. Defective expression of hematopoietic cell protein tyrosine phosphatase (HCP) in lymphoid cells blocks Fas-mediated apoptosis. *Immunity* 1995; **2**: 353–362.

146. Itoh N, Tsujimoto Y and Nagata S. Effect of bcl-2 on Fas antigen-mediated cell death. *J. Immunol.* 1993; **151**: 621–627.

147. Chiou SK, Tseng CC, Rao L and White E. Functional complementation of the adenovirus E1B 19-kilodalton protein with Bcl-2 in the inhibition of apoptosis in infected cells. *J. Virol.* 1994; **68**: 6553–6566.

148. Takayama S, Sato T, Krajewski S *et al.* Cloning and functional analysis of BAG-1: a novel Bcl-2-binding protein with anti-cell death activity. *Cell* 1995; **80**: 279–284.

149. Chiu VK, Walsh CM, Liu C-C, Reed JC and Clark WB. Bcl-2 blocks degranulation but not fas-based cell-mediated cytotoxicity. *J. Immunol.* 1995; **154**: 2023–2032.

150. Green DR, Mahboubi A, Nishioka W *et al.* Promotion and inhibition of activation-induced apoptosis in T-cell hybridomas by oncogenes and related signals. *Immunol. Rev.* 1994; **142**: 321–342.

151. Brunner T, Modil RJ, LaFace D *et al.* Cell-autonomous Fas (CD95)/Fas-ligand interaction mediates activation-induced apoptosis in T-cell hybridomas. *Nature* 1995; **373**: 441–444.

152. Ju S-T, Panka DJ, Cui H *et al.* Fas (CD95)/FasL interactions required for programmed cell death after T-cell activation. *Nature* 1995; **373**: 444–448.

153. Yang Y, Mercep M, Ware CF and Ashwell JD. Fas and activation-induced Fas ligand mediate apoptosis of T cell hybridomas: inhibition of Fas ligand expression by retinoic acid and glucocorticoids. *J. Exp. Med.* 1995; **181**: 1673–1682.

154. Drappa J, Brot N and Elkon KB. The Fas protein is expressed at high levels on CD4+CD8+ thymocytes and activated mature lymphocytes in normal mice but not in the lupus-prone strain, MRL 1pr/1pr. *Proc. Natl. Acad. Sci. USA* 1993; **90**: 10340–10344.

155. Evan G and Littlewood T. The role of c-myc in cell growth. *Curr. Opin. Genet. Dev.* 1993; **3**: 44–49.

156. DePinho RA, Schreiber-Agus N and Alt FW. myc family oncogenes in the development of normal and neoplastic cells. *Adv. Cancer Res.* 1991; **57**: 1–46.

157. Blackwood EM and Eisenman RN. Max: a helix-loop-helix zipper protein that forms a sequence-specific DNA-binding complex with Myc. *Science.* 1991; **251**: 1211–1217.

158. Amati B, Dalton S, Brooks MW, Littlewood TD, Evan GI and Land H. Transcriptional activation by the human c-Myc oncoprotein n yeast requires interaction with Max. *Nature* 1992; **359**: 423–426.

159. Rustgi AK, Dyson N and Bernards R. Amino-terminal domains of c-myc and N-myc proteins mediate binding to the retinoblastoma gene product. *Nature* 1991; **352**: 541–544.

160. Roy AL, Carruthers C, Gutjahr T and Roeder RG. Direct role for myc in transcription initiation mediated by interactions with TFII-I. *Nature* 1993; **365**: 359–361.

161. Askew D, Ashmun R, Simmons B and Cleveland J. Constitutive c-myc expression in IL-3-dependent myeloid cell

line suppresses cycle arrest and accelerates apoptosis. *Oncogene* 1991; **6**: 1915–1922.

162. Evan GI, Wyllie AH, Gilbert CS *et al.* Induction of apoptosis in fibroblasts by c-myc protein. *Cell* 1992; **69**: 119–128.

163. Shi Y, Glynn JM, Guilbert LJ, Cotter TG, Bissonnette RP and Green DR. Role for c-myc in activation-induced apoptotic cell death in T cell hybridomas. *Science* 1992; **257**: 212–214.

164. Amati B, Littlewood T, Evan GI and Land H. The c-Myc protein induces cell cycle progression and apoptosis through dimerization with Max. *EMBO J.* 1994; **12**: 5083–5087.

165. Bissonnette RP, McGahon A, Mahboubi A and Green DR. Functional Myc–Max heterodimer is required for activation-induced apoptosis in T cell hybridomas. *J. Exp. Med.* 1994; **180**: 2413–2418.

166. Bissonnette RF, Echeverri F, Mahboubi A and Green DR. Apoptotic cell death induced by c-myc is inhibited by bcl-2. *Nature* 1992; **359**: 552–554.

167. Fanidi A, Harrington EA and Evan GI. Cooperative interaction between *c-myc* and *bcl-2* proto-oncogenes. *Nature* 1992; **359**: 554–556.

168. Hermeking H and Eick D. Mediation of c-myc-induced apoptosis by p53. *Science* 1994; **265**: 2091–2093.

169. Wagner AJ, Meyers C, Laimins LA and Hay N. c-Myc induces the expression and activity of ornithine decarboxylase. *Cell Growth Differ.* 1994; **11**: 879.

170. Packham G and Cleveland JL. Ornithine decarboxylase is a mediator of c-Myc-induced apoptosis. *Mol. Cell. Biol.* 1994; **14**: 5741–5747.

171. Brüne B, Hartzell P, Nicotera P and Orrenius S. Spermine prevents endonuclease activation and apoptosis in thymocytes. *Exp. Cell. Res.* 1991; **195**: 323–329.

172. Dypbukt JM, Ankarcrona M, Burkitt M *et al.* Different prooxidant levels stimulate growth, trigger apoptosis, or produce necrosis of insulin-secreting RINm5F cells. The role of intracellular polyamines. *J. Biol. Chem.* 1994; **269**: 30553–30560.

173. Tsujimoto Y, Gorham J, Cossman J, Jaffe E and Croce CM. The t(14;18) chromosome translocations involved in B-cell neoplasms result from mistakes in VDJ joining. *Science* 1985; **229**: 1390–1393.

174. Bakhshi A, Jensen JP, Goldman P *et al.* Cloning the chromosomal breakpoint of t(14;18) human lymphomas: clustering around JH on chromosomal 14 and near a transcriptional unit on 18. *Cell* 1985; **41**: 889–906.

175. Cleary ML and Sklar J. Nucleotide sequence of a t(14;18) chromosomal breakpoint in follicular lymphoma and demonstration of a breakpoint cluster region near a transcriptionally active locus on chromosome 18. *Proc. Natl. Acad. Sci. USA.* 1985; **82**: 7439–7443.

176. Hockenbery D, Nuñez G, Milliman C, Schreiber RD and Korsmeyer SJ. Bcl-2 is an inner mitochondrial membrane protein that blocks programmed cell death. *Nature* 1990; **348**: 334–336.

177. de Jong D, Prins FA, Mason DY, Reed JC, van Ommen GB and Kluin PM. Subcellular localization of the bcl-2 protein in malignant and normal lymphoid cells. *Cancer Res.* 1994; **54**: 256–260.

178. Lithgow T, van Driel R, Bertram JF and Strasser A. The protein product of the oncogene bcl-2 is a component of the nuclear envelope, the endoplasmic reticulum, and the outer mitochondrial membrane. *Cell Growth Differ.* 1994; **5**: 411–417.

179. Hockenbery DM, Zutter M, Hickey W, Nahm M and Korsmeyer SJ. Bcl-2 protein is topographically restricted in tissues characterized by apoptotic cell death. *Proc. Natl. Acad. Sci. USA* 1991; **88**: 6961–6965.

180. Vaux DL, Cory S and Adams JM. *Bcl-2* gene promotes haemopoietic cell survival and cooperates with *c-myc* to immortalize pre-B cells. *Nature* 1988; **335**: 440–442.

181. Nuñez G, London L, Hockenbery D, Alexander M, McKearn J and Korsmeyer SJ. Deregulated Bcl-2 gene expression selectively prolongs survival of growth factor-deprived hemopoietic cell lines. *J. Immunol.* 1990; **144**: 3602–3610.

182. Miyashita T and Reed J. bcl-2 gene transfer increases relative resistance of S49.1 and WEHI7.2 lymphoid cells to cell death and DNA fragmentation induced by glucocorticoids and multiple chemotherapeutic drugs. *Cancer Res.* 1992; **52**: 5407–5411.

183. Miura M, Zhu H, Rotello R, Hartwieg EA and Yuan J. Induction of apoptosis in fibroblasts by IL-1 β-converting enzyme, a mammalian homolog of the C. elegans cell death gene ced-3. *Cell* 1993; **75**: 653–660.

184. Rao L, Debbas M, Sabbatini P, Hockenbery D, Korsmeyer S and White E. The adenovirus E1A proteins induce apoptosis, which is inhibited by the E1B 19-kDa and Bcl-2 proteins. *Proc. Natl. Acad. Sci. USA* 1992; **89**: 7742–7746.

185. Torigoe T, Millan JA, Takayama S, Taichman R, Miyashita T and Reed JC. Bcl-2 inhibits T-cell-mediated cytolysis of a leukemia cell line. *Cancer Res.* 1994; **54**: 4851–4854.

186. McDonnell TJ, Deane N, Platt FM *et al. bcl-2*-immunoglobulin transgenic mice demonstrate extended B cell survival and follicular lymphoproliferation. *Cell* 1989; **57**: 79–88.

187. Sentman CL, Shutter JR, Hockenbery D, Kanagawa O and Korsmeyer SJ. bcl-2 inhibits multiple forms of apoptosis but not negative selection in thymocytes. *Cell* 1991; **67**: 879–888.

188. Strasser A, Harris AW and Cory S. *bcl-2* transgene inhibits T cell death and perturbs thymic self-censorship. *Cell* 1991; **67**: 889–899.

189. Siegel RM, Katsumata M, Miyashita T, Louie DC, Greene MI and Reed JC. Inhibition of thymocyte apoptosis and negative antigenic selection in bcl-2 transgenic mice. *Proc. Natl. Acad. Sci. USA* 1992; **89**: 7003–7007.

190. Korsmeyer SJ. Regulators of cell death. *Trends Geneti.* 1995; **11**: 101–105.

191. Boise LH, Gonzalez-Garcia M, Postema CE *et al.* bcl-x, a bcl-2 related gene that functions as a dominant regulator of apoptotic cell death. *Cell* 1993; **74**: 597–608.

192. Gonzalez-Garcia M, Perez-Ballestero R, Ding L *et al.* bcl-xL is the major bcl-x mRNA form expressed during murine development and its product localizes to mitochondria. *Development* 1994; **120**: 3033–3042.

193. Oltvai ZN, Milliman CL and Korsmeyer SJ. Bcl-2 hetero-dimerizes in vivo with a conserved homolog, Bax, that accelerates programmed cell death. *Cell* 1993; **74**: 609–619.

194. Yin XM, Oltval ZN and Korsmeyer SJ. BH1 and BH2 domains of Bcl-2 are required for inhibition of apoptosis and heterodimerization with Bax. *Nature* 1994; **369**: 321–323.

195. Yang E, Zha J, Jockel J, Boise LH, Thompson CB and Korsmeyer SJ. Bad, a heterodimeric partner for Bcl-XL and Bcl-2, displaces Bax and promotes cell death. *Cell* 1995; **80**: 285–291.

196. Chittenden T, Harrington EA, O'Connor R *et al.* Induction of apoptosis by the Bcl-2 homologue Bak. *Nature* 1995; **374**: 733–736.

197. Farrow SN, White JH, Martinou I *et al.* Cloning of a bcl-2 homologue by interaction with adenovirus E1B 19K. *Nature* 1995; **374**: 731–733.

198. Kiefer MC, Brauer MJ, Powers VC *et al.* Modulation of apoptosis by the widely distributed Bcl-2 homologue Bak. *Nature* 1995; **374**: 736–739.

199. Henderson S, Huen D, Rowe M, Dawson C, Johnson G and Rickinson A. Epstein–Barr virus-coded BHRF1 protein, a viral homolog of bcl-2, protects human B-cells from programmed cell death. *Proc. Natl. Acad. Sci. USA* 1993; **90**: 8479–8483.

200. Neilan J, Lu Z, Afonso C, Kutish G, Sussman M and Rock D. An African swine fever virus gene with similarity to bcl-2 and the Epstein–Barr virus gene BHRF1. *J. Virol.* 1993; **67**: 4391–4394.

201. Hengartner MO and Horvitz HR. *C. elegans* cell survival gene ced-9 encodes a functional homolog of the mammalian protooncogene bcl-2. *Cell* 1995; **76**: 665 676.

202. Lin EY, Orlofsky A, Berger MS and Prystowsky MB. Characterization of A1, a novel hemopoietic-specific early-response gene with sequence similarity to bcl-2. *J. Immunol.* 1993; **151**: 1979–1988.

203. Kozopas KM, Yang T, Buchan HL, Zhou P and Craig RW. MCL1, a gene expressed in programmed myeloid cell differentiation, has sequence similarity ͑to BCL2. *Proc. Natl. Acad. Sci. USA* 1993; **90**: 3516–3520.

204. Zhong LT, Sarafian T, Kane DJ *et al.* bcl-2 inhibits death of central neural cells induced by multiple agents. *Proc. Natl. Acad. Sci. USA* 1993; **90**: 4533–4537.

205. Hockenbery DM, Oltvai ZN, Yin X-M, Milliman CL and Korsmeyer SJ. Bcl-2 functions in an antioxidant pathway to prevent apoptosis. *Cell* 1993; **75**: 241–251.

206. Kane DJ, Sarafian TA, Anton R *et al.* Bcl-2 inhibition of neural death: decreased generation of oxygen species. *Science* 1993; **262**: 1274–1277.

207. Jacobson MD and Raff MC. Programmed cell death and Bcl-2 protection in very low oxygen. *Nature* 1995; **374**: 814–816.

208. Shimizu S, Eguchi Y, Kosaka H, Kamiike W, Matsuda H and Tsujimoto Y. Prevention of hypoxia-induced cell death by Bcl-2 and Bcl-x. *Nature* 1995; **374**: 811–813.

209. Lam M, Dubyak G, Chen L, Nunez G, Miesfeld RL and Distelhorst CW. Evidence that BCL-2 represses apoptosis by regulating endoplasmic reticulum-associated Ca2+ fluxes. *Proc. Natl. Acad. Sci. USA* 1994; **91**: 6569–6573.

210. Fernandez-Sarabia M and Bischoff JR. Bcl-2 associates with the ras-related protein R-ras p23. *Nature* 1993; **366**: 274–275.

211. Wang HG, Miyashita T, Takayama S *et al.* Apoptosis regulation by interaction of bcl-2 protein and raf-1 kinase. *Oncogene* 1994; **9**: 2751–2756.

212. Newmeyer DD, Farschon DM and Reed JC. Cell-free apoptosis in *Xenopus* egg extracts: inhibition by bcl-2 and requirement for an organelle fraction enriched in mitochondria. *Cell* 1994; **79**: 353–364.

213. Donehower LA, Harvey M, Slagle BL *et al.* Mice deficient for p53 are developmentally normal but are susceptible to spontaneous tumours. *Nature* 1992; **356**: 215–221.

214. Jacks T, Remington L, Williams BO *et al.* Tumor spectrum analysis in p53-mutant mice. *Curr. Biol.* 1994; **4**: 1–7.

215. Purdie CA, Harrison DJ, Peter A *et al.* Tumour incidence, spectrum and ploidy in mice with a large deletion in the p53 gene. *Oncogene* 1994; **9**: 603–609.

216. El-Deiry WS, Tokino T, Velculescu VE *et al.* WAF1, a potential mediator of p53 tumor suppression. *Cell* 1993; **75**: 817–825.

217. Miyashita T and Reed JC. Tumor suppressor p53 is a direct transcriptional activator of the human bax gene. *Cell* 1995; **80**: 293–299.

218. Donehower LA and Bradley A. The tumor suppressor p53. *Biochim. Biophys. Acta* 1993; **1155**: 181–205.

219. Levine AJ, Momand J and Finlay C. The p53 tumour suppressor gene. *Nature* 1991; **351**: 453–456.

220. Momand J, Zambetti GP, Olson DC, George D and Levine AJ. The mdm-2 oncogene product forms a complex with the p53 protein and inhibits p53-mediated transactivation. *Cell* 1992; **69**: 1237–1245.

221. White E. Regulation of apoptosis by the transforming genes of the DNA tumor virus adenovirus. *Proc. Soc. Exp. Biol. Med.* 1993; **204**: 30–39.

222. Truant R, Xiao H, Ingles CJ and Greenblatt J. Direct interaction between the transcriptional activation domain of human p53 and the TATA box-binding protein. *J. Biol. Chem.* 1993; **268**: 2284–2287.

223. Kuerbitz SJ, Plunkett BS, Walsh VW and Kastan MB. Wild type p53 is a cell cycle checkpoint determinant following irradiation. *Proc. Natl. Acad. Sci. USA* 1992; **89**: 7491–7495.

224. Lane DP. p53, guardian of the genome. *Nature* 1992; **358**: 15–16.

225. Michalovitz D, Halevy O and Oren M. Conditional inhibition of transformation and of cell proliferation by a temperature-sensitive mutant of p53. *Cell* 1990; **62**: 671–680.

226. Yonish-Rouach E, Resnitzky D, Lotem J, Sachs L, Kimchi A and Oren M. Wild-type p53 induces apoptosis of myeloid leukaemic cells that is inhibited by interleukin-6. *Nature* 1991; **352**: 345–347.

227. Shaw P, Bovey R, Tardy S, Sahli R, Sordat B and Costa J. Induction of apoptosis by wild-type p53 in a human colon tumor-derived cell line. *Proc. Natl. Acad. Sci. USA* 1992; **89**: 4495–4499.

228. Ramqvist T, Magnusson KP, Wang Y, Szekely L, Klein G and Wiman K. Wild-type p53 induces apoptosis in a Burkitt lymphoma (BL) cell line that carries mutant p53. *Oncogene* 1993; **8**: 1495–1500.

229. Wang Y, Ramqvist T, Szekely L, Axelson H, Klein G and Wiman K. Reconstitution of wild-type p53 expression triggers apoptosis in a p53-negative v-myc retrovirus-induced T-cell lymphoma line. *Cell Growth Differ* 1993; **4**: 467–473.

230. Ryan JJ, Danish R, Gottlieb CA and Clarke MF. Cell cycle analysis of p53-induced cell death in murine erythroleukemia cells. *Mol. Cell. Biol.* 1993; **13**: 711–719.

231. Johnson P, Chung S and Benchimol S. Growth suppression of Friend virus-transformed erythroleukemia cells by p53 protein is accompanied by hemoglobin production and is sensitive to erythropoietin. *Mol. Cell. Biol.* 1993; **13**: 1456–1463.

232. Lowe SW, Schmitt EM, Smith SW, Osborne BA and Jacks T. p53 is required for radiation-induced apoptosis in mouse thymocytes. *Nature* 1993; **362**: 847–849.

233. Clarke AR, Purdie CA, Harrison DJ *et al.* Thymocyte apoptosis induced by p53-dependent and independent pathways. *Nature* 1993; **362**: 849–852.

234. Strasser A, Harris AW, Jacks T and Cory S. DNA damage can induce apoptosis in proliferating lymphoid cells via p53-independent mechanisms inhibitable by bcl-2. *Cell* 1994; **79**: 329–339.

235. Lowe SW and Ruley HE. Stabilization of the p53 tumor suppressor is induced by adenovirus 5 E1A and accompanies apoptosis. *Genes Dev.* 1993; **7**: 535–545.

236. Debbas M and White E. Wild type p53 mediates apoptosis by E1A, which is inhibited by E1B. *Genes Dev.* 1993; **7**: 546–554.

237. Pan H and Griep AE. Altered cell cycle regulation in the lens of HPV-16 E6 or E7 transgenic mice: implications for tumor suppressor gene function in development. *Genes Dev.* 1994; **8**: 1285–1299.

238. Wu X and Levine A. p53 and E2F-1 cooperate to mediate apoptosis. *Proc. Natl. Acad. Sci. USA* 1994; **91**: 3602–3606.

239. Morgenbesser SD, Williams BO, Jacks T and DePinho RA. p53-dependent apoptosis produced by Rb-deficiency in the developing mouse lens. *Nature.* 1994; **371**: 72–74.

240. Scher CD and Siegler R. Direct tranformation of 3T3 cells by abelsone murine leukemia virus. *Nature* 1975; **253**: 729–731.

241. Konopka JB and Witte ON. Detection of a c-abl tyrosine kinase activity in vitro permits direct comparison of normal and altered abl gene products. *Mol. Cell. Biol.* 1985; **5**: 3116–3123.

242. Chen YY and Rosenberg N. Lymphoid cells transformed by abelson virus require the v-abl protein-tyrosine kinase only during early G1. *Proc. Natl. Acad. Sci. USA* 1992; **89**: 6683–6687.

243. McGahon A, Bissonnette R, Schmitt M, Cotter KM, Green DR and Cotter TG. BCL-ABL maintains resistance of chronic myelogenous leukemia cells to apoptotic cell death. *Blood* 1994; **83**: 1179–1187.

244. Chapman RS, Whetton AD and Dive C. The suppression of drug-induced apoptosis by activation of v-ABL protein tyrosine kinase. *Cancer Res.* 1994; **54**: 5131–5137.

245. Groffen J, Stephenson JR, Heisterkamp N, De Klein A, Bartram CB and Grosveld G. Philadelphia chromosomal breakpoints are clustered within a limited region, bcr, on chromosome 22. *Cell* 1984; **36**: 93–99.

246. Smetsers TF, Skorski T, van de Locht LT *et al.* Antisense BCR-ABL oligonucleotides induce apoptosis in the philadelphia chromosomal positive cell line BV173. *Leukemia* 1994; **8**: 129–140.

247. Bedi A, Zehnbauer BA, Barber JP, Sharkis SJ and Jones RJ. Inhibition of apoptosis by BCL-ABL in chronic myeloid leukemia. *Blood* 1994; **83**: 2038–2044.

248. Tybulewicz VLJ, Crawford CE, Jackson PK, Bronson RT and Mulligan RC. Neonatal lethality and lymphopenia in mice with a homozygous disruption of the c-abl proto-oncogene. *Cell* 1991; **65**: 1153–1163.

249. Yuan J, Shaham S, Ledoux S, Ellis HM and Horvitz HR. The *C. elegans* cell death gene ced-3 encodes a protein similar to mammalian interleukin-1 β-converting enzyme. *Cell* 1993; **75**: 641–652.

250. Kumar S, Tomooka Y and Noda M. Identification of a set of genes with developmentally down-regulated expression in the mouse brain. *Biochem. Biophys. Res. Commun.* 1992; **185**: 1155–1161.

251. Wang L, Miura M, Bergeron L, Zhu H and Yuan J. Ich-1, an Ice/ced-3-related gene, encodes both positive and negative regulators of programmed cell death. *Cell* 1994; **78**: 739–750.

252. Kumar S, Kinoshita M, Noda M, Copeland NG and Jenkins NA. Induction of apoptosis by the mouse Nedd2 gene, which encodes a protein similar to the product of the *Caenorhabditis elegans* cell death gene ced-3 and the mammalian IL-1β-converting enzyme. *Genes Dev.* 1994; **8**: 1613–1626.

253. Shiver JW, Su L and Henkart P. Cytotoxicity with target DNA breakdown by rat basophilic leukemia cells expressing both cytolysin and granzyme A. *Cell* 1992; **71**: 315–322.

254. Heusel JW, Wesselschmidt RL, Shresta S, Russell JH and Ley TJ. Cytotoxic lymphocytes require granzyme B for the rapid induction of DNA fragmentation and apoptosis in allogeneic target cells. *Cell* 1994; **76**: 977–987.

255. Shi L, Nishioka WK, Th'ng J, Bradbury EM, Litchfield DW and Greenberg AH. Premature p34cdc2 activation required for apoptosis. *Science* 1994; **263**: 1143–1145.

256. Kaufmann SH, Desnoyers S, Ottaviano Y, Davidson NE and Poirier GG. Specific proteolytic cleavage of poly (ADP-ribose) polymerase: an early marker of chemotherapy-induced apoptosis. *Cancer Res.* 1993; **53**: 3976–3985.

257. Lazebnik YA, Kaufmann SH, Desnoyers S, Poirier GG and Earnshaw WC. Cleavage of poly(ADP-ribose) polymerase by a proteinase with properties like ICE. *Nature* 1994; **371**: 346–347.

258. Casciola-Rosen LA, Miller DK, Anhalt GJ and Rosen A. Specific cleavage of the 70 kDa ptotrin component of the U1 small nuclear ribonucleoprotein is a characteristic biochemical feature of apoptotic cell death. *J. Biol. Chem.* 1994; **269**: 30757–30760.

259. Bruno S, Lassota P, Giaretti W and Dazynkiewicz Z. Apoptosis of rat thymocytes triggered by prednisolone, camptothecin, or teniposide is selective to Go cells and is prevented by inhibitors of proteases. *Oncol. Res.* 1992; **4**: 29–35.

260. Sarin A, Adams DH and Henkart PA. Protease inhibitors selectively block T cell receptor-triggered programmed cell

death in a murine T cell hybridoma and activated peripheral T cells. *J. Exp. Med.* 1993; **178**: 1693–1700.

261. Gagliardini V, Fernandez P-A, Lee RKK *et al.* Prevention of vertebrate neuron death by the crmA gene. *Science* 1994; **263**: 826–828.

262. Tewari M and Dixit VM. Fas-and tumor necrosis factor-induced apoptosis is inhibited by the poxvirus crmA gene product. *J. Biol. Chem.* 1995; **270**: 3255–3260.

263. Enari M, Hug H and Nagata S. Involvement of an ICE-like protease in Fas-mediated apoptosis. *Nature* 1995; **375**: 78–80.

264. Los M, Van de Craen M, Penning LC *et al.* Requirement of an ICE/CED-3 protease for Fas/Apo-1-mediated apoptosis. *Nature* 1995; **375**: 81–83.

265. Li P, Allen H, Banerjee S *et al.* Mice-deficient in IL-1β-converting enzyme are defective in production of mature IL-1b and resistant to endotoxic shock. *Cell* 1995; **80**: 401–411.

266. Kuida K, Lippke JA, Ku G *et al.* Altered cytokine export and apoptosis in mice deficient in interleukin-1β converting enzyme. *Science* 1995; **267**: 2000–2003.

267. Zhivotovsky B, Wade D, Gahm A, Orrenius S and Nicotera P. Formation of 50 kbp chromatin fragments in isolated liver nuclei is mediated by protease and endonuclease activation. *FEBS Letts.* 1994; **351**: 150–154.

268. Brown DG, Sun X-M and Cohen GM. Dexamethasone-induced apoptosis involves cleavage of DNA to large fragments prior to internucleosomal fragmentation. *J. Biol. Chem.* 1993; **268**: 3037–3039.

269. Cohen GM, Sun X-M, Fearnhead H *et al.* Formation of large molecular weight fragments of DNA is a key committed step of apoptosis in thymocytes. *J. Immunol.* 1994; **153**: 507–516.

270. Sun DY, Jiang S, Zheng L-M, Ojcius DM and Young JD-E. Separate metabolic pathways leading to DNA fragmentation and apoptotic chromatin condensation. *J. Exp. Med.* 1994; **179**: 559–568.

271. Gaido ML and Cidlowski JA. Identification, purification, and characterization of a calcium-dependent endonuclease (NUC18) from apoptotic rat thymocytes: NUC18 is not histone H2B. *J. Biol. Chem.* 1991; **266**: 18580–18585.

272. Hughes Jr, FM and Cidlowski JA. Apoptotic DNA degradation: evidence for novel enzymes. *Cell Death Differ.* 1994; **1**: 11–17.

273. Montague JW, Gaido ML, Frye C and Cidlowski JA. A calcium-dependent nuclease from apoptotic rat thymocytes is homologous with cyclophilin. Recombinant cyclophilins A, B, and C have nuclease activity. *J. Biol. Chem.* 1994; **269**: 18877–18880.

274. Ucker DS, Obermiller PS, Eckhart W, Apgar JR, Berger NA and Meyers J. Genome digestion is a dispensable consequence of physiological cell death mediated by cytotoxic T lymphocytes. *Mol. Cell. Biol.* 1992; **12**: 3060–3069.

275. Peitsch MC, Polzar B, Stephan H *et al.* Characterization of the endogenous deoxyribonuclease involved in nuclear DNA degradation during apoptosis (programmed cell death). *EMBO J.* 1993; **12**: 371–377.

276. Noll M. DNA folding in the nucleosome. *J. Mol. Biol.* 1977; **116**: 49–71.

277. Peitsch MC, Mannherz HG and Tschopp J. The apoptosis endonucleases: cleaning up after cell death? *Trends Cell. Biol.* 1994; **4**: 37–41.

278. Barry MA and Eastman A. Identification of deoxyribonuclease II as an endonuclease involved in apoptosis. *Arch. Biochem. Biophys.* 1993; **300**: 440–450.

279. Bernardi G. Mechanism of action and structure of acid deoxyribonuclease. *Adv. Enzymol.* 1968; **31**: 1–49.

280. Mogil RJ, Shi Y, Bissonnette RP, Bromley P, Yamaguchi I and Green DR. Role of DNA fragmentation in T cell activation-induced apoptosis *in vitro* and *in vivo*. *J. Immunol.* 1994; **152**: 1674–1683.

281. Dipasquale B and Youle RJ. Programmed cell death in heterokaryons: a study of the transfer of apoptosis between nuclei. *Am. J. Pathol.* 1992; **141**: 1471–1479.

282. Hengartner MO and Horvitz HR. The ins and outs of programmed cell death during *C. elegans* development. *Phil. Trans. Roy. Soc. Lond. B.* 1994; **345**: 243–246.

283. Hedgecock EM, Sulston JE and Thomson JN. Mutations affecting programmed cell deaths in the nematode *Caenorhabditis elegans*. *Science* 1983; **220**: 1277–1279.

284. Jacobson MD, Burne JF and Raff MC. Programmed cell death and Bcl-2 protection in the absence of a nucleus. *EMBO J.* 1994; **13**: 1899–1910.

285. Schulze-Osthoff K, Walczak H, Droge W and Krammer PH. Cell nucleus and DNA fragmentation are not required for apoptosis. *J. Cell. Biol.* 1994; **127**: 15–20.

286. Nakajima H, Golstein P and Henkart P. The target cell nucleus is not required for cell-mediated granzyme- or Fas-based cytotoxicity. *J. Exp. Med.* 1995; **181**: 1905–1909.

287. Bansal N, Houle A and Mclnykovych G. Apoptosis: mode of cell death induced in T cell leukemia lines by dexamethasone and other agents. *FASEB J.* 1991; **5**: 211–216.

288. Abbas AK, Lichtman AH and Pober JS. *Cellular and Molecular Immunology*. Philadelphia: W. B. Saunders Company, 1991.

289. Snodgrass HR, Kisielow P, Kiefer M, Steinmetz M and von Boehmer H. Ontogeny of the T-cell antigen receptor within the thymus. *Nature* 1985; **313**: 592–595.

290. Raulet DH, Garman RD, Saito H and Tonegawa S. Developmental regulation of T-cell receptor gene expression. *Nature* 1985; **314**: 103–107.

291. Mombaerts P, Clarke AR, Rudnicki MA *et al.* Mutation in T cell antigen receptor genes α and β block thymocyte development at different stages. *Nature* 1992; **360**: 225–231.

292. Shinkai Y, Koyasu S, Nakayama K *et al.* Restoration of T cell development in RAG-2-deficient mice by functional TCR transgene. *Science* 1993; **259**: 822–825.

293. Groettrup M, Ungewiss K, Azogui O *et al.* A novel disulfide-linked heterodimer on pre-T cells consists of the T cell receptor β chain and a 33 kd glycoprotein. *Cell.* 1993; **75**: 283–294.

294. Saint-Ruf C, Ungewiss K, Groettrup M, Bruno L, Fehling HJ and von Boehmer H. Analysis and expression of a cloned pre-T cell receptor gene. *Science* 1994; **266**: 1208–1212.

295. Levelt CN, Ehrfeld A and Eichmann K. Regulation of thymocyte development through CD3. I. Timepoint of ligation of CD3 determines clonal deletion or induction

of developmental program. *J. Exp Med.* 1993; **177**: 707–716.

296. Huesmann M, Scott B, Kisielow P and von Boehmer H. Kinetics and efficacy of positive selection in the thymus of normal and T cell receptor transgenic mice. *Cell* 1991; **66**: 533–540.

297. von Boehmer H. Positive selection of lymphocytes. *Cell* 1994; **76**: 219–228.

298. Nossal GJV. Negative selection of lymphocytes. *Cell* 1994; **76**: 229–239.

299. Ahn NG, Seger R and Krebs EG. The mitogen-activated protein kinase activator. *Curr. Opin. Cell. Biol.* 1992; **4**: 992–999.

300. Cowley S, Paterson H, Kemp P and Marshall CJ. Activation of MAP kinase is necessary and sufficient for PC12 differentiation and for transformation of NIH 3T3 cells. *Cell* 1994; **77**: 841–852.

301. Alberola-Ila J, Forbush KA, Seger R, Krebs EG and Perlmutter RM. Selective requirement for MAP kinase activation in thymocyte differentiation. *Nature* 1995; **373**: 620–623.

302. Swan KA, Alberola-lla J, Gross JA *et al.* Involvement of p21 ras distinguishes positive and negative selection in thymocytes. *EMBO J.* 1995; **14**: 276–285.

303. Jameson SC and Bevan MJ. T cell receptor antagonists and partial agonists. *Immunity.* 1995; **2**: 1–11.

304. Evavold BD and Allen PM. Separation of IL-4 production from Th cell proliferation by an altered T cell receptor ligand. *Science* 1991; **252**: 1308–1310.

305. Madrenas J, Wange RL, Wang JL, Isakov N, Samelson LE and Germain RN. ζ phosphorylation without ZAP-70 activation induced by TCR antagonists or partial agonists. *Science* 1995; **267**: 515–518.

306. Hogquist KA, Jameson SC, Heath WR, Howard JL, Bevan MJ and Carbone FR. T cell receptor antagonist peptides induce positive selection. *Cell* 1994; **76**: 17–27.

307. Sebzda E, Wallace VA, Mayer J, Yeung RS, Mak TW and Ohashi PS. Positive and negative thymocyte selection induced by different concentrations of a single peptide. *Science* 1994; **263**: 1615–1618.

308. Ashton-Rickardt PG, Banderia A, Delaney JR *et al.* Evidence for a differential avidity model of T cell selection in the thymus. *Cell* 1994; **76**: 651–663.

309. Zacharchuk CM, Mercep M and Ashwell JD. Thymocyte activation and death: a mechanism for molding the T cell repertoire. In: Edelson RL (ed.) *Antigen and Clone-specific Immunoregulation.* New York: *Ann. New York Acad. Sci.*, 1991: 52–70.

310. Vacchio MS, Yang Y and Ashwell JD. Nuclear receptors and thymocyte apoptosis: shaping the immune repertoire. In: Mihich E and Schimke RT (eds), New York: Plenum Press, 1994: 179–194.

311. Andjelic S, Jain N and Nikolic-Zugic J. Immature thymocytes become sensitive to calcium-mediated apoptosis with the onset of CD8, CD4, and the T cell receptor expression: a role for bcl-2? *J. Exp. Med.* 1993; **178**: 1745–1751.

312. Tao W, Teh SJ, Melhado I, Jirik F, Korsmeyer SJ and Teh HS. The T cell receptor repertoire of CD4CD4$^-$8$^+$ thymocytes is altered by overexpression of the BCL-2 protooncogene in the thymus. *J. Exp. Med.* 1994; **179**: 145–153.

313. Linette GP, Grusby MJ, Hedrick SM, Hansen TH, Glimcher LH and Korsmeyer SJ. Bcl-2 is upregulated at the CD4$^+$CD8$^+$ stage during positive selection and promotes thymocyte differentiation at several control points. *Immunity* 1994; **1**: 197–205.

314. Strasser A, Harris AW, von Boehmer H and Cory S. Positive and negative selection of T cells in T-cell receptor transgenic mice expressing a bcl-2 transgene. *Proc. Natl. Acad. Sci. USA* 1994; **91**: 1376–1380.

315. Veis DJ, Sorenson CM, Shutter JR and Korsmeyer SJ. Bcl-2-deficient mice demonstrate fulminant lymphoid apoptosis, polycystic kidneys, and hypopigmented hair. *Cell* 1993; **75**: 229–240.

316. Motoyama N, Wang F, Roth KA *et al.* Massive cell death of immature hematopoietic cells and neurons in Bcl-x-deficient mice. *Science* 1995; **267**: 1506–1510.

317. Russell JH and Wang R. Autoimmune gld mutation uncouples suicide and cytokine/proliferation pathways in activated, mature T cells. *Eur. J. Immunol.* 1993; **23**: 2379–2382.

The Role of Tyrosine Kinases in T Cell Development

Nigel Sharfe and Chaim M. Roifman

Division of Immunology and Allergy, Hospital for Sick Children, University of Toronto, Toronto, Canada

INTRODUCTION

With developments in the field of signal transduction research, the central role of tyrosine phosphorylation in developmental signalling events has been revealed. Only since the late 1980s has the full importance of the protein tyrosine kinases been realized, as various families of these kinases have been shown to control events ranging from cell proliferation to differentiation and mature physiological function. These kinases play a crucial role in determining progression along diverging pathways, transducing signals from a myriad of membrane bound receptors destined to influence gene transcription, in some cases gene arrangement, and ultimately cellular morphology and phenotype.

In particular, our understanding of lymphocyte differentiation and signalling has greatly increased as the function of the tyrosine kinases has become apparent. From the earliest stages of B and T lymphocyte development, tyrosine kinases appear to play an essential role in determining differentiation pathways. While regulating expression of the antigen receptors and their accessory molecules, they also play a central role in determining the recognition repertoires arriving at the periphery and provide a mechanism of turning signals from the antigen receptors of mature functional cells into physiological reponses. Both receptor-derived and cytoplasmic tyrosine kinase activities contribute to this picture of cellular tyrosine phosphorylation, which is daily growing in complexity as our understanding of these proteins increases. Significant contributions to our understanding have been made by homologous recombination "knock-out" and transgenic mice, by transfection experiments and ultimately, by the description of human SCID patients.

This chapter will attempt to present the contribution of the various tyrosine kinases, as presently understood, to the development of T lymphocytes in their differentiation from progenitors to mature T cells. Progressing from those expressed in the earliest T cell progenitors, to those participating in the later events of differentiation, we shall describe the continuous role of tyrosine kinases in the T cell differentiation process.

THYMOCYTE DIFFERENTIATION—AN OVERVIEW

Most work on T cell differentiation has been performed in mice, due to the experimental difficulties of investigating this process in humans. Therefore, much of the data available originates from descriptions of the murine system. Unless otherwise stated the material discussed in this chapter was derived from this murine work. $\alpha\beta$ thymocyte differentiation is most easily viewed by dividing the process into a series of steps defined by the expression of plasma membrane proteins. Haematopoietic progenitor cells characterized by their expression of the CD34 cell surface marker contain a subpopulation of cells destined to migrate from the bone marrow of the adult (or from fetal liver) into the thymus [1]. These prethymic cells, as well as several other subpopulations of CD34+ progenitor cells, already express low levels of the T cell antigen receptor (TCR) coreceptor CD4 [2], although at levels approximately five-fold lower than mature peripheral T lymphocytes.

These precursor cells move through the bloodstream and into the thymic environment [3]. Migration to the thymus is assumed to be directed by chemoattractants derived from thymic epithelium; although the precise nature of these signals, suggested to be soluble β-microglobulin, has yet to be defined. While there are no obvious structural boundaries, thymus architecture can be divided into regions populated by thymocytes of distinct maturational stages. Precursor cells arrive at the cortico-medullary junction or subcapsule, and begin differentiating under the influence of cellular contacts supplied by this tissue and factors secreted from these cells. A multitude of stromal elements contribute to create a series of specialized microenvironments through which the thymocyte will pass during its development. As maturation progresses, thymocytes move through the cortical regions of the thymus undergoing negative and positive selection pressures; until mature self-tolerant T cells exit the thymic medulla into the periphery. The basis for thymocyte movement within the thymus is currently undertermined but may be mediated by contractile proteins such as actin (which is expressed at sites of cell–cell contact in both lymphocytic and stromal cells) [3].

Thymocyte maturation can be represented in terms of progression from an immature CD4CD8 double-negative phenotype to a CD4CD8 double-positive cell (Figure 13.1). Investigations of the murine system suggest that double-negative cells can be further divided on the basis of the surface expression of CD44 (phagocytic glycoprotein 1, pgp1) and CD25 (IL2 receptor α chain). Precursor cells arriving in the thymic environment progress through stages defined by the presence and absence of these markers; initially displaying a CD44$^+$CD25$^-$ phenotype, progressing to CD44$^+$CD25$^+$ and subsequently CD44$^-$CD25$^+$; and finally to cells which have down-regulated expression of both markers. In humans thymocyte progression is marked by expression of CD7 in the subcortical regions, with development of CD1 expression occurring in the cortex. The first step to mature TCR expression occurs in CD44$^-$CD25$^+$ cells with initiation of rearrangement of the germline V,D and J segments of the TCRβ gene to form a clonotypic chain and its subsequent expression.

Expression of CD25 (IL2 receptor) is preceded by receptors to Interleukin 7, which are found on

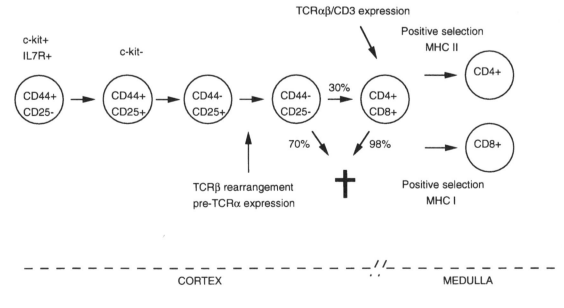

Figure 13.1   Model of T cell differentiation indicating thymocyte progression through the cortical and medullary areas of the thymus

pro-thymocytes in the blood [1]. Activation of this cytokine receptor appears critical to CD25 expression, as deletion of the IL7 receptor expression in mice prevents the generation of these cells [4]. However, the more clearly defined role of the IL7 receptor at this stage occurs in response to Interleukin 7 secreted by thymic stromal cells which initiates expression of the recombinase genes RAG-1 and RAG-2 [5–7]. These gene products are essential to the process rearrangement, as deletion of either results in an arrest of thymocyte maturation at a $CD44^-CD25^+$ phenotype, retaining the $TCR\beta$ gene in germline configuration [8–9]. Progression to a $CD44^-CD25^-$ phenotype occurs when the $TCR\beta$ chain is correctly expressed, initially associated with a pre-α chain [10, 11], and latterly with a mature $TCR\alpha$ protein expressed from the rearranged gene. Expression of a productively rearranged $TCR\beta$ gene product inhibits further gene rearrangement by an allelic exclusion mechanism. While progression from prethymic CD4 expressing triple negative cells to a $CD44^-CD25^-$ phenotype occurs under the influence of the stromal cells of the outer cortical microenvironment, further differentiation into $CD4^+CD8^+$ thymocytes occurs independently of thymic influence.

Concomitant with rearrangement of the germline $TCR\alpha$ and β genes and expression of the products, there is development of CD4 and CD8 expression; while induction of CD3 expression appears to precede TCR rearrangement. Maturing thymocytes arrive at a triple positive ($CD4^+CD8^+CD3^+$) phenotype displaying surface expression of a mature TCR heterodimer. These cells are now subject to TCR-mediated selective pressures, both negative and positive, which will determine the TCR repertoire expressed in the periphery. The TCR and coreceptor chains upon immature thymocytes interact with class I and class II major histocompatibility proteins expressed by the thymic epithelium [3]. Failure to receive the correct signals, or signals of the appropriate intensity, from the TCR and its coreceptors throughout this process will result in deletion of the thymocyte. High affinity interactions between MHC displaying self-peptides and thymocyte TCR result in deletion of those TCR clones, a necessary requirement to maintain self-tolerance [12]. Whether this functional deletion occurs by clonal

apoptosis or induction of anergy may depend upon the precise degree of maturation that the thymocyte has achieved. In addition to negative selection, thymocytes are subject to a positive selection process [13]. Cells expressing a suitable TCR, that in combination with CD4 or CD8 coreceptor, can bind MHC with appropriate affinity are expanded; while simultaneously down-regulating either CD4 or CD8 coreceptor expression. CD4 expression is known to be controlled by a lineage-specific silencer element which represses CD4 expression in both immature double-negative cells and $CD8^+$ lymphocytes [14]. Ultimately, $CD4^+$ and $CD8^+$ cells emerge into the periphery as mature T lymphocytes expressing a functional $TCR\alpha\beta$ complex and CD3 coreceptor chains.

T CELL SIGNALLING—AN OVERVIEW

The T cell antigen receptor (TCR) and the closely associated proteins $CD3\gamma,\delta$, ϵ and ζ are all members of the immunoglobulin superfamily. The TCR itself consists of disulphide linked α and β chains capable of recognizing and binding to antigen peptide bound to major histocompatibility complex (MHC) molecules. Both receptor chains possess glycosylated extracellular immunoglobulin-like domains, but contain little cytoplasmic sequence; transducing signals via interaction with the non-covalently associated CD3 γ,δ and ϵ coreceptor chains [15]. The fourth CD3 protein, ζ, is expressed as a disulphide-linked homodimer and unlike the other CD3 proteins contains little extracellular sequence (nine amino acids) and has a large cytoplasmic domain containing repeated SH2 domain binding motifs responsible for the recruitment of signal transducing proteins. $CD3\zeta$ RNA also undergoes alternative splicing to generate a protein with an alternative cytoplasmic carboxy terminal, η [16]. Recognition of peptide–MHC complex by the TCR complex is regulated by CD4 and CD8: these single chain transmembrane coreceptors recognizing MHC class I and class II respectively.

The CD3 complex appears to consist of two relatively independent signal transduction units [17]. Chimeras consisting of the cytoplasmic domains of $CD3\zeta$ linked to CD8 extracellular and transmembrane domains, transduce signals indistinguishable from those generated by TCR cross-linking (CD8 does not normally activate these signalling events). This coupling occurs in the absence of other CD3 subunits in mutant Jurkat lines. Similarly, a 22 amino acid sequence in the cytoplasmic tail of CD3ϵ is also capable of independently signalling [18]. Activation via this pathway results in a pattern of events distinct from complete TCR or CD3ζ-mediated signalling [19].

The initiating signals derived from these receptors are primarily tyrosine phosphorylation cascades [20, 21]. Upon antigen-receptor-induced activation, multiple tyrosine phosphorylation events occur rapidly resulting in hydrolysis of phosphatidylinositol-bisphosphate [22], liberating inositol phosphate and diacylglycerol messengers and temporarily increasing cytosolic calcium levels. Coupling of the antigen receptor to tyrosine phosphorylation occurs primarily through interaction with members of the src family of non-receptor tyrosine kinases [23]. These kinases are attached to the plasma membrane via an amino terminal myristoylation and possess src homology SH2 and SH3 interaction domains. While expression of pp60src itself is widespread, a number of related proteins are preferentially expressed in lymphoid cells. In T cells p56lck is found coupled to CD4 and CD8 [24], while p59fyn is associated with the TCR itself [25]. p56lck associates noncovalently with CD4 and CD8 via charged cysteine-rich motifs [26] utilizing a unique amino terminal interaction sequence. While CD3ϵ coprecipitates p59fyn, CD3ζ becomes closely associated with a cytosolic tyrosine kinase, Zap-70, in a tyrosine-phosphorylation-dependent manner upon activation [27, 28]. ZAP-70 differs from the src-kinases in lacking a myristoylation site and possessing tandem SH2 domains, while lacking SH3. ZAP-70 also displays expression limited to lymphoid cells, predominantly the T cell lineage.

In addition to TCR and coreceptor-mediated events, T cell differentiation is heavily influenced by cytokines. A variety of cytokine receptors are expressed from the earliest stages of T cell differentiation, such as the Interleukin 7 receptor [29], present on pro-thymocytes in fetal blood [1], and the Interleukin 2 receptor, which is up-regulated upon CD44[4] triple-negative thymocytes. These receptors belong within a highly homologous superfamily [30] (particularly within the extracel-

lular domain) and generally consist of multiple chains. These can be divided into private "ligand specific" chains and public class chains such as gp130, IL2Rγ and β_c which are shared by smaller subfamilies [31]. Not possessing intrinsic tyrosine kinase activity, these receptor complexes transduce signals primarily by activation of the src-like and Jak tyrosine kinase families; some receptors activating multiple members of both families. Through activation of these cytoplasmic tyrosine kinases the cytokine receptors are linked to a variety of signalling pathways including PI3 kinase and p21ras activation and Shc and MAPK phosphorylation [32–35]. Evidence is also emerging that the two branches of tyrosine kinase activity may regulate distinct transcriptional events [36,37]. The cytoplasmic domains of the cytokine receptors share varying degrees of homology (although generally quite low) and contain multiple SH2 domain binding motifs. Perhaps the most detailed analysis has been performed upon the IL2 receptor, where distinct portions of the IL2Rβ cytoplasmic domain appear to be responsible for the binding of the src (acidic region) [38, 39] and the Jak (serine-rich region) [40,41] kinases. Although the mechanism of these associations is not well understood, it does not appear to be dependent upon either SH2-or SH3-mediated interactions, as is generally the case in TCR complex interactions.

While a number of cytoplasmic tyrosine kinases perform well-detailed roles in T cell differentiation, the participation of receptor tyrosine kinases is more poorly documented. However, one, c-kit, is suspected to play an important role starting in precursor T cells of the bone marrow. Starting at c-kit this chapter will present the various tyrosine kinases approximately in the order they are believed first to participate in the differentiation process.

C-KIT

One of the earliest tyrosine kinase activities to be expressed in developing thymocytes is associated with the transmembrane protein, c-kit; although its actual role in T cell differentiation remains to be determined. c-kit belongs to the PDGF family of receptors, possessing intrinsic tyrosine kinase activity. The extracellular domain of c-kit contains five immunoglobulin-like domains, while the cytoplasmic region contains a kinase domain divided by an insert sequence [42]. c-kit is expressed upon haematopoietic stem cells from the earliest stages of development. The c-kit ligand is known variously as Steel Factor, KL (kit ligand) and SI (Steel) and is known to promote cell proliferation, survival, adhesion, migration and differentiation in a variety of early cells. Interestingly, both human and murine heterozygous carriers of c-kit mutations manifest the condition piebaldism while remaining otherwise normal. c-kit maps to the white spotting (W) locus of the mouse genome [43]; deletion resulting in defects in haematopoiesis, mast cells and primordial germ cells [44]. Although deletion has a more profound effect upon erythroid and mast cell development than the lymphoid lineage, c-kit expression is probably important for the maintenance of T precursor cells; as treatment of $c - Kit^+$ bone marrow cells with anti-c-kit antibodies blocks T cell generation upon transfer to irradiated mice. c-kit expression is highest in very immature thymocytes [45,46] and is diminished as the cells progress to up-regulate CD25(IL2Rα) expression. While the precise role of c-kit within the T cell lineage remains undescribed, c-kit kinase activation is known to induce several signal transduction pathways [47]; including PI3 kinase [48], which is necessary to induce proliferation, p21ras activation [49], p95vav [50] and p50Shc [51] phosphorylation and stimulation of MAPK activity. c-kit ligation also results in association with the haematopoietic phosphatase PTP1C, with coincident activation of the phosphatase activity [52]. PTP1C can dephosphorylate c-kit *in vitro*, suggesting that this phosphatase may regulate c-kit activity.

THE JAK KINASES

Recent advances in the field of cytokine receptor signal transduction have revealed the potential of a newly described family of tyrosine kinases to influence T cell development. The Jak family of cytoplasmic tyrosine kinases [53] are activated by a variety of receptors, as diverse as those for Interleukin 2 [54], growth hormone [55] and erythropoietin [56]. These rapidly activate members of the Jak family with resultant tyrosine phosphorylation

of the activated Jak kinases. Jak activation appears to be required for the activation of a cytoplasmic family of transcriptional activators, the Stat family [57–59]. The Stat factors homo and hetero-dimerize upon tyrosine phosphorylation before migration into the nucleus, binding specific DNA elements to initiate gene transcription [60]. Jak kinases appear to serve solely as signal transducers of the cytokine receptor superfamily (although Jak activation by G protein coupled receptor has recently been described) [61] and are not activated in response to antigen-receptor-induced stimulation of T cells [62]. Most receptors appear simultaneously to activate multiple Jak kinases. The interferon receptors (α, β and γ) each activate two Jak kinases, both of which are essential to signal transduction. Experiments in fibroblasts displaying a deficient range of Jak kinase expression have demonstrated the absolute requirement for activation of both kinases in IFN signal transduction [63, 64]. Both of the cytokine receptors with a well-characterized role in the development of T cells, IL2 [65] and IL7 [66], utilize coincident activation of Jak1 and Jak3 to transduce their signal.

The Jak family currently contains four members, Jaks 1 to 3 and Tyk2 [67]. While Jak 1, Jak 2 and Tyk2 are widely expressed, including the thymus and T cells, Jak3 displays highly restricted expression [68,69]. Initial descriptions of the Jak3 kinase, both murine and human, found expression restricted to cells of the lymphoid and myeloid lineages. Extremely low level Jak3 expression was detected in immature haematopoietic cells which was dramatically up-regulated in cells representing terminally differentiated or mature, activated stages. Jak3 up-regulation has been described in PHA-activated peripheral blood T lymphocytes (PBL-T), in granulocyte colony stimulating factor differentiated myeloid cells and in cell lines representative of B lymphocyte developmental stages [68,69].

However, a more recent study of Jak3 expression in primary human T cells has detected Jak3 expression at all stages of differentiation including a high level in thymocytes and substantial, albeit relatively reduced, expression in peripheral T lymphocytes [70]. Moreover, despite their immaturity CD4$^-$CD8$^-$ thymocytes also strongly express the Jak3 message and protein. This expression pattern differs significantly from that described in myeloid cells and the B lymphocyte lineage, suggesting Jak3 plays a unique role in T cell differentiation. Based upon the relative levels of Jak3 expression, Jak3 kinase would appear to play a highly important role in both double-negative thymocytes and mature, activated T lymphocytes. The reduction in Jak3 expression in peripheral T lymphocyte may be linked to a requirement for tighter regulation of Jak3 kinase activity within these periphery cells, possibly preventing premature activation by erroneous stimuli. Regulation of Jak3 expression may be similar to the src-like kinase p56lck; which is expressed under the control of distinct promoter elements in thymocytes and mature T lymphocytes [71]. This may reflect changes in the stimuli responsible for the induction of Jak3 expression.

While the activation of Jak3, in concert with Jak1, by Interleukin 2, Interleukin 4 and Interleukin 7 has been reported in mature and activated T lymphocytes; Jak3 is also activated by IL7 in peripheral T cells and in thymocytes [70], and activation has been specifically demonstrated as early in development as double-negative thymocytes. The importance of IL7 to the differentiation of CD4$^-$CD8$^-$ cells [72,73] is already well established. Not yet subject to selection, having recently entered the thymic environment, these cells have yet to express TCR$\alpha\beta$ or CD3 coreceptor chains. Of all the factors examined, IL7 alone was capable of promoting V(D)J rearrangement of the T cell receptor β chain gene [5], required for expression of the TCR complex and clonal expansion. IL7 secreted by thymic epithelial cells acts upon these thymocytes to induce and sustain expression of the recombinase genes RAG-1 and RAG-2; both necessary for productive VDJ rearrangement of the TCRβ gene.

In T cells and thymocytes initiation of IL7 signal transduction appears to be primarily dependent upon tyrosine phosphorylation events. Three cytoplasmic kinases capable of associating with the receptor have been identified that are activated by IL7; Jak1, Jak3 and p56lck [74]. However, the murine models provided by the p56lck knockout [75] and p56lck dominant negative [76] transgenic mice (see page 233, p56lck), while displaying early thymic maturational arrest at a CD3$^-$ CD4$^-$ CD8$^-$ phenotype, possess the TCRβ gene in a rearranged configuration. This implies that while p56lck is required for the progression of CD4$^-$CD8$^-$TCRβ^+ thymocytes, it is not essential to the process of TCRβ gene rearrangement. Jak1 and Jak3 expression and activation are therefore likely to be important to this process, possibly inducing RAG-1 and RAG-2 expression via Stat-mediated transcriptional activation of these genes.

The presence of Jak3 in CD4$^-$CD8$^-$ cells has implications for understanding the signalling of cytokines other than IL7 in thymocytes, in particular for IL2 which also activates the pairing of Jak1 and Jak3 in mature activated T lymphocytes. CD25 (IL2Rα chain) expression is up-regulated before IL7-induced TCRβ rearrangement commences, in CD44$^+$CD25$^+$CD4$^-$CD8$^-$ thymocytes. This chain, in conjunction with the IL2Rγ chain, may also transduce signals via Jak3. It has been suggested that mutation of the IL2R$_\gamma$ chain such that Jak3 association is decreased or absent, may contribute to X-linked combined immunodeficiency (XCID) [77] in humans. Mutation of the IL2Rγ chain is observed in patients displaying X-linked severe combined immunodeficiency (XSCID) with a resultant absence, or substantial reduction of T cells. Using mutated IL2Rγ chains corresponding to those found in XSCID patients, Russell et al. (1994) demonstrated a substantial reduction in IL2-induced association of Jak3 in transfected COS cells. The high level of Jak3 expression observed in the earliest stages of thymocyte development suggests this failure to associate with IL2Rγ may be truly relevant to the clinical development of XSCID. As both the IL2 and IL7 receptor complexes utilize the IL2Rγ chain, failure of Jak3 to interact with the IL2Rγ chain in early double-negative thymocytes would result in a complete failure of the T lineage to develop; as observed in the XSCID condition.

Two other Jak activating members of the cytokine receptor superfamily, the prolactin and LIF receptors, are also expressed in the very early stages of T cell differentiation. Evidence is increasingly accumulating which suggests that the pituitary gland hormone prolactin may influence the immune system. Not only is a high level of prolactin receptor expression observed on double-negative thymocytes [78], but chicken embryos lacking a pituitary gland (partially decapitated Dcx embryos) display delayed T cell development with an accumulation of double-negative thymocytes and a concomitant reduction in the number of double-positive cells [79]. This defect can essentially be rescued by the application of exogenous prolactin. Although $\alpha\beta$ thymocyte development is profoundly retarded by the lack of prolactin, $\gamma\delta$ differentiation procedes normally. The prolactin receptor has recently been shown to be capable of binding Jak2 and activating Stat factors by tyrosine phosphorylation [80]. Jak2 activation may be of importance to T cell development, as treatment of T lymphocytes and T cell lines with prolactin up-regulates the IL2 receptor [81]. Without prolactin induced Jak2 mediated transcriptional activation, it is possible that early thymocytes may fail to fully up-regulate IL2 receptor expression and become refractory to stimulation by IL2 secreted from thymic epithelial cells.

Leukaemia inhibitory factor (LIF) receptor signalling also involves Jak2 kinase activation and this may also play an important role in the development of T cells. Overexpression of LIF specifically in the T cell compartment of mice resulted in severe dysfunction of both the B and T lineages [82], with $CD4^+CD8^+$ cells present in the lymph nodes while reduced numbers are present within the thymus. LIF overexpression appears to result in premature maturation of the thymocytes, possibly before the advent of selective pressures, resulting in premature exit from the thymus. These experiments tentatively suggest that LIF is required to regulate maturational processes and highlight once again the importance of the Jak family in mediating the early signalling events of T cell differentiation.

p56LCK

Of all the tyrosine kinases that play a role in the differentiation of T cells, perhaps the most thoroughly investigated is the lymphocyte-specific src-family kinase p56lck. Not only is p56lck associated with the CD4 and CD8 TCR coreceptor chains and transducing signals to the CD3 coreceptor chains, but it is also involved in signalling from a number of cytokine receptors including Interleukins 2 and 7. This combination of roles places p56lck central to signal transduction events in both thymocyte differentiation and mature T lymphocyte activation.

Expression of lck is itself developmentally regulated, distinct distal and proximal promoters achieving differential expression in immature thymocytes and mature T lymphocytes [71]. Little sequence homology is seen between the promoter elements, both of which are responsible for expression in thymocytes, while only the distal promoter drives expression in mature lymphocytes. This suggests that expression is regulated by different assemblies of transcription factors, presumably formed in response to differing stimuli at progressive stages of development. Transcription from the single distal promoter in mature T lymphocytes may also reflect a necessity for a tighter control of lck activity in these cells.

The importance of the lck gene product to development was fully realized with the deletion of the gene in mice, utilizing homologous recombination techniques [75]. While heterozygous mice functioned normally, with normal thymocytes and peripheral T lymphocytes, a minor decrease in the peripheral CD4 positive T lymphoctye population was noted, suggesting that lck was required for positive selection of $CD4^+$ cells. However, upon cross-breeding to create a true homozygous knockout, the true importance of lck was revealed. Lck null mice display considerable thymic atrophy with a significant decrease in thymocyte populations. While a few T lymphocytes are present in the periphery, mature single-positive thymocytes cannot be detected. The total thymocyte population is reduced 10-to 100- fold, with a significant decrease in the proportion of $CD4^+CD8^+$ cells. Of the remaining thymocytes 20–40% are of a double-negative phenotype, a phenotype usually contributing to only 2% of the thymic population. As the total number of double-negative cells is similar to wildtype mice, the decrease in thymocyte numbers is primarily due to a depletion of the $CD4^+CD8^+$ population. The few T cells detectable in the periphery display proliferative response to $TCR\alpha\beta$ and CD3 crosslinking, albeit reduced, suggesting that lck is not absolutely essential to signal transduction within these cells. Alternatively, these cells may represent a normally rare population utilizing an altered pathway of signal transduction from the TCR complex, and may not be indicative of the true importance of lck within the T cells of the periphery. This is supported by studies of human lck-deficient T cell lines which manifest gross abnormalities in TCR and CD3 mediated signalling [83].

Data obtained from lck knockout mice were confirmed and strengthened by transgenic mice expressing a dominant negative lck [76]. These also display a defective expansion of CD4$^-$CD8$^-$ cells to CD4$^+$CD8$^+$ thymocytes. These mice, expressing a catalytically inactive form of p56lck (K237R), under the transcriptional control of the proximal lck promoter, exhibit a decrease in thymus size and total thymocyte number directly correlated with the level of transgene overexpression. A dose-dependent reduction in T cell populations was observed, except of the highly immature CD4$^-$CD8$^-$ thymocytes. The generation of $\gamma\delta$ lymphocytes, however, appears to be unaffected by the lck dominant negative transgene, indicating that the differentiation pathway of this lineage diverges from the $\alpha\beta$ lineage at an earlier stage.

The majority of cells within the thymus of these transgenic animals are large, a characteristic of thymoblasts, expressing the heat stable antigen (HSA) and the IL2α receptor (CD25), while having already lost expression of Pgp1 (CD44); CD44$^-$ CD25$^+$HSA$^+$. Lck expression and activity appear crucial for progression to the next stage of development, loss of IL2Rα and up-regulation of CD4 and CD8 expression. The CD44$^-$CD25$^+$HSA$^+$ thymoblasts of these mice are arrested at the GO/G1 cell cycle boundary suggesting that an lck-mediated signal is required to transit this point. Further analysis of this population demonstrated that these cells had undergone germline TCR beta gene rearrangement, but had yet to rearrange the TCR alpha gene. V(D)J recombination events of the TCRβ gene were detected by PCR at levels equivalent to those in wildtype cells demonstrating the independence of this particular step from lck activity.

The T cell populations of lck knockout and dominant negative transgenic mice are phenotypically similar to those observed in TCRβ null mice [84]. These animals display a similar arrest of double-negative expansion to CD4$^+$CD8$^+$ thymocytes demonstrating that productive rearrangement and expression of the TCRβ gene is crucial to progression beyond this point. It has therefore been hypothesized that the TCRβ chain delivers a signal required to initatiate TCRβ allelic exclusion, re-arrangement of the TCRα locus and proliferation

via lck. This hypothesis is supported by experiments demonstrating that overexpression of constitutively activated lck overcomes maturational arrest in TCRβ null cells, permitting progression to TCRα rearrangement, up-regulation of CD4 and CD8 expression and proliferation. In contrast, overexpression of a TCRβ chain transgene in lck-deficient mice does not overcome the block in maturation [85], demonstrating that TCRβ expression *per se* is not sufficient.

Similarly, although expression of a TCRβ transgene in mice blocks rearrangement of the endogenous β locus almost completely by an allelic exclusion process, this does not occur in mice co-expressing the dominant negative lck transgene [85]. Expression of the dominant negative lck overrides allelic exclusion mechanisms, even in the presence of the prerearranged TCRβ transgene [85]. Furthermore, introduction of activated lck as a transgene into a RAG-1-deficient background (required for the rearrangement of the TCRβ gene), rescues these thymocytes from the block in proliferation, without rearrangement and expression of the endogenous TCRβ locus occurring [85]. Conversely, overexpression of dominant negative lck in a TCRβ^+ RAG-1 null background, displaying almost normal levels of thymocytes and development due to TCRβ chain rescue of the RAG-1 defect, causes a reversion of phenotype to that of the TCRβ or lck knockout [86].

These studies have all produced data consistent with the theory that p56lck mediates signals originating from a rearranged and expressed TCRβ gene. This signal is responsible for the GO/G1 transition of thymoblasts expressing low levels of CD4/CD8 into proliferating cells with high expression levels of CD4 and CD8, which will progress to expression of the mature TCR upon rearrangement of the TCRα locus.

While these experiments demonstrate that p56lck is essential to this step, the precise mechanism of signalling is undetermined. The rearranged TCRβ chain is believed to be expressed on the surface in association with a pre-TCRα chain, a transmembrane protein containing a cytoplasmic SH3 recognition domain [10,11]. This pre-T cell receptor is assumed to associate with p56lck, although the nature of the signal transmitted is unknown. Some evidence exists suggesting that crosslinking of the TCRβ chain expressed in the absence of TCRα chain upon double-positive thymocytes (CD4$^+$CD8$^+$CD3$^+$) generates an increase in cytoplasmic Ca^{2+} levels [87], perhaps inferring that p56lck induces phosphorylation of PLCγ or similar effectors. The signal derived from the TCRβ-preα complex appears to be responsible for inducing both clonal expansion and allellic exclusion. As the requirement for TCRβ chain signalling can be bypassed by crossinking CD3 [88], it is believed that the pre-TCR complex may function by associating with the CD3 proteins.

The contribution of p56lck to T cell differentiation in its association with the TCR coreceptors CD4 and CD8 [89–91] and the phosphatase CD45 [89,93] are more difficult to assess. Association with CD4 and CD8 is not required for p56lck to function in the early events of maturation, as activated lck transgenes with the CD4/CD8 interaction domain deleted still promote expansion of the TCRβ^+ thymocyte population. While little is known of signalling from these complexes, CD4-induced p56lck activity has been demonstrated to regulate TCR expression on immature CD4$^+$CD8$^+$ thymocytes by influencing assembly and transport of the nascent TCR chains [94–96]. Chronic engagment of CD4 by interaction with MHC class II proteins results in aggregation and activation of p56lck, lck disengagement from CD4 and degradation of the "empty" CD4 proteins, resulting in the continued association of p56lck with the remaining CD4. Although more highly expressed, CD8 does not appear to activate lck as efficiently due to a lower degree of association and does not seem to influence TCR expression. The importance of these interactions to differentiation remains to be determined but perhaps suggests the basis for a difference in the CD4 and CD8 selection pathways.

p59FYN

p59fyn, like p56lck, exhibits limited expression but differs in being expressed as two isoforms, as the result of alternative splicing, predominantly expressed in the brain (B) and haematopoietic system (T). The p59fyn isoforms are essentially identical apart from an alternative region of 50 amino acids at the junction of the SH2 and kinase domains [23]. The function of this hinge region is undetermined, overexpression of the two isoforms independently in the thymocytye compartment of mice resulting in identical phenotypic responses [97].

p59fyn functions to transduce signals from both the antigen receptor and cytokine superfamily receptors. Association with the TCR was detected by co-immunoprecipitation and cocapping experiments, p59fyn preferentially associating with the CD3ϵ complex rather than the CD3ζ chain. This pattern of association differs significantly from p56lck, suggesting the two kinases perform distinct functions in response to TCR ligation. p59fyn expression appears to be developmentally regulated, increasing significantly upon the transition from double-positive to single-positive status [98], suggesting its contribution to differentiation may be limited to events in mature DP thymocytes. Surprisingly, mice in which p59fyn(T) expression is deleted by homologous recombination techniques (97,99), while retaining expression of the brain specific isoform, are essentially normal. Lymphoid development continues in the absence of p59fyn and mature T lymphocytes are found in circulation. However, while mature T lymphocytes of p59fyn(T) null mice display significant (although reduced) responses to mitogens, thymocyte responses are virtually ablated. Fyn(T) null thymocytes do not essentially display either calcium or proliferative responses to antigen receptor stimulation or to mitogens such as concanavalin A. Deletion of fyn also affects signalling in more mature cells as although splenic T cells mobilize calcium, albeit at a reduced level compared to wildtype, and proliferate; they fail to produce IL2 following anti-CD3 stimulation. After 24 hour culture with anti-CD3 and PMA fyn null T cells secret little, if any, IL2. Thymocyte anti-Thy-1 responses are also impaired, while those of peripheral T cells remain normal.

Although thymocyte responses are significantly altered, the inability to respond to various stimuli does not appear to impair the maturation process as deletion of p59fyn does not appear to have a significant effect upon thymocyte differentiation, as witnessed in p56lck knockout mice. A normal progression from CD4$^-$ CD8$^-$ to CD4$^+$CD8$^+$ cells occurs in the thymus, with mature single-positive cells emerging in the periphery; although the ability to delete thymocytes expressing TCR clones to certain superantigens appears to be compromised [98], particularly those undergoing a significant degree of late deletion. The inability to proliferate in response to signals derived from the TCR does not appear to significantly hinder negative or positive selection of thymocytes, nor the ability of splenic T cells to provide B lymphocyte help or to function in a mixed lymphocyte reaction [98]. Thymocytes appear to reacquire responsiveness once they have moved from the thymus into the periphery. As single-positive CD4 and CD8 thymocytes display the same impairment in proliferative response demonstrated by the total thymocyte population, these functions must be reacquired at a very late stage of development. Thus while not playing a significant role in directing thymocyte development, p59fyn(T) is required for TCR and CD3-mediated signalling events in these cells. This is supported by p59fyn overexpression studies, where thymocytes demonstrate enhanced signalling responses in terms of tyrosine phosphorylation, calcium release, IL2 secretion and proliferation [97]. This amplification is only achieved by p59fyn overexpression and is not observed in mice or cells overexpressing other src-like kinases, including p56lck. Overexpression of catalytically inactive p59fyn produced a similar phenotype to fyn knockout mice, substantially reducing TCR-mediated proliferative responses in thymocytes [97]. While p59fyn expression appears to be essential to the initiation of TCR-and CD3-induced signalling in thymocytes, its role in mature T lymphocytes may be more in the nature of an amplification of signals derived from another kinase, such as p56lck.

A significant role of p59fyn in mice highlighted by both overexpression and deletion experiments is its role in mediating Thy-1 signalling. While TCR/CD3ϵ signalling appears to consist of two modules, based around the CD3ζ and CD3ϵ chains, Thy-1 can only utilize the ζ chain module [17]. As Thy-1-mediated proliferative responses were essentially ablated in fyn null thymocytes, fyn may play a major role in transducing this signal, perhaps by activation of ZAP-70 and other secondary kinases. The importance of fyn to this pathway is reinforced by the high level of association with Thy-1 detected in normal thymocytes [100].

p50CSK

Both p56lck and p59fyn, like other members of the src family, are negatively regulated by phosphorylation of a carboxy terminal tyrosine residue [101]. This residue is believed, when phosphorylated, to bind the SH2 domain; folding the protein into a conformation denying substrates access to the kinase active site. p50csk [102,103] appears to phosphorylate this residue in all src-like kinases and is the only kinase identified capable of doing so *in vitro*.

Csk expression is ubiquitous and essential, as deletion of expression by homologous recombination in mice is lethal [104]. Within T cells p50csk is rapidly, albeit transiently, activated after TCR/CD3 stimulation, associating with a 70 kDa phosphoprotein [105] of undetermined function. In agreement with its regulation of p56lck and p59fyn *in vitro* [106], overexpression of p50csk in T cells results in a marked reduction of the tyrosine phosphorylation induced by TCR/CD3 ligation and ultimately of IL2 production [107]. An even more complete inhibition is observed when the overexpressed csk is targeted to the plasma membrane, presumably due to a more efficient interaction with the TCR/CD3 associated kinases p56lck and p59fyn. Furthermore, overexpression of p50csk is capable of inhibiting signalling from a constitutively activated form of p56lck.

p50csk is also capable of phosphorylating the CD45 phosphatase *in vitro*, creating a p56lck binding site [108] and with a resultant stimulation of CD45 activity. Similarly, cotransfection of p50csk and CD45 in COS cells results in tyrosine phosphorylation of CD45. The presence and activity of various CD45 isoforms is essential to TCR-mediated signalling, as is the activation of p56lck and p59fyn. Thus p50csk is well placed to have a significant role in regulating the contribution of various tyrosine kinases and phosphatases to T cell differentiation. There is a degree of circularity to this regulation, as lck and fyn are also believed to be substrates of CD45 [109–111]. Because of the lethality of csk deletion in mice, further investigation of its role in T cell differentiation will require overexpression of wildtype and dominant negative forms of the kinase restricted to the T cell compartment.

A kinase with a high degree of homology to p50csk has recently been cloned, ntk [112], suggesting that there may be a family of these regulatory kinases. Ntk, unlike csk, is expressed in two forms (52 and 56 kDa) as the result of alternative RNA splicing of exon 2; introducing an alternative initatiation site for translation. The relative expression levels of these forms varies in haematopoietic cells; with normal B and T lymphocytes expressing predominantly exon-2-deficient transcripts while immature B cells (WEHI-231) express proportionately more ntk RNA retaining the exon. This suggests that the alternative forms play specific roles at distinct stages of differentiation.

ZAP-70

ZAP-70 is a member of the second distinct family of tyrosine kinases to be involved in transduction of signals from the antigen receptor of the T cell. This family currently consists of two kinases: Syk [113], which is most highly expressed in platelets and B lymphocytes; and ZAP-70 [114], which is expressed in T lymphocytes and natural killer cells. ZAP-70 was initially isolated as a ATP binding phosphoprotein associating with the CD3ζ chain after TCR stimulation [27,28]. Association between the 70 kDa zeta-associated protein (ZAP-70) and the CD3ζ chain is dependent upon prior tyrosine phosphorylation of the CD3ζ chain and requires the two phosphotyrosine binding SH2 domains of ZAP-70. Activation of ZAP-70 therefore appears to be downstream of the primary tyrosine kinase activity stimulated by ligation of the TCR. Cotransfection experiments in non-lymphoid cells demonstrate an enhancement in ZAP-70 association with CD3ζ when either of the p56lck or p59fyn src-like kinases is cotransfected [114]. This appears to be primarily due to increased phosphorylation of the transfected zeta chains. Originally this interaction was described in the Jurkat T cell line, where stimulation via the TCR or CD3 complex was required to observe association. However, a high level of zeta phosphorylation had already been described in purified thymocytes [115], suggesting that this interaction may occur at a higher basal level in primary cells. Subsequently a portion of the ZAP-70 expressed was found to be constitutively associated with tyrosine phosphorylated zeta chain in thymocytes and in T lymphocytes from lymph nodes [116].

However, while this association was constitutive, the ZAP-70 kinase domain was only activated after stimulation via the TCR or CD3 complex. Thus CD3ζ–ZAP-70 complexes are preformed to a degree, primed to respond to activating stimuli.

The full importance of ZAP-70 to T lineage development was only realized with the description of immunodeficient patients displaying a selective failure to express ZAP-70 [117,118]. While lacking ZAP-70, these patients display an otherwise normal pattern of tyrosine kinase expression. The selective deletion of this protein results in severe combined immunodeficiency (SCID), with the complete disappearance of CD8 single-positive lymphocytes from the periphery. While double-positive thymocytes can be detected in the thymus and CD4$^+$ cells develop normally, there is a failure to positively select CD8 lymphocytes. Although CD4$^+$ cells are present in the periphery of ZAP–70–deficient patients, these cells exhibit abnormal signalling patterns and are unable to secrete IL2 or proliferate in response to mitogenic stimulation. ZAP-70 is therefore also required for TCR and CD3 signal transduction to occur normally in mature T cells. While some tyrosine phosphorylation does occur within these cells, the lack of a ZAP-70 contribution may serve to reduce these signals below a desirable threshold. The combination of activation-deficient CD4$^+$ cells and the absence of CD8$^+$ cells combine in these patients to create the SCID condition.

The failure to generate CD8 single-positive cells suggests that ZAP-70 plays a unique role within this particular lineage, crucial to positive selection from the population of CD4CD8 expressing thymocytes within the thymic medulla. The basis for the difference between CD4 and CD8 selection is unclear. Studies of the link between CD4/CD8 and the p56lck kinase suggest that a greater proportion of cell surface CD4 is associated with p56lck than CD8 in immature CD4$^+$CD8$^+$ thymocytes. Subsequently, CD4 crosslinking results in a higher degree of p56lck activation than that induced by CD8. During the selection process this may impact on the relative degree of ZAP-70 activation induced by TCR/CD4 ligation compared to TCR/CD8, or may alternatively suggest the basis for different selection pathways. Further evidence for an lck-independent CD8 signalling pathway is presented by the expression of CD8α (mutated to prevent p56lck association) in a CD8 null background in the context of a transgenic TCR. These mice develop a normal $CD8^+$ lineage despite the inability of CD8 to signal via p56lck [119]. Therefore, while CD8 selection signalling appears to be dependent upon ZAP-70, CD4 selection signals may be mediated by lck.

Little is understood of the process involved in positive selection, let alone of the signal transduction events necessary for its occurence. It is possible that the contribution of ZAP-70 may be as simple as reinducing the silencing element controlling CD4 expression [14]. This silencer element is normally active in immature double-negative thymocytes, repressed in double-positive cells, and active once more in mature $CD8^+$ thymocytes. Alternatively, ZAP-70 may contribute more complex signals at various levels of the selection process.

The precise role of Zap-70 in TCR-mediated signalling remains difficult to determine. Although ZAP-70 activation is required for $CD8^+$ maturation, double-positive thymocytes expressing signalling-disabled versions of CD3ζ (by mutation of SH2 domain recognition sites) [120] mature into single-positive cells of both CD4 and CD8 lineages. While the zeta chain appears to be necessary for the stable expression and association of the TCR/CD3 complex [121,122], its signalling capacity does not appear absolutely necessary [120]. Despite a reduction in thymocyte numbers in mice expressing signalling-disabled zeta chains, the ratio of CD4 to CD8 T cells emerging into the periphery remains unaltered. This is consistent with an equivalent effect upon both lineages rather than a specific inhibition of $CD8^+$ development. ZAP-70 therefore appears to be capable of exerting its influence upon selection independently of the zeta chain. This view is supported by the demonstration of ZAP-70 binding to each member of the CD3 complex (γ, δ, ϵ) [123]. These interactions also require tyrosine phosphorylation of the CD3 chains and participation of the ZAP-70 SH2 domains [124]. Mutation of the SH2 binding motifs of CD3ϵ specifically inhibit its association with ZAP-70, while p59fyn binding to this chain remains unaltered [123]. It is therefore possible that ZAP-70 association with these chains is essential to the selection process, while its role in association with CD3ζ contributes to signal

amplification–essential to normal TCR/CD3-mediated signal transduction, but of less importance to thymocyte commitment.

Activation of the other family member, Syk, has also been detected in T cells in reponse to TCR/CD3 ligation [125,126]. However, the contribution of this kinase to TCR signalling remains to be determined. Deletion of Syk by homologous recombination in mice is lethal due to platelet dysfunction, causing death 10–12 days after birth as the result of tissue damage caused by haemolysis [127]. However, transfer of Syk null fetal liver cells into a RAG2 null mouse background (where endogenous thymocyte development arrests at $CD4^- CD8^- CD3^-$) results in the generation of normal double-positive thymocytes and, furthermore, the appearance of both CD4 and CD8 single-positive thymocyte populations. In contrast, B lymphocyte development is completely arrested. This suggests that Syk does not play a role in determining T cell differentiation pathways, CD4 or CD8, despite its relatively high level of expression in thymocytes [127]. However, it has been suggested that in ZAP-70-deficient patients, Syk may be capable of functionally substituting for ZAP-70 to a sufficient degree to allow positive selection of CD4 but not CD8 cells. This is unlikely to be physiological as adoptive transfer of Syk null fetal liver cells into RAG2 null mice results in the generation of normal $CD4^+ : CD8^+$ ratios [127]. Furthermore, substitution of ZAP-70 with Syk would seem unlikely to provide a phosphorylation signal of identical intensity. The importance of the relative intensity of phosphorylation signals to physiological response has been clearly demonstrated in T lymphocytes, where minor alterations in activation stimuli can result in dramatically altered degrees of activation of the cytoplasmic tyrosine kinases including ZAP-70; this in turn resulting in responses as apposite as T cell activation and anergy [128]. However, as Syk may potentially correct ZAP-70 deficiency to some degree, a role for ZAP-70 in CD4 single-positive selection cannot be completely excluded.

EMT (ITK,TSK)

Recently a novel T cell specific tyrosine kinase known variously as emt (expressed mainly in T

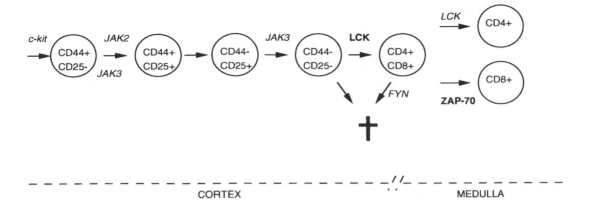

Figure 13.2 The role of tyrosine kinases in T cell development. The points indicated in the differentiation process are, or may be (italics), dependent upon tyrosine kinase activity

cells), itk (inducible tyrosine kinase) and tsk (T cell specific kinase) [129–132] has been independently identified and cloned by several groups. With the closely related Tec I and II kinases and the B cell specific atk (emb, btk), emt forms a new subfamily of non-receptor cytoplasmic tyrosine kinases. Although possessing a similar structure to the src kinases with single SH2 and SH3 and kinase domains, they possess larger amino terminal unique domains and lack both myristoylation and the negative regulatory phosphorylation sites. While lack of myristoylation may potentially indicate a difference in subcellular location, the regulation of kinase activity appears to be significantly different to the src family. Emt expression is restricted to T and NK cells and may be induced in T lymphocytes by activation stimuli such as PHA and IL2. While strongly expressed in activated T lymphocytes, high levels of emt are also detected in thymocytes [132], suggestive of a role in development. The potential importance of emt to T cell differentiation is suggested by studies of the homologous B cell specific tyrosine kinase atk. Deficient atk expression in humans results in the condition X-linked agammaglobulinaemia [133,134]. This inherited humoral immuno-deficiency arises from profoundly decreased levels of circulating immunoglobulins. Atk deficiency arrests pre-B cell differentiation and expansion, resulting in a severe reduction in circulating B lymphocytes, while T cell development and cellular immunity remain normal. The precise signals pro-

vided by atk and its actual role have yet to be ascertained.

TYROSINE KINASES AND T CELL DIFFERENTIATION: THE EVOLVING PICTURE

The current picture of the tyrosine kinase contribution to T cell developmental events is far from complete. While several of the major players have been identified, their precise roles remain unclear and other tyrosine kinases undoubtedly remain to be identified. As the data presented here demonstrate, tyrosine kinase activites are present and necessary for progression from the earliest stages of differentiation. In their turn, the Jak, the Src and the ZAP families of kinases play essential roles in determining the fate of the maturing cell. Combining the data obtained from *in vitro* manipulation, homologous recombination knockouts and transgenes of these kinases enables us to draw an overall scheme attesting to the importance of this group of signal transducers to T cell differentiation (Figure 13.2).

Even before T precursor cells enter the maturation-promoting environment of the thymus, receptors for the cytokine Interleukin 7 are being expressed; suggesting that at this early stage, or immediately upon entering the thymus, Jak1 and Jak3 kinase activities are required for growth and survival. Failure to receive this signal may arrest development at this stage (CD44$^+$CD25$^-$) as deletion of IL7 receptor expression results in an apparent failure to up-regulate CD25, while HSA induction occurs normally. Signals derived from the CD25 IL2 receptor may also require participation of the Jak kinases, although this assumption is based upon our knowledge of the $\beta\gamma$ heterodimer IL2 receptor. However, the potential importance of Jak3 to signalling at this point is highlighted by the demonstration that XSCID IL2Rγ chain mutations can result in reduced Jak3 activation. XSCID is characterized by the absence or severe impairment of T cell development. Therefore it is probable that most of the Jak family kinases are involved in the earliest of developmental stages; immediately following, if not preceding, migration of the progenitor cells to the thymus. c-kit, the receptor tyrosine kinase, would also appear to be involved during these stages of T cell differentiation, although its precise role is currently difficult to ascertain.

While events during the initial stages of development remain relatively unclear, the ensuing events, and the role of tyrosine kinases, are beginning to be more clearly defined. Following expression of CD25 and subsequent CD44 down-regulation, thymocytes proceed to initiate rearrangement of the germline TCRβ gene. As discussed earlier, this appears to be induced by Interleukin 7 secreted from thymic epithelial cells and is likely to be dependent upon the activation of the Jak1 and Jak3 kinases. Interleukin-7-activated p56lck may also play a role in this process, although deletion studies suggest it is not absolutely required. Only a small proportion of the double-negative cells proceed to successfully rearrange the TCRβ gene, most failing and succumbing to programmed cell death at this stage. However, approximately 30% [135] express a productively rearranged TCRβ chain and signal through this protein complexed with pre-α chain and CD3 proteins. This signal serves to initiate proliferation and subsequent expansion of these clones. The ability to negotiate this point has clearly been demonstrated to be dependent upon p56lck activity. Deletion of p56lck or overexpression of a dominant negative transgene arrests development at this stage, CD44$^-$CD25$^-$TCRβ^+. p56lck does not appear to be required to induce the expression of CD4 and CD8, as lck activity deficient G0/G1 arrested cells

express very low levels of these markers, but rather for the expansion of this small population of TCRβ^+ cells. Expression of an activated p56lck transgene alone is sufficient to bypass the requirement for TCRβ rearrangement and permit premature expansion of the thymocytes. These experiments define the earliest point at which lck is absolutely essential to maturation. Subsequent to TCRβ expression and signalling the thymocytes proceed to rearrangement of the TCRα locus and expression of a mature TCR$\alpha\beta$ heterodimer at the cell surface. Selective forces can now act upon the thymocyte population, eliminating self-reactive clones while positively expanding other more appropriate TCR-bearing cells.

Signals responsible for negative and positive selection decisions are believed to be delivered by a combination of TCR/CD3 and CD4/CD8 ligation events. Deletion of p59fyn expression revealed that it plays a role in transducing signals from the TCR of CD4$^+$ CD8$^+$ thymocytes, but curiously the absence of p59fyn has little effect upon the generation of single-positive mature lymphocytes. An antigen-specific defect in clonal deletion is apparent, but these mice do not develop auto-immune conditions, suggesting that self-tolerance mechanisms are still operational. As selective influences are believed to act through engagement of the TCR$\alpha\beta$ chains, this ability to respond correctly to selection pressures in the face of impaired signalling ability is puzzling. However, increased expression of fyn coincides with the transition form DP to SP thymocytes, suggesting that p59fyn is not required for negative selection events occurring early in DP thymocyte maturation. Deletion of clones relatively late in this progression may be more dependent upon p59fyn. As thymocytes up-regulate 59fyn expression they also acquire a much greater sensitivity to stimulation via the TCR, suggesting fyn may act to amplify these signals.

A more definite role in the selection process can be assigned to the ZAP-70 kinase, as the development of mature CD8$^+$ lymphocyte is ablated in ZAP-70-deficient humans while CD4$^+$ lymphocytes are selected normally. Therefore, the activation signal delivered by coincident ligation of TCR and CD4 presumably impacts differentially upon ZAP-70 from TCR and CD8 engagement. Signalling through CD4 and CD8 is unlikely to be equivalent due to the differential coupling of

CD4 and CD8 to p56lck. While CD8 appears to be less tightly associated with the kinase, the relative degree of p56lck activation obtained through the two coreceptor molecules *in vivo* is unclear. The ability to generate $CD8^+$ cells in the absence of CD8–p56lck interaction does suggest that this kinase is not required to deliver signals in this pathway. However, as overexpression of a CD4 transgene also affects the positive selection of MHC class I restricted TCR [136], presumably by sequestering p56lck, its participation cannot be ruled out. Current evidence suggests that ZAP-70 is activated as part of a phosphorylation cascade, presumably initiated by p56lck or p59fyn, rather than by auto-phosphorylation. This may strengthen the argument for the involvement of lck, as deletion of p59fyn did not impact significantly upon the maturation of single-positive thymocytes, suggesting that p59fyn is unlikely to be responsible for transducing either the CD4 or CD8 pathways.

In conclusion, it can be seen that each progressive stage along the pathway of T cell differentiation relies heavily upon the contribution of different tyrosine kinase activities with little redundancy. Both cytokine receptor and T cell receptor induced events call upon a changing selection of these cytoplasmic kinases to accomplish their functions. The full complexity of these tyrosine phosphorylation events has probably yet to be appreciated as many of the substrates of the kinases remain undefined. Furthermore, the regulation of these tyrosine kinases is poorly detailed and further investigation will undoubtedly reveal an equally complex web of tyrosine phosphatase activities, which is working to maintain tight control over the activities of these essential signal transduction elements.

REFERENCES

1. Rodewald H-R, Kretzschamar K, Takeda S, Hohl C and Dessing M. Identification of pro-thymocytes in murine fetal blood: T lineage commitment can precede thymus colonization. *EMBO J.* 1994; **13**: 4229–4240.
2. Louache F, Debili N, Marandin A, Coulombel L and Vainchenker W. Expression of CD4 by human hematopoietic progenitors. *Blood* 1994; **84**: 3344–3355.
3. Boyd RL, Tucek CL, Godfrey DI, Izon DJ, Wilson TJ, Davidson NJ, Bean AGD, Ladyman HM, Ritter MA and Hugo P. The thymic microenvironment. *Immunol. Today* 1993; **14**: 445–459.
4. Peschon JJ, Morrissey PJ, Grabstein KH, Ramsdell FJ, Maraskovsky E, Gliniak BC, Park LS, Ziegler SF, Williams DE, Ware CB, Meyer JF and Davidson BL. Early lymphocyte expansion is severely impaired in interleukin 7 receptor deficient mice. *J. Exp. Med.* 1994; **180**: 1955–1960.
5. Muegge K, Vila, MP and Durum SK. Interleukin-7: a cofactor for V(D)J rearrangement of the T cell receptor β gene. *Science* 1993; **261**: 93–95.
6. Schatz DG, Oettinger MA and Baltimore D. The V(D)J recombination activating gene. *Cell* 1989; **59**: 1035–1048.
7. Oettinger MA, Schatz DG, Gorka C and Baltimore D. RAG-1 and RAG-2, adjacent genes that synergistically activate V(D)J recombination. *Science* 1990; **248**: 1517–1523.
8. Mombaerts P, Iacomini J, Johnson RS, Herrap K, Tonegawa S and Papaioannou VE. RAG-1 deficient mice have no mature B and T lymphocytes. *Cell* 1992; **68**: 869–877.
9. Shinkai Y, Rathbun G, Lam K-P, Oltz EM, Stewart V, Mendelsohn M, Charron J, Datta M, Young F, Stall AM and Alt FW. RAG-2 deficient mice lack mature lymphocytes owing to inability to initiate V(D)J rearrangement. *Cell* 1992; **68**: 855–867.
10. Saint-Ruf C, Ungewiss K, Groettrup M, Bruno L, Fehling HJ and von Boehmer H. Analysis and expression of a pre-T cell receptor gene. *Science* 1994; **266**: 1208–1212.
11. Groettrup M, Baron A, Griffiths G, Palacios R and von Boehmer H. A novel disulfide-linked heterodimer on pre-T cells consists of the T cell receptor β chain and a 33 kd glycoprotein. *Cell* 1993; **75** 283–294.
12. Nossal GJV. Negative selection of lymphocytes. *Cell* 1994; **76**: 229–239.
13. von Boehmer H. Positive selection of lymphocytes. *Cell* 1994; **76**: 219–228.
14. Sawada S, Scarbourgh JD, Killeenn N and Littman DR. A lineage-specific transcriptional silencer regulates CD4 gene expression during T lymphocyte development. *Cell* 1994; **77**: 917–929.
15. Ashwell JD. Genetic and mutational analysis of the T-cell antigen receptor. *Ann. Rev. Immunol.* 1990; **8**: 139–167.
16. Jin Y-J, Clayton LK, Howard FD, Koyasu S, Sieh M, Steinbrich R, Tarr GE and Reinherz EL. Molecular cloning of the CD3ζ subunit identifies a CD3ζ-related product in thymus derived cells. *Proc.Natl.Acad.Sci.USA* 1990; **87**: 3319–3323.
17. Wegner A-MK, Letourneur F, Hoeveler A, Brocker T, Luton F and Mallisen. The T cell receptor/CD3 complex is composed of at least two autonomous transduction modules. *Cell* 1992; **68**: 83–95.
18. Letourner F and Klausner RD. Activation of T cells by a tyrosine kinase activation domain in the cytoplasmic tail of CD3ε. *Science* 1992; **255**: 79–82.
19. Bauer A, McConkey DJ, Howard FD, Clayton LK, Novick D, Koyasu S and Reinherz EL. Differential signal transduction via T cell receptor CD3₂, CD3ζ-n and CD3₂ isoforms. *Proc.Natl.Acad.Sci.* USA 1991; **88**: 3842–3846.
20. Isakov N. Tyrosine phosphorylation and dephosphorylation in lymphocyte activation. *Mol. Immunol.* 1993; **30**: 197–210.

21. June CH, Fletcher MC, Ledbetter JA, Schieven GL, Siegel JN, Phillips AF and Samelson LE. Inhibition of tyrosine phosphorylation prevents T cell receptor mediated signal transduction. *Proc.Natl.Acad.Sci.USA* 1990; **87**: 7722–7726.

22. Mustelin T, Coggeshall KM, Isakov N and Altman A. T cell antigen receptor-mediated activation of phospholipase C requires tyrosine phosphorylation. *Science* 1990; **247**: 1584–1587.

23. Bolen JB, Thompson PA, Eiseman E and Horak ID. Expression and interactions of the Src family of tyrosine protein kinases in T lymphocytes. *Adv Cancer Res.* 1991; **57**: 103.

24. Shaw AS, Amrein KE, Hammond C, Stern DF, Sefton BM and Rose JK. The lck tyrosine protein kinase interacts with the cytoplasmic tail of the CD4 glycoprotein through its unique amino terinal domain. *Cell* 1989; **59**: 627–636.

25. Samelson LE, Phillips AF and Luong Etand Klausner RD. Association of the fyn protein tyrosine kinase with the T cell antigen receptor. *Proc.Natl.Acad.Sci.USA*, 1990; **87**: 4358–4362.

26. Turner JM, Brodsky MH, Irving BA, Levin SD, Perlmutter RM and Littman DR. Interaction of the unique *N*-terminal region of tyrosine kinase p56lck with cytoplasmic domains of CD4 and CD8 is mediated by cysteine motifs. *Cell* 1990; **60**: 755–765.

27. Chan AC, Irving BA, Fraser JD and Weiss A. The ζ chain is associated with a tyrosine kinase and upon T cell antigen receptor stimulation associates with ZAP-70, a 70 kDa tyrosine phosphoprotein. *Proc.Natl.Acad.Sci.USA* 1991; **88**: 9166–9170.

28. Wange RL, Kong A-NT and Samelson LE. A tyrosine-phosphorylated 70-kDa protein binds a photoaffinity analogue of ATP and associates with both the ζ chain and CD3 components of the activated T cell antigen receptor. *J.Biol.-Chem.* 1992; **267**: 11685–11688.

29. Suda T and Zlotnik A. IL-7 maintains the T cell precursor potential of CD^{3-} thymocytes. *J. Immunol.* 1991; **146**: 3068–3073.

30. Bazan JF. Structural design and molecular evolution of a cytokine receptor superfamily. *Proc.Natl.Acad.Sci.USA.* 1990; **87**: 6934–6938.

31. Kishimoto T, Taga T and Akira S. Cytokine signal transduction. *Cell* 1994; **76**: 253–262.

32. Minami Y, Kono T, Miyazaki T and Taniguchi T. The IL-2 receptor complex: its structure, function, and target genes. *Ann.Rev.Immunol.* 1993; **11**: 245–267.

33. Miyajima A, Mui AL-F, Ogorochi T and Sakamaki K. Receptors for granulocyte-macrophage colony stimulating factor, interleukin-3 and interleukin-5. *Blood* 1993; **82**: 1960–1974.

34. Dadi HK and Roifman CM. Activation of phosphatidyli-nositol-3 kinase by ligation of the interleukin 7 receptor on human thymocytes. *J. Clin. Invest.* 1993; **92**: 1559–1563.

35. Dadi H, Ke S and Roifman CM. Activation of phosphati-dylinositol-3 kinase by ligation of the interleukin 7 receptor is dependent on protein tyrosine kinase activity. *Blood* 1994; **84**: 1579–1586.

36. Asao H, Tanaka N, Ishii N, Higuchi M, Takeshita T, Nakamura M, Shirasawa T and Sugamura K. Interleukin

2-induced activation of JAK3: possible involvement in signal transduction for *c-myc* induction and cell proliferation. *FEBS.Letts.* 1994; **351**: 201–206.

37. Shibuya H, Kohu K, Yamada K, Barsoumian EL, Perlmutter RM and Tanigucgi T. Functional dissection of p56lck, a protein tyrosine kinase which mediates interleukin-2 induced activation of the c-fos gene. *Mol.Cell.Biol.* 1994; **14**: 5812–5819.

38. Minami Y, Kono T, Yamada K, Kobayashi N, Kawahara A, Perlmutter RM and Taniguchi T. Association of p56lck. with the IL2 receptor chain is critical for the IL2 induced activation of p56lck. *EMBO J.* 1993; **12**: 759–768.

39. Hatakeyama M, Kono T, Kobayashi N, Kawahara A, Levin SD, Perlmutter RM and Taniguchi T. Interaction of the IL-2 receptor with the *src*-family kinase p56lck. Identification of novel intermolecular association. *Science* 1991; **252**: 1523–1528.

40. Miyazaki T, Kawahara A, Fujii H, Nakagawa Y, Minami Y, Liu Z-J, Oishi I, Silvennoinen O, Witthuhn BA, Ihle JN and Taniguchi T. Functional activation of Jak1 and Jak3 by selective association with IL2 receptor subunits. *Science* 1994; **266**: 1045–1047.

41. Merida I, Williamson P, Kuziel WA, Greene WC and Gaulton GN. The serine rich cytoplasmic domain of the interleukin 2 receptor beta chain is essential for interleukin 2 dependent tyrosine protein kinase and phosphatidylinositol-3-kinase activation. *J.Biol.Chem.* 1993; **268**: 6765–6770.

42. Majumder S, Brown K, Qiu F-H and Besmer. *c-kit* protein, a transmembrane kinase: identification in tissues and characterization. *Mol.Cell.Biol.* 1988; **8**: 4896–4903.

43. Chabot B, Stephenson DA, Chapman VM, Besmer P and Bernstein A. The proto-oncogene c-kit encoding a transmembrane tyrosine kinase receptor maps to the mouse W locus. *Nature* 1988; **335**: 88–89.

44. Rusell ES. Hereditary anemias of the mouse. *Adv. Genet.* 1979; **20**: 357–459.

45. Godfrey DI, Zlotnik A and Suda T. Phenotypical and functional characterization of c-kit expression during T cell development. *J. Immunol.* 1992; **149**: 2281–2285.

46. Matsuzaki Y, Gyotuku JI, Ogawa M, Nishikawa GI, Katsura Y, Gachelin G and Nakauchi H. Characterization of c-kit positive intrathymic stem cells that are restricted to lymphoid differentiation. *J.Exp.Med.* 1993; **178**: 1283–1292.

47. Lev S, Givol D and Yarden Y. A specific combination of substrates is involved in signal transduction by the kit-encoded receptor. *EMBO J.* 1991; **10**: 647–654.

48. Serve H, Yee NS, Stella G, Sepp-lorenzino L, Tan JC and Besmer P. Differential roles of PI3 kinase and Kit tyrosine 821 in Kit receptor-mediated proliferation, survival and cell adhesion in mast cells. *EMBO J.* 1995; **14**: 473–483.

49. Duronio V, Welham MJ, Abraham S, Dryden P and Schrader JW. p21^{ras} activation via hematopoietin receptors and c-kit requires tyrosine kinase activity but not tyrosine phosphorylation of p21^{ras} GTPase-activating protein. *Proc.Natl.Acad.Sci.USA.* 1992; **89**: 1587–1591.

50. Alai M, Mui AL, Cutler RL, Bustelo XR, Barbacid M and Krystal G. Steel factor stimulates the tyrosine phosphorylation of the proto-oncogene p95^{vav}, in human hematopoietic cells. *J. Biol. Chem.* 1992; **267**: 18021–18025.

51. Cutler RL, Liu L, Damen JE and Krystal G. Multiple cytokines induce the tyrosine phosphorylation of Shc and its association with Grb2 in hematopoietic cells. *J.Biol.-Chem.* 1993; **268**: 21463–21465.

52. Yi T and Ihle JN. Association of hematopoietic cell phosphatase with c-kit after stimulation with c-kit ligand. *Mol.-Cell.Biol.* 1993; **13**: 3350–3358.

53. Wilks AF, Harpur AG, Kurban RR, Ralph SJ, Zurcher G and Ziemiecki A. Two novel protein-tyrosine kinases, each with a second phosphotransferase-related catalytic domain, define a new class of protein kinase. *Mol.Cell.Biol.* 1991; **11**: 2057–2065.

54. Witthuhn BA, Silvennoinen O, Miura O, Lai, KS, Cwik C, Liu ET and Ihle JN. Involvement of the Jak-3 janus kinase in signalling by Interleukins 2 and 4 in lymphoid and myeloid cells. *Nature* 1994; **370**: 153–157.

55. Argetsinger LS, Campbell GS, Yang X, Witthuhn BA, Silvennoinen O, Ihle JN and Carter-Su C. Identification of Jak2 as a growth hormone receptor associated tyrosine kinase. *Cell* 1993; **74**: 237–244.

56. Witthuhn BA, Quelle FW, Silvennoinen O, Yi T, Tang B, Miura O and Ihle JN. Jak2 associates with the erythropoietin receptor and is tyrosine phosphorylated and activated following stimulation with erythropoietin. *Cell* 1993; **74**: 227–236.

57. Shaul K, Ziemiecki A, Wilks AF, Harpur AG, Sadowski HB, Gilma MZ and Darnell JE. Polypeptide signalling to the nucleus through tyrosine phosphorylation of Jak and Stat proteins. *Nature* 1993; **366**: 580–583.

58. Silvennoinen O, Ihle JN, Schlessinger J and Levy DE. Interferon-induced nuclear signalling by Jak protein tyrosine kinases. *Nature* 1993; **366**: 583–585.

59. Zhong Z, Wen Z and Darnell Jr, JE. Stat3: a STAT family member activated by tyrosine phosphorylation in response to epidermal growth factor and Interleukin 6. *Science* 1994; **264**: 95–98.

60. Darnell Jr, JE, Kerr IM and Stark GR. Jak–STAT pathways and transcriptional activation in response to IFNs and other extracellular signaling proteins. *Science* 1994; **264**: 1415–1421.

61. Marrero MB, Schieffer B, Paxton WG, Heerdt L, Berk BC, Dalafontaine P and Bernstein KE. Direct stimulation of Jak/STAT pathway by angiotensin II AT1 receptor. *Nature* 1995; **375**: 247–250.

62. Beadling C, Guschin D, Witthuhn BA, Ziemiecki A, Ihle JN, Kerr IM and Cantrell DA. Activation of JAK kinases and STAT proteins by interleukin-2 and interferon α, but not the T cell antigen receptor, in human T lymphocytes. *EMBO J.* 1994; **13**: 5605–5615.

63. Muller M, Briscoc J, Laxton C, Guschin D, Ziemiecki A, Silvennionen O, Harpur AG, Barbieri G, Witthuhn BA, Schindler C, Pelligrini S, Wilks AF, Ihle JN, Stark GR and Kerr IM. The protein tyrosine kinase JAK1 complements defects in interferon $-\alpha/\beta$ and $-\gamma$ signal transduction. *Nature* 1993; **366**: 129–135.

64. Watling D, Guschin D, Muller M, Silvennoinen O, Witthuhn BA, Quelle FW, Rogers NC, Schindler C, Stark GR, Ihle JN and Kerr IM. Complementation by the protein tyrosine kinase JAK2 of a mutant cell line deficiency in the interferon $-\gamma$ signal transduction pathway. *Nature* 1993; **366**: 166–170.

65. Johnston JA, Kawamura M, Kirken RA, Chen Y-Q, Blake TB, Shibuyas K., Ortaldo JR, McVicar DW and O'Shea JJ. Phosphorylation and activation of the Jak-3 kinase in response to interleukin-2. *Nature* 1994; **370**: 151–153.

66. Boussiotis VA, Barber DL, Nakarai T, Freeman GJ, Gribben JG, Bernstein GM, D'Andrea AD, Ritz J and Nadler LM. Prevention of T cell anergy by signalling through the γ chain of the IL2 receptor. *Science* 1994; **266**: 1039–1041.

67. Velazquez L, Fellous M, Stark GR and Pelligrini S. A protein tyrosine kinase in the interferon alpha/beta signaling pathway. *Cell* 1992; **70**: 713–722.

68. Kawamura M, McVicar DW, Johnston JA, Blake TB, Chen ZY-Q, Lal, BK, Lloyd AR, Kelvin DJ, Staples JE, Ortaldo JR and O'Shea, JJ. Molecular cloning of L-Jak, a Janus kinase family protein-tyrosine kinase expressed in natural killer cells and activated leukocytes. *Proc. Natl.Acad.Sci.USA* 1994; **91**: 6374–6378.

69. Rane SG and Reddy EP. Jak3: a novel Jak kinase associated with terminal differentiation of hematopoietic cells. *Oncogene* 1994; **9**: 2415–2423.

70. Sharfe N, Dadi H, O'Shea JJ and Roifman CM. Interleukin 7 signalling in double negative thymocytes involves Jak3. Submitted for publication, 1995.

71. Wildin RS, Garvin AM, Pawar S, Lewis DB, Abraham KM, Forbush KA, Ziegler SF, Allean JM and Perlmutter RM. Developmental regulation of lck gene expression in T lymphocytes. *J.Exp.Med.* 1991; **173**: 383–393.

72. Plum J, De Smedt M and Leclerq G. Exogeneous IL-7 promotes the growth of $CD4^-CD8^-CD44^-CD25^{+/-}$ precursor cells and blocks the differentiation pathway of $TCR\alpha\beta$ cells in fetal thymus organ culture. *J. Immunol.* 1993; **150**: 2706–2716.

73. Watson JD, Morrissey PJ, Namen AE, Conlon PJ and Widmer MB. Effect of IL-7 on the growth of fetal thymocytes in culture. *J. Immunol.* 1989; **143**: 1215–1222.

74. Sharfe N, Dadi H and Roifman CM. Jak3 protein tyrosine kinase mediates IL7 induced activation of phosphatidylinoistol 3'-kinase. *Blood* 1995; **86**: 2077–2085.

75. Molina TK, Kishihara K, Siderovski DP, van Ewijk W, Narendran A, Timms E, Wakeham A, Paige CJ, Hartmann K-U, Veillette A, Davidson D and Mak TW. Profound block in thymocyte development in mice lacking p56lck. *Nature* 1992; **357**: 161–164.

76. Levin SD, Anderson SJ, Forbush KA and Perlmutter RM. A dominant negative transgene defines a role for p65lck. *EMBOJ.* 1993; **12**: 1671–1680.

77. Russell SM, Johnston JA, Noguchi M, Kawamura M, Bacon CM, Friedmann M, Berg M, McVicar DW, Witthuhn BA, Silvennoinen O, Goldman AS, Schmalstieg FC, Ihle JN, O'Shea JJ and Leonard WJ. Interaction of IL-2Rβ and γc chains with Jak1 and Jak3: Implications for XSCID and XCID. *Science* 1994; **266**: 1042–1045.

78. Jeurissen SH, Janse EM, Ekinao S, Nieuwenhuis P, Koch G and De Boer GF. Monoclonal antibodies as probes for defining cellular subsets in the bone marrow, thymus, bursa of fabricius and spleen of the chicken. *Vet. Immunol. Immunopath.* 1988; **19**: 225–238.

79. Moreno J, Vicente A, Heijnen I and Zapata AG. Prolactin and early T-cell development in embryonic chicken. *Immunol. Today*. 1994; **15**: 524–526.

80. Gouilleux F, Wakao H, Mundt M and Groner B. Prolactin induces phosphorylation of Tyr694 of Stat5(MGF), a prerequisite for DNA binding and induction of transcription. *EMBOJ*. 1994; **13**: 4361–4369.

81. Mukherjee P, Mastro AM and Hymer WC. Prolactin induction of interleukin-2 receptors on rat splenic lymphocytes. *Endocrinology* 1990; **126**: 270–273.

82. Shen MM, Skoda RC, Cardiff RD, Campos-Torres J, Leder P and Ornitz DM. Expression of LIF in transgenic mice results in altered thymc epithelium and apparent interconversion of thymic and lymph node morphologies. *EMBOJ*. 1994; **13**: 1375–1385.

83. Strauss DB and Weiss A. Genetic evidence for the involvement of the Lck tyrosine kinase in signal transduction through the T cell antigen receptor. *Cell* 1992; **70**: 585–593.

84. Mombaerts P, Clarke AR, Rudnicki MA, Iacomini J, Itohara S, Lafaille JJ, Wang L, Ichikawa Y, Jaenisch R, Hooper ML and Tonegawa S. Mutations in the T cell antigen receptor genes α and β block thymocyte development at different stages. *Nature* 1992; **360**: 225–231.

85. Anderson SJ, Levin SD and Perlmutter RM. Protein tyrosine kinase p56[lck] controls allelic exclusion of T-cell receptor β-chain genes. *Nature* 1993; **365**: 552–554.

86. Mombaerts P, Anderson SJ, Perlmutter RM, Mak TW and Tonegawa S. An activated lck transgene promotes thymocyte development in RAG-1 mutant mice. *Immunity* 1994; **1**: 261–267.

87. Groettup M, Baron A, Griffiths G, Palacios R and von Boehmer H. T cell receptor (TCR) β chain homodimers on the surface of immature but not mature α, β, γ chain deficient T cell lines. *EMBOJ*. 1992; **11**: 2735–2746.

88. Levelt CN, Mombaerts P, Iglesias A, Tonegawa S and Eichmann K. Restoration of early thymocyte differentiation in T-cell receptor β-chain-deficient mutant mice by transmembrane signaling through CD3. *Proc. Natl. Acad. Sci. USA* 1993; **90**: 11401–11405.

89. Glaichenhaus N, Shastri N, Littman DR and Turner JM. Requirement for association of p56[lck] with CD4 in antigen specific signal transduction in T cells. *Cell* 1991; **64**: 511–520.

90. Shaw AS, Amrein KE, Hammond C, Stern DF, Sefton BM and Rose JK. The Lck tyrosine protein kinase interacts with the cytoplasmic tail of the CD4 glycoprotein through its unique amino-terminal domain. *Cell* 1989; **59**: 627–636.

91. Turner JM, Brodsky MH, Irving BA, Levin SD, Perlmutter RM and Littman DR. Interaction of the unique *N*-terminal region of tyrosine kinase p56[lck] with cytoplasmic domains of CD4 and CD8 is mediated by cysteine motifs. *Cell* 1990; **60**: 755–765.

92. Mustelin T, Coggeshall KM and Altamn A. Rapid activation of the T cell kinase pp56[lck] by the CD45 phosphotyrosine phosphatase. 1989. *Proc. Natl. Acad. Sci. USA* **86**, 6302–6306.

93. Ross SE, Schraven B, Goldman FD, Crabtree J and Koretzky GA. The association between CD45 and LCK does not require CD4 or CD8 and is independent of T cell receptor stimulation. *Biochem. Biophys. Res. Commun.* 1994; **198**: 88–96.

94. Kearse KP, Roberts JL, Munitz TI, Weist DL, Nakayama T and Singer A. Developmental regulation of α β T cell antigen receptor expression results from differential stability of nascent TCRα proteins within the endoplasmic reticulum of immature and mature T cells. *EMBOJ*. 1994; **13**: 4505–4514.

95. Bonifacino JS, McCarthy SA, Maguire JE, Nakayama T, Singer DS, Klausner RD and Singer A. Novel post-translational regulation of TCR expression in CD4[+]CD8[+] thymocytes influenced by CD4. *Nature* 1990; **344**: 247–251.

96. Wiest DL, Yuan L, Jefferson J, Benveniste P, Tsokos M, Klausner RD, Glimcher LH, Samelson LE and Singer A. Regulation of T cell receptor expression in immature CD4[+]CD8[+] thymocytes by p56[lck] tyrosine kinase: Basis for differential signaling by CD4 and CD8 in immature thymocytes expressing both coreceptor molecules. *J. Exp. Med*. 1993; **178**: 1701–1712.

97. Cooke MP, Abraham KM, Forbush KA and Perlmutter RM. Regulation of T cell receptor signalling by a src family protein tyrosine kinase (p59[fyn]). *Cell* 1991; **65**: 281–291.

98. Stein PL, Lee H-M, Rich S and Soriano P. pp59[fyn] mutant mice display differential signalling in thymocytes and peripheral T cells. *Cell* 1992; **70**: 741–750.

99. Appleby MW, Gross JA, Cooke MP, Levin SD, Qian X and Perlmutter RM. Defective T cell receptor signalling in mice lacking the thymic isoform of pp59[fyn]. *Cell* 1992; **70**: 751–763.

100. Thomas PM and Samelson LE. The glycophosphatidylinositol-anchored Thy-1 molecule interacts with the p60[fyn] protein tyrosine kinase in T cells. *J. Biol. Chem*. 1992; **267**: 11685–11688.

101. Mustelin T and Burn P. Regulation of *src* family tyrosine kinases in lymphocytes. *Trends Biochem*. 1993; **18**: 215–220.

102. Partanen J, Armstrong E, Bergman M, Makela TP, Hirvonen H, Huebner K and Alitalo K. *Cyl* encodes a putative cytoplasmic tyrosine kinase lacking the conserved tyrosine autophosphorylation site (Y416[arc]). *Oncogene* 1991; **6**: 2013–2018.

103. Nada S, OkadaM, MacAuley A, Cooper JA and Nakagawa H. Cloning of a complementary DNA for a protein-tyrosine kinase that specifically phosphorylates a negative regulatory site of p60[c−arc]. *Nature* 1991; **351**: 69–72.

104. Imamoto A and Soriano P. Disruption of the csk gene, encoding a negative regulator of Src family tyrosine kinases, leads to neutral tube defects and embryonic lethality in mice. *Cell* 1993; **73**: 1117–1124.

105. Oetken C, Couture C, Bergman M, Bonnefoy-Berard N, Williams S, Alitalo K, Burn P and Mustelin T. TCR/CD3 triggering causes increased activity of the p50[cak] tyrosine kinase and engagement of its SH2 domain. *Oncogene* 1994; **9**: 1625–1631.

106. Bergman M, Mustelin T, Oetken C, Partanen J, Flint NA, Amrein KE, Autero M, Burn P and Alitalo K. The human p50[cak]tyrosine kinase phosphorylates p50[lck] at Tyr-505

and down regulates its catalytic activity. *EMBOJ.* 1992; **11**: 2919–2924.

107. Chow LM, Fournel M, Davidson D and Veillette A. Negative regulation of T-cell receptor signalling by tyrosine protein kinase p50^{cak}. *Nature* 1993; **365**: 156–160.

108. Autero M, Saharinen J, Pessa-Morikawa T, Soula-Rothhut M, Oetken C, Gassmann M, Bergman M, Alitalo K, Burn P, Gahmberg CG and Mustelin T. Tyrosine phosphorylation of CD45 phosphotyrosine phosphatase by p50^{lck} creates a binding site for p50^{lck} tyrosine kinase and activates the phosphatase. *Mol. Cell. Biol.* 1994; **14**: 1308–1321.

109. Hurley TR, Hyman R and Sefton BM. Differential effects of expression of the CD45 tyrosine protein phosphatase on the tyrosine phosphorylation of the *Lck, fyn* and *c-src* tyrosine protein kinases. *Mol. Cell. Biol.* 1993; **13**: 1651–1656.

110. Schraven B, Kirchgessner H, Gaber Y, Samstag Y and Meur S. A functional complex is formed in human T lymphocytes between the protein tyrosine phosphatase CD45 and the protein tyrosine kinase p56^{lck} and pp32, a possible common substrate. *Eur. J. Immunol.* 1991; **21**: 2469–2477.

111. Mustelin T, Pessa-Morikawa T, Autero M, Gassman M, Gahmberg CG, Andersson LC, and Burn P. Regulation of the p59^{fyn} protein tyrosine kinase by the CD45 phosphotyrosine phophatase. *Eur. J. Immunol.* 1992; **22**: 1173–1178.

112. Chow LML, Davidson D, Fournel M, Gosselin P, Lemieux S, Lyu MS, Kozak CA, Matis LA and Veillette A. Two distinct protein isoforms are encoded by *ntk*, a *csk*-related tyrosine protein kinase gene. *Oncogene* 1994; **9**: 3437–3448.

113. Taniguchi T, Kobayashi T, Kondo J, Takahashi K, Nakamura H, Suzuki J, Nagi K, Yamada T, Nakamura S and Yamamura H. Molecular cloning of a porcine gene *syk* that encodes a 72kDa protein tyrosine kinase showing high susceptibility to proteolysis. *J. Biol. Chem.* 1991; **266**: 15790–15796.

114. Chan AC, Iwashima M, Turck CW and Weiss A. ZAP-70: a 70kDa protein tyrosine kinase that associates with the TCRζ chain. *Cell* 1992; **71**: 649–662.

115. Nakayama T, Singer A, Hsi ED and Samelson LE. Intrathymic signalling in immature CD4$^+$CD8$^+$ thymocytes results in tyrosine phosphorylation of the T-cell receptor zeta chain. *Nature* 1989; **341**: 651–654.

116. van Oers NSC, Killeen N and Weiss A. ZAP-70 is constitutively associated with tyrosine phosphorylated TCR ζ in murine thymocytes and lymph node T cells. *Immunity* 1994; **1**: 675–685.

117. Arpaia E, Sharar M, Dadi H, Cohen A and Roifman CM. Defective T cell receptor signalling and CD8$^+$ thymic selection in humans lacking ZAP-70 kinase. *Cell* 1994; **76**: 947–958.

118. Chan AC, Kadlecek TA, Elder ME, Filipovich AH, Kuo W-L, Iwasima M, Parslow TG and Weiss A. ZAP-70 deficiency in an autosomal recessive form of severe immunodeficiency. *Science* 1994; **264**: 1599–1601.

119. Chan IT, Limmer A, Louie MC, Bullock ED, Fung LWP, Mak TW and Loh DY. Thymic selection of cytotoxic T cells independent of CD8α-lck association. *Science* 1993; **261**: 1581–1584.

120. Shores EW, Huang K, Tran T, Lee E, Grinberg A and Love PE. Role of TCR ζ chain in T cell development and selection. *Science* 1994; **266**: 1047–1050.

121. Liu C-P, Ueda R, She J, Sancho J, Wang B, Weddell G, Loring J, Kurahara C, Dudley EC, Hayday A, Terhorst C and Huang M. Abnormal T cell development in CD3 – ζ$^{-/-}$ mutant mice and identification of a novel T cell population in the intestine. *EMBOJ.* 1993; **12**: 4863–4875.

122. Love PE, Shores EW, Johnson MD, Tremblay ML, Lee EJ, Grinberg A, Huang SP, Singer A and Westphal H. T cell development in mice that lack the ζ chain of the T cell antigen receptor complex. *Science* 1993; **261**: 918–921.

123. Timson Gauen LK, Zhu Y, Letourneur F, Hu Q, Bolen JB, Matis LA, Klausner RD and Shaw AS. Interactions of p59^{fyn} and ZAP-70 with T cell receptor activation motifs: Defining the nature of a signalling motif. *Mol. Cell. Biol.* 1994; **14**: 3729–3741.

124. Wange RL, Malek SN, Desiderio S and Samelson LE. The tandem SH2 domains of ZAP-70 bind the TCRζ and CD3ε from activated Jurkat T cells. *J. Biol. Chem.* 1993; **268**: 19797–19801.

125. Kolanus W, Romeo C and Seed B. T cell activation by clustered tyrosine kinases. *Cell* 1993; **74**: 171–183.

126. Chan AC, van Oers NSC, Tran A, Turka L, Law C-L, Ryan JC, Clark AE and Weiss A. Differential expression of ZAP-70 and Syk protein tyrosine kinases, and the role of this family of protein tyrosine kinases in T cell antigen receptor signalling. *J. Immunol.* 1994; **152**: 4758–4766.

127. Cheng A, Rowley B, Pao W, Hayday A, Bolen J and Pawson T. The protein tyrosine kinase Syk regulates developmental selection of murine B lymphocytes. Signal Transduction in Normal and Tumor Cells. AACR Conference in Cancer Research, Banff. April 1995.

128. Sloan-Lancaster J, Shaw AS, Rothbard JB and Allen PM. Partial T cell signalling: Altered phospho-ζ and lack of ZAP-70 recruitment in APL-induced T cell anergy. *Cell* 1994; **79**: 912–922.

129. Tanaka N, Asao H, Ohtani K, Nakamura M and Sugamura K. A novel human tyrosine kinase gene inducible in T cells by Interleukin 2. *FEBS Letts.* 1993; **324**: 1–5.

130. Yamada N, Kawakami Y, Kimura H, Fukamachi H, Baier G, Altman A, Kato T, Inagak Y and Kawakami T. Structure and expression of novel protein-tyrosine kinases, Emb and Emt, in hematopoietic cells. *Biochem Biophys. Res. Commun.* 1993; **192**: 231–240.

131. Siliciano JD, Morrow TA and Desiderio SV. itk, a T-cellspecific tyrosine kinase gene inducible by interleukin 2. *Proc. Natl. Acad. Sci. USA* 1992; **89**: 11194–11198.

132. Heyeck SD and Berg LJ. Developmental regulation of a murine T-cell-specific tyrosine kinase gene, Tsk. *Proc. Natl. Acad. Sci. USA* 1993; **90**: 669–673.

133. Vetrie D, Vorechovsky I, Sideras P, Holland J, Davies A, Flinter F, Hammarstrom L, Kinnon C, Levinsky R, Bobrow M, Smith C and Bentley D. The gene involved in X-linked agammaglobulinemia is a member of the src family of protein tyrosine kinases. *Nature* 1993; **361**: 226–233.

134. Tsukada S, Saffran DC, Rawlings DJ, Parolini O, Allen RC, Klisak I, Sparkes RS, Kubagawa H, Mohandas T, Quan S, Belmont JW, Cooper MD, Conley ME and Witte ON. Deficient expression of a B cell cytoplasmic tyrosine kinase in human X-linked agamaglobulinemia. *Cell* 1993; **72**: 279–290.

135. Penit C, Lucas B and Vasseur F. Cell expansion and growth arrest phases during the transition from precursor CD4⁻8⁻ to immature CD4⁺8⁺ thymocytes in normal and genetically modified mice. *J. Immunol.* 1995; **154**: 5103–5113.

136. Van Oers NSC, Garvin AM, Davis CB, Forbush KA, Littman DR, Perlmutter RM and The H-S. Disruption of CD8-dependent negative and positive selection of thymocytes is correlated with a decreased association between CD8 and the protein tyrosine kinase, p56^lck. *Eur. J. Immunol.* 1992; **22**: 735–743.

14

Cyclic Nucleotide Signalling Throughout T Cell Maturation

Alison M. Michie, Margaret M. Harnett and Miles D. Houslay

Division of Biochemistry and Molecular Biology, Institute of Biomedical and Life Sciences, University of Glasgow, Glasgow, Scotland, UK

INTRODUCTION

The cyclic nucleotides, adenosine-3′,5′-cyclic monophosphate (cAMP) and guanosine-3′,5′-cyclic monophosphate (cGMP) play central roles in signal transduction processes [1,2]. The precise function of these cyclic nucleotides in the immune response is still poorly understood despite having been under investigation for many years. Early research suggested that while cAMP negatively modulated lymphocyte proliferation [3], cGMP could promote lymphocyte responses. However, it has become increasingly evident that this picture is oversimplified. Thus, while it is accepted that cAMP generally inhibits lymphocyte responses [3], there is also a body of research that supports more complex roles for cAMP and cGMP in lymphocyte function [4,5].

THE CYCLIC NUCLEOTIDE SIGNALLING SYSTEM

The intracellular second messengers, cAMP and cGMP are generated by adenylyl cyclase and guanylyl cyclase, respectively [6,7], in response to external stimuli, such as hormones, neurotransmitters or cytokines. Cyclic nucleotides can then elicit their effects via the activation of specific protein kinases, causing the phosphorylation of target proteins within the cell and leading to a physiological response [8–10]. cAMP has been shown to bind with high affinity to the regulatory subunits of a protein kinase (PKA), thereby activating the enzyme [8]. Binding of cAMP to the PKA-holoenzyme causes the dissociation of the two regulatory subunits from the two active catalytic subunits [8,11]. The catalytic subunits are then free to phosphorylate serine/threonine residues of target proteins, either in the nucleus, where they are believed to bind to DNA and alter gene transcription, or in the cytoplasm, where the subunits may alter post-transcriptional events [3,11]. Indeed, the modulation of transcription factor activities, such as the cAMP-responsive element binding protein (CREB) or the activation transcription factor (ATF) families, is thought to be under the control of PKA-mediated phosphorylation (Figure 14.1). [12] This regulation can control either the stimulation or repression of specific gene expression [13, 14]. By comparison, substantially less is known about how cGMP mediates its actions within the cell: it has, however, been shown to interact with several intracellular receptor proteins, including protein kinases [15,16], cyclic nucleotide phosphodiesterases [17] and cGMP gated ion channels [18, 19]. Nevertheless, cGMP is considered to be of great importance in cellular signalling, especially as it has recently been found to be elevated in response to nitric oxide (NO) [20]. Clearly then, the regulation of intracellular levels of cyclic nucleotides is critical and, thus, cellular cyclic nucleotide homeostasis is maintained not only by regulating their synthesis by adenylyl cyclase or guanylyl cyclase [21, 22], but also by control of their degradation, through the action of the cyclic nucleotide phosphodiesterases (PDEs) [23, 24].

Lymphocyte Signalling: Mechanisms, Subversion and Manipulation. Edited by M. M. Harnett and K. P. Rigley © 1997 John Wiley & Sons Ltd.

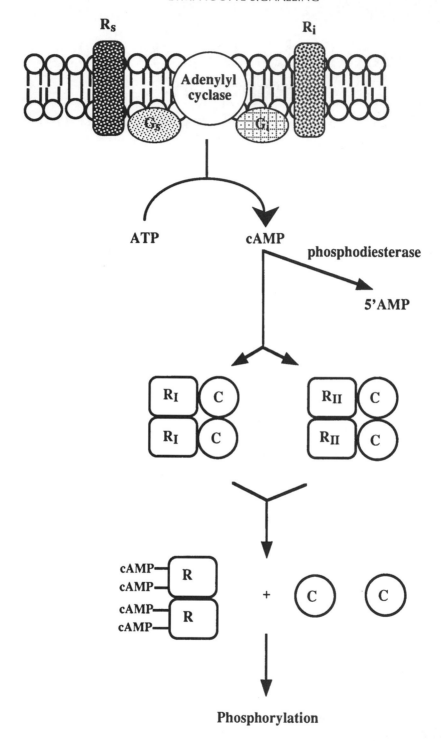

Figure 14.1   cAMP metabolism. cAMP levels can be regulated at the level of: (i) adenylyl cyclase, via receptor coupling through inhibitory G proteins (G_i) or stimulatory G proteins (G_s); (ii) cAMP phosphodiesterase, by regulating the level of breakdown of cAMP

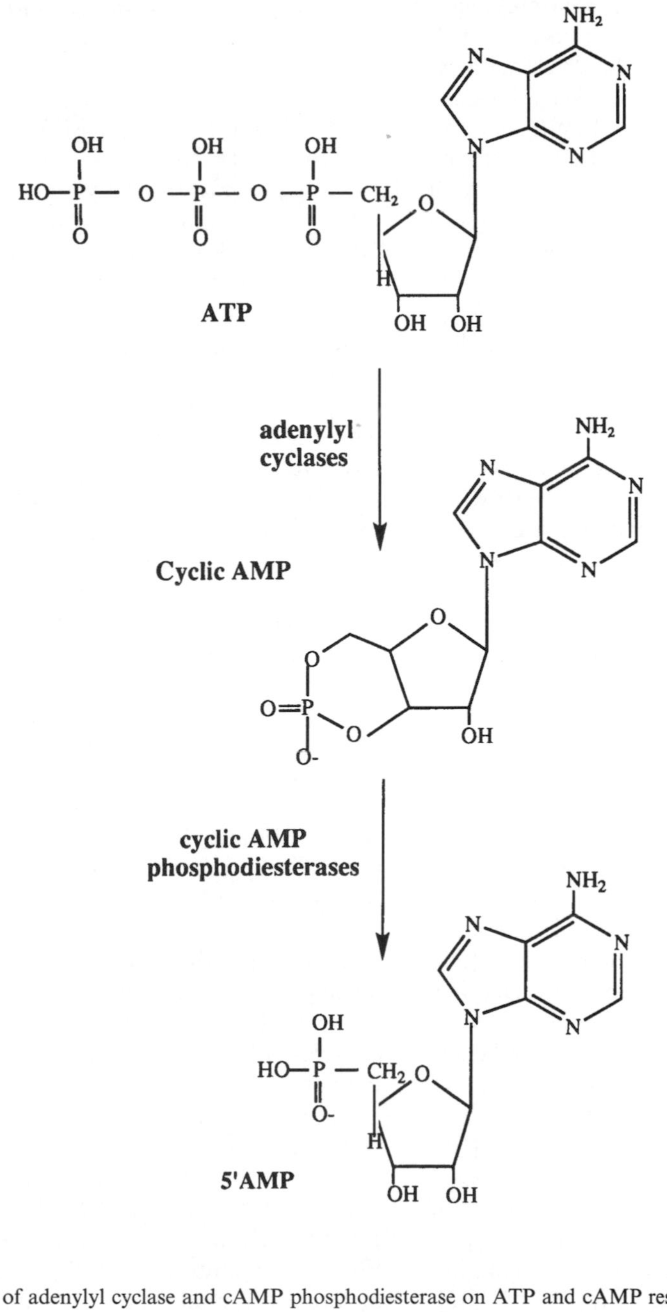

Figure 14.2   Action of adenylyl cyclase and cAMP phosphodiesterase on ATP and cAMP respectively

Adenylyl Cyclases

Adenylyl cyclases are a group of membrane-bound proteins with molecular weights ranging from 120 to 150 kDa which are responsible for catalysing the generation of the second messenger cAMP from ATP (Figure 14.2). [25, 26]. The initial isolation of a cDNA encoding a polypeptide with adenylyl cyclase activity [27], termed type I, was followed by the isolation of multiple other forms of adenylyl cyclase [28–31]. There are now more than eight genetically distinct mammalian adenylyl cyclase isoforms originating from separate genes. In addition, alternatively spliced transcripts [32] and partial sequence identity of novel adenylyl cyclases (type VII) [33] are evident, yet the extent of molecular diversity within this family is unresolved as yet. However, the conservation of primary sequence between species is very high ($> 90\%$, but dissimilar between subtypes (overall homology among the various isoforms is about 50%). Therefore it appears that these adenylyl cyclase isoforms have evolved differently. Adenylyl cyclase consists of a single polypeptide, which is predicted to have six transmembrane helices followed by a large cytoplasmic loop, another six transmembrane helices and a long intracellular tail. This structure is similar to that predicted for transporter species and ion channels [27]. Although each isoenzyme family displays different properties of regulation by calcium, protein kinase C (PKC), PKA, G proteins ($G_{i\alpha 2}$ and $G_{i\alpha 3}$ and $\beta\gamma$ subunits) [21], all the adenylyl cyclase activities are stimulated via G proteins G_s [34] and by the diterpene forskolin [35].

Guanylyl Cyclases

It is now known that the regulation of synthesis of cGMP can occur via two routes. Some agonists cause calcium-dependent synthesis of membrane-permeant free radicals such as NO, which are able to stimulate a cytoplasmic guanylyl cyclase [36–38]. A cytoplasmic form of guanylyl cyclase has been known to exist for some years as a heterodimer [39]. The activity of this guanylyl cyclase is dependent on the co-expression of α (70 kDa) and β (82 kDa) subunits as no detectable cyclase activity is found associated with the monomers [40]. Interestingly, from cloning experiments, it appears that each subunit possesses a catalytic domain [41]. In contrast, other agonists interact directly with an extracellular domain of a membrane-spanning guanylyl cyclase, hence increasing its activity [42]. Three mammalian isoforms of plasma membrane guanylyl cyclases have now been cloned. They have been designated GC-A, GC-B and GC-C [43, 44]. Further analysis has revealed that two of these cyclases are the natriuretic peptide receptors GC-A and GC-B, found expressed in the brain [45].

Cyclic-nucleotide-dependent Protein Kinases

PKA exists as two major isoforms, type I and type II, which differ with respect to their regulatory (R) subunits, as identified by their distinct molecular weights, antigenicity, amino acid sequence and affinity for cAMP analogues. Moreover, isotypes of each of these regulatory subunits, RI (α and β) and RII (α and β), as well as of the catalytic (C) subunit (α, β and γ) have now also been identified. Expression of these isoforms varies between tissues, with type II PKA being present in all cells, whereas the tissue distribution of type I PKA is more restricted [46, 47]. Intracellular localization also varies, with RI isoforms found primarily in the cytosolic compartment of the cell, although a fraction has been found to be located to the membranes of erythrocytes [47] and associated with the T cell receptor of activated lymphocytes [48]. In contrast, the RII isoforms are generally particulate, with up to 75% of the total intracellular RII pool being associated with the plasma membrane and cytoskeletal compartments [49, 50]. In all cases, the inactive holoenzyme form of PKA is a tetrameric entity, consisting of two R subunits and two C subunits (R_2C_2) (Figure 14.1) [8]. However, upon binding of four molecules of cAMP to the regulatory subunits (two per R monomer), the active catalytic subunits are released and can diffuse through the cytoplasm and even into the nucleus [51]. The catalytic subunits of PKA are then able to phosphorylate a wide range of physiological substrates containing the target serine in the context X–R–R–X–S–X.

As with PKA, there are two classes of cGMP-dependent protein kinase (PKG, type I and type II forms) expressed in mammalian cells: PKG appears to be predominantly localized in the cytosol although the enzyme in platelets and the type II PKG from intestinal brush border appear to be membrane bound. PKG is found to be expressed to particularly high levels in cerebellar Purkinje cells, heart and lung tissue [52, 53]. Interestingly, PKG is absent from many regions of the brain where cGMP is produced, which might indicate that it is not the sole modulator of cGMP effects [18]. While the type I PKG is widely distributed and is a homodimer consisting of subunits of approximately 78 kDa, the type II PKG is a monomer with a mass of 86 kDa and is found only in intestinal epithelial cells [54]. The type I class of PKG consists of two closely related isoforms termed type Iα and 1β [15], which differ in their N-terminal region, indicating that they may arise from alternatively spliced transcripts [55]. Both isoforms can be activated by cAMP and cGMP, although greater concentrations of cyclic nucleotides are needed to activate the β isoform [56]. Binding of cGMP to PKG results in enzyme activation resulting in the phosphorylation of proteins containing the consensus sequence, R–L–R–S–R–L–G. This site is very similar to that of PKA and, indeed, these protein kinases have even been shown to phosphorylate the same sites in certain proteins [57]. Nevertheless, there is some specificity between these kinase activities: indeed, where phenylalanine residues are C-terminal to the target serine residues, this enhances the selectivity for PKG compared with PKA [58].

Cyclic Nucleotide Phosphodiesterases

The degradation of the second messengers cAMP and cGMP is achieved by cyclic nucleotide phosphodiesterase (PDE) activities which elicit the hydrolysis of the 3′ phosphodiester bond to give the corresponding 5′ nucleoside monophosphate $(5' - AMP; 5' - GMP)$ (Figure 14.2) [59, 60]. This activity is provided by a diverse and complex group of enzymes generated by multiple genes together with alternative splicing mechanisms (reviewed in references [23, 24, 61–63]). Cyclic nucleotide phosphodiesterases have been shown to regulate various physiological responses in many different cells and tissues including platelet aggregation [64, 65], vascular relaxation [66, 67], cardiac muscle contraction [68, 69] and inflammation [70, 71].

Cyclic nucleotide hydrolysing PDEs exhibit distinct biochemical characteristics and can be divided into at least seven functional classes (Table 14.1): (i) PDE1, which can hydrolyse both cGMP and cAMP and whose activity is stimulated by calcium/calmodulin (Ca^{2+}/CaM); (ii) PDE2 hydrolyses cAMP and cGMP, with the activity being stimulated by micromolar concentrations of cGMP; (iii) PDE3 specifically hydrolyses cAMP in

a manner which is inhibited by micromolar concentrations of cGMP; (iv) PDE4 enzymes are cAMP specific PDEs which are insensitive to cGMP and are selectively inhibited by rolipram; (v) PDE5 and PDE6 enzymes specifically hydrolyse cGMP, but differ in their structure and tissue distribution; (vi) PDE7 specifically hydrolyses cAMP and is insensitive not only to cGMP but to all known PDE inhibitors including the nonselective PDE inhibitor isobutylmethylxanthine (IBMX).

The PDE enzyme family shows a diverse range of properties that are exemplified by their differential sensitivity to phosphorylation by various kinases [62, 72–77], their inhibitor specificities and cofactor requirements [62, 63, 73, 75, 78, 79], their hormonal regulation [23, 80, 81], their differential expression patterns [82–84] and their intracellular localization [82, 85]. Members of this diverse family thus have the potential to modulate cAMP metabolism to the specific requirements of the cell, including being able to integrate regulatory signals from lipid, tyrosyl and NO signalling pathways.

MODULATION OF T CELL ACTIVATION BY CYCLIC NUCLEOTIDE SIGNALLING

The vast majority of pharmacological evidence in which intracellular levels of cAMP in lymphocytes have been manipulated, for example by the addition of: (i) agonists to receptors which are coupled to G_s; (ii) the diterpene forskolin which activates the catalytic unit of adenylate cyclase directly; (iii) cholera toxin which elicits the ADP-ribosylation and activation of G_s subunits; (iv) cell-permeable cAMP analogues; and (v) cAMP phosphodiesterase inhibitors [86–88], supports the view that elevation of intracellular levels of cAMP provides an anti-proliferative signal to lymphocytes [3]. Indeed, sustained elevation of cAMP levels has even been shown to induce programmed cell death (apoptosis) in T cell hybridomas and thymocytes [89, 90]. Taken together with the findings that the level of PDE activities in disease states, such as leukaemia (high PDE, uncontrolled proliferation) and sarcoidosis (low PDE, functional unresponsiveness), can influence the proliferation status of lymphocytes [91, 92], the general conclusion has been that cAMP plays an anti-proliferative role in the immune system.

PKA Inhibition of TCR-mediated Signalling Pathways

It is well established that following ligation of the TCR/CD3 complex, the activation of protein tyrosyl kinases (PTKs) such as $p59^{fyn(T)}$, $p56^{lck}$ and $p70^{zap(T)}$ results in the rapid tyrosine phosphorylation, recruitment and activation of key signal transducing enzymes such as the phosphatidylinositol 4,5-bisphosphate (PPIP$_2$)-specific phospholipase (PLC)γ − 1 [93–95]. Hydrolysis of PIP$_2$ generates the second messengers, inositol trisphosphate (IP$_3$) and diacylglycerol (DAG), leading to a release of calcium from intracellular stores and the activation of protein kinase C (PKC). Other signalling events, downstream of PTK activation, include the recruitment and activation of the PI-3-kinase [96] and Ras

[97]-MAP kinase [91, 98] signalling cascades (for review see Chapter 2 of this volume).

cAMP has been shown to be inhibitory at a number of sites on these signalling pathways leading to uncoupling of the TCR/CD3 complex from its downstream signalling mediators. Targets of cAMP action include the inhibition of tyrosine phosphorylation [99–101], PIP$_2$ breakdown [86–88, 99, 102, 103], [$^{2+}$]$_i$ elevations, PKC activation [104–107], c-raf-1 activation [108] and K$^+$ conductance [109, 110]. The mechanisms underlying such modulation of T cell signalling are as yet poorly understood, but there are some interesting indications of crosstalk between cAMP and TCR-coupled mitogenic signalling pathways which may provide insight into the role(s) of cAMP pathways in T cell signalling. cAMP appears to exert its negative effects on lymphocyte proliferation via PKA-I [111]. Indeed, recent work [48] has demonstrated that in activated T lymphocytes, the PKA-Iα isoform is found associated with the TCR/CD3 complex. and that following phosphorylation of Ser1248 by PKA, PLCγ1 becomes inactivated,

Table 14.1 Summary of cyclic nucleotide phosphodiesterase isoforms

Isoenzyme type	Effectors	Substrate specificity	Selective inhibitors
PDE1 formerly type I, Ca^{2+}-stimulated PDE	Stimulated by calcium/calmodulin	Differing K_m for cAMP and cGMP depends on the tissue	Vinpocetine
PDE2 formerly type II, cAMP-stimulated PDE	cAMP activity is stimulated by low (μM) [cGMP]	High K_m for both cAMP and cGMP	EHNA
PDE3 formerly type III, cAMP-inhibited PDE	cAMP activity is inhibited by low (μM) [cAMP]	Low K_m for both cAMP and cGMP	Milrinone, cilostamide, amrinone
PDE4 formerly type IV, cAMP-specific PDE	Specific for cAMP	Low K_m for cAMP only	Rolipram, Ro20–1724
PDE5 formerly type V, cGMP-binding cGMP-specific PDE	Specific for cGMP located in the periphery, e.g. heart or lung	Isoforms with high and low K_m for cGMP only	Zaprinast, dipyridamole
PDE6 formerly type V, photoreceptor PDE	Specific for cGMP Located in rods and cones of visual system	Micromolar K_m for cGMP	Dipyridamole, M&B22 948
PDE7 formerly type VII, cAMP-specific rolipram-insensitive PDE	Specific for cAMP	Very low K_m for cAMP only	None reported. Insensitive to rolipram and IBMX

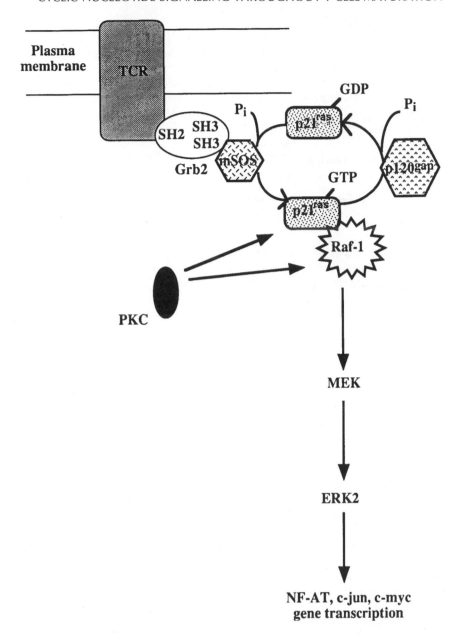

Figure 14.3   Model of p21^{ras}/Raf − 1/MAPK signalling pathway upon ligation of the TCR/CD3 complex

thus preventing PIP$_2$ hydrolysis and the activation of PKC [86, 99, 102, 103]. Moreover, cAMP elevation suppresses Raf-1 activity, which has also been reported to be associated with the TCR/CD3 complex in T cells [112], through causing its phosphorylation by PKA, thus enabling cAMP to regulate negatively the MAP kinase signalling cascade. A

downstream effector of Ras, Raf-1 (a serine/threonine kinase) associates with GTP-bound Ras via its *N*-terminal regulatory domain, causing Raf-1 activation [113, 114]. Raf-1 is responsible for activating MAPKK (MEK), which upon phosphorylation stimulates ERK2 form of MAP kinase in T cells. Activation of Ras–MAP kinase pathway in T cells appears to lead to the activation of the transcription factors NF-AT, c-Jun, and c-Myc and ultimately to the transcriptional activation of the IL-2 gene [115–118]. In addition to activation by Ras, Raf-1 has also been shown to be phosphorylated and activated by PKC in T lymphocytes [98, 108], suggesting the possibility of Ras-independent activation of the MAP kinase cascade in T lymphocytes (Figure 14.3). Recent studies have shown that cAMP, presumably via interaction of PKA and Raf-1 localized at the T cell receptor, appears to modulate the Ras-independent mechanism of Raf-1 activation in T cells [108]. Interestingly, in fibroblasts, cAMP is able to inhibit Raf-1 activation via both routes [119, 120].

As discussed above, ligation of the TCR by antigen may ultimately lead to the production of interleukin-2 (IL-2), and the proliferation of specific T cell clones [121]. In addition to modulating early TCR-coupled signalling events, cAMP can also impact on these later events leading to T cell proliferation by regulating transcription of the IL-2 gene [100, 106, 122–125]. The regulatory region of the IL-2 gene consists of compact, clustered binding sites for a range of transcription factors that are differentially regulated. It is clear that some factors are more critical than others in the regulation of IL-2 gene transcription: for example, AP-1 and NF-AT appear to be rate limiting for IL-2 expression. The precise mechanism(s) of cAMP modulation of IL-2 levels is, at present, not fully understood, but IL-2 promoter activity is found to be reduced by elevations in cAMP [106], and such inhibition can be blocked by the addition of a PKA inhibitor [100, 125]. In addition, PKA activation has been shown to reduce the half-life of IL-2 mRNA transcripts by > 50% [100], block induction of c-*jun* encoding Jun/AP1 transcription factors [103] and modulate DNA binding activities of nuclear factor κB (NF-κB) and a TGGGC binding factor [126]. Cumulatively, these effects abrogate stable protein/DNA contacts over the entire IL-2 enhancer region and are specific for IL-2 production as transcription of the IL-2R, c-*fos*, c-*myc* and c-*myb* genes is unaffected [127–131]. Moreover, analysis of cytokine production from T helper subsets showed that while increases in cAMP appeared to inhibit IL-2 production from T_{h1} cells there was no effect on IL-4 production from T_{h2} cells [86, 132, 133]. Interestingly, T_{h2}

clones not only have a higher cAMP content per cell than T_{h1} cell lines [133] but also, elevation of intracellular cAMP appears to up-regulate T_{h2} cytokine production and hence cAMP levels may perhaps serve to bias the helper response to particular antigens.

Positive Roles of Cyclic Nucleotides in T Cell Activation

It was shown over fifteen years ago that lymphocyte proliferation could not proceed without fluctuations in the levels of cyclic nucleotides [134]: cAMP was shown to be transiently elevated upon stimulation of lymphocytes with mitogenic lectins and treatment of the cells with indomethacin reversibly prevented the increase in cAMP and the onset of S phase. However, sustained elevation of cAMP levels was found to be inhibitory to proliferation [134]. Since then, several studies have demonstrated that elevation of intracellular levels of cAMP/cGMP and calcium may work in concert to promote mitogen-stimulated proliferation of lymphocytes [5, 135, 136]. In addition, suppression of adenylyl cyclase activity with 2,3-dideoxyadenosine, was found to lead to a marked inhibition of the level of lymphocyte proliferation observed in response to stimulation with phorbol ester and ionomycin. Furthermore, this inhibition could be reversed with the addition of micromolar concentrations of dibutyryl cAMP leading to an enhancement of the phorbol ester/calcium ionophore-induced lymphocyte proliferation by some 25–50% [137].

The transient increases in cAMP observed following ligation of the TCR/CD3 complex [138], result in a rapid activation of the PKAI isoform [139], and may be involved in aggregation of receptors as crosslinking of CD3, CD4 and CD8 activates a cAMP-dependent pathway that regulates the mobility and directional polarization of these molecules [140]. Investigation of the role of PKA in T cell activation by treatment of cytotoxic T lymphocytes (CTLs) with antisense mRNA for the catalytic subunit of PKA (Cα), has identified a role for PKA in the regulation of nuclear and/or cytoplasmic events in CTL activation, such as cytokine synthesis and secretion. In addition,

TCR-triggered protein-synthesis-independent responses, such as cytotoxicity and exocytosis, were down-regulated by addition of exogeneous PKA subunits suggesting a dual regulation model of CTL function by cAMP [141].

Effects of cAMP on Thymocyte Differentiation and Cell Death

The elevation of cAMP levels has been implicated both in the differentiation of early thymocytes [142, 143] and in the stimulation of thymic apoptosis during negative selection [89, 144–146]. These findings highlight the potential for differential responses to cAMP signalling at different developmental stages of lymphopoiesis. For example, the Thy-1 antigen appears to play a major role in the transduction of early thymic-differentiation signals, and it has been found that crosslinking of Thy-1 on early thymocytes results in increases of cAMP and receptor mobility [147]. Moreover, Thy-1 expression appears to be under the control of cAMP at $CD4^-/CD8^-$ stage of thymocyte development and such cells are induced to enter accelerated differentiation in response to prostaglandin E2 or cAMP analogues, protecting them from apoptosis [142, 143, 148, 149]. Thus, these results may suggest that cAMP may play a role in the transduction of survival signals during the early stages of thymocyte maturation.

Elevation of cAMP levels following stimulation of $CD4^+CD8^+$ thymocytes with PHA or anti-CD3 antibodies has also been reported [150, 151]. Whereas ligation of thymocytes with anti-CD3 abs caused a slow, sustained increase in cAMP, PHA stimulation of thymocytes only led to a transient elevation of cAMP levels. Furthermore, when IBMX was used to inhibit all PDE activity in the cells, allowing changes in cAMP level to be regarded as an index of adenylyl cyclase activity, the increase in cAMP levels observed upon anti-CD3 or PHA treatment of thymocytes indicated a pronounced activation of adenylyl cyclase activity, suggesting that a complex balance of adenylyl cyclase and PDE activities occurred following TCR activation [151]. The use of PDE isoform-selective inhibitors then indicated that PDE4 activity was likely to play a central role in the regulation of cAMP degradation in such thymocytes as the CD3-coupled elevation in cAMP observed in the presence of the PDE4 inhibitor, rolipram, was similar to that seen in the presence of IBMX. In contrast, addition of the PDE3-selective inhibitor, milrinone, did not potentiate cAMP concentrations, implying that the PDE3 isoform was not involved in the regulation of cAMP levels in murine thymocytes.

The differential cAMP signalling responses resulting from stimulation of thymocytes with PHA and anti-CD3 may be due, at least in part, to the complement of cell surface receptors recruited by these agents. Whereas anti-CD3 acts solely through the TCR/CD3 complex, PHA is known to crosslink multiple cell surface receptors, including the TCR/CD3 complex and CD2 [152], an accessory molecule which is expressed from the early stages of thymic development [153] and thought to trigger differentiation and IL-2-dependent proliferation signals in immature thymocytes via interaction with its complementary molecule, LFA-3, on thymic epithelial cells [153–155]. In addition, PHA and anti-CD3 may be able to trigger differential functional responses in thymocytes: thus, while anti-CD3-stimulated thymocytes apoptose even in the presence of PMA or IL-2 [156], thymocytes can respond mitogenically to costimulation with lectins and PMA [157], suggesting that PHA might prime thymocytes to be rescued from programmed cell death by coligating (at least) CD3 and CD2. McConkey et al. have shown that a sustained elevation of cAMP in thymocytes leads to apoptosis [89]. Thus, the above finding that anti-CD3 induced sustained cAMP generation, while PHA only elicited a transient cAMP signal, might explain why thymocytes undergo apoptosis, rather than proliferation in response to anti-CD3 [158–160]. Further supporting evidence for a role for the sustained cAMP generation in transducing CD3-mediated apoptotic signals is provided by the finding that CD3 is only coupled to transient increases in cAMP levels in mature T cells under conditions in which costimulation with PMA leads to T cell proliferation [138, 140, 161, 162]. Furthermore, although thymocytes have been shown to have lower basal concentrations of cAMP than mature lymphocytes [163], they are more sensitive to stimulation via isoproterenol, leading to more substantial cAMP signals in thymocytes relative to mature T cells [163, 164].

Regulation of cAMP Phosphodiesterases in T Cells

It is has been proposed that cAMP may play a pivotal role in regulating thymocyte maturation and/or cell death [89, 143]. As such, differential regulation of cAMP levels by cAMP PDE isoforms in a maturation-dependent manner may provide a molecular mechanism for the varying biological responses induced by ligation of the TCR on immature and mature T cells. Indeed, there is some evidence which suggests that differential activation of these enzymes may play a central role in regulating cAMP levels during lymphocyte activation and cell cycle progression [165–168].

Our own studies using isoform-selective inhibitors have identified that PDE4 and PDE2 isoforms are the predominant activities contributing to cAMP homeostasis in murine thymocytes and that these PDE isoforms are differentially regulated in response to stimulation with PHA or anti-CD3 antibodies [150]. For example, while PHA mediates a rapid decrease in PDE4 activity (2–3 minutes) followed by a slow recovery towards basal levels (20 minutes), anti-CD3 or anti-TCR mAbs mediate an immediate small and transient decrease in PDE4 activity followed by a slow, but sustained, increase to levels approximately two-fold above basal by 20 minutes. In addition, PHA, but not anti-CD3, caused a dramatic, but transient, loss in PDE2 activity providing further evidence that PHA and CD3 are differentially coupled to cyclic nucleotide metabolism in thymocytes [150].

Pharmacological studies have indicated that anti-CD3-stimulated induction of PDE4 activity appears to be mediated by a PKC-dependent mechanism, as selective inhibitors of PKC block this induction and the phorbol ester, PMA, replicates the elevation of PDE4 activity observed following stimulation with anti-CD3 antibodies (AMM, unpublished data). In addition, elevation of PDE4 activity was also abrogated by pre-incubation with PTK inhibitors, a finding consistent with earlier reports that tyrosine phosphorylation leads to activation of the peripheral plasma membrane PDE species in intact hepatocytes following stimulation of the cells with insulin [77]. At this stage, it is tempting to suggest that early activation of PDE4 may be due at least partially to the PTK-mediated activation of PLC-γ1 and the subsequent activation of PKC [94]. Although no pharmacological studies on the mechanisms involved in mediating the rapid changes in PDE2 and PDE4 activity following stimulation of thymocytes with PHA have been carried out, they are also likely to involve alteration of phosphorylation status of these enzymes. Consistent with this, increased activity correlating with PDE phosphorylation has been noted for particular splice variants of PDE4D [81] and for a PDE4 activity found in hepatocytes [77, 169]. No studies, however, have thus far characterized any phosphorylation of PDE2, although it has been suggested that insulin may lead to transient activation of this activity in hepatocytes [170]. Likewise, dephosphorylation of these isoforms would provide a ready mechanism for the the transient decreases in PDE2 and PDE4 activities observed in PHA-stimulated cells.

Further studies have indicated that at least some of the CD3-mediated increase in PDE4 activity may be due to induction of additional splice variants of PDE4 in an analogous manner to that reported by Sette *et al.* [171] where PKA was shown to mediate the expression of certain splice variants of PDE4D (PDE4D3.1 and PDE4D3.2), while increasing activity of PDE4D3.3 by phosphorylation in response to thyroid-stimulating hormone. This investigation demonstrated that splice variants from the same locus could be differentially regulated on stimulation of the cell by a particular hormone. Similarly, an elevation of cAMP has also been shown to cause to an increase in PDE4D expression in immature Sertoli cells [80, 171, 172], in cardiac myoblasts [173] and in Jurkat cells [83b]. The induction of the PDE4 is extremely fast, indicating that its gene expression, like that of c-*jun* [174] may be controlled by a *cis*-acting TRE to which AP-1 transcription factors bind. This rapid induction bears comparison with work carried out on the induction of PDE1 in CHO cells [175]. The PDE1 induction process occurred within 30 minutes of phorbol ester treatment of the cells, and could also be achieved by overexpression of selected isoforms of PKC (α and) within the cells. Both of these PKC isoforms are found to be present in T cells, and have been implicated in the early events of T cell activation [176, 177].

cGMP and Implications for NO Signalling

The finding that the cGMP-stimulated PDE2 activity plays a major role in maintaining cAMP homeostasis in murine thymocytes suggests that cGMP may act as "molecular switch" during signalling in these cells. At concentrations of cGMP which are too low to stimulate PDE2 activity, the predominant PDE species involved in the hydrolysis of cAMP in thymocytes/lymphocytes would be the PDE4 activity. However, under conditions where cGMP levels become elevated, such as through the stimulation of soluble guanylyl cyclase via NO [20], then a dramatic increase in PDE2 activity may become apparent. Moreover, cGMP has been documented to increase following stimulation of thymocytes and lymphocytes in the absence of adherent cells with PHA, Con A, PMA and calcium ionophore, A23187 [135, 178, 179]. In the case of PHA, this may be due to its ability to activate membrane and soluble forms of lymphocyte guanylyl cyclase [5, 136].

cGMP may be elevated through the stimulation of soluble guanylyl cyclase by NO [38]. NO is a freely diffusible molecule synthesized from the amino acid L-arginine by NO synthase, and can function as an intercellular as well as an intracellular messenger [20, 180, 181]. It appears to have a role in mediating cytotoxicity in the immune system, being responsible for stimulating the phagocytosis of pathogens by cytokine-activated macrophages [182]. However, NO has recently been shown to inhibit apoptosis in B lymphocytes [183], indicating that nitric oxide may also have a role as a "survival" signal in the immune system.

It has been established that sustained elevations of cAMP by agents such as cAMP analogues or E series prostaglandins, as well as glucocorticoid treatment, lead to apoptosis in thymocytes [89, 90, 144, 184–186]. However, these cell death pathways can be linked, as cAMP can potentiate apoptosis caused by glucocorticoids [187]. Interestingly, glucorticoids have been shown to inhibit the induction, but not the activity, of calcium independent NO synthase in response to LPS alone or in combination with IFN-γ *in vitro* [188–190] and *in vivo* [191]. It is tempting to suggest that one of the actions of glucocorticoids may be to stimulate apoptosis by raising cAMP levels by removing cGMP regulation of cAMP hydrolysis.

The connection between cyclic nucleotides and NO signalling pathways is exciting, especially in the context of the murine thymus, where it is apparent from these studies that cAMP hydrolysis may be greatly influenced by the intracellular levels of cGMP.

Cyclic Nucleotide Phosphodiesterases in Lymphoproliferative Disorders

Much of the interest in the roles that specific PDE isoforms may play in the regulation of maturation, proliferation and cell death of lymphocytes has arisen from findings documenting that lymphocytes from patients with lymphoproliferative disorders have aberrant levels of PDE activities relative to normal controls. For example, leukaemic T cells have elevated cAMP and cGMP PDE activities compared to the levels observed in normal T cells [192–195]. Moreover, while cGMP has been shown to play a regulatory role in cAMP homeostasis in normal resting lymphocytes by reducing cAMP PDE activities by 80% at micromolar concentrations [196, 197], this inhibition is diminished in transformed human B and T cell lines and in lymphocytes from patients with lymphocytic leukemia [197]. Consistent with this, Takemoto *et al.* showed that [196] within four hours of mitogenic stimulation, activated human peripheral blood lymphocytes stopped expressing the cGMP-inhibited PDE isoform (PDE3). The resultant elevation of cAMP PDE activity and lower cAMP levels present in these cells may therefore suggest that removal of cGMP regulation could provide a molecular mechanism for the loss of negative growth control in activated and/or proliferating lymphocytes. In contrast, PDE4 activity may be important in controlling the levels of cAMP in activated lymphocytes as the use of the PDE4-specific inhibitor, rolipram, in mitogenically stimulated thymocytes leads to an inhibition of proliferation by 60% [168]. Interestingly, elevation of cGMP via the stimulation of guanylyl cyclase synergized with the rolipram-mediated inhibition of PDE4 activity implying that cGMP may also play a central role in the negative regulation of rat thymocyte proliferation [168].

Sarcoidosis is a chronic granulomatous disorder which is characterized by low proliferative responses of peripheral blood lymphocytes to mitogenic lectins. Interestingly, cAMP and cGMP PDE activities in lymphocytes from such patients are decreased by 26% compared with normal controls [198]. Thus, the associated sustained elevation in cyclic nucleotide levels observed in these patients may go some way to explaining the poor proliferative responses of these cells. Together, these contrasting lymphoproliferative disorders have shown not only how elevated cAMP levels may exert negative effects on lymphocyte proliferation but also how distinct PDE activities may regulate differential lymphocyte responses: although such studies illustrate the deleterious effects that dysfunctions in discrete components of the cyclic nucleotide homeostasis system can have on the immune response, they also highlight the potential for development of strategies for therapeutic intervention with respect to specific PDE activities

FUTURE PROSPECTS

The control of cAMP PDEs appears to play as an important role in regulating the metabolism of cAMP as does regulation of adenylate cyclase [23, 62]. Recently, it has become clear that at least PDE4 subfamilies have the ability to be specifically targeted to specific membranes by virtue of an alternatively spliced *N*-terminal domain [199]. Members of the PDE4A family which have a proline/arginine rich *N*-terminal splice region appear to bind to SH3 domains of *src*-related non-receptor tyrosine kinases such as $p59^{fyn}$, $p56^{lck}$ and $p50^{csk}$[200]. The importance of these kinases is well appreciated in T cell activation which may imply that cAMP PDEs may play a central role in the regulation of T cell signalling. Indeed, PDE4 activity may be regulated by a PTK-mediated pathway (Michie, Erdogan, Harnett and Houslay, unpublished results).

Recent research has rekindled interest in cAMP signalling pathways with the observation that cAMP can inhibit Raf-1 kinase activity in fibroblasts by eliciting its PKA-mediated phosphorylation [119, 201, 202]. Raf-1 is a central player in the activation of the MAP kinase signalling cascade [203] and has recently been found associated with the TCR/CD3 complex [112] as has PKA [48].

PKA is known to mediate stimulatory as well as inhibitory effects on lymphocyte proliferation and appears to be involved in early signalling events of T cell activation [48, 139]. Although the precise role of PKA type I and type II is not known in lymphoid cells, it has been shown that a balance between the two protein kinases is required to maintain normal cell functions [204]. It has been suggested that cAMP-mediated stimulation of PKA-I may release this kinase activity from the TCR/CD3 complex and, through phosphorylation of specific proteins (e.g. Raf-1) uncouple the TCR from intracellular signalling pathways [48]. The finding that PDE4 isoforms may be localized to this area of the cell by their association with *src*-family non-receptor PTKs, may indicate that cAMP PDEs could play a crucial role in T cell proliferation pathways. These findings may provide the basis of a molecular mechanism for the modulatory effects of cAMP in T cell biology.

REFERENCES

1. Sutherland EW and Rall TW. Fractionation and characterisation of a cyclic adenine ribonucleotide formed by tissue particles. *J. Biol. Chem.* 1958; **232**: 1077–1091.
2. Sutherland EW. Studies on the mechanism of hormone action. *Science* 1972; **177**: 401–408.
3. Kammer GM. The adenylate cyclase-cAMP-protein kinase pathway and regulation of the immune response. *Immunol. Today* 1988; **9**: 222–229.
4. Wang T, Sheppard JR and Foker JE. Rise and fall of cyclic AMP required for onset of lymphocyte DNA synthesis. *Science* 1978; **201**: 155–157.
5. Rochette-Egly C and Kempf J. Cyclic nucleotides and calcium in human lymphocytes induced to divide. *J. Physiol. (Paris)* 1981; **77**: 721–725.
6. Sutherland EW, Robinson GA and Butcher RW. Some aspects of the biological role of adenosine 3′, 5′-monophosphate (cyclic AMP). *Circulation* 1968; **37**: 279–306.
7. Goldberg ND and Haddox MK. Cyclic GMP metabolism and involvement in biological regulation. *Ann. Rev. Biochem.* 1977; **46**: 823–896.
8. Taylor SS. cAMP-dependent protein kinase. *J. Biol. Chem.* 1989; **264**(15): 8443–8446.
9. Kuo JF and Greengard P. Cyclic nucleotide-dependent protein kinases, IV. Widespread occurrence of adenosine 3′, 5′ monophosphate-dependent protein kinase in various tissue and phyla of the animal kingdom. *Proc. Natl. Acad. Sci. USA* 1969; **64**: 1349–1355.
10. Edelman AM, Bluminthal DK and Krebs EG. Protein serine/threonine kinases. *Ann. Rev. Biochem.* 1987; **56**: 567–613.
11. Krebs EG and Beavo JA. Phosphorylation–dephosphorylation of enzymes. *Ann. Rev. Biochem.* 1979; **48**: 923–959.

12. Meek DW, Street AJ. Nuclear protein phosphorylation and growth control. *Biochem. J.* 1992; **287**: 1–15.

13. Montminy MR, Sevarino KA, Wagner JA, Mandel G and Goodman RH. Identification of a cyclic-AMP-responsive element within the rat somatostain gene. *Proc. Natl. Acad. Sci. USA* 1986; **83**: 6682–6686.

14. Comb M, Birnberg NC, Seasholtz A, Herbert E and Goodman HM. A cyclic AMP-and phorbol ester-inducible DNA element. *Nature* 1986; **323**: 353–356.

15. Lincoln TM, Thompson M and Cornwell TL. Purification and characterisation of two forms of cyclic GMP-dependent protein kinase from bovine aorta. *J. Biol. Chem.* 1988; **263**: 17632–17637.

16. Francis SH, Noblett BD, Todd BW, Wells JN and Corbin JD. Relaxation of vascular and tracheal smooth muscle by cyclic nuleotide analogs that preferentially activate purified cGMP-dependent protein kinase. *Mol. Pharmacol.* 1988; **504**: 506–517.

17. Charbonneau H, Beier N, Walsh KA and Beavo JA. Identification of a conserved domain among cyclic nucleotide phosphodiesterases from diverse species. *Proc. Natl. Acad. Sci. USA* 1986; **83**: 9308–9312.

18. Lincoln TM and Cornwell TL. Intracellular cyclic GMP receptor proteins. *FASEB J.* 1993; **7**: 328–338.

19. Kaupp UB, Niidome T, Tanabe T *et al.* Primary structure and functional expression form complementary DNA of the rod photoreceptor ctclic GMP-gated channel. *Nature* 1989; **342**: 762–766.

20. Moncada S, Palmer RMJ and Higgins EA. Nitric oxide: physiology, pathophysiology, and pharmacology. *Pharmacol. Rev.* 1991; **43**: 109–142.

21. Taussig R and Gilman AG. Mammalian membrane-bound adenylyl cyclases. *J. Biol. Chem.* 1995; **270**(1): 1–4.

22. Cooper DMF, Mons N and Fagan K. Ca2+-sensitive adenylyl cyclases. *Cell. Signal.* 1994; **6**: 823–840.

23. Conti M, Nemoz G, Sette C and Vincini E. Recent progress in understanding the hormonal regulation of phosphodiesterases. *Endocrine Rev.* 1995; **16**(3): 370–389.

24. Beavo JA, Conti M and Heaslip RJ. Multiple cyclic nucleotide phosphodiesterases. *Mol. Pharmacol.* 1994; **46**: 399–405.

25. Gilman AG. G proteins: Transducers of receptor-generated signals. *Ann. Rev. Biochem.* 1987; **56**: 615–649.

26. Casey PJ and Gilman AG. G protein involvement in receptor–effector coupling. *J. Biol. Chem.* 1988; **263**: 2577–2580.

27. Krupinski J, Coussen F, Bakalyar HA *et al.* Adenylyl cyclase amino acid sequence: possible channel or transporter like structure. *Science* 1989; **244**: 1558–1564.

28. Feinstein PG, Schrader KA, Bakalyar HA *et al.* Molecular cloning and characterisation of a Ca2+/calmodulin insensitive adenylate cyclase from rat brain. *Proc. Natl. Acad. Sci. USA* 1991; **88**: 10173–10177.

29. Bakalyar HA and Reed RR. Identification of a specialised adenylyl cyclase that may mediate odorant detection. *Science* 1990; **250**: 1403–1406.

30. Premont RT, Chen JQ, Ma HW, Ponnapalli M and Ivengar R. 2 members of a widely expressed subfamily of hormone-stimulated adenylyl cyclases. *Proc. Natl. Acad. Sci. USA* 1992; **89**: 9809–9813.

31. Cali JJ, Zwaagstra JC, Mons N, Cooper DMF and Krupinski J. Type VIII adenylyl cyclase. *J. Biol. Chem.* 1994; **269**(16): 12190–12195.

32. Wallach J, Droste M, Kluxen FW, Pfeuffer T and Frank R. Molecular cloning and expression of a novel type V adenylyl cyclase from rabbit myocardium. *FEBS Letts.* 1994; **338**: 257–263.

33. Krupinski J, Lehman TC, Frankenfield CD, Zwaagstra JC and Watson PA. Molecular diversity of the adenylyl cyclase family. *J. Biol. Chem.* 1992; **267**: 24858–24862.

34. Pieroni JP, Jacobwitz O, Chen J and Iyengar R. Signal recognition and integration by Gs-stimulated adenylyl cyclases. *Curr. Opin. Neurobiol.* 1993; **3**: 345–351.

35. Seamon KB, Padgett W and Daly JW. Forskolin: unique diterpene activator of adenylate cyclase in membranes and in intact cells. *Proc. Natl. Acad. Sci. USA* 1981; **78**(6): 3363–3367.

36. Rapoport RM and Murad F. Agonist induced endothelium-derived relaxation in rat thoracic aorta may be mediated through cyclic GMP. *Circ. Res.* 1983; **52**: 352–357.

37. Mellion BT, Ignarro LJ, Ohlstein EH, Pontecorvo EG, Hyman AL and Kadowitz PJ. Evidence for the inhibitory role of guanosine 3′ 5′-monophosphate in ADP-induced human platelet aggregation in the presence of nitric oxide and related vasodilators. *Blood* 1981; **57**: 946–955.

38. Furchgott RF and Zawadski JV. The obligatory role of endothelial cells in the relaxation of arterial smooth muscle by acetylcholine. *Nature* 1980; **288**: 373–376.

39. Kamisaki Y, Saheki S, Nakane M *et al.* Soluble guanylate cyclase from rat lung exists as a heterodimer. *J. Biol. Chem.* 1986; **261**: 7236–7241.

40. Buechler WA, Nakane M and Murad F. Expression of guanylate cyclase activity requires both enzyme subunits. *Biochem. Biophys. Res. Commun.* 1991; **174**: 351–357.

41. Nakane M, Arai K, Saheki S, Kuno T, Buechler W and Murad F. Molecular cloning and expression of cDNAs coding for soluble guanylate cyclase from rat lung. *J. Biol. Chem.* 1990; **265**: 16841–16845.

42. Garbers DL. Guanylyl cyclase receptors and their ligands. *Adv. Second Messenger Phosphoprotein Res.* 1993; **28**: 91–95.

43. Lowe DG, Chang MS, Hellmiss R *et al.* Human atrial natriuretic peptide receptor defines a new paradigm for second messenger signal transduction. *EMBO J.* 1989; **8**: 1377–1384.

44. Schulz S, Singh S, Bellet RA *et al.* The primary structure of a plasma membrane guanylate cyclase demonstrates diversity within this new receptor family. *Cell* 1989; **58**: 1155–1162.

45. Chinkers M, Garbers DL, Chang M-S, Lowe DG, Goeddel DV and Schulz S. Molecular cloning of a new type of cell surface receptor: a membrane form of guanylate cyclase is an atrial netriuretic peptide receptor. *Nature* 1989; **338**: 78–83.

46. Corbin JD, Sugden PH, Lincoln TM and Keely SL. Compartmentalisation of adenosine 3′ : 5′-monophosphate and adenosine 3′ : 5′-monophosphate-dependent protein kinase in heart tissue. *J. Biol. Chem.* 1977; **252**: 3854–3861.

47. Rubin CS, Erlichman J and Rosen OR. Cyclic adenosine 3′, 5′-monophosphate-dependent protein kinase of human

erythrocyte membranes. *J. Biol. Chem.* 1972; **247**: 6135–6139.

48. Skalhegg BS, Tasken K, Hansson V, Huitfeldt HS, Jahnsen T and Lea T. Location of cAMP-dependent protein kinase type I with the TCR-CD3 complex. *Science* 1994; **263**: 84–87.

49. Salavatori S, Damiani E, Barhanin J, Furlan S, Giovanni S and Margreth A. Co-localisation of the dihydropyridine receptor and the cyclic AMP-binding subunit of an intrinsic protein kinase to the junctional membrane of the transverse tubules of skeletal muscle. *Biochem. J.* 1990; **267**: 679–687.

50. Nigg EA, Schafer G, Hilz H and Eppenberger H. Cyclic AMP-dependent protein kinase type-II is associated with the Golgi complex and with centrosomes. *Cell* 1985; **41**: 1039–1051.

51. Meinkoth JL, Ji Y, Taylor SS and Feramisco JR. Dynamics of the distribution of cyclic AMP-dependent protein kinase in living cells. *Proc. Natl. Acad. Sci. USA* 1990; **87**: 9595–9599.

52. Lincoln TM and Keely SL. Regulation of cardiac cGMP-dependent protein kinase. *Biochim. Biophys. Acta* 1980; **676**: 230–244.

53. Lohmann SM, Walter U, Miller PE, Greengard P and De-Cammilli P. Immunohistochemical localization of cyclic GMP-dependent protein kinase in mammalian brain. *Proc. Natl. Acad. Sci. USA* 1981; **78**: 653–657.

54. deJonge HR. Cyclic GMP-dependent protein kinase in intestinal brush borders. *Adv. Cycl. Nucleot. Res.* 1981; **14**: 315–323.

55. Francis SH, Woodford TA, Wolfe L and Corbin JD. Types Ia and Ib isozymes of cGMP-dependent protein kinase: alternative mRNA splicing may produce different inhibitory domains. *Second Mess. Phosphoproteins* 1989; **12**: 301–310.

56. Landgraf WRP, Keilbach A, May B, Welling A and Hofmann F. Cyclic GMP-dependent protein kinase and smooth muscle relaxation. *J. Cardiovasc. Pharm.* 1992; **20**(S1): S18–S22.

57. Lincoln TM and Corbin JD. Adenosine 3′: 5′;-cyclic monophosphate and guanosine 3′:5′-cyclic monophosphate-dependent protein kinases: possible homologous proteins. *Proc. Natl. Acad. Sci. USA* 1977; **74**: 3239–3244.

58. Colbran JL, Francis SH, Leach AB *et al.* A phenylalanine in peptide substrates provides for selectivity between cGMP-dependent and cAMP-dependent protein kinases. *J. Biol. Chem.* 1992; **267**: 9589–9594.

59. Drummond GI and Perrot-Yee S. Enzymatic hydrolysis of adenosine 3′, 5′ phosphoric acid. *J. Biol. Chem.* 1961; **236**: 1126–1129.

60. Butcher RW and Sutherland EW. Adenosine 3′5′-phosphate in biological materials. *J. Biol. Chem.* 1962; **237**: 1244–1250.

61. Bolger G. Molecular biology of the cAMP-specific cyclic nucleotide phosphodiesterases. *Cell. Signal.* 1994; **6**: 851–859.

62. Houslay MD and Kilgour E. Cyclic nucleotide phosphodiesterases in liver—a review of their characterization, regulation by insulin and glucagon and their role in controlling cyclic AMP concentrations. In: Beavo JA and Houslay MD (eds) *Molecular Pharmacology of Cell Regulation*, Vol. 2. Chichester: John Wiley, 1990: 185–226.

63. Reeves ML and England PJ. Cardiac phosphodiesterases and the functional effects of selective inhibitors. In: Beavo JA and Houslay MD (eds) *Molecular Pharmacology of Cell Regulation*, Vol. 2. Chichester: John Wiley, 1990: 299–316.

64. Hidaka H, Hayashi H, Kohri H *et al.* Selective inhibitors of platelet cyclic adenosine monophosphate phosphodiesterase, cilostamide, inhibits platelet aggregation. *J. Pharm. Exp. Ther.* 1979; **211**: 26–30.

65. Simpson AWM, Reeves ML and Rink TJ. Effects of SK&F94120, an inhibitor of cyclic nucleotide phosphodiesterase type-III, on human platelets. *Biochem. Pharmacol.* 1988; **37**: 2315–2320.

66. Kauffman RF, Schenck KW, Utterbeck BG, Crowe VG and Cohen ML. In vitro vascular relaxation by new inotropic agents: relationship to phosphodiesterase inhibition and cyclic nucleotides. *J. Pharmacol. Exp. Ther.* 1987; **242**: 864–872.

67. Tanaka T, Ishikawa T, Hagiwara M, Onoda K, Itoh K and Hidaka H. Effects of cilostazol, a selective cAMP phosphodiesterase inhibitor on the contraction of vascular smooth muscle. *Pharmacology* 1988; **36**: 313–320.

68. Farah AE, Alousi AA and Schwarz RP. Positive inotropic agents. *Annl. Rev. Pharmacol. Toxicol.* 1984; **24**: 275–328.

69. Weishaar RE, Kobylarz-Singer DC, Steffen RP and Kaplan HR. Subclasses of cyclic AMP-specific phosphodiesterase in left ventricular muscle and their involvement in regulating myocardial contractility. *Circ. Res.* 1987; **61**: 529–537.

70. Torphy TJ, Livi GP, Balcarek JM, White JR, Chilton FH and Undem BJ. Therapeutic potential of isozyme-selective phosphodiesterase inhibitors in the treatment of asthma. *Adv. Second Messenger Phosphoprotein Res.* 1992; **25**: 289–305.

71. Plaut M, Marone G, Thomas LL and Lichtenstein LM. Review: cyclic nucleotides in immune responses and allergy. In: Sands PHH (ed.) *Advances in Cyclic Nucleotide Research*, Vol. 12. New York: Raven Press, 1980; 161–172.

72. Beltman J, Sonnenberg WK and Beavo JA. The role of protein phosphorylation in the regulation of cyclic nucleotide phosphodiesterases. *Mol. Cell. Biol.* 1993; **127/128**: 239–253.

73. Swinnen JV, Joseph DR and Conti M. The mRNA encoding a high affinity cAMP phosphodiesterase is regulated by hormones and cAMP. *Proc. Natl. Acad. Sci. USA* 1989; **86**: 8197–8201.

74. Manganiello VC, Smith CJ, Degerman E and Belfrage P. Cyclic GMP-inhibited cyclic nucleotide phosphodiesterases. *Mol. Pharmacol. Cell Reg.* 1990; **2**: 87–116.

75. Degerman E, Smith CJ, Tornqvist H, Vasta V, Belfrage P and Manganiello VC. Evidence that insulin and isoprenaline activate the cyclic GMP-inhibited low Km cyclic AMP phosphodiesterase in rat fat cells by phosphorylation. *Proc. Natl. Acad. Sci. USA* 1990; **87**: 533–537.

76. Kilgour E, Anderson NG and Houslay MD. Activation and phosphorylation of the 'dense-vesicle' high-affinity cyclic AMP phosphodiesterase by cyclic AMP-dependent protein kinase. *Biochem. J.* 1989; **260**: 27–39.

77. Pyne NJ, Cushley W, Nimmo HG and Houslay MD. Insulin stimulates the tyrosyl phosphorylation and activation of the 52 kDa peripheral plasma-membrane cyclic

AMP phosphodiesterase in intact hepatocytes. *Biochem. J.* 1989; **261**(3): 897–901.

78. Hall IP and Hill SJ. Effects of isozyme selective phosphodiesterase inhibitors on bovine tracheal smooth muscle tone. *Biochem. Pharmacol.* 1992; **43**(1): 15–17.

79. Moore JB, Combs DW and Tobia AJ. Bemoradan—a novel inhibitor of rolipram-insensitive cyclic AMP phosphodiesterase from canine heart tissue. *Biochem. Pharmacol.* 1991; **42**: 679–683.

80. Swinnen JV, Tsikalas KE and Conti M. Properties and hormonal regulation of two structurally related cAMP phosphodiesterases from rat sertoli cell. *J. Biol. Chem.* 1991; **266**(27): 18370–18377.

81. Sette C, Iona S and Conti M. The short-term activation of a rolipram-sensitive, cAMP-specific phosphodiesterase by thyroid-stimulating hormone in thyroid FRTL-5 cells is mediated by a cAMP-dependent phosphorylation. *J. Biol. Chem.* 1994; **269**: 9245–9252.

82. Lobban M, Shakur Y, Beattie J and Houslay MD. Identification of two splice variant forms of type IVB cyclic AMP phosphodiesterase, DPD (rPDE-IVB1 and PDE-4(rPDE-IVB2 in brain: selective localization in membrane and cytosolic compartments and differential expression in various brain regions. *Biochem. J.* 1994; **304**: 399–406.

83a. Engels P, Fichtel K and Lubbert H. Expression and regulation of human and rat phosphodiesterase type IV isogene. *FEBS Letts* 1994; **350**: 291–295.

83b. Erdogan S, and Houslay MD. Challenge of human Jurkat T cells with the adenylate cyclase activator forskolin elicits major changes in cAMP PDE expression by upregulating PDE3 and inducing PDE4D1 and PDE4D2 splice variants as well as downregulating a novel PDE4A splice variant. *Biochem. J.* 1997 (in press).

84. Engels P, Abdel' Al S, Hulley P and Lubbert H. Brain distribution of four rat homologues of the *Drosophila* dunce phosphodiesterase. *J. Neurosci. Res.* 1995; **41**: 169–178.

85. Shakur Y, Pryde JG and Houslay MD. Engineered deletion of the unique *N*-terminal domain of the cyclic AMP-specific phosphodiesterase RD1 prevents plasma membrane association and the attainment of enhanced thermostability without altering its sensitivity to inhibition by rolipram. *Biochem. J.* 1993; **292**: 677–686.

86. Alava MA, Debell KE, Conti A, Hoffman T and Bonvini E. Increased intracellular cyclic AMP inhibits inositol phospholipid hydrolysis induced by perturbation of the T-cell receptor CD3 complex but not by G-protein stimulation—association with protein kinase-A-mediated phosphorylation of phospholipase C-γ-1. *Biochem. J.* 1992; **284**: 189–199.

87. O'Shea JJ, Urdahl KB, Luong HT, Chused TM, Samelson LE and Klausner RD. Aluminium fluoride induces phosphatidylinositol turnover, elevation of cytoplasmic free calcium, and phosphorylation of T cell antigen receptor in murine T cells. *J. Immunol.* 1987; **139**(10): 3463–3469.

88. Lerner A, Jacobson B and Miller RA. Cyclic AMP concentrations modulate both calcium flux and hydrolysis of phosphatidylinositol phosphates in mouse T lymphocytes. *J. Immunol.* 1988; **140**(3): 936–940.

89. McConkey DJ, Orrenius S and Jondal M. Agents that elevate cAMP stimulate DNA fragmentation in thymocytes. *J. Immunol.* 1990; **145**(4): 1227–1230.

90. Dowd DR and Miesfeld RL. Evidence that glucocorticoid-induced and cyclic AMP-induced apoptotic pathways in lymphocytes share distal events. *Mol. Cell. Biol.* 1992; **12**: 3600–3608.

91. Nel AE, Ledbetter JA, Ho P, Akerley B, Franklin K and Katz R. Activation of MAP-2 kinase activity by the CD2 receptor in Jurkat T cells can be reversed by CD45 phosphatase. *Immunology* 1991; **73**: 129–133.

92. Satoh T, Nafakuku M and Kaziro Y. Function of ras as a molecular switch in signal transduction. *J. Biol. Chem.* 1992; **267**: 24149–24152.

93. Weiss A and Littman DR. Signal transduction by lymphocyte antigen receptors. *Cell* 1994; **76**: 263–274.

94. Secrist JP, Karnitz L and Abraham RT. T-cell antigen receptor ligation induces tyrosine phosphorylation of phospholipase C-γ1. *J. Biol. Chem.* 1991; **266**(19): 12135–12139.

95. June CH, Fletcher MC, Ledbetter JA and Samelson LE. Increases in tyrosine phosphorylation are detectable before phospholiase C activation after T cell receptor stimulation. *J. Immunol.* 1990; **144**(5): 1591–1599.

96. Rudd CE, Janssen O, Cai Y-C, da Silva AJ, Raab M and Prasad KVS. Two step TCRζ/CD3-CD4 and CD28 signalling in T cells: SH2/SH3 domains. protein tyrosine and lipid kinases. *Immunol. Today* 1994; **15**(5): 225–234.

97. Downward J, Graves J and Cantrell D. The regulation and function of p21^{ras} in T cells. *Immunol. Today* 1992; **13**: 89–92.

98. Siegel JN, Klausner RD, Rapp UR and Samelson LE. T cell antigen receptor engagement stimulates c-*raf* phosphorylation and induces c-*raf*-associated kinase activity via a protein kinase C-dependent pathway. *J. Biol. Chem.* 1990; **265**(30): 18472–18480.

99. Park DJ, Min HK and Rhee SG. Inhibition of CD3-linked phospholipase C by phorbol ester and by cAMP is associated with decreased phosphotyrosine and increased phosphoserine contents of PLC-γ1. *J. Biol. Chem.* 1992; **267**(3): 1496–1501.

100. Anastassiou ED, Paliogianni F, Balow JP, Yamada H and Boumpas DT. Prostaglandin-E2 and other cyclic AMP-elevating agents modulate IL-2 and IL-2R-alpha gene-expression at multiple levels. *J. Immunol.* 1992; **148**: 2845–2852.

101. Klausner RD, O'Shea JJ, Luong H, Ross P, Bluestone JA and Samelson LE. T cell receptor tyrosine phosphorylation. *J. Biol. Chem.* 1987; **262**(26): 12654–12659.

102. Takayama H, Trenn G and Sitkovsky MV. Locus of inhibitory action of cAMP-dependent protein kinase in the antigen receptor-triggered cytotoxic T lymphocyte activation pathway. *J. Biol. Chem.* 1988; **263**(5): 2330–2336.

103. Tamir A and Isakov N. Increased intracellular cyclic AMP levels block PKC-mediated T cell activation by inhibition of c-jun transcription. *Immunol. Letts.* 1991; **27**: 95–100.

104. Chouaib S, Robb RJ, Welte K and Dupont B. Analysis of prostaglandin E2 effect on T lymphocyte activation. *J. Clin. Invest.* 1987; **80**: 339–346.

105. Rothman BL, Kennure N, Kelley KA, Katz M and Aune TM. Elevation of intracellular cAMP in human T

lymphocytes by an anti-CD44 mAb. *J. Immunol.* 1993; **151**(11): 6036–6042.

106. Paliogianni F, Kincaid RL and Boumpas DT. Prostaglandin E2 and other cyclic AMP elevating agents inhibit interleukin 2 gene transcription by counteracting calcineurin-dependent pathways. *J. Exp. Med.* 1993; **178**: 1813–1817.

107. Gray LS, Gnarra J, Hewlett EL and Engelhard VH. Increased intracellular cyclic adenosine monophosphate inhibits T lymphocyte-mediated cytolysis by two distinct mechanisms. *J. Exp. Med.* 1988; **167**: 1963–1968.

108. Whitehurst CE, Owaki H, Bruder JT, Rapp UR and Geppert TD. The MEK kinase activity of the catalytic domain of Raf-1 is regulated independently of ras binding in T cells. *J. Biol. Chem.* 1995; **270**(10): 5594–5599.

109. Payet MD and Dupuis G. Dual regulation of the n type K+ channel in Jurkat T lymphocytes by protein kinases A and C. *J. Biol. Chem.* 1992; **267**(26): 18270–18273.

110. Bastin B, Payet MD and Dupuis G. Effects of modulators of adenylyl cyclase on interleukin 2 production, cytosolic calcium elevation and K+ channel activity in Jurkat T cells. *Cell. Immunol.* 1990; **128**: 385–399.

111. Skalhegg BS, Landmark BF, Doskeland SO, Hansson V, Lea T and Jahnsen T. Cyclic AMP dependent protein kinase Type I mediates the inhibitory effects of cyclic AMP on cell replication in human T lymphocytes. *J. Biol. Chem.* 1992; **267**(22): 15707–15714.

112. Loh C, Romeo C, Seed B, Bruder JT, Rapp U and Rao A. Association of raf with the CD3 δ and γ chains of the T cell receptor-CD3 complex. *J. Biol. Chem.* 1994; **269**(12): 8817–8825.

113. Hallberg B, Rayter S and Downward J. Interaction of ras and raf in intact mammalian cells upon extracellular stimulation. *J. Biol. Chem.* 1994; **269**: 3913–3916.

114. Avruch J, Zhang X-F and Kyriakis JM. Raf meets ras: completing the framework of a signal transduction pathway. *TIBS* 1994; **19**: 279–283.

115. Woodrow M, Rayter S, Downward J and Cantrell DA. p21ras function is important for T cell antigen receptor and protein kinase C regulation of nuclear factor of activated cells. *J. Immunol.* 1993; **150**: 1.

116. Northrop JP, Ullman KS and Crabtree GR. Characterisation of the nuclear and cytoplasmic components of the lymphoid specific nuclear factor of activated T cells (NFAT) complex. *J. Biol. Chem.* 1993; **268**: 22917.

117. Pulverer BJ, Kyriakis JM, Avruch J, Nikolakaki E and Woodgett JR. Phosphorylation of c-jun mediated by MAP kinases. *Nature* 1991; **353**: 670–674.

118. Seth A, Gonzalez FA, Gupta S, Raden DL and Davis RJ. Signal transduction within the nucleus by mitogen activated protein kinase. *J. Biol. Chem.* 1992; **34**: 24796.

119. Cook SJ and McCormick F. Inhibition by cAMP of ras-dependent activation of raf. *Science* 1993; **262**: 1069–1072.

120. Lange Carter CA and Johnson GL. Ras-dependent growth factor regulation of MEK kinase in PC12 cells. *Science* 1994; **265**: 1458–1461.

121. Altman A, Mustelin T and Coggeshall KM. T lymphocyte activation: a biological model of signal transduction. *Crit. Revs. Immunol.* 1990; **10**(4): 347–391.

122. Aussel C, Mary D, Peyron J-F, Pelassy C, Ferrua B and Fehlmann M. Inhibition and activation of interleukin 2 synthesis by direct modification of guanosine triphosphate-binding proteins. *J. Immunol.* 1988; **140**(1): 215–220.

123. Novogrodsky A, Patya M, Rubin AL and Stenzel KH. Agents that increase cellular cAMP inhibit production of interleukin-2, but not its activity. *Biochem. Biophys. Res. Commun.* 1983; **114**(1): 93–98.

124. Mary D, Aussel C, Ferrua B and Fehlmann M. Regulation of interleukin 2 synthesis by cAMP in human T cells. *J. Immunol.* 1987; **139**(4): 1179–1184.

125. Averill LE, Stein RL and Kammer GM. Control of human T lymphocyte interleukin 2 production by a cAMP-dependent pathway. *Cell. Immunol.* 1988; **115**: 88–99.

126. Chen D and Rothenberg EV. Interleukin 2 transcription factors as molecular targets of cAMP inhibition: delayed inhibition kinetics and combinatorial transcription roles. *J. Exp. Med.* 1994; **179**: 931–942.

127. Chouaib S, Welte K, Mertelsmann R and Dupont B. Prostaglandin E2 acts at two distinct pathways of T lymphocyte: inhibition of interleukin 2 production and down-regulation of transferrin receptor expression. *J. Immunol.* 1985; **135**(2): 1172–1179.

128. Scholz W and Altman A. Synergistic induction of interleukin 2 receptor (Tac) expression on YT cells by interleukin 1 of tumor necrosis factor in combination with cAMP inducing agents. *Cell. Signal.* 1989; **1**(4): 367–375.

129. Farrar WL, Evans SW, Rapp UR and Cleveland JL. Effects of anti-proliferative cyclic AMP on interleukin 2-stimulated gene expression. *J. Immunol.* 1987; **139**(6): 2075–2080.

130. Natumiya S, Hirata M, Nanba T *et al.* Activation of IL2 receptor gene by forskolin and cyclic AMP analogues. *Biochem. Biophys. Res. Commun.* 1987; **143**(2): 753–760.

131. Ramarli D, Fox DA and Reinherz EL. Selective inhibition of interleukin 2 gene function following thymocyte antigen/major histocompatibility complex receptor crosslinking: possible thymic selection mechanism. *Proc. Natl. Acad. Sci. USA* 1987; **84**: 8598–8602.

132. Galjewski TF, Schell SR and Fitch FW. Evidence implicating utilisation of different T cell receptor-associated signalling pathways by Th1 and Th2 clones. *J. Immunol.* 1990; **144**(11): 4110–4120.

133. Novak TJ and Rothenburg EV. Cyclic AMP inhibits induction of interleukin 2 but not interleukin 4 in T cells. *Proc. Natl. Acad. Sci. USA* 1990; **87**: 9353–9357.

134. Wang JL, McClain DA and Edelman GM. Modulation of lymphocyte mitogenesis. *Proc. Natl. Acad. Sci. USA* 1975; **72**(5): 1917–1921.

135. Coffey RG, Hadden EM and Hadden JW. Evidence for cyclic GMP and calcium mediation of lymphocyte activation by mitogens. *J. Immunol.* 1977; **119**(4): 1387–1394.

136. Coffey RG, Hadden EM and Hadden JW. Phytohemagglutinin stimulation of guanylate cyclase in human lymphocytes. *J. Biol. Chem.* 1981; **256**(9): 4418–4424.

137. Koh WS, Yang KH and Kaminski NE. Cyclic-AMP is an essential factor in immune-responses. *Biochem. Biophys. Res. Commun.* 1995; **206**: 703–709.

138. Ledbetter JA, Parsons M, Martin PJ, Hansen JA, Rabinovitch PS and June CH. Antibody binding to CD5 (Tp67) and Tp44 T cell surface molecules: effects in cyclic nucleotides, cytoplasmic free calcium, and cAMP mediated suppression. *J. Immunol.* 1986; **137**: 3299–3305.

139. Laxminarayana D, Berrada A and Kammer GM. Early events of human T lymphocyte activation are associated with Type I protein kinase A activity. *J. Clin. Invest.* 1993; **92**: 2207–2214.

140. Kammer GM, Bochm CA, Rudolph SA and Schultz LA. Mobility of the human T lymphocyte surface molecules CD3, CD4, and cD8: regulation by a cAMP-dependent pathway. *Proc. Natl. Acad. Sci. USA* 1988; **85**: 792–796.

141. Sugiyama H, Chen P, Hunter M, Taffs R and Sitkovsky M. The dual role of the cAMP-dependent protein kinase Ca subunit in T-cell receptor-triggered T-lymphocytes effector functions. *J. Biol. Chem.* 1992; **267**(35): 25256–25263.

142. Scheid MP, Goldstein G, Hammerling U and Boyse EA. Lymphocytes differentiation from precursor cells *in vitro*. *Ann. NY Acad. Sci.* 1975; **249**: 531–540.

143. Bach M-A, Fournier C and Bach J-F. Regulation of θ-antigen expression by agents altering cyclic AMP level and by thymic factor. *Ann. NY Acad. Sci.* 1975; 249: 316–327.

144. Kaye J and Ellenberger DL. Differentiation of an immature T cell line: A model of thymic positive selection. *Cell* 1992; **71**: 423–435.

145. Swat W, Ignatowicz L, Von Boehmer H and Kisielow P. Clonal deletion of immature CD4+CD8+ thymocytes in suspension culture by extrathmic antigen-presenting cells. *Nature* 1991; **351**: 150–153.

146. Murphy KM, Heimberger AB and Loh DY. Induction by antigen of intrathymic apoptosis of CD4+CD8+ TCRlo thymocytes *in vivo*. *Science* 1990; **250**: 1720.

147. Butman BT, Jacobsen T, Cabatu OG and Bourguignon LYW. The involvement of cAMP in lymphocyte capping. *Cell. Immunol.* 1981; **61**: 397–403.

148. Singh U and Owen JJT. Studies on the effect of various agents on the maturation of thymus stem cells. *Eur. J. Immunol.* 1975; **5**: 286–288.

149. Goetzl EJ, An SZ and Zeng L. Specific suppression by prostaglandin-E(2) of activation-induced apoptosis of human CD4(+)CD8(+) T-lymphoblasts. *J. Immunol.* 1995; **154**: 1041–1047.

150. Michie AM, Lobban MD, Muller T, Harnett MM and Houslay MD. Rapid regulation of PDE-2 and PDE-4 cyclic AMP phosphodiesterase activity following ligation of the T cell antigen receptor on thymocytes: analysis using the selective inhibitors erythro-9-(2-hydroxy-3-nonyl)-adenine (EHNA) and rolipram. *Cell. Signal.* 1996: **8**: 97–110.

151. Michie AM. The differential regulation of cyclic AMP phospphodiesterases in T lymphocytes. PhD Thesis 1996: University of Glasgow.

152. O'Flynn K, Krensky AM, Beverley PCL, Burakoff SJ and Linch DC. Phytohaemagglutinin activation of T cells through the sheep red blood cell receptor. *Nature* 1985; **313**: 686–687.

153. Fox DA, Hussey RE, Fitzgerald KA *et al.* Activation of human thymocytes via the 50 kDa T11 sheep erythrocyte binding protein induces the expression of interleukin 2 receptors on both T3+ and T3-populations. *J. Immunol.* 1985; **134**(1): 330–335.

154. Denning SM, Tuck DT, Vollager LW, Springer TA, Singer KH and Haynes BF. Monoclonal antibodies to CD2 and lymphocyte function-associated antigen 3 inhibit human thymic epithelial cell-dependent mature thymocyte activation. *J. Immunol.* 1987; **139**(8): 2573–2578.

155. Reem GH, Carding S and Reinherz EL. Lymphokine synthesis is induced in human thymocytes by activation of the CD2 (T11) pathway. *J. Immunol.* 1987; **139**(1): 130–134.

156. Sancho J, Silverman LB, Castigli E *et al.* Developmental regulation of transmembrane signalling via the T cell antigen receptor/CD3 complex in human T lymphocytes. *J. Immunol.* 1992; **148**(5): 1315–1321.

157. Taylor MV, Metcalf JC, Hesketh TR, Smith GA and Moore JP. Mitogens increase phosphorylation of phosphoinositides in thymocytes. *Nature* 1984; **312**: 462–464.

158. Smith CA, Williams GT, Kingston R, Jenkinson EJ and Owen JJT. Antibodies to CD3/TcR complex induce death by apoptosis in immature T cells in thymic cultures. *Nature* 1989; **337**: 181–184.

159. Jondal M, Okret S and McConkey D. Killing of immature CD4+CD8+ thymocytes in vivo by anti-CD3 or 5'-(N-ethyl)-carboxamido-adenosine is blocked by glucocorticoid receptor antagonist Ru-486. *Eur. J. Immunol.* 1993; **23**: 1246–1250.

160. Shi YF, Bissonnette RP, Parfrey N, Szalay M, Kubo RT and Green DR. In vivo administration of monoclonal antibodies to the CD3 T cell receptor complex induces cell death (apoptosis) in immature thymocytes. *J. Immunol.* 1991; **146**: 3340–3346.

161. Carrera AC, Rincon M, De Landazuri MO and Lopez-Botet M. CD2 is involved in regulating cyclic AMP levels in T cells. *Eur. J. Immunol.* 1988; **18**: 961–964.

162. Hahn WC, Rosenstein Y, Burakoff SJ and Bierer BE. Interaction of CD2 with its ligand lymphocyte function-associated antigen-3 induces adenosine 3'5'-cyclic monophospate production in T lymphocytes. *J. Immunol.* 1991; **147**(1): 14–21.

163. Niaudet P, Beaurain G and Bach M-A. Differences in effect of isoprenaline stimulation on levels of cyclic AMP in human B and T lymphocytes. *Eur. J. Immunol.* 1976; **6**: 834–836.

164. Bach M-A. Differences in cyclic AMP changes after stimulation by prostagladins and isopreterenol in lymphocyte subpopulations. *J. Clin. Invest.* 1975; **55**: 1074–1081.

165. Meskini N, Hosni M, Nemoz G, Lagarde M and Pringent A-F. Early increase in lymphocyte cyclic nucleotide phosphodiesterase activity upon mitogenic activation of human peripheral blood mononuclear cells. *J. Cell. Physiol.* 1992; **150**: 140–148.

166. Epstein PM, Mills JS, Hersh EM, Strada SJ and Thompson WJ. Activation of cyclic nucleotide PDE from isolated human peripheral blood lymphocytes by mitogenic agents. *Cancer Res.* 1980; **40**: 379–386.

167. Hurwitz RL, Hirsch KM, Clark DJ, Holcombe VN and Hurwitz MY. Induction of a Ca/CaM PDE during PHA stimulated lymphocyte mitogenesis. *J. Biol. Chem.* 1990; **265**(15): 8901–8907.

168. Marcoz P, Prigent AF, Lagarde M and Nemoz G. Modulation of rat thymocyte proliferative response through the inhibition of different cyclic nucleotide phosphodiesterase isoforms by means of selective inhibitors and cGMP-elevating agents. *Mol. Pharmacol.* 1993; **44**: 1027–1035.

169. Marchmont RJ, Ayad SR and Houslay MD. Purification and properties of the insulin-stimulated cyclic AMP phosphodiesterase from rat liver plasma membranes. *Biochem. J.* 1981; **195**: 645–652.

170. Pyne NJ and Houslay MD. An insulin mediator preparation serves to stimulate cyclic GMP activated cyclic AMP phosphodiesterase rather than other purified insulin-activated cyclic AMP phosphodiesterases. *Biochem. Biophys. Res. Commun.* 1988; **156**: 290–296.

171. Sette C, Vincini E and Conti M. The rat PDE3/IVD phosphodiesterase gene codes for multiple proteins differentially activated by cAMP-dependent protein kinase. *J. Biol. Chem.* 1994; **269**(28): 18271–18274.

172. Monaco I, Vicini E and Conti M. Structure of 2 rat genes-coding for closely related rolipram-sensitive cAMP phosphodiesterases-multiple messenger-RNA variants originate from alternative splicing and multiple start sites. *J. Biol. Chem.* 1994; **269**(1): 347–357.

173. Kovala T, Lorimer IAJ, Brickenden AM, Ball EH and Sanwall BD. Protein kinase A regulation of cAMP phosphodiesterase expression in rat skeletal myoblasts. *J. Biol. Chem.* 1994; **269**: 8680–8685.

174. Angel P, Hattori K, Smeal T and Karin M. The jun proto-oncogene is positively autoregulated by its product Jun/AP-1. *Cell* 1988; **55**: 875–887.

175. Spence S, Rena G, Sweeney G, Houslay MD. Induction of Ca2+/calmodulin-stimulated cyclic AMP phosphodiesterase (PDE1) activity in Chinese hamster ovary cells (CHO) by phorbol 12-myristate 13-acetate and by the selective overexpression of protein kinase C isoforms. *Biochem. J.* 1995; **310**: 975–982.

176. Szamel M and Resch K. Signalling pathways leading to activation of protein kinase C isoenzymes in T lymphocytes. *Biol. Chem. Hopp-Seyler* 1992; **373**: 828.

177. Strulovici B, Daniel-Issakani S, Baxter G *et al.* Distinct mechanisms of regulation of protein kinase C by hormones and phorbol diesters. *J. Biol. Chem.* 1991; **266**: 168–173.

178. Hadden JW, Hadden EM, Haddox MK and Goldberg NG. *Proc. Natl. Acad. Sci. USA* 1972; **69**: 3024.

179. Hadden JW and Coffey RG. Cyclic nucleotides in mitogen-induced lymphocyte proliferation. *Immunol. Today* 1982; **3**: 299–304.

180. Lee DHK, Faunce D, Henry D, Sturm R and Rimele T. Rat polymorphonuclear leukocytes (PMN) increase cGMP levels in the rat aorta. *FASEB J.* 1988; **2**(A518): 1296.

181. Southam E and Garthwaite J. The nitric oxide–cGMP signalling pathway in rat brain. *J. Neuropharmacol.* 1993; **32**: 1267–1277.

182. Nathan C and Hibbs JB. Role of nitric oxide synthesis in macrophage antimicrobial activity. *Curr. Opin. Immunol.* 1991; **3**: 65–70.

183. Genaro AM, Hortelano S, Alverez A and Martinez-A. C. Splenic B lymphocyte programmed cell death is prevented by nitric oxide release through mechanisms involving sustained Bcl-2 levels. *J. Clin. Invest.* 1995; **95**: 1884–1890.

184. McConkey DJ, Nicotera P, Hartzell P, Bellomo G and Wyllie AH. Glucocorticoids activate a suicide process in thymocytes through an elevation of cytosolic calium concentration. *Arch. Biochem. Biophys.* 1989; **269**(1): 365–370.

185. Wyllie AH. Glucocorticoid-induced thymocyte apoptosis is associated with endogenous endonuclease activation. *Nature* 1980; **284**: 555.

186. Cohen JJ and Duke RC. Glucocorticoid activation of a calcium-dependent endonuclease in thymocyte nuclei leads to cell death. *J. Immunol.* 1984; **132**(1): 38–42.

187. McConkey DJ, Orrenius S, Okret S and Jondal M. Cyclic AMP potentiates glucocorticoid-induced endogenous endonuclease activation in thymocytes. *FASEB J.* 1993; **7**: 580–585.

188. Radomski MW, Palmer RMJ and Moncada S. Modulation of platelet aggregation by an L-arginine–nitric oxide pathway. *Proc. Natl. Acad. Sci. USA* 1990; **87**: 10043–10047.

189. Rees DD, Cellek S, Palmer RMJ and Moncada S. Dexamethasone prevents the induction by endotoxin of a nitric oxide synthase and the associated effects on vascular tone: an insight into endotoxic shock. *Biochem. Biophys. Res. Commun.* 1990; **173**: 541–547.

190. Di Rosa M, Radomski M, Carnuccio R and Moncada S. Glucocorticoids inhibit the induction of nitric oxide synthase in macrophages. *Biochem. Biophys. Res. Commun.* 1990; **172**: 1246–1252.

191. Knowles RG, Salter M, Brooks SL and Moncada S. Anti-inflammatory glucocorticoids inhibit the induction of endotoxin of NO synthase in the lung, liver and aorta of the rat. *Biochem. Biophys. Res. Commun.* 1990; **172**: 1042–1048.

192. Epstein PM, Mills JS, Strada SJ, Hersh EM and Thompson WJ. Increased cyclic nucleotide phosphodiesterase activity associated with proliferation and cancer in human and murine lymphoid cells. *Cancer Res.* 1977; 37: 4016–4023.

193. Hait WN and Weiss B. Increased cyclic nucleotide phosphodiesterase activity in leukaemic lymphocytes. *Nature* 1976; **259**: 321–323.

194. Hait WN and Weiss B. Characteristics of the cyclic nucleotide phosphodiesterases of normal and leukemic lymphocytes. *Biochim. Biophys. Acta* 1977; **497**: 86–100.

195. Epstein PM and Hachisu R. Cyclic nucleotide phosphodiesterase in normal and leukemic human lymphocytes and lymphoblasts. In: Thompson, WJ and Strada, SJ (eds) *Advances in Cyclic Nucleotide and Protein Phosphorylation Research*, Vol. 16. New York: Raven Press, 1984: 303–323.

196. Takemoto DJ, Kaplan SA and Appleman MM. Cyclic GMP and phosphodiesterase activity in mitogen-stimulated human lymphocytes. *Biochem. Biophys. Res. Commun.* 1979; **90**(2): 491–497.

197. Takemoto DJ, Lee W-NP, Kaplan SA and Appleman MM. Cyclic AMP phosphodiesterase in human lymphocytes and lymphoblasts. *J. Cyc. Nucl. Res.* 1978; **4**: 123–132.

198. Nemoz G, Prigent AF, Aloui R *et al.* Impaired G proteins and cyclic nucleotide phosphodiesterase activity in T lymphocytes from patients with sarcoidosis. *Eur. J. Clin. Invest.* 1993; **23**: 18–27.

199. Shakur Y, Wilson M, Pooley L *et al.* Identification and characterisation of the type-IVA cyclic AMP-specific phosphodiesterase RD1 as a membrane-bound protein expressed in cerebellum. *Biochem. J.* 1995; **306**: 801–809.

200. O'Connell JC, McCallum JF, McPhee I, Bolger G, Frame M and Houslay MD. The cyclic AMP phosphodiesterase RPDE-6 (RNPDE4A5) binds to the SH3 domain of v-src through its *N*-terminal splice region. *Biochem. J.* 1996: in press.

201. Wu J, Dent P, Jelinek T, Wolfman A, Weber MJ and Sturgill TW. Inhibition of the EGF-activated MAP kinase signaling pathway by adenosine-3′, 5′-monophosphate. *Science* 1993; **262**: 1065–1069.

202. Hafner S, Adler HS, Mischak H *et al.* Mechanism of inhibition of Raf-1 by protein kinase A. *Mol. Cell. Biol.* 1994; **14**: 6696–6703.

203. Gupta S, Weiss A, Kumar G, Wang S and Nel A. The T cell antigen receptor utilises lck, raf-1, MEK-1 for activating mitogen-activated protein kinase. *J. Biol. Chem.* 1994; **269**(25): 17349–17357.

204. Tortora G, Pepe S, Bianco C *et al.* Differential effects of protein kinase A subunits on Chinese hamster ovary cell cycle and proliferation. *Int. J. Cancer* 1994; **59**(5): 712–716.

15

Cytokines of the Immunoglobulin Superfamily: Signalling Through the Interleukin-1 Receptor

Pauline R.M. Dobson[1], David R. Jones[2] and Barry L. Brown[3]

[1]Institute for Cancer Studies University of Sheffield Medical School, Sheffield, UK; [2]Instituto de Investigaciones Biomedicas, Madrid, Spain and [3]Institute of Endocrinology, University of Sheffield Medical School Sheffield, UK

CYTOKINES OF THE IMMUNOGLOBULIN SUPERFAMILY

As will be apparent from other chapters in this book, many of the receptors for cytokines have been characterized and cloned. It has become clear that these receptors can be grouped into distinct superfamilies which, though not necessarily functionally linked, are derived from a common primordial gene encoding a basic domain structure. The largest of these families is the immunoglobulin superfamily (Ig-SF), which comprises approximately 40% of the leukocyte membrane polypeptides, all containing such a conserved domain. Immunoglobulin heavy and light chains are the prototype members of this superfamily. Each member possesses a characteristic region of approximately 100 amino acids forming a fold—the "Ig fold"—caused by two sandwiched β sheets, composed of antiparallel β strands which are held in position usually (but not always) by a disulphide bond. The sequence similarity occurs mainly in the inwardly pointing residues of the β strands. It is thought that these domains may be required for homologous and heterologous interactions between members of this family.

The Ig superfamily includes the poly-Ig receptors for IgA and IgM, MHC class I and II, adhesion molecules (ICAM-1 and LFA-3), the T cell receptor and domains of T cell accessory proteins (CD2, 3, 4 and 8). Most of the members are membrane bound and do not bind antigen. The cytokine receptors of this superfamily include receptors for platelet derived growth factor (PDGFR), fibro-blast growth factor (FGFR), colony stimulating factor-1 (CSF-1R; M-CSFR; c-fms), stem cell factor (SCFR; or "Steel" factor R; c-kit), and Interleukin-1 (IL-1). More distant relatives include cytokine receptors such as granulocyte macrophage colony stimulating factor R (GM-CSFR), granulocyte-CSFR (G-CSFR), ciliary neurotrophic factor R (CNTFR), oncostatin-MR and those for several interleukins—IL-3R, IL-5R, IL-6R, and the soluble IL-12R — each of which has Ig-like domains. For example, in the IL-6R, one Ig-like domain exists in the α chain (the IL-6 binding portion), and one on the β chain (gp130) which does not bind IL-6, but once associated with the α chain is responsible for signal transduction (see Figure 15.1). In addition, X-ray crystallography of the growth hormone (GH)/GH-soluble receptor complex has revealed that the sequence repeats of haemopoietin domain fold in a manner in some ways similar to the Ig-fold [1]. These cytokines, however, are more closely related to the haemopoietin or cytokine type 1 receptor family (the largest of the cytokine families), now more commonly referred to as the cytokine receptor family, than they are to the Ig family.

As indicated earlier, the common theme within the receptors for CSF-1, PDGF, FGF, SCF, and IL-1 is that within the extracellular domain of their receptors multiple Ig-like domains are located (see Figure 15.1). In the case of CSF-1R, SCFR and PDGFR, five Ig-like domains are present; in FGFR and IL-1R type I, there are three [2–4]. It is interesting to note that, in addition to multiple ligands within each family (e.g. at least

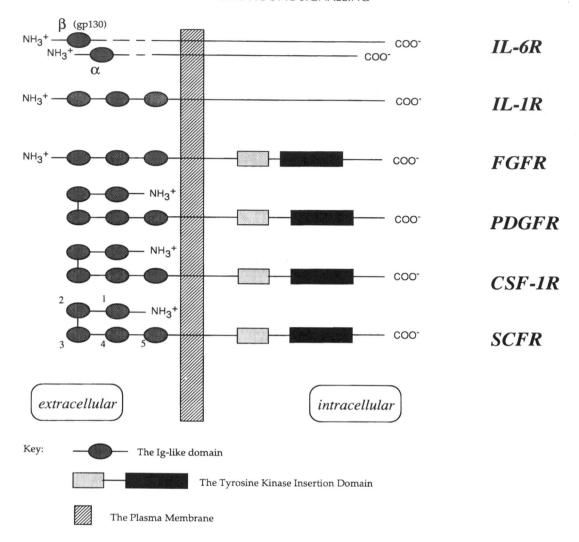

Figure 15.1 Cytokine receptors of the immunoglobulin superfamily

nine FGFs), distinct subreceptor types occur. In considering the actions of a "single" ligand, these complexities increase both the potential variety of signalling pathways and likely specificity, and also complicate the interpretation of experimental data; the knowledge of these intricacies has not always been available at the time of the study.

The signalling profiles activated by cytokines via this family of receptors are very varied. However, again we can sub-group these cytokines into those where the receptors contain intrinsic protein tyrosine kinase domains within their structures

(PDGF, FGF, CSF-1 and SCF), or where the receptor is known to associate with and activate protein tyrosine kinases (for example IL-6R and several others in the cytokine family involve the JAnus Kinases; JAKs), and finally, IL-1 which acts in an apparently unique way, and where involvement with tyrosine kinases is unclear.

The mechanisms relating to the cytokine receptor family and receptor tyrosine kinases (RTKs) are described in separate chapters in this book. In this chapter, therefore, we will concentrate on IL-1 signalling, and its relationship to other members of this and related groups, as appropriate (see other relevant reviews [5–7]). We will, however, briefly mention signalling via receptors involving tyrosine kinases in order to place the growth factors/cytokines of this grouping into context.

THE Ig SF CYTOKINE RECEPTOR TYROSINE KINASES (RTKS)

A key feature evident within the intracellular domains of the CSF-1, SCF, PDGF and FGF receptors is the intrinsic protein tyrosine kinase activity which they possess. This is a prerequisite for placing these receptors within the family of receptor tyrosine kinases (RTKs), which also includes the receptors for epidermal growth factor (EGF) and insulin. The downstream signalling mechanisms of these receptors involve the trans/autophosphorylation of dimerized growth factor receptors which leads to the recognition of specific phosphorylated tyrosine residues within the receptor by cytoplasmic proteins containing src-homology-2 (SH2) and src-homology-3 (SH3) domains. These proteins either act as adaptors for the activation of other enzymes or themselves contain intrinsic enzyme activity [8,9]. An interesting feature of the family of RTKs is that individual subclasses exist. One such distinguishing characteristic is that the receptors for CSF-1, SCF, PDGF and FGF are distinct from those for EGF and insulin in that the intracellular tyrosine kinase domain is divided into two by what is termed a kinase insertion sequence (Figure 15.1).

Dimerization of the RTKs is recognized to be central to signalling through these receptors. A neat demonstration of how this may occur was recently given by Blechman and colleagues [10]. SCFR (or c-kit), like many cytokine receptors including IL-1R, has both membrane associated and soluble forms of the receptor, though only the former are capable of signal transduction. These authors developed a monoclonal antibody which blocked dimerization of either the transmembrane or the soluble receptor. Using truncated mutant forms of the receptors, they were able to ascertain that the antibody recognized the fourth Ig-like domain of the receptor (i.e. the domain second from the transmembrane leaflet in the case of the signalling receptor, there being five in total—see

Figure 15.1). Deletion of this domain in either the soluble or the signalling form of the receptor prevented dimerization. In addition, the antibody blocked signal transduction through the signalling receptor. This suggests that the fourth domain is required for dimerization and signalling, and perhaps implies that a conformational change occurs at this locus on ligand binding. This may represent an important clue to the mechanisms involved in dimerization. Dimerization is more usually represented as a single ligand interacting with two receptors and in so doing imparting stability to the dimeric receptor complex. However, the work of Blechman and colleagues suggests the concept of the soluble extracellular portion of the receptor being sufficient by itself for the dimerization to occur. This may turn out to be the case not only for the Ig superfamily. It will be interesting to see whether other receptor types which dimerize such as the growth hormone receptors (which lack intrinsic RTK and Ig domains) have an equivalent mechanism.

For IL-1R, the situation is quite different from the Ig family cytokines mentioned above in that no protein serine, threonine or tyrosine kinase activity is intrinsic to the receptor. Enhanced protein kinase activity is observed, though, as a post IL-1 receptor–ligation event (see below). Nonetheless, although it has not been clearly demonstrated so far, dimerization of the IL-1R may occur and, if so, the means might be similar to that for SCFR.

BACKGROUND TO IL-1 SIGNALLING

The history of investigations into IL-1 action brims with confusion. It is important, however, to remember that many of the early studies were performed on impure preparations of the cytokine leading to effects of contaminating substances or even synergistic responses of the various contaminants. Also, a number of the early studies did not take sufficient account of the dynamics of the responses and many of the reported results were from experiments conducted over long time intervals (often 24 hours or more), and these were frequently extrapolated to form conclusions relating to the initial events following receptor activation, where second(s) or minute(s) time points are more relevant. Also, long-term studies suffered since activated cells may, and often do, release other substances in response to the initial activator and may act back on the same cells or associated cells in an autocrine or paracrine way. Many of the early longer term studies also relied on similarities between the action of IL-1 and some drugs such as phorbol 12-myristate 13-acetate (PMA) and ionomycin, leading to what are now thought to be erroneous conclusions. For example, the use of PMA in this way led to the belief by many that IL-1 signalling involved acute activation of protein kinase C, which has since been shown not to be the case.

INTERLEUKIN-1 AND ITS RECEPTOR

Little over a decade ago, there was much excitement over novel and potent paracrine activities between leukocytes in the immune system. The so-called monokine, Interleukin-1 (IL-1), was recognized as a molecule responsible for several previously described and apparently separate activities including lymphocyte activating factor, endogenous pyrogen, mononuclear cell factor, leukocyte endogenous mediator, catabolin and osteodast activating factor. We now know that the effects of IL-1 are pleiotropic and are able to be manifest in numerous cell types [11, 12]. Alongside this, activities such as T cell growth factor, and B cell growth factor were shown to be attributable to IL-2. This trend of generic categorization of seemingly separate activities has continued in respect of nomenclature of similar agents. With this knowledge came the awareness that the terms "monokine" or "lymphokine" represented too narrow a definition, as other cell types were shown to release and/or be affected by these activities, and "cytokine" became the more appropriate term to be adopted.

The biochemical nature of cytokines escaped elucidation primarily due to the extremely low concentrations present in biological fluids, albeit that the activities were potent. Molecular cloning was to provide the means whereby these activities could be discerned. In the same year that the cDNA for the mouse IL-1β was cloned from an established cell line [13], human IL-1β was cloned from the blood of donors [14], which revealed IL-1 to be an 18 kDa molecule with a precursor form of

31 kDa. IL-1α, the product of a separate gene and sharing 26% homology with IL-1β was cloned and characterized at approximately the same time. Although there are a lot of similarities in the biological properties of IL-1α and IL-1β, there are some important differences between these two agents. For example, whereas the precursor molecule for IL-1β is inactive, that for IL-1α is active and is found in association with the plasma membrane. Also, IL-1α is rarely found in the circulation or in inflammatory body fluids and an intracrine role has been speculated upon.

In terms of the signal transduction mechanisms, perhaps the greater excitement related to the characterization of the cellular receptor for IL-1, where it became clear that a single transmembrane protein was involved. An anti-IL-1 receptor antibody was used to characterize the gene product from molecular cloning. Initial progress was hampered by the inability to radioiodine-label IL-1β successfully on tyrosine residues since biological activity was lost on iodination — a problem which beset many laboratories internationally at the time, and which was later found not to be the case for radiolabelling IL-1α. Radiolabelled IL-1α in fact became the means of investigation of the IL-1R. Saturation binding of IL-1R was successfully achieved [15], but perhaps the major importance of this pioneering work was that it highlighted that in LBRM-33-1 A5 (a cell line developed by Steve Gillis and Steve Mizel), receptor numbers were exceptionally low as compared to other known receptor types for endocrine and neuronal ligands. LBRM-33-1 A5 expressed only 100–200 IL-1R per cell as compared to thousands in the case of other cellular receptor populations known, but nonetheless, signals could be effectively transduced. Later this was found not to be a unique circumstance; many cell types express this relatively low number of functional receptors for IL-1, or sometimes 10-fold above or below this number.

Two receptors for IL-1 were described, one of molecular mass 86 kDa (IL-1RI), later found to be the signalling receptor for both IL-1α and IL-1β [16–21] and the second, a smaller mass 65 kDa receptor (IL-1RII) lacking an intracellular domain and capacity for signalling [22]. Binding studies revealed that there is preferential binding of IL-1β over IL-1α to the type I IL-1R and in the case of the type II IL-1R this is more so [23]. A third member of the IL-1 family exists, the IL-1R antagonist (IL-1RA) [24, 25]. IL-1RA binds to the type I receptor without effect in terms of signal transduction other than to prevent occupation of the receptor by the agonist ligands, IL-1α and IL-1β[26]. The discovery of soluble extracellular IL-1Rs [27, 28] was followed by the conclusion that the type II receptor exists as a 'decoy' able to dissociate from the cell surface on IL-1 binding [20, 21], providing yet another means of regulating the amount of IL-1 accessible to the type I signalling receptor.

Very recently, Greenfeder and colleagues [29, 30] cloned a possible third IL-1R, which although having a molecular weight close to IL-1RII (66 kDa) is distinct from it and is called the IL-1 receptor accessory protein (IL-1RAP or IL-1R AcP). The function of this receptor (if indeed it is a receptor) is as yet unknown, however, it does not appear to signal. The receptors for IL-1 have structural similarities with a variety of other proteins, prompting the suggestion that there may be a family of related signalling proteins and this is reviewed by Sims and Dower [31].

It is not the intention of the authors of the current chapter to present a detailed analysis of the Interleukin-1 receptors, and so we refer the reader to an excellent review [31]. Rather, we will concentrate on the signal transduction mechanisms *per se* which relate to IL-1 cellular action.

INTERLEUKIN-1 SIGNAL TRANSDUCTION

The elucidation of the IL-1 receptor marked a major breakthrough in the understanding of the diversities of likely receptor structures. Perhaps, therefore, there was a naive expectancy that knowledge of the receptor structure for IL-1 would provide evidence for how it acted in terms of signal transduction. In fact this was not to be the case: prior knowledge of receptor structure and signalling processes for hormonal and other described agents, when compared with IL-1R, did not provide special insight to the cellular mechanism of action of IL-1. On the basis of the structure, though, some potential aspects could be eliminated: IL-1R was not an intrinsic tyrosine kinase

receptor and, similarly, direct association with heterotrimeric G proteins seemed unlikely (discussed more fully below). Almost all known mediators of signal transduction have been investigated in terms of IL-1 action, often with largely conflicting results.

Cyclic Nucleotides in IL-1 Action

Adenosine 3', 5'-Cyclic Monophosphate (cAMP)

Perhaps the singularly most controversial area of Interleukin-1 action is whether or not adenosine 3', 5'-cyclic monophosphate (cAMP) is involved in IL-1 signal transduction. Table 15.1 indicates the state of play: in essence several groups have found that IL-1 does or does not affect cAMP in various cellular systems, and sometimes these apparently contrasting findings are in the same cell type, and presumably through the same single signalling IL-1 receptor type. It should be noted that not all are direct measurements. The situation is also as confused in respect of IL-1 stimulated protein kinase A (PKA), the major kinase activated by cAMP leading to a cascade of enzyme activity within the cell. However, Brooks and Mizel [32] have speculated that PKA offers a "permissive" step in IL-1 action in cells (EL4) which do not respond to IL-1 in terms of cAMP accumulation.

Only recently has evidence been obtained for the existence of isoforms of adenylyl cyclase (AC), the enzyme which forms cAMP. The functions of these individual isoforms are not yet resolved [33]. The differences in the literature in terms of whether cAMP changes have been observed in response to IL-1 might reflect the expression of particular isoforms of AC already present in cellular populations. If IL-1 caused elevation of cAMP via only one or two isoforms of AC, for example, or the affected isoforms were expressed in low amounts in a particular cell type, it is possible that the level of cAMP produced could be outwith the sensitivity limits of conventional assay procedures which reveal quantities in the low femtomolar region. It may be that cyclic AMP imaging may offer an insight, as markedly lower quantities of cAMP can be accurately determined and in the physiological state by this technique

[34]. Alternatively, effects of IL-1 on cAMP could be as a result of crosstalk mechanisms rather than direct receptor mediated activation by IL-1. Again the expression of key enzymes, not only cyclase itself, but of those having the potential to interact with cyclase, may be a ruling parameter. Our earlier work in a somewhat different set of systems suggested that crosstalk mechanisms were highly cell specific. For example, we observed that vasoactive intestinal polypeptide (VIP), which acts through cAMP in both GH3 and UMR106 cells, was only subject to down-regulation by tumour promoting phorbol ester in the former cell type, whereas PGE2 stimulated cAMP accumulation was down-regulated by tumour promoting phorbol ester in the UMR106 cells [35, 36]. The importance of these findings is that, as well as the receptor specificity, the intracellular phenotype in terms of expression of isoenzymes, for example, may be a determinant in the specificity of signal transduction pathways and crosstalk mechanisms. If, on the other hand, there are truly differing responses in different cell systems in terms of IL-1 and cAMP (i.e. that in some cell types the cAMP pathway is not affected by IL-1 and in others it is), it implies that cAMP is probably not a major messenger in IL-1 action.

Guanosine 3', 5'-Cyclic Monophosphate (cGMP)

Guanosine 3' 5' cyclic monophosphate (cGMP) has also been implicated in IL-1 signalling, and although less work has been done in terms of this potential mediator, the results have been confusing. In rat smooth muscle cells, IL-1 has been shown to cause the accumulation of intracellular cGMP and nitrite [60, 61]. Contrary to these reports, evidence has suggested that IL-1 does not affect the cGMP content of human T lymphocytes [50] nor human Detroit fibroblasts (Dobson and Brown, unpublished observations). At this time there is not enough information to be able to draw firm conclusions in terms of the importance of cGMP in IL-1 signalling.

G Proteins in IL-1 Action

The role of heterotrimeric guanine nucleotide-binding proteins (G proteins) during IL-1 transmembrane signalling has also been investigated. Some evidence for G protein involvement in IL-1 action has come from the use of pertussis toxin (PTX) which abrogates IL-1-stimulated production of diradylglycerol (DRG) in murine T lymphocytes [51], cAMP accumulation in 70Z/3 cells, YT cells, fibroblasts and synovial cells [41], DRG in 70Z/3 pre-B cells (Dobson and Brown, unpublished observation), release of IL-2 and prostaglandin production in murine T lymphocytes and human gingival fibroblasts respectively [62], and stimulated phosphatidic acid phosphohydrolase activity in rat mesangial cells [63]. Taken together, these results suggest that there may be G protein involvement in IL-1R-mediated events.

However, experiments using PTX have been placed in some doubt as two independent groups

Table 15.1 Studies relating to cAMP as a second messenger in IL-1 signal transduction

Reports of IL-1 stimulated cAMP	Reports of no effect of IL-1 on cAMP
Giri et al., 1984 [37]	Abraham et al., 1987 [49]
Edelman et al., 1987 [38]	Didier et al., 1988 [50]
Shirakawa et al., 1988 [39]	Dobson et al., 1989 [51]
Zhang et al., 1988 [40]	Scholz and Altman, 1989 [52]
Chedid et al., 1989 [41]	Dobson et al., 1990 [53]
Shirakawa et al., 1989 [42]	Kasahara et al., 1990 [54]
Chedid and Mizel, 1990 [43]	Schlegel-Hauter and Aebischer, 1990 [55]
Munoz et al., 1992 [44]	O'Neill et al., 1990 [56]
Renkonen et al., 1990 [45]	Pfielschifter et al., 1991 [57]
Turunen et al., 1990 [46]	Yamamoto et al., 1992 [58]
Zeki et al., 1991 [47]	Garcia et al., 1992 [59]
Wietzmann and Savage, 1993 [48]	Dobson and Brown (70Z/3; unpublished)

have shown that the B oligomer of PTX, which binds the toxin to membranes, but does not cause ADP-ribosylation, was itself able to exert similar effects of PTX in T cells [64] and inhibit IL-1-induced IL-2 release [62]. Also, neither the active subunit of PTX nor the B oligomer altered IL-1/IL-1R binding or IL-1-induced activation of the transcription factor, NF-κB [62]. These findings argue against the direct coupling of a G protein to IL-1R, as one would expect to see a shift in binding to higher concentrations of IL-1 and inhibition or blockade of NF-κB activity by PTX, were a G protein involved. On the other hand effects of IL-1 on GTP hydrolysis have been noted [41, 56]. Also, Mitcham and Sims [65] have identified a novel protein (IIP1) of 110 kDa with sequence homology to GTPase activating proteins (GAPs) which interacts with the cytoplasmic domain of the type I IL-1 receptor at the region required for signal transduction. It remains to be seen whether or not there is a G protein mediating IL-1 signalling. It is possible that a non-heterotrimeric or small molecular weight G protein is involved in IL-1 action, as has been shown for IGF-1. This would not explain the effects of PTX on responses to IL-1, but might account for differences observed in response to IL-1 in terms of GTP hydrolysis. It may be that the effect of PTX is artefactual, at least in the T cell, and the inhibition of IL-1 stimulated IL-2 release by the B oligomer [62] suggests this to be so. The evidence for PTX causing ADP ribosylation is very substantial, but in the case of IL-1 actions it cannot be ruled out that PTX and the B oligomer cause the blockade of a secondary response initiated by IL-1, perhaps by an agent acting in an autocrine or intracrine manner. Considering the time courses that have been investigated in terms of blockade of IL-1 effects by PTX, this secondary response would have to be as a result of a rapidly generated mediator, which is fast acting. Such a premise is not unprecedented. Sphingosine 1-phosphate, for example, is rapidly and transiently formed in response to PDGF [66]. It has also been demonstrated to have effects when added exogenously to cells which are sensitive to PTX [67]. It is unlikely that sphingosine 1-phosphate is released from the cell [68], therefore, it could be exerting an intracrine PTX sensitive effect on G proteins. A similar IL-1 induced stimulation of a secondary mediator which acts extracellularly or intracellularly, and whose action is mediated by a PTX-sensitive G protein, may provide the explanation for the effects of IL-1 relating to GTP hydrolysis.

Ionic Changes in IL-1 Action

Intracellular calcium $Ca^{2+}{}_i$

Increases in intracellular calcium levels ($Ca^{2+}{}_i$) in response to IL-1 have been observed in murine fibroblasts [69], human fibroblasts [69, 70] and in rabbit osteoclasts [71], although there are as many reports where no change in ($Ca^{2+}{}_i$) has been observed [49, 72–74]. In addition, a decrease in ($Ca^{2+}{}_i$) concentration has been observed in murine pre-B lymphocytes stimulated with IL-1 [75]. The reasons for the apparent variations in responses to IL-1 observed by different groups is not clear and secondary and/or crosstalk mechanisms cannot be ruled out.

Intracellular pH (the Na^+/H^+ Antiporter)

The Na^+/H^+ antiporter exchange system within cells is the key mechanism for the control of cytosolic pH, leading to alkalinization preceding DNA replication. In terms of IL-1 action, in murine pre-B lymphocytes, murine T lymphocytes and isolated rat pancreatic islet cells, IL-1 was shown to increase intracellular pH_i in an amiloride-sensitive manner [75–77], contrasting with the observations of Kester and colleagues [78] who observed no change in pH_i in rat mesangial cells and our observations in Swiss 3T3 fibroblasts, where there was no direct effect of IL-1 on the antiporter, as cellular pH was unchanged in response to IL-1 when measured directly using Quene-1.

In conclusion, the situation in relation to effects of IL-1 on ionic changes is by no means clear. However, in line with other parameters which have been tested (see below), such as those relating to PI hydrolysis (which can generate calcium transients) and protein kinase C activation (which is a means of altering Na^+/H^+ channels), the evidence is against these being essential components of IL-1 action.

Putative Lipid Mediators and Hydrolysing Enzymes in IL-1 Action

Diradylglycerol (DRG)

Agonist-stimulated phosphodiesteric hydrolysis of phosphatidylinositol-4,5-bisphosphate by a phospholipase C is a key pathway for the generation of inositol polyphosphates and diradylglycerol (DRG) in many systems. However, we and others found that exogenous addition of IL-1 does not generate inositol phosphates [49,51,74], indicating that phosphatidylinositol (PI) turnover is not involved in IL-1 action. However, there is a contrasting report of breakdown of inositol-containing phospholipids following exposure of cells to IL-1 [79], and a separate report that hydrolysis of PI occurred within the cell nucleus [80].

Despite the lack of effect on PI metabolism, we observed a rapid, pertussis-toxin-sensitive, generation of DRG in response to IL-1β, but this did not cause protein kinase C (PKC) activation [51,53]. In experiments over extensive time courses (1 second to 96 hours) we did not detect any effect of IL-1 on apparent PKC activity that could not be blocked by cycloheximide [53]. This lack of effect of IL-1 on acute apparent PKC stimulation has been confirmed by other groups [81]. This situation is also similar to our observations with recombinant basic FGF, which was without an acute effect on PI metabolism or DRG formation and PKC activation in 3T3 cells [82,83]. However, there are reports of activation of PKC by FGF [84] and activation of PKC in the absence of calcium changes or PI metabolism [85], and effects on DRG formation without PI hydrolysis [86]. Also, the involvement of PI metabolism in FGF action [87, 88] has been reported, albeit in different tissues. These differences may be due to receptor subtypes of the FGFR. The situation for IL-1 and FGF action is rather different from that of PDGF, however, where there is good agreement that PI metabolism and PKC activation are influenced.

Several other authors have also reported the formation of DRG in response to IL-1, however, the source of this lipid remains an enigma. Phosphatidylcholine has been implicated as one putative source of the DRG [72, 89, 90]. In addition, phosphatidylethanolamine [78] and glycosylphosphatidylinositol [53, 74] have been suggested as hydrolysable substrates during IL-1 signalling, and potential sources of DRG. It has also been suggested that IL-1 causes phosphorylcholine to be released extracellularly, with a correlating decrease in cellular PC content. This apparent PC-specific phospholipase C activity has been further defined and has been suggested to be calcium dependent and held in position on the outer leaflet of the plasma membrane through a glycosylphosphatidylinositol (GPI) anchor.

It is of interest that Musial and colleagues [91] have recently shown that IL-1 stimulates the production of alkyl-acyl- and alkenyl-acyl-glycerols and that these had no effect on PKC in a cell-free system. These data accord well with studies from our group and others showing IL-1 stimulated DRG accumulation in the absence of PKC activation.

There is little evidence for involvement of phospholipase D activation by IL-1, although, Garcia and colleagues [59] have suggested that it may be stimulated by IL-1. More work needs to be done to draw a firm conclusion with respect to the involvement of PLD in IL-1 action.

Phospholipase A2

There have been numerous reports indicating that incubation of cells with IL-1 leads to the mobilization of polyunsaturated fatty acids (FAs) from phospholipids, via activation of phospholipase A_2 (PLA$_2$). The source of the polyunsaturated fatty acid has been reported less widely with differing results in distinct cell systems. In murine macrophages, phosphatidylcholine (PC) and phosphatidylinositol (PI) were reported to be substrates for a PLA$_2$ in IL-1-stimulated cells [92]; in murine T lymphocytes, PC was reported as the sole substrate [93]; in human synoviocytes PC, PI and phosphatidylethanolamine (PE) were found to be hydrolysed to form FAs as a result of IL-1 stimulation [94], whereas in human myometrial cells, PC and phosphatidic acid were reported to be the substrates for the IL-1-stimulated PLA$_2$ activity [95].

At the present time, however, the role of PLA$_2$ activity in IL-1 action is not clear. The fate of the released polyunsaturated fatty acid is usually thought to be either the direct secretion from the cell (possibly to act in a paracrine manner), or its metabolism to form eicosanoids (which in turn might be released from the cell to fulfil a paracrine role). The effect of IL-1 on eicosanoids, however, has been suggested not to be as a result of a direct effect on PLA$_2$ activity, but rather one involving protein synthesis of, for example, cyclooxygenase [96, 97] (discussed below).

The immediate effects of PlA$_2$ activity outwith those directly relating to eicosanoid production, comprise an exciting new and uncharted area of investigation. There are indications that we are perhaps not recognizing in general the potential of membrane phospholipids in signalling mechanisms, both intracellular and extracellular. In our studies with rbFGF, for example, we found that arachidonic acid (AA) was generated and secreted in response to this stimulus [82], an effect which was very rapid and which we attributed to PLA$_2$ activity. Abrogation of calcium entry prevented this action. Bombesin, an agent acting through the PI cycle and known to release calcium from intracellular stores, was as sensitive to blockade of calcium entry in terms of AA release as was bFGF, which did not affect PI hydrolysis acutely. Quite separately, the calcium status in terms of PLA$_2$ activity is important, though elevated calcium is not a requirement for activation of all isoforms of PLA$_2$ and we did not see changes in calcium in response to bFGF. In contrast to the situation for bFGF, however, we have neither observed calcium transients nor AA formation or release following exogenous addition of IL-1 to Swiss 3T3 fibroblasts.

Linoleic acid has also been suggested to be formed after exposure of cells to IL-1. There is a report by Bomalaski and colleagues [93] indicating that linoleic acid was preferentially released, as compared to arachidonic acid, and the postulated role of the released linoleic acid was the augmentation of protein kinase C (PKC)-mediated IL-2 release.

Ceramides

Another major group of lipids possibly involved in IL-1 signalling are the sphingolipids. It has been proposed that ceramide, produced from sphingolipid hydrolysis acts as a second messenger in IL-1 action. The possible role of the sphingomyelin cycle in the action of a variety of extracellular agents has been extensively reviewed recently [98, 99]. Classically, ceramide production has been thought to occur via the hydrolysis of sphingomyelin catalysed by sphingomyelinases (the so-called "sphingomyelin cycle"). Acidic sphingomyelinase [100, 101] is activated in vitro by diradylglycerols, and Schutze and colleagues [102] have suggested that the release of DRG in response to TNFα activates this enzyme. There are also two neutral sphingomyelinases whose mode of activation is unknown. While the evidence for a role for ceramide in the action of TNFα (and some other agents) is quite compelling, albeit in the face of some opposing views, its role in IL-1 action remains controversial and has indeed been disputed. Ballou and colleagues [103] reported that sphingomyelinase and C$_2$-ceramide (a membrane-soluble ceramide analogue) reproduced some of the effects of IL-1 stimulation, such as an increase in PGE$_2$ release and induction of the cyclooxygenase gene. They also reported an IL-1 stimulated breakdown of sphingomyelin and production of ceramide. Mathias and coworkers [104] also showed that IL-1 stimulated sphingomyelin turnover in EL4 cells. It was also reported that IL-1 induced ceramide production in cell-free extracts and that IL-1 action was associated with a ceramide-activated protein kinase activity. However, it appears that C$_2$-ceramide does not mimic IL-1 activation of AP-1 [32]. Furthermore, although we have observed similar patterns of kinase activity in response to lipid extracts of IL-1 treated cells and C$_6$-ceramide [105] we have been unable to confirm IL-1 induced production of ceramide in IL-1 responsive T47D breast cancer cells.

Ceramide has been shown to activate a number of intracellular targets. It activates a 97 kDa membrane associated serine/threonine kinase [104] which is reported to phosphorylate the EGF receptor [106] and Raf-1 [107]. It has also been reported that PKCζ is activated following treatment of 3T3 cells with sphingomyelinase and, in vitro, by ceramide. Ceramide has also been shown to activate a phosphatase—"ceramide-activated protein phosphatase" (CAPP), which is of the protein phosphatase 2A type [108, 109] and a MAP kinase [110]. The reported activation of PPA2 is intriguing since it has been shown that IL-1 inactivates this phosphatase [111, 112].

Most attention has been given to the role of stimulated sphingomyelin turnover via sphingomyelinase action. However, Kolesnick and colleagues [113] have recently shown that an anti-cancer drug, daunorubicin, influences the activity of ceramide synthase. It will be interesting to determine the relative roles of sphingomyelinases and ceramide synthase in the action of extracellular agents. Clearly much remains to be done to resolve the controversy surrounding the possible role of ceramide in IL-1 action. Perhaps a related lipid mimics the action of ceramide on intracellular enzymes and/or we are witnessing another possible example of some of the problems that have beset investigations of IL-1 action, viz., cellular diversity and, possibly, differential crosstalk mechanisms.

Phosphatidylinositol 3-Kinase (PI3K)

The enzyme which is responsible for the generation of 3'-phosphorylated inositol-containing phospholipids is phosphatidylinositol 3-kinase (PI3K) [114]. PI3K is activated by PTKs and as might be expected, therefore, is known to be involved in the action of several cytokine members of the Ig superfamily including PDGF, CSF-1 and FGF (see review by Poyner and colleagues [115]). PI3K consists of a heterodimer of 85 kDa and 110 kDa subunits, in which the former contains SH2 and SH3 domains and the latter has catalytic activity yielding phosphatidylinositol-3,4,5-trisphosphate (PIP_3) and possibly other 3'-phosphorylated inositol-containing phospholipids. Four isoforms of PI3K have been discovered. The precise role of PI3K in cellular function is uncertain, however it seems to be intrinsically linked to tyrosine kinase activity. Activation of selective isomers of protein kinase C (PKC) [116], action filament rearrangements [117] and membrane ruffling events [118] have been reported to be influenced by PI3K. In the case of PDGF, PI3K binds to tyrosine phosphate residues within the kinase insertion sequence of the tyrosine kinase domain of the PDGF receptor [119]. For GM-CSF the position seems to be quite different and more complex. In this case, activation of PI3K appears to involve a multiprotein complex consisting of the activated receptor, the activated *src* type protein (non-receptor) tyrosine kinases *lyn* and c-*yes* and, of course, PI3K [120]. All the FGF receptors contain a consensus sequence for the binding of PI3K located near the C-terminus of the receptor. In addition, the activation of PI3K by FGF has been observed [121].

Our recent findings indicate that an acute activation of PI3K occurs following exogenous addition of IL-1 to T47D cells and also that the mitogenic affect of IL-1 can be inhibited by drugs which block PI3K activity ([122]; also manuscript in preparation). An implication of these results is that IL-1 signalling may, indeed, involve a tyrosine kinase.

Kinases and Phosphatases in IL-1 Action

The involvement of kinases and phosphatases in IL-1 action has been given much attention in recent years and, although patterns seem to be beginning to emerge, the role of protein phosphorylation in the mechanism of action of IL-1 remains ill-defined. There have been a number of reports of IL-1-stimulated phosphorylation of cellular proteins [123–125], some of which have been identified whereas the nature of others remains unknown. Clearly, there are many potential ways by which protein phosphorylation could be increased by IL-1. For example, receptor-associated proteins might initiate kinase cascades or putative second messengers may activate kinases or modulate phosphatases leading to changes in phosphorylation patterns within the cell. Intrinsic receptor phosphatases have been described for other receptors (though this is not the case for IL-1R) and it is possible that receptor-associated phosphatases may also exist. In terms of second messenger activation of kinase pathways, if, as suggested, cyclic AMP (see section on cyclic nucleotides) is induced in some cell types, then the cyclic-AMP-dependent protein kinases (PKAs) may be activated in response to IL-1. However, it seems clear that IL-1 does not affect cyclic AMP in all cells. Another possible second messenger that may influence protein phosphorylation in response to IL-1 is ceramide or related lipids. As indicated above, the potential role of ceramide in IL-1 signalling is controversial, but our results suggest that a lipid produced on IL-1 stimulation has similar effects to ceramide on the activation of several, apparently novel, kinases in breast cancer cells.

Moreover, ceramide has been shown to activate a membrane-associated kinase (p97) that phosphorylates the EGF receptor [106], protein phosphatase 2A [108, 109] and a MAP kinase [110]. How other putative second messengers, such as diradylglycerol(s), may influence protein phosphorylation is not yet known. Although there have been a few reports indicating that PKC may be activated by IL-1, by far the majority of evidence indicates that there is no acute stimulation of PKC by IL-1 and, therefore, it seems very unlikely that PKC plays a role in the initial signalling events induced by IL-1. It is clear that even in those situations where phorbol esters and IL-1 have similar effects, e.g. NF-κB activation, the IL-1 effects are PKC independent [126–128]. Since IL-1 may induce the synthesis of at least some of the PKC isoforms (see below), this enzyme may feed into signalling pathways at longer time intervals.

A protein that has been shown to be phosphorylated in response to IL-1 is the EGF receptor at Thr 669 [129, 130]. There is evidence that the kinase responsible for this phosphorylation is a member of the MAP kinase family and there are also reports of the activation of members of this family by IL-1 in a variety of cell types [131–135]. Saklatvala and colleagues [123, 134, 136, 137] have described IL-1 stimulation of the phosphorylation of the small heat shock protein, hsp27, via activation of an hsp27 kinase. This kinase, although similar in molecular mass to MAP kinase, is not activated, unlike MAP kinases, by phorbol esters via activation of PKC.

The MAPK family comprises a number of different protein kinases all of which are activated by dual phosphorylation on both threonine and tyrosine residues. The prototype MAP kinases are the p42/p44 MAP kinases (also known as ERK2 and ERK1). Other members of the family include the p54 stress-activated protein kinases (SAPKs)[138] or jun N-terminal kinase (JNK) [139] and the more recently identified p38 MAPK which is a homologue of the yeast kinase, HOG1 [140]. While the evidence for activation of MAPK(s) by IL-1 is quite substantial, it is not clear what the upstream pathways are, nor which MAPKs are activated in different cell types. Indeed here again with IL-1, there appear to be cell-specific pathways activated. For example, our own studies have revealed that p42/44 MAPKs are activated in T47D human

breast cancer cells, but not in EL4 rodent lymphoma cells [141]. There is also evidence that ERK1 and ERK2 are activated in other cell types, e.g. KB epidermoid cells [142]. Interestingly, we found that IL-1 stimulated T669 phosphorylation was similar in the T47D and EL4 cells suggesting that increased activity of p42/44 MAPK was not essential for T669 kinase activation and that other routes for this effect are possible. There is evidence that IL-1 activates p54 MAPK (SAPK/JNK) in preference to p42/44 MAPK, at least in some cell types [135]. Initially the activation of T669 in the EGF receptor (T669 kinase) was attributed to p42/44 MAPKs, but it has since been suggested to be due to SAPK/JNK activity [143, 144] and this would accord with our own results. Recently, the p38 (RK) kinase cascade has been implicated in IL-1 action [140, 145, 146]. The p38 kinase (RK) is probably the major activator of MAPKAPK2 which is responsible for the rapid increase in phosphorylation of hsp27 seen upon stimulation by IL-1 [145, 147, 148].

Thus, it would appear that IL-1 does activate members of the MAPK family, but that the pattern of activation is cell specific, IL-1 activating a range of MAPKs in different cells, but also activating several of these cascades simultaneously in at least some cell types. MAP kinase activation requires phosphorylation on threonine and tyrosine residues and this dual phosphorylation is catalysed by MAP kinase kinase. IL-1 has been shown to cause both the activation [142] and the de novo synthesis [149] of this enzyme. As stated at the beginning of this chapter, the receptor for IL-1 contains no intrinsic tyrosine kinase activity, however protein tyrosine phosphorylation in response to IL-1 has been observed [44, 124, 150–153]. Studies using tyrosine kinase inhibitors have implicated protein tyrosine phosphorylation in IL-1-induced gene expression [44, 152] and IL-1-induced nitric oxide release [151, 153]. The possibility exists that some tyrosine kinase activity separate from the tyrosine/threonine phosphorylation of the MAPK family is involved in IL-1 signalling, but which tyrosine kinases are involved is unknown. However, one recent report suggests the activation of a p60[src]-type kinase by IL-1 [154].

Reports in the literature also point to the existence and role in IL-1 signalling of a number of potentially novel kinases. Rachie and colleagues

[155] reported an IL-1 activated, 85 kDa Ser/Thr kinase in the nucleus, while Kracht and coworkers [144] provided evidence for a 45 kDa kinase that phosphorylated the EGF receptor peptide containing T669, which was not p44 MAPK. There has also been a report of an IL-1-activated, and potentially novel, 65 kDa β casein kinase [134]. A very recent study suggests that for IL-1 activation of NF-κB, the serine and threonine phosphorylation and subsequent degradation of MAD3 IκB is a necessary event[156].

Much interest has recently been aroused by the discovery of IL-1 type I receptor associated kinases (IRAKs). Erickson and colleagues have described two serine/threonine kinases associated with the receptor; one has histone kinase activity while the other phosphorylates a 60 kDa endogenous substrate which is also associated with the receptor[157]. Dower and colleagues have suggested that the endogenous substrate(s) may be cell-type specific[158]. It appears that IRAK activity, which is detectable within 30 seconds, is only observed after IL-1 stimulation and not in cells which are unresponsive to IL-1[159].

The possible role of protein phosphatases in IL-1 action has received less attention, although there have been a few studies employing enzyme inhibitors. Okadaic acid, a protein serine/threonine phosphatase (type I and 2A) inhibitor, has been reported to have some similar effects on cells compared to those treated with IL-1 [160, 161] and these include a rapid increase in phosphorylation of the heat shock protein, hsp27, strathmin and the EGF receptor. These findings combined with the suggestion of an IL-1-dependent inactivation of a redox-sensitive protein phosphatase [111] suggest a possible crucial role of protein phosphatases in the mechanism of action of IL-1. It is clearly possible that ceramide-activated protein phosphatase(s) may also have a role to play.

Despite kinase/phosphatase effects being the aspect of IL-1 signalling where arguably most progress has been made, clearly we are a long way from defining the kinase cascades and the influence of phosphatases in IL-1 signalling. We know very little of the way in which kinase cascades are influenced by second messengers and, although we do know some of the kinases which are activated, how the various potential pathways are integrated remains much of a mystery. In addition, the whole question of crosstalk between pathways has hardly been addressed at all. Moreover, the problem of (apparent) differences in IL-1 activation of pathways in different cell types remains to be addressed, although if the activated kinase cascades converge then the possibility of linking all of the pathways from the IL-1 receptor at the membrane to induction of specific genes within the nucleus is truly within our grasp.

Transcription Factors

One transcription factor relevant to this discussion is the nuclear factor NF-κB (NF-κB). The NF-κB binding site was first described as an enhancer element which bound a nuclear factor in B lymphocytes that increased transcription of genes encoding the kappa chain of immunoglobulins [162]. Interleukin-1 activates NF-κB [127, 163, 164] which comprises at least three subunits: these are p50 (which is responsible for DNA binding), p65 and an inhibitory protein, IκB. Release of IκB following phosphorylation and/or proteolysis leads to formation of a heterotetramer of two p50 and two p65 units, which apparently translocates to the nucleus [165]. Both of these subunits belong to a large family of *rel* related factors [166, 167]. Indeed, both the p50 and p65 subunits share substantial homology with the protooncogenes *v-rel*, its viral counterpart, *v-rel*, and the *Drosophila* morphogen, *dorsal*. IκB shares sequence homology with MAD3 [168]. In their excellent review, Brooks and Mizel [32] indicate that activation of both PKA and PKC can convert NF-κB into a DNA binding form which translocates to the nucleus. However, the lack of acute stimulation of PKA or PKC by IL-1 in some cell types suggests alternative routes to this effect. Interestingly, Ostrowski and colleagues [169] isolated a 65 kDa κB-associated protein which was phosphorylated by a nuclear ser/thr kinase that was activated in response to IL-1. They suggested that the p65 protein was the IL-1 stimulated kinase.

Studies using mutant forms of the IL-1 receptor have revealed interesting results in terms of the functional portions of the IL-1R [170, 171]. In a report by Croston and colleagues [170], a region of the IL-1R type I cytoplasmic domain was identified to be required for both NF-κB activation and

IL-1 receptor associated kinase (IRAK) activity. None of the IL-1R mutants were found to be capable of activating NF-κB in the absence of IRAK activity. The authors concluded that IRAK activity was essential for NF-κB activity. In a separate study by Dower and colleagues [171], the most N-terminal region of the cytoplasmic domain of the type I receptor was found to be critically important for the function of NF-κB and IRAK, but also SAPK activities.

Another IL-1 responsive transcription factor is AP-1 which regulates a number of genes via binding to a DNA sequence known as TRE (the TPA [same as PMA] response element). AP-1 comprises the products of *jun* and *fos* oncogenes. In studies of the costimulatory effects of IL-1, Muegge and colleagues [172, 173] reported that IL-1 stimulated expression of c-jun and that this formed an active dimer (i.e. AP-1) with c-*fos* activated by mitogen. In some cells it appears that IL-1 can increase the expression of both c-*jun* and c-*fos* [32]. As indicated earlier, it is now known that *jun* is activated via N-terminal phosphorylation catalysed by SAPK/JNK which is activated by IL-1, at least in some cell types.

A third IL-1 responsive transcription factor is NF-IL-6. Messenger RNA for NF-IL-6 is induced by IL-1 as well as by LPS, TNF and IL-6. NF-IL-6 regulates some acute phase protein and cytokine genes as well as IL-6 expression. It has been reported that *ras* and MAP kinases cause phosphorylation and activation of NF-IL-6 [174]. How this translates to an action of IL-1 is unclear at the present time since the profile of MAPK-type activation in response to IL-1 in different cell types is not yet clarified, neither is the possible role of *ras* in IL-1 action.

SYNERGISTIC ACTIONS OF IL-1 AT A CELLULAR LEVEL

Taking a simplistic view of cytokine action at a cellular level, cytokines act potently, pleiotropically and in synergy. There are many examples of IL-1 acting in synergy with either other cytokines or with other agents such as hormones, neurotransmitters or neuromodulators. The consequences are that a highly interactive network is established. An interesting question relating to these intercellular communications is how do the synergistic actions occur at a cellular level?

Regulation of signalling events by virtue of up- or down-regulation of receptors (either homologous or heterologous) is well established. In the present case the receptor is a potentially important target, as cytokines on binding to their specific receptors are known to induce expression of other cytokine receptors and sometimes, as in the case of IL-1, their own receptors. The classical example is in the T cell where IL-1 induces IL-2 release. This is accompanied by both IL-1-induced IL-1R expression and IL-1-induced IL-2R expression. In one move, an effective autocrine loop of IL-2 action is established and IL-2 may in addition act on neighbouring cells. Also, further signalling through the IL-1 receptor to the T cell may be enhanced. This, however, probably only represents one possible level of mediation of synergistic action at a cellular level and others (see below) may occur. Lastly, negative regulatory components which eventually turn off the signal are mobilized such as IL-1-induced IL-1RA release to antagonize IL-1R activity and IL-1-induced release of soluble receptors which compete for IL-1 binding.

In terms of mediation of IL-1 synergies at a cellular level, it is possible that intracellular sites may also be recruited. These may be at the level of key enzymes involved in signal transduction. There are a growing number of examples in the literature of IL-1 causing the synthesis of enzymes involved in signal transduction pathways. In these cases, IL-1 does not necessarily activate the enzyme in question whose synthesis it induces, but rather it seems that IL-1 may be priming the cell for a subsequent activation by an agent recruiting the signal transduction pathway of which the enzyme is a component.

Burch and colleagues [175] demonstrated that IL-1 was capable of causing the synthesis of PLA$_2$ and cyclooxygenase, but not the synthesis of PLC, in 3T3 fibroblasts. This could account for the synergy in terms of PGE$_2$ production known to occur as a result of bradykinin (BK) activation following IL-1 incubation (each on its own able to stimulate PGE$_2$ production) [175]. This was an important paper not only because a likely locus of synergy between IL-1 and BK at the level of signal transduction pathways was highlighted, but also a degree of specificity in terms of the induction by IL-1 of signalling enzyme was indicated, as the expression of PLC was unaffected by prior incubation with IL-1 in the same set of experiments. Later TNF was found to augment IL-1 and BK stimulated arachidonic acid (AA) production [176], implicating an effect at the level of PLA$_2$. We later observed that BK could synergize with IL-1 in terms of PGE$_2$ production from human osteoblast-like cells in a dramatic and rapid manner [177] and more recently Bathon and colleagues [178] have also observed synergy between BK and IL-1 in terms of PGE$_2$ production, which may be attributable to the same type of mechanism which Burch and colleagues identified.

We have observed a similar potential for synergy in respect of the effect of IL-1 on PKC synthesis. While, we could not detect an acute effect of IL-1 on the activation of PKC [53], the apparent sustained increase in PKC activity in both cytosol and membrane fractions observed after 6–48 hours of incubation with IL-1 could be blocked by simultaneous incubation with cycloheximide. We concluded that the effect of IL-1β on PKC was solely due to synthesis of the PKC enzyme and not an acute activation of PKC in response to IL-1.

A similar situation may also pertain to adenylyl cyclase. Hertelendy and colleagues[179] indicated that, in human myometrial cells, both IL-1 and TNFα caused the induction of the synthesis of adenylyl cyclase, an effect which they attributed to be at the level of the catalytic unit of cyclase as forskolin-stimulated cAMP was augmented after IL-1 incubation. Our findings in Swiss 3T3 fibroblasts and a human osteoblast-like line, MG63, essentially agree with these findings. We observed that pre-incubation of IL-1 led to augmentation of NaF and forskolin-stimulated cAMP [180], and that basal and NaF-stimulated cAMP could be prevented by pre-incubation with cycloheximide. However, in a recent publication by Beasley and colleagues, in which apparently similar findings were reported in vascular smooth muscle cells [181], IL-1 stimulated a delayed increase in cAMP production, which could be blocked by indomethacin, implying that the effect was due to IL-1-induced PG release. These systems need to be investigated further for a clear indication of the level of influence by IL-1, as an effect

other than directly on cyclase synthesis itself cannot be excluded without further experimentation.

Finally, as indicated above there are several reports of acute effects of IL-1 on the MAPK pathway, which appears to occur in many, but not all, cell types. In addition, however, it has recently been shown that IL-1 also causes the induction of synthesis of MAP kinase kinase [149], providing perhaps another intracellular site for potential synergy.

FUTURE PROSPECTS

Despite the growing body of information accumulated so far, the mechanism by which IL-1 mediates its biological effects is a long way from being completely understood. From the numerous reports acknowledged above (and others not mentioned) several points are worth noting. Because the distribution of IL-1Rs is thought to be widespread, many different cell types have been used to study transmembrane signalling by IL-1, which makes for a demanding task to review and highlight specific points of interest.

The use of cell lines for IL-1 signalling studies has been criticized on the basis that they are not "normal" cells and that this may be responsible for the diversity in signals apparently generated by IL-1. However, these lines are living systems which respond to IL-1 and other agents and, in that context give an indication of how IL-1 signals in a cell of that particular type. Using cell lines, therefore, can provide the information on the potential effects of IL-1 at a cellular level. This is just as important in primary cultures of normal cells where the populations are likely to be contaminated by adjacent populations, and even if 100% purification of a normal cell type were ever achieved, cells of different maturity or cellular subtypes of the same apparent population are likely to be still present in the culture. Single cell analysis *in situ* is not yet at the stage where we can address most of the questions that need to be answered. In the meantime, cell lines offer a good model and the fact that IL-1 acts differently in different cell lines and sub-clones may well be physiologically relevant rather than artefactual. An important point which comes out in the section above on induction of key enzymes by IL-1 is that not only is it likely that a cell is set up in a particular way to receive and tune signals, but gene expression can be modified by an external agent, such that subsequent signalling from other agents may be modified, enhanced or reduced. It should be remembered that the immediate history of the cell being studied is, therefore, very important in terms of drawing conclusions about cellular mechanisms.

Following on from this, another important feature of cellular signalling relates to the possible paracrine/autocrine mechanisms which may be operating concurrently or sequentially to the initial signalling pathways. These paracrine mechanisms are likely to be fairly specific in different cell types and even cell sub-clones, and are particularly relevant to studies of cytokine action, where initiation of paracrine actions may be implicit to the signal from the cytokine. In effect, the paracrine action(s) may be an accessory function of the signal from the initiating cytokine, but perhaps only in that cell type or sub-clone. The dose response and time course of these paracrine effects to the initiating cytokine is crucial to the interpretation of the results obtained in terms of mechanisms: the question is in essence, which changes in signalling patterns observed are due to the initiating cytokine and which to a subsequent agent produced in response to the initiating cytokine? Also, could the signals from the initiating cytokine and the induced agent(s) crosstalk? Again, the information points to the cell having its own agenda, perhaps due to its phenotype, differentiation state or level of senescence.

IL-1 has probably been the most well studied cytokine *in vivo* and *in vitro*. Despite this, there is little or no consensus on the primary intracellular signals which mediate its actions. The literature cited in this review is not exhaustive in terms of what is known about IL-1 signalling, instead specific publications have been chosen to emphasize specific areas of interest and questions that have been raised. One thing is for sure, almost as many questions remain as have been answered, which should lead to an especially exciting decade of research to come.

ACKNOWLEDGEMENTS

We gratefully acknowledge the Yorkshire Cancer Research Campaign and the Arthritis and Rheumatism Council for financial support towards the studies cited from the laboratories (Sheffield) of the authors.

NOTE ADDED IN PROOF

The extracellular portion of IL-1RI has been shown to be released in a similar fashion to IL-1RII [182]. Recent work by O'Neill and colleagues indicates that the A (active) oligomer of PTX is present as a contaminant in the B (inactive) oligomer preparation in sufficient quantities to account for the effects attributed to the B oligomer (L. O'Neill, personal communication). Goeddel and colleagues have identified a new TRAF protein. TRAF 6 and have demonstrated that IL-1 induces association of TRAF 6 with IRAK leading to activation of NF-κB [183].

REFERENCES

1. de Vos AM, Ultsch M and Kossiakoff AA. Human growth hormone and the extracellular domain of its receptor: crystal structure of the complex. *Science* 1992; **255**: 306–312.
2. Williams AF. A year in the life of the immunoglobulin superfamily. *Immunol. Today* 1987; **8**, 298.
3. Sims JE, March CJ, Cosman D, Widmer MB, Mac Donald HR, McMahan CJ, Grubin C E, Wignall JM, Jackson JL, Call SM, Friend D, Alpert AR, Gillis S, Urdal DL and Dower SK. cDNA expression cloning of the IL-1 receptor, a family of the immunoglobulin superfamily. *Science* 1988; **241**: 585–589.
4. Ullrich A and Schlessinger J. Signal transduction by receptors with tyrosine kinase activity. *Cell* 1990; **61**: 203–212.
5. Miyajima A, Kitamura T, Harada N, Yokada T and Arai K. Cytokine receptors and signal transduction. *Ann. Rev. Immunol.* 1992; **10**: 295–331.
6. Kishimoto T, Taga T and Akira S. Cytokine signal transduction. *Cell* 1994; **76**: 253–262.
7. Taniguchi T. Cytokine signalling through nonreceptor protein-tyrosine kinases. *Science* 1995; **268**: 251–255.
8. Kazlauskas RJ. Elucidating structure–mechanism relationships in lipases—prospects for predicting and engineering catalytic properties. 1994; **12**: 464–472.
9. Schlessinger J. SH2/SH3 signalling proteins. *Curr. Opin. Genet. Devel.* 1994; **4**: 25–30.
10. Blechman JM, Lev S, Barg J, Eisenstein M, Vaks B, Vogel Z, Givol D and Yarden Y. The 4th immunoglobulin domain of the stem cell factor–receptor couples ligand binding to signal transduction. *Cell* 1995; **80**: 103–113.
11. Dinarello CA. The biology of interleukin-1. In: Kishimoto K. (ed.) *Chemical Immunology*, Vol. 51, 1992: 1–32.
12. Oppenheim JJ and Gery I. From lymphodrek to interleukin 1 (IL-1). *Immunol. Today* 1993; **14**: 232–234.

13. Lomedico PT, Gubler R, Hellmann CP, Dukovich M, Giri J, Pan YE, Collier K, Semionow R, Chua AO and Mizel SB. Cloning and expression of the murine interleukin-1 cDNA in *Escherichia coli*. *Nature* 1984; **312**: 458–000

14. Auron PE, Webb AC, Rosenwasser LJ, Mucci SF, Rich A, Wolff, SM and Dinarello CA. Nucleotide sequence of human monocyte interleukin-1 precursor cDNA. *Proc. Natl. Acad. Sci. USA* 1984; **81**: 7909–7911.

15. Dower SK, Kronheim SR, March CJ, Conlon PJ, Hopp TP and Urdal DL. High affinity plasma membrane receptors for human interleukin-1—detection and characterisation. *J. Leuk. Biol.* 1985; **37**: 698.

16. Dower SK, Kronheim SR, Hopp TP, Cantrell M, Deeley M, Gillis S, Henney CS and Urdal DL. The cell surface receptors for interleukin-1α and interleukin-1β are identical. *Nature* 1986; **324**: 266–268.

17. Kilian PL, Kaffka KL, Stern AS, Woehle D, Benjamin WR, Dechiara TM, Gubler U, Farrar JJ, Mizel SB and Lomedico PT. Interleukin 1α and interleukin 1β bind to the ame receptor on T cells. *J. Immunol.* 1986; **136**: 4509–4514.

18. Matsushima K, Akahoshi T, Yamada M, Furtani Y and Oppenheim JJ. Properties of a specific interleukin-1 receptor on human Epstein–Barr virus-transformed B lymphocytes: Identity of the receptor for IL-1α and IL-1β. *J. Immunol.* 1986; **36**: 4496–4502.

19. Chin J, Cameron PM, Rupp EA and Schmidt JA. Identification of a high affinity receptor for native human Interleukin-1α and Interleukin-1β on normal human lung fibroblasts. *J. Exp. Med.* 1987; **165**: 70–86.

20. Sims JE, Gayle MA, Slack JL, Alderston MR, Bird TA, Giri JG, Colotta F, Re F, Mantovani A, Shanebeck K, Grabstein KH and Dower SK. Interleukin 1 signalling occurs exclusively via the type I receptor. *Proc. Natl. Acad. Sci., USA* 1993; **90**: 6155–6159.

21. Colotta F, Re F, Muzio M, Bertini R, Polentarutti N, Sironi M, Giri JG, Dower SK, Sims JE and Mantovani A. Interleukin-1 type II receptor-a decoy target for IL-1 that is regulated by IL4. *Science* 1993; **261**: 472–475.

22. Mc Mahan CJ, Slack JL, Mosely B, Cosman D, Lupton SD, Brunton LL, Grubin CE, Wignall JM, Jenkins NA, Brannan CI, Copeland NG, Huebner K, Croce CM, Cannizzarro LA, Benjamin D, Dower SK, Spriggs MK and Sims JE. A novel IL-1 receptor, cloned from B cells by mammalian expression, is expressed in many cell types. *EMBO J.* 1991; **10**: 2821–2832.

23. Dower SK, Sims JE, Cerretti DP and Bird TA. The interleukin-1 system: Receptors, ligands and signals. In: Kishimoto K. (ed.) *Chemical Immunology*, Vol. 51. 1992: 33–64.

24. Arend WP, Joslin FP and Massoni RJ. Effects of immune complexes on production by human monocytes of interleukin 1 or an interleukin-1 inhibitor. *J. Immunol.* 1985; **134**: 3868–3875.

25. Dinarello CA. The biology of interleukin-1. In Kishimoto K. (ed.) *Interleukins: Molecular Biology and Immunology*, Vol 51. 1992: 1–32.

26. Dripps DJ, Brandhuber BJ, Thompson RC and Eisenberg SP. Interleukin-1 (IL-1) receptor antagonist binds to the 80-kDa IL-1 receptor but does not initiate IL-1 signal transduction. *J. Biol. Chem.* 1991; **266**: 10331–10336.

27. Symonds JA, Eastgate JA and Duff GW. Purification and characterisation of a novel soluble receptor for interleukin 1. *J. Exp. Med.* 1991; **174**: 1251–1254.

28. Giri JG, Newton RC and Horuk R. Identification of soluble interleukin-1 binding proteins in cell-free supernatants. *J. Biol. Chem.* 1990; **265**: 17416–17419.

29. Greenfeder SA, Chizzonite R and Ju G. Expression cloning of a cDNA encoding a novel murine interleukin-1 receptor accessory protein. *J. Cell. Biochem.* (Suppl.) 1994; **18B**: 1124.

30. Greenfeder SA, Nunes P, Kwee L, Labow M, Chizzonite PA and Ju G. Molecular cloning and characterisation of a second subunit of the interleukin-1 receptor complex. *J. Biol. Chem.* 1995; **270**: 13757–13765.

31. Sims JE and Dower SK. Interleukin-1 receptors. *Eur. Cytokine Network* 1994; **5**: 539–546.

32. Brooks JW and Mizel SB. Interleukin-1 signal transduction. *Eur. Cytokine Network* 1994; **5**: 547–561.

33. Iyengar R. Multiple families of G_s-regulated adenylyl cyclases. In: Brown BL and Dobson PRM (eds) *Advances in Second Messenger and Phosphoprotein Research*, Vol. 28. Raven Press, 1993: 27–36.

34. Adams S, Bacskai B, Harootunian AT, Mahaut-Smith M, Sammak PJ, Taylor SS and Tsien RY. Imaging of cAMP signals and A-kinase translocation in single living cells. In: Brown BL and Dobson PRM (eds) *Advances in Second Messenger and Phosphoprotein Research*, Vol. 28. Raven Press, 1993: 167–170.

35. Quilliam LA, Dobson PRM and Brown BL. Regulation of GH_3 pituitary tumour cell adenylate cyclase activity by activators of protein kinase C. *Biochem. J.* 1989; **262**: 829–834.

36. Dobson PRM, Brown BL, Michelangeli VP, Moseley JM and Martin TJ. Interactive regulation of signalling pathways in bone cells; possible modulation of PGE_2 stimulated adenylate cyclase activity by protein kinase C. *Biochim. Biophys. Acta* 1990; **1065**: 323–326.

37. Giri JG, Kincade PW and Mizel SB. Interleukin-1-mediated induction of κ-light chain synthesis and surface immunoglobulin expression of pre-B cells. *J. Immunol.* 1984; **132**: 223–228.

38. Edelman AM, Blumenthal DK and Krebs EG. Protein serine/threonine kinases. *Ann. Rev. Biochem.* 1987; **56**: 567–613.

39. Shirakawa F, Yamashita U, Chedid M and Mizel SB. Cyclic AMP—an intracellular second messenger for interleukin-1. *Proc. Natl. Acad. Sci. USA* 1988; **85**: 8201–8205.

40. Zhang Y, Lin J-X, Yip YK and Vilcek J. Enhancement of cAMP levels and of protein kinase activity by tumor necrosis factor and interleukin 1 in human fibroblasts: role in the induction of interleukin 6. *Proc. Natl. Acad. Sci. USA* 1988; **85**: 6802–6805.

41. Chedid M, Shirakawa F, Naylor P and Mizel SB. Signal transduction pathway for IL-1-involvement of pertussis toxin-sensitive GTP-binding protein in the activation of adenylate cyclase. *J. Immunol.* 1989; **142**: 4301–4306.

42. Shirakawa F, Chedid M, Suttles J, Pollock BA and Mizel SB. Interleukin-1 and cyclic AMP induce κ-immunoglobulin light chain expression via activation of an NF-κ-B-like DNA binding protein. *Mol. Cell. Biol.* 1989; **9**: 959–964.

43. Chedid M and Mizel SB. Involvement of cyclic AMP-dependent protein kinases in the signal transduction pathway for interleukin-1β. *Mol. Cell. Biol.* 1990; **10**: 3824–3827.

44. Muñoz E, Zubiaga A, Huang C and Huber B. T. Interleukin-1 induces protein tyrosine phosphorylation in T-cells. *Eur. J. Immunol.* 1992; **22**: 1391–1396.

45. Renkonen R, Mattila P, Hayry P and Ustiov J. Interleukin 1-induced lymphocyte binding to endothelial cells. Role of cAMP as a second messenger. *Eur. J. Immunol.* 1990; **20**: 1563–1567.

46. Turunen JP, Mattila P and Renkonen R. Cyclic AMP mediates IL-1-induced lymphocyte penetration through endothelial monolayers. *J. Immunol.* 1990; **145**: 4192–4197.

47. Zeki K, Azuma H, Suzuki H, Morimoto I and Eto S. Effects of interleukin 1 on growth and 3', 5'-monophosphate generation of the rat thyroid cell line, FRTL-5 cells. *Acta Endocrin.* 1991; **124**: 60–66.

48. Weitzmann MN and Savage N. Cyclic adenosine 3' 5'-monophosphate, a second messenger in interleukin-1 mediated K562 cytostasis. *Biochem. Biophys. Res. Communi.* 1993; **190**: 564–570.

49. Abraham RT, Ho SN, Barna TJ and McKean DJ. Transmembrane signalling during interleukin 1-dependent T cell activation. *J. Biol. Chem.* 1987; **262**: 2719–2728.

50. Didier M, Aussel C, Pelassy C and Fehlman M. IL-1 signalling for IL-2 production in T cells involves a rise in phosphatidylserine synthesis. *J. Immunol.* 1988; **141**: 3078–3080.

51. Dobson PRM, Plested CP, Jones DR, Barks T and Brown BL. Interleukin-1 induces a pertussis-toxin sensitive increase in diacylglycerol accumulation in mouse thymona cells. *J. Mol. Endocrin.* 1989; **2**: R5–R8.

52. Scholz W and Altman A. Synergistic induction of interleukin-2 receptor (TAC) expression on YT cells by interleukin-1 or tumour necrosis factor alpha in combination with cyclic AMP inducing agents. *Cell. Signal.* 1989; **1**: 367.

53. Dobson PRM, Skjodt H, Plested CP, Short AD, Virdee K, Russell RGG and Brown BL. Interleukin-1 stimulates diglyceride accumulation in the absence of protein kinase C activation. *Reg. Peptides* 1990; **29**: 109–116.

54. Kasahara T, Yagisawa H, Yamashita K, Yamaguchi Y and Akiyama Y. IL1 induces proliferation and IL6 mRNA expression in a human astrocytoma cell line: positive and negative modulation by cholera toxin and cAMP. *Biochem. Biophys. Res. Commun.* 1990; **167**: 1242–1248.

55. Schlegel-Haueter SE and Aebischer F. Cyclic AMP modulates interleukin-1 action in a cytotoxic T-cell hybridoma. *Cell. Signal.* 1990; **2**: 489–496.

56. O'Neill LAJ, Bird TA, Gearing AJ H and Saklatvala J. Interleukin-1 signal transduction: Increased GTP binding and hydrolysis in membranes of murine thymoma line (EL4). *J. Biol. Chem.* 1990; **265**: 3146–3152.

57. Pfeilschifter J, Leighton J, Pignat W, Marki F and Vosbeck K. Cyclic AMP mimics, but does not mediate, interleukin-1 and tumour-necrosis-factor-stimulated phospholipase A2 secretion from rat renal mesangial cells. *Biochem. J.* 1991; **273**: 199–204.

58. Yamamoto M, Yasuda M, Shiokawa S and Nobunga M. Intracellular signal transduction in proliferation of synovial cells. *Clin. Rheum.* 1992; **11**: 92–96.

59. Garcia JGN, Stasek JE, Bahler C and Natarajan V. Interleukin-1-stimulated prostacyclin synthesis in endothelium: Lack of phospholipase C, phospholipase D, or protein kinase C involvement in early signal transduction. *J. Lab. Clin. Med.* 1992; **120**: 929–940.

60. Beasely D. Interleukin 1 and endotoxin activate soluble guanylate cyclase in vascular smooth muscle. *Am. J. Physiol.* 1990; **259**: R38–R44.

61. Beasely D, Schwartz JH and Brenner BR. Interleukin 1 induces prolonged L-arginine-dependent cyclic guanosine monophosphate and nitrite production in rat vascular smooth muscle cells. *J. Clin. Invest.* 1991; **87**: 602–608.

62. O'Neill LAJ, Ikebe T, Sarsfield SJ and Saklatvala J. The binding subunit of pertussis toxin inhibits IL-1 induction of IL-2 and prostaglandin production. *J. Immunol.* 1992; **148**: 474–479.

63. Bursten SL, Harris WE, Bomztyk K and Lovett D. Interleukin-1 rapidly stimulates lysophosphatidate acyltransferase and phosphatidate phosphohydrolase activities in human mesangial cells. *J. Biol. Chem.* 1991; **266**: 20732–20743.

64. Gray LS, Huber KS, Gray MC, Hewlett EL and Engelhard VH. Pertussis toxin effects on T lymphocytes are mediated through CD3 and not by pertussis-toxin catalysed modification of a G protein. *J. Immunol.* 1989; **142**: 1631–1638

65. Mitcham JL and Sims JE. A novel human protein that interacts with the IL-1 receptor. *Cytokine* 1995; **7**(6): 595.

66. Olivera A and Spiegel S. Sphingosine-1-phosphate as 2nd messenger in cell proliferation induced by PDGF and FCS mitogens. *Nature* 1993; **365**: 557–560.

67. Wu J, Spiegel S and Sturgill TW. Sphingosine-1-phosphate rapidly activates the mitogen-activated protein kinase pathway by a G-protein dependent mechanism. *J. Biol. Chem.* 1995; **270**: 11484–11488.

68. Michell RH and Wakelam MJO. Sphingolipid signalling. *Curr. Biol.* 1994; **4**: 370–373.

69. Bouchelouche PN, Reimert C and Bendtzen K. Effects of natural and recombinant interleukin-1α and β on cytosolic free calcium in human and murine fibroblasts. *Leukemia* 1988; **2**: 691–696.

70. Corkey BE, Geschwind JF, Deeney JT, Hale DE, Douglas SD and Kilpatrick L. Ca²⁺ responses to interleukin 1 and tumor necrosis factor in cultured human skin fibroblasts. *J. Clin. Invest.* 1991; **87**: 778–786.

71. Yu HS and Ferrier J. Interleukin-1α induces a sustained increase in cytosolic free calcium in cultured rabbit osteoclasts. *Biochem. Biophys. Res. Commun.* 1993; **191**: 343–350.

72. Rosoff PM, Savage N and Dinarello CA. Interleukin-1 stimulates diacylglycerol production in T lymphocytes by a novel mechanism. *Cell* 1988; **54**: 73–81.

73. Schettini G, Florio T, Meucci O, Scala G, Landolfi E and Grimaldi M. Effect of interleukin-1β on transducing mechanisms in 235-1 cells—1. Modulation of adenylate cyclase activity. *Biochem. Biophys. Res. Commun.* 1988; **155**: 1089–1096.

74. Dobson PRM, Plested CP, Jones DR and Brown BL. Studies on the mechanism of action of interleukin-1. In: Oppenheim JJ, Powanda MC, Kluger MJ and Dinarello CA (eds) *Progress in Leukocyte Biology*, Vol. 10. wiley. *Molecular and Cellular Biology of Cytokines*. 1990: 209–214.

75. Stanton TH, Maynard M and Bomsztyk PT. Effect of interleukin-1 on intracellular concentration of sodium, calcium and potassium in 70Z/3 cells. *J. Biol. Chem.* 1986; **261**: 5699–5701.

76. Civitelli R, Teitelbaum SL, Hruska KA and Lacey DL. IL-1 activates the Na+/H+ antiport in a murine T cell. *J. Immunol.* 1989; **143**: 4000–4008.

77. Helqvist S, Bouchelouche PN, Johannesen J and Nerup J. Interleukin 1β increases the cytosolic free sodium concentration in isolated rat Islets of Langerhans. *Scand. J. Immunol.* 1990; **32**: 53–58.

78. Kester M, Simonson MS, Mene P and Sedor JR. Interleukin-1 generates transmembrane signals from phospholipids through novel pathways in cultured rat mesangial cells. *J. Clini. Invest.* 1989; **83**: 718–723.

79. Wijelath ES, Kardasz AM, Drummond R and Watson J. Interleukin-one induced inositol phospholipid breakdown in murine macrophages: possible mechanism of receptor activation. *Biochem. Biophys. Res. Commun.* 1988; **152**: 392–397.

80. Marmiroli N, Ognibene A, Bavelloni Cinti C, Cocco L and Maraldi NM. Interleukin-1α stimulates nuclear phospholipase C in human ostreosarcoma SaO5-2 cells. *J. Biol. Chem.* 1994; **269**: 13–16.

81. Hulkower KI, Georgescu HI and Evans CH. Evidence that responses of articular chondrocytes to interleukin-1 and basic fibroblast growth factor are not mediated by protein kinase C. *Biochem. J.* 1991; **276**: 157–162.

82. Virdee K, Brown BL and Dobson PRM. The mitogenic action of recombinant bFGF in Swiss 3T3 cells is independent of early diradylglycerol production and downregulatable PKC activity. *Biochim. Biophys. Acta* 1994; 1224; 4889–494.

83. Virdee K, Brown BL and Dobson PRM. Stimulation of arachidonic acid release from Swiss 3T3 cells by recombinant bFGF: independence from phosphoinositide turnover. *Biochem. Biophys. Acta* 1994; **1220**: 171–180.

84. Presta M, Maier JAM and Ragnotti G. The mitogenic signalling pathway but not the plasminogen activator inducing pathway of basic fibroblast growth factor is mediated through protein kinase C in fetal bovine aortic endothelial cells. *J. Cell. Biol.* 1989; **109**: 1977–1884.

85. Nanberg E, Morris C, Higgins T, Vara F and Rozengurt E. Fibroblast growth factor stimulates protein kinase C in quiescent 3T3 cells without calcium mobilisation or inositol phosphate accumulation. *J. Cell Physiol.* 1990; **143**: 232–242.

86. Magnaldo I, L'Allemain G, Chambard JC, Moenner M, Barritault D and Pouyseggur J. The mitogenic signalling pathway of fibroblast growth factor is not mediated through polyphosphoinositide hydrolysis and protein kinase C activation in hamster fibroblasts. *J. Biol. Chem.* 1986; **261**: 16916–16922.

87. Tsuda T, Kaibuchi K, Kwahara Y, Fukuzaki H and Takai Y. Induction of protein kinase C activation and calcium mobilisation by fibroblast growth factor in Swiss 3T3 cells. *FEBS Letts.* 1985; **191**: 205–210.

88. Brown KD, Blakeley DM and Brigstock DR. Stimulation of polyphosphoinositide hydrolysis in Swiss 3T3 cells by recombinant fibroblast growth factors. *FEBS Letts* 1989; **247**: 227–231.

89. Rosoff PM. Characterisation of the interleukin-1-stimulated phospholipase C activity in human T lymphocytes. *Lymphokine Res.* 1989; **8**: 407–413.

90. Galella G, Medini L, Stragliotto E, Stefanini P, Rise P, Tremoli E and Galli C. In human monocytes interleukin-1 stimulates a phospholipase C active on phosphatidylcholine and inactive on phosphatidylinositol. *Biochem. Pharmacol.* 1992; **44**: 715–720.

91. Musial A, Mandal A, Coroneos E and Kester M. Interleukin-1 and endothelin stimulate distinct species of diglycerides that differentially regulate protein kinase C in mesangial cells. *J. Biol. Chem.* 1995; **270**: 21632–21638.

92. Watson J and Wijelath ES. Interleukin-one induced arachidonic acid turnover in macrophages. *Autoimmunity* 1990; **8**: 71–76.

93. Bomalaski JS, Steiner MR, Simon PL and Clark MA. IL-1 increases phospholipase A2 activity, expression of phospholipase A2-activating protein and release of linoleic acid from the murine T helper cell line EL4. *J. Immunol.* 1992; **148**: 155–160.

94. Angel J, Colard O, Chevy F and Fournier C. Interleukin-1-mediated phospholipid breakdown and arachidonic acid release in human synovial cells. *Arth. Rheum.* 1993; **36**: 158–167.

95. Molnar M, Romero R and Hertelendy F. Interleukin-1 and tumour necrosis factor stimulate arachidonic acid release and phospholipid metabolism in human myometrial cells. *Am. J. Obstet. Gynecol.* 1993; **169**: 825–829.

96. Mizel SB, Dayer JM, Krane SM and Mergenhagen SE. Stimulation of rheumatoid synovial cell collagenase and prostaglandin production by partially purified lymphocyte activating factor (interleukin-1). *Proc. Natl. Acad. Sci. USA* 1981; **78**: 2474–2477.

97. Dukovich M, Severin JM, White SJ, Yamazaki S and Mizel SB. Stimulation of fibroblast proliferation and prostaglandin production by purified recombinant murine interleukin-1. *Clin. Immunol. Immunopathol.* 1986; **38**: 381–389.

98. Hannun YA and Obeid LM. Ceramide—an intracellular signal for apoptosis. *Trends Biochem. Sci.* 1995; **20**: 73–77.

99. Kolesnick R. Signal transduction through the sphingomyelin pathway. *Mol. Chem. Neuropathol.* 1994; **21**: 287–297.

100. Spence MW. Sphingomyelinases. *Adv. Lipid Res.* 1993; **26**: 3–23.

101. Merrill AH, Hannun YA and Bell RM. Sphingolipids and their metabolites in cell regulation. *Adv. Lipid Res.* 1993; **25**: 1–24.

102. Schutze S, Potthoff K, Machleidt T, Berkovic D, Weigmann K and Kronke M. TNF activates NF-κB by phosphatidylcholine-specific phospholipase C-induced acidic sphingomyelin breakdown. *Cell* 1992; **71**: 765–776.

103. Ballou LR, Chao CP, Holness MA, Barker SC and Raghow R. Interleukin-1-mediated PGE2 production and sphingomyelin metabolism. *J. Biol. Chem.* 1992; **267**: 20044–20050.

104. Mathias S, Younes A, Kan C-C, Orlow I, Joseph C and Kolesnick RN. Activation of the sphingomyelin signalling

pathway in intact EL4 cells and in a cell free-system by IL-1β. *Science* 1993; **259**: 519–522.

105. Corbett NR, Brown BL and Dobson PRM. The signal transduction mechanism of interleukin-1β may involve a release of ceramide, with subsequent activation of protein kinases. *Cytokine* 1995; **7**: A51.

106. Mathias S, Dressler KA and Kolesnick RN. Characterisation of a ceramide-activated protein kinase-stimulation by tumour necrosis factor alpha. *Proc. Natl. Acad. Sci. USA* 1991; **88**: 10009–10013.

107. Yao B, Zhang YH, Delikat S, Mathias S, Basu S and Kolesnick R. Phosphorylation of Raf by ceramide-activated protein kinase. *Nature* 1995; **378**: 307–310.

108. Dobrowsky RT and Hannun YA. Ceramide activates a cytosolic protein phosphatase. *J. Biol. Chem.* 1992; **267**: 5048–5051.

109. Dobrowsky RT, Kamibayashi C, Mumby MC and Hannun YA. *J. Biol. Chem.* 1993; **268**: 15523–15530.

110. Raines MA, Kolesnick RM and Golde DW. Sphingomyelinase and ceramide activate mitogen-activated protein kinase in myeloid HL-60 cells. *J. Biol. Chem.* 1993; **268**: 14572–14575.

111. Guy GR, Cairns J, Siew Bee Ng and Tan YH. Inactivation of a redox-sensitive protein phosphatase during the early events of tumour necrosis factor/interleukin-1 signal transduction. *J. Biol. Chem.* 1993; **268**: 2141–2148.

112. Guy GR, Philp R and Tan YH. Activation of protein kinases and the inactivation of protein phosphatase 2A intumour necrosis factor and interleukin 1 signal transduction pathways. *Eur. J. Biochem.* 1995; **229**: 503–511.

113. Bose R, Verheij M, Haimovitzfriedman A, Scotto K, Fuks Z and Kolesnick R. Ceramide synthase mediates daunorubicin-induced apoptosis—an alternative mechanism for generating death signals. *Cell* 1995; **82**: 405–414.

114. Stephens LR, Jackson TR and Hawkins PT. Agonist stimulated synthesis of phosphatidylinositol (3,4,5)-triphosphate—a new intracellular signalling system. *Biochem. Biophys. Acta* 1993; **1179**: 27–75.

115. Poyner DR, Hanley MR, Jackson TR and Hawkins PT. Receptor regulation of phosphoinositide 3-hydroxykinase in the NG115-401L-C3 neuronal cell line—stimulation by insulin-like growth factor-1. *Biochem. J.* 1993; **290**: 901–905.

116. Nakanishi H, Brewer KA and Exton JH. Activation of the zeta-isoenzyme of protein kinase C by phosphatidylinositol 3,4,5-triphosphate. *J. Biol. Chem.* 1993; **268**: 13–16.

117. Wyman M and Arcaro A. Platelet-derived growth factor-induced phosphatidylinositol 3-kinase activation mediates actin rearrangements in fibroblasts. *Biochem. J.* 1994; **298**: 517–520.

118. Wennstrm S, Hawkins PT, Cooke F, Hara K, Yonezawa K, Kasuga M, Jackson T, Claessonwelsh L and Stephens L. Activation of phosphoinositide 3-kinase is required for PDGF-stimulated membrane ruffling. *Curr. Biol.* 1994; **4**: 385–393.

119. Cantley LC, Auger KR, Carpenter C, Duckworth B, Graziani A, Kapeller R and Soltoff S. Oncogenes and signal transduction. *Cell* 1991; **64**: 281–302.

120. Corey S, Equinoa A, Puyantheatt K, Bolen JB, Cantley L, Molinedo F, Jackson TR, Hawkins PT and Stephens LR. Granulocyte macrophage colony stimulating factor stimulates both association and activation of phosphoinositide 3OH-kinase and src-related tyrosine kinase(s) in human myeloid derived cells. *EMBOJ.* 1993; **12**: 2681–2690.

121. Jackson TR, Stephens LR and Hawkins PT. Receptor specificity of growth factor-stimulated synthesis of 3-phosphorylated lipids in Swiss 3T3 cells. *J. Biol. Chem.* 1992; **267**: 16627–16636.

122. Ratcliff H, Al-Sakkaf KA, Dobson PRM and Brown BL. A possible role for phosphatidylinositol 3′-kinase in IL-1β and prolactin-induced T47D cell proliferation. *Cytokine* 1995; **7**: A339.

123. Kaur P and Saklatvala J. Interleukin-1 and tumour necrosis factor increase phosphorylation of fibroblast proteins. *FEBS Letts.* 1988; **241**: 6–10.

124. Lovett DH, Martin M, Bursten S, Szamel M, Gemsa D and Resch K. Interleukin 1 and the glomerular mesangium III. IL-1-dependent stimulation of mesangial cell protein kinase activity. *Kidney Int.* 1988; **34**: 26–35.

125. Shiroo M and Matsushima K. Enhanced phosphorylation of 65 and 74 kDa proteins by tumor necrosis factor and interleukin-1 in human peripheral blood mononuclear cells. *Cytokine* 1990; **2**: 13–20.

126. Dornand J, Sekkat C, Mani JC and Gerber M. Lipoxygenase inhibitors suppress IL-2 synthesis—relationship with rise of $[Ca_i^{2+}]$ and the events dependent on protein kinase-C activation. *Immunol. Letts.* 1987; **16**: 101–106.

127. Bomsztyk K, Rooney JW, Iwasaki T, Rachie NA, Dower SK and Sibley CH. Evidence that interleukin-1 and phorbol esters activate NF-Kappa-B by different pathways—role of protein-kinase-C. *Cell Reg.* 1991; **2**: 329–335.

128. O'Neill LAJ. Towards an understanding of the signal transduction pathways for interleukin-1. *Biochim. Biophys. Acta* 1995; **1266**: 31–44.

129. Bird TA and Saklatvala J. IL-1 and TNF transmodulate epidermal growth factor receptors by a protein kinase C-independent mechanism. *J. Immunol.* 1989; **142**: 126–133.

130. Bird TA and Saklatvala J. Down modulation of epidermal growth factor receptor affinity in fibroblasts treated with interleukin-1 or tumour necrosis factor is associated with phosphorylation at a site other than threonine 654. *J. Biol. Chem.* 1990; **265**: 235–240.

131. Bird TA, Sleath PR, DeRoos PC, Dower SK and Virca GD. Interleukin-1 represents a new modality for the activation of extracellular signal-regulated kinases/microtubule-associated protein-2 kinases. *J. Biol. Chem.* 1991; **266**: 22661–22670.

132. Bird TA, Schule HD, Delaney PB, Sims JE, Thoma B and Dower SK. Evidence that MAP (mitogen-activated protein) kinase activation may be a necessary but not sufficient signal for a restricted subset of responses in IL-1-treated epidermoid cells. *Cytokine* 1992; **4**: 429–440.

133. Guy GR, Chua SP, Wong NS, Ng SB and Tan YH. Interleukin 1 and tumor necrosis factor activate common multiple protein kinases in human fibroblasts. *J. Biol. Chem.* 1991; **266**: 14343–14352.

134. Guesdon F, Freshney N, Waller RJ, Rawlinson L and Saklatvala J. Interleukin 1 and tumor necrosis factor stimulate two novel protein kinases that phosphorylate the

heat shock protein hsp27 and β-casein. *J. Biol. Chem.* 1993; **268**: 4236–4243.

135. Bird TA, Schule HD, Delaney PB, De Roos P, Sleath P, Dower SK and Virca GD. The interleukin-1-stimulated protein kinase that phosphorylates heat shock protein hsp27 is activated by MAP kinase. *FEBS Letts.* 1994; **338**: 31–36.

136. Saklatvala J, Kaur P and Guesdon F. Phosphorylation of the small heat-shock protein is regulated by interleukin 1, tumor necrosis factor, growth factors, bradykinin and ATP. *Biochem. J.* 1991; **277**: 635–642.

137. Guesdon F and Saklatvala J. Identification of a cytoplasmic protein-kinase regulated by IL-1 that phosphorylates the small heat-shock protein, hsp27. *J. Immunol.* 1991; **147**: 3402–3407.

138. Kyriakis JM and Avruch J. PP54 microtubule-associated protein-2-kinase—a novel serine threonine protein kinase regulated by phosphorylation and stimulated by poly-L-lysine. *J. Biol. Chem.* 1990; **265**: 17355–17363.

139. Hibi M, Lin AN, Smeal T, Minden A and Karin M. Identification of an oncoprotein-responsive and UV-responsive protein kinase that binds and potentiates the c-jun activation domain. *Genes Dev.* 1993; **7**: 2135–2148.

140. Han J, Lee JD, Bibbs L and Ulevitch RJ. A MAP kinase targetted by endotoxin and hyperosmolarity in mammalian cells. *Science* 1994; **265**: 808–811.

141. Dunford JE, Brown BL and Dobson PRM. Multiple kinases in interleukin-1 action. *Cytokine* 1994; **6**: A115a.

142. Saklatvala J, Rawlinson LM, Marshall CJ and Kracht M. Interleukin 1 and tumour factor activate the mitogen-activated protein (MAP) kinase in cultured cells. *FEBS Letts.* 1993; **334**: 189–192.

143. Kracht M, Truong O, Totty NF, Shiroo M and Saklatvala J. Interleukin-1 alpha activates two forms of p54-alpha MAP kinase in rabbit liver. *J. Exp. Med.* 1994; **180**: 2017–2025.

144. Kracht M, Shiroo M, Marshall CJ, Hsuan JJ and Saklatvala J. Interleukin-1 activates a novel protein-kinase that phosphorylates the epidermal growth-factor receptor peptide T669. *Biochem. J.* 1994; **302**: 897–905.

145. Freshney NW, Rawlinson L, Guesdon F, Jones E, Cowley S, Hsuan J and Saklatvala J. Interleukin-1 activates a novel protein-kinase cascade that results in the phosphorylation of hsp27. *Cell* 1994; **78**: 1039–1049.

146. Cuenda A, Rouse J, Doza YN, Meier R, Cohen P, Gallagher TF, Young PR and Lee JC. SB-203580 is a specific inhibitor of a MAP kinase homologue which is stimulated by cellular stresses and interleukin-1. *FEBS Letts.* 1995; **364**: 229–233.

147. Rouse J, Cohen P, Trigon S, Morange M, Alonso-Llamazares A, Zamanillo D, Hunt T and Nebreda AR. A novel kinase cascade triggered by stress and heat shock that stimulates MAPKAP kinase-2 and phosphorylation of the small heat shock proteins. *Cell* 1994; **78**: 1027–1037.

148. Huot AE, Lambert H, Lavoie JN, Guimond A, Houle F and Landry J. Characterisation of a 45 kDa 54 kDa hsp27 kinase, a stress sensitive kinase which may activate the phosphorylation-dependent protective function of mammalian 27 kDa heat shock protein hsp27. *Eur. J. Biochem.* 1995; **227**: 416–427.

149. Huwiler A and Pfeilschifter J. Interleukin-1 stimulates de-novo synthesis of mitogen activated protein kinase in glomerular mesangial cells. *FEBS Letts.* 1994; **350**: 135–138.

150. Iwasaki T, Uehara Y, Graves L, Rachie N and Bomsztyk K. Herbimycin A blocks IL-1-induced NF-κB DNA-binding in lymphoid cell lines. *FEBS Letts.* 1992; **298**: 240–244.

151. Corbett JA, Wang JL, Hughes JH, Wolf BA, Sweetland MA, Lancaster JR and McDaniel ML. Nitric oxide and cyclic GMP formation induced by interleukin 1β in Islets of Langerhans. *Biochem. J.* 1992; **287**: 229–235.

152. Joshi-Barve SS, Rangnekar VV, Sells SF and Rangnekar VM. Interleukin-1-inducible expression of gro-β via NF-kB activation is dependent upon tyrosine kinase signalling. *J. Biol. Chem.* 1993; **268**: 18018–18029.

153. Marczin N, Papapetropoulos A and Catravas JD. Tyrosine kinase inhibitors suppress endotoxin-and IL-1β-induced NO synthesis in aortic smooth muscle cells. *Heart Circ. Physiol.* 1993; **34**: H1014–H1018.

154. Guy GR, Philp R and Tan YH. Activation of protein-kinases and the inactivation of protein phosphatase 2A in tumour-necrosis factor and interleukin-1 signal-transduction pathways. *Eur. J. Biochem.* 1995; **229**: 503–511.

155. Rachie NA, Seger R, Valentine MA, Ostrowski J and Bomsztyk, K. Identification of an inducible 85 kDa nuclear protein kinase. *J. Biol. Chem.* 1993; **268**: 22143–22149.

156. Guesdon F, Ikebe T, Stylianou E, WarwickDavis J, Haskill S and Saklatvala J. Interleukin-1-induced phosphorylation of MAD3, the major inhibitor of nuclear factor kappa B of HELA cells—interference in signalling by the proteinase inhibitors, 2,4-dichloroisocoumarin and tosylphenylalenyl chloromethyleketone. *Biochem. J.* 1995; **307**: 287–295.

157. Eriksson A, Bird T, Virca GD and Dower SK. Biochemical characterisation of Type I IL-1 receptor associated kinase (IRAK) activities. *Cytokine* 1995; **7**: A331.

158. Dower SK, Eriksson A, Mitcham J, Silber D, Bird T, Gayle M, Gayle RB, Sims J and Slack J. Analysis of Type I IL-1 receptor function by alanine scanning mutagenesis. *Cytokine* 1995; **7**: A15.

159. Martin MU, Wesche H, Falk W and Resch K. The Type I IL-1 receptor signal transduction complex contains a protein kinase. *Cytokine* 1995; **7**: A273.

160. Guy GR, Cao X, Chua SP and Tan YH. Okadaic acid mimics multiple changes in early protein phosphorylation and gene expression induced by tumor necrosis factor or interleukin-1. *J. Biol. Chem.* 1992; **267**: 1846–1852.

161. Kracht M, Heiner A, Resch K and Szamel M. Interleukin-1-induced signaling in T-cells. Evidence for the involvement of phosphatases PP1 and PP2A in regulating protein kinase-C-mediated protein phosphorylation and interleukin-2 synthesis. *J. Biol. Chem.* 1993; **268**: 21066–21072.

162. Sen R and Baltimore D. Multiple nuclear factors interact with the immunoglobulin enhancer sequences. *Cell* 1986; **46**: 705–716.

163. Bomsztyk K, Toivola B, Emery DW, Rooney JW, Dower SK, Rachie NA and Sibley CH. Role of cyclic AMP in interleukin-1-induced kappa light chain gene expression in murine B-cell line. *J. Biol. Chem.* 1990; **265**: 9413–9417.

164. Osborn L, Kunkel S and Nabel GJ. Tumour necrosis factor-alpha and interleukin-1 stimulate the human immunodeficiency virus enhancer by activation of the nuclear factor KAPPA-B. *Proc. Natl. Acad. Sci. USA* 1989; **86**: 2336–2340.

165. Beg AA and Baldwin Jr AS. The IκB proteins: multifunctional regulators of Rel/NF-κB transcription factors. *Genes Dev.* 1993; **7**: 2064–2070.

166. Kieran M, Blank V, Logeat F, Vandekerckhove J, Lottspeich F, Le Bail O, Urban MB, Kourilsky P, Baeuerle PA and Israel A. The DNA subunit of NF-κB is identical to factor KBF1 and homologous to the rel oncogene product. *Cell* 1990; **62**: 1007–1018.

167. Nolan GP, Ghosh S, Liou H-C, Tempst P and Baltimore D. DNA binding and IκB inhibition of the cloned p65 subunit of NF-κB, and a rel-related polypeptide. *Cell* 1991; **64**: 961–969.

168. Haskill S, Beg AA, Tompkins SM, Morris JS, Yurochko AD, SampsonJohannes A, Mondal K, Ralph P and Baldwin AS. Characterisation of an immediate-early gene induced in adherent monocytes that encodes I-κB-like activity. *Cell* 1991; **65**: 1281–1289.

169. Ostrowski J, Sims JE, Sibley CH, Valentine MA, Dower SK, Meier KE and Bomsztyk K. A serine/threonine kinase activity is closely associated with a 65 kDa phosphoprotein specifically recognised by the κB enhancer element. *J. Biol. Chem.* 1991; **266**: 12722–12733.

170. Croston GE, Cao ZD and Goeddel DV. NF-κB activation by interleukin-1 (IL-1) requires an IL-1 receptor-associated protein kinase activity. *J. Biol. Chem.* 1995; **270**: 16514–16517.

171. Slack J, Ericksson A, Mitcham J, Silber D, Bird T, Gayle M, Gayle RB, Sims J and Dower SK. Analysis of Type I IL-1 receptor cytoplasmic region function by site-directed mutagenesis. *Cytokine* 1995; **7**: A336.

172. Muegge K, Williams TM, Kant J, Karin M, Chiu R, Schmidt A, Siebenlist U, Young HA and Durum SK. Interleukin-1 costimulatory activity on the interleukin-2 promoter via AP-1. *Science* 1989; **246**: 249–251.

173. Muegge K, Vila M, Gusella GL, Musso T, Herrlich P, Stein B and Durum SK. Interleukin-1 induction of the c-jun promoter. *Proc. Natl. Acad. Sci. USA* 1993; **90**: 7054–7058.

174. Nakajima T, Kinoshita S, Sasagawa T, Sasaki K, Naruto M, Kishimoto T and Akira S. Phosphorylation at threonine-235 by a RAS-dependent mitogen-activated protein-kinase cascade is essential for transcription factor NF-IL6. *Proc. Natl. Acad. Sci. USA* 1993; **90**: 2207–2211.

175. Burch RM, Connor JR and Axelrod J. Interleukin-1 amplifies receptor-mediated activation of phospholipase-A2 in 3T3-fibroblasts. *Proc. Natl. Acad. Sci. USA* 1988; **85**: 6306–6309.

176. Burch RM and Tiffany CW. Tumour necrosis factor causes amplification of arachidonic-acid metabolism in response to interleukin-1, bradykinin, and other agonists. *J. Cell. Physiol.* 1989; **141**: 85–89.

177. Rahman S, Bunning RAD, Dobson PRM, Evans DB, Chapman K, Jones TH, Brown BL and Russell RGG. Bradykinin stimulates the production of prostaglandin E2 and interleukin-6 in human osteoblast-like cells. *Biochim. Biophys. Acta* 1992; **1135**: 97–102.

178. Bathon JM, Croghan JC, MacGlashan Jr DW and Proud D. Bradykinin is a potent and relatively selective stimulus for cytosolic calcium elevation in human synovial cells. *J. Immunol.* 1994; **153**: 2600–2608.

179. Hertelendy F, Romero R, Miklos M, Todd H and Baldassare JJ. Cytokine-initiated signal transduction in human myometrial cells. *Am. J. Repr. Immunol.* 1993; **30**: 49–57.

180. Armour KJ, Smith NWP, Brown BL and Dobson PRM. Interleukin-1β induces the synthesis of adenylyl cyclase in Swiss 3T3 fibroblasts and MG-63 osteosarcoma cells. *Biochem. Biophys. Res. Commun.* 1995; **212**: 293–299.

181. Beasley D and McGuiggin ME. Interleukin 1 induces prostacyclin-dependent increases in cyclic AMP production and does not affect cyclic GMP production in human vascular smooth muscle cells. *Cytokine* 1995; **7**: 417–426.

182. Arend WP, Malyak M, Smith MF, Whisenand TD, Slack JL, Sims J, Giri JG, Dower SK. Binding of IL-la, IL-1b and IL-1 receptor antagonist by soluble IL-1 receptors and levels of soluble IL-1 receptors in synovial fluids. *J. Immunol.* 1994, **153**, 4767–4774.

183. Cao Z, Xiong J, Takeuchi M, Kurama T, Goeddel DV. TRAF 6 is a signal transducer for interleukin-1. *Nature.* 1996, **383**, 443–446.

16

Haemopoietic Growth Factors and Their Receptors

Caroline A. Evans[1], Clare M. Heyworth[2] and Anthony D. Whetton[1]

[1] Leukaemia Research Fund Cellular Development Unit, Department of Biochemistry and Applied Molecular Biology, UMIST, Manchester, UK and [2] Cancer Research Campaign Department of Experimental Haematology, Paterson Institute for Cancer Research, Christie Hospital NHS Trust, Manchester, UK

INTRODUCTION

Myeloid and lymphoid cells of the haemopoietic system arise in the bone marrow. Many of the mature peripheral blood cells have extremely short lifetimes, for example, neutrophils have a half-life of about 7 hours. There is therefore a constant demand for new myeloid and lymphoid blood cells but this demand can vary in both a quantitative and qualitative sense: during an infection more neutrophils, monocytes and lymphocytes are required to mount an effective host response against invading pathogenic (often bacterial) organisms; during parasitic infections eosinophils are required to engage and destroy organisms such as roundworms; anoxia or haemorrhage leads to a greater requirement for erythrocyte production. Blood cell production must therefore be constant and flexible. How are these targets met day to day and how can production be varied to meet the needs of the organism? How are these mechanisms disrupted in diseases such as aplastic anaemia, myelodysplastic syndromes or leukaemias? Basic studies have yielded some important information in these areas which in part answers these questions. Indeed, while it is beyond the scope of this review, describing the mechanisms of blood cell production in the bone marrow has been a key to developing ways in which the rate of blood cell production can be altered by the clinician for the benefit of patients. The rapid development and clinical application of two regulators of blood cell production for clinical use, granulocyte colony stimulating factor, (G-CSF) and erythroprotein (EPO), have shown the immense benefits that can be derived from manipulating blood cell production [1,2].

The Haemopoietic System

All the different types of blood cell (see Figure 16.1) are derived from the common ancestral haemopoietic stem cell. Haemopoietic stem cells can undergo self-renewal, and differentiation to produce more developmentally restricted cells which proliferate and develop to form the various types of lymphoid and myeloid cells that are present in peripheral blood and tissues. Stem cells can be defined by their ability to establish long-term constitution of haemopoiesis when transferred to an appropriate recipient animal [3]. Committed progenitor cells are derived from stem cells which have a restricted developmental capacity, e.g. granulocyte-macrophage colony forming cells (GM-CFC) or erythroid colony forming cells (BFU-e). In the bone marrow, proliferation and differentiation of haemopoietic stem and progenitor cells occur in intimate contact with the marrow stromal cells and the associated extracellular matrix (ECM) [4,5]. We are a long way from understanding the specific molecular interactions that are involved in the association between haemopoietic cells, stromal cells and ECM but certain soluble and membrane-bound cytokines are known to be able to stimulate the survival, proliferation and development of progenitor cells. It has now been established that stromal cells and

Lymphocyte Signalling: Mechanisms, Subversion and Manipulation. Edited by M. M. Harnett and K. P. Rigley © 1997 John Wiley & Sons Ltd.

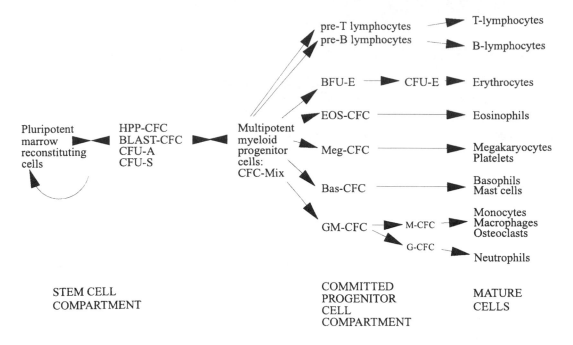

Figure 16.1   Schematic representation of haemopoiesis. Pluripotential haemopoietic stem cells have the potential to self-renew or proliferate and differentiate into a range of mature blood cell types. The stem cell compartment comprises the pluripotent marrow reconstituting cells together with high proliferative potential colony forming cells (HPP-CFC) and blast colony forming cells (Blast-CFC) and spleen and A colony forming units (CFU-S and CFU-A respectively) which have been defined in *in vitro* and *in vivo* assays. Stem cells give rise to lineage restricted, committed progenitor cells which develop to form mature myeloid and lymphoid cells

resident bone marrow macrophages produce a variety of cytokines such as stem cell factor (SCF), G-CSF and granulocyte-macrophage colony stimulating factor (GM-CSF) [6]. In some instances, these cytokines can be produced as both soluble and membrane-bound forms by stromal cells and may have a dual role to play, acting as adhesion molecules as well as growth stimulating molecules [7,8]. The levels and types of cytokines produced by stromal cells can be potentiated by cytokines—such as interleukin-1 (IL-1)—which are associated with the inflammatory response. Many of the cytokines produced from the stromal cells in a soluble from can specifically associated with components of the ECM and in particular by binding heparan sulphate proteoglycans [9,10]. Such binding of cytokines to the ECM may well be important for limiting proteolysis. Transforming growth factor β1 (TGF β) and macrophage inflammatory protein, cytokines that are also produced by bone marrow cells can similarly inhibit progenitor cell

proliferation [11,12]. Any assessment of the influence of the stromal cell environment on the regulation of stem cell proliferation must therefore take into account the balance of signals from growth inhibitors and growth promoters. However, few data are available to date on the molecular mode of action of growth inhibitors such as TGFβ and macrophage inflammatory protein-1α. The following review will therefore concentrate on an area where a great deal more information is available: the actions of growth promoters on haemopoietic cells.

MODE OF ACTION OF HAEMOPOIETIC GROWTH FACTORS

Haemopoietic growth factors (HGFs) elicit their biological effects by interaction with specific receptors expressed on the surface of target cells, resulting in activation of intracellular signalling events

[13,14] and ultimately regulation of gene transcription promoting survival, proliferation lineage commitment and development. In recent years, the availability of recombinant HGF and the molecular cloning of HGF receptors (HGF R) have facilitated the study of their mode of action. HGF R can be subdivided into two major sub-groups: members of the tyrosine kinase receptor [15] and cytokine receptor [16] superfamilies, based on the presence or absence of an intrinsic tyrosine kinase activity respectively (reviewed in reference [17]). TGF-β receptors form a unique family with an intrinsic serine/threonine kinase activity. TGFβ can act as a growth promoter as well as an inhibitor of haemopoietic cells. The reader is referred to a recent review on TGFβ R function [18].

Tyrosine Kinase HGF R

Tyrosine kinase receptors are composed of a cytoplasmic tyrosine kinase domain, a single membrane spanning domain and a large glycosylated extracellular ligand binding domain [19]. There are currently four subfamilies, class I, II, III and IV, based on structural characteristics. The SCF R, M-CSF R (encoded by *c-kit* and *c-fms* respectively) and the novel *flk1* and *flk2/flt3* ligand receptors are members of this receptor family and are designated class III (reviewed in reference [20]). They each possess five immunoglobulin (Ig)-like repeats, a single transmembrane domain and a cytoplasmic tyrosine kinase domain divided by a kinase insert region. The platelet-derived growth factor (PDGF) receptor is the prototype receptor of this family and much of the work on signal transduction of class III PTK receptors has been performed in this receptor [21].

Cytokine HGF R

The functional characteristics of the cytokine receptor superfamily have been mainly determined for the erythroprotein (EPO) and growth hormone (GH) receptors. The HGF R for EPO, G-CSF, GM-CSF and IL-2, IL-3, IL-4, IL-5, IL-6 and IL-7 are members of this superfamily and share common structural features, including a modified fibronectin type III repeat in the extracellular ligand binding domain. This comprises the cysteine motif at the *N*-terminal end and a C-terminal WS*x*WS box (where *x* is any amino acid) which exists in a degenerate form in the GH receptor. The WS*x*WS box is predicted to be situated at the base of the ligand binding site adjacent to the membrane spanning region [22], its precise function is unknown but the integrity of the WS*x*WS motif has been demonstrated to be essential for EPO receptor processing, ligand binding and activation [23,24].

Many of these receptors have low binding affinities for their respective ligands; the formation of functional high affinity receptors requires interaction with an additional subunit(s). Several members of the cytokine receptor superfamily function as homodimers and these include the receptors for GH prolactin and EPO. Other cytokine receptors require a second type of subunit to form a heterodimeric high affinity receptor capable of eliciting a signal. The subunit responsible for low affinity binding associates with a common subunit, following ligand binding, to form a high affinity receptor which mediates signal transduction [25]. A common receptor subunit, βc was first described for the IL-3, GM-CSF and IL-5 receptor complexes [16] and other receptor subunits, for example gp130 (IL-6 R subunit), have since been identified [26].

A detailed description of all the cytokine HGF Rs and their signalling pathways is outwith the scope of this chapter: thus this review will focus on the structure–functional relationships of the most well characterized receptors of the cytokine superfamily (e.g. the M-CSF, EPO, IL-3 and IL-6 families) and highlight features (e.g. receptor oligomerization, distinct ligand binding and signalling subunits and common receptor chains shared among cytokine receptor subfamilies) which are a paradigm for cytokine receptor signalling. Moreover, signalling associated with other cytokine receptor families is discussed extensively in Chapters 12 (IL-7), 15 (IL-1), 17 (e.g. IFN, IL-2, IL-4 and IL-12) and 19 (IL-2 and four receptor families) of this volume.

Signal Transduction Pathways

Agonist-stimulated tyrosine phosphorylation of receptors is a critical event in signal transduction and leads to recruitment of signalling proteins by the occupied receptor and activation of signal transduction pathways as shown in Figure 16.2. These interactions may occur via the src homology-2 (SH2) domain of the target protein and phosphorylated tyrosine residues on the receptor [27,28] and are further discussed on page 297.

HGFs stimulate several cellular signalling events, including activation of phospholipid hydrolysis and activation of the low molecular weight G protein p21^{ras}, which lead to activation of a number of downstream signalling proteins, such as raf protein kinase and mitogen-activated protein kinase (MAPK) [29]. Protein kinases such as pro-

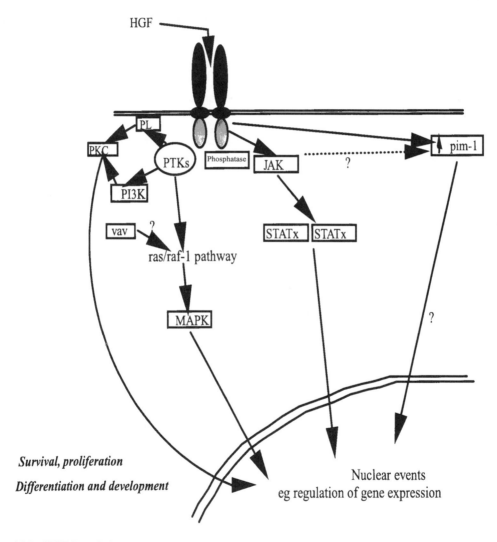

Figure 16.2 HGF R and signal transduction. A representative receptor is shown which may function as either homo-or heterodimers upon ligand binding. Receptor activation results in tyrosine phosphorylation by the intrinsic tyrosine kinase and/or association with cytosolic protein tyrosine kinase(s). These non-receptor tyrosine kinases are of two main types: members of the src family of tyrosine kinases (denoted PTK) or Janus kinases (JAKS). A range of signalling proteins are recruited by phosphotyrosine–SH2 domain interactions leading to signalling events such as formation of STAT homo/heterodimers, activation of the ras/MAP kinase pathway, activation of protein kinase C following PI-3 kinase or phospholipase activation. Expression of pim-1 kinase may also increase. Receptor signalling is subject to positive and negative feedback by tyrosine phosphatases. Receptor mediated signalling events occur at the level of the cytoplasm and nucleus to regulate survival, proliferation, differentiation and development

tein kinase C [30,31] and *pim-1* kinase may also be activated [32]. In addition to these pathways, HGF R of the cytokine receptor superfamily activate the novel JAK/STAT signalling pathways [33]. These signals converge at the level of the nucleus to regulate expression of transcription factors, including c-myc, c-fos c-jun and cell cycle regulators [34]. The next section will discuss the structure–function relationship of HGF R with reference to their interaction with specific signalling proteins. The signal transduction pathways associated with HGF R signalling are then discussed in detail.

HGF-R: STRUCTURE–FUNCTION

M-CSF, SCF-R and the Novel flk1 and flk2/flk3 PTK

These receptors are members of the tyrosine kinase receptor superfamily of which the PDGF R (α and β forms) is the most well characterized. These HGF R resemble the PDGF R in dimerizing upon ligand binding and activation leading to receptor autophosphorylation by the intrinsic tyrosine kinase and recruitment of signalling proteins via phosphotyrosine–src homology region 2 (SH2) domain interactions.

The *c-fms* proto-oncogene encodes the receptor for M-SCF (also known as CSF-1). Single point mutations in the extracellular domain of the M-CSF R are sufficient to promote constitutive tyrosine kinase activity, possibly by promoting MCSF R aggregation [35]. The viral homologue of *c-fms* is constitutively active and induces feline fibrosarcoma, the transforming potential resulting from the replacement of 40 amino acids at the *C*-terminal by 11 encoded by the virus [36]. M-CSF promotes association of the p85 subunit of P13-kinase [37–39] with Tyr^{723} (human) and Tyr^{721} (murine) autophosphorylation sites in the kinase insert domain [40]. the SHC, Grb2, Sos1 proteins and a 150 kDa tyrosine phosphorylated protein form a complex with the M-CSF R [41]. Grb2 binds Tyr^{697} of the murine M-CSF R. Mutation of this tyrosine residue and Tyr^{721}, a binding site for the p85 subunit of PI3-kinase, generates signalling defective M-CSF R [42]. The M-CSF

R also binds and activates members of the *src* family of non-receptor tyrosine kinases. This interaction is mediated in part by one of the autophosphorylation sites, Tyr^{809} (human), independently of the kinase insert domain [43]. A series of proliferation signalling defective M-CSF R have been generated, one of which promotes proliferation when the cells are cotransfected with c-myc. Further details of this and other features of M-CSF R function are reviewed by Roussel, 1994 [44].

The receptor for SCF is the product of the *c-kit* proto-oncogene. The formation of *c-kit* receptor homodimers is due to binding of the dimeric ligand and receptor–receptor interaction mediated by the extracellular domain [45,46]. The ligand for the *c-kit* receptor was identified subsequent to the description of the *c-kit* protein. Before SCF was widely available, experiments on signal transduction by the *c-kit* receptor were performed using a chimeric EGF/c-KIT receptor composed of the extracellular domain of EGF R and the transmembrane and cytoplasmic domain of C-KIT. EGF stimulated autophosphorylation of the chimeric receptor on the cytoplasmic domain of C-KIT and generation of a mitogenic signal which was associated with P13-kinase and phosphorylation of phospholipase Cγ (PI-PLCγ) and raf-1 kinase [47,48]. These results are confirmed by the demonstration that SCF stimulates autophosphorylation of c-KIT and binding to P13-kinase and PI-PLCγ [49,50]. The interkinase domain of the *c-kit* receptor binds the p85 subunit of P13-kinase and this association is dependent on phosphorylation of tyrosine residues in the kinase insert domain [51]. The phosphorylation of PI-PLCγ was not associated with increases in inositol phosphates, a result confirmed in haemopoietic cells [52]. c-KIT forms a multiprotein signal transfer complex composed of non-catalytic and enzymatically active subunits of PI-3-kinase, PI-PLCγ and GAP. PI-PLCγ and P13-kinase (p85 subunit) compete for association with c-kit binding sites *in vitro*. GAP and P13-kinase bind cooperatively which may enhance formation of the signalling complex [53].

The novel receptor tyrosine kinases encoded by *flk1* and *flk2* were identified in primitive haemopoietic cells from fetal liver by screening for protein tyrosine kinase domain sequence homology [54,55]. *Flk2* was also cloned independently from the placenta and described as *flt3* [56]. The ligand for the orphan *flk2/flt3* receptor has recently been identified and this, together with the use of chimeric *fms/flk2(flt3)* receptor has allowed further characterization of its biological role [57–59]. The *flk2/flt3* receptor is expressed exclusively on a stem cell subpopulation and acts synergistically with other HGF to expand the number of primitive progenitor cells and to promote myeloid differentiation [57,60,61]. The synergistic effects of the FLK2/FLT3 ligand were greater for HGFs binding members of the cytokine receptor superfamily than for tyrosine kinase receptors [61]. Experiments performed on the chimeric *fms/flk2(flt3)* receptor transfected into IL-3 dependent Ba/F3 cell line indicate that the chimeric receptor promoted proliferation and abrogated the requirement for IL-3. These effects were associated with ligand-binding-stimulated receptor autophosphorylation and association with PI-PLCγ, p21[ras], P13-kinase, SHC, Grb2, vav, Fyn and Src. Of these, PI-PLCγ, the p85 subunit of P13-kinase, SHC, Grb2, Src and Fyn were physically associated with the cytoplasmic domain of FLK2/FLT3 [62]. The p85 P13-kinase subunit binds a region outside of the kinase insert in the *C*-terminal end of this receptor [63]. STK1 is the human homologue of FLK2/FLT3, is selectively expressed on CD34[+] bone marrow cells and plays a role in proliferation of stem and early progenitor cells [64].

EPO R

The EPO R is a member of the cytokine receptor superfamily and, as such, lacks a tyrosine kinase domain. This receptor is relatively well characterized in terms of the relationship between structure and function and its ability to activate signalling pathways [65]. EPO-induced receptor homodimerization is thought to be an initial step in EPO-R-mediated signalling. This hypothesis is supported by data from experiments examining the effect of site directed mutagenesis of amino acid residues predicted to be critical for homodimer formation. These residues were identified by analysis of the crystal structure of ligand-bound homodimeric GH R and alignment with EPO R sequence. Substitution of Arg[129], Glu[132] or Glu[133] with cysteine resulted in constitutive receptor activation due to

formation of disulphide-linked homodimers. The cytoplasmic domain of the EPO R appears to be critical for signal transduction since co-expression of wild-type EPO R together with inactive mutant receptors possessing truncated cytoplasmic domains generates dominant negative receptors [66].

The cytoplasmic domain contains non-overlapping positive and negative growth regulatory domains [67]. This domain is composed of 236 amino acids, the membrane proximal domain (≤ amino acids) is sufficient for signal transduction and promotes EPO R association with JAK2 kinase, tyrosine phosphorylation of *vav* protein and expression of *pim-1* serine/threonine kinase. The membrane proximal region has sequence homology with the β subunit of the IL-2 R and gp130 of the IL-6/LIF/oncostatin M receptor family [68], and contains the common box 1 and 2 motifs which are essential for JAK2 activation [69]. Mutation of Trp[282] to Arg, in the membrane proximal region, abrogates the interaction with JAK2 [70]. The C-terminal region of the EPO R has been suggested to be critical for EPO mediated activation of MAP kinase via the p21[ras] pathway [71]. A serine-rich portion of the *C*-terminal region (approximately 40 amino acids) is subject to negative control and is proposed to interact with the SH-PTP1 tyrosine phosphatase [72]. The EPO R is also proposed to interact with the *syp* tyrosine phosphatase (see page 298).

The function of the WSxWS motif in the extracellular domain, close to the transmembrane region, has been investigated by use of deletion and insertion receptor mutants. The resulting mutants were compared to the wild-type EPO R following transfection into the IL-3-dependent Ba/F3 cell line. In contrast to cells expressing wild-type EPO-R, cells expressing EPO-R mutated in WSxWS motif retained EPO R in the endoplasmic reticulum and were thus unable to bind EPO or proliferate in response to this growth factor [24]. The loss of function mutation in the WSxWS motif cannot be abolished by the conversion of Arg[129] to a cysteine (a mutation which leads to constitutive activation of the wildtype EPO R), neither can it be complemented by gp55, a membrane glycoprotein encoded by Friend spleen focus forming virus which associates with the EPO R to stimulate EPO-independent proliferation [73,74].

Thus, while the data indicate that the WSxWS motif is required for EPO R activity, its precise role remains to be determined.

GM-CSF, IL-3 and IL-5 R

The GM-CSGF, IL-3 and IL-5 R are composed of two subunits, α and β. The α subunit is unique for each receptor. The α_{IL-5}; α_{GM-CSF} and α_{IL-5} subunits bind the appropriate cytokine only and, while each is unique, they share a common structural organization and bind their cognate ligand with low affinity. Functional high affinity receptors result from interaction of the α with the 135 kDa βc subunit which are proposed to form a heterodimeric receptor complex [16]. The situation in the murine system is more complex than that in man due to the existence of an IL-3-specific β subunit (β_{IL-3} or AIC2A), in addition to the βc [75]. These two β subunits form functionally equivalent high affinity IL-3 receptors [76], the major difference between the IL-3-specific AIC2A and the murine βc (also known as AIC2B) and human βc is that AIC2A can bind IL-3 with low affinity while βc is unable to bind IL-3, GM-CSF or IL-5 in the absence of the cognate α subunit [16]. The role of the IL-3 Rα subunit in IL-3 signalling has been clarified by experiments using A/J mice which have an impaired IL-3 response. This is due to aberrant splicing of the IL-3 Rα gene to generate a deletion mutant of IL-3 Rα which remains inside the cell rather than being expressed on the cell surface. The IL-3 Rβ is normally expressed thus the defect in IL-3 signalling is at the level of IL-3 Rα [77,78].

There is some evidence that the unoccupied GM-CSF Rα is constitutively phosphorylated and not subject to further phosphorylation upon binding of GM-CSF [79]. The GM-CSF Rα subunit also exists in soluble forms which lack the transmembrane and cytoplasmic domains [80]. These soluble receptor subunits antagonize GM-CSF and compete for GM-CSF binding with the transmembrane GM-CSF Rα and high affinity native GMCSF R$\alpha\beta$c complex [81]. Experiments using mutant α subunits indicate that the cytoplasmic domain of the α subunit is required for signal transduction [82–84]. Furthermore, the GM-CSF Rα subunit has been demonstrated, in the absence

of βc, to activate glucose transport in oocytes [85]—an event associated with the suppression of apoptosis in haemopoietic cells [86]. Two conserved, membrane proximal, cytoplasmic sequences designated box 1 and box 2 have been identified as candidates for signalling since deletion of this portion of the intracellular domain results in inhibition of proliferation with no change in ligand binding affinity [87,88]. The cytoplasmic domains of the GM-CSF and IL-5 receptors contain conserved proline-rich clusters which may be involved in the binding of additional receptor subunits or signalling proteins via proline-rich region–SH3 domain interactions [89].

The βc subunit also plays a role in signal transduction events stimulated by GM-CSF, IL-3 and IL-5. The cytoplasmic domain is extensively phosphorylated on threonine and tyrosine residues upon activation and associates with a number of signalling proteins. Analysis of chimeric IL-5 R composed of the extracellular domain of IL-5 Rα and the cytoplasmic domain of βc subunit have been suggested to interact with specific signal transduction pathways based on studies performed using a number of βc subunits with truncated cytoplasmic domains [90,91]. The membrane proximal region upstream of Glu^{517} is required for activation of JAK2 kinase and STAT5 [92] and for induction of *pim-1* kinase and the transcription factor, c-myc (see below): biochemical events associated with DNA synthesis and cell cycle progression [91,93]. The distal cytoplasmic region of 140 amino acid residues between Leu^{626} and Ser^{765} appears to be essential for activation of p21^{ras}, raf and MAPK to promote the suppression of apoptotic cell death [93]. Truncation of the C-terminus enhances the tyrosine phosphorylation signal [90,91] and this region has been suggested to be subject to negative regulation by interacting with SH-PTP1 (also known as haemopoietic cell phosphatase) [94].

GM-CSF, IL-3 and IL-5 promote distinct as well as common biological responses in their target cell populations. This may reflect differences in signal transduction pathways activated and/or receptor expression on the target cell. A study investigating this has been performed recently using transgenic mice constitutively expressing the IL-5Rα subunit in haemopoietic cells. The IL-5 responses of bone marrow cells from these animals

JAK concentration allows transphosphorylation and activation of JAKs which phosphorylate the receptor and signalling protein substrates which are recruited to the receptor–JAK complex [120].

The GM-CSF, IL-3 and IL-5 receptors form heterodimers between the ligand-specific α subunit and common β_c. These heterodimers activate JAK2 which binds to the β_c subunit [84, 121]. the membrane proximal region of the α subunit containing the box 1 and box 2 motifs is also required. Based on the model of EPO R interaction with JAKs, it is thought that JAK2 associates with β_c following ligand binding. The $\alpha\beta_c$ heterodimers further oligomerize and bring JAKs into close proximity. The α subunit is thought not to be required for JAK binding *per se* but for formation of $\alpha\beta_c$ receptor dimers. IL-6 receptors interact with the gp130 subunit, which is functionally equivalent to β_c, and JAK2 is activated by gp130 [122]. In common with the other cytokine receptors, activation is dependent upon the presence of the membrane proximal region containing box 1 and box 2 motifs. JAK activation does not appear to be limited to members of the cytokine receptor superfamily; the tyrosine kinase receptor for SCF, the *c-kit* protooncogene product, has also been proposed to activate the JAK2 kinase pathway, in common with IL-3 and GM-CSF [121].

JAKs phosphorylate a novel family of transcription factors, STATs (signal transducers and activators of transcription), which provide a direct link between receptor activation and gene transcription [33]. STATs are a family of proteins found predominantly in the cytosol and seven STAT proteins have been identified to date, STAT1–4, STAT5A, STAT5B and STAT6 each of which is restricted to a subset of HGF R. For example, G-CSF and IL-6 activate STAT3 [123, 124] while EPO activates STAT5 [125]. IL-3, GM-CSF and IL-5 all activate STAT5 which exists as two highly homologous STAT5A and STAT5B proteins [92, 126]. It has also been recently demonstrated that IL-3, GM-CSF and IL-5 stimulate the formation of at least two DNA-binding complexes, one of which was STAT1α in human eosinophils [127].

STATs exist in a latent form and are activated upon JAK-mediated tyrosine phosphorylation. This induces the formation of STAT homo/heterodimers, via a conserved SH2 domain, and their translocation to the nucleus via an undefined mechanism. The STAT transcription factors bind promoters of specific genes to regulate gene transcription. The STAT transcription factors bind promoters of specific genes to regulate gene transcription. In general, the DNA sequences bound by STATs resemble the gamma interferon (IFNγ) activated site (GAS) which is a regulatory element in the promoter of IFNγ inducible genes [33].

Tyrosine Phosphatases

HGFs elicit rapid but transient tyrosine phosphorylation of a range of signalling proteins. The termination of the tyrosine phosphorylation signals probably results from dephosphorylation by phosphatases but little is known about the specific proteins involved. The identification of a cytoplasmic protein tyrosine phosphatase, SH-PTP1 [128], as the product of the "moth-eaten" locus, mutations which give rise to a number of haemopoietic abnormalities [129], suggest a critical role for this phosphatase in negative regulation of haemopoiesis. SH-PTP1 (also called PTPIC, HCP and SHP) possesses a *C*-terminal phosphatase domain and two *N*-terminal SH2 domains and binds the phosphorylated EPO receptor [72], c-KIT [130] and the IL-3 receptor β_c subunit [94] via the *N*-terminal SH2 domain. *In vitro* binding studies suggest Tyr^{429}, in the cytoplasmic domain of EPO-R, as the binding site for SH-PTP1. Mutant EPO R which lack this residue cannot bind SH-PTP1 and Ba/F3 cells expressing this mutant receptor are hypersensitive to EPO which promotes autophosphorylation of JAK2 kinase over a longer time course relative to the wildtype receptor. On the basis of these results, SH-PTP1 is suggested to be specifically recruited and activated by EPO R and to play a role in inactivation of JAK2 kinase [72]. This is the first identification of a substrate for SH-PTP1 and a model for regulation of members of the JAK family of PTks has been proposed based on studies using the EPO-R. The model suggests that SH-PTP1 exists in an inactive conformation in the cytosol until EPO binds to its receptor and activates JAK2 kinase to phosphorylate Tyr^{429}. Recruitment of SH-PTP1 via SH2 domain–Tyr429 interaction is suggested to mediate both activation and translocation of the phosphatase to the vicinity of its substrate(s).

The *syp* tyrosine phosphatase (also called SHPTP2, PTP1D and PTP2C) has also been implicated in HGF R signalling, but as a positive rather than negative regulator, with a role distinct from that of SHPTP1 [131]. This phosphatase is expressed in all mammalian tissues (unlike SHPTP1 which is restricted to haemopoietic cells) and is homologous to the *corkscrew* gene product of *Drosophila* which is required for signalling downstream of the *Torso* receptor tyrosine kinase [132]. Syp is phosphorylated on tyrosine and threonine residues in response to a range of growth factors and binds, for example, to activated PDGF R [133], p21$^{brc-abl}$ fusion protein [134] and c-KIT [135] and is auto-dephosphorylated when activated [136]. The role of syp phosphatase in EPO R function has recently been investigated using Mo7e cells transfected with EPO R [137]. Syp was tyrosine phosphorylated upon EPO stimulation, bound to the phosphorylated EPOR and complexed with the Grb2 protein which is a component of the ras/raf-1 kinase signalling pathway. These data suggest that syp may act as an adaptor protein between tyrosine phosphorylated EPO R and Grb2 to propagate cellular signalling. IL-3 and GM-CSF also induce tyrosine phosphorylation and activation of the human homologue of syp, SHPTP2, which associates with Grb2 and PI3-kinase and is proposed to play a role in the integration of signals from the IL-3 and GM-CSF receptors to the p21ras and PI3-kinase pathways [138]. The dynamic relationship between protein tyrosine phosphorylation/dephosphorylation by kinases/phosphatases appears to be an important determinant in HGF-R-mediated regulation of haemopoiesis.

Phospholipid Signalling Pathways

Phospholipid hydrolysis by phospholipases is an early event following receptor–ligand interaction and results in the generation of second messenger molecules which activate downstream effector signalling proteins such as the serine/threonine kinase, protein kinase C (see page 300). Activation of the lipid kinase, phosphatidylinositol 3-kinase also occurs in response to several G-HGFs.

The inositol phospholipid, phosphatidylinositol 4,5-bisphosphate, is subject to receptor-mediated cleavage by phosphoinositide-specific phospholipase C to generate the second messengers sn-1, 2-diacylglycerol (DAG) and inositol 1,4,5-trisphosphate. There are at least several isozymes of PI-PLC enzymes [139], of which PI-PLCβ and PI-PLCγ are differentially regulated by G proteins and tyrosine kinases respectively. SCF R and the flk2/flt3 receptor phosphorylate PI-PLCγ but the functional significance of this event remains to be determined [48, 49, 62]. To date, there is no evidence for inositol phospholipid hydrolysis by M-CSF, IL-3, GM-CSF, IL-5, G-CSF, IL-6 or EPO. IL-3 and M-CSF can stimulate phosphatidylcholine (PtdCho) breakdown [52, 140–143].

P13-kinase is a kinase which phosphorylates the D3 position of phosphatidylinositol, phosphatidylinositol-4-phosphate and phosphatidylinositol-4,5-bisphosphate [144] and is involved in mitogenic signalling. The enzyme is composed of two subunits: an 85 kDa regulatory subunit and a 110 kDa catalytic subunit which is activated upon binding of the 85 kDa subunit to a phosphorylated receptor via the SH2 domain. IL-5 has been demonstrated to tyrosine phosphorylate PI3-kinase in the IL-5 dependent, early B cell line Y16 [84]. GM-CSF, IL-3, M-CSF and SCF and EPO also stimulate tyrosine phosphorylation and activation of PI3-kinase [37, 49, 50, 113, 145]. The novel *Flk2/Flt3* receptor has also been suggested to activate PI3-kinase [62]. The precise role of PI3-kinase activation in HGF R action remains to be defined but in the case of the PDGF R, its activation is critical for mitogenic signalling by the PDGF R [146] and mediates 70 kDa S6 kinase activation [147]. Furthermore, phosphatidylinositol 3, 4, 5-trisphosphate, the product of PI3-kinase, can activate the PKCζ isozyme [148].

Activation of Serine/Threonine Kinase Pathways

p21^{ras}/raf/MAP Kinase Pathway

p21^{ras} is activated by a number of HGF including IL-3, GM-CSF, G-CSF and the *flk2/flt3* ligand [62, 107, 149–151] but not IL-4 [152]. p21^{ras} is subject to regulation by the 120 kDa p21^{ras}

GTPase activating protein (GAP) [153, 154] which negatively regulates p21^{ras} by enhancing the rate of GTP hydrolysis by its intrinsic GTPase (this is low compared to Gα of heterotrimeric G proteins). The tyrosine phosphorylation of GAP is stimulated in response to M-CSF but the functional significance of this is unclear since there is no effect on activity [38]. It has recently been reported that IL-3, GM-CSF, IL-5 and SCF activate p21^{ras} in the absence of tyrosine phosphorylation of GAP suggesting that GAP phosphorylation is not a major mechanism for HGF regulation of p21^{ras} [150]. Furthermore, SCF activation of p21^{ras} in Mo7e and R6X myeloid cells is accompanied by minimal phosphorylation of GAP [155].

HGF-mediated activation of p21^{ras} can occur via tyrosine phosphorylation of the SHC adaptor protein [62, 152, 156] which functions as a bridge or adaptor molecule in association with the GRB2 adaptor protein. Overexpression of SHC has been demonstrated to potentiate proliferation in response to GM-CSF and recruit the GRB2 and Son of Sevenless (SoS) proteins [157]. Cytokines stimulate binding of Grb2 to SHC and SoS proteins via its SH2 and SH3 domains respectively. SoS was first identified in *Drosophila* and is a GDP nucleotide exchange protein (GNEP) which activates p21^{ras} by promoting dissociation of GDP. Activated p21^{ras} promotes translocation and activation of raf kinase at the plasma membrane and activation of mitogen-activated protein (MAP) kinase leading to transcriptional activation. Several other GNEP proteins have been identified including the novel vav protein (see page 301).

The Raf-1 proto-oncogene product is a serine/threonine kinase and is activated in response to a range of growth factors [158]. Serine hyperphosphorylation and activation of raf-1 kinase have been detected in response to a number of HGF including IL-3, GM-CSF, M-CSF, SCF and EPO [155, 159–161]. The precise mechanism of raf-1 activation and regulation is unclear but possibilities include phosphorylation by PKC [162, 163] and protein kinase A [164]. In addition, raf-1 kinase can bind p21^{ras}, an interaction which requires the effector domain of p21^{ras} (residues 32–40) and two distinct regions in the *N*-terminal region of raf-1 protein [165]. Raf-1 kinase is recruited to the plasma membrane by ras for activation by tyrosine phosphorylation [166] and is

currently considered to function downstream of p21^{ras} and MAP kinase kinase [167].

MAP kinase is a cytoplasmic serine/threonine kinase which phosphorylates and regulates a range of cytosolic and nuclear substrates. These include the cytoskeletal protein, microtubule-activated-protein-2, and nuclear proteins such as lamins and transcription factors. Nuclear translocation of MAP kinase occurs after prolonged activation and leads to phosphorylation of nuclear proteins providing a potential link between cytoplasmic signalling and transcriptional regulation [168]. The activation of MAP kinase occurs upon phosphorylation of both threonine and tyrosine residues and is mediated by MAP kinase kinase (MAPKK or MEK), a dual specificity threonine/tyrosine kinase which is in turn regulated by phosphorylation by raf-1 kinase and MAP kinase kinase kinase [169].

To date, several HGFs, including IL-3, GM-CSF, IL-5, G-CSF, IL-6 and SCF but not IL-4, have all been demonstrated to activate the Raf-1/MAP kinase pathway [29, 107, 155, 156, 160, 170, 171].

Protein Kinase C

PKC is a family of structurally related cytoplasmic serine/threonine kinases whose activation is required for the regulation of many key cellular events including survival, proliferation and differentiation [172]. PKC was initially described as a Ca^{2+}/PtdSer/DAG dependent protein kinase but multiple isozymes of PKC have since been identified which have distinct differences in their biochemical properties. These include classical Ca^{2+}-dependent cPKCs α, β_I, β_{II}, γ, novel Ca^{2+}-independent nPKCs δ, , η and θ and atypical phorbol-ester-insensitive isozymes PKCζ and λ which have different substrate specificities, cofactor sensitivities and are subject to cell-specific expression, dependent on cell type and differentiation status [173]. While sn-1,2 DAG is the major physiological activator of PKC, other lipids such as lysophosphatidylcholine and free fatty acids, e.g. arachidonic acid, can activate PKC; PKCζ may be selectively activated by PI 3, 4, 5P$_3$ a product of PI3-kinase [148] (see page 299).

PKCs are composed of a single polypeptide chain of approximately 80 kDa with two domains: an *N*-terminal regulatory domain and a *C*-terminal of catalytic domain. PKC activation plays a role in signal transduction by a number of HGF R including GM-CSF, IL-3, M-CSF and EPO to promote cell survival, proliferation and lineage determination. For example, PKC activation is involved in determining lineage commitment of GM-CFC, in which M-CSF-stimulated macrophage development is associated with nuclear localization and activation of PKCα [52].

PKC can interface with other signalling pathways by phosphorylation of signalling proteins. In the case of HGF R action this can occur at the level of MAP kinase pathway via phosphorylation of raf-1 serine/threonine kinase (discussed on page 000).

pim-1 Kinase

The pim-1 proto-oncogene encodes a cytoplasmic serine/threonine protein kinase(s) which is expressed predominantly in haemopoietic cells [174, 175]. Pim-1 kinase has been implicated in mast cell IL-3 signalling using pim-1 transgenic mice [176] and is associated with the induction of proliferation by the cytokine receptors GM-CSF, IL-3, IL-6 and G-CSF. No such activation was observed in GM-CSF stimulated neutrophils which are terminally differentiated suggesting that pim-1 kinase activation depends on cell phenotype and cellular response [32]. The pim-1 gene is suggested to be constitutively activated and its levels regulated by transcriptional attenuation [177]. Expression of pim-1 kinase and tyrosine phosphorylation of vav are associated with JAK2-mediated EPO-stimulation proliferation [178]. Activation of JAKs appears to be a common event in signalling by cytokine receptors which activate pim-1 kinase.

Vav

The vav protooncogene encodes a 95 kDa protein which is expressed exclusively in haemopoietic cells [179]. Vav protein possesses sequences commonly found in transcription factors: helix–loop–helix/leucine zipper-like and zinc finger and nuclear localization signals [180]. Deletion of the helix–loop–helix/leucine-like motif results in the formation of an oncogenic form of vav. Vav is tyrosine phosphorylated in response to a range of cytokines including the HGFs. For example, SCF (but not IL-3 or GM-CSF) stimulates the tyrosine phosphorylation of vav in the Mo7e and TF-1 cell lines [181]. Vav is also implicated in EPO R and FLK-2/FLT-3 signalling pathways and is tyrosine phosphorylated in response to EPO and FLK-2/FLT-3 ligand respectively [62, 178]. The vav protein has therefore been described as a novel signalling molecule with potential to transduce tyrosine phosphorylation into transcriptional events [179]. In this respect, the fact that vav contains an SH2 domain flanked by two SH3 domains, a pleckstrin homology domain, a cysteine-rich region with homology to PKC and a region structurally related to proteins such as ras GPTase which have guanine nucleotide exchange activity, increases the likelihood that vav plays a role in HGF signalling. The central domain of vav shows homology to the human dbl oncogene which encodes a GDP–GTP exchange factor for the rho/CDC42 family of low molecular weight G proteins [182, 183]. Following on from this observation, vav has been suggested to play a role in haemopoietic cell signalling by coupling tyrosine kinase pathways (via the SH3–SH2–SH3 domain) to ras-like GTPases through regulation of guanine nucleotide exchange [184, 185]. The SH2 domain of vav thus appears to play a critical role in its function since mutation of conserved residues in the vav SH2 domain reduces its transforming potential [186]. The precise role of vav in haemopoiesis is unknown but vav has been shown to play a role in T and B cell development using vav-deficient embryonic stem cells [187, 188].

The yeast two-hybrid system, which detects in vivo interactions via reconstitution of a functional transcriptional activator, has recently identified human heterogeneous nuclear ribonucleoprotein K as an additional potential partner for the SH3–SH2–SH3 domain of vav in mouse fibroblasts and the haemopoietic Jurkat T cell line. A novel 45 kDa poly(rC) binding protein associates with vav in vitro. Based on these results, vav has been proposed to play a role in the regulation of RNA biogenesis by modulating the function of poly(rC)-specific ribonucleoproteins [189].

Gene transcription

HGFs promote the survival, proliferation and differentiation of haemopoietic progenitor cells and regulate mature cell function. These responses are dependent on transcriptional events, subject to repression or activation by transcription factors located in the nucleus which bind DNA and interact with the transcription apparatus. HGF R signals to promote these responses converge at the level of transcription. The mechanism by which the specificity of the transcriptional response is generated is complex but certain transcription factors are known to be associated with specific cellular events, e.g. GATA is associated with erythroid development while the immediate early response genes, such as c-myc, c-fos, c-jun and egr-1 mediate cell cycle progression. The reader is referred to two recent reviews for detailed discussions of transcriptional regulation and the role of transcription factors in haemopoiesis respectively [190, 191].

ACKNOWLEDGEMENTS

The authors are supported by the BBSRC and the Leukaemic Research Fund. Thanks to Sian Nicholls for help with Figure 16.1.

REFERENCES

1. Maurer AB, Ganser A, Seipelt G *et al*. Changes in erythroid progenitor cell and accessory cell compartments in patients with myelodysplatic syndromes during treatment with all-trans retinoic acid and haemopoietic growth factors. *Br. J. Haematol.* 1995; **89**: 449–456.
2. Frampton JE, Lee CR, Faulds D *et al*. Filgrastim. A review of its pharmacological properties and therapeutic efficacy in neutropenia. *Drugs* 1994; **48**: 731–760.
3. Till JE and McCulloch EA. A direct measurement of the radiation sensitivity of normal mouse bone marrow cells. *Rad. Res.* 1961; **14**: 213–222.
4. Adams JC and Watt FM. Regulation of development and differentiation by the extracellular matrix. *Development* 1993; **117**: 1183–1198.
5. Yoder MC and Williams DA. Matrix molecular interactions with the haemopoietic stem cell. *Exp. Haem.* 1995; **23**: 961–967.
6. Gualtieri RJ, Shadduck RK, Baker DG and Quesenberry PJ. Hematopoietic regulatory factors produced in long-term murine bone marrow cultures and the effect of in vitro irradiation. *Blood* 1984; **64**: 516–525.
7. Rettenmier CW and Roussel MF. Differential processing of colony-stimulating factor 1 procursors encoded by two human cDNAs. *Mol. Cell. Biol.* 1988; **8**: 5026–5034.
8. Williams DE. The steel factor. *Dev. Biol.* 1992; **151**: 368–376.
9. Goron MY, Riley GP, Watt SM and Greaves MF. Compartmentalization of a haematopoietic growth factor (GM-CSF) by glycosaminoglycans in the bone marrow microenvironment. *Nature* 1987; **326**: 403–405.
10. Roberts R, Gallagher J, Spooncer E, Allen TD, Bloomfield F and Dexter TM. Heparan sulphate bound growth factors: a mechanism for stromal cell mediated haemopoiesis. *Nature* 1988; **332**: 376–378.
11. Eaves CJ, Cashman JD, Kay RJ *et al*. Mechanisms that regulate the cell cycle status of very primitive hematopoietic cells in long-term human marrow cultures. II. Analysis of positive and negative regulators produced by stromal cells within the adherent layer. *Blood* 1991; **78**: 110–117.
12. Keller JR, Bartelmez SH, Sitnicka E *et al*. Distinct and overlapping effects of macrophage inflammatory protein 1 alpha and transforming growth factor beta on hematopoietic progenitor/stem cell growth. *Blood* 1994; **84**: 2175–2181.
13. Ihle JN, Witthuhn B, Tang B, Yi T and Quelle FW. Cytokine receptors and signal transduction. *Baillière's Clin. Haematol.* 1994, **7**: 17–48.
14. Pearson MA, Heyworth CM., Owen-Lynch PJ, Dexter TM and Whetton AD. Molecular mechanisms of signal transduction in haemopoietic lineages. In: Testa NG, Lord BI and Dexter T.M (eds), *Haemopoietic Cell Lineages*. New York: Marcel Dekker (in press).
15. Ullrich A and Schlessinger J. Signal transduction by receptors with tyrosine kinase activity. *Cell* 1990; **61**: 203–212.
16. Miyajima A. Molecular structure of the IL-3, GM-CSF and IL-5 receptors. *Int. J. Cell Cloning* 1992; **10**: 126–134.
17. O'Farrell A-M, Kinoshita T and Miyajima A. The haemopoetic cytokine receptors. In: Whetton AD (ed.) *Blood Cell Biochemistry*. 1996; **7**: 1–40.
18. Attisano L, Wrana JL, Lopez CF and Massague J. TGF-beta receptors and actions. *Biochim. Biophys. Acta* 1994; **1222**: 71–80.
19. Hunter T and Cooper JA. Protein-tyrosine kinases. *Ann. Rev. Biochem.* 1985; **54**: 897–930.
20. Rosnet O and Birnbaum D. Hematopoietic receptors of class III receptor-type tyrosine kinases. *Crit. Rev. Oncog.* 1993; **4**: 595–613.
21. van der Geer P, Hunter T and Lindberg RA. Receptor protein-tyrosine kinases and their signal transduction pathways. *Ann. Rev. Cell Biol.* 1994; **10**: 251–337.
22. Bazan, J.F. Structural design and molecular evolution of a cytokine receptor superfamily. *Proc. Natl. Acad. Sci. USA* 1990; **87**: 6934–6938.
23. Watowich SS, Yoshimura A, Longmore GD, Hilton DJ, Yoshimura Y and Lodish HF. Homodimerization and constitutive activation of the erythropoietin receptor. *Proc. Natl. Acad. Sci. USA* 1992; **89**: 2140–2144.
24. Yoshimura A, Zimmers T, Neumann D, Longmore G, Yoshimura Y and Lodish HF. Mutations in the Trp-Ser-X-Trp-Ser motif of the erythropoietin receptor abolish processing, ligand binding, and activation of the receptor. *J. Biol. Chem.* 1992; **267**: 11619–11625.
25. Cosman, D. The hematopoietin receptor superfamily. *Cytokine* 1993; **5**: 95–106.

26. Kitamura T, Ogorochi T and Miyajima A. Multimeric cytokine receptors. *Trends Endocrinol. Metab.* 1994; **5**: 8–14.

27. Fantl WJ, Escobedo JA, Martin GA *et al*. Distinct phosphotyrosines on a growth factor receptor bind to specific molecules that mediate different signalling pathways. *Cell* 1992; **69**: 413–423.

28. Marengere LE and Pawson T. Structure and function of SH2 domains. *J. Cell Sci.* (Suppl.) 1994; **18**: 97–104.

29. Welham MJ, Duronio V, Sanghera JS, Pelech SL and Schrader JW. Multiple hemopoietic growth factors stimulate activation of mitogen-activated protein kinase family members. *J. Immunol.* 1992b; **149**: 1683–1693.

30. Fields AP, Pincus SM, Kraft AS and May WS. Interleukin-3 and bryostatin 1 mediate rapid nuclear envelope protein phosphorylation in growth factor-dependent FDC-P1 hematopoietic cells. A possible role for nuclear protein kinase C. *J. Biol. Chem.* 1989; **264**: 21896–21901.

31. Shearman MS, Heyworth CM, Dexter TM, Haefner B, Owen PJ and Whetton AD. Haemopoietic stem cell development to neutrophils is associated with subcellular redistribution and differential expression of protein kinase C subspecies. *J. Cell Sci.* 1993; **104**: 173–180.

32. Lilly M, Le T, Holland P and Dendrickson SL. Sustained expression of the pim-1 kinase is specifically induced in myeloid cells by cytokines whose receptors are structurally related. *Oncogene* 1992; **7**: 727–732.

33. Ihle JN and Kerr IM. Jaks and Stats in signaling by the cytokine receptor superfamily. *Trends Genet.* 1995; **11**: 69–74.

34. Sherr CJ. Growth factor-regulated G1 cyclins. *Stem Cells Dayt* 1994; **1**: 47–55.

35. van Daalen Wetters T, Hawkins SA, Roussel MF and Sherr CJ. Random mutagenesis of CSF-1 receptor (FMS) reveals multiple sites for activating mutations within the extracellular domain. *EMBO J.* 1992; **11**: 551–557.

36. Browning PJ, Bunn HF, Coline A, Shuman M and Nienhuis AW. "Replacement" of COOH-terminal truncation of v-fms with c-fms sequences markedly reduces transformation potential. *Proc. Natl. Acad. Sci. USA* 1986; **83**: 7800–7805.

37. Varticovski L, Druker B, Morrison D, Cantley L and Roberts T. The colony stimulating factor-1 receptor associates with and activates phosphatidylinositol-3 kinase. *Nature* 1989; **342**: 699–702.

38. Reedijk M, Liu XQ and Pawson T. Interaction of phosphatidylinositol kinase, GTPase-activating protein (GAP), and GAP-associated proteins with the colony-stimulating factor 1 receptor. *Mol. Cell Biol.* 1990; **10**: 5601–5608.

39. Shurtleff SA, Downing JR, Rock CO, Hawkins SA, Roussel MF and Sherr CJ. Structural features of the colony-stimulating factor 1 receptor that affect its association with phosphatidylinositol 3-kinase. *EMBO J.* 1990; **9**: 2415–2421.

40. Reedijk M, Liu X, van der Geer P, Letwin K, Waterfield MD, Hunter T and Pawson T. Tyr721 regulates specific binding of the CSF-1 receptor kinase insert to PI3′-kinase SH2 domains: a model for SH2-mediated receptor-target interactions. *EMBO J.* 1992; **11**: 1365–1372.

41. Lioubin MN, Myles GM, Carlberg K, Bowtell D and Rohrschneider LR. Shc, Grb2, Sos1, and a 150-kilodalton tyrosine- phosphorylated protein form complexes with

Fms in hematopoietic cells. *Mol. Cell. Biol.* 1994; **14**: 5682–5691.

42. van der Geer P and Hunter T. Mutation of Tyr697, a GRB2-binding site, and Tyr721, a PI3-kinase binding site, abrogates signal transduction by the murine CSF-1 receptor expressed in Rat-2 fibroblasts. *EMBO J.* 1993; **12**: 5161–5172.

43. Courtneidge SA, Dhand R, Pilat D, Twamley GM, Waterfield MD and Roussel MD. Activation of Src family kinases by colony stimulating factor-1, and their association with its receptor. *EMBO J.* 1993; **12**: 943–950.

44. Roussel MF. Signal transduction by the macrophage-colony-stimulating factor receptor (CSF-1R). *J. Cell Sci.* (Suppl.) 1994; **18**: 105–108.

45. Lev S, Yarden Y and Givol D. Dimerization and activation of the kit receptor by monovalent and bivalent binding of the stem cell factor. *J. Biol. Chem.* 1992a; **267**: 15970–15977.

46. Lev S, Yarden Y and Givol D. A recombinant ectodomain of the receptor for the stem cell factor (SCF) retains ligand-induced receptor dimerization and antagonizes SCF-stimulated cellular responses. *J. Biol. Chem.* 1992b; **267**: 10866–10873.

47. Lev S, Yarden Y and Givol D. Receptor functions and ligand-dependent transforming potential of a chimeric kit proto-oncogene. *Mol. Cell. Biol.* 1990; **10**: 6064–6068.

48. Lev S, Givol D and Yarden Y. A specific combination of substrates is involved in signal transduction by the kit-encoded receptor. *EMBO J.* 1991; **10**: 647–654.

49. Rottapel R, Reedijk M, Williams DE *et al*. The Steel/W transduction pathway: kit autophosphorylation and its association with a unique subset of cytoplasmic signaling proteins is induced by the Steel factor. *Mol. Cell. Biol.* 1991; **11**: 3043–3051.

50. Welham MJ and Schrader JW. Steel factor-induced tyrosine phosphorylation in murine mast cells. Common elements with IL-3-induced signal transduction pathways. *J. Immunol.* 1992a; **149**: 2772–2783.

51. Lev S, Givol D and Yarden Y. Interkinase domain of kit contains the binding site for phosphatidylinositol 3′kinase. *Proc. Natl. Acad. Sci. USA* 1992c; **89**: 678–682.

52. Whetton AD, Heyworth CM, Nicholls SE *et al*. Cytokine-mediated protein kinase C activation is a signal for lineage determination in bipotential granulocyte macrophage colony-forming cells. *J. Cell. Biol.* 1994; **125**: 6551–6559.

53. Herbst R, Shearman MS, Jallal B, Schlessinger J and Ullrich A. Formation of signal transfer complexes between stem cell and platelet-derived growth factor receptors and SH2 domain proteins in vitro. *Biochemistry* 1995; **34**: 5971–5979.

54. Matthews W, Jordan CT, Wiegand GW, Pardoll D and Lemischka IR. A receptor tyrosine kinase specific to hematopoietic stem and progenitor cell-enriched populations. *Cell* 1991a; **65**: 1143–1152.

55. Matthews W, Jordan CT, Gavin M, Jenkins NA, Copeland NG and Lemischka IR. A receptor tyrosine kinase cDNA isolated from a population of enriched primitive hematopoietic cells and exhibiting close genetic linkage to c-kit. *Proc. Natl. Acad. Sci. USA* 1991b; **88**: 9026–9030.

56. Rosnet O, Marchetto S, deLapeyriere O and Birmbaum D. Murine Flt3, a gene encoding a novel tyrosine kinase

receptor of the PDGFR/CSF1R family. *Oncogene* 1991; **6**: 1641–1650.

57. Hannum C, Culpepper J, Campbell D *et al.* Ligand for FLT3/FLK2 receptor tyrosine kinase regulates growth of haematopoietic stem cells and is encoded by variant RNAs. *Nature* 1994; **368**: 643–648.

58. Lyman SD, Brasel K, Rousseau AM and Williams DE. The flt3 ligand: a hematopoietic stem cell factor whose activities are distinct from steel factor. *Stem Cells Dayt* 1994: **1**: 99–107.

59. Rossner MT, McArthur GA, Allen JD and Metcalf D. Fms- like tyrosine kinase 3 catalytic domains can transduce a proliferative signal in FDC-P1 cells that is qualitatively similar to the signal delivered by c-Fms. *Cell Growth Differ.* 1994; **5**: 549–555.

60. Zeigler FC, Bennett BD, Jordan CT *et al.* Cellular and molecular characterization of the role of the flk-2/flt-3 receptor tyrosine kinase in hematopoietic stem cells. *Blood* 1994; **84**: 2422–2430.

61. Jacobsen SE, Okkenhaug C, Myklebust J, Veiby OP and Lyman SD. The FLT3 ligand potently and directly stimulates the growth and expansion of primitive murine bone marrow progenitor cells in vitro: synergistic interactions with interleukin (IL) 11, IL-12, and other hematopoietic growth factors. *J. Exp. Med.* 1995; **181**: 1357–1363.

62. Dosil M, Wang S and Lemischka IR. Mitogenic signalling and substrate specificity of the Flk/Flt3 receptor tyrosine kinase in fibroblasts and interleukin 3-dependent hematopoietic cells. *Mol. Cell. Biol.* 1993; **13**: 6572–6585.

63. Rottapel R, Turck CW, Casteran N *et al.* Substrate specificites and identification of a putative binding site for PI3K in the carboxy tail of the murine Flt3 receptor tyrosine kinase. *Oncogene* 1994; **9**: 1755–1765.

64. Small D, Levenstein M, Kim E *et al.* STK-1, the human homolog of Flk-2/Flt-3, is selectively expressed in CD34+ human bone marrow cells and is involved in the proliferation of early progenitor/stem cells. *Proc. Natl. Acad. Sci. USA* 1994; **91**: 459–463.

65. Youssoufian H, Longmore G, Neumann D, Yoshimura A and Lodish HF. Structure, function, and activation of the erythropoietin receptor. *Blood* 1993; **81**: 2223–2236.

66. Watowich SS, Hilton DJ and Lodish HF. Activation and inhibition of erythropoietin receptor function: role of receptor dimerization. *Mol. Cell. Biol.* 1994; **14**: 3535–3549.

67. D'Andrea AD, Yoshimura A, Youssoufian H, Zon LI, Koo JW and Lodish HF. The cytoplasmic region of the erythropoietin receptor contains non-overlapping positive and negative growth- regulatory domains. *Mol. Cell. Biol.* 1991; **11**: 1980–1987.

68. D'Andrea AD, Fasman GD and Lodish HF. Erythropoietin receptor and interleukin-2 receptor beta chain: a new receptor family. *Cell* 1989; **58**: 1023–1024.

69. Witthuhn BA, Quelle FW, Silvennoinen O *et al.* JAK2 associates with the erythropoietin receptor and is tyrosine phosphorylated and activated following stimulation with erythropoietin. *Cell* 1993; **74**: 227–236.

70. Miura O, Nakamura N, Quelle FW, Witthuhn BA, Ihle JN and Aoki N. Erythropoietin induces association of the JAK2 protein kinase with the erythropoietin receptor in vivo. *Blood* 1994; **84**: 1501–1507.

71. Miura Y, Miura O, Ihle JN and Aoki N. Activation of the mitogen-activated protein kinase pathway by the erythropoietin receptor. *J. Biol. Chem.* 1994; **269**: 29962–29969.

72. Klingmuller U, Lorenz U, Cantley LC, Neel BG and Modish HF. Specific recruitment of SH-PTP1 to the erythropoietin receptor causes inactivation of JAK2 and termination of proliferative signals. *Cell* 1995; **80**: 729–738.

73. Li JP, D'Andrea AD, Lodish HF and Baltimore D. Activation of cell growth by binding of Friend spleen focus-forming virus gp55 glycoprotein to the erythropoietin receptor. *Nature* 1990; **343**: 762–764.

74. Li JP, Hu HO, Niu QT and Fang C. Cell surface activation of the erythropoietin receptor by Friend spleen focus-forming virus gp55. *J. Virol.* 1995; **69**: 1714–1719.

75. Gorman DM, Itoh N, Kitamura T *et al.* Cloning and expression of a gene encoding an interleukin 3 receptor-like protein: identification of another member of the cytokine receptor gene family. *Proc. Natl. Acad. Sci. USA* 1990; **87**: 5459–5463.

76. Hara T and Miyajima A. Two distinct functional high affinity receptors for mouse interleukin-3 (IL-3). *EMBO J.* 1992; **11**: 1875–1884.

77. Hara T, Ichihara M, Takagi M and Miyajima, A. Interleukin-3 (IL-3) poor-responsive inbred mouse strains carry the identical deletion of a branch point in the IL-3 receptor alpha subunit gene. *Blood* 1995; **85**: 2331–2336.

78. Ichihara M, Hara T, Takagi M, Cho LC, Gorman DM and Myiajima A. Impaired interleukin-3 (IL-3) response of the A/J mouse is caused by a branch point deletion in the IL-3 receptor alpha subunit gene. *EMBO J.* 1995; **14**: 939–950.

79. Jubinsky PT, Laurie AS, Nathan DG, Yetz AJ and Sieff CA. Expression and function of the human granulocyte-macrophage colony-stimulating factor receptor alpha subunit. *Blood* 1994; **84**: 4174–4185.

80. Raines MA, Liu L, Quan SG, Joe V, DiPersio JF and Golde DW. Identification and molecular cloning of a soluble human granulocyte-macrophage colony-stimulating factor receptor. *Proc. Natl. Acad. Sci. USA* 1991; **88**: 8203–8207.

81. Brown CB, Beaudry P, Laing TD, Showmaker S and Kaushansky K. In vitro characterization of the human recombinant soluble granulocyte-macrophage colony-stimulating factor receptor. *Blood* 1995; **85**: 1488–1495.

82. Polotakaya A, Zhao Y, Lilly ML and Kraft AS. A critical role for the cytoplasmic domain of the granulocyte-macrophage colony-stimulating factor alpha receptor in mediating cell growth. *Cell Growth Differ.* 1993; **4**: 523–531.

83. Takaki S, Murata Y, Kitamura T, Miyajima A, Tominaga A and Takatsu K. Reconstitution of the functional receptors for murine and human interleukin 5. *J. Exp. Med.* 1993; **177**: 1523–1529.

84. Sato S, Katagiri T, Takaki J *et al.* IL-5 receptor- mediated tyrosine phosphorylation of SH2/SH3-containing proteins and activation of Bruton's tyrosine and Janus 2 kinases. *J. Exp. Med.* 1994; **180**: 2101–2111.

85. Ding DX, Rivas CI, Heaney ML, Raines MA, Vera JC and Golde DW. The alpha subunit of the human granulocyte-macrophage colony-stimulating factor receptor signals for glucose transport via a phosphorylation-independent pathway. *Proc. Natl. Acad. Sci. USA* 1994; **91**: 21537–21541.

86. Kan O, Baldwin SA and Whetton AD. Apoptosis is regulated by the rate of glucose transport in an IL-3-dependent haemopoietic cell line. *Biochem. Soc. Trans.* 1994; **22**.

87. Weiss M, Yokoyama C, Shikama Y, Naugle C, Druker B and Sieff CA. Human granulocyte-macrophage colony-stimulating factor receptor signal transduction requires the proximal cytoplasmic domains of the alpha and beta subunits. *Blood* 1993; **82**: 3298–3306.

88. Ronco LV, Silverman SL, Wong SG, Slamon DJ, Park LS and Gasson JC. Identification of conserved amino acids in the human granulocyte-macrophage colony-stimulating factor receptor alpha subunit critical for function. Evidence for formation of a heterodimeric receptor complex prior to ligand binding. *J. Biol. Chem.* 1994; **269**: 277–283.

89. Takaki S, Kanazawa H, Shiiba M and Takatsu K. A critical cytoplasmic domain of the interleukin-5 (IL-5) receptor alpha chain and its function in IL-5-mediated growth signal transduction. *Mol. Cell. Biol.* 1994; **14**: 7404–7413.

90. Sakamaki K, Miyajima I, Kitamura T and Miyohima A. Critical cytoplasmic domains of the common beta subunit of the human GM-CSF, IL-3 and IL-5 receptors for growth signal transduction and tyrosine phosphorylation. *EMBO J.* 1992; **11**: 3541–3549.

91. Sato N, Sakamaki K, Terada N, Arai K and Miyajima, A. Signal transduction by the high-affinity GM-CSF receptor: two distinct cytoplasmic regions of the common beta subunit responsible for different signaling. *EMBO J.* 1993; **12**: 4181–4189.

92. Mui AL, Wakao H, O'Farrell AM, Harada N and Miyajima A. Interleukin-3 granulocyte-macrophage colony stimulating factor and interleukin-5 transduce signals through two STAT5 homologs. *EMBO J.* 1995; **14**: 1166–1175.

93. Kinoshita T, Yokota T, Arai K and Miyajima A. Suppression of apoptotic death in hematopoietic cells by signalling through the IL-3/GM-CSF receptors. *EMBO J.* 1995; **14**: 266–275.

94. Yi T, Mui AL, Krystal G and Ihle JN. Hematopoietic cell phosphatase associates with the interleukin-3 (IL-3) receptor beta chain and down-regulates IL-3-induced tyrosine phosphorylation and mitogenesis. *Mol. Cell. Biol.* 1993a; **13**: 7577–7586.

95. Takagi M, Hara T, Ichihara M, Takatsu K and Miyajima A. Multi-colony stimulating activity of interleukin 5 (IL-5) on hematopoietic progenitors from transgenic mice that express IL-5 receptor alpha subunit constitutively. *J. Exp. Med.* 1995; **181**: 889–899.

96. Kishimoto T, Akira S and Taga T. Interleukin-6 and its receptor: a paradigm for cytokines. *Science* 192; **258**: 593–597.

97. Hibi M, Murakami M, Saito M, Hirano T, Taga T and Kishimoto T. Molecular cloning and expression of an IL-6 signal transducer, gp130. *Cell* 1990; **63**: 1149–1157.

98. Murakami M, Narazaki M, Hibi M *et al.* Critical cytoplasmic region of the interleukin 6 signal transducer gp130 is conserved in the cytokine receptor family. *Proc. Natl. Acad. Sci. USA* 1991; **88**: 11349–11353.

99. Murakami M, Hibi M, Nakagawa N *et al.* IL-6-induced homodimerization of gp130 and associated activation of a tyrosine kinase. *Science* 1993; **260**: 1808–1810.

100. Painessa G, Graziani R, De SA *et al.* Two distinct and independent sites on IL-6 trigger gp130 dimer formation and signalling. *EMBO J.* 1995; **14**: 1942–1951.

101. Demetri GD and Griffin JD. Granulocyte colony-stimulating factor and its receptor. *Blood* 1991; **78**: 2791–2808.

102. Baumann H, Gearing D and Ziegler SF. Signaling by the cytoplasmic domain of hematopoietin receptors involves two distinguishable mechanisms in hepatic cells. *J. Biol. Chem.* 1994; **269**: 16297–16304.

103. Dong F, van Buitenen C, Pouwels K, Hoefsloot LH, Lowenberg B and Touw IP. Distinct cytoplasmic regions of the human granulocyte colony-stimulating factor receptor involved in induction of proliferation and maturation. *Mol. Cell. Biol.* 1993; **13**: 7774–7781.

104. Fukunaga R, Ishizaka I and Nagata S. Growth and differentiation signals mediated by different regions in the cytoplasmic domain of granulocyte colony-stimulating factor receptor. *Cell* 1993; **74**: 1079–1087.

105. Ziegler SF, Bird TA, Morella KK, Mosley B, Gearing DP and Baumann H. Distinct regions of the human granulocyte-colony-stimulating factor receptor cytoplasmic domain are required for proliferation and gene induction. *Mol. Cell. Biol.* 1993; **13**: 2384–2390.

106. Avalos BR, Hunter MG, Parker JM *et al.* Point mutations in the conserved Box 1 region inactivate the human granulocyte colony-stimulating factor receptor for growth signal transduction and tyrosine phosphorylation of p75c-rel. *Blood* 1995; **85**: 3117–3126.

107. Bashey A, Healy L and Marshall C. Proliferative but not nonproliferative responses to granulocyte colony-stimulating factor are associated with rapid activation of the p21ras/MAP kinase signalling pathway. *Blood* 1994; **83**: 949–957.

108. Dong F, van Paassen M, van Buitenen C, Hoefsloot LH, Lowenberg B and Touw IP. A point mutation in the granulocyte colony-stimulating factor receptor (G-CSF-R) gene in a case of acute myeloid leukemia results in the overexpression of a novel G-CSF-R isoform. *Blood* 1995; **85**: 902–911.

109. Stahl N, Farruggella TJ, Boulton TG, Zhong Z, Darnell JJ and Yancopoulos GD. Choice of STATs and other substrates specified by modular tyrosine-based motifs in cytokine receptors. *Science* 1995; **267**: 1349–1353.

110. Schlessinger J. SH2/SH3 signaling proteins. *Curr. Opin. Genet. Devel.* 1994; **4**: 25–30.

111. Kishimoto T, Taga T and Akira S. Cytokine signal transduction. *Cell* 1994; **76**: 253–262.

112. Torigoe T, O'Connor R, Santoli D and Reed JC. Interleukin-3 regulates the activity of the LYN protein-tyrosine kinase in myeloid-committed leukemic cell lines. *Blood* 1992; **80**: 617–624.

113. Corey S, Eguinoa A, Puyana-Theall K *et al.* Granulocyte macrophage-colony stimulating factor stimulates both association and activation of phosphoinositide 30H-kinase and src-related tyrosine kinase(s) in human myeloid derived cells. *EMBO J.* 1993; **12**: 2681–2690.

114. Hanazono Y, Chiba S, Sasaki K *et al.* c-fps/fes protein-tyrosine kinase is implicated in a signaling pathway

triggered by granulocyte-macrophage colony-stimulating factor and interleukin-3. *EMBO J.* 1993a; **12**: 1641–1646.

115. Hanazono Y, Chiba S, Sasaki K, Mano H, Yazaki Y and Hirai H. Erythropoietin induces tyrosine phosphorylation and kinase activity of the c-fps/fes proto-oncogene product in human erythropoietin-responsive cells. *Blood* 1993b; **81**: 3193–3196.

116. Matsuda T, Takahashi TM, Fukada T *et al.* Association and activation of Btk and Tec tyrosine kinases by gp130, a signal transducer of the interleukin-6 family of cytokines. *Blood* 1995; **85**: 627–633.

117. Desiderio S and Siliciano JD. The Itk/Btk/Tec family of protein-tyrosine kinases. *Chem. Immunol.* 1994; **59**: 191–210.

118. Mano H, Yamashita Y, Sato K, Yazaki, Y. and Hirai H. Tec protein kinase is involved in interleukin-3 signalling pathway. *Blood* 1995; **85**: 343–350.

119. Tang B, Mano H, Yi T and Ihle JN. Tec kinase associates with c-kit and is tyrosine phosphorylated and activated following stem cell factor binding. *Mol. Cell. Biol.* 1994; **14**: 8432–8437.

120. Ihle JN, Witthuhn BA, Quelle FW *et al.* Signaling by the cytokine receptor superfamily: JAKs and STATs. *Trends Biochem. Sci.* 1994; **19**: 222–227.

121. Brizzi MF, Zini MG, Aronica MG, Blechman JM, Yarden Y and Pegoraro L. Convergence of signaling by interleukin-3, granulocyte-macrophage colony-stimulating factor, and mast cell growth factor on JAK2 tyrosine kinase. *J. Biol. Chem.* 1994; **269**: 1680–1684.

122. Narazaki M, Witthuhn BA, Yoshida K, Silvennoinen O, Yasukawa Kishimoto T and Taga T. Activation of JAK2 kinase mediated by the interleukin 6 gp130. *Proc. Natl. Acad. Sci. USA* 1994; **91**: 2285.

123. Tian SS, Lamb P, Seidel HM, Stein RB and Rosen J. Rapid activation of the STAT3 transcription factor by granulocyte colony-stimulating factor. *Blood* 1994; **84**: 1760–1764.

124. Zhong Z, Wen Z and Darnell JJ. Stat3: a STAT family member activated by tyrosine phosphorylation in response to epidermal growth factor and interleukin-6. *Science* 1994; **264**: 95–98.

125. Gouilleux F, Pallard C and Dunsanter FI *et al.* Prolactin growth hormone, erythropoietin and granulocyte-macrophage colony stimulating factor induce MGF-Stat5 DNA binding activity. *EMBO J.* 1995; **14**: 2005–2013.

126. Azam M, Erdjument BH, Kreider BL *et al.* Interleukin-3 signals through multiple isoforms of Stat5. *EMBO J.* 1995; **14**: 1402–1411.

127. van der Bruggen T, Caldenhoven E, Kanters D *et al.* Interleukin-5 signaling in human eosinophils involves JAK2 tyrosine kinase and Stat1 alpha. *Blood* 1995; **85**: 1442–1448.

128. Plutzky J, Neel BG and Rosenberg RD. Isolation of a src homology 2-containing tyrosine phosphatase. *Proc. Natl. Acad. Sci. USA* 1992; **89**: 1123–1127.

129. Schultz LD, Schweitzer PA, Rajan TV, Yi T, Ihle JN and Matthews D. Mutations at the murine motheaten locus are within the protein-tyrosine phosphatase (Hcph) gene. *Cell* 1993; **73**: 1445–1454.

130. Yi T and Ihle JN. Association of hematopoietic cell phosphatase with C-Kit after stimulation with c-Kit ligand. *Mol. Cell. Biol.* 1993b; **13**: 3350–3358.

131. Vogel W, Lammers R, Huang J and Ullrich A. Activation of a phosphotyrosine phosphatase by tyrosine phosphorylation. *Science* 1993; **259**: 1611–1614.

132. Freeman RJ, Plutzky J and Neel BG. Identification of a human src homology 2-containing protein-tyrosine-phosphatase: putative homolog of *Drosophila* corkscrew. *Proc. Natl. Acad. Sci. USA* 1992; **89**: 11239–11243.

133. Feng GS, Hui CC and Pawson T. SH2-containing phosphotyrosine phosphatase as a target of protein-tyrosine kinases. *Science* 1993; **259**: 1607–1611.

134. Tauchi T, Feng GS, Shen R *et al.* SH2-containing phosphotyrosine phosphatase Syp is a target of p210bcr-abl tyrosine kinase. *J. Biol. Chem.* 1994a; **269**: 15381–15387.

135. Tauchi T, Feng GS, Marshall MS *et al.* The ubiquitously expressed Syp phosphatase interacts with c-kit and Grb2 in hematopoietic cells. *J. Biol. Chem* 1994b; **269**: 25206–25211.

136. Lechleider RJ, Sugimoto S, Bennet AM *et al.* Activation of the SH2-containing phosphotyrosine phosphatase SH-PTP2 by its binding site, phosphotyrosine 1009, on the human platelet-derived growth factor receptor, *J. Biol. Chem.* 1993; **268**: 21478–21481.

137. Tauchi T, Feng GS, Shen R *et al.* Involvement of SH2-containing phosphotyrosine phosphatase Syp in erythropoietin receptor signal transduction pathways. *J. Biol. Chem.* 1995; **270**: 5631–5635.

138. Welham MJ, Dechert U, Leslie KB, Jirik F and Schrader, J.W. Interleukin (IL)-3 and granulocyte/macrophage colony-stimulating factor, but not IL-4, induce tyrosine phosphorylation, activation and association of SHPTP2 with Grb2 and phosphatidylinositol 3'-kinase. *J. Biol. Chem.* 1994a; **269**: 23764–23768.

139. Cockcroft S and Thomas GM. Inositol-lipid-specific phospholipase C isoenzymes and their differential regulation by receptors. *Biochem. J.* 1992; **288**: 1–14.

140. Whetton AD, Monk PN, Consalvey SD and Downes CP. The haemopoietic growth factors interleukin-3 and colony stimulating-1 stimulate proliferation but do not induce inositol lipid breakdown in murine bone-marrow-derived macrophages. *EMBO J.* 1986; **5**: 3281–3286.

141. Duronio B, Nip L and Pelech SL. Interleukin 3 stimulates phosphatidylcholine turnover in a mast/megakaryocyte cell line. *Biochem. Biophys. Res. Commun.* 1989; **164**: 804–808.

142. Hartmann T, Seuwen K, Roussel MF, Sherr CJ and Pouyssegur J. Functional expression of the human receptor for colony-stimulating factor 1 (CSF-1) in hamster fibroblasts: CSF-1 stimulates Na+/H+ exchange and DNA-synthesis in the absence of phosphoinositide breakdown. *Growth Factors* 1990; **2**: 289–300.

143. Imamura K, Dianoux A, Nakamura T and Kufe D. Colony-stimulating factor 1 activates protein kinase C in human monocytes. *EMBO J.* 1990; **9**: 2423–2428.

144. Downes CP and Carter AN. Phosphoinositide 3-kinase: a new effector in signal transduction? *Cell Signal.* 1991; **3**: 501–513

145. Damen JE, Mui AL, Puil L, Pawson T and Krystal G. Phosphatidylinositol 3-kinase associates, via its Src

homology 2 domains, with the activated erythropoietin receptor. *Blood* 1993; **81**: 3204–3210.

146. Valius M and Kazlauskas A. Phospholipase C-gamma 1 and phosphatidylinositol 3 kinase are the downstream mediators of the PDGF receptor's mitogenic signal. *Cell* 1993; **73**: 321–334.

147. Chung J, Grammer TC, Lemon KP, Kazlauskas A and Blenis J. PDGF-and insulin-dependent pp70S6k activation mediated by phosphatidylinositol-3-OH kinase. *Nature* 1994; **370**: 71–75.

148. Nakanishi H, Brewer KA and Exton JH. Activation of the zeta isozyme of protein kinase C by phosphatidylinositol 3,4,5-triphosphate. *J. Biol. Chem.* 1993; **268**: 13–16.

149. Satoh T, Nakafuku M, Miyajima A and Kaziro Y. Involvement of ras p21 protein in signal-transduction pathways from interleukin 2, interleukin 3, and granulocyte/macrophage colony-stimulating factor, but not from interleukin 4. *Proc. Natl. Acad. Sci. USA* 1991; **88**: 3314–3318.

150. Duronio V, Welham MJ, Abraham S, Dryden P and Schrader JW. p21ras activation via hemopoietin receptors and c-kit requires tyrosine kinase activity but not tyrosine phosphorylation of p21ras GTPase-activating protein. *Proc. Natl. Acad. Sci. USA* 1992; **89**: 1587–1591.

151. Heidaran MA, Molloy CJ, Pangelinan M *et al.* Activation of the colony-stimulating factor 1 receptor leads to the rapid tyrosine phosphorylation of GTPase-activating protein and activation of cellular p21ras. *Oncogene* 1992; **7**: 147–152.

152. Welham MJ, Duronio V, Leslie KB, Bowtell D and Schrader JW. Multiple hemopoietins, with the exception of interleukin-4, induce modifications of Shc and mSos1, but not their translocation. *J. Biol. Chem.* 1994b; **269**: 21165–21176.

153. McCormick F. ras GTPase activating protein: signal transmitter and signal terminator. *Cell* 1989; **56**: 5–8.

154. McCormick F. Activators and effectors of ras p21 proteins. *Curr. Opin. Genet. Devel.* 1994; **4**: 76–76.

155. Miyazawa K, Hendrie PC, Mantel C, Wood K, Ashman LK and Broxmeyer HE. Comparative analysis of signalling pathways between mast cell growth factor (c-kit ligand) and granulocyte-macrophage colony-stimulating factor in a human factor-dependent myeloid cell line involves phosphorylation of Raf-1, GTPase-activating protein and mitogen-activated protein kinase. *Exp. Hematol.* 1991; **19**: 1110–1123.

156. Owen LP, Wong AK and Whetton AD. v-Abl-mediated apoptotic suppression is associated with SHC phosphorylation without concomitant mitogen-activated protein kinase activation. *J. Biol. Chem.* 1995; **270**: 5956–5962.

157. Lanfrancone L, Pellici G, Brizzi MF *et al.* Overexpression of Shc protein potentiates the proliferative response to the granulocyte-macrophage colony-stimulating factor and recruitment of Grb2/SoS and Grb2/p140 complexes to the beta receptor subunit. *Oncogene* 1995; **19**: 907–917.

158. Magnuson NS, Beck T, Vahidi H, Hahn H, Smola U and Rapp UR. The Raf-1 serine/threonine protein kinase. *Semin. Cancer Biol.* 1994; **5**: 247–253.

159. Baccarini M, Sabatini DM, Rapp UR and Stanley ER. Colony stimulating factor-1 (CSF-1) stimulates temperature dependent phosphorylation and activation of the RAF-1 proto-oncogene product. *EMBO J.* 1990; **9**: 3649–3657.

160. Kanakura Y, Druker B, Cannistra SA, Furukawa Y, Torimoto Y and Griffin JD. Signal transduction of the human granulocyte-macrophage colony-stimulating factor and interleukin-3 receptors involves tyrosine phosphorylation of a common set of cytoplasmic proteins. *Blood* 1990; **76**: 706–715.

161. Carroll MP, Spivak JL, McMahon M, Weich N, Rapp UR and May WS. Erythropoietin induces Raf-1 activation and Raf-1 is required for erythropoietin-mediated proliferation. *J. Biol. Chem.* 1991; **266**: 14964–14969.

162. Kolch W, Heidecker G, Kochs G *et al.* Protein kinase C alpha activates RAF-1 by direct phosphorylation. *Nature* 1993; **364**: 249–252.

163. Carroll MP and May WS. Protein kinase C-mediated serine phosphorylation directly activates Raf-1 in murine hematopoietic cells. *J. Biol. Chem.* 1994; **269**: 1249–1265.

164. Schramm K, Niehof M, Radziwill G, Rommel C and Moelling, K. Phosphorylation of c-Raf1 by protein kinase A interferes with activation. *Biochem. Biophys. Res. Commun.* 1994; **201**: 740–747.

165. Brtva TR, Drugan JK, Ghosh S *et al.* Two distinct Raf domains mediate interaction with Ras. *J. Biol. Chem.* 1995; **270**: 9809–9812.

166. Marais R, Light Y, Paterson HF and Marshall CJ. Ras recruits raf-1 to the plasma membrane for activation by tyrosine phosphorylation. *EMBO J.* 1995; **14**: 3136–3145.

167. Moodie SA, Willumsen BM, Weber MJ and Wolfman, A. Complexes of Ras. GTP with Raf-1 and mitogen-activated protein kinase kinase. *Science* 1993; **260**: 1658–1661.

168. Guan KL. The mitogen activated protein kinase signal transduction pathway: from the cell surface to the nucleus. *Cell Signal.* 1994; **6**: 581–589.

169. Marshall CJ. MAP kinase kinase kinase, MAP kinase kinase and MAP kinase. *Curr. Opin. Genet. Devel.* 1994; **4**: 82–89.

170. Okuda K, Sanghera JS, Pelech SL *et al.* Granulocyte-macrophage colony-stimulating factor, interleukin-3, and steel factor induce rapid tyrosine phosphorylation of p42 and p44 MAP kinase. *Blood* 1992; **79**: 2880–2887.

171. Kumar G, Gupta S, Wang S and Nel AE. Involvement of Janus kinases, p52shc, Raf-1, and MEK-1 in the IL-6 induced mitogen-activated protein kinase cascade of a growth responsive B cell line. *J. Immunol.* 1994; **153**: 4436–4447.

172. Hug H and Sarre TF. Protein kinase C isoenzymes: divergence in signal transduction? *Biochem. J.* 1993; **291**: 329–343.

173. Dekker LV and Parker PJ. Protein kinase C–a question of specificity. *Trends Biochem. Sci.* 1994; **19**: 73–77.

174. Meeker TC, Nagarajan L, Ar-Rushdi A and Croce CM. Cloning and characterization of the human PIM-1 gene: a putative oncogene related to the protein kinases. *J. Cell. Biochem.* 1987; **35**: 105–112.

175. Meeker TC, Nagarajan L, Ar-Rushdi A, Rovera G, Huebner K and Croce CM. Characterization of the human PIM-1 gene: a putative proto-oncogene coding for a tissue specific member of the protein kinase family. *Oncogene Res.* 1987; **1**: 87–101.

176. Domen J, van der Lugt LN, Laird PW *et al.* Impaired interleukin-3 response in Pim-1-deficient bone marrow-derived mast cells. *Blood* 1993; **82**: 1445–1452.

177. Nagaranjan L and Narayana L. Transcriptional attenuation of PIM-1 gene. *Biochem. Biophys. Res. Commun.* 1993; **190**: 435–439.

178. Miura O, Miura Y, Nakamura N *et al.* Induction of tyrosine phosphorylation of Vav and expression of Pim-1 correlates with Jak2-mediated growth signaling from the erythropoietin receptor. *Blood* 1994; **84**: 4135–4141.

179. Bustelo XR, Ledbetter JA and Barbacid M. Product of vav proto-oncogene defines a new class of tyrosine protein kinase substrates [see comments]. *Nature* 1992; **356**: 68–71.

180. Margolis B, Hu P, Katzav S *et al.* Tyrosine phosphorylation of vav proto-oncogene product comtaining SH2 domain and transcription factor motifs [see comments]. *Nature* 1992; **356**: 71–74.

181. Alai M, Mui AL, Cutler RL, Bustelo XR, Barbacid M and Krystal G. Steel factor stimulates the tyrosine phosphorylation of the proto-oncogene product, p95vav, in human hemopoietic cells. *J. Biol. Chem.* 1992; **267**: 18021–18025.

182. Hart MJ, Eva A, Evans T, Aaronson SA and Cerione RA. Catalysis of guanine nucleotide exchange on the CDC42Hs protein by the dbl oncogene product. *Nature* 1991; **354**: 311–314.

183. Hart MJ, Eva A, Zangrilli D *et al.* Cellular transformation and guanine nucleotide exchange activity are catalyzed by a common domain on the dbl oncogene product. *J. Biol. Chem.* 1994; **269**: 62–65.

184. Gulbins E, Coggeshall KM, Baier G, Katzav S, Burn P and Altman A. Tyrosine kinase-stimulated guanine nucleotide exchange activity of Vav in T cell activation. *Science* 1993; **260**: 822–825.

185. Gulbins E, Coggeshall KM, Baier G *et al.* Direct stimulation of Vav guanine nucleotide exchange activity for Ras by phorbol esters and diglycerides. *Mol. Cell. Biol.* 1994; **14**: 4749–4758.

186. Katzav S. Single point mutations in the SH2 domain impair the transforming potential of vav and fail to activate proto-vav. *Oncogene* 1993; **8**: 1757–1763.

187. Fuscher KD, Zmuldzinas A, Gardner S, Barbacid M, Bernstein A and Guidos C. Defective T-cell receptor signalling and positive selection of Vav-deficient CD4+ CD8+ thymocytes. *Nature* 1995; **374**: 474–477.

188. Zhang R, Alt FW, Davidson L, Orkin SH and Swat W. Defective signalling through the T-and B-cell antigen receptors in lymphoid cells lacking the vav proto-oncogene. *Nature* 1995; **374**: 470–473.

189. Bustelo XR, Suen KL, Michael WM, Dreyfuss G and Barbacid M. Association of the vav proto-oncogene product with poly(rC)-specific RNA-binding proteins. *Mol. Cell. Biol.* 1995; **15**: 1324–1332.

190. Hill CS and Treisman R. Transcriptional regulation by extracellular signals: mechanisms and specificity. *Cell* 1995; **80**: 199–211.

191. Orkin SH. Transcription factors and hematopoietic development. *J. Biol. Chem.* 1995; **270**: 4955–4958.

17

Cytokines and Transcription Factors in Regulation of Lymphocyte Function

Ellen L. Burlinson[1], Bradford W. Ozanne[2] and William Cushley[1]

[1]*Division of Biochemistry and Molecular Biology, Institute of Biomedical and Life Sciences, University of Glasgow, Glasgow Scotland, UK and* [2]*CRC Beatson Laboratories, Switchback Road, Glasgow Scotland, UK*

INTRODUCTION

All aspects of cellular function depend upon the accurate expression of genes required for performance of particular biological functions. Important genes in lymphocytes range from those involved in somatic rearrangement of antigen receptor genes via others necessary for regulation of cell death or survival, to large groups encoding cytokines and their receptors. For many immunological processes, expression of particular genes must occur at a defined window in the differentiation programme of the cell and the mechanisms which govern temporal patterns of gene expression are therefore of great importance.

Since promoter and enhancer elements governing the expression of individual genes are identical in all cell types in an individual organism, with the exception of modifications at the levels of methylation and translocation, it is axiomatic that control must be exerted via factors which interact with these elements and so confer transcriptional activity upon them and their associated genes. Consequently, transcription factors are excellent candidates for proteins which can be subject to regulation at various levels, including developmental regulation of expression, reversible modification and protein–protein interactions necessary for stimulation or inhibition of DNA binding activity. In this chapter, we review and contrast the properties of several well-characterized transcription factor systems which are important in regulation of lymphoid function, overview others involved in expression of cytokine genes, and consider transcription factors which are essential for accurate lymphopoiesis. The functional linkage of these transcription factor systems to stimuli originating in the extracellular environment is also discussed.

THE NF-AT PROTEIN FAMILY: REGULATION OF CYTOKINE GENE EXPRESSION IN T LYMPHOCYTES

Stimulation of T cells via their antigen receptors leads not only to cellular proliferation but also to expression of cytokine genes. While many of the transcription factors which bind to *cis* elements in the promoters of cytokine genes will be common to many systems, it is not surprising that T lymphocytes possess a group of factors which are targeted following stimulation of the T cell receptors and which facilitate cytokine gene expression. This group of transcriptional regulatory factors are the nuclear factors of activated T lymphocytes or NF-AT proteins.

Structural Features and Tissue Distribution of NF-AT Proteins

There are two principal forms of NF-AT proteins present in T cells, NF-ATc (cytosolic) and NF-ATp (pre-existing), which are the products of two distinct genes [1–3]. NF-ATc is a protein of 716 amino acids, with a predicted molecular weight of approximately 78 kDa; the migration of NF-ATc, like many transcription factors, is somewhat higher on SDS–PAGE ($\sim 100\,kDa$) than is predicted by conceptual translation of cDNA sequences [3]. The molecule is divisible into three broad regions, with a proline-rich N-terminal region of 418 residues followed by a central region of some 290 amino acids which shows homology to the *rel* (NF-κB) transcription factors and to the *dorsal* factor of *Drosophila*; there is a small C-terminal tail region. NF-ATp has a predicted size of 890 amino acids and displays the same three homology regions as those noted above for NF-ATc [2]; NF-ATp is a 120 kDa protein as determined by SDS–PAGE analysis.

The weak homology to the *rel / dorsal* families has led to the identification of two further human NF-AT family members, NF-AT3 and NF-AT4 [4]. The sequence of the NF-AT3 cDNA encodes a protein of 902 amino acids, while three species of NF-AT4 cDNA were isolated, each of which differed at their 3′-terminal ends. NF-AT4a encodes a protein of 708 amino acids, while its NF-AT4b and NF-AT4c partners are proteins of 739 and 1068 residues respectively. All of the NF-AT3 and NF-AT4 proteins contain a central 290 amino acid domain with homology to *rel* transcription factors [4].

Given the likely role of NF-AT proteins in regulation of cytokine expression in activated T cells, it is surprising that expression of NF-AT proteins is widespread throughout the tissues. Thus, NF-ATp appears to be expressed at very low levels in most human cell types [4], but a range of tissues, including skeletal muscle, pancreas, placenta and lymphoid tissue, display strong levels of NF-ATp RNA. NF-ATc appears to be more restricted in its pattern of expression, being found principally in thymus and spleen, but with skeletal muscle also giving a strong hybridization signal for one of the two RNA species detected in the Northern blot analysis [4]. NF-ATc RNA levels are inducible in T cells and in T cell lines, and are more abundant in extracts of activated spleen and thymus than resting lymphoid tissue; in transgenic animals bearing an NF-AT-dependent transgene, NF-ATc RNA was detectable in skin [3]. By contrast, NF-AT3 appeared to be most highly expressed outwith haematopoietic lineages, and lung, kidney, testes and ovary had high levels [4]. NF-AT4 RNA is readily detectable in kidney, skeletal muscle and thymus, and the size of the RNA species is consistent with NF-AT4c being the most abundant NF-AT4 isoform expressed at these sites [4].

NF-AT protein binding sites in DNA appear to possess the minimal motif GGAAA [5]. NF-ATp is known to bind and activate several genes bearing the GGAAA motif as part of their promoters, including those for IL-2, IL-4, the CD40 ligand and TNFα, although the latter NF-ATp site has the form GGAGAA [5, 6]. Integrity of NF-AT binding motifs in the promoters of the target genes is also exquisitely sensitive to the state of methylation, with the presence of methyl-guanyl residues in the site leading to abolition of NF-AT binding and transactivation [5]. Direct comparison of all four NF-AT family members showed that all could bind a decanucleotide from the IL-4 promoter (GGAAAATTTT) with similar affinity, although NF-AT4 bound the oligonucleotide less well than the others. However, comparison of binding to a very similar sequence from the IL-2 promoter (GGAAAAACTG) revealed that NF-ATp bound well to this site, NF-ATc and NF-AT 3 associated with lower affinity, and NF-AT4 was unable to bind the decanucleotide [4]. Thus, NF-AT family members may have slightly different specificities for DNA binding sites. In fact, the majority of functional NF-AT sites in the regulatory elements of cytokine genes are linked to sites for AP-1 transcription factors and it appears that NF-AT and AP-1 family members interact cooperatively to bind to DNA at composite NF-AT / AP-1 binding sites [7]. Finally, in sharp contrast to the *rel*, AP-1 and STAT families of transcription factors, at least NF-ATp and NF-ATc appear to be able to bind to target sequences in the DNA as monomers [4].

Co-operative Interactions of NF-AT and AP-1 Transcription Factors

The interaction of NF-AT proteins with elements of the IL-2 promoter in human T cells has been well characterized as a model for the regulation of NF-AT activity. Activation of the IL-2 gene requires delivery of signals via two distinct transduction pathways, a protein-kinase-C-dependent pathway and a cyclosporin-sensitive, calcium-dependent pathway. Maximal induction of cytokine genes can be obtained by treating cells with calcium ionophore and phorbol ester to mimic the signals delivered via the T cell receptor (reviewed in reference [5]).

It is now clear that all NF-AT proteins require to interact with AP-1 components to show maximal DNA binding and transactivation [8]. The IL-2 promoter region contains two NF-AT sites, one located some 150 bp upstream of the cap site (proximal site), while the second, higher affinity, distal NF-AT site resides at −255 to −285 bp from the cap site (see Figure 17.4). While neither c-*fos* nor c-*jun* proteins can bind to the distal site in their own right, they bind cooperatively with NF-ATp, indicating that the latter factor confers specificity for DNA binding [7]. The AP-1 proteins, either as c-*jun* dimers or c-*fos* / c-*jun* heterodimers, stabilize the binding of NF-ATp to DNA [1] and available data indicate that all *jun* or *fos* family members are capable of interacting with NF-ATp [4]. The importance of the cooperative binding of NF-AT and AP-1 elements is underscored by the observation that mutation of the AP-1 site abolishes NF-ATp-mediated transcription of reporter constructs [6]. In a comparative study of the capacity of *jun* homodimers and *jun* / *fra1* heterodimers to cooperate with each of the four members of the NF-AT protein family in binding to a distal IL-2 promoter oligonucleotide, it was noted that the *jun* homodimers were less effective than the *jun* / *fra1* heterodimers in stabilizing NF-AT binding, but that, in each case, the AP-1 components facilitated DNA binding to the same extent [4].

The IL-2 promoter is complex in the sense that it possesses two NF-AT sites, plus sites which are regulated by other incoming signals to the T cell (e.g. NF-κB, CD28). Studies of the NF-AT / AP-1 composite site of the IL-4 promoter have allowed a finer dissection of the roles of NF-AT and AP-1 components in DNA binding to, and transactivation of, these sites [9]. Thus, NF-ATp/NF-ATc and AP-1 components bind cooperatively to their sites in the IL-4 promoter but, in contrast to the IL-2 promoter, the *fra1*, *fra2* and *jun*B and *jun*D members of the AP-1 family predominate in the AP-1 complexes from anti-CD3 stimulated cells; moreover, *jun* homodimers appeared to bind to the AP-1 site with high affinity. Binding of both NF-AT and AP-1 components to the composite site is necessary for transactivation of the site in reporter constructs [9].

Regulation of NF-AT Activity

In common with a number of transcription factor families, NF-AT proteins are found in an inactive state in the cytosol which, following stimulation of the cell, translocate to the nucleus where their transcriptional activation potential is expressed [1, 10, 11]. However, in contrast to systems, such as the STATs where phosphorylation of factors is required for activation, it is dephosphorylation which is the critical event in NF-AT protein activation. *In vivo*, conversion of inactive phospho-NF-AT to active dephospho-NF-AT is mediated by the protein phosphatase calcineurin, and the immunosuppressive agents cyclosporin and FK506 inhibit calcineurin activity and so attenuate NF-AT protein activation and cytokine gene expression [7, 11]. The activity of calcineurin is dependent upon calcium and calmodulin, and it is known that perturbation of the T cell receptor promotes a rise in intracellular calcium which, in turn, will lead to activation of calcineurin.

Expression of the IL-2 gene requires input from two cellular signalling pathways. One pathway is calcium dependent and cyclosporin sensitive, while the second is calcium independent and protein kinase C dependent; these data are usually interpreted in terms of the target of the latter pathway as being the AP-1 components of the transcriptional network [8], and NF-AT proteins being the targets of the former [3]. Similarly, ionomycin promotes NF-AT activity in a cyclosporin-inhibitable manner, suggesting activation of calcineurin by this stimulation, while PKC activation by phorbol ester leads to elevated AP-1 activity. However, activation of the AP-1 and NF-AT transcription factor systems by the two signalling pathways may not be mutually exclusive. Thus, while NF-ATp is generally accepted to be a direct substrate for calcineurin, there are data which indicate that NF-ATc activity can be modulated by phorbol ester, suggesting that NF-ATc may be subject to regulation by PKC-regulated pathways in addition to those under the control of Ca^{2+} and calcineurin [3].

THE NF-κB FAMILY OF TRANSCRIPTION FACTORS

Characterization of NF-κB

NF-κB is an ubiquitous transcription factor, which was originally characterized as a nuclear factor which binds to the immunoglobulin κ light-chain transcriptional enhancer, active within B cells [12], although the decameric κB motif of the consensus 5'-GGGPuNNPyPyCC-3' has since been found within the promoter/enhancer regions of many genes [13]. Although NF-κB is widely distributed, its main effects are exerted within the immune system in immune, inflammatory and acute phase responses, and can be activated by a wide range of stimuli including viruses, parasites, cytokines, endotoxins and stress. The apparent lineage-specific effect of NF-κB within immune cells is gained through both structural and functional interactions with other transcriptional activator proteins which recognize adjacent sequences within the promoter regions of responsive genes. Primary transcription factors such as NF-κB, pre-exist within the cytoplasm of the cell, unlike many inducible transcription factors which are synthesized de novo in response to a range of extracellular signals. Such primary factors are rapidly activated within the cytoplasm, in response to an appropriate extracellular signal, and then translocated to the nuclear compartment, where they can interact with their appropriate recognition sequences and activate gene expression (Figure 17.1).

To date, five independent genes have been identified which encode transcription activator proteins which recognize the κB motif: these proteins are NF-κB1 (p105), NF-κB-2 (p100), Rel-A (p65), Rel-B and c-Rel, and all are classified as members of the rel family of transcription factors. Both NF-κB1 and NF-κB2, unlike the other Rel family members, are initially synthesized as precursor proteins which are proteolytically processed to generate mature NF-κB1 (p50) and NF-κB2 (p52) proteins [14]. The X-ray crystallographic structure of the p50 dimer of NF-κB has recently been elucidated [15, 16] with each of the p50 subunits consisting of two, flexibly linked domains, which resemble that of immunoglobulin. The structure of p50 NF-κB is unrelated to any of the previously identified DNA binding proteins, in that it does not interact with the DNA via short α-helices, but instead through loops which wrap themselves around the DNA, leaving only the minor groove free.

While the p50 and p65 subunits of NF-κB exist within virtually every cell type, RelB is preferentially expressed within cells of the immune system. Considering the primary amino acid structure of the NF-κB subunit proteins, they share a 300 amino acid Rel homology domain which contains both DNA-binding and dimerization domains, and also a nuclear localization signal [17, 18]. Proteins of the Rel family can associate as homo-or heterodimeric complexes, with the exception of RelB which cannot homodimerize, but can form heterodimers with p50 or p52. Thus, the dimerization of Rel family members generates a wide variety of trans-acting factors, each of which preferentially binds to distinct κB-binding site elements, resulting in a range of transcriptional activities [18]. Purification studies suggest that p50, RelA and c-Rel are the major components of NF-κB complexes which bind to most identified cis-acting κB sites, with the most frequently occurring NF-κB complex being that of p50–RelA. The differential activity of the dimeric complexes is demonstrated in transgenic mice expressing a gene with a minimal promoter containing three copies of the binding site for this protein complex, p50-RelA heterodimers were constitutively active in lymphoid tissues, whereas in organs containing p50 homodimers no transgene expression was seen [19]. Evidence suggests that p50 homodimers may act as negative regulators of κB-dependent transcription by displacing p50–RelA heterodimers from binding sites, although whether p50 homodimers' sole function is in negative regulation remains to be resolved, as they do have some transcriptional activity in cell-free assays and in yeast cells [20–22]. The promoter element for the major histocompatibility complex class II invariant chain contains two NF-κB binding sites which, depending on the cell type, can act as either positive or negative regulators of transcription. In B cell lines and in the invariant chain-expressing T cell line, H9, the κB sites act as a positive regulatory element, whereas, in contrast, in myelomonocytic and glial cells such regulatory elements act negatively [23]. Interestingly, it seems that the

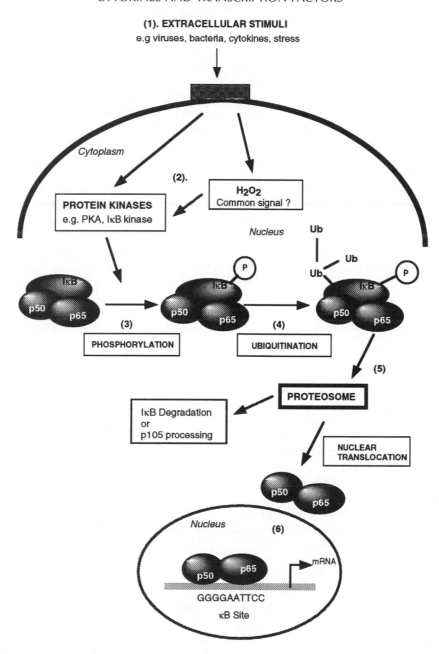

Figure 17.1 Activation pathway for NF-κB. A wide range of extracellular stimuli can activate NF-κB via association with their high affinity binding sites at the cell surface (*Stage 1*). This receptor engagement causes activation of protein kinases, including PKA and I-κB kinase, and also the production of reactive oxygen intermediates, the latter possibly serving as a common signalling mechanism for the wide range of extracellular stimuli (*Stage 2*). Heterodimers of p50–p65 are retained in an inactive form within the cytoplasm by association with inhibitory subunits, such as I-κB or p105, which are substrates for protein kinases (*Stage 3*).

Phosphorylation of I-κB facilitates its subsequent ubiquitination by ubiquitin conjugating enzyme (*Stage 4*), and the ubiquitin conjugated complex is then degraded by the multicatalytic cytoplasmic protease, the proteasome, thereby releasing p50–p65 heterodimers, which are free to translocate to the nucleus (*Stage 5*). The released, transcriptionally- active p50–p65 heterodimers bind to specific sequences within the promoter regions of responsive genes and, in conjunction with other transcription factors, initiate gene expression (*Stage 6*) presence of p50, p52, p65 and c-rel correlate with positive regulation, whereas the presence of p50 alone correlates with negative regulation, thus lending support to the idea that p50 homodimers can act as negative regulators of κB motif-containing promoters. Although c-rel-RelA dimers have been identified through both *in vitro* studies and in intact cells, their functional relevance *in vivo* remains to be determined [24–26]. RelB, which was cloned from mouse fibroblasts, is an immediate early response gene [27] which, in combination with p50, can bind to κB sites. The human homologue of RelB, termed ReII, lacks transactivating ability itself, instead, it attenuates p50 and p50-RelB binding activity and supresses NF-κB transactivation. It would appear, therefore, that different combinations of NF-κB subunits allow for fine control over DNA-binding specificity.

Regulation of NF-κB

Rel family members are sequestered within the cytoplasm in a number of ways. Some homo-or heterodimers (e.g. p50 and p65) are bound to a member of the I-κB family of inhibitor proteins (I-κB-α, I-κB-β, I-κB-γ, Bcl-3 and the *Drosophila* protein *cactus*), while others are bound to the p105 rel family member. Such inhibitor proteins contain a 30 amino acid ankyrin repeat motif, required for interactions with Rel proteins, and a *C*-terminal PEST sequence, which may be involved in protein degradation [28]. For activation of NF-κB, the I-κB inhibitory subunit must be released from the cytoplasmic complex, thus allowing transcriptionally active NF-κB to translocate to the nucleus. Although I-κB-α and I-κB-β bind to the same range of Rel proteins, it appears that while I-κB-α is degraded in response to a wide range of NF-κB inducing agents, I-κB-β responds to a limited number of these inducers [29]. Studies by Henkel and colleagues suggested that the degradation of I-κB after stimulation of cells with phorbol ester, IL-1, LPS or TNF-α, is facilitated by a cytoplasmic chymotrypsin-like protease [30]. In studies by Traenckner and coworkers, a peptide which specifically inhibits the chymotrypsin-like activity of the multicatalytic cytosolic protease, the proteasome, was found to inhibit the TNF or okadaic-acid-mediated activation of NF-κB, by

blocking the degradation of I-κB-α [31]. The results of this study also suggested that oxidants, such as H_2O_2, may be involved in the phosphorylation of I-κB-α, as treatment of cell with the antioxidant pyrrolidine dithiocarbamate, prevented the phosphorylation of I-κB-α and subsequent activation of NF-κB. These findings contrasted with those gained from cell-free systems, which suggested that phosphorylation of I-κB by protein kinase A and other kinases was sufficient to cause release of I-κB and subsequent activation of NF-κB. Recent evidence suggests that the ubiquitin–proteasome pathway is involved in the activation of Rel protein complexes and that the phosphorylation of I-κB or p105, rather that facilitating degradation directly, instead allows for their recognition by ubiquitin-conjugating enzymes and their subsequent ubiquitination, followed by degradation and processing [29]. It remains to be determined whether all NF-κB inducers activate a common I-κB kinase or discrete I-κB kinases, although one such candidate kinase (*pelle*) has recently been identified in *Drosophila* [32]. In addition, studies by Frantz and colleagues [33] provided *in vivo* evidence that activation of calcineurin, a Ca^{2+}/calmodulin-dependent protein phosphatase, can inactivate I-κB, thus releasing NF-κB for binding to its appropriate recognition sites. I-κB is also regulated by Rel family proteins, in that overexpression of p65 in carcinoma and lymphoid cells increases I-κB protein levels, both by stabilization of I-κB and also by increasing transcription of I-κB mRNA [34]. p65 and I-κB are therefore linked in an autoregulatory loop.

A role for phosphorylation of NF-κB1-p50 has been proposed in both NF-κB activation and in its stable DNA binding [35]. After stimulation of Jurkat T cells with PMA and PHA, p50 was hyperphosphorylated, and this phosphorylated form of p50 was translocated to the nucleus, corresponding with the appearance of an active κB binding DNA-complex within the nuclear compartment.

Genes which are Regulated by NF-κB

Recognition sequences for NF-κB complexes have been identified within the promoter / enhancer regions of a large number of cytokine genes, including those of interleukins 1, 2, 3, 4, 6, 8,

GM-CSF, M-CSF, G-CSF, IFN-γ and IFNβ and also within regulatory regions of other immunologically important genes, such as immunoglobulin, c-fos and the IL-2 receptor [36]. NF-κB may also be a target for anti-inflammatory agents such as glucocorticoids, which down-regulate the expression of a number of cytokine genes, such as IL-6, which are involved in inflammatory responses. Activation of the IL-6 promoter, by a combination of NF-IL-6 and the p65 subunit of NF-κB, is inhibited by dexamethasone-activated glucocorticoid receptor, possibly through a direct physical association between p65 and the glucocorticoid receptor, as shown through protein crosslinking and co-immunoprecipitation experiments [37]. Both sodium salicylate and aspirin, which are potent anti-inflammatory drugs, also inhibit the activation of NF-κB, by preventing the degradation of I-κB [38].

To identify the particular Rel family members which recognize the κB site within the IL-8 promoter region, Kunsch and colleagues stimulated Jurkat T cells with PMA and then analysed protein binding to the κB site, in the presence of antibodies against the various members of the Rel homology group [39]. The p65 (RelA) protein was identified as the predominant component within the NF-κB complexes, as only anti-p65 antibodies inhibited the formation of nuclear complexes, and antisense oligonucleotides to RelA, but not p50, inhibited the PMA induction of the IL-8 gene in Jurkat T cells. Further gel shift assays using in vitro translated products or purified proteins revealed that the IL-8 κB site can bind RelA, c-rel and NF-κB2 homodimers, but not the more commonly found p50–p65 heterodimeric NF-κB complex. In more recent studies, a role for NF-IL-6 in IL-8 gene activation was also defined [40]. Within the IL-8 promoter the binding sites for NF-κB and NF-IL-6 are adjacent, and studies on Jurkat T cells revealed that for maximal gene expression both sites must be intact. Furthermore, they found that NF-IL-6 and RelA form a ternary complex, which synergistically activates the IL-8 gene. This induction of the IL-8 gene by TNF is inhibited at the transcriptional level by IFN-β, through the κB site [40]. IFN-β does not block either the activation or binding of NF-κB complexes in response to TNF, but rather it causes additional complexes to be formed, which appear to contain components from the NF-κB and NF-IL-6 family of proteins [41].

The gene for the α chain of the interleukin-2 receptor, is rapidly expressed on both T cells and B cells in response to a wide range of mitogenic stimuli [42] and its promoter region is known to contain an enhancer region which contains binding sites for NF-κB and serum response factor [43]. These two cis-acting sites are of importance in the activation of the IL-2Rα gene in T cells in response to stimulation with PMA, TNFα, or the transactivator protein Tax from the human T cell lymphotropic virus type I. More detailed analysis of the promoter region revealed the presence of binding sites for Elf-1 (an Ets family protein) and HMG-1 (non histone chromatin associated proteins) downstream from the NF-κB enhancer region, and also a physical interaction between Elf-1 and NF-κB and HMG-1 proteins was defined. Such protein–protein and protein–DNA interactions may be functionally significant in both cell-specific and inducible expression of the IL-2Rα gene [43].

Control of NF-κB Expression

A number of cytokines and other receptor–ligand interactions are known to induce NF-κB activation thereby allowing this transcription factor to fulfil its role in the regulation of cytokine and immunoglobulin gene expression. For example, mature B cells produce IL-1 and TNFα, both of which are potent activators of NF-κB, which, in turn, acts at the transcriptional level in the regulation of these two cytokine genes [44, 45]. Within T cells, TNFα can also induce NF-κB expression in a T cell receptor independent manner [46]. A TCR-dependent activation of NF-κB has also been described in primary human T cells and in the Jurkat T cell line, in response to Ab mediated ligation of the TCR/CD3 complex [47, 48]. Indeed, such stimulation of primary T cells resulted in an immediate ($\sim$ 1 hour) nuclear accumulation of p50–p65 NF-κB complexes, recognizing an oligonucleotide containing the κB site of the IL-2Rα promoter, with such heteodimeric complexes replacing the p50 homodimers predominantly found in resting T cells. At a later time point ($\sim$ 7 to 16 hours), p50-c-rel binding activity was also detected.

For optimal T cell activation both antigen-specific signals, as delivered through the TCR and costimulatory signals, which can be delivered through the CD28 molecule, are required. Signalling through CD28 accelerates the rate of translocation of p50, RelA and c-Rel into the nuclear compartment in human peripheral blood T cells, and c-rel translocation in the Jurkat T cell line. This CD28-mediated enhancement of transcriptionally active NF-κB correlates with an increased production of IL-2 by, and proliferation of, the stimulated T cells [49]. The elevation in transcriptionally active NF-κB complexes, as driven by CD28 ligation, appears to be due to an increased rate of inactivation of I-κB [49, 50]. Furthermore, the observation that I-κB degradation is prevented by the immunosuppressant rapamycin, suggests that I-κB is a the downstream signalling target of both CD28 and rapamycin [50].

The interaction between T cell CD40 ligand (a member of the TNF family) and B cell CD40 (a member of the TNF receptor superfamily) is of crucial importance in many aspects of B cell immunological function, including their growth, differentiation and survival from apoptosis. Although the functional effects of CD40 ligation on B cells have been fairly well characterized [48], the downstream signalling events triggered by CD40 ligation are poorly understood. At the level of transcription factors, binding activity for a κB site containing oligonucleotide was increased in a time-dependent manner in B cells, after stimulation with fixed, activated T cells. This T cell mediated induction of NF-κB was inhibited by an anti-CD40L antibody, suggesting that the activation of NF-κB within this system is dependent upon CD40 ligation [51]. A role for CD40 induced NF-κB activation was additionally supported by the studies of Berberich and coworkers who showed that antibody-mediated crosslinking of B cell CD40 also led to the production of p50, RelA and c-Rel containing complexes within tonsillar B cells [52]. Furthermore, this induction of NF-κB seemed to be independent of PKC, as prior treatment of the B cells with phorbol ester did not diminish NF-κB activation [51]. T cell independent activation of B cells, as mimicked by BCR ligation with anti-Ig in human cells, or LPS in murine B cells also leads to NF-κB activation [53].

OCTAMER BINDING PROTEINS

Characterization of Octamer-motif Binding Proteins

The octamer binding element of sequence 5'-ATGCAAAT-3', was originally described as a conserved motif within the promoter region of the histone H2b gene [54], and was subsequently characterized as a functional element within the immunoglobulin heavy and light chain promoters [55, 56]. All octamer DNA-binding proteins share a common structural feature, the POU-domain motif, which is comparable with other structural motifs, such as zinc fingers or leucine zippers, which define other families of structurally related proteins. Biophysical and biochemical evidence suggests that POU-domains are bipartite in structure, consisting of a 75 residue amino terminal POU-specific sub-domain and a 60 amino acid carboxy terminal variant homeodomain, connected by a proteolytically sensitive linker region of variable length and sequence [57]. Both the sub-domains appear necessary for the high affinity interaction of the octamer DNA-binding proteins with DNA, via helix–turn–helix motifs.

The two best characterized octamer binding proteins are Oct-1, a 90–100 kDa ubiquitously expressed DNA-binding protein, and Oct-2, a 60 kDa protein, which is only expressed in lymphocytes and neuronal cells [58]. Oct-1 is thought to be responsible for ubiquitous octamer motif activities, whereas in B cells and neuronal cells, Oct-2 can facilitate cell-specific transcriptional activity. Additional members of the POU-domain family of DNA-binding proteins include Oct-3 [59], Pit-1 [60], [61] and SCIP/Tst-1 [62, 63].

Oct-2

As Oct-2 is believed to mediate lymphocyte-specific octamer motif activities, then the main focus of this section will be this factor and not the ubiquitously acting factor, Oct-1. Although Oct-2 is encoded by a single gene, alternative splicing yields six protein isoforms with differing C-terminal domains, termed Oct-2.1 to Oct-2.6 [64]. A number of studies have sought to identify whether the transactivating potential of Oct-2 is mediated via its amino or carboxy-terminal domains. Indeed, through mutational analysis of the two domains, Annweiler and associates [65] showed that the *C*-terminal domain is critical for enhancer activation, while the glutamine-rich *N*-terminal POU-domain only minimally contributes to transcriptional activation from a remote position. In subsequent studies, Annweiler and colleagues [66] assessed the transcriptional activity of chimeric DNA-binding proteins, which consisted of the DNA-binding domain of the yeast GAL4 transcription factor in conjunction with individual Oct-2 domains. Analysis of such constructs revealed comparable results to those gained through mutational analysis in that, while the *N*-and *C*-terminal domains of Oct-2 can independently stimulate transcription from a promoter proximal position, only the *C*-terminal domain of Oct-2 can stimulate transcription from a distal position, in the presence of an additional B cell restricted factor [66]. One candidate B cell specific co-activator, is OCA-B [67], which facilitates transcription from an Ig heavy chain promoter, but has no significant effect on other octamer-dependent genes, such as that of histone H2b, suggesting that it is a tissue, promoter and factor-specific co-activator protein. A second tissue-specific co-activator protein of octamer binding proteins was identified by Gstaiger and colleagues [68] using the yeast one-hybrid assay. This B cell specific Oct-binding protein, termed Bob1 can associate with Oct-1 and Oct-2, via an interaction with their POU-domains. A third 40 kDa octamer co-activator protein, OAP40, has also been identified as acting in association with Oct-1 to induce IL-2 gene expression [69]. OAP40 shows homology with JunD and c-jun, and appears to function in an AP-1-independent manner, to form a complex with Oct-1, which can contribute to IL-2 gene activation.

Differential Effects of Oct-2 Isoforms

In B cells, octamer motifs exert positive effects on transcription, whereas in neuronal cells their effect is negative. This differential transcriptional effect was shown to be dependent on a number of factors, including the precise sequence of the octamer motif, its position within the promoter and also the cell type [64]. Analysis of the Oct-2 isoforms within B cells and neuronal cells, revealed that while in B cells the Oct-2.1 and 2.3 isoforms predominate, neuronal cells display elevated levels of the Oct-2.2, 2.4 and 2.5 isoforms. The regulatory capacity of the different Oct-2 isoforms was investigated by introducing them into cells which lack endogenous Oct-2 [64]. Such studies revealed that while Oct-2.1, 2.2 and 2.3 had stimulatory effects on promoters, Oct-2.4 and 2.5 stimulated some promoters, and yet repressed others [64]. The transcriptional activation capacities of different Oct-2 isoforms were also studied in a cell-free system, using HeLa cell nuclear extracts, which had been depleted of their Oct-1 activity [70]. In this system Oct-2.3 facilitated a two-fold increase in transcription, while Oct-2.5 stimulated transcription twenty-fold. The experiments of Annweiler and colleagues defined the *N*-terminal domain of Oct-2 as being of importance in determining whether the Oct-2 isoform acts as a positive or negative regulator of transcription [66]. Furthermore, phosphorylation of a negative regulatory domain within Oct-2.3 may be important in controlling the function of this isoform.

Transcriptional activation by Oct-2 seems to be gained through its interaction with the basal transcription factor TFIID, thus allowing formation of a pre-initiation complex, and subsequent transcription by RNA polymerase II. This physical interaction of Oct-2 with basal transcription factors is dependent on its *N*-terminal, glutamine-rich, activation domain [71].

Genes which are Regulated by Octamer-binding Proteins

Octamer motif DNA-binding proteins are known to interact functionally with the 5' regulatory regions of a number of immunologically important genes, including immunoglobulin heavy and light chain genes (reviewed in reference [72]), IL-2, IL-4 [73], CD20 and CD21 [74]. Although Oct-2 is exclusively considered to act as a developmental and tissue specific regulator of Ig gene transcription, the findings of Feldhaus and co-workers [75] suggest that Oct-1 can activate transcription of Ig genes as efficiently as Oct-2. Thus, targeted disruption of the Oct-2 gene in the murine B cell line

WEHI-231, with the consequential 20-fold depletion in Oct-2 levels, did not alter the expression of endogenous Ig genes, or the activity of transfected Ig promoters. Additional evidence in support of Oct-1 transcriptional activity in B cells was provided by Pfisterer and colleagues [76]. They found that in B cell lines which lack endogenous Oct-2 activity, Oct-1 acts in conjunction with a B cell restricted activity, stimulating octamer-containing promoters to a level comparable to that of Oct-2. Furthermore they suggest that in non-B-cells the levels of Oct-2 are insufficient to facilitate octamer-dependent promoter activity.

Three other T cell derived cytokines of importance during an inflammatory reaction are GM-CSF, IL-3 and IL-5. Kaushansky and colleagues found that the basal transcription of each of these cytokine genes is controlled by Oct-1 [77]. In addition, two other T cell specific proteins of 45 kDa and 43 kDa were identified in stimulated T cells, which are immunologically distinct from c-jun or OAP40 (jun-D), and appear to be involved in the coordinated regulation of GM-CSF, IL-3 and IL-5 gene expression.

Octamer Binding Proteins in B-cell Development

As Oct-2 appears to be involved in the B cell restricted expression of Ig genes in B cells, a number of studies have sought to determine whether this factor is necessary for B cell development or function. Miller and coworkers found that the Oct-2 gene is expressed at low levels in a range of transformed pre-B-cell lines, being induced in such cells by LPS stimulation [78]. This induction of Oct-2 in pre-B-cells correlates with the inducible expression of the kappa-light chain immunoglobulin gene, as facilitated *in vivo* by B cell ligation by a stromal cell antigen. The synthesis of both Oct-2 and the kappa-light chain are also induced by IL-1 and inhibited by TGF$_\beta$, lending further support to the idea that Oct-2 is involved in the stage-specific expression of the kappa gene and hence in the expression of Ig molecules on the B cell surface. Disruption of the Oct-2 gene by homologous recombination resulted in mutants which have normal numbers of B cell precursors, but relatively low numbers of IgM$^+$ B cells, with defec-

tive Ig secretion in response to antigenic stimulation [79], suggesting that Oct-2 is of importance in the maturation of Ig-secreting cells.

Other immunologically important genes have also been demonstrated to be developmentally controlled by Oct-2. Indeed, CD20(B1) a B cell specific protein which is involved in the regulation of B cell proliferation and differentiation, is positively regulated by octamer binding proteins. The octamer motif seems to influence both the constitutive expression of CD20 in mature B cells and its induction in pre-B-cells [74]. In more mature B cells, terminal differentiation into plasma cells is characterized by Ig secretion and ablation of MHC class II antigen expression. *In vitro* studies have suggested that IL-6 can induce both B cell lines and primary B cells to undergo this differentiation process by initiating signalling events which regulate Oct-2 DNA-binding activity [80].

STATS AND CYTOKINE SIGNALLING

Cytokines are potent modifiers of cellular function [81], and considerable interest has surrounded the mechanisms of signal transduction for the cytokine receptors [82]. While a number of cytokines, such as IL-8, appear to deliver signals via classical, seven transmembrane loop, G-protein coupled receptors, a much larger group appears to bind to receptors which utilize a quite different intracellular signalling pathway. Thus, the interferons, prolactin, growth hormone and a majority of the interleukins, bind to receptors which activate members of the *Janus* (or JAK) family of protein tyrosine kinases which, in their turn, phosphorylate and activate latent cytosolic transcription factors; this group of transcription factors are the signal transducers and activators of transcription (STATs) [82, 83]. Cytokine receptors of the haematopoietin receptor superfamily (HRS) can be grouped into subfamilies on the basis of shared receptor subunits [82]. For example, cytokines including IL-6, leukaemia inhibitory factor (LIF), oncostatin M and IL-11 act via receptors which, in addition to possessing a chain specific for individual cytokines, share a 130 kDa signal transducer (and are therefore referred to as the gp 130 family). Receptors for IL-3, IL-5 and GM-CSF share a common β chain (β_c or myeloid family), while the receptors for IL-2, IL-4, IL-7, IL-9 and IL-15 possess a common γ chain (γc or lymphoid family of cytokine receptors).

In general terms, STATs are cytoplasmic proteins which play key roles in translating acute signals generated at occupied receptors into biological responses at the transcriptional level. The STAT family members are typically in the 80–100 kDa molecular weight range, and have varying degrees of homology to each other (25–50% overall). The STATs contain three main areas of substantial homology: a 50 amino acid region adjacent to the N-terminus of the molecule; and two domains in the centre of the STAT which have striking homology to SH2 and SH3 domains (Figure 17.2); finally, there is a C-terminal tail region [83]. STATs exist as inactive monomeric molecules in the cytoplasm, and require tyrosine phosphorylation for dimerization, translocation and expression of transactivating activity. STATs bind to specific sequences related to the γ-activating sequences (GAS) found in the promoters of many cytokine sensitive genes [83]. While there is considerable variability in the precise sequences bound by individual STATs, the elements are palindromic and of the general form $T - T - (N)_5 - A - A$ [84]. However, the size of the spacer between the pairs of TT and AA motifs, the palindromic half-sites, has a considerable influence upon which STATs can bind to the GAS-like element. Thus, a 5 bp spacer favours general STAT binding with little selectivity, while a 4 bp spacer element favours complexes activated by, for example, IL-6; a 6 bp spacer between the palindromic half-sites favours STATs triggered by IL-4 [84]. However, while the spacer confers a general restriction in terms of specificity of binding of a STAT to a promoter element, the precise sequence of nucleotides within the spacer is pivotal in determining whether transcriptional activation actually takes place.

STAT1 and STAT2: IFNα and IFNγ Signalling

The pioneering work on STATs was performed in studies of interferon-driven induction of transcription [83]. For IFNα signalling, a 15 bp interferon stimulated response element (ISRE) was found in the 5' promoter region of many IFN-sensitive genes, and insertion of ISRE into reporter plasmids led to IFNα-regulated reporter gene expression [85, 86]. Electrophoretic mobility shift assays demonstrated the presence of ISRE-specific binding protein(s) and the rapid kinetics of appearance of binding activity suggested that the DNA binding protein pre-exisited in an inactive form [86–88]. The protein initially identified as a positive regulator of IFNα-sensitive genes was designated interferon stimulated gene factor-3 (ISGF-3), and a variety of studies identified four proteins in the ISGF-3 complex of molecular weights 48 kDa, 84 kDa, 91 kDa and 113 kDa [89]. The 48 kDa DNA-binding protein (p48) was found to be a member of the interferon regulatory factor (IRF) family [90, 91], and sequence analysis of the remaining three proteins showed them to be highly homologous, but with no obvious homology to any existing protein family. The 91 kDa and 84 kDa proteins are identical, with the sole exception of a 38 amino acid C-terminal region unique to the former protein, and are designated STAT1α and STAT1β, respectively [92]. STAT1α and STAT1β are generated by alternate splicing of exons at the 3' end of the single STAT1 gene. The 113 kDa protein is designated STAT2 and appears to be the sole product of the single STAT2 gene [83, 92].

In contrast to IFNα, IFNγ stimulates transcription of different genes with both immediate [93, 94] and delayed [95] kinetics. Thus, MHC genes are expressed several hours after IFNγ binding, implying a need for de novo protein synthesis, whereas

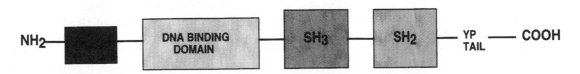

Figure 17.2 Schematic diagram of a STAT family member. A typical STAT family member is shown. In addition to the amino terminal domain and the tyrosine phosphorylated (YP) domain, the SH2 and SH3 domains are shown as individual boxes, as is the separate domain essential for DNA binding activity

other genes are transcribed immediately upon IFNγ binding. As with ISRE in the IFNα model, a consensus γ-activated site (GAS) has been defined for those genes subject to immediate transcriptional activation by IFNγ. The GAS element has a 9 bp core, but it is clear that considerable variation is possible within this 9 bp element before transcriptional activatory potential is abolished [83]. Analysis of proteins binding to the GAS element revealed the presence of 91 kDa STAT1α [96].

A range of somatic cell mutants, deficient in various aspects of IFN signalling, have revealed the complex roles of STAT1 and STAT2 proteins, and the JAK kinases which regulate their activities, in regulating the expression of IFNα and IFNγ sensitive genes in the same cell. Thus, in cells where both STAT1α and STAT1β activities are absent, no response to either IFNα or IFNγ is observed [97]. Introduction of cDNA for either STAT1α or STAT1β complements the defect and restores IFNα responsiveness. However, only STAT1α was capable of restoring IFNγ responsiveness in the cells, despite the fact that STAT1β protein was phosphorylated (Y^{701}), translocated to the nucleus and bound the GAS element [97]. These studies indicate that the 38 amino acid C-terminal region is of importance for GAS-element mediated transcriptional activation. The data indicate that STAT1α alone is sufficient for IFNγ-stimulated transcription, but that STAT2 is necessary for IFNα-stimulated transcription to proceed; that is, STAT2 can associate with either STAT1α or STAT1β to form a transcriptionally competent ISGF-3 complex. STAT2 also shows a single site for tyrosine kinase action (Y^{690}) and phosphorylation of Y^{690} appears to occur independently of STAT1α or STAT1β phosphorylation, implying that a different kinase is involved [98]. However, translocation of phosphorylated STAT2 proceeds with greater efficiency in the presence of STAT1α or STAT1β.

Biochemical studies have indicated that, in resting cells, STAT1α and STAT1β exist as monomers which are converted to dimers following receptor activation. Moreover, monomers and dimers can be isolated from cytosolic fractions, indicating that STAT homo-and heterodimers form in the cytoplasm prior to translocation to the nucleus [99]. A number of mutations have proved informative in

assessing the importance of tyrosine phosphorylation to STAT function. Treatment of cells with IFNα or IFNγ leads to tyrosine phosphorylation of each of STAT1α and STAT1β at a single identical site, tyrosine 701 (Y^{701}) [100]. The Y701F mutation abolishes sensitivity to both IFNα and IFNγ, and blocks phosphorylation of STAT1α, dimer formation, assembly of the ISGF-3 complex and transcriptional activation [96]. Both STAT1α and STAT1β possess SH2 domains in which arginine 602 (R^{602}) is critical in recognizing the tyrosine phosphorylated peptide present in activated, ligand-occupied receptors [101]. The R602L substitution abrogates binding to tyrosine phosphate and also dimerization of STAT molecules, a result which suggests that activated STATs dimerize via their SH2 domains [96]. Dimers can be driven to slowly dissociate and re-form by incubation with a phosphorylated Y^{701}-containing peptide [83]. Building upon the theme of peptide inhibition of STAT function, Greenlund and colleagues demonstrated that a peptide derived from the region around Y^{440} in the IFNγ receptor in which Y^{440} was phosphorylated could inhibit STAT1β activation [102]. Thus, phosphotyrosyl-containing peptides from both the activated receptor and the STAT itself can inhibit STAT function by blocking either receptor association and/or STAT dimerization.

STAT3: Regulation of the Acute Phase Response

The sequence analysis of STAT1 and STAT2, together with the observation that ligands other than the interferons could activate JAK family protein tyrosine kinases, suggested that the STATs may be a large family of related molecules responsible for linking cell surface ligand binding events to alterations in gene expression. PCR-based approaches using conserved sequences led to the isolation and characterization of the STAT3 and STAT4 (see below) family members [103].

Murine STAT3 appears to comprise 770 amino acids (92 kDa) and is widely expressed in murine tissues [103]. It was initially defined as acute phase response factor and, as such, it is unsurprising that STAT3 is involved in signalling via those cytokine receptors which comprise the gp130 signalling component (i.e. IL-6, LIF, oncostatin M and IL-11) [104,105]. STAT3 can also be tyrosine phosphorylated and activated by epidermal growth factor or lipopolysaccharide treatment of mouse liver [104,106], and by granulocyte-colony stimulating factor [107]. Finally, in the CMK megakaryocytic cell line, thrombopoietin promotes tyrosine phosphorylation and activation of both STAT1 and STAT3 [108].

STAT4: Signalling via IL-12 receptors

STAT4 was cloned by PCR-based methods using oligonucleotides which reflected conserved sequences in the human STAT1α and STAT1β proteins [103]. The STAT4 cDNA sequence is predicted to encode a protein of 749 amino acids and immunoprecipitation studies suggest that STAT4 is a protein of 89 kDa. The pattern of tissue expression of STAT4 is restricted, being found principally in spleen, testis and thymus [103]. Although initially thought to be an "orphan" transcription factor, recent data indicate that STAT4 can bind to GAS elements following tyrosine phosphorylation by JAKs [109]. Moreover, two groups have demonstrated independently that IL-12 can induce tyrosine phosphorylation of STAT4 in human T lymphoid cells [110,111].

STAT5: Signal Coupling in Murine Myeloid Cells and Human T Lymphocytes

As noted earlier, cytokine receptors can be grouped into three groups based on common receptor subunits. Thus, two main receptor groups are defined on the basis of the common γ subunit or by gp130, while the final subset, the myeloid family, utilizes a common β chain and includes the receptors for IL-3, IL-5 and GM-CSF. As with other cytokine families, stimulation of cells with these ligands leads to activation of JAK tyrosine kinases [82].

Studies of signal transducing factors (STF) binding to GAS-like elements in myeloid cells stimulated with IL-3 revealed that two complexes could be defined. STF-IL-3a was found in immature myeloid cells and contained two GAS-binding proteins of 77 kDa and 80 kDa. STF-IL3a was

biochemically distinct from STF-IL3b, which possessed GAS-binding proteins of 94 kDa and 96 kDa and was found mainly in more mature cell lines [112]. Immunochemical data suggest that the 80 kDa and 77 kDa molecules are generated by proteolytic processing of the C-terminal region of the 96 kDa and 94 kDa molecules, respectively [112]. The cloning of cDNAs corresponding to the 77 kDa and 80 kDa species has now been accomplished by two independent groups [112,113]. The p80 cDNA clone encodes a protein predicted to contain 786 amino acids, and this is designated STAT5B; STAT5A encodes the p77 component of STF-IL3a. While STAT5A and STAT5B show significant homology to each other, they show striking homology to ovine mammary gland factor, a prolactin-sensitive factor present in lactating sheep mammary gland [112,113]. Analysis of the sequence data for STAT5A and STAT5B suggests that, unlike the STAT1α and STAT1β proteins which are alternative splice variant products of a single gene, the STAT5A and STAT5B proteins are products of distinct genes [112, 113].

STAT5 expression is widespread, being found in spleen, thymus and bone marrow, as well as in a range of other tissues and cell lines. By RNAase protection analysis it appears that STAT5A and STAT5B are expressed coordinately, with broadly equivalent levels of the two RNA species being found in tissues which express STAT5; there were no differences based upon whether cells were IL-3, IL-5 or GM-CSF-dependent [113]. However, Northern bolt analysis of tissue expression suggested the presence of only a single species of STAT5 mRNA [112].

Consistent with the mechanism of action of other STATs, treatment of a range of cell lines with IL-3 leads to phosphorylation of STAT5 molecules and to formation of STAT5 dimers. Analysis of peptides from STAT5A and STAT5B indicates that Y^{699} is the key residue in STAT5 activation [112]; this tyrosine is in the equivalent position to those involved in regulating activation of STAT1α, STAT1β and STAT2. In terms of activation, IL-3 leads to tyrosine phosphorylation and activation of both STAT5A and STAT5B, and it is clear that both homodimeric and heterodimeric complexes can be formed [113]. Consistent with overlaps in biological activities and sharing of the βc subunit in their respective receptors, IL-5 and GM-CSF can elicit identical STAT5 responses to those driven by IL-3.

In the murine models, STAT5 isoforms appear to be regulated principally by IL-3, IL-5 or GM-CSF. However, human STAT5, which shows homology both to the murine STAT5 molecules and to ovine mammary gland factor, is activated by IL-2 [114]. Like other STAT family members, human STAT5 activation is regulated by tyrosine phosphorylation, in this case by JAK3, a result which is entirely consistent with the finding that the C-terminal region of the IL-2 receptor β chain is required for STAT5 activation [114].

STAT6: an IL-4-regulated STAT

A human DNA binding protein regulated by JAK kinases and IL-4 was isolated from a HUVEC cDNA library [115]. STAT6 is predicted to be an 848 amino acid protein, giving rise to a 94 kDa protein; in fact, the mobility of STAT6 by SDS–PAGE is 100 kDa. Northern blotting analyses revealed that expression of STAT6 mRNA was widespread, and also that several distinct RNA species were detectable in different tissues, raising the possibility of multiple STAT6 variants whose expression may be developmentally regulated [115].

STAT6 activation can be inhibited by phospho-peptides derived from either the IL-4 receptor or STAT6 [115]. A group of five IL-4 receptor penta-decapeptides containing phosphotyrosine as their central residues were tested for inhibition of STAT6 dimerization, and only two related sequences with conserved residues flanking the central tyrosine (Y^{578} or Y^{606}) were active; non-phosphorylated peptides and those derived from the IFNγ receptor were unable to inhibit STAT6 dimerization. The same two peptides which blocked STAT6 dimerization were also capable of abolishing STAT6 DNA-binding activity, even although STAT6 had been correctly phosphory-lated, suggesting that the peptides perturb the STAT6 dimer giving rise to monomers which are not able to interact with STAT6 target DNA [115].

Murine STAT6 has recently been cloned [111], and displays 83% homology to its human counter-part; a tyrosine residue (Y^{641}) has been identified which is likely to serve the equivalent function to Y^{701} in STAT1 species. Like human STAT6, mur-ine STAT6 mRNA is expressed in a range of tissues and at least two hybridzation signals are evident in northern blots. By SDS–PAGE analysis, human and murine STAT6 molecules migrate at 100 kDa and 102 kDa, respectively, but minor species of 94 kDa and 84 kDa are also evident which appear to be products of processing of the N-terminal regions of the molecules [111]. Given that IL-4, but not IL-2, activates STAT6 it seems that the α chain of the IL-4 receptor confers spe-cificity for STAT6 activation. Studies of trunca-tion and point mutants of the IL-4 receptor α chain show that removal of the C-terminus from either amino acid 557 or from residue 437 leads to loss of STAT6 activation whereas the Y497F point mutation, which leads to loss of mitogenic responses to IL-4, is without noticeable effect upon tyrosine phosphorylation of STAT6 [111].

Entraining Specificity; Signalling Connections to the *Janus* System

The *Janus* (or JAK) family of non-receptor protein tyrosine kinases provides the link between ligand activation of the receptor and initiation of down-stream signalling. Thus, essentially all of the cyto-kines binding to HRS members activate JAKs and STATs.

As noted earlier, IL-6 appears to utilize the STAT1 and STAT3 pathways. The contribution of individual JAKs to STAT activation was stu-died in mutant human fibrosarcoma cells lacking JAK1, JAK2 or Tyk2 activities [116]. IL-6 stimulates tyrosine phosphorylation of each of these three JAK family members in wildtype cells and, in cells lacking any one of these JAKs, tyrosine phosphorylation of the remaining two family members is normal; thus, absence of one JAK member does not abolish IL-6-driven tyro-sine phosphorylation of the others. However, the absence of JAK1, but not JAK2 or Tyk2, resulted in a striking absence of tyrosine phosphorylation of the gp130 molecule upon IL-6 stimulation. This dominant effect of JAK1 was again evident in terms of STAT activation; IL-6-driven activation of STAT1 and STAT3 was greatly compromised in JAK1-cells. This study demonstrated the essen-tial role of the JAK–STAT system in the response to IL-6, and the dominant role played by JAK1 in driving the response [116]. It remains to be determined if other ligands which utilize gp130 have a similar reliance on JAK1 for mobilization of the biological response or whether JAK2 and/or Tyk 2 have more dominant roles in these responses. However, it is noteworthy that thrombopoietin, like IL-6, IL-11 and LIF, has effects upon megakaryoctye proliferation but acts via c-*mpl*, a non-gp130 receptor, and mediates its effects upon STAT1 and STAT3 activation via JAK2 [108].

IL-2 is important in regulation of lymphocyte growth. The high affinity IL-2 receptor complex comprises a 55 kDa α chain, a 75 kDa β chain and the 64 kDa γc subunit. In T lymphocytes, addition of IL-2 results in activation of tyrosine phosphorylation, and this can be largely accounted for by JAK1 and JAK3 activation. JAK3 has been shown to be associated with the γc subunit of the IL-2 receptor complex and JAK1 with the β chain [117]. IFNα induces JAK and Tyk2 phosphorylation in T cells and, in terms of STAT activation, IFNα induced STAT1, STAT2 and STAT3 activation in T cells, whereas IL-2 failed to activate any of these STATs [118]. In this case, STAT5 is activated by IL-2 in human T cells [114], and STAT6 also appears to be a substrate for IL-4-driven tyrosine phosphorylation in T cells [111]. Since the γc subunit is common to the IL-2 and IL-4 receptor complexes [119], and this selectively associates with JAK3 [117], it is possible that STAT5 and STAT6 activation in T cells is mediated by JAK3, while IFN-mediated activation of STAT1, STAT2 and STAT3 is dependent upon JAK2 or Tyk2 activity [118].

The data from several JAK–STAT systems makes it possible to develop a theme, with variations, for signalling via cytokine receptors [83]. The key to initiation of signal transduction is, of course, ligand binding. Ligand binding leads to dimerization of receptor components, recruitment of JAK family members to the occupied receptor complex and phosphorylation of the receptor at specific tyrosine residues. Monomeric STAT molecules with specificity for particular tyrosine phosphorylated motifs on the receptor bind via their SH2 domains and themselves become phosphorylated upon tyrosine residues, leading to formation of STAT homo-or heterodimers. The activated STAT dimers are now capable of migrating to the nucleus where, in conjunction with other elements of the transcriptional machinery, they bind to consensus elements in the promoter elements of genes sensitive to the influence of the initial stimulatory ligand (Figure 17.3). While there are only a limited number of JAKs and STATs, there is considerable scope for elaboration within this basic framework. Thus, recruitment of a particular JAK to an activated receptor complex can lead to phosphorylation of several different STAT monomers; once phosphorylated, these are able to associate to form a range of dimeric forms, each of which may have distinctive specificity for promoter elements and characteristic transcriptional activation potential, either stimulatory or inhibitory, once the complexes arrive at the promoter element and interact with other transcription factors.

REGULATION OF IL-2 GENE EXPRESSION

Interleukin-2 is a T cell derived cytokine which plays a pivotal role in both T cell and B cell biology. In T cells it is required for their activation, proliferation and differentiation, whereas in B cells, it is necessary for IL-2-dependent B cell growth, and also for their differentiation into antibody secreting plasma cells. IL-2 gene transcription within T cells is developmentally regulated, in that IL-2 is not produced by resting T cells, but is produced by CD4$^+$ T cells following their activation by a range of stimuli, including antigens, mitogens and costimulatory signals, resulting in the synthesis and maximal cytokine secretion at approximately 24 hours post-stimulation [36]. The T cell restricted nature of IL-2 gene transcription, and its developmental regulation within T lineage cells, have resulted in this gene serving as a model system for studying the mechanisms of inducible gene transcription in T cells.

IL-2 Promoter Region

Analysis of the 5' flanking region of the IL-2 gene by deletion analysis [120], gel shift assays [121] and *in vivo* PCR genomic footprinting [122] has identified a number of functional regions which are necessary for maximal induction of the gene by

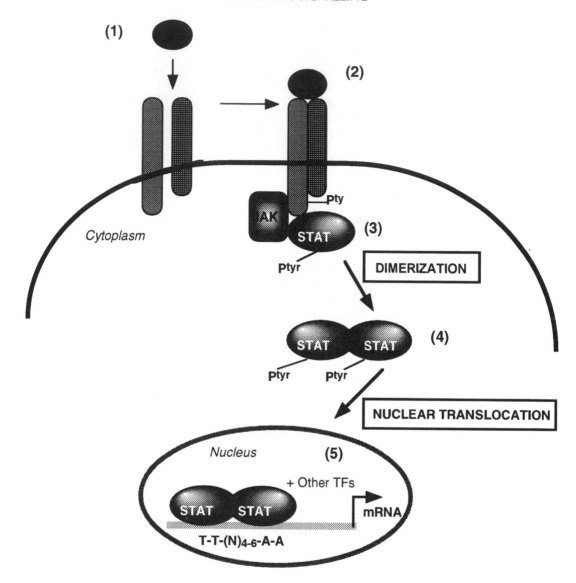

Figure 17.3 The JAK–STAT cytokine receptor signalling system. Cytokine binds to its cognate receptor resulting in oligomerization of receptor components (*Stage 1*). One or more JAKs associated with the oligomerized receptor tyrosine phosphorylate specific residues within the cytoplasmic tail of the cytokine receptor, and latent STAT proteins bind to the phosphorylated receptor via their SH$_2$ domains; the bound STATs, in turn, become tyrosine phosphorylated by JAK kinases (*Stage 2*).

The tyrosine phosphorylated STAT proteins are released from the receptor (*Stage 3*), and form homo-or heterodimers, probably via their SH$_2$ domains, prior to translocating to the nucleus (*Stage 4*). The STAT dimers bind to defined sequences within the promoter regions of cytokine-responsive genes and, in combination with other transcription factors, initiate transcription (*Stage 5*).

mitogens, both in the Jurkat T cell line and in normal primary T cells (Figure 17.4). These regions include binding sites for NF-κB [123]; NF-AT, AP1, AP3, Ets and Oct-1/NF-IL-2A [124], CD28RC and NF-IL2B (a non-classical NF-κB site), which appears to be distinct from

AP1 [125] (reviewed in reference [36]). Comparable studies on the 5′ flanking region of the mouse IL-2 gene, which shares approximately 86% identity with the human gene, suggested that such regulatory sites are conserved both in sequence and function [126]. Analysis of the sequences of the

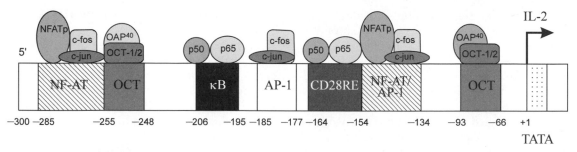

Figure 17.4 The IL-2 promoter region. The main *cis* elements of the promoter region of the IL-2 gene are illustrated as boxes with distinct styles, and the locations of the elements relative to the transcription initiation site are shown as numbers along the bottom edge of the figure. The complexes of transcriptional regulatory factors which interact with each *cis* motif are shown on the top of the figure

regulatory elements within both the human and the murine IL-2 gene revealed that they are all non-canonical, and therefore do not optimally bind their associated transcription factors. Even so, such weak binding sites are of functional importance, as their removal decreases the inducibility of the IL-2 promoter, and their conversion to canonical sites increases the activity of the promoter in T cells and also facilitates IL-2 gene transcription in non-T cells [127]. Through *in vivo* footprinting studies [128] it has been demonstrated that only activated T cells showed full occupancy of DNA-binding sites within the IL-2 promoter region, implying that both the T cell restricted nature and developmental control of IL-2 gene transcription may be attributable to the stable binding of such transcription factors. Based on a number of observations [127,128], it was proposed that the differential binding affinities of ubiquitous transcription factors, in conjunction with the activities of T cell specific and signal dependent binding factors, may contribute to the cell specific activity of the IL-2 promoter.

NF-AT

A binding site for NF-AT has been identified within the 5' regulatory region of the IL-2 gene, between −255 and −285 bp, which shows binding activity for two polypeptides, of 45 kDa and 90 kDa [129]. It appears that following engagement of the TCR, NF-AT is activated by a CsA-sensitive calcium-dependent signalling pathway, involving cyclophilin, a cyclophilin binding protein, termed calcium-signal modulating cyclophilin ligand

(CAMT) and calcineurin [130]. Although the primary initiation of IL-2 gene transcription appears to be NF-AT dependent and CsA sensitive, the subsequent T cell proliferation as supported by phorbol ester stimulation, is independent of NF-AT and CsA resistant [131].

ETS1

Two potential Ets binding sites, termed EBS1 and EBS2 have been identified within the promoter regions of both the murine and human IL-2 genes [132]. Both the EBS sites are located within the area of the NF-AT and NFIL-2B motifs, and are essential for the formation of NF-AT-1 and NFIL-2B nuclear protein complexes. The EBS1 and EBS2 associated binding activities were not attributable to the previously characterized Ets-1 or Ets-2 family members, but to a new family member, termed Elf-1 [132].

AP-1

AP-1 transcriptional activity is induced in T cells following TCR ligation and is implicated in IL-2 gene transcriptional events. The importance of AP-1 in the production of IL-2 by T cells was highlighted through transient transfections of Jurkat cells with TAM-67, a dominant negative mutant of the c-jun protein [133]. In such cells, TAM-67 expression inhibited the endogenous transcriptional activity of both AP-1 and NF-AT, resulting in the blockade of IL-2 gene activity, but had no effect on the NF-κB, Oct-1/NF-IL2A or proximal TRE-like sites. Additional evidence supporting a role for AP-1 in IL-2 gene regulation was provided by studies on splenic T cells from c-fos transgenic mice [134]. Following anti-CD3 treatment of such cells, nuclear AP-1 levels were elevated in comparison to that of control cells, but no significant differences in NF-κB, NF-AT or Oct-1 levels were detected. This elevation in AP-1 was concomitant with both enhanced and prolonged IL-2 gene expression and cytokine production. IL-1, which acts as a costimulator of T cell activation by antigen, enhances levels of c-jun mRNA, thus complementing the elevation in c-fos mRNA seen after antigenic stimulation [135]. Other T cell costimulatory signals, such as CD28 ligation by B7-1 or B7-2, are also necessary for optimal AP-1 transcriptional function, and hence for IL-2 gene activity [136].

Oct-1 and Oct-2

Two octamer binding sites have been defined within the 5' flanking region of the IL-2 gene, a proximal site at −75 to −66 bp, and a distal site at −256 to −248 bp, relative to the transcription initiation site at −1 bp, and each of these two sites can be regulated by Oct-1 or Oct-2 binding [137]. Subsequent studies by Ullman and coworkers (124), defined a role for both Oct-1 and OAP40 (an octamer associated binding protein) in antigen mediated stimulation of IL-2 gene activity in T cells. In addition, it is suggested that the close proximity of the distal octamer and AP-1 binding sites results in a functional interaction between bound transcription factors, which appears to be crucial for optimal IL-2 gene transcription [73].

NF-κB

The principal κB response element is located at between −206 and −195 bp in the promoter region, with a secondary non-classical site at −159 to −134 bp (sometimes called the NF-IL-2B site). Elevation in cAMP levels can cause a gene specific inhibition in IL-2 expression, by inhibiting of the binding of NF-κB family members to their regulatory elements. However, in contrast, cAMP elevation does not affect NF-AT or AP-1 binding [138].

Negative Regulation of the IL-2 Gene

In recent experiments, Romano-Spica and coworkers suggested that the ETS1 gene, which is expressed at high levels in resting T cells, but at very low levels in activated T cells, may act as a negative regulator of IL-2 gene transcription [139]. Indeed, Jurkat T cells transfected with antisense ETS1 produce higher levels of IL-2 after stimulation, than cells transfected with sense ETS1, and the activity of the IL-2 promoter, as measured via reporter gene activity, was higher in antisense transfectants than their sense ETS1 counterparts. These two observations support the idea that the ETS1 gene product suppresses IL-2 gene expression. Further negative regulation of the IL-2 gene expression is provided by a 735 amino acid protein, NIL-2A, which binds to a negative response element (NRE-A) located at −110 to −101 bp in the 5' promoter of the IL-2 gene [140]. The NIL-2A activity is present in resting and mitogen-activated T cells, and is CsA insensitive [140]. Finally, other negative regulatory regions have been proposed such as the Pud element at position −292 to −264 bp, which silences the IL-2 gene in resting T cells, being displaced by a positively acting transcription factor following T-cell activation events [141].

The differential regulation of IL-2 gene expression is clearly evident in both murine and human Th1 and Th2 subsets, in that Th1 cells secrete IL-2, whereas Th2 cells do not. This differential gene activity was also demonstrated *in vitro* by Barve and colleagues [142] using Th1 or Th2 T cell clones transfected with the IL-2 promoter linked to a reporter gene. In Th2 T cells, IL-2 promoter activity could be initiated if the cells were cotransfected with a vector containing the eukaryotic initiation factor 4E (eIF-4E), which is known to be rate limiting for protein synthesis. This overexpression of eIF-4E was concomitant with an increase in NF-AT levels, raising the possibility that differences in the concentrations of transcription factors within Th1 and Th2 cells may be of importance in determining whether or not they express the IL-2 gene. Previous studies by Schwartz and coworkers suggested that the mutually exclusive patterns of cytokine expression by Th1 and Th2 cells may be achieved at the level of transcription factors, in that IL-4, a Th2 derived cytokine, inhibits NF-

IL2B binding activity, thus potentially disrupting transcriptional IL-2 gene activation in Th1 T cells [125].

A distinction can also be made between anergic and activated T cells based on their IL-2 secretion patterns. Thus, antigenic activation of T cells in the absence of an appropriate costimulatory signal, renders such cells anergic, possibly due to their failure to activate NF-AT and NF-κB. This is in contrast to their functionally activated counterparts, where such transcription factors are activated thus allowing IL-2 gene activity [143].

T Cell Development

During intrathymic T cell development of pre T cells three stages are demarcated by their competence to switch on the IL-2 gene in response to acute stimulation. Thus, IL-2 inducibility is acquired prior to T cell receptor gene rearrangement, is lost during positive and negative selection stages, and re-appears in the mature, positively selected thymocytes. [144]. In an attempt to assign a molecular basis for these developmentally regulated transitions, Chen and Rothenberg examined DNA-binding protein activity within the different subpopulations of T cells, after stimulation with phorbol ester and calcium ionophore. Their results indicated that while Oct-1 and Sp1 are constitutively expressed in all thymocytes, and NF-κB could be induced within all thymocytes, NF-AT and AP-1 could only be activated in cells not undergoing selection processes; finally, a factor containing CREB (cyclic AMP response element binding protein) could be activated in all thymocytes in the absence of specific induction. Thus, NF-AT and AP-1 seem to be particularly sensitive to changes in thymocytes undergoing the positive or negative selection processes [144].

TRANSCRIPTION FACTORS AND LYMPHOID DEVELOPMENT

Haematopoiesis is a complex and dynamic process, giving rise to cells of erythroid, myeloid and lymphoid lineages. Progression from the pluripotent stem cell to a mature cell phenotype of any lineage requires programmed expression of particular genes in the differentiating cells. In view of this, it is likely that included in the genes expressed as part of the developmental programme will be transcription factor genes which can, in their turn, direct the expression of other genes as the appropriate developmental stimuli are encountered. Recent work has identified genes encoding distinct groups of transcription factors, including E2A, *Pax*-5 and *Ikaros*, which are required for accurate lymphopoiesis and where targeted disruption of the genes leads to increasingly severe effects upon development of mature lymphocytes.

Ikaros and Lymphoid Development

The *Ikaros* gene is a candidate for a gene with overall control of early steps in lymphopoiesis, and was initially defined in terms of its ability to bind to and activate the enhancer of the CD3δ gene [145]. The *Ikaros* gene encodes a family of five proteins (*Ik1–Ik5*) via alternate splicing of primary transcripts (Molnar in reference [146]). All *Ikaros* gene products have two zinc-finger DNA-binding domains at their N-and C-termini, with the precise structure and specificity of the N-terminal domain being determined by the alternate splicing. As well as differing in DNA binding capacity, the *Ikaros* gene products all have different transcriptional activation potential, with some behaving as strong activators of transcription while others are suppressors. Binding sites for *Ikaros* gene products have the general form GGGAA/T, and are found in the regulatory sites for a number of genes critical for lymphoid function including *Igh*, *Igk*, *Igl*, CD3, CD25, RAG-1, TdT and *mb*-1 (Molnar in reference [146]). The importance of these genes in B lymphoid development has been recently reviewed [147, 148].

Generation of *Ikaros*$^{-/-}$ mice revealed that this gene exerts its influence upon lymphopoiesis at a very early stage. Thus, *Ikaros*$^{-/-}$ mice lack B and T lymphocytes, natural killer cells and all precursors of these cell types, but have normal erythroid and myeloid lineages [146]. Not surprisingly, *Ikaros*$^{-/-}$ mice succumb rapidly to opportunistic infections. The fact that the myeloid and erythroid lineages are intact in the *Ikaros*$^{-/-}$ mice, while the lymphoid lineages are ablated, indicates that the

Ikaros gene acts close to the level of the pluripotent stem cell and prior to the stage where committed lymphoid progenitors appear [145, 146]. Thus, the *Ikaros* gene products may act to drive early haematopoeitic precursors towards commitment to lymphoid cells and so function as a master regulator for early lymphoid development [146]. In the absence of functional *Ikaros* proteins, haematopoietic precursors flow towards the myeloid and erythroid compartments.

E2A Genes and B Lymphopoiesis

The E2A genes encode transcription factors of the helix–loop–helix (HLH) type which have a pair of amphipathic helices separated by a loop region which can be of different lengths [149]. Some HLH factors have an additional conserved basic domain adjacent to the helix–loop–helix motif which is necessary for DNA binding; these are designated bHLH proteins. The E protein genes, including those for E2A and E2-2 proteins, are widely expressed and, via the HLH structure, are capable of forming homodimers or heterodimers with tissue-specific HLH proteins, with the latter being more efficient in transcriptional activation [149]. The E2A gene gives rise to two distinct proteins, E12 and E47, by alternate splicing of E12 or E47-specific bHLH exons [150]. These proteins are widely expressed, but it is only in B lineage cells that they appear to form dimeric complexes which bind to E sites (e.g. μE2, μE5, κE1) in immunoglobulin gene enhancers [151, 152]. E2-2 proteins are pre-dominant over E2A proteins in the pre-B-cell compartment but this feature is reversed as the cells develop towards the mature B cell [151].

Targeted disruption of the E2A gene results in knockout mice (E2A$^{-/-}$) with an unusual defined phenotype [153, 154]. Thus, although E2A mRNA and protein levels are widespread in normal mice, the principal phenotype of E2A$^{-/-}$ mice is a lack of B lymphoid cells; the levels of all other haematopoietic lineages, including T cells, are essentially normal. Moreover, the levels of mRNA for several genes critical for B cell development are absent in the E2A$^{-/-}$ animals, including those for *RAG-1*, *mb-1*, CD19, *pax-5* and λ_5 [153, 154]. The mice lack rearranged immunoglobulin heavy chain genes, suggesting that the block in differentiation

caused by the lack of E2A proteins occurs extremely early in the B cell developmental pathway, perhaps at the early pro-B-cell stage. Moreover, it is known that two transcripts, Iμ and μ°, are formed from *un*rearranged heavy chain genes, and it is proposed that these reflect the availability of the locus for rearrangement [148]. In the E2A$^{-/-}$ animals, both Iμ and μ° transcripts are depressed, particularly those of Iμ, and this may signal inaccessibility of the *Igh* locus for rearrangement, leading ultimately to a failure in B lymphopoiesis [148, 153, 154].

In addition to the E2A proteins themselves, B cells also express inhibitors of E2A protein activity, the *Id* gene products *Id*1 and *Id*2 [155], which are expressed at high levels at early stages of B cell development and are coordinately down-regulated as the cells reach the pre-B-cell stage. The *Id* proteins block E2A activity by binding to their HLH domains, forming inactive heterodimers [155]. While E2A proteins are expressed throughout B cell development [151], DNA binding activity can only be detected in extracts of cells at the pre-B and later stages, which correlates with loss of *Id* protein. The proposal that *Id* proteins function to restrain E2A activity is supported by the finding that mice transgenic for constitutive expression of *Id*1 show impaired B cell development which is related to the extent of *Id*1 transgene expression [156]. The mice show blockade of transition from the pro-B to pre-B compartments and display reduced rearrangement of immunoglobulin heavy and κ light chain loci. The transcription of several B cell-specific genes is diminished and, in parallel with data from the E2A$^{-/-}$ mice [153, 154], levels of Iμ transcripts were very low (μ° transcripts were only slightly reduced [156]). Thus, overexpression of E2A inhibitory proteins can lead to broadly similar effects to those noted in E2A$^{-/-}$ mice, demonstrating the importance of E2A proteins in B cell development.

Pax-5 (BSAP)

Pax-5 encodes the paired domain transcription factor B cell lineage-specific activator protein (BSAP) [157]. *Pax*-5 is expressed in all B lineage cells with the exception of plasma cells and has been shown to regulate the expression of CD19

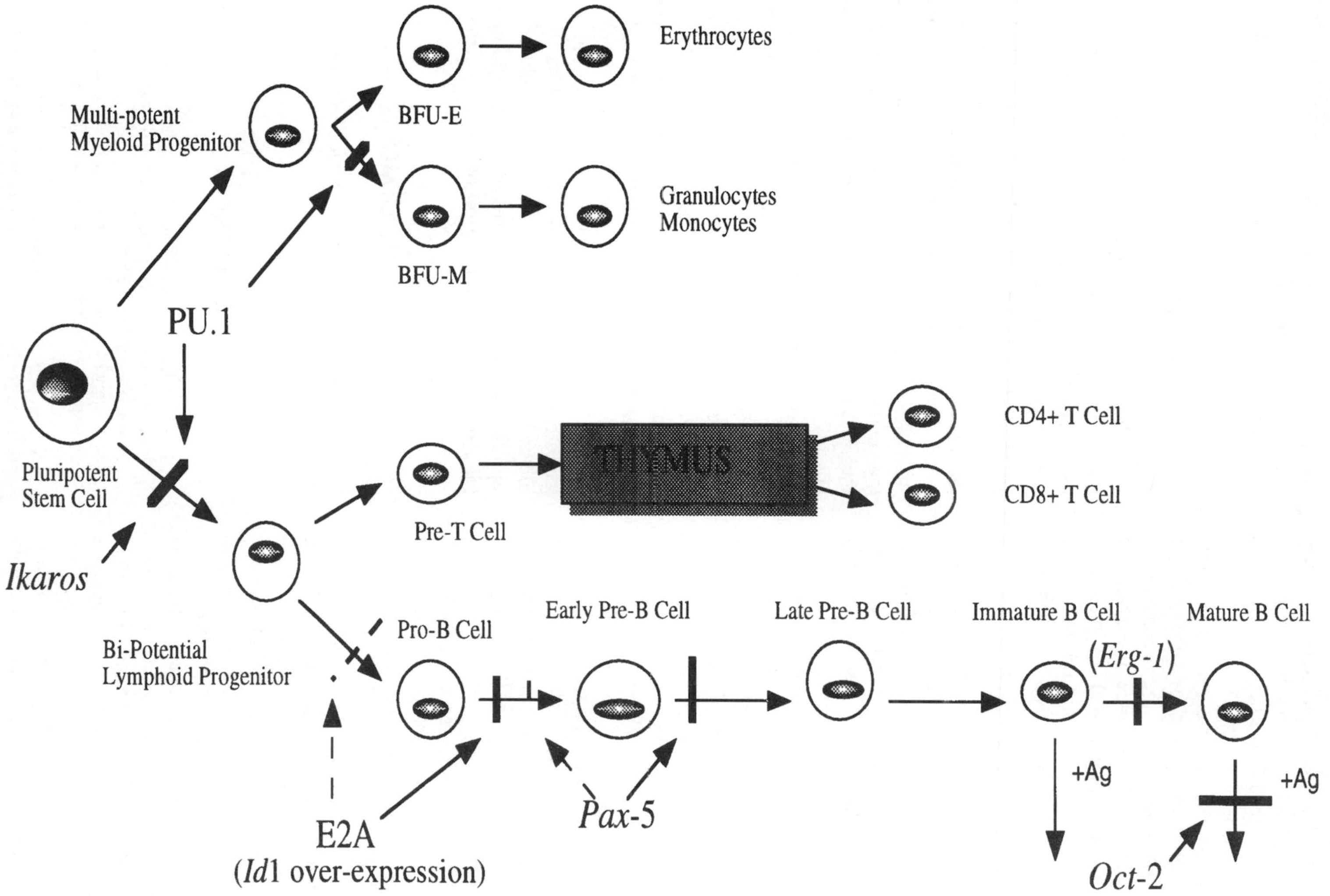

Figure 17.5 Transcription factors in lymphoid development. The figure shows the principal haematopoietic lineages, and the transcription factors which regulate particular steps are shown in bold type. The points where transcription factors act are given by double solid lines (or broken lines where the point of action is tentative). *Id*1 data are based upon transgenic mice which constitutively overexpress the inhibitory factor. Only the B cell differentiation pathway has been given in detail

[158]. A number of other B-cell-specific genes appear to have *Pax*-5 binding sites in their promoters including the C_ϵ gene of the heavy chain locus [159] and the V_{pre-B1} gene [160]. That *Pax*-5 has a role in B cell function is underscored by the finding that B cell proliferation can be suppressed if *Pax*-5-antisense oligonucleotides are present in stimulated cells [161]. The importance of *Pax*-5 in B cell development was demonstrated by generation of *Pax*-5$^{-/-}$ mice [162]. In addition to altered patterning in the midbrain, *Pax*-5$^{-/-}$ animals showed a severe block in B cell differentiation. Thus, although early pro-B cell precursors were present in the animals, there was a persistent failure to develop small pre-B cells, leading to a lack of B cells and plasma cells in all lymphoid organs of the mice, although T cell levels were normal; serum immunoglobulins were also essentially undetectable [162]. At the level of immunoglobulin gene rearrangements, a small fraction of cells from the *Pax*-5$^{-/-}$ mice showed heavy chain rearrangements, but the κ light chains were in the germline configuration in all of the mice. Finally, in addition to effects on conventional B-2 B cells, the *Pax*-5$^{-/-}$ animals showed a lack of Ly-1$^+$(CD5$^+$) B cells implying that *Pax*-5 is important for development of B-1 B cells [162]; this may reflect a failure to express V_{pre-B1}. Thus, in contrast to the more severe effects of loss of *Ikaros* or E2A function, *Pax*-5 appears to regulate the pro-B to pre-B transition in B cell development.

It is therefore clear that transcription factors have a significant role to play in both the antigen-independent and antigen-dependent phases of lymphoid development (Figure 17.5). The available data suggest that *Ikaros* may act as a committing factor for lymphoid development [146], acting after PU.1, inactivation of which leaves only erythrocytes and megakaryocytes intact but destroys all of the lymphoid, myeloid and granulocytic compartments [163]. After action of *Ikaros* and commitment to the lymphoid lineages, expression of E2A proteins is required for progression beyond the early stages of B cell development [153, 154], and *Pax*-5 must act after the pre-B-cell stage is reached to ensure continuation towards the antigen-dependent compartment [162]. Oct-2 family transcription factors are then required for mobilization of responses to antigen or mitogens in B

cells [79] and, on a wider canvass, intact NF-κB is required for accurate performance of a number of immunological functions in the periphery [164].

FUTURE PROSPECTS

It is clear that transcriptional regulatory factors have a central role in the development, activation and regulation of lymphoid function. Thus, it is clear that lack of expression of individual transcription factors can lead to immunodeficiency either via lack of lymphocytes themselves or an inability of the cells to respond to antigenic challenge. While many DNA binding proteins and their target sequences are defined, and understanding of the involvement of certain factors such as the STATs in the signalling networks of the cell are advancing rapidly, much remains to be discovered about the precise regulation of transcription factor activation and de-activation and how lineage-specific factors interact with other proteins to generate transcriptionally competent complexes. Equally, the mechanisms by which transcription factors can act to activate gene expression at one gene but suppress transcription at another await detailed characterization. Finally, the developmental controls which govern expression of the genes encoding critical transcription factors themselves is not well appreciated and this field offers significant promise for a greater understanding of the genetic basis of regulation of lymphopoiesis.

ACKNOWLEDGEMENTS

ELB is supported by the BBSRC *Intracellular Signalling* Programme.

REFERENCES

1. Jain J, Miner Z and Rao A. Analysis of the pre-existing and nuclear forms of nuclear factor of activated T cells. *J. Immunol.* 1993; **151**: 837–848.
2. McCaffrey PG, Luo C, Kerppola TK *et al.* Isolation of the cyclosporin-sensitive transcription factor NF-ATp. *Science* 1993; **262**: 750–754.
3. Northrop JP, Ho SN, Chen L *et al.* NF-AT components define a family of transcription factors targeted in T cell activation. *Nature* 1994; **369**: 497–502.

4. Hoey T, Sun Y-L, Williamson K and Xu X. Isolation of two new members of the NF-AT gene family and funtional characterization of the NF-AT proteins. *Immunity* 1995; **2**: 461–472.

5. Rao A. NF-ATp: a transcription factor required for the co-ordinate induction of several cytokine genes. *Immunol. Today* 1994; **15**: 274–281.

6. Boise LH, Petryniak B, Moa X *et al*. The NFAT-1 DNA binding complex in activated T-cells contains Fra-1 and Jun B. *Mol. Cell. Biol.* 1993; **13**: 1911–1919.

7. Jain J, McCaffrey PG, Miner Z *et al*. The T cell transcription factor NF-ATp is a substrate for calcineurin and interacts with fos and jun. *Nature* 1993; **365**: 352–355.

8. Jain J, Burgeon E, Badalian TM, Hogan PG and Rao A. A similar DNA binding motif in the NF-AT family protens and the Rel homology region. *J. Biol. Chem.* 1992; **270**: 4138–4145.

9. Rooney JW, Hoey T and Glimcher LH. Coordinate and cooperative roles for NF-AT and AP-1 in the regulation of the murine IL-4 gene. *Immunity* 1995; **2**: 473–483.

10. Flanagan WM, Corthesy B, Bram RJ and Crabtree GR. Nuclear association of a T cell transcription factor blocked by FK-506 and cyclosporin A. *Nature* 1991; **352**: 803–807.

11. Clipstone NA and Crabtree GR. Identification of calci-neurin as a key signalling enzyme in T lymphocyte activation. *Nature* 1992; **357**: 695–697.

12. Sen R and Baltimore D. Multiple nuclear factors interact with the immunoglobulin enhancer sequences. *Cell* 1986; **46**: 705–716.

13. Grimm S and Baeuerle PA. The inducible transcription factor NFκB: structure function relationship of its protein subunits. *Biochem. J.* 1993; **290**: 297–308.

14. Fan CM and Maniatis T. Generation of p50 the subunit of NF-κB by processing of p105 through an ATP-dependent pathway. *Nature* 1991; **354**: 395–398.

15. Ghosh G, Van Duyne G, Ghosh S and Sigler PB. Structure of NF-κB p50 homodimer bound to a κB site. *Nature* 1995; **373**: 303–310.

16. Muller CWR FA, Sodeoka M, Verdine GL and Harrison SC. Structure of the NF-κB p50 homodimer bound to DNA. *Nature* 1995; **373**: 311–317.

17. Blank V, Kourilsky P and Israel A. NF-κB and related proteins: Rel/dorsal homologies meet ankyrin-like repeats. *Trends Biochem. Sci.* 1992; **17**: 135–140.

18. Nolan GP and Baltimore D. The inhibitory ankyrin and activator Rel proteins. *Curr. Opin. Genet. Devel.* 1992; **2**: 211–220.

19. Lernbecher T, Muller U and Wirth T. Distinct NF-kappa B/Rel transcription factors are responsible for tissue-specific and inducible gene activation. *Nature* 1993; **365**: 767–770.

20. Moore PA, Ruben SM and Rosen CA. Conservation of transcriptional activation functions of the NF-κB p50 and p65 subunits in mammalian cells and in *Saccharomyces cerevisiae*. *Mol. Cell. Biol.* 1993; **13**: 1666–1674.

21. Kretzshcmar M, Meisterernst M, Scheidereit C, Li G and Roeder RG. Transcriptional regulation of the HIV-I pro-moter by NF-κB in vitro. *Genes Dev.* 1992; **6**: 761–774.

22. Fujita T, Nolan GP, Ghosh S and Baltimore D. Indepen-dent modes of transcriptional activation by the p50 and p65 subunits of NF-κB. *Genes Dev.* 1992; **6**: 775–787.

23. Brown AM, Linhoff MW, Stein B *et al*. Function of NF-kappa B/Rel binding sites in the major histocompatibility complex class II invariant chain promoter is dependent on cell-specific binding of different NF-kappa B/Rel subunits. *Mol. Cell. Biol.* 1994; **14**: 2926–2935.

24. Ballard DW, Walker WH, Doerre S *et al*. The v-rel onco-gene encodes a κB enhancer binding protein that inhibits NF-κB function. *Cell* 1990; **63**: 803–814.

25. Hansen SK, Nerlov C, Zabel U *et al*. A novel complex between the p65 subunit of NFκB and c-Rel binds to a DNA element involved in the phorbol ester induction of the human urokinase gene. *EMBO J.* 1992; **11**: 205–213.

26. Kieran M, Blank V, Logeat F *et al*. The DNA binding subunit of NF-κB is identical to factor KBF1 and homo-logous to the rel oncogene product. *Cell* 1990; **62**: 1007–1018.

27. Ryseck R-P, Bull P, Takamiya M *et al*. RelB, a new family transcriptional activator that can interact with p50 NF-κB. *Mol. Cell. Biol.* 1992; **12**: 674–684.

28. Beg AA and Baldwin AS. The I-κB proteins: multi-func-tional regulators of Rel/NF-κB transcription factors. *Genes. Dev.* 1993; **7**: 2064–2070.

29. Thanos D and Maniatis T. NF-κB: a lesson in family values. *Cell* 1995; **80**: 529–532.

30. Henkel T, Machleidt T, Alkalay I, Kronke M, Ben-Neriah Y and Baeuerle PA. Rapid proteolysis of I kappa B-alpha is necessary for activation of transcription factor NF-kappa B. *Nature* 1993; **365**: 182–185.

31. Traenckner EB-M, Wilk S and Baeuerle A. A proteasome inhibitor prevents activation of NF-κB and stabilizes a newly phosphorylated form of IκB-α that is still bound to NF-κB. *EMBO J.* 1994; **13**: 5433–5441.

32. Wasserman SA. A conserved signal transduction pathway regulating the activity of the rel-like proteins *dorsal* and NF-κB. *Mol. Biol. Cell* 1993; **4**: 767–771.

33. Frantz B, Nordby EC, Bren G *et al*. Calcineurin acts in synergy with PMA to inactivate IκB/MAD3, an inhibitor of NF-κB. *EMBO J.* 1994; **13**: 861–870.

34. Scott ML, Fujita T, Liou HC, Nolan GP and Baltimore D. The p65 subunit of NF-kappa B regulates I kappa B by two distinct mechanisms. *Genes Dev.* 1993; **7**: 1266–1276.

35. Li C-CH, Dai R-M, Chen E and Longo DL. Phosphoryla-tion of NF-κB1-p50 is involved in NF-κB activation and stable DNA binding. *J. Biol. Chem.* 1994; **269**: 30089–30092.

36. Ullman KS, Northrop JP, Verweij CL and Crabtree GR. Transmission of signals from the T lymphocyte antigen receptor to the genes responsible for cell proliferation and immune function: the missing link. *Ann. Rev. Immunol.* 1990; **8**: 421–452.

37. Ray A and Prefontaine KE. Physical association and func-tional antagonism between the p65 subunit of transcription factor NF-kappa B and the glucocorticoid receptor. *Proc. Natl. Acad. Sci. USA* 1994; **18**: 752–756.

38. Kopp E and Ghosh S. Inhibition of NF-kappa B by sodium salicylate and aspirin. *Science* 1994; **265**: 956–959.

39. Kunsch C and Rosen CA. NF-kappa B subunit-specific regulation of the interleukin-8 promoter. *Mol. Cell. Biol.* 1993; **13**: 6137–6146.

40. Kunsch C, Lang RK, Rosen CA and Shannon MF. Synergistic transcriptional activation of the IL-8 gene by NF-kappa B p65 (RelA) and NF-IL-6. *J. Immunol.* 1994; **153**: 153–164.

41. Oliveira IC, Mukaida N, Matsushima K and Vilcek J. Transcriptional inhibition of the interleukin-8 gene by interferon is mediated by the NF-κB site. *Mol. Cell. Biol.* 1994; **14**: 5300–5308.

42. Cushley W and Harnett MM. Regulation of B lymphocyte growth and differentiation by soluble mediators. In: Snow EC (ed.) *Handbook of B and T Lymphocytes*. London: Academic Press, 1994.

43. John SJ, Reeves RB, Lin J-X *et al.* Regulation of cell-type-specific interleukin-2 receptor α-chain gene expression: Potential role of physical interactions between Elf-1, HMG-I(Y), and NF-κB family proteins. *Mol. Cell. Biol.* 1995; **15**: 1786–1796.

44. Sung S-S, Jung LK, Walters JA, Chen W, Wang CY and Fu SM. Production of tumour necrosis factor/cachetin by human B cell lines and tonsillar B cells. *J. Exp. Med.* 1988; **168**: 1539–1544.

45. Matsushima K, Procopio A, Abe H, Scala G, Ortaldo JR and Oppenheim JJ. Production of interleukin 1 activity by normal human peripheral blood B lymphocytes. *J. Immunol.* 1985; **135**: 1132–1139.

46. Grilli M, Chen-Tran A and Lenardo MJ. Tumour necrosis factor alpha mediates a T cell receptor-independent induction of the gene regulatory factor NF-kappa B in T lymphocytes. *Mol. Immunol.* 1993; **30**: 1287–1294.

47. Pimentel-Muinos FX, Mazana J and Fresno M. Biphasic control of nuclear factor-κB activation by the T cell receptor complex: role of tumour necrosis factor α. *Eur. J. Immunol.* 1995; **25**: 179–186.

48. Baeuerle PA and Henkel T. Function and activation of NF-κB in the immune system. *Ann. Rev. Immunol.* 1994; **12**: 141–179.

49. Bryan RG, Li Y, Lai JH *et al.* Effect of CD28 signal transduction on c-rel in human peripheral blood T-cells. *Mol. Cell. Biol.* 1994; **14**: 7933–7942.

50. Lai JH and Tan TH. CD28 signalling causes a sustained down-regulation of I kappa B alpha which can be prevented by the immunosuppressant rapamycin. *J. Biol. Chem.* 1994; **269**: 30077–30080.

51. Lamanach-Girard AC, Chiles TC, Parker DC and Rothstein TL. T cell-dependent induction of NF-kappa B in B cells. *J. Exp. Med.* 1993; **177**: 1215–1219.

52. Berberich I, Shu GL and Clark EA. Cross-linking CD40 on B cells rapidly activates nuclear factor-kappa B. *J. Immunol.* 1994; **153**: 4357–4366.

53. Muller JM, Ziegler-Heitbrock HW and Baeuerle PA. Nuclear factor kappa B, a mediator of lipopolysaccharide effects. *Immunobiology* 1993; **187**: 233–256.

54. Harvey RP, Robins AJ and Wells JRE. Independently evolving chicken histone H2B genes: identification of a ubiquitous H2B-specific 5' element. *Nucl. Acid Res.* 1982; **10**: 7851–7863.

55. Falkner F and Zachau Hca. Correct transcription of an immunoglobulin gene requires an upstream fragment containing conserved sequence elements. *Nature* 1984; **310**: 71–74.

56. Parslow T, Blair D, Murphy W and Granner D. Structure of 5' ends of immunoglobulin genes: a novel conserved sequence. *Proc. Natl. Acad. Sci. USA* 1984; **81**: 2650–2654.

57. Botfield MC, Jancso A and Weiss MA. Biochemical characterisation of the Oct-2 POU domain with implications for bipartite DNA recognition. *Biochemistry* 1992; **31**: 5841–5848.

58. Dent CL, Lillycrop KA, Estridge JK, Thomas NS and Latchman DS. The B-cell and neuronal forms of the octamer-binding protein Oct-2 differ in DNA-binding specificity and functional activity. *Mol. Cell. Biol.* 1991; **11**: 3925–3930.

59. Leonardo MJ, Staudt L, Robbins P, Kuang A, Mulligan RC and Baltimore D. Repression of the IgH enhancer in tetracarcinoma cells associated with a novel octamer factor. *Science* 1989; **243**: 544–546.

60. Bodner M, Castrillo JL, Theill LE, Deerinck T, Ellisman M and Karin M. The pituitary-specific transcription factor GHF-1 is a homebox-containing protein. *Cell* 1988; **55**: 505–518.

61. Ingraham HA, Chen R, Mangalam HJ *et al.* A tissue specific transcription factor containing a homedomain specifies a pituitary phenotype. *Cell* 1988; **55**: 519–529.

62. Monuki ES, Weinmaster G, Kuhn R and Lemke G. A glial POU domain gene regulated by cAMP. *Neuron* 1989; **3**: 783–793.

63. Xi H, Treacy MN, Simmons DM, Ingraham HA, Swanson LW and Rosenfeld MG. Expression of a large family of POU-domain regulatory genes in mammalian brain development. *Nature* 1989; **340**: 35–42.

64. Lillycrop KA and Latchman DS. Alternative splicing of the Oct-2 transcription factor RNA is differentially regulated in neuronal cells and B-cells and results in protein isoforms with opposite effects on the activity of octamer/TAAT-GARAT-containing promoters. *J. Biol. Chem.* 1992; **267**: 24960–24965.

65. Annweiler A, Muller-Immergluck M and Wirth T. Oct2 transactivation from a remote enhancer position requires a B-cell-restricted activity. *Mol. Cell. Biol.* 1992; **12**: 3107–3116.

66. Annweiler A, Zwilling S and Wirth T. Functional differences between the Oct2 transactivation domains determine the transactivation potential of individual Oct2 isoforms. *Nucl. Acid Res.* 1994; **11**: 4250–4258.

67. Luo Y, Fujii H, Gerster T and Roeder RG. A novel B cell derived coactivator potentiates the activation of immunoglobulin promoters by octamer-binding transcription factors. *Cell* 1992; **71**: 231–241.

68. Gstaiger M, Knoepfel L, Georgiev O, Schaffner W and Hovens CM. A B-cell coactivator of octamer-binding transcription factors. *Nature* 1995; **373**: 360–362.

69. Uckun FM, Schieven GL, Dibirdik I, Chandan-Langlie M, Tuel-Ahlgren L and Ledbetter JA. Stimulation of protein tyrosine phosphorylation, phosphoinositide turnover, and multiple previously unidentified serine, threonine-specific protein kinases by the pan B-cell receptor CD40/Bp50 at discrete developmental stages of human B-cell ontogeny. *J. Biol. Chem.* 1991; **266**: 17478–17485.

70. Annweiler A, Zwilling S, Hipskind RA and Wirth T. Analysis of transcriptional stimulation by recombinant Oct proteins in a cell-free system. *J. Biol. Chem.* 1993; **268**: 2525–2534.

71. Arnosti DN, Merino A, Reinberg D and Schaffner W. Oct-2 facilitates functional preinitiation complex assembly and is continuously required at the promoter for multiple rounds of transcription. *EMBO J.* 1993; **12**: 157–166.

72. Staudt LM. Immunoglobulin gene transcription. *Ann. Rev. Immunol.* 1991; **9**: 373–398.

73. Pfeuffer T, Klein-Hessling S, Heinfling A *et al.* Octamer factors exert a dual effect on the IL-2 and IL-4 promoters. *J. Immunol.* 1994; **153**: 5572–5585.

74. Thevenin C, Lucas BP, Kozlow EJ and Kehrl JH. Cell type-and stage-specific expression of the CD20/B1 antigen correlates with the activity of a diverged octamer DNA motif present in its promoter. *J. Biol. Chemi.* 1993; **268**: 5949–5956.

75. Feldhaus AL, Klug CA, Arvin KL and Singh H. Targeted disruption of the Oct-2 locus in a B-cell provides genetic evidence for two distinct cell type-specific pathways of octamer element-mediated gene activation. *EMBO J.* 1993; **12**: 2763–2772.

76. Pfisterer P, Annweiler A, Ullmer C, Corcoran LM and Wirth T. Differential transactivation potential of Oct1 and Oct2 is determined by additional B cell-specific activities. *EMBO J.* 1994; **13**: 1655–1663.

77. Kaushansky K, Shoemaker SG, O'Rork CA and McCarty JM. Coordinate regulation of multiple human lymphokine genes by Oct-1 and potentially novel 45 and 43 kDa polypeptides. *J. Immunol.* 1994; **152**: 1812–1820.

78. Miller CL, Feldhaus AL, Rooney JW, Rhodes LD, Sibley CH and Singh H. Regulation and a possible stage-specific function of Oct-2 during B-cell differentiation. *Mol. Cell. Biol.* 1991; **11**: 4885–4894.

79. Corcoran LM, Karelas M, Nossal GJ, Ye ZS, Jacks T and Baltimore D. Oct-2, although not required for early B-cell development, is critical for later B-cell maturation and for postnatal survival. *Genes Dev.* 1993; **7**: 570–582.

80. Natkunam Y, Zhang X, Liu Z and Chen-Kiang S. Simultaneous activation of Ig and Oct-2 synthesis and reduction of surface MHC class II exxpression by IL-6. *J. Immunol.* 1994; **153**: 3476–3484.

81. Callard RE (ed.). *Cytokines and B Lymphocytes.* London: Academic Press, 1990.

82. Kishimoto T, Taga T and Akira S. Cytokine signal transduction. *Cell* 1994; **76**: 253–262.

83. Darnell JE Jr, Kerr IM and Stark GR. Jak–STAT pathways and transcriptional activation in response to IFNs and other extracellular signalling proteins. *Science* 1994; **264**: 1415–1420.

84. Seidel HM, Milocco LH, Lamb P, Darnell JE, Stein RB and Rosen J. Spacing of palindromic half sites as a determinant of selective STAT (signal transducers and activator of transcription) DNA binding and transcriptional activity. *PNAS* 1995; **92**: 3041–3045.

85. Kessler DS, Levy DE and Darnell JE, Jr. Two interferon induced nuclear factors bind a single promoter element in interferon-stimulated genes. *Proc. Natl. Acad. Sci. USA* 1988; **85**: 8521–8525.

86. Levy DE, Kessler DS, Pine R, Reich N and Darnell JE, Jr. Interferon-induced nuclear factors that bind a shared promoter element correlate with positive and negative transcriptional control. *Genes Dev.* 1988; **2**: 383–393.

87. Dale TC, Rosen JM, Guille MJ *et al.* Overlapping sites for constitutive and induced DNA binding factors involved in interferon-stimulated transcription. *EMBO J.* 1989; **8**: 831–839.

88. Porter AGC, Chernajovsky Y, Dale TC, Gilbert CS, Stark GR and Kerr IM. Interferon response element of the human gene 6–16. *EMBO J.* 1988; **7**: 85–92.

89. Fu X-Y, Kessler SD, Veals SA, Levy DE and Darnell JE, Jr. ISGF3, the transcriptional activator induced by interferon α consists of multiple interacting polypeptide chains. *Proc. Natl. Acad. Sci. USA* 1990; **87**: 8555–8559.

90. Miyamoto M, Fujita T, Kimura Y *et al.* Regulated expression of a gene encoding a nuclear factor, IRF-1, that specifically binds to IFN-β gene regulatory elements. *Cell* 1988; **54**: 903–913.

91. Pine R, Decker T, Kessler DS, Levy DE, Darnell JE, Jr. Purification and cloning of interferon-stimulated gene factor 2 (ISGF2): ISGF2 (IRF-1) can bind to the promoters of both beta interferon-and interferon-stimulated genes but is not a primary transcriptional activator of either. *Mol. Cell. Biol.* 1990; **10**: 2448–2457.

92. Fu X-Y, Schindler C, Improta T, Aebersold R and Darnell JE, Jr. The proteins of ISGF3, the interferon α-induced transcriptional activator, define a gene family involved in signal transduction. *Proc. Natl. Acad. Sci. USA* 1992; **89**: 7836–7841.

93. Decker T, Lew DJ, Mircovitch J and Darnell JE, Jr. Cytoplasmic activation of GAF, an IFN-gamma-cegulated DNA-binding factor. *EMBO J.* 1991; **10**: 927–932.

94. Pearse RN, Feinman R, Shuai K and Darnell JE, Jr. Intcfcron α-induced transcription of the high affinity Fc receptor for IgG requires assembly of a complex that includes the 91 kDa subunit of transcription factor ISGF3. *Proc. Natl. Acad. Sci. USA* 1993; **90**: 4314–4318.

95. Blanar MA, Boettger EC and Flavell RA. Transcriptional activation of HLA-DRα by interferon γ requires a trans acting protein. *Proc. Natl. Acad. Sci. USA* 1988; **85**: 4672–4676.

96. Shuai K, Stark GR, Kerr IM and Darnell JE, Jr. A single phosphotyrosine residue of Stat91 required for gene activation by interferon gamma. *Science* 1993; **261**: 1744–1746.

97. John J, McKendry R, Pellegrini S, Flavel D, Kerr IM and Stark GR. Isolation and characterisation of a new mutant human cell line unresponsive to alpha and beta interferons. *Mol. Cell. Biol.* 1991; **11**: 4189–4195.

98. Improta T, Schindler C, Horvath CM, Kerr IM, Stark GR and Darnell JEJ. Transcription factor ISGF-3 formation requires phosphorylated Stat91 protein, but Stat113 protein is phosphorylated independently of Stat91 protein. *PNAS* 1994; **91**: 4776–4780.

99. Schindler C, Shuai K, Prezioso VR and Darnell JE, Jr. Interferon-dependent phosphorylation of a latent cytoplasmic transcription factor. *Science* 1992; **257**: 809–813.

100. Shuai K, Ziemiecki A, Wilks AF *et al.* Polypeptide signalling to the nucleus through tyrosine phosphorylation of Jak and Stat proteins. *Nature* 1993; **366**: 580–583.

101. Overduin M, Rios CB, Mayer BJ and Baltimore D. Three-dimensional solution structure of the src homology 2 domain of c-abl. *Cell* 1992; **70**: 697–704.

102. Greenlund AC, Farrar MA, Viviano BL and Schreiber RD. Ligand-induced IFNγ receptor tyrosine phosphorylation couples the receptor to its signal transduction system (p91). *EMBO J.* 1994; **13**: 1591–1600.

103. Zhong Z, Wen Z and Darnell JEJ. Stat3 and Stat4: Members of the family of signal transducers and activators of transcription. *PNAS* 1994; **91**: 4806–4810.

104. Zhong Z, Wen Z and Darnell JEJ. Stat3: a new family member that is activated through tyrosine phosphorylation in response to EGF and IL-6. *Science* 1994; **264**: 95–98.

105. Feldman GM, Petricoin III EF, David M, Larner AC and Finbloom DS. Cytokines that associate with the signal transducer gp130 activate the interferon-induced transcription factor p91 by tyrosine phosphorylation. *Biol. Chem.* 194; **269**: 10747–10752.

106. Ruff-Jamison S, Zhong Z, Wen Z, Chen C, Darnell JE, Jr and Cohen S. Epidermal growth factor and lipopolysaccharide activate Stat3 transcription factor in mouse liver. *J. Biol. Chem.* 1994; **269**: 21933–21935.

107. Tian S, Lamb P, Seidel HM, Stein RB and Rosen J. Rapid activation of the Stat3 transcription factor by granulocyte colony-stimulating factor. *Blood* 1994; **84**: 1760–1764.

108. Gurney AL, Wong SC, Henzel WJ, de Sauvage FJ. Distinct regions of c-Mpl cytoplasmic domain are coupled to the JAK–STAT signal transduction pathway and Shc phosphorylation. *PNAS* 1995; **92**: 5292–5296.

109. Yamamoto K, Quelle FW, Thiefelder WE *et al.* Stat4: a novel gamma interferon activation binding site protein expressed in early myeloid differentiation. *Mol. Cell. Biol.* 1994; **14**: 4342–4349.

110. Jacobson NG, Szabo S, Weber-Nordt RM *et al.* Interleukin 12 activates Stat3 and Stat4 by tyrosine phosphorylation in T cells. *J. Exp. Med.* 1995: in press.

111. Quelle FW, Shimoda K, Thierfelder W *et al.* Cloning of murine Stat6 and human Stat6, Stat proteins that are tyrosine phosphorylated in responses to IL-4 and IL-3 but are not required for mitogenesis. *Mol. Cell. Biol.* 1995; **15**: 3336–3343.

112. Azam M, Erdjument-Bromage H, Kreider BL *et al.* Interleukin-3 signals through multiple isoforms of Stat5. *EMBO J.* 1995; **14**: 1402–1411.

113. Mui A-LF, Wakao H, O'Farrell A-M, Harada N and Miyajima A. Interleukin-3, granulocyte-macrophage colony stimulating factor and interleukin-5 transduce signals through two STAT5 homologs. *EMBO J.* 1995; **14**: 1166–1175.

114. Fujii H, Nakagawa Y, Schindler U *et al.* Activation of Stat5 by interleukin 2 requires a carboxyl-terminal region of the interleukin 2 receptor β chain but it is not essential for the proliferative signal transmission. *PNAS* 1995; **92**: 5482–5486.

115. Hou J, Schindler U, Henzel WJ, Ho TC, Brasseur M and McKnight SL. An interleukin-4-induced transcription factor: IL-4 Stat. *Science* 1994; **265**: 1701–1705.

116. Guschin D, Rogers N, Briscoe J *et al.* A major role for the protein tyrosine kinase JAK1 in the JAK/STAT signal transduction pathway in response to interleukin-6. *EMBO J.* 1995; **14**: 1421–1429.

117. Miyazaki T, Kawahara A, Fujii H *et al.* Functional activation of Jak1 and Jak3 by selective association with IL-2 receptor subunits. *Science* 1994; **266**: 1045–1047.

118. Beadling C, Guschin D, Witthuhn BA *et al.* Activation of JAK kinases and STAT proteins by interleukin-2 and interferon α, but not the T cell antigen receptor, in human T lymphocytes. *EMBO J.* 1994; **13**: 5605–5615.

119. Russell SM, Keegan AD, Harada N *et al.* The interleukin 2 receptor γ chain is a functional component of the interleukin 4 receptor. *Science* 1993; **262**: 1880–1883.

120. Shibuya H and Taniguchi T. Identification of multiple cis-elements and trans-acting factors involved in the induced expression of the IL-2 gene. *Nucl. Acid Res.* 1989; **17**: 9173–9184.

121. Granelli-Piperno A and Nolan P. Nuclear transcription factors that bind to elements of the IL-2 promoter. Induction requirements in primary human T-cells. *J. Immunol.* 1991; **15**: 2734–2739.

122. Brunvand MW, Krumm A and Groudine M. In vivo footprinting of the human IL-2 gene reveals a nuclear factor bound to the transcription factor start site in T cells. *Nucl. Acid Res.* 1993; **21**: 4824–4829.

123. Hoyos B, Ballard DW, Bohlein E, Siekevitz M and Greene WC. Kappa B-specific DNA binding proteins: role in the regulation of human interleukin-2 gene expression. *Science* 1989; **244**: 457–460.

124. Ullman KS, Flanagan WM, Edwards CA and Crabtree GR. Activation of early gene expression in T lymphocytes by Oct-1 and an inducible protein, OAP40. *Science* 1991; 254: 558–562.

125. Schwartz EM, Salgame P and Bloom BR. Molecular regulation of human interleukin 2 and T-cell function by interleukin 4. *Proc. Natl. Acad. Sci. USA* 1993; **90**: 7734–7738.

126. Novak TJ, White PM and Rothenberg EV. Regulatory anatomy of the murine interleukin-2 gene. *Nucl. Acid Res.* 1990; **18**: 4523–4533.

127. Hentsch B, Mouzaki A, Pfeuffer T, Rungger D and Serfling E. The weak, fine-tuned binding of ubiquitous transcription factors to the IL-2 enhancer contributes to its T cell restricted activity. *Nucl. Acid Res.* 1992; **20**: 2657–2665.

128. Garrity PA, Chen D, Rothenberg EV and Wold BJ. Interleukin-2 transcription is regulated in vivo at the level of coordinated binding of both constitutive and regulated factors. *Mol. Cell. Biol.* 1994; **14**: 2158–2169.

129. Corthesy B and Kao PN. Purification by DNA affinity chromatography of two polypeptides that contact the NF-AT DNA binding site in the interleukin 2 promoter. *J. Biol. Chem.* 1994; **269**: 20682–20690.

130. Bram RJ and Crabtree GR. Calcium signalling in T cells stimulated by cyclophilin B-binding protein. *Nature* 1994; **371**: 355–358.

131. Terada N, Or R, Weinberg K, Domenico J, Lucas JJ and Gelfand EW. Transcription of IL-2 and IL-4 genes is not inhibited by cyclosporin A in competant T-cells. *J. Biol. Chem.* 1992; **267**: 21207–21210.

132. Thompson CB, Wang CY, Ho TC *et al.* Cis-acting sequences required for inducible interleukin-2 enhancer

function bind a novel Ets-related protein, Elf-1. *Mol. Cell. Biol.* 1992; **12**: 1043–1053.

133. Petrak D, Memon SA, Birrer MJ, Ashwell JD and Zacharchuk CM. Dominant negative mutant of c-jun inhibits NF-AT transcriptional activity and prevents IL-2 gene transcription. *J. Immunol.* 1994; **153**: 2046–2051.

134. Ochi Y, Koizumi T, Kobayashi S *et al.* Analysis of IL-2 gene regulation in c-fos transgenic mice. Evidence for an enhancement of IL-2 expression in splenic T cells stimulated via TCR/CD3 complex. *J. Immunol.* 1994; **153**: 3485–3490.

135. Muegge K, Williams TM, Kant J *et al.* Interleukin-1 costimulatory activity on the interleukin-2 promoter via AP-1. *Science* 1989; **246**: 249–251.

136. Rincon M and Flavell RA. AP-1 transcriptional activity requires both T-cell receptor-mediated and co-stimulatory signals in primary T lymphocytes. *EMBO J.* 1994; **13**: 4370–4381.

137. Kamps MP, Corcoran T, LeBowitz JH and Baltimore D. The promoter of the human interleukin-2 gene contains two octamer-binding sites and is partially activated by the expression of Oct-2. *Mol. Cell. Biol.* 1990; **10**: 5464–5472.

138. Chen D and Rothenberg EV. Interleukin 2 transcription factors as molecular targets of cAMP inhibition: delayed inhibition kinetics and combinatorial roles. *J. Exp. Med.* 1994; **179**: 931–942.

139. Romano-Spica V, Georgiou P, Suzuki H, Papas TS and Bhat NK. Role of ETS1 in IL-2 gene expression. *J. Immunol.* 1995; **154**: 2724–2732.

140. Williams TM, Moolten D, Burlein J *et al.* Identification of a zinc finger protein that inhibits IL-2 gene expression. *Science* 1991; **254**: 1791–1794.

141. Mouzaki A, Dai Y, Weil R and Rungger D. Cyclosporin A and FK506 prevent the derepression of the IL-2 gene in mitogen-induced primary T lymphocytes. *Cytokine* 1992; **4**: 151–160.

142. Barve SS, Cohen DA, De-Benedetti A, Rhodas RE and Kaplan AM. Mechanism of differential regulation of IL-2 in murine Th1 and Th2 T cell subsets. 1. Induction of IL-2 transcription in Th2 cells by up-regulation of transcription factors with the protein synthesis initiation factor 4E. *J. Immunol.* 1994; **152**: 1171–1181.

143. Go C and Miller J. Differential induction of transcription factors that regulate the interleukin 2 gene during anergy induction and restimulation. *J. Exp. Med.* 1992; **175**: 1327–1336.

144. Chen D and Rothenberg EV. Molecular basis for developmental changes in interleukin-2 gene inducibility. *Mol. Cell. Biol.* 1993; **13**: 228–237.

145. Georgopoulos K, Moore DD and Deefler B. *Ikaros*, an early lymphoid restricted transcription factor, a putative mediator for T cell commitment. *Science* 1992; **258**: 808–812.

146. Georgopoulos K, Bigby M, Wang J-H *et al.* The Ikaros gene is required for the development of all lymphoid lineages. *Cell* 1994; **79**: 143–156.

147. Dorshkind K. Transcriptional control points during lymphopoiesis. *Cell* 1994; **79**: 751–753.

148. Desiderio S. Transcription factors controlling B-cell development. *Curr. Biol.* 1995; **5**: 605–608.

149. Murre C, Bain G, van Dijk MA *et al.* Structure and function of helix-loop-helix proteins. *Biochim. Biophys. Acta* 1994; **1218**: 129–135.

150. Murre C, McCaw PS and Baltimore D. A new DNA binding and dimerisation motif in immunoglobulin enhancer binding, *daughterless, MyoD* and *myc* proteins. *Cell* 1989; **56**: 777–783.

151. Bain G, Gruenwald S and Murre C. E2A and E2-2 are subunits of B-cell specific E2 box DNA-binding proteins. *Mol. Cell. Biol.* 1993; **13**: 3522–3529.

152. Murre C, Voronova A and Baltimore D. B cell and myocyte-specific E2 box binding factors contain E12/E47-like subunits. *Mol. Cell. Biol.* 1991; **11**: 1156–1160.

153. Bain G, Robanus Maandag EC, Izon DJ *et al.* E2A proteins are required for proper B cell development and initiation of immunoglobulin gene rearrangements. *Cell* 1994; **79**: 885–892.

154. Zhuang Y, Soriano P and Weintraub H. The helix-loop helix gene E2A is required for B cell formation. *Cell* 1994; **79**: 875–884.

155. Sun X-H, Copeland NG, Jenkins NA and Baltimore D. Id proteins, Id1 and Id2, selectively inhibit DNA binding by one class of helix-loop-helix proteins. *Mol. Cell. Biol.* 1991; **11**: 5603–5611.

156. Sun X-H. Constitutive expression of the *Id*1 gene impairs mouse B cell development. *Cell* 1994; **79**: 893–900.

157. Adams B, Dorfler P, Aguzzi A *et al.* Pax-5 encodes the transcription factor BSAP and is expressed in B lymphocytes, the developing CNS and adult testis. *Genes Dev.* 1992; **6**: 1589–1607.

158. Kozmik Z, Wang S, Dorfler P, Adams B and Busslinger M. The promoter of the CD19 gene is a target for the B cell-specific transcription factor BSAP. *Mol. Cell. Biol.* 1992; **12**: 2662–2672.

159. Rothman P, Li SC, Gorham B, Glimcher L, Alt FW and Boothby M. Identification of a conserved lipopolysaccharide-plus-interleukin-4-responsive element located at the promoter of the germ-line transcript. *Mol. Cell. Biol.* 1991; **11**: 5551–5561.

160. Okabe T, Watanabe T and Kudo A. A pre-B and B cell specific DNA-binding protein EBB-1, which binds to the promoter of the Vpre-B1 gene. *Eur. J. Immunol.* 1992; **22**: 37–43.

161. Wakatsuki Y, Neurath MF, Max EE and Strober W. The B cell specific transcription factor BSAP regulates B cell proliferation. *J. Exp. Med.* 1994; **179**: 1099–1108.

162. Urbanek P, Wang Z-Q, Fetka I, Wagner EF and Busslinger M. Complete block of early B cell differentiation and altered patterning of the posterior midbrain in mice lacking *Pax*-5 / BSAP. *Cell* 1994; **79**: 901–912.

163. Scott EW, Simon EC, Anastasi J and Singh H. Requirement of transcription factor PU. 1 in the development of multiple haematopoietic lineages. *Science* 1994; **265**: 1573–1577.

164. Sha WC, Liou HC, Tuomanen EI and Baltimore D. Targetted disruption of the p50 subunit of NF-κB leads to multi-focal defects in immune responses. *Cell* 1995; **80**: 321–330.

Part IV

Disease and Disruption of Lymphocyte Signalling—Immunopharmacology and Potential Sites of Therapeutic Intervention

18

X-linked Agammaglobulinaemia, an Inherited Deficiency of Antibody Production

Ruth C. Lovering, Steve Hinshelwood and Christine Kinnon

Molecular Immunology Unit, Institute of Child Health, 30 Guilford Street, London, UK

INTRODUCTION

Inherited immunodeficiencies arise in the population as the result of naturally occurring mutations in genetic components of the immune system. With developments in gene mapping and gene cloning techniques many of these disorders can now be investigated at the molecular level. Those disease genes which lie on the X chromosome have been particularly amenable to this kind of investigation. As a result, several of the genes responsible for these X-linked disorders have recently been identified (reviewed in reference [1]).

The gene for one of these disorders, X-linked agammaglobulinaemia (XLA), was identified as Bruton's tyrosine kinase (*Btk*). The protein product of the *Btk* gene is required for normal B cell differentiation and a lack of this protein results in a severe B cell deficiency and consequently antibody deficiency. The Btk protein is a non-receptor tyrosine kinase and, although its precise function is unknown, it is thought to be involved in B cell signalling pathways. In this chapter we discuss the nature of the disease, the *Btk* gene and its protein product and what is known about its function in both the normal and diseased immune system.

THE DISEASE

XLA, originally described by Bruton [2], was the first disorder of humoral immunity to be recognized. Affected boys have drastically reduced numbers of circulating B lymphocytes and, consequently, very little or no immunoglobulin of any isotype is produced. Although it was initially suggested that the truncated immunoglobulin proteins, resulting from partial gene rearrangements (DJ) and observed in B cell lines derived from XLA patients, were one of the characteristic abnormalities associated with the disease [3], these are now accepted as being a characteristic of normal pre-B cells and do not arise solely as a result of XLA. The majority of immunoglobulin that is present in these patients appears to be expressed from complete functional gene rearrangements (VDJ) [4,5]. There are normal or elevated numbers of pre-B cells in the bone marrow of XLA patients, indicating that the XLA defect does not prevent entry of bone marrow progenitor cells into the B cell lineage but does appear to cause a block in B cell differentiation [6–8]. The defect in this disease was shown to be intrinsic to B cells by X-inactivation studies. Such studies identified an apparently non-random pattern of X-inactivation in the mature B cells of carrier women and a random pattern in cells of other haematopoietic lineages [9, 10]. Such results indicate that the B cell population in these carrier women is derived solely from progenitor cells with the active X chromosome expressing the normal *Btk* gene.

As a consequence of a lack of B cells and immunoglobulin these boys suffer a high frequency of severe bacterial infections, particularly in the upper and lower respiratory tract. Once the immunodeficiency is recognized as XLA, often when the patient is still quite young, the usual treatment is a lifetime of regular intravenous immunoglobulin infusions. This significantly reduces the episodes of infection and, with good management, the life expectancy of most patients is extended to a near normal length. Unfortunately, intravenous immunoglobulin does not protect these patients against rare neurotropic viruses and such infections can be fatal. Furthermore, despite treatment, some of the patients succumb to the long-term consequences of bacterial infection.

The Genetics of XLA

XLA is a single gene disorder which was originally mapped to the Xq21.3–Xq22 region on the long arm of the X chromosome [11]. Subsequent genetic linkage studies refined its localization within Xq22 [12]. Several families have been reported in which XLA apparently cosegregates with an associated growth hormone deficiency [13, 14], however, in many cases this may be accounted for by growth hormone insufficiency due to delayed growth and puberty [15]. There is still controversy as to whether or not there are autosomal recessive forms of this disease, since females with an apparently identical clinical phenotype have been described [16]. The current estimate of the incidence of XLA is about 1 in 150 000 live births.

Isolation of the Causative Gene

The *Btk* gene was identified and isolated in 1993 by two groups working independently using two quite different approaches [17, 18]. Genetic linkage studies had mapped the XLA locus to a 2 to 4 centi morgan region of Xq22 and no recombinations had been observed between the polymorphic marker DXS178 and the XLA locus in over 70 affected chromosomes [12, 18]. Genomic DNA from this region, in the form of a yeast artificial chromosome (YAC) encompassing the DXS178 locus and surrounding DNA, was screened for B-cell-specific cDNAs. One of the genes isolated in this way was found to be expressed in B cells and mutated in eight XLA patients, suggesting that defects in this gene are causative of XLA [18]. Sequence analysis indicated that the product of this gene is a protein tyrosine kinase (PTK) and it was therefore named *Bruton's tyrosine kinase* (*Btk*, also known as *Atk* or *Bpk*). At the same time the mouse homologue of *Btk* was identified by screening a mouse lymphoid progenitor cDNA

library with a DNA probe specific for the kinase domain of the human *Ltk* gene, with the intention of isolating genes involved in B-lymphocyte development [17]. The restricted expression of one of the genes identified, its position on the human X chromosome, and its lack of expression in B cell lines derived from patients with XLA, suggested it was involved in the XLA phenotype [17].

The *xid* mouse

In the mouse the most well studied primary immunodeficiency disorder is the *xid* mutation of CBA/N mice, characterized by a lack of response to certain polysaccharide antigens [19]. The *xid* locus lies in a region of the mouse X chromosome which is syntenic with that which contains the XLA locus in humans. It was, therefore, rapidly shown that not only did the murine *Btk* gene locus cosegregate with the *xid* locus, but that a defect in this gene, resulting in an amino acid substitution R28C, is responsible for the *xid* phenotype [20, 21]. There are, however, a number of distinct differences between the disease phenotypes of *xid* and XLA which bring into question the value of *xid* mice as an animal model of XLA (see later).

BRUTON'S TYROSINE KINASE

The *Btk* Gene

The genomic structures of the human and mouse *Btk* genes have recently been established by several groups [22–24]. The gene is encoded in 19 exons and in the human is contained within 37.5 kb of genomic DNA, whereas the mouse gene has been found to cover approximately 43.5 kb [24]. The *Btk* mRNA is only 2.5 kb, with a 1.9 kb open reading frame beginning at an initiation codon at nucleotide 133 [18]. In the human the exons vary in size from 55 bp to 503 bp and the intronic regions vary from 179 bp to approximately 9 kb.

The Btk Protein

The *Btk* mRNA is translated into a 659 amino acid protein with a molecular weight of approximately 77 kDa. Sequence analysis has shown that Btk shares considerable homology with members of the Src-related non-receptor tyrosine kinase family of proteins. This family includes many of the PTKs thought to be involved in signal transduction in haematopoietic cells including Fyn, Lck and Lyn (reviewed in references [25, 26]). As a result of sequence analysis, these PTKs can be viewed as consisting of a number of modular globular domains, each associated with a particular function and which can be arranged in different orders in the various proteins. Indeed, X-ray crystallographic studies of these isolated domains have demonstrated that they are capable of assuming a globular conformation, thought to be similar to the native structure.

In analysing the Btk protein sequence and comparing it to previously characterized domains, it is apparent that the Btk protein can be subdivided into five modular domains (Figure 18.1). The most well characterized domains are the Src homology (SH) domains SH1, SH2 and SH3. The tyrosine kinase activity of the protein is provided by the SH1 domain, the SH2 and SH3 domains have been shown to mediate protein–protein interactions through their ability to recognize and bind specific peptide motifs. The N-terminal of Btk was originally described as having a unique sequence, however recently it has been recognized as being composed of two distinct regions. The extreme N-terminal of Btk is homologous to the recently characterized Pleckstrin homology (PH) domain. The function of this domain is still unclear, although there is some evidence to suggest that it is involved in membrane localization [27]. There is some conservation of the sequence in the region separating the PH and SH3 domains between Btk and other closely related proteins, and this region has been described as the Tec homology (TH) domain.

The Btk/Tec/Itk Subfamily of Tyrosine Kinases

Btk shares a high degree of homology with several other non-receptor PTKs, including Bmx, Dsrc28c, Itk, Tec and Txk [28–32]. These have been suggested to form a Src-related subfamily based on the high level of similarity between their protein sequences and also because of specific features that they have in common (Table 18.1). These proteins all lack the N-terminal myristoylation site, common to other Src-related kinases and used for membrane localization. In addition they lack the C-terminal tyrosine (Y527 in Src) which has been shown to regulate the kinase activity of several Src related PTKs, through its interaction with the SH2 domain (see later). The lack of this residue in members of this family suggests that their activity may be regulated in a novel way.

Distinct features of the N-terminus of Btk are also shared with other members of this subfamily (Table 18.1). The PH domain is present in most members as is the TH domain, which has been subdivided into the Btk motif and the proline rich motif, on the basis of sequence analysis [33]. All the members of this subfamily appear to be expressed predominantly in haematopoietic cells.

Expression of Btk

Btk expression has been shown to be limited to hematopoietic cell lineages, in particular to B cells, but it is also expressed in myeloid cell lineages; it is not present in T cell lineages [17, 18, 34–36]. As expected the product of the Btk gene is present in progenitor B cell lines, where its expression appears to be essential for the cells to differentiate. The early block in B cell differentiation in patients with XLA, which results in low or absent numbers of recirculating B cells, does not preclude function(s) for the Btk protein in more mature B cells or myeloid cells. Indeed, the Btk protein is expressed in all B cell lineage lines investigated to date, with the exception of plasma cell lines [35,36]. This suggests that the Btk protein may have a functional role in mature B cells, in addition to that in early B cell maturation, but that Btk is not required for immunoglobulin production in terminally differentiated cells.

Although expression of Btk in B cells was expected, its expression in myeloid cells was surprising. No defects in the myeloid lineages of affected boys have been observed and, furthermore, there is no evidence from X-inactivation studies that Btk expression is essential in these cells. It can only be assumed that the function of

Table 18.1 A comparison of the Btk/Itk/Tec family members

Protein	Distribution of expression	N-terminal site myristoylation site	PH domain	Btk motif	Proline-rich motif	C-terminal tyrosine	Homology to Btk
Btk	B cell and myeloid cell lineages	-	+	+	+	-	100%
Bmx	Bone marrow, endothelial and several human tissues	-	+	+	-	-	61%
DSrc28c	-	-	+	-	+	-	58%
Itk/Tsk	IL-2-stimulated T cells	-	+	+	+	-	51%
Tec	All hematopoietic lineages and liver tissue	-	+	+	+	-	57%
Txk	T cell and myeloid cell lineages	-	-	-	+	-	57% (SH1-3)
Src	ubiquitous	+	-	-	-	+	39%

Btk in these cells is redundant or is not essential for their development.

MUTATIONS IN THE *BTK* GENE

The XLA phenotype was the first reported disease associated with the loss of function of a tyrosine kinase. This contrasts with the oncogenic potential of other PTKs in which specific mutations result in constitutive activation of the protein (e.g. v-Src). Mutations within the *Btk* gene which result in a change in the activity of the protein, and subsequently cause human disease, can provide an insight into the structure–function relationship of the different domains of Btk and these may be

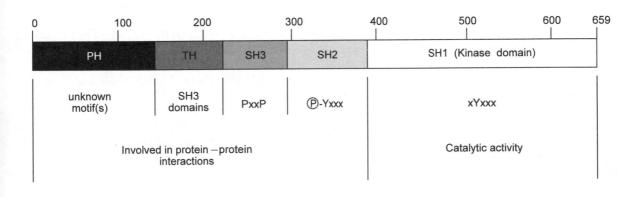

Figure 18.1 Schematic representation of the Btk protein, highlighting the functional domains. An approximate scale is given in amino acid residues

extrapolated to the homologous domains in this important family of proteins.

In the clinical context, confirmation of mutations in the *Btk* gene in patients with atypical forms of XLA has enabled the disease to be diagnosed with confidence. This has implications for the management of the disease in the patient and also for genetic counselling of the whole family. Mutation analysis for all forms of XLA enables carrier status determination and prenatal diagnosis to be made with absolute certainty based on the identification of the mutation in relevant family members.

Mutation Analysis for *Btk*

Mutation analysis of patients with XLA has confirmed that amino acid substitutions can severely disrupt the function of this protein. As many of the substitutions are at residue positions which are highly conserved between proteins, the functional relevance of these substitutions can be extrapolated to other homologous domains which are present in a wide range of different proteins. Amino acid substitutions have been observed in all of the different domains of Btk, with the exception of the TH and SH3 domains. Since such mutations all give rise to the same XLA phenotype it is likely that each of these domains is essential for the correct function of the Btk protein. The presence of amino acid substitutions in two of the domains thought to mediate protein-protein interactions, PH and SH2, suggests that multiple protein–protein interactions may be involved in its functioning. The fact that no amino acid substitutions have been identified in the TH and SH3 domains in any study so far, may be ascribed to chance or it may be because many of the amino acid substitutions in these domains do not severely disrupt the essential function(s) of the protein.

Over a hundred mutations associated with XLA have been identified in the *Btk* gene to date [18, 22, 23, 37–46]. Mutations are present throughout the gene and include deletions, insertions, nonsense, missense and splice site mutations. These mutations have been detected through the use of a number of different techniques including sequencing, SSCP analysis, chemical mismatch and Southern blot analyses of the *Btk* cDNA or genomic DNA. Of the 119 mutations characterized, 52 are missense mutations (which result in an amino acid substitution). However, as the same mutations have been identified by different groups there are, to date, only 32 different missense mutations (Table 18.2). In general, these mutations appear to be randomly distributed throughout the gene, however, some mutational 'hotspots' may exist [22, 38, 45]. One mutation in particular, a G to A transition at nucleotide 1691, resulting in R520Q, has been identified in eight different apparently unrelated patients and may reflect a mutation "hotspot" for this gene [22, 38, 44, 45]. This mutation is at a CpG dinucleotide and such sites are often hypermutable [47]. There is a run of seven adenosines at nucleotide positions 341 to 347 and so far four unrelated patients have been found to have an extra A in this run, resulting in a frameshift and premature termination at nucleotide positions 383 to 385 [45]. This mutation most likely arises relatively frequently as a result of replication slippage [48].

Analysis of the Btk protein expressed in B cell lines derived from XLA patients has confirmed that some mutations result in unstable or inactive protein being produced [17, 42, 43]. The expression levels of the Btk protein in the progenitor B cells of these XLA patients is likely to be similar to that in their EBV transformed B cell lines and would be consistent with the hypothesis that a lack of functional Btk protein prevents normal B cell development. However, in the majority of cases, it can only be inferred what effect these mutations will have on protein expression.

Phenotypic Variation

Characterization of mutations within the *Btk* gene has enabled some patients with less severe symptoms to be diagnosed as having atypical forms of XLA. A patient previously diagnosed as having common variable immunodeficiency was found to have a genomic deletion of the *Btk* gene [49], illustrating that patients diagnosed as having immunodeficiencies other than XLA may have mutations in Btk. In order to evaluate the relationship between mutations in the *Btk* gene and atypical XLA phenotypes it is necessary for a standard classification of the XLA phenotype to be

Table 18.2 Substitution, insertion and deletion mutations in the Btk protein which cause XLA

Amino acid change	Region	XLA phenotype	Source
M1V	Initiation	Classical	37, 38
M1T	Initiation	Classical	37
F25S	PH	Classical	46
R28H	PH	Moderate	44
		Leaky	23
T33P	PH	Classical	44,45
Q103 7 aa insertion	PH	Classical	37
V113D	PH	Classical	38
V131 43 aa deletion	PH	Classical	45
Q260 21 aa deletion	SH3	Not known	45
		Slightly leaky	44
R288W	SH2	Delayed onset	46
		Leaky	34
G302 1 aa deletion	SH2	Classical	45
R307G	SH2	Classical	37
Y334S	SH2	Classical	22
Y361C	SH2	Leaky/mild	38,42
I370M	SH2	Classical	46
L408P	Kinase	Moderate	44
K430E	Kinase	Classical	18
C506R	Kinase	Classical	22
M509V	Kinase	Classical	46
R520Q	Kinase	Classical	23,38,43,44,45
R525Q	Kinase	Classical	44,18
		Leaky	23
R525P	Kinase	Classical	46
N526K	Kinase	Classical	46
L542P	Kinase	Classical with GHD	38
R562W	Kinase	Classical	22,38
		Delayed onset	46
R562P	Kinase	Classical	43
W581R	Kinase	Classical	38
A582V	Kinase	Delayed onset	46
		Classical	43
G584 61aa deletion	Kinase	Classical with GHD	40
E589G	Kinase	Moderate	44
		Classical	43
G594R	Kinase	Classical	43
G594E	Kinase	Leaky with GHD	43
A607D	Kinase	Leaky	37
G613D	Kinase	Leaky	43,44
M630K	Kinase	Classical	22,38
R641C	Kinase	Classical	45
L652P	Kinase	Classical	38

employed. Fortunately, many authors have described the B cell numbers and immunoglobulin levels of their patients at presentation which does enable some comparisons between patients to be made. Twenty mutations have been described as being associated with a "non-classical" XLA phenotype (variously described as leaky, slightly leaky, mild, moderate and delayed onset) [23, 37–39, 42–44, 46]. Of these twenty patients, fourteen have missense mutations, two have nonsense mutations, three have splice site mutations and one has an insertion. The proportion of missense mutations (70%) in these non-classical XLA patients is significantly higher $p(0.01)$ than the proportion of missense mutations in patients with classical XLA (38%). This is consistent with the idea that some of the amino acid substitutions may only partially inactivate the Btk protein and, in these cases, B cell differentiation is only partially blocked. Mild forms of XLA are also associated with mutations which result in the truncation of the Btk protein [38, 44, 50], as these mutations are likely to totally inactivate the protein it appears that occasionally some B cell differentiation can occur in the absence of a functional Btk protein. However, in the majority of cases the lack of a functional Btk protein blocks B cell differentiation and, in fact, there are only a handful of cases where B cell differentiation occurs in the absence of a functional protein.

Within these atypical XLA families the severity of the XLA phenotype can be quite variable and this has led to the suggestion that modifying or secondary factors may influence the severity of the XLA phenotype [23, 37–39, 42, 44]. Such factors could include past infections and treatments received as well as the T cell population and the genetic background of the patient.

Genetic Counselling

Identification of *Btk* as the causative gene in XLA has made it possible to give a more accurate diagnosis of XLA, based on direct mutation detection [18, 37, 41]. In families where detectable mutations have been identified in affected boys, carrier status diagnosis for their female relatives can now be made with absolute certainty and, in many cases, prenatal diagnosis can be offered. Previously it was only possible to provide an estimate for the likelihood of inheritance of XLA based on indirect assays such as genetic linkage, using closely linked informative polymorphic markers, or X-inactivation studies of B cells. Both of these analyses can give ambiguous results and the risk could only be estimated with varying degrees of confidence.

In cases where it has been possible to confirm the genetic defect in patients with less severe forms of XLA there will be implications for the future management of the disease. There is also a possibility that very mild B cell abnormalities, such as an inability to raise antisera to particular types of antigen, may be associated with mutations within Btk, analogous to the phenotype of the *xid* mouse.

STRUCTURE AND FUNCTION OF THE BTK DOMAINS

The identification of Btk as a member of a family of PTKs already well established as being involved in haematopoietic signal transduction has given some very useful indicators as to possible functions for Btk. In many cases the various structures and functions of these related PTKs have been well worked out. Because of their modular nature it is easier to assess, and extrapolate, potential functions for Btk on a domain by domain basis.

The Pleckstrin-homology Domain

The PH domain is a recently defined protein motif which was originally recognized as a repeated sequence in pleckstrin and has since been found in a wide range of proteins, including Btk, p120GAP, PLCγ and βARK-1 [27, 51–53]. PH domains are modules of approximately 100 amino acids in length (137 in Btk) and have been implicated in binding to a wide variety of substrates including protein kinase C (PKC) [54,55] $\beta\gamma$ subunits of the heterotrimeric G proteins [56], phosphatidylinositol-4,5-biphosphate (PtdIns(4,5)P$_2$) [57], as well as to phosphorylated serine or threonine residues [27]. There are some suggestions that localization to the membrane may be mediated by the PH domain [27]. However, there appear to be too many different types of associations mediated by PH domains to enable a generalization to be made about its function.

At least two substrates for the Btk PH domain have been identified, Yao *et al.* [55] have shown that the isolated PH domain of Btk, as a glutathione-*S*-transferase (GST) fusion protein, will bind to several isoforms of PKC from mouse bone marrow derived mast cells, and that PKCβ1 is co-immunoprecipitated with Btk from these cells. They also provide some evidence that Btk kinase activity may be decreased by PKC serine phosphorylation, which would suggest that the signals transmitted by these two different classes of protein kinases can converge. The use of GST fusion proteins has also enabled the demonstration of an association between recombinant $\beta_1\gamma_2$ subunits of heterotrimeric G proteins and the *N*-terminal portion of the PH domain of Btk [58]. This association was confirmed by a cotransfection assay. However, in the cases of Btk and other proteins (e.g. βARK1 and PLCγ) [56], the $\beta\gamma$-binding appears to be mediated by only the *C*-terminal portion of the PH domain and seems to require sequences from the adjacent region [56,58]. An interaction between Btk and heterotrimeric G proteins would enable receptors that activate G proteins to regulate Btk function.

A variety of mutations within the PH domain of Btk have been detected in patients with XLA, several of which are amino acid substitutions and there is also a 7 amino acid insertion. These mutations suggest that the PH domain is important in Btk function and have implications for the structure–function relationship of this domain in other proteins. It has been suggested that the causative mutation in the *xid* mouse, R28C, has functional implications for the PH domain of Btk [20,21]. Using nuclear magnetic resonance techniques the structure of the PH domain has been determined and from this the likely ligand binding surface has been predicted [59,60]. In Btk R28 is present on an exposed surface within the predicted functional surface where it is likely to interact with a specific ligand [59,60]. As the R28C mutation affects B cell differentiation but has no effect on the stability or activity of the protein, it is likely that this mutation disrupts the ligand binding of the PH domain [59,60]. This mutation therefore provides additional evidence for the position of the functional ligand binding surface of the PH domain.

The mutations identified in the PH domain of Btk which are associated with XLA, and which result in amino acid substitutions, are all within the β strands of the domain (Table 18.2). In particular, βb (nomenclature as described in reference [27]) has three different substitutions, F25S, R28H and T33P (Table 18.2), including one at the same position as the *xid* mouse mutation. The β strands are thought to provide a potential substrate binding cleft and therefore amino acid substitutions within this cleft may be expected to influence its binding capacity. However, in a cell line established from a classical phenotype XLA patient with a T33P mutation the protein itself is undetectable, suggesting that it is the lack of Btk protein which is responsible for the XLA phenotype [35]. This cell line appears to have a phenotype characteristic of a very immature B cell, with no immunoglobulin genes rearranged. The insertion of 7 amino acids into β strand f has been predicted to result in a catastrophic structural perturbation [27] which would be consistent with the classical XLA phenotype observed. However, it is only in the case of the R28C substitution of the *xid* mouse that the stability and kinase activity of the mutated Btk protein is known to be normal [20,21] and therefore the effect of these XLA mutations on the Btk protein can only be guessed at.

In one case an activating mutation has been described in the PH domain of Btk [61]. The E41K amino acid substitution was generated using a random mutagenesis approach and results in transforming activity in NIH 3T3 cells plated in soft agar. The results suggest that transformation activation and regulation of Btk are critically dependent on the PH domain and clearly implicate Btk in signalling in a cell proliferation pathway.

The Tec Homology Region

The region separating the PH and the SH3 domains of Btk, Itk and Tec has recently been described as the TH domain [33]. The major feature of this region (amino acids 138 to 214 in Btk) is the presence of either one or two KKPLPPT/EP motifs. This short proline-rich sequence is typical of the core recognition sequence of SH3 domains, which suggests that the role of this region in Btk may be to facilitate the binding of other SH3 containing proteins. Indeed, using a yeast hybrid trapping method and a fusion protein system the SH3 domains of Fyn, Lyn and Hck have been shown to bind to the proline-rich motifs of the Btk TH domain [62]. As Fyn, Lyn and Hck are Src-related PTKs known to be involved in signal transduction pathways in B cells, their association with Btk provides some evidence to suggest that Btk is also involved in these pathways.

A second motif has been described within the TH region. The sequence kyHPxfwxdG-xyxCCxqxxkxapGC has been called the Btk motif [33] and is present in four of the Btk/Itk/Tec family members (Table 18.1) as well as two non-tyrosine kinase proteins. As described above, there is a suggestion that part of the TH region is necessary for the binding of $\beta\gamma$ subunits of heterotrimeric G proteins by the PH domain [58], the region required includes the Btk motif.

No amino acid substitutions associated with XLA have been identified in the TH region, although a deletion has been found in one patient within the PH domain which does extend into this region (see Table 18.2).

Src-homology 3 Domain

SH3 domains are domains of 60 amino acids present in a variety of proteins, many of which are known to be involved in transduction of extracellular signals [63]. There is some evidence that the cellular localization of some proteins may be directed by their SH3 domain [64]. SH3 domains mediate protein–protein interactions via recognition of specific proline-rich peptide sequences [65,66] and to date, although the sequence PXXP is always present, most SH3 domain ligands contain multiple prolines. Structural analysis suggests that SH3 domains bind up to ten amino acid peptides adopting a poly-proline II helix conformation. The presence of multiple prolines stabilizes this conformation [65–67].

The Btk SH3 domain, expressed as a GST fusion protein in bacterial cells, has been shown to bind *in vitro* to several polypeptides in B cell lysates [68]. In particular, we have identified an *in vitro* association between the Btk SH3 domain and the protooncogene c-Cbl. c-Cbl is a 120 kDa proline-rich protein that was originally characterized as the cellular homologue of a murine viral oncogene that causes pre-B-cell lymphomas [69]. This protein is expressed in haematopoietic cells and the SH3 domains from a large number of proteins, including those of Fyn, Grb2, Lck and Nck, have already been shown to bind c-Cbl [69–71]. c-Cbl is expressed in B cell and T cell lineages and is rapidly phosphorylated following stimulation of the B cell receptor (BCR) and T cell receptor (TCR), respectively—it is therefore likely to be involved in both of these signalling pathways (68,70). Further investigation will establish the exact nature of the interaction between the Btk SH3 domain and c-Cbl and identify the nature of other proteins which are also apparently associated with the Btk SH3 domain.

No amino acid substitutions within the Btk SH3 domain have been detected, so far, and this may reflect a redundancy of function in this domain or may relate to its small size. A 21 amino acid deletion which removes part of the SH3 domain and most of the sequences between the SH3 and SH2 domains, has been identified independently in two unrelated patients [45,72]. This deletion does not appear to affect the stability or kinase activity of the mutated Btk protein, when compared to the normal protein [72]. From computer modelling studies it was suggested that this mutated protein will be lacking the majority of the SH3 ligand binding site which may account for the severe XLA phenotype of the patients in these families [72]. However, there is also a possibility that the deletion of amino acids separating the SH3 and SH2 domains affects the secondary structure of the SH2 domain and that it is the disruption of this domain which is responsible for the XLA phenotype of these patients.

Src-homology 2 Domain

The structure and function of the SH2 domain has been extensively studied and well characterized. SH2 domains, modular structures of approximately 100 amino acids with no catalytic function, have been shown to be involved in both inter-and intramolecular interactions (reviewed in reference [73]). These interactions are mediated through specific tyrosine residues which have been phosphorylated by activated PTKs [74]. This domain recognizes these conserved phosphorylated tyrosine residues within a specific peptide motif. These motifs, known as a TAMs (tyrosine-containing activation motifs) or ARAMs (antigen receptor activation motifs), are characteristically based on the sequence YXXI/L. They are present in a wide range of molecules, including the Igα and Ig chains of the BCR. They are implicated in recruitment of specific SH2-containing proteins, such as PTKs, to receptor complexes at the cell surface. In many cases this interaction activates the kinase activity of the SH2-containing PTK.

A model for the structure of the SH2 domain of Btk can be predicted on the basis of its close homology to Src [75]. This was used to predict a recognition sequence, YEXI/L, in the target protein, however attempts to identify such target protein(s) have so far been unsuccessful ([76]; see also our unpublished observations). Possibly the system used in both cases, expression of the Btk SH2 domain as a bacterial GST fusion protein, is not suitable for assessing the interactions of this particular domain.

Two of the amino acid substitutions we have identified in Btk as causing XLA fall within the SH2 domain of the protein and presumably in these patients the activity of the Btk protein has been reduced as a consequence of the disruption in the protein–protein interactions between Btk and its targets. X-ray crystallography and modelling of the Src SH2 domain suggests that this domain forms a two holed socket which can bind specific peptide sequences in a two "pronged" conformation, with the phosphorylated tyrosine residue providing one of the "prongs" [77]. The substitution of the arginine at residue 288 (Figure 18.2, aA2, notation as in reference [78]) for a tryptophan results in the affected boy having a mild form of XLA [37,39]. The substituted arginine, R288, has been suggested to lie on the rim of the first pocket

and is suggested to reduce, but not abolish, the activity of the Btk protein (Figure 18.2) [37,39]. In contrast, the disease in a patient with classical XLA was shown to be due to a substitution of a glycine for arginine βB5 (Figure 18.2, notation as above) at residue 307 [37]. Arginine is positively charged and at position βB5 it is involved directly in binding the phosphotyrosine peptide in other SH2 domains, this residue, R307, is the only invariant residue present in all SH2 domains known. The change to a neutral glycine residue appears to disrupt the binding potential of this region and completely abolish the functioning of Btk, in this patient. This is consistent with the observation that substitution of the equivalent residue in the Abl SH2 domain, R171, disrupts phosphotyrosine binding and reduces the transforming activity of this protein [79]. Four other substitutions within the SH2 domain have been identified in patients with XLA [22,42,45,46]. One of these involves a tyrosine residue, Y334, likely to be involved in binding the peptide in the hydrophobic pocket, which is substituted with a serine in one classical XLA patient [22]. Another of the substitutions, Y361C, is adjacent to a site that probably participates in the hydrophobic peptide binding. However, although the protein does have kinase activity, it is unstable and it may be the low amount of Btk protein which accounts for the associated atypical XLA phenotype [42].

Kinase Domain

Many signalling pathways are mediated through the activation of protein kinases (reviewed in references [80, 81]). Activated PTKs phosphorylate specific tyrosine residues in target molecules. The activation of a PTK initiates the formation of multiprotein complexes through the phosphorylation

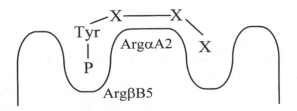

Figure 18.2 SH2 domains bind to phosphorylated tyrosine motifs

and subsequent binding of these tyrosines to SH2 domains. Catalytic domains of protein kinases are responsible for the ability of the kinase to phosphorylate its substrate and usually comprise about 250 to 300 amino acids. The region is known as either the kinase domain or SH1 (with reference to PTKs) and its functional and structural boundaries have been defined by the analysis of conserved sequences and by measuring the activity of truncated proteins [82]. Two classes of protein kinases have been defined with respect to their substrate specificity, serine/threonine-specific (PSKs) and tyrosine-specific (PTKs). Although there is considerable homology between catalytic domains, specific motifs have been identified which are characteristic of each class. These motifs can be used to predict what the substrate specificity of the kinase will be. In the case of Btk this was predicted to be tyrosine and this has now been confirmed experimentally [39].

Due to the sequence homology between different protein kinases it has been possible to extrapolate functional and structural data from one protein kinase to another. The first crystal structure determined for a protein kinase was that of a serine/threonine kinase, cAMP-dependent protein kinase (cAPK), and this has been used to construct a three-dimensional model for the Btk kinase domain [43]. Since then the crystal structure of the catalytic domain of a PTK, human insulin receptor (IRK), has been determined [83]. These structural analyses have identified which residues are important for the correct folding of the kinase domain, for the catalytic activity of the protein and for its substrate binding. Furthermore, a comparison between the structures of cAPK and IRK has shown how the structure of the catalytic site determines the substrate preference of the protein. From these models it is possible to predict how mutations in the Btk kinase domain which lead to the XLA disease affect the structure and function of this domain [43].

Twenty-one amino acid substitutions have been identified in the kinase domain of Btk. It is assumed that these mutations decrease the kinase activity of Btk, although this has only been demonstrated in one instance, R525Q [43]. Nine of these mutations are at highly conserved residue positions and it is possible to predict their functional consequences. Vihinen *et al.* [43] have

mapped eight substitutions to one face of the Btk kinase domain, suggesting a clustering of functionally important residues. Four substitution sites identified in the Btk kinase domain, R520Q, L542P, R562P/W, W581R, are also mutated in the homologous region of the IRK, and in these cases result in non-insulin-dependent diabetes mellitus (NIDDM) [84]. From the crystal structure of the IRK kinase domain it has been predicted that three of these NIDDM mutations, R1131Q, R1174Q and W1193L, would destabilize the kinase domain [83].

From crystallographic studies of cAPK, a PSK, roles for many of the highly conserved residues in the protein kinase family have been elucidated [85,86]. Superimposition of these catalytic loops reveals that the catalytic loop of cAPK, characterized by the sequence YRDLKPEN, shows remarkable conformational similarities to the analogous sequence, HRDLAARN, in IRK and Btk, despite the sequence differences (Figure 18.3) [83]. In the phosphotransfer reaction D166 of cAPK (D521 in Btk) is the catalytic base, K168 provides charge neutralization, N171 (N526 in Btk) is hydrogen bonded to D166 and is involved in Mg^{2+} coordination (Figure 18.3). R1136 in IRK is the counterpart to K168 in cAPK and serves a dual purpose providing charge neutralization at the phosphotransfer site and making a hydrogen bond with the catalytic base. Two substitutions have been detected in Btk at R525 (R525Q and R525P), this is the counterpart to R1136 in IRK. As the majority of patients with substitutions at this site are described as having classical XLA (Table 18.2) it would be predicted that these mutations dramatically decrease the kinase activity of the protein. Indeed, analysis of the Btk protein isolated from a patient with the R525Q substitution showed that this protein has no kinase activity [43] suggesting that glutamine at this position is unable to functionally replace arginine.

There appear to be at least two ways in which the activity of a PTK can be regulated. As mentioned previously, the SH2 domain of Src-related PTKs in the inactive conformation are bound to a *C*-terminal phosphotyrosine (Y527 in Src) and it is thought that in this conformation the active site of the kinase domain is hidden. Recognition of a phosphotyrosine motif substrate by the SH2 domain results in the concomitant release of the

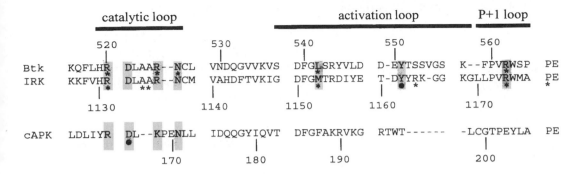

Figure 18.3 Alignment of the active sites of the Btk, IRK and cAPK kinase domains. The shaded residues are discussed in the text and residue numbers are indicated.*, Substitution site associated with XLA or NIDDM; Y, autophosphorylation site; D, catalytic base

C-terminal phosphotyrosine and the exposure of the kinase domain to possible tyrosine substrates. Btk does not have this *C*-terminal tyrosine residue and so its activity is unlikely to be regulated in the same way as Src.

A novel autoinhibition mechanism of PTK activity was revealed by the X-ray crystal structure of IRK [83] and it is possible that this regulatory method could be employed by Btk. Hubbard *et al.* [83] have shown that the hydroxyl group of Y1162 of IRK is bound in the active site. This tyrosine lies within the activation loop (Figure 18.3) and is known to be autophosphorylated in response to insulin, however, it appears likely that this auto-phosphorylation occurs via a mechanism operating *in trans*. As the ATP-binding site is blocked in the unphosphorylated IRK structure, Hubbard *et al.* [83] conclude that the binding of ATP and of "self" Y1162 in the active site are mutually exclusive, *cis*-autophosphorylation at this site cannot occur to any appreciable extent. *Trans*-autophosphorylation of this tyrosine stabilizes the non-inhibiting conformation of the activation loop through electrostatic interactions between the phosphotyrosine, Y1162, and positively charged residues, such as R1131. R1131 is thought to be particularly important in the stabilization of the active conformation of IRK [83] and an amino acid substitution of this residue (R1131Q) has been observed which is associated with NIDDM. As Btk has a tyrosine (Y551) at the equivalent residue position as the Y1162 autophosphorylation site of IRK, it seems likely that autophosphorylation may be important in the regulation of the kinase activ-

ity of Btk. Furthermore, several XLA patients have been identified with a substitution at R520 (Figure 18.3, Table 18.2), which is at the equivalent residue position R1131 in IRK. Presumably in patients with R520Q substitutions the active conformation of Btk cannot be stabilized.

From crystal structure analysis of kinase domains, comparisons between different kinase domains and mutation analysis a clearer picture is emerging about the key residues within this catalytic domain and the roles they play.

THE ROLE OF BTK AS A SIGNALLING MOLECULE

Btk appears to have a pivotal role in B cell differentiation and signalling. Signalling through this tyrosine kinase may provide a bottleneck in B cell maturation, without this signal the pre-B cells fail to differentiate. However, it is not known what external stimuli lead to Btk activation, nor what the substrates of Btk might be. Since expression of the BCR is specific to B cells, and critical to the development of mature B lymphocytes, it was considered possible that Btk might play a role in signal transduction through either the mature BCR or the pre-B-cell surrogate light chain complex. These molecules, however, are not the only candidates for initiating signalling pathways involving Btk in B cell development. Cells respond to a variety of stimuli during B cell development, including soluble cytokines such as IL-7, IL-4 and

IL-6 and other molecules present on the surface of bone marrow stromal cells.

Stimulation through the B Cell Antigen Receptor

Antibody stimulated crosslinking of the BCR leads to increased cellular protein tyrosine phosphorylation [87,88] and to increased activity of several non-receptor PTKs including Lyn, Blk and Syk [89–91]. In addition, several groups have shown increased *in vivo* phosphorylation of Btk in B cell lines and increased *in vitro* kinase activity in Btk immunoprecipitates following stimulation of B cell lines by BCR crosslinking. This has been seen in human B cell lymphoma lines, Ramos [92] and Daudi [93], and in the murine immature B line WEHI 231 [94,95]. Constitutive Btk tyrosine phosphorylation has also been reported in one murine pre-B-cell line [95]. Furthermore Btk has been shown to be capable of phosphorylating recombinant Ig-α in *in vitro* kinase assays [94]. However, while these reports demonstrate that Btk activity may be stimulated by crosslinking the BCR, they do not show that Btk is an essential, or even important, component of signal transduction through the BCR.

Crosslinking Fc∈RI

Stimulation of *in vitro* kinase activity of Btk immunocomplexes has been demonstrated in several other systems. The high affinity immunoglobulin E receptor (Fc∈RI) on mast cells is a multimeric complex associated with cytoplasmic non-receptor PTKs, analogous to the BCR in B cells. The receptor is a tetrameric complex of an IgE binding α chain, β chain and homodimer of two γ chains. Both the β and γ chains contain TAMs and have been shown to be associated with Lyn and Syk PTKs, respectively, and crosslinking of the receptor leads to protein tyrosine phosphorylation. Kawakami *et al.* [96] have shown that crosslinking of Fc∈RI also leads to activation of Btk in a rat mast cell line, although they do not see the concomitant increase in *in vivo* phosphorylation of Btk that is seen in B cell systems. A role

for Btk in signalling through the Fc∈RI receptor was unexpected since defects have not been observed in the mast cells of XLA patients, nor has any abnormality in this cell lineage been reported in *xid* mice. This may reflect other proteins expressed in mast cells which can compensate for the loss of Btk function.

Cytokine Stimulation of Btk

Recently two reports have suggested a role for Btk in signal transduction of cytokine stimulated pathways. Matsuda *et al.* [97] have shown that in a murine pro-B cell line stably transfected with gp130, Btk *in vitro* immune complex kinase activity is increased following crosslinking of the transfected gp130—which is the common receptor subunit of the interleukin (IL)-6, IL-11, oncostatin M, leukaemia inhibitory factor and ciliary neurotrophin factor receptors. Stimulation with IL-6 normally leads to increased tyrosine phosphorylation of several cellular proteins, although none of the components of the receptor contain intrinsic PTK activity. While it is likely that Janus kinases (JAKs) play a role in signal transduction through the IL-6 receptor, the further demonstration by Matsuda *et al.* (97) that Btk is associated with gp130 also implicates Btk in this signalling pathway.

Treatment of an IL-5-dependent murine pre-B-cell line, Y16, results in increased cytoplasmic protein tyrosine phosphorylation. IL-5 is a T cell derived cytokine and is an eosinophil differentiation factor in humans and mice, it is also a growth and differentiation factor for mouse though not human B cells. The IL-5 receptor consists of an IL-5 binding α chain and a 130 kDa β chain that is common to the IL-3 receptor and the GM-CSF receptor. Neither the α nor the β chain contains catalytic kinase domains. Sato *et al.* [98] have shown that in Y16 cells stimulation with IL-5 leads to an activation of JAK2 and Btk PTKs, and that phosphatidyl-inositol 3-kinase, shc, vav and HS1 become tyrosine phosphorylated. Whether or not Btk plays a direct or indirect role in phosphorylation of these substrates is unclear, although it is interesting to note that B cells from *xid* mice do not respond to IL-5 [99]. Furthermore, Li *et al.* [61] have shown that transfection of Y16 cells with an activating mutation of Btk relieved the IL-5 dependency of these cells.

Abnormal Signalling in *xid* Mouse B Cells

The CBA/N strain of mice [19] expresses a Btk protein with an amino acid substitution, R28C, in the PH domain which results in a selective immunodeficiency [20,21]. As discussed, this mutation does not appear to disrupt the stability or kinase activity of the Btk protein. These *xid* mice may be considered as an animal model of XLA. There are however several crucial differences between the phenotype of these *xid* mice and human XLA patients.

The immunodeficiency in *xid* mice is less severe than that seen in XLA, B cells are present in *xid* mice, and these mice are defective only in so far as they are unable to produce immunoglobulin in response to inoculation with type II T cell independent antigens. Although, unlike XLA in man, *xid* mice do have some B cell development they do lack a major compartment of phenotypically mature sIgm^{lo}/sIgD^{hi} B cells in their peripheral organs and these B cells do not respond normally in several *in vitro* assays.

B cells from CBA/N mice do not proliferate following ligation of CD40 [100], an otherwise mitogenic stimulus, and perhaps more significantly, *xid* B cells show abnormal response to crosslinking of surface IgM [101,102]. *xid* B cells are also deficient in response to the crosslinking of a recently characterized 105 kDa antigen that is capable of delivering a signal protecting B cells from radiation-induced apoptosis [103]. These cells also show a lack of response to stimulation with an anti-CD38 antibody that stimulates tyrosine phosphorylation and cell proliferation in normal B cells [104,105]. In addition, they respond abnormally to IL-5 [99] and IL-10 [106]. IL-10 stimulation of resting B cells leads to up-regulation of class II MHC antigens but this response is absent in *xid* mice [106].

It is possible that Btk plays a role in some aspects of signal transduction from all of these surface molecules. However, it is important when interpreting such data to be aware that abnormal signalling in B cells from *xid* mice may not necessarily reflect merely a molecular lesion in these cells, but may possibly reflect the development of different, or abnormal, B cell lineages as a result of the effect of the *xid* mutation on B cell ontogeny of the whole animal. It may be possible to address some of these questions by attempting to reconstitute expression of "wildtype" Btk in these mice.

Subcellular Localization of Btk

For Btk to play an effective role in signal transduction, it must interact with other molecules involved in this process at the cell surface, either directly or indirectly. Btk has been found to be predominantly located in the cell cytoplasm in resting cells [17]. This reflects the absence of a myristoylation site that is generally a characteristic of the Src-related non-receptor tyrosine kinase [17,18]. A small amount of Btk has been found to be translocated to the particulate membrane fractions following stimulation of surface antigen in B cells [92] and FcRI in mast cells [96]. The amount of Btk associated with the membrane fraction in transformed NIH 3T3 cells was found to increase when the E41K activating mutation is introduced [61]. Localization to a particulate fraction would reflect association with membrane or cytoskeletal proteins on activation. Such interactions may be short-lived, as evidenced by the difficulty in seeing appreciable amounts of Btk associated with these fractions.

FUTURE PROSPECTS

It has now been shown that Btk is indeed activated following stimulation of the BCR surface antigen receptor, and probably some cytokine receptors, on B cells. In other cell types Btk plays a role in additional signalling pathways. The next step in this process will be to fully determine the substrates for Btk and how they fit into the signal transduction pathways. Although substantial progress has been made in elucidating the role played by Btk in mature B lineage and other cell types, there is still a long way to go before the function of Btk in the development of B cells in the whole organism is fully understood. It may then be possible to suggest alternative strategies for treatment of XLA, which is still a potentially fatal disorder.

One possibility of developing a "cure" for this disorder is through the use of somatic gene therapy techniques, wherein a normal copy of gene would be introduced efficiently and stably into the pluripotential stem cells of the bone marrow. Once restored to the patient, these transduced bone marrow cells should be able to generate mature immunocompetent B cells for the remaining lifetime of the individual. Obviously there are numerous technical obstacles to be overcome before such a possibility is feasible. Furthermore, attractive as this possibility is, it will first be necessary to eliminate any likelihood of an oncogenic role for Btk. Such experiments are already underway in a number of laboratories.

ACKNOWLEDGEMENTS

We would like to thank all our colleagues at the Institute of Child Health for their help in preparing this chapter and particularly members of the Molecular Immunology Unit for providing us with their unpublished results. We are also grateful to the Wellcome Trust for its continued support.

REFERENCES

1. Kinnon C. Inherited immunodeficiencies. In: Dickson G (ed.) *Molecular and Cell Biology of Human Gene Therapeutics*. London, UK: Chapman & Hall, 1995 (in press).
2. Bruton OC. Agammaglobulinemia. *Pediatrics* 1952; **9**: 722–727.
3. Schwaber J, Molgaard H, Orkin SH, Gould HJ and Rosen FS. Early pre-B cells from normal and X-linked agammaglobulinaemia produce C mu without an attached VH region. *Nature* 1983; **304**: 355–358.
4. Mensink EJ, Schuurman RK, Schot JD, Thompson A and Alt FW. Immunoglobulin heavy chain rearrangements in X-linked agammaglobulinemia. *Eur. J. Immunol.* 1986; **16**: 963–967.
5. Milili M, Le Deist F, de Saint-Basile G, Fischer A, Fougereau M and Schiff C. Bone marrow cells in X-linked agammaglobulinemia express pre-B-specific genes (lambda-like and V pre-B) and present immunoglobulin V-D-J gene usage strongly biased to a fetal-like repertoire. *J. Clin. Invest.* 1993; **91**: 1616–1629.
6. Pearl ER, Vogler LB, Okos AJ, Crist WM, Lawton AR and Cooper MD. B lymphocyte precursors in human bone marrow: an analysis of normal individuals and patients with antibody-deficiency status. *J. Immunol.* 1978; **120**: 1169–1175.
7. Conley ME. B cells in patients with X-linked agammaglobulinemia. *J. Immunol.* 1985; **134**: 3070–3074.
8. Campana D, Farrant J, Inamdar N, Webster AD and Janossy G. Phenotypic features and proliferative activity of B cell progenitors in X-linked agammaglobulinemia. *J. Immunol.* 1990; **145**: 1675–1680.
9. Conley ME, Brown P, Pickard AR, Buckley RH, Miller DS, Raskind WH *et al.* Expression of the gene defect in X-linked agammaglobulinemia. *N. Engl. J. Med.* 1986; **315**: 564–567.
10. Fearon ER, Winkelstein JA, Civin CI, Pardoll DM and Vogelstein B. Carrier detection in X-linked agammaglobulinemia by analysis of X-chromosome inactivation. *N. Engl. J. Med.* 1987; **316**: 427–431.
11. Kwan S-P, Kunkel L, Bruns G, Wedgewood RJ, Latt S and Rosen FS. Mapping of the X-linked agammaglobulinemia locus by use of restriction fragment-length polymorphism. *J. Clin. Invest.* 1986; **77**: 649–652.
12. Lovering R, Middleton Price HR, O'Reilly MA, Genet SA, Parkar M, Sweatman AK *et al.* Genetic linkage analysis identifies new proximal and distal flanking markers for the X-linked agammaglobulinemia gene locus, refining its localization in Xq22. *Hum. Mol. Genet.* 1993; **2**: 139–141.
13. Fleisher TA, White RM, Broder S, Nissley SP, Blaese RM, Mulvihill JJ *et al.* X-linked hypogammaglobulinemia and isolated growth hormone deficiency. *N. Engl. J. Med.* 1980; **302**: 1429–1434.
14. Conley ME, Burks AW, Herrod HG and Puck JM. Molecular analysis of X-linked agammaglobulinemia with growth hormone deficiency. *J. Pediatr.* 1991; **119**: 392–397.
15. Buzi F, Notarangelo LD, Plebani A, Duse M, Parolini O, Monteleone M and Ugazio AG. X-linked agammaglobulinemia, growth hormone deficiency and delay of growth and puberty. *Acta Paediatr.* 1994; **83**: 99–102.
16. Conley ME and Sweinberg SK. Females with a disorder phenotypically identical to X-linked agammaglobulinemia. *J. Clin. Immunol.* 1992; **12**: 139–143.
17. Tsukada S, Saffran DC, Rawlings DJ, Parolini O, Allen RC, Klisak I *et al.* Deficient expression of a B cell cytoplasmic tyrosine kinase in human X-linked agammaglobulinemia. *Cell* 1993; **72**: 279–290.
18. Vetrie D, Vorechovsky I, Sideras P, Holland J, Davies A, Flinter F *et al.* The gene involved in X-linked agammaglobulinaemia is a member of the src family of protein-tyrosine kinases. *Nature* 1993; **361**: 226–233.
19. Scher I. CBA/N immune defective mice; evidence for the failure of a B cell subpopulation to be expressed. *Immunol. Rev.* 1982; **64**: 117–136.
20. Rawlings DJ, Saffran DC, Tsukada S, Largaespada DA, Grimaldi JC, Cohen L *et al.* Mutation of unique region of Bruton's tyrosine kinase in immunodeficient XID mice. *Science* 1993; **261**: 358–361.
21. Thomas JD, Sideras P, Smith CI, Vorechovsky I, Chapman V and Paul WE. Colocalization of X-linked agammaglobulinemia and X-linked immunodeficiency genes. *Science* 1993; **261**: 355–358.
22. Hagemann TL, Chen Y, Rosen FS and Kwan S-P. Genomic organization of the *Btk* gene and exon scanning for mutations in patients with X-linked agammaglobulinemia. *Hum. Mol. Genet.* 1994; **3**: 1743–1749.
23. Ohta Y, Haire RN, Litman RT, Fu SM, Nelson RP, Kratz J *et al.* Genomic organization and structure of Bruton

agammaglobulinemia tyrosine kinase: Localization of mutations associated with varied clinical presentations and course in X-chromosome linked agammaglobulinemia. *Proc. Natl. Acad. Sci. USA* 1994; **91**: 9062–9066.

24. Sideras P, Müller S, Shiels H, Jin H, Khan WN, Nilsson L *et al.* Genomic organisation of mouse and human Bruton's agammaglobulinemia tyrosine kinase (*Btk*) loci. *J. Immunol.* 1994; **153**: 5607–5617.

25. Cambier JC, Pleiman CM and Clark MR. Signal transduction by the B cell receptor and its coreceptors. *Ann. Rev. Immunol.* 1994; **12**: 457–486.

26. Harnett MM. Antigen receptor signalling: from the membrane to the nucleus. *Immunol. Today* 1994; **15**: P1–P3.

27. Gibson TJ, Hyvönen M, Musacchio A and Saraste M. PH domain: the first anniverary. *TIBS* 1994; **19**: 349–353.

28. Tamagnone L, Lahtinen I, Mustonen T, Virtaneva K, Francis F, Muscatelli F *et al.* BMX, a novel nonreceptor tyrosine kinase gene of the *BTK/ITK/TEC/TXK* family located in chromosome Xp22.2. *Oncogene* 1994; **9**: 3683–3688.

29. Gregory RJ, Kammermeyer KL, Vincent WS and Wadsworth SG. Primary sequence and developmental expression of a novel *Drosophila melanogaster src* gene. *Mol. Cell. Biol.* 1987; **7**: 2119–2127.

30. Siliciano JD, Morrow TA and Desiderio SV. *itk*, a T-cell-specific tyrosine kinase gene inducible by interleukin 2. *Proc. Natl. Acad. Sci. USA* 1992; **89**: 11194–11198.

31. Mano H, Mano K, Tang B, Koehler M, Yi T, Gilbert DJ *et al.* Expression of a novel form of Tec kinase in hematopoietic cells and mapping of the gene to chromosome 5 near Kit. *Oncogene* 1993; **8**: 417–424.

32. Haire RN, Ohta Y, Lewis JE, Fu SM, Kroisel P and Litman GW. *TXK*, a novel human tyrosine kinase expressed in T cells shares sequence identity with *Tec* family kinases and maps to 4p12. *Hum. Mol. Genet.* 1994; **3**: 897–901.

33. Vihinen M. Nilsson L and Smith CI. Tec homology (TH) adjacent to the PH domain. *FEBS Letts.* 1994; **350**: 263–265.

34. de Weers M, Verschuren MCM, Kraakman MEM, Mensink RGJ, Schuurman RKB, van Dongen JJM and Hendriks RW. The Bruton's tyrosine kinase gene is expressed throughout B cell differentiation, from early precursor B cell stages preceding immunoglobulin gene rearrangement up to mature B cell stages. *Eur. J. Immunol.* 1993; **23**: 3109–3114.

35. Genevier HC, Hinshelwood S, Gaspar HB, Rigley KP, Brown D, Saeland S *et al.* Expression of Bruton's tyrosine kinase protein within the B cell lineage. *Eur. J. Immunol.* 1994; **24**: 3100–3105.

36. Smith CIE, Baskin B, Humire-Greiff P, Zhou J-N, Olsson PG, Maniar HS *et al.* Expression of Bruton's agammaglobulinemia tyrosine kinase gene, BTK, is selectively down-regulated in T lymphocytes and plasma cells. *J. Immunol.* 1994; **152**: 557–565.

37. Bradley LAD, Sweatman AK, Lovering RC, Jones AM, Morgan G, Levinsky RJ and Kinnon C. Mutation detection in the X-linked agammaglobulinemia gene, BTK, using single strand conformation polymorphism analysis. *Hum. Mol. Genet.* 1994; **3**: 79–83.

38. Conley ME, Fitch-Hilgenberg ME, Cleveland JL, Parolini O and Rohrer J. Screening of genomic DNA to identify mutations in the gene for Bruton's tyrosine kinase. *Hum. Mol. Genet.* 1994; **3**: 1751–1756.

39. de Weers M, Mensink RGJ, Kraakman MEM, Schuurman RKB and Hendriks RW. Mutation analysis of the Bruton's tyrosine kinase gene in X-linked agammaglobulinemia: identification of a mutation which affects the same codon as is altered in immunodeficient *xid* mice. *Hum. Mol. Genet.* 1994; **3**: 161–166.

40. Duriez B, Duquesnoy P, Dastot F, Bougneres P, Amselem S and Goossens M. An exon-skipping mutation in the *btk* gene of a patient with X-linked agammaglobulinemia and isolated growth hormone deficiency. *FEBS Letts.* 1994; **346**: 165–170.

41. Lovering RC, Sweatman A, Genet SA, Middleton-Price HR, Vetrie D, Vorechovsky I *et al.* Identification of deletions in the *btk* gene allows unambiguous assessment of carrier status in families with X-linked agammaglobulinemia. *Hum. Genet.* 1994; **94**: 77–79.

42. Saffran DC, Parolini O, Fitch-Hilgenberg ME, Rawlings DJ, Afar DEH, Witte ON and Conley ME. Brief report: a point mutation in the SH2 domain of Bruton's tyrosine kinase in atypical X-linked agammaglobulinemia. *N. Engl. J. Med.* 1994; **330**: 1488–1491.

43. Vihinen M, Vetrie D, Maniar HS, Ochs HD, Zhu Q, Vorechovsky I *et al.* Structural basis for chromosome X-linked agammaglobulinemia: a tyrosine kinase disease. *Proc. Natl. Acad. Sci. USA* 1994; **91**: 12803–12807.

44. Zhu Q, Zhang M, Winkelstein J, Chen S-H and Ochs HD. Unique mutations of Bruton's tyrosine kinase in fourteen unrelated X-linked agammaglobulinemia families. *Hum. Mol. Genet.* 1994; **3**: 1899–1900.

45. Gaspar HB, Bradley LAD, Katz F, Lovering RC, Roifman CM, Morgan G, Levinsky RJ and Kinnon C. Mutation analysis in Bruton's tyrosine kinase, the X-linked agammaglobulinaemia gene, including identification of an insertional hotspot. *Hum. Mol. Genet.* 1995; **4**: 755–757.

46. Vorechovsky I, Vihinen M, de Saint Basile G, Honsová S, Hammarström L, Müller S *et al.* DNA-based mutation analysis of Bruton's tyrosine kinase gene in patients with X-linked agammaglobulinaemia. *Hum. Mol. Genet.* 1995; **4**: 51–58.

47. Cooper DN and Youssoufian H. The CpG dinucleotide and human genetic disease. *Hum. Genet.* 1988; **78**: 151–155.

48. Cooper DN and Krawczak M. Mechanisms of insertional mutagenesis in human genes causing genetic disease. *Hum. Genet.* 1991; **87**: 409–415.

49. Vorechovsky I, Zhou J-N, Vetrie D, Bentley D, Björkander J, Hammerström L and Smith CIE. Molecular diagnosis of X-linked agammaglobulinemia. *Lancet* 1993; **341**: 1153.

50. Kornfeld SJ, Good RA and Litman GW. Atypical X-linked agammaglobulinemia. *N. Engl. J. Med.* 1994; **331**: 949–950.

51. Haslam RJ, Koide HB and Hemmings BA. Pleckstrin domain homology. *Nature* 1993; **363**: 309–310.

52. Mayer BJ, Ren R, Clark KL and Baltimore D. A putative modular domain present in diverse signaling proteins. *Cell* 1993; **73**: 629–630.

53. Musacchio A, Gibson T, Rice P, Thompson J and Saraste M. The PH domain: a common piece in the structural patchwork of signalling proteins. *TIBS* 1993; **18**: 343–348.

54. Konishi H, Kuroda S and Kikkawa U. The pleckstrin homology domain of Rac protein kinase associates with the regulatory domain of protein kinase C zeta. *Biochem. Biophys. Res. Commun.* 1994; **205**: 1770–1775.

55. Yao L, Kawakami Y and Kawakami T. The pleckstrin homology domain of Bruton tyrosine kinase interacts with protein kinase C. *Proc. Natl. Acad. Sci. USA* 1994; **91**: 9175–9179.

56. Touhara K, Inglese J, Pitcher JA, Shaw G and Lefkowitz RJ. Binding of G protein $\beta\gamma$-subunits to pleckstrin homology domains. *J. Biol. Chem.* 1994; **269**: 10217–10220.

57. Harlan JE, Hajduk PJ, Yoon HS and Fesik SW. Pleckstrin homology domains bind to phosphatidylinositol-4,5-bisphosphate. *Nature* 1994; **371**: 168–170.

58. Tsukada S, Simon MI, Witte ON and Katz A. Binding of $\beta\gamma$ subunits of heterotrimeric G proteins to the PH domain of Bruton tyrosine kinase. *Proc. Natl. Acad. Sci. USA* 1994; **91**: 11256–11260.

59. Macias MJ, Musacchio A, Ponstingl H, Nilges M, Saraste M and Oschkinat H. Structure of the pleckstrin homology domain from β-spectrin. *Nature* 1994; **369**: 675–677.

60. Yoon HS, Hajduk PJ, Petros AM, Olejniczak ET, Meadows RP and Fesik SW. Solution structure of a pleckstrin-homology domain. *Nature* 1994; **369**: 672–675.

61. Li T, Tsukada S, Satterthwaite A, Havlik MH, Park H, Takatsu K and Witte ON. Activation of Bruton's tyrosine kinase (BTK) by a point mutation in its pleckstrin homology (PH) domain. *Immunity* 1995; **2**: 451–460.

62. Cheng G, Ye Z-S and Baltimore D. Binding of Bruton's tyrosine kinase to Fyn, Lyn, or Hck through a Src homology 3 domain-mediated interaction. *Proc. Natl. Acad. Sci. USA* 1994; **91**: 8152–8155.

63. Pawson T and Gish GD. SH2 and SH3 domains: from structure to function. *Cell* 1992; **71**: 359–362.

64. Bar-Sagi D, Rotin D, Batzer A, Mandiyan V and Schlessinger J. SH3 domains direct cellular localization of signaling molecules. *Cell* 1993; **74**: 83–91.

65. Ren R, Mayer BJ, Cicchetti P and Baltimore D. Identification of a ten-amino acid proline-rich SH3 binding site. *Science* 1993; **259**: 1157–1161.

66. Yu H, Chen JK, Feng S, Dalgarno DC, Brauer AW, Schreiber SL. Structural basis for the binding of proline-rich peptides to SH3 domains. *Cell* 1994; **76**: 933–945.

67. Feng S, Chen JK, Yu H, Simon JA, Schreiber SL. Two binding orientations for peptides to the Src SH3 domain: development of a general model for SH3-ligand interactions. *Science* 1994; **266**: 1241–1247.

68. Cory GOC, Lovering RC, Hinshelwood S, MacCarthy-Morrogh L, Levinsky RJ and Kinnon C. The protein product of the c-*cbl* protooncogene is phosphorylated after B cell receptor stimulation and binds the SH3 domain of Bruton's tyrosine kinase. *J. Exp. Med.* 1995; **182**: 611–615.

69. Blake TJ, Shapiro M, Morse HC and Langdon WY. The sequences of the human and mouse c-*cbl* proto-oncogenes show v-*cbl* was generated by a large truncation encompassing a proline-rich domain and a leucine zipper-like motif. *Oncogene* 1991; **6**: 653–657.

70. Donovan JA, Wange RL, Langdon WY and Samelson LE. The protein product of the c-cbl protooncogene is the 120-kDa tyrosine-phosphorylated protein in Jurkat cells acti-

vated via the T cell antigen receptor. *J. Biol. Chem.* 1994; **269**: 22921–22924.

71. Rivero-Lezcano OM, Sameshima JH, Marcilla A and Robbins KC. Physical association between Src homology 3 elements and the protein product of the c-cbl proto-oncogene. *J. Biol. Chem.* 1994; **269**: 17363–17366.

72. Zhu Q, Zhang M, Rawlings DJ, Vihinen M, Hagemann T, Saffran DC *et al.* Deletion within the Src homology domain 3 of Bruton's tyrosine kinase resulting in X-linked agammaglobulinemia. *J. Exp. Med.* 1994; **180**: 461–470.

73. Cohen GB, Ren R and Baltimore D. Modular binding domains in signal transduction pathways. *Cell* 1995; **80**: 237–248.

74. Songyang Z, Shoelson SE, Chaudhuri M, Gish G, Pawson T, Haser WG *et al.* SH2 domains recognize specific phosphopeptide sequences. *Cell* 1993; **72**: 767–778.

75. Vihinen M, Nilsson L and Smith CIE. Structural basis of SH2 domain mutations in X-linked agammaglobulinemia. *Biochem. Biophys. Res. Commun.* 1994; **205**: 1270–1277.

76. Aoki Y, Isselbacher KJ, Cherayil BJ and Pillai S. Tyrosine phosphorylation of Blk and Fyn Src homology 2 domain-binding proteins occurs in response to antigen-receptor ligation in B cells and constitutively in pre-B cells. *Proc. Natl. Acad. Sci. USA* 1994; **91**: 4204–4208.

77. Waksman G, Shoelson SE, Pant N, Cowburn D and Kuriyan J. Binding of a high affinity phosphotyrosyl peptide to the Src SH2 domain: crystal structures of the complexed and peptide-free forms. *Cell* 1993; **72**: 779–790.

78. Eck MJ, Shoelson SE and Harrison SC. Recognition of a high-affinity phosphotyrosyl peptide by the Src homology-2 domain of p56^{lck}. *Nature* 1993; **362**: 87–91.

79. Mayer BJ, Jackson PK, Van Etten RA and Baltimore D. Point mutations in the abl SH2 domain coordinately impair phosphotyrosine binding in vitro and transforming activity in vivo. *Mol. Cell. Biol.* 1992; **12**: 609–618.

80. Johnson GL and Vaillancourt RR. Sequential protein kinase reactions controlling cell growth and differentiation. *Curr. Opin. Cell. Biol.* 1994; **6**: 230–238.

81. Sun H and Tonks NK. The coordinated action of protein tyrosine phosphatases and kinases in cell signaling. *TIBS* 1994; **19**: 480–485.

82. Hanks SK, Quinn AM and Hunter T. The protein kinase family: conserved features and deduced phylogeny of the catalytic domains. *Science* 1988; **241**: 42–52.

83. Hubbard SR, Wei L, Ellis L and Hendrickson WA. Crystal structure of the tyrosine kinase domain of the human insulin receptor. *Nature* 1994; **372**: 746–754.

84. Accili D, Cama A, Barbetti F, Kadowaki H, Kadowaki T and Taylor SI. Insulin resistance due to mutations of the insulin receptor gene: an overview. *J. Endocrinol. Invest.* 1992; **15**: 857–864.

85. Zheng J, Knighton DR, ten Eyck LF, Karlsson R, Xuong N, Taylor SS and Sowadski JM. Crystal structure of the catalytic subunit of cAMP-dependent protein kinase complexed with MgATP and peptide inhibitor. *Biochemistry* 1993; **32**: 2154–2161.

86. Bossemeyer D, Engh RA, Kinzel V, Ponstingl H and Huber R. Phosphotransferase and substrate binding mechanism of the cAMP-dependent protein kinase catalytic subunit from porcine heart as deduced from the 2.0 A structure of the

complex with Mn^{2+} adenylyl imidodiphosphate and inhibitor peptide PKI(5–24). *EMBO J.* 1993; **12**: 849–859.

87. Campbell M-A and Sefton BM. Protein tyrosine phosphorylation is induced in murine B lymphocytes in response to stimulation with anti-immunoglobulin. *EMBO J.* 1990; **9**: 2125–2131.

88. Gold MR, Law DA and DeFrance AL. Stimulation of protein tyrosine phosphorylation by the B-lymphocyte antigen receptor. *Nature* 1990; **345**: 810–813.

89. Burkhardt AL, Brunswick M, Bolen JB and Mond JJ. Anti-immunoglobulin stimulation of B lymphocytes activates src-related protein-tyrosine kinases. *Proc. Natl. Acad. Sci. USA* 1991; **88**: 7410–7414.

90. Yamanashi Y, Fukui Y, Wongsasant B, Kinoshita Y, Ichimori Y, Toyoshima K and Yamamoto T. Activation of Src-like protein-tyrosine kinase Lyn and its association with phosphatidylinositol 3-kinase upon B-cell antigen receptor- mediated signaling. *Proc. Natl. Acad. Sci. USA* 1992; **89**: 1118–1122.

91. Kurosaki T, Takata M, Yamanashi Y, Inazu T, Taniguchi T, Yamamoto T and Yamamura H. Syk activation by the Src-family tyrosine kinase in the B cell receptor signaling. *J. Exp. Med.* 1994; **179**: 1725–1729.

92. de Weers M, Brouns GS, Hinshelwood S, Kinnon C, Schuurman RKB, Hendriks RW and Borst J. B-cell antigen receptor stimulation activates the human Bruton's tyrosine kinase, which is deficient in X-linked agammaglobulinemia. *J. Biol. Chem.* 1994; **269**: 23857–23860.

93. Hinshelwood S, Lovering RC, Genevier HC, Levinsky RJ and Kinnon C. The protein defective in X-linked agammaglobulinemia, Bruton's tyrosine kinase, shows increased autophosphorylation activity *in vitro* when isolated from cells in which the B cell receptor has been cross-linked. *Eur. J. Immunol.* 1995; **25**: 1113–1116.

94. Saouaf SJ, Mahajan S, Rowley RB, Kut SA, Fargnoli J, Burkhardt AL *et al.* Temporal differences in the activation of three classes of non-transmembrane protein tyrosine kinases following B-cell antigen receptor surface engagement. *Proc. Natl. Acad. Sci. USA* 1994; **91**: 9524–9528.

95. Aoki Y, Isselbacher KJ and Pillai S. Bruton tyrosine kinase is tyrosine phosphorylated and activated in pre-B lymphocytes and receptor-ligated B cells. *Proc. Natl. Acad. Sci. USA* 1994; **91**: 10606–10609.

96. Kawakami Y, Yao L, Miura T, Tsukada S, Witte ON and Kawakami T. Tyrosine phosphorylation and activation of Bruton tyrosine kinase upon FcRI cross-linking. *Mol. Cell. Biol.* 1994; **14**: 5108–5113.

97. Matsuda T, Takahashi-Tezuka M, Fukada T, Okuyama Y, Fujitani Y, Tsukada S *et al.* Association and activation of Btk and Tec tyrosine kinases by gp130, a signal transducer of the interleukin-6 family of cytokines. *Blood* 1995; **85**: 627–633.

98. Sato S, Katagiri T, Takaki S, Kikuchi Y, Hitoshi Y, Yonehara S *et al.* IL-5 receptor-mediated tyrosine phosphorylation of SH2/SH3-containing proteins and activation of Bruton's tyrosine and Janus 2 kinases. *J. Exp. Med.* 1994; **180**: 2101–2111.

99. Koike M, Kikuchi Y, Tominaga A, Takaki S, Akagi K, Miyazaki J *et al.* Defective IL-5-receptor-mediated signaling in B cells of X-linked immunodeficient mice. *Int. Immunol.* 1995; **7**: 21–30.

100. Hasbold J and Klaus GGB. B cells from CBA/N mice do not proliferate following ligation of CD40. *Eur. J. Immunol.* 1994; **24**: 152–157.

101. Rigley KP, Harnett MM, Phillips RJ and Klaus GGB. Analysis of signaling via surface immunoglobulin receptors on B cells from CBA/N mice. *Eur. J. Immunol.* 1989; **19**: 2081–2086.

102. Lindsberg M-L, Brunswick M, Yamada H, Lees A, Inman J, June CH and Mond JJ. Biochemical analysis of the immune B cell defect in xid mice. *J. Immunol.* 1991; **147**: 3774–3779.

103. Miyake K, Yamashita Y, Hitoshi Y, Takatsu K and Kimoto M. Murine B cell proliferation and protection from apoptosis with an antibody against a 105 kDa molecule: unresponsiveness of X-linked immunodeficient B cells. *J. Exp. Med.* 1994; **180**: 1217–1224.

104. Kirkham PA, Santos-Argumedo L, Harnett MM and Parkhouse RME. Murine B-cell activation via CD38 and protein tyrosine phosphorylation. *Immunology* 1994; **83**: 513–516.

105. Santos-Argumedo L, Lund FE, Heath AW, Solvason N, Wu WW, Grimaldi JC, Parkhouse RME and Howard M. CD38 unresponsiveness of *xid* B cells implicates Bruton's tyrosine kinase (*btk*) as a regulator of CD38 induced signal transduction. *Int. Immunol.* 1995; **7**: 163–170.

106. Go NF, Castle BE, Barrett R, Kastlein R, Dang W, Mosmann TR *et al.* Interleukin 10, a novel B cell stimulatory factor: unresponsiveness of X chromosome-linked immunodeficiency B cells. *J. Exp. Med.* 1990; **172**: 1625–1631.

19

The IL-2Rγ Chain in X-linked Severe Combined Immunodeficiency (SCIDX1)

Robin E. Callard[1], David J. Matthews[1] and Paula A. Clark[2]

[1]*Immunobiology Unit, Institute of Child Health, 30 Guilford Street, London, UK and*
[2]*Molecular Immunology Unit, Institute of Child Health, 30 Guilford Street, London, UK*

INTRODUCTION

In the past five years the genes responsible for six of the seven known X-linked immunodeficiencies have been identified [1]. These are X-linked severe combined immunodeficiency (SCIDX1), X-linked agammaglobulinaemia (XLA), X-linked hyper IgM syndrome (HIGM-1), Wiskott–Aldrich syndrome (WAS), X-linked chronic granulomatous disease (CGD), and properdin deficiency. The gene responsible for X-linked lymphoproliferative disease (Duncan's syndrome) has yet to be identified.

Each of these X-linked immunodeficiencies can be considered as a naturally occurring gene inactivation model with a well-characterized disease phenotype. The identification and characterization of the proteins encoded by the genes responsible has led to important advances in our understanding of both the immunodeficiency itself and the basic cellular and molecular mechanisms that determine normal immune function. Ultimately, these studies of basic immune mechanisms in immunodeficiency and their clinical sequelae will lead to new forms of treatment including gene replacement therapy.

X-LINKED SEVERE COMBINED IMMUNODEFICIENCY (SCIDX1)

There is no better example of how the molecular characterization of an immunodeficiency has had a major impact on basic immunology than SCIDX1.

This is a devastating immunodeficiency characterized by profoundly defective cellular and humoral immunity resulting in greatly increased susceptibility to infection and is uniformly fatal by 1–2 years of age unless treated by bone marrow transplantation [2,3]. Affected boys have markedly reduced or absent T cells, but B cells are often present in normal or even increased numbers [3]. B cells from SCIDX1 patients can secrete immunoglobulin on stimulation with pokeweed mitogen in the presence of normal T cells, but other *in vitro* assays have uncovered defective responses to mitogens and cytokines [3]. In obligate carriers of SCIDX1, the X chromosome is non-randomly inactivated in T cells, and may be non-randomly inactivated in B cells, neutrophils and monocytes consistent with a gene expressed in most if not all haematopoietic cell lineages [4,5]. The gene responsible for SCIDX1 has been mapped to Xq13 [6–8]. The gene coding for the IL-2 receptor γ chain (IL-2Rγ), now called the common γ chain (γc chain) also maps to this locus [9,10] and mutations in the γc-chain gene are responsible for SCIDX1 [9,11].

THE IL-2 RECEPTOR

The IL-2 receptor is a complex of three distinct polypeptide subunits (Figure 19.1) [12]. The α chain (Tac, p55 or CD25) binds IL-2 with low affinity (kd 10^{-8} M) and a short dissociation half-life of 1.7 s [13]. The larger β chain (p75, CD122) also binds IL-2 with low affinity (kd 10^{-7} M) whereas the γ chain (γc chain, p64) does not bind IL-2. The heterotrimeric complex consisting

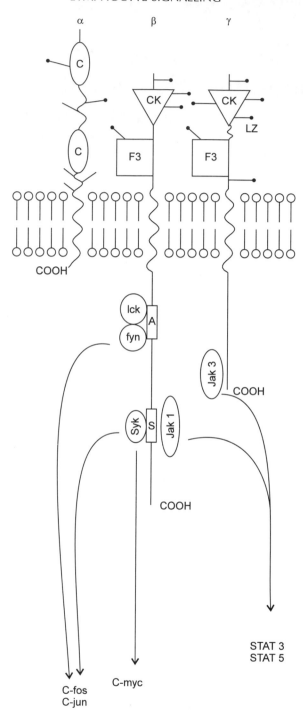

Figure 19.1   The IL-2 receptor complex showing CKR-SF structure of IL-2Rβ and γ chains and PTK association with cytoplasmic domains of the β and γc chains

of the $\alpha/\beta/\gamma$ chains binds IL-2 with high affinity (kd 10^{-11} M). Receptor complexes of the α/β chain also bind IL-2 with high affinity (kd 10^{-10} M) but may not deliver a mitogenic signal whereas complexes of the β/γ chain bind IL-2 (kd 10^{-9} M) and can deliver a mitogenic signal [13]. The α chain has two extracellular domains with homology to the complement control protein (CCP) also known as the "sushi domain" or GP-I motif. It has a very short intracellular region and does not signal. The IL-15Rα chain has homology with the IL-2Rα chain (CD25) and together they define a new cytokine receptor family [14]. The IL-2Rβ and γc chains are both members of the cytokine receptor superfamily (CKR-SF) (Figure 19.1). They both have an extracellular segment containing a CK domain of about 100 amino acids with a characteristic Cys–X–Trp motif and three other conserved Cys residues, and an FN III domain containing the WS–X–WS motif required for ligand binding and signal transduction [15,16]. The γc chain also has a short leucine zipper motif between the CK and FNIII domains. Heterodimerization of the IL-2Rβ and γ chains is essential for high affinity IL-2 binding and signal transduction [12,17].

Mutational analysis has identified three sites on the IL-2 molecule that are important for interaction with the IL-2R [18]. One site formed by residues around Lys57 on α-helix A and residues around Glu76 on α-helix B is important for interaction with IL-2Rα. A second site defined by Asp34 on α-helix A and Asn 103 on α-helix C is required for interaction with IL-2Rβ and a third site defined by Gln141 on α-helix D is required for interaction with IL-2Rγ. Computer modelling of the IL-2R has confirmed the importance of these sites for IL-2 binding and identified loop regions on the IL-2Rβ and γ chains as the sites which interact with IL-2 [19].

SIGNALLING BY THE IL-2 RECEPTOR

Binding of IL-2 to its receptor triggers a series of signalling events including activation of protein kinases such as p56^{lck} [20], p53/56^{lyn} [21], Raf-1 [22], and S6 kinase [23]; phosphorylation of cellular proteins including the IL-2Rβ and γc chains [24,25]; activation of p21 ras [26,27]; activation of phosphatidylinositol 3-kinase [28,29]; and hydrolysis of glycosyl-phosphatidylinositol [30]. These signalling pathways induce expression of c-fos, c-jun, c-myb and c-myc [31–34] culminating in lymphocyte proliferation and differentiation.

Neither the IL-2Rβ or γ chains have intrinsic kinase activity and the phosphorylation events triggered by IL-2 binding and receptor heterodimerization depend on activation of cellular protein kinases. Mutational analysis of the cytoplasmic domains of the receptor subunits has begun to elucidate the molecular events involved in IL-2 signal transduction. The cytoplasmic segment of IL-2Rβ includes an acidic region which was shown in a series of mutagenesis experiments to be a binding site for the Src family tyrosine kinases, Lck and Fyn [35] and is required for increased expression of c-fos and c-jun but not c-myc [35,36]. Signal transduction and mitogenic activity of the IL-2Rβ also depends on a more membrane proximal serine-rich region that binds and activates Syk. This region is required for expression of c-myc as well as c-fos and c-jun. Both the serine-rich and acidic regions are required for activation of p56^{lck} [35] and Ras [27]. In the absence of p56^{lck}, the src-PTKs, Fyn and Lyn may be activated by IL-2 binding but there is evidence only for association of Fyn with the β chain [37]. The janus tyrosine kinase, Jak1, also binds to the serine-rich region of the IL-2Rβ chain [27,38,39] and Jak3 binds to the C-terminal region of the γc chain [39]. Jak1 and Jak3 are also activated by IL-4, IL-7 and IL-9 whose receptors include the γc chain [40,41].

Heterodimerization of the IL-2Rβ and γc chains required for signal transduction is thought to bring Jak1 and Jak3 into close proximity allowing cross-phosphorylation and activation of their catalytic domains followed by phosphorylation of STAT3 and STAT5 [42,43]. The IL-2R reconstituted with a mutant form of the γc chain, lacking the C-terminal 68 amino acids, was unable to induce expression of c-fos, c-jun or c-myc, but IL-2 binding and internalization were not affected [44]. In contrast, reconstitution with a mutant γc chain, lacking only the C-terminal 30 amino acids, was able to activate a kinase and induce c-myc expression but not c-fos or c-jun expression [45]. This shows that the γc chain is essential for at least two signalling pathways, but the exact associations with signalling proteins has not yet been fully determined. Truncations of the γc chain gene and point mutations causing loss of Jak3 association with the γc chain have been found in SCIDX1 [41]. Jak3 is also required for induction of c-fos but not *c-myc*. More recently, mutations in the Jak3 gene have been found in an autosomal form of SCID [46]. These findings demonstrate the importance of Jak3 association and activation by the γc chain for responses to IL-2 and other cytokines whose receptors use the γc chain, and explains the need for receptor heterodimerization for functional IL-2 signalling.

OTHER CYTOKINE RECEPTORS THAT USE THE γc CHAIN

The γc chain is known to be a functional component of the receptors for IL-2 [47], IL-4 [48,49], IL-7 [50], IL-9 [51] and IL-15 [52]. The importance of the γc chain for signal transduction by IL-2 has been established in fibroblast transfection experiments showing that its presence in the IL-2R complex is required for protein tyrosine phosphorylation and induction of *c-myc*, *c-fos* and *c-jun* [45], and activation of Jak-3 [39]. Receptor internalization following IL-2 binding also requires a functional γc chain [47]. Other studies have shown that the γc chain is required for tyrosine phosphorylation of the insulin receptor substrate-1 (IRS-1) and IRS-2 (4PS) in response to IL-4 [49], and efficient internalization of the IL-7R on binding of IL-7 [53]. In other transfection experiments, proliferation of mouse F7 cells to IL-4 or IL-7 required expression of the γc chain in addition to the IL-4R or IL-7R [44]. An antibody to the murine γc chain blocks binding and inhibits IL-9 responses by IL-9-dependent cell lines [51]. These experiments clearly show that the γc chain is a signalling component of the receptors for IL-2, IL-4, IL-7 and IL-9.

The fact that the γc chain is a functional component of the receptors for IL-4 [44,48,49], IL-7 [44] IL-9 [51] and IL-15 [54] suggests that signalling elements dependent on the γc chain may be common to all of them. IL-2, IL-4, IL-7, IL-9 and IL-15 all activate Jak1 and Jak3 [42,55], but IL-2, IL-15 and IL-7 activate different STAT proteins than do IL-4 and IL-13 [56]. IL-2, IL-15 and IL-7 activate STAT3 and STAT5 [42,43,56]. The activation of the same STATs by IL-2, IL-7 and IL-15, even though they bind to different receptors, can be explained by the presence of similar tyrosine phosphorylated motifs in the cytoplasmic domain

of the β chain of the IL-7R and the common β chain of the IL-2R and IL-15R [52] that serves as a STAT5 docking site [56]. In contrast, the other two cytokines IL-4 and IL-13 activate STAT6 and not STAT5 [42,57]. STAT6 binds to a docking site on the IL-4Rα chain which is now known also to be a component of the IL-13R [58]. Significantly, although IL-13 activates STAT6, it does not activate Jak3 [59] suggesting that the specificity of STAT6 activation is due to its docking site on the IL-4Rα chain and not the activation of Jak3. It will be interesting to determine whether the recently identified IL-13 binding chain [60,61], which associates with the IL-4R α chain to form the IL-13 receptor complex, is able to bind and activate an as yet unidentified Jak protein or some other kinase which can activate STAT6. These findings suggest that specificity of STAT activation is determined by the STAT docking sites on the receptor subunits rather than which Jaks are activated. Specificity of cellular responses defined by STAT association with receptor subunits rather than Jak activation has also been shown for other cytokines [42,62,63].

γc CHAIN MUTATIONS IN SCIDX1

More than 30 different γc chain gene mutations in unrelated SCIDX1 patients have been described (Table 19.1). These include point mutations giving rise to single amino acid substitutions, deletions which may cause frame shifts with premature stop codons giving rise to a predicted truncated protein, and splice site mutations. The different mutations may compromise γc chain function by altering or destroying sites on the molecule important for cytokine binding, anchoring in the membrane, and/or interaction with other signalling proteins. Others may prevent receptor subunit association (e.g. IL-2Rβ and IL-2Rγ heterodimerization) essential for signal transduction [17,64]. In other examples, no mRNA and/or γc chain protein can be detected. In a recent study of γc chain gene mutations in six X-SCID patients by Di Santo *et al.* [65], loss of most or all high affinity IL-2 binding sites in four of them was due to point mutations in the extracellular domain resulting in loss of tertiary structure or ligand binding (P1, P3, P4, P6 in Table 19.1). Another mutation consisting of

Table 19.1 γc Chain mutations in SCIDX1

Patient	cDNA position (affected exon)	Base change	Protein sequence change	Comments
Pt[a]	115 (1)	a. G→A b. Insertion of 27 bp of intronic sequence	a. D_{39}-N b. Aberrant mRNA and premature stop	Atypical/mild SCID 2 mRNA species, one functional less abundant form with point mutation and an abundant non-functional form
1[b]	186 (2)	T→A	Premature stop at C_{62}	
B[c]	202 (2)	G→A	E_{68}→K	Gives 78 amino acid protein
P11[d]	205 (2)	1 bp insertion (T)	Frameshift after E_{68}	
T[c]	241 (2)	C→T	Premature stop at Q_{81}	No binding, no γc-chain staining
P3 (YY)[e]	Deletion of exon 2 with frameshift and abnormal amino acid sequence			
Pt[f]	intron 2	A→G	—	Splice acceptor site created, unstable mRNA
7[b]	341 (3)	G→A	G_{114}→D	
P1[g]	344 (3)	G→A	C_{115}→F	Very low level of high affinity IL-2 binding
1[h]	355 (3)	A→T	Premature stop at K_{119}	
C[c]	430 (3)	C→T	Premature stop at Q_{144}	
5[b]	intron 3	G→A	—	Disrupts splice donor site
6[b]	458 (4)	T→A	I_{153}→N	
P1 (YO)[e]	467 (4)	C→T	A_{156}→V	Loss of IL-2 binding
P10[d]	548 (4)	T→C	L_{183}→S	
P4, P12[d]	intron 4	G→A	—	Disrupts splice donor site, no γc-chain staining
P9[d]	664 (5)	C→T	R_{222}→C	
P2[d]	665 (5)	1 bp deletion	Frameshift at F_{221} with truncation at amino acid 271	
P7[d]	671–686 (5)	16 bp deletion	Frameshift at R_{224} with truncation at amino acid 266	
P6[g]	676 (5)	C→T	R_{226}→C	Very low level of high affinity IL-2 binding
G[c]	697 (5)	1 bp deletion	Frameshift at S_{233} with truncation at amino acid 272	
P6[d]	703 (5)	C→T	Premature stop at Q_{235}	
III-2[i]	702/703 (5)	9 bp duplication and insertion	Duplication and insertion of 3 amino acids (QHW) between A_{234} and Q_{235}	Associated with female germline mosaicism
P3[g]	720 (5)	G→C	W_{240}→C	Loss of high affinity IL-2 binding
P4[g]	722 (5)	G→T	S_{241}→I	Loss of high affinity IL-2 binding
P2[g]	812–815 or 816–819 (6)	GATT deletion	Frameshift at L_{272}	Loss of high affinity IL-2 binding
AA[d]	854 (6)	G→A	R_{285}→Q	
2[h]; P5, P8[d]	865 (7)	C→T	Premature stop at R_{289}	
P[j]	878 (7)	T→A	L_{293}→Q	Mild SCIDX1 phenotype
P2 (AY)[e]	between 954/960 (8)	2 bp deletion (GA)	Frameshift at S_{320} in SH2 domain with truncated protein with 6 different aa after E_{319}	Normal binding but no signalling
3[h]	923 (7)	C→A	Premature stop at S_{308}	
P5[g]	No detectable γc-chain mRNA and no protein			
P1[d]	Deletions of exons 6, 7, 8 and all or part of exon 5			No detectable γc-chain

Sources: a = DiSanto (70); b = Puck (10); c = Markiewicz (97); d = Clark (98); e = Ishii (55); f = Tassara (99); g = Di Santo (64); h = Noguchi (9); i = Puck (100); j = Schmalstieg (68).

a 4 bp deletion in the transmembrane domain of the γc chain also resulted in loss of cytokine binding which may be explained by a loss of structural integrity affecting membrane anchoring (P2 in Table 19.1). Loss of binding in the sixth example was due to absence of γc chain mRNA and protein expression [65]. In a similar study, Ishii *et al.* described three γc chain mutations in SCIDX1 [66]. One of these was an A $\rightarrow$ V$_{156}$ substitution in the loop joining the CK and FIII domains just distal to the leucine zipper motif and resulted in loss of high affinity IL-2 binding sites (YO in Table 19.1). This residue is close to the site which binds the D helix of IL-2 [19] and the A $\rightarrow$ V mutation may disrupt cytokine binding although a loss of structural integrity may also prevent association with the other receptor subunit components. A second mutation consisting of a 2 bp deletion, causing a frame shift of the coding region in the SH2 domain of the cytoplasmic domain, resulted in a loss of ability to signal but did not affect cytokine binding (AY in Table 19.1). The third mutation was a deletion of exon 2 with a frame shift and no detectable γc chain expression (YY in Table 19.1). An E $\rightarrow$ K$_{68}$ mutation described by Markiewicz also results in loss of high affinity binding sites without loss of γc chain mRNA. Although γc chain protein expression may occur in this patient, no γc chain expression could be detected by antibody staining in one of our patients with the same mutation. Three other nonsense mutations resulted in reduced mRNA expression with abnormal splicing in one with no functional protein.

Some γc chain gene mutations may result in only partial loss of γc chain function giving rise to a less severe or variable disease phenotype [67]. In one large family, affected boys had chronic bacterial and viral infections but were not affected with *Pneumocystis carinii* typical of SCIDX1, and some lived into late childhood or early adulthood [68]. Serum Ig levels were normal. CD4$^+$ and CD8$^+$ numbers were decreased but not as low as usually seen in SCIDX1. X-inactivation studies of obligate carriers showed non-random inactivation in T cells and B cells consistent with findings in SCIDX1. A point mutation giving rise to a Leu$_{293}$ $\rightarrow$ Gln substitution in the cytoplasmic domain of the γc chain was found in affected boys of this family (Table 19.1) [69]. The effect of this mutation on γc chain association

with other components of the receptor complex or the effects on signalling have not yet been determined. It will be interesting to see whether the mutation results in partial receptor dysfunction for IL-2, IL-4, IL-7, IL-9 and IL-15, or loss of activity in only some of these receptors. The latter is a possibility as it is known that the epitope on the γc chain defined by monoclonal antibodies required for binding IL-2 and IL-7 is distinct from the epitope required for binding of IL-4 [70].

Another family with a mild form of SCIDX1 has been described by DiSanto *et al.* [71]. Affected boys in this family suffered from recurrent infection, severe diarrhoea and failure to thrive [72]. Serum Ig levels and peripheral T cell, B cell and NK cell numbers were normal but specific antibody responses were defective [65,71]. T cell receptor β chain expression was oligoclonal suggesting that the defect allowed the generation of a limited number peripheral T cell clones [71]. Analysis of γc chain cDNA and genomic sequences revealed a splice site mutation which gave rise to two γc chain transcripts [71]. One of these was an abundant non-functional form with an insertion of a small intronic sequence. The other much less abundant form was functional and produced a γc chain with a single D$_{39}$ amino acid substitution. High affinity IL-2R expression was five-fold lower than in normal controls—consistent with the abundancy of the non-functional transcript. Loss of function may be due to loss of high affinity receptors and/or the point mutation in the γc chain expressed in the residual high affinity receptors.

These examples will no doubt encourage other investigators to consider γc chain mutations in atypical combined immunodeficiencies with an X-linked pedigree. Characterization of γc chain gene mutations that give rise to milder forms of SCIDX1 will prove invaluable for determining the functional sites on the γc chain and the role of these sites in cytokine binding and signalling.

LYMPHOID DEVELOPMENT IN MICE WITH A TARGETED DELETION OF THE γc CHAIN GENE

Targeted deletion of the γc chain gene in mice has been reported by two groups. DiSanto *et al.* [73]

used Cre/loxP–mediated recombination to delete a region of the γc chain gene encompassing exons 2–6 encoding extracellular and transmembrane domains. Male mice deficient in γc chain gene expression (γc⁻) had severe abnormalities in lymphoid development. Thymuses were hypoplastic and peripheral T cells numbers were low but not absent. B cell development was also abnormal and peripheral B cell numbers were greatly reduced in contrast to normal B cell numbers seen in human SCIDX1. NK cells were also very low or absent. The spleens in the γc⁻ mice were small with greatly reduced numbers of lymphoid cells compared to normal littermates but T cell areas and primary follicles were present, although there were no germinal centres. Development of gut associated intraepithelial lymphocytes was also severely reduced and Peyer's patches were not detected. A very similar phenotype was reported by Cao et al. in γc chain gene knockout mice lacking part of exon 3 and all of exons 4–8 with no expression of γc chain mRNA or protein [74]. The presence of some functional T cells, and the reduced numbers of peripheral B cells in γc⁻ male mice contrast with the very low numbers or absence of T cells and normal numbers of B cells in SCIDX1. This species-specific difference may be due to different roles for γc⁻chain-dependent cytokines in lymphoid development and/or the existence of alternative (redundant) cytokine signalling pathways. For example, the loss of B cells in male γc⁻ mice suggests that IL-7 is more important for murine than human B cell development.

THE EFFECT OF γc CHAIN GENE MUTATIONS ON LYMPHOCYTE FUNCTION IN SCIDX1

SCIDX1 males typically have profoundly defective cell mediated and humoral immunity with very low numbers or no T cells but normal numbers of B cells. In contrast, autosomal SCID patients unable to make IL-2 [75], and mice in which the IL-2 gene [76], IL-4 gene [77], or both IL-2 and IL-4 genes [78] have been inactivated by targeted gene disruption, have normal numbers of T cells and do not show the same immunological abnormalities as SCIDX1. The profound cell mediated and humoral immunodeficiency in SCIDX1 must therefore be due to the combined inability to respond to those cytokines (IL-2, IL-4, IL-7, IL-9 and IL-15) whose receptors include the γc chain as a functional component. This conclusion itself raises some problematical questions. For example, IL-7 and IL-9 are involved in neuronal differentiation in the hippocampus yet abnormal brain development or function does not appear to be a feature of SCIDX1 [79]. Similarly, although IL-7 is an important cytokine for B cell development, and IL-4 is a significant B cell growth and differentiation factor, development and numbers of B cells in SCIDX1 appear to be normal.

These apparent paradoxes may be explained by cytokine binding to alternative receptors that do not include the γc chain. Some evidence for alternative IL-7 and IL-4 receptors has been obtained. For example, we have found recently that IL-7 can elicit production of IL-6 from normal and SCIDX1 monocytes (Paula Clark et al., unpublished observations). Although this finding conflicts with the known functional dependence of the IL-7R on the γc chain [44,50,53], it may be explained by IL-7 acting through an alternative receptor that does not include the γc chain. A low affinity IL-7 receptor has been identified but not cloned or fully characterized [80]. In addition, a novel IL-7-like cytokine called thymic stromal cell derived lymphopoietin (TSLP) binds to a receptor which includes the IL-7Rα chain but may not include the γc chain [81]. It is possible that IL-7 acts on B cell precursors (BCP) through this second receptor. A second functional IL-7 receptor that also binds TSLP would explain why B cell defects are more severe in IL-7R-deficient than in γc-chain-deficient mice [81].

In another study, a requirement for the γc chain in B cell responses to IL-2, IL-4, IL-13 and IL-15 was investigated. In these experiments, B cells from two SCIDX1 patients with defined mutations in the γc chain gene and no detectable γc chain expression were unable to respond to IL-2 or IL-15 confirming the functional importance of the γc chain in the IL-2R and IL-15R. In contrast, SCIDX1 B cells responded normally to IL-4 and as well or better than control B cells to IL-13 in assays for B cell activation (CD23 expression), proliferation, and IgE secretion [54]. These experiments showed either that the γc chain is not required for signal transduction by IL-4 or IL-13 or that IL-4 and IL-13 can act through an alternative receptor which does not have the γc chain as a component.

Together, these observations suggest that the phenotype in SCIDX1 and in γc-chain-deficient mice may be ameliorated to some extent by alternative receptors for IL-4 and IL-7 (at least) that do not use the γc chain.

THE ROLE OF THE γc CHAIN IN THE RECEPTORS FOR IL-4 AND IL-13: EVIDENCE FOR A SECOND IL-4 RECEPTOR THAT DOES NOT INCLUDE THE γc CHAIN

The relationship between the receptors for IL-4 and IL-13 and the functional role of the γc chain has been examined in more detail in experiments with IL-4 mutant proteins. Functional analysis of IL-4 mutant proteins has identified two sites on the IL-4 protein (site I and site II) important for receptor binding and biological activity respectively. Site II mutants such as IL-4$_{Y124D}$ bind with almost normal affinity to the IL-4R α chain but are unable to signal. Recent biochemical experiments and computer modelling of receptor binding have shown that Tyr124 on IL-4 is required for interaction with the γc chain consistent with an important role for this residue in signal transduction [19,82]. The IL-2 mutant protein IL-2$_{Q141D}$ is analogous to IL-4$_{Y124D}$ and also binds to its receptor with an affinity comparable to wildtype IL-2 but has defective signalling properties. The Gln residue at position 141 in IL-2$_{Q141D}$ is known to be important for interactions with the γc chain of the IL-2 receptor which is disrupted by the Asp substitution [83,84]. The IL-4$_{Y124D}$ mutant has been shown to inhibit responses to IL-4 and IL-13 by B cells [85], and the TF1 erythroleukaemic cell line [86] consistent with antagonistic binding to the IL-4Rα chain. IL-4$_{Y124D}$ also inhibits SCIDX1 B cell responses to IL-4 and IL-13 showing that expression of the γc chain is not required for the antagonistic activity of this mutant [58,87]. Confirmation that the IL-4Rα chain is a component of the IL-13R has been obtained by showing that B cell responses to IL-4 and IL-13 are inhibited by monoclonal antibodies specific for the IL-4Rα chain [56,88].

A model for the IL-4 and IL-13 receptors suggested by these experiments is shown in Figure 19.2. According to this model, IL-4Rα is a component of the IL-4 receptor and the IL-13 receptor. A similar conclusion has been reached recently by Zurawski et al. [89] and Lin et al. [56]. The classical (type I) IL-4 receptor is formed by association of the IL-4Rα chain with the γc chain and does not bind IL-13 [86]. In addition, the IL-4R α chain can associate with the Mr 70K IL-13 binding protein that has been recently described and found to be expressed on several cell types [60,61] to form a functional IL-13R which does not include the γc chain. Recent experiments showing that an antibody to the mouse γc chain inhibits responses to IL-4 but not to IL-13 also show that the γc chain is not shared by the IL-13 receptor [90].

The proposed structures for the IL-4 and IL-13 receptors have several important functional implications. The most significant of these is that IL-4 can bind to and activate B cells through two different receptors. One of these is the high affinity IL-4R$\alpha/\gamma c$ chain (type I) receptor and the other is the IL-13 receptor complex made up of the IL-4Rα and the IL-13 binding chain (the type II IL-4 receptor). There is no indication that an IL-4Rα dimer could be a functional receptor for IL-4R. Previous studies have indicated the existence of more than one type of IL-4 receptor on human lymphocytes. For example, high and low affinity IL-4 binding sites on human lymphoid cells have been described [91]. Furthermore, functional experiments have shown that 50-fold higher concentrations of IL-4 are required for increased expression of surface CD23 than surface IgM on human B cells consistent with differential binding and activation of high and low affinity receptors [92]. Two signalling pathways activated by IL-4 have also been described [93]. In these experiments, high concentrations of IL-4 were found to activate a unique signal transduction pathway in human B cells characterized by rapid and transitory hydrolysis of phosphatidylinositol bisphosphate (PIP$_2$) and calcium mobilization followed after a short lag period by an increase in cytoplasmic cAMP [93]. A recent study has shown that IL-13 activation of human monocytes also triggers this signalling pathway [94], and we have found

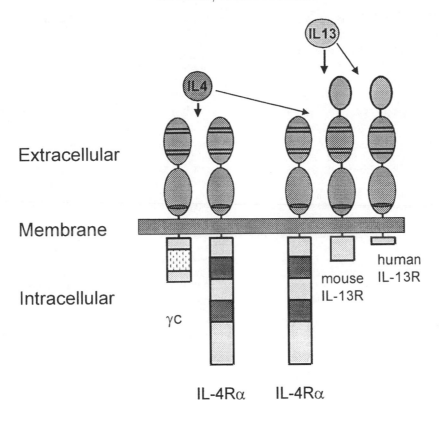

Extracellular

Membrane

Intracellular

γc

human
IL-13R

mouse
IL-13R

IL-4Rα   IL-4Rα

key   cytokine receptor domain   acidic region

fibronectin III domain   SH2 like homology domain

immunoglobulin like domain

Figure 19.2 A model for the IL-4R and IL-13R complexes

that IL-13 stimulates IP$_3$ production and increased intracellular cAMP in B cells (Hibbert *et al.*, in preparation). On the other hand, IL-4 binding to the high affinity receptor (IL-4Rα/γc chain) activates protein tyrosine kinases including Jak1 and Jak3 and phosphorylation of 4PS (IRS-2) [40,59,95–97]. JAK1 forms complexes with the IL-4R α chain and the 4PS protein [40] which has recently been cloned and identified as IRS-2

[96] whereas Jak3 is associated with the γc chain [39,41,97]. The recent finding that IL-13 activates JAK1 kinase but not JAK3 [59] and that both IL-4 and IL-13 induce phosphorylation of IL-4Rα [98] and activation of STAT6 [57,59] also implicates IL-4Rα but not the γc chain in the IL-13R. This model for the IL-13 receptor explains the many overlapping biological properties of IL-4 and IL-13. It also has major implications for the immuno-

deficiency in SCIDX1 and in construction of receptor antagonists to control IL-4 and IL-13 mediated responses such as IgE production in asthma and allergy.

REFERENCES

1. Kinnon C. Prospects for gene therapy of X-linked immunodeficiency diseases. In: Latchman DS (ed.) *From Genetics to Gene Therapy*. Oxford: Bios Scientific Publishers, 1994: 25–44.
2. Fischer A. Severe combined immunodeficiencies. *Immunodef. Rev.* 1992; **3**: 83–100.
3. Conley ME. Molecular approaches to analysis of X-linked immunodeficiencies. *Ann. Rev. Immunol.* 1992; **10**: 215–238.
4. Conley ME, Lavoie A, Briggs C, Brown P, Guerra C and Puck JM. Nonrandom X chromosome inactivation in B cells from carriers of X chromosome-linked severe combined immunodeficiency. *Proc. Natl. Acad. Sci. USA* 1988; **85**: 3090–3094.
5. Goodship J, Malcolm S, Lau YL, Pembrey ME, and Levinsky RJ. Use of X-chromosome inactivation analysis to establish carrier status for X-linked severe combined immunodeficiency. *Lancet* 1988; **1**: 729–732.
6. de Saint Basile G, Arveiler B, Oberle I, Malcolm S, Levinsky RJ, Lau YL, Hofker M, Debre M, Fischer A, Griscelli C and Mandel JL. Close linkage of the locus for X-linked severe combined immunodeficiency to polymorphic DNA markers in Xq11-q13. *Proc. Natl. Acad. Sci. USA* 1987; **84**: 7576–7579.
7. Puck JM, Conley ME and Bailey LC. Refinement of linkage of human severe combined immunodeficiency (SCIDX1) to polymorphic markers in Xq13. *Am. J. Hum. Genet.* 1993; **53**: 176–184.
8. Markiewicz S, DiSanto JP, Chelly J, Fairweather N, Le Marec B, Griscelli C, Graeber MB, Muller U, Fischer A, Monaco AP *et al.* Fine mapping of the human SCIDX1 locus at Xq12–13.1. *Hum. Mol. Genet.* 1993; **2**: 651–654.
9. Noguchi M, Yi H, Rosenblatt HM, Filipovich AH, Adelstein S, Modi WS, McBride OW and Leonard WJ. Interleukin-2 receptor gamma chain mutation results in X-linked severe combined immunodeficiency in humans. *Cell* 1993; **73**: 147–157.
10. Puck JM, Deschenes SM, Porter JC, Dutra AS, Brown CJ, Willard HF and Henthorn PS. The interleukin 2 receptor gamma chain maps to Xq13.1 and is mutated in X-linked severe combined immunodeficiency, SCIDX1. *Hum. Mol. Genet.* 1993; **2**: 1099–1104.
11. Voss SD, Hong R and Sondel PM. Severe combined immunodeficiency, interleukin 2 (IL-2) and the IL-2 receptor: experiments of nature continue to point the way. *Blood* 1994; **83**: 626–635.
12. Nakamura M, Asao H, Takeshita T and Sugamura K. The interleukin 2 receptor heterotrimer complex and intracellular signalling. *Semin. Immunol.* 1993; **5**: 309–317.
13. Minami Y, Kono T, Miyazaki T and Taniguchi T. The IL-2 receptor complex: its structure, function and target genes. *Ann. Rev. Immunol.* 1993; **11**: 245–267.
14. Giri JG, Kumaki S, Ahdieh M, Friend DJ, Loomis A, Shanebeck K, DuBose R, Cosman D, Park LS and Anderson DM. Identification and cloning of a novel IL-15 binding protein that is structurally related to the alpha chain of the IL-2 receptor. *EMBO J.* 1995; **14**: 3654–3663.
15. Eck MJ, Ultsch M, Rinderknecht E, De-Vos AM and Sprang SR. The structure of human lymphotoxin (tumour necrosis factor beta) at 1.9 A resolution. *J. Biol. Chem.* 1992; **267**: 2119–2122.
16. de Vos AM, Ultsch M and Kossiakoff AA. Human growth hormone and extracellular domain of its receptor: crystal structure of the complex. *Science* 1992; **255**: 306–312.
17. Nakamura Y, Russell SM, Mess SA, Friedmann M, Erdos M, Francois C, Jacques Y, Adelstein S and Leonard WJ. Heterodimerisation of the IL-2 receptor beta and gamma cytoplasmic domains is required for signalling. *Nature* 1994; **369**: 330–333.
18. Zurawski SM, Vega F, Doyle EL, Huyghe B, Flaherty K, McKay DB and Zurawski G. Definition and spatial location of mouse interleukin-2 residues that interact with its heterotrimeric receptor. *EMBO J.* 1993; **12**: 5113–5119.
19. Bamborough P, Hedgecock CJR and Richards WG. The interleukin-2 and interleukin-4 receptors studied by molecular modelling. *Structure* 1994; **2**: 839–851.
20. Horak ID, Gress RE, Lucas PJ, Horak EM, Waldmann TA and Bolen JB. T lymphocyte interleukin 2 dependent tyrosine protein kinase signal transduction involves the activation of p56lck. *Proc. Natl. Acad. Sci. USA* 1991; **88**: 1996–2000.
21. Torigoe T, Saragovi HU and Reed JC. Interleukin 2 regulates the activity of the Lyn protein tyrosine kinase in a B cell line. *Proc. Natl. Acad. Sci. USA* 1992; **89**: 2674–2678.
22. Turner B, Rapp U, App H, Greene M, Dobashi K and Reed J. Interleukin 2 induces tyrosine phosphorylation and activation of p72–74 Raf-1 kinase in a T cell line. *Proc. Natl. Acad. Sci. USA* 1991; **88**: 1227–1231.
23. Evans SW and Farrar WL. Interleukin 2 and diacylglycerol stimulate phosphorylation of 40S ribosomal S6 protein. Correlation with increased protein synthesis and S6 kinase activation. *J. Biol. Chem.* 1987; **262**: 4624–4630.
24. Asao H, Takeshita T, Nakamura M, Nagata K and Sugamura K. Interleukin 2 (IL-2)-induced tyrosine phosphorylation of IL-2 receptor p75. *J. Exp. Med.* 1990; **171**: 637–644.
25. Asao H, Kumaki S, Takeshita T, Nakamura M and Sugamura K. IL-2 dependent in vivo and in vitro tyrosine phosphorylation of IL-2 receptor gamma chain. *FEBS Letts.* 1992; **304**: 141–145.
26. Satoh T, Nakafuku M, Miyajima A and Kaziro Y. Involvement of ras p21 protein in signal-transduction pathways from interleukin 2, interleukin 3, and granulocyte/macrophage colony-stimulating factor, but not from interleukin 4. *Proc. Natl. Acad. Sci. USA* 1991; **88**(8): 3314–3318.
27. Satoh T, Minami Y, Kono T, Yamada K, Kawahara A, Taniguchi T and Kaziro Y. Interleukin 2 induced activation of Ras requires two domains of interleukin 2 receptor beta subunit, the essential region for growth stimulation and

Lck binding domain. *J. Biol. Chem.* 1992; **267**: 25423–25427.

28. Merida M, Diez E and Gaulton GN. IL-2 binding activates a tyrosine-phosphorylated phosphatidylinositol-3-kinase. *J. Immunol.* 1991; **147**: 2202–2207.

29. Remillard B, Petrillo R, Maslinski W, Tsudo M, Strom TB, Cantley L and Varticovski L. Interleukin-2 receptor regulates activation of phosphatidylinositol 3-kinase. *J. Biol. Chem.* 1991; **266**(22): 14167–14170.

30. Eardley DD and Koshland ME. Glycosylphosphatidylinositol: a candidate system for interleukin-2 signal transduction. *Science* 1991; **251**: 78–81.

31. Reed JC, Sabath DE, Hoover RG, and Prystowsky MB. Recombinant interleukin 2 regulates levels of c-myc mRNA in a cloned murine T lymphocyte. *Mol. Cell. Biol.* 1985; **5**: 3361–3368.

32. Stern JB and Smith KA. Interleukin 2 induction of T cell G1 progression and c-myb expression. *Science* 1986; **233**: 203–206.

33. Farrar WL, Cleveland JL, Beckner SK, Bonvini E and Evans SW. Biochemical and molecular events associated with interleukin 2 regulation of lymphocyte proliferation. *Immunol. Rev.* 1986; **92**: 49–65.

34. Shibuya H, Yoneyama M, Ninomiya-Tsuji J, Matsumoto K and Taniguchi T. IL-2 and EGF receptors stimulate the haematopoietic cell cycle via different signalling pathways: demonstration of a novel role for c-myc. *Cell* 1992; **70**: 57–67.

35. Minami Y, Kono T, Yamada K, Kobayashi N, Kawahara A, Perlmutter RM and Taniguchi T. Association of p56^lck with IL-2 receptor beta chain is critical for the IL-2 induced activation of p56^lck. *EMBO J.* 1993; **12**: 759–768.

36. Hatakeyama M, Kono T, Kobayashi N, Kawahara A, Levin SD, Perlmutter RM and Taniguchi T. Interaction of the IL-2 receptor with the src-family kinase p56lck: identification of novel intermolecular association. *Science* 1991; **252**(5012): 1523–1528.

37. Kobayashi N, Kono T, Hatakeyama M, Minami Y, Miyazaki T, Perlmutter RM and Taniguchi T. Functional coupling of the src- family protein tyrosine kinases p59fyn and p53/56lyn with the interleukin 2 receptor: implications for redundancy and pleiotropism in cytokine signal transduction. *Proc. Natl. Acad. Sci. USA* 1993; **90**: 4201–4205.

38. Hatakeyama M, Mori H, Doi T and Taniguchi T. A restricted cytoplasmic region of IL-2 receptor beta chain is essential for growth signal transduction but not for ligand binding and internalisation. *Cell* 1989; **59**: 837–845.

39. Miyazaki T, Kawahara A, Fujii H, Nakagawa Y, Minami Y, Liu Z-J, Oishi I, Silvennoinen O, Witthuhn BA, Ihle JN and Taniguchi T. Functional activation of Jak1 and Jak3 by selective association with IL-2 receptor subunits *Science* 1994; **266**: 1045–1047.

40. Yin T, Tsang ML and Yang YC. JAK1 kinase forms complexes with interleukin 4 receptor and 4PS/insulin receptor substrate 1 like protein and is activated by interleukin 4 and interleukin 9 in T lymphocytes. *J. Biol. Chem.* 1994; **269**: 26614–26617.

41. Russell SM, Johnston JA, Noguchi M, Kawamura M, Bacon CM, Friedmann M, Berg M, McVicar DW, Witthuhn BA, Silvennoinen O, Goldman AS, Schmalsteig FC, Ihle JN, O'Shea JJ and Leonard WJ. Interactions of IL-2 beta and gamma chains with Jak1 and Jak3: Implications for XSCID and XCID. *Science* 1994; **266**: 1042–1045.

42. Ihle JN and Kerr IM. Jaks and Stats in signalling by the cytokine receptor superfamily. *Trends Genet.* 1995; **11**: 69–74.

43. Johnston JA, Bacon CM, Finbloom DS, Rees RC, Kaplan D, Shibuya K, Ortaldo JR, Gupta S, Chen YQ, Giri JG and O'Shea JJ. Tyrosine phosphorylation and activation of STAT5, STAT3 and Janus kinases by interleukins 2 and 15. *Proc. Natl. Acad. Sci. USA* 1995; **92**: 8705–8709.

44. Kawahara A, Minami Y and Taniguchi T. Evidence for a critical role for the cytoplasmic region of the interleukin 2 (IL-2) receptor gamma-chain in IL-2, IL-4 and IL-7 signalling. *Mol. Cell. Biol.* 1994; **14**: 5433–5440.

45. Asao H, Takeshita T, Ishii N, Kumaki S, Nakamura M and Sugamura K. Reconstitution of functional interleukin 2 receptor complexes on fibroblastoid cells: involvement of the cytoplasmic domain of the gamma chain in two distinct signalling pathways. *Proc. Natl. Acad. Sci. USA* 1993; **90**: 4127–4131.

46. Macchi P, Villa A, Giliani S, Sacco MG, Frattini A, Porta F, Ugazio AG, Johnston JA, Candotti F, O'Shea JJ, Vezzoni P and Notarangelo L. Mutations of Jak3 gene in patients with autosomal severe combined immunodeficiency (SCID). *Nature* 1995; **377**: 65–68.

47. Takeshita T, Asao H, Ohtani K, Ishii N, Kumaki S, Tanaka N, Munakata H, Nakamura M and Sugamura K. Cloning of the gamma chain of the human IL-2 receptor. *Science* 1992; **257**: 379–382.

48. Kondo M, Takeshita T, Ishii N, Nakamura M, Watanabe S, Arai K and Sugamura K. Sharing of the interleukin-2 (IL-2) receptor gamma chain between receptors for IL-2 and IL-4. *Science* 1993; **262**: 1874–1877.

49. Russell SM, Keegan AD, Harada N, Nakamura Y, Noguchi M, Leland P, Friedmann MC, Miyajima A, Puri RK, Paul WE and Leonard WJ. Interleukin-2 receptor gamma chain: a functional component of the interleukin-4 receptor. *Science* 1993; **262**: 1880–1883.

50. Kondo M, Takeshita T, Higuchi M, Nakamura M, Sudo T, Nishikawa S-I and Sugamura K. Functional participation of the IL-2 receptor gamma-chain in IL-7 receptor complexes. *Science* 1994; **263**: 1453–1454.

51. Kimura Y, Takeshita T, Kondo M, Ishii N, Nakamura M, Van Snick J and Sugamura K. Sharing of the IL-2 receptor gamma chain with the functional IL-9 receptor complex. *Int. Immunol.* 1995; **7**: 115–120.

52. Giri JG, Ahdieh M, Eisenman J, Shanebeck K, Grabstein K, Kumaki S, Namen A, Park LS, Cosman D and Anderson D. Utilization of the beta and gamma chains of the IL-2 receptor by the novel cytokine IL-15. *EMBO J.* 1994; **13**: 2822–2830.

53. Noguchi M, Nakamura Y, Russell SM, Ziegler SF, Tsang M, Cao X and Leonard WJ. Interleukin-2 receptor gamma chain: a functional component of the interleukin-7 receptor. *Science* 1993; **262**: 1877–1880.

54. Matthews DJ, Clark PA, Herbert J, Morgan G, Armitage RJ, Kinnon C, Minty A, Grabstein KH, Caput D, Ferrara P and Callard RE. Function of the IL-2 receptor gamma

chain in biological responses of X-SCID B cells to IL-2, Il-4, IL-13 and IL-15. *Blood* 1995; **85**: 38–42.

55. Yin T, Yang L and Yang Y-C. Tyrosine phoshorylation and activation of Jak family tyrosine kinases by interleukin 9 in MO7E cells. *Blood* 1995; **85**: 3101–3106.

56. Lin J-X, Migone T-S, Tsang M, Friedmann M, Weatherbee JA, Zhou L, Yamauchi A, Bloom ET, Mietz J, John S and Leonard WJ. The role of the shared receptor motifs and common Stat proteins in the generation of cytokine pleiotropy and redundancy by IL-2, IL-4, IL-7, IL-13 and IL-15. *Immunity* 1995; **2**: 331–339.

57. Hou J, Schindler U, Henzel WJ, Ho TC, Brasseur M and McKnight SL. An interleukin 4 induced transcription factor: IL-4 stat. *Science* 1994; **265**: 1701–1706.

58. Callard RE, Matthews DJ and Hibbert L. *Immunol. Today* (in press).

59. Welham M, Learmonth L, Bone H and Schrader JW. Interleukin 13 signal transduction in lymphohaematopoietic cells. *J. Biol. Chem.* 1995; **270**: (in press).

60. Obiri NI, Debinski W, Leonard WJ and Puri RK. Receptor for interleukin 13: interaction with interleukin 4 by a mechanism that does not involve the common gamma chain shared by receptors for interleukins 2,4,7,9 and 15. *J. Biol. Chem.* 1995; **270**: 8797–8804.

61. Vita N, Lefort S, Laurent P, Caput D and Ferrara P. Characterisation and comparison of the interleukin 13 receptor with the interleukin 4 receptor on several cell types. *J. Biol. Chem.* 1995; **270**: 3512–3517.

62. Stahl N, Farruggella TJ, Boulton TG, Zhong Z, Darnell JE and Yancopoulos GD. Choice of STATs and other substrates specified by modular tyrosine-based motifs in cytokine receptors. *Science* 1995; **267**: 1349–1353.

63. Heim MH, Kerr IM, Stark GR and Darnell JE. Contribution of STAT SH2 groups to specific interferon signalling by the Jak-STAT pathway. *Science* 1995; **267**: 1347–1349.

64. Nelson BH, Lord JD and Greenberg PD. Cytoplasmic domains of the interleukin 2 receptor beta and gamma chains mediate the signal for T cell proliferation. *Nature* 1994; **369**: 333–336.

65. DiSanto JP, Dautry-Varsat A, Certain S, Fischer A and de Saint Basile G. Interleukin-2 (IL-2) receptor gamma chain mutations in X-linked severe combined immunodeficiency disease result in the loss of high affinity IL-2 receptor binding. *Eur. J. Immunol.* 1994; **24**: 475–479.

66. Ishii N, Asao H, Kimura Y, Takeshita T, Nakamura M, Tsuchiya S, Konno T, Maeda M, Uchiyama T and Sugamura K. Impairment of ligand binding and growth signalling of mutant IL-2 receptor gamma-chains in patients with X-linked severe combined immunodeficiency. *J. Immunol.* 1994; **153**: 1310–1317.

67. DiSanto JP, Le Deist F, Caniglia M, Markiewicz S, Lebranchu Y, Griscelli C, Fischer A and de Saint Basile G. Variant forms of X-linked severe combined immunodeficiency disease: one or many genes. *Immunodeficiency* 1993; **4**: 253–258.

68. Brooks EG, Schmalstieg FC, Wirt DP, Rosenblatt HM, Adkins LT, Lookingbill DP, Rudloff HE, Rakusan TA and Goldman AS. A novel X-linked combined immunodeficiency disease. *J. Clin. Invest.* 1990; **86**: 1623–1631.

69. Schmalstieg FC, Leonard WJ, Noguchi M, Berg M, Rudloff HE, Denney RM, Dave SK, Brooks EG and Goldman AS. Missense mutation in exon 7 of the common gamma chain gene causes a moderate form of X-linked combined immunodeficiency. *J. Clin. Invest.* 1995; **95**: 1169–1173.

70. He Y-W, Adkins B, Furse RK and Malek TR. Expression and function of the common gamma chain subunit of the IL-2, IL-4 and IL-7 receptors. *J. Immunol.* 1995; **154**: 1596–1605.

71. DiSanto JP, Rieux-Laucat F, Dautry-Varsat A, Fischer A and de Saint Basile G. Defective human interleukin 2 receptor gamma chain in an atypical X chromosome linked severe combined immunodeficiency with peripheral T cells. *Proc. Natl. Acad. Sci. USA* 1994; **91**: 9466–9470.

72. de Saint Basile G, Le Deist F, Caniglia M, Lebranchu Y, Griscelli C and Fischer A. Genetic study of a new X-linked recessive immunodeficiency syndrome. *J. Clin. Invest.* 1992; **89**: 861–866.

73. DiSanto JP, Muller W, Guy-Grand D, Fischer A and Rajewsky K. Lymphoid cell development in mice with a targeted deletion of the interleukin 2 receptor gamma chain. *Proc. Natl. Acad. Sci. USA* 1995; **92**: 377–381.

74. Cao X, Shores EW, Hu-Li J, Anver MR, Kelsall BL, Russell SM, Drago J, Noguchi M, Grinberg A, Bloom ET, Paul WE, Katz SI, Love PE and Leonard WJ. Defective lymphoid development in mice lacking expression of the common cytokine receptor gamma chain. *Immunity* 1995; **2**: 223–238.

75. DiSanto JP, Keever CA, Small TN, Nichols GL, O'Reilly RJ and Flomenberg N. Absence of interleukin 2 production in a severe combined immunodeficiency disease syndrome with T cells. *J. Exp. Med.* 1990; **171**: 1697–1704.

76. Schorle H, Holtschke T, Hunig T, Schimpl A and Horak I. Development and function of T cells in mice rendered interleukin-2 deficient by gene targeting. *Nature* 1991; **352**: 621–624.

77. Kuhn R, Rajewsky K. and Muller W. Generation and analysis of interleukin 4 deficient mice. *Science* 1991; **254**: 707–710.

78. Sadlack B, Kuhn R, Schorle H, Rajewsky K, Muller W and Horak I. Development and proliferation of lymphocytes in mice deficient in both interleukins 2 and 4. *Eur. J. Immunol.* 1994; **24**: 281–284.

79. Mehler MF, Rozental R, Dougherty M, Spray DC and Kessler JA. Cytokine regulation of neuronal differentiation of hippocampal progenitor cells. *Nature* 1993 **362**: 62–65.

80. Armitage RJ, Ziegler SF, Friend DJ, Park LS and Fanslow WC. Identification of a novel low-affinity receptor for human interleukin 7. *Blood* 1992; **79**: 1738–1745.

81. Peschon JJ, Morrissey PJ, Grabstein KH, Ramsdell FJ, Maraskovsky E, Gliniak BC, Park LS, Ziegler SF, Williams DE, Ware CB, Meyer JD and Davison BL. Early lymphocyte expansion is severely impaired in interleukin 7 receptor deficient mice. *J. Exp. Med.* 1994; **180**: 1955–1960.

82. Duschl A. An antagonistic mutant of interleukin 4 fails to recruit the common gamma chain into the receptor complex. *Eur. J. Biochem.* 1995; **228**: 305–310.

83. Zurawski SM and Zurawski G. Receptor antagonist and selective agonist derivatives of mouse interleukin 2. *EMBO J.* 1992; **11**: 3905–3910.

84. Zurawski SM, Imler J-L and Zurawski G. Partial agonist/antagonist mouse interleukin-2 proteins indicate that a third component of the receptor complex functions in signal transduction. *EMBO J.* 1990; **9**: 3899–3905.

85. Aversa G, Punnonen J, Cocks BG, de Waal Malefyt R, Vega F, Zurawski SM, Zurawski G and De Vries JE. An interleukin 4 (IL- 4) mutant protein inhibits both IL-4 or IL-13 induced human immunoglobulin G4 (IgG4) and IgE synthesis and B cell proliferation: support for a common component shared by IL-4 and IL-13 receptors. *J. Exp. Med.* 1993; **178**: 2213–2218.

86. Zurawski SM, Vega F, Huyghe B and Zurawski G. Receptors for interleukin 13 and interleukin 4 are complex and share a novel component that functions in signal transduction. *EMBO J.* 1993; **12**: 2663–2670.

87. Matthews DJ *et al*. Submitted for publication.

88. Renard N, Duvert V, Banchereau J and Saeland S. Interleukin-13 inhibits the proliferation of normal and leukaemic human B cell precursors. *Blood* 1994; **84**: 2253–2260.

89. Zurawski SM, Chomarat P, Djossou O, Bidaud C, McKenzie ANJ, Miossec P, Banchereau J and Zurawski G. The primary binding subunit of the human interleukin 4 receptor is also a component of the interleukin 13 receptor. *J. Biol. Chem.* 1995; **270**: 13869–13878.

90. He Y-W and Malek TR. The IL-2 receptor common gamma chain does not function as a subunit shared by the IL-4 and IL-13 receptors. *J. Immunol.* 1995; **155**: 9–12.

91. Foxwell BMJ, Woerly G and Ryffel B. Identification of interleukin 4 receptor-associated proteins and expression of both high and low affinity binding on human lymphoid cells. *Eur. J. Immunol.* 1989; **19**; 1637–1641.

92. Shields JG, Armitage RJ, Jamieson BN, Beverley PCL and Callard RE. Increased expression of surface IgM but not IgD or IgG on human B cells in response to interleukin 4. *Immunology*, 1989; **66**: 224–227.

93. Rigley KP, Thurstan SM and Callard RE. Independent regulation of interleukin 4 (IL-4) induced expression of human B cell surface CD23 and IgM: functional evidence for two IL-4 receptors. *Int. Immunol.* 1991; **3**: 197–203.

94. Sozzani P, Cambon C, Vita N, Seguelas M-H, Caput D, Ferrara P and Pipy B. Interleukin 13 inhibits protein kinase C triggered respiratory burst in human monocytes: role of calcium and cyclic AMP. *J. Biol. Chem.* 1995; **270**: 5084–5088.

95. Wang L, Keegan AD, Li W, Lienhard GE, Pacini S, Gutkind JS, Myers MG, Sun X, White MF, Aaronson SA, Paul WE and Pierce JH. Common elements in interleukin 4 and insulin signalling pathways in factor dependent haematopoietic cells. *Proc. Natl. Acad. Sci. USA* 1993; **90**: 4032–4036.

96. Sun XJ, Wang L-M, Zhang Y, Yenush L, Myers MG, Glasheen E, Lane WS, Pierce JH, and White MF. Role of IRS-2 in insulin and cytokine signalling. *Nature* 1995; **377**: 173–177.

97. Musso T, Johnston JA, Linnekin D, Varesio L, Rowe TK, O'Shea JJ and McVicar DW. Regulation of JAK3 expression in human monocytes: phosphorylation in response to interleukins 2, 4, and 7. *J. Exp. Med.* 1995; **181**: 1425–1431.

98. Smerz-Bertling C and Duschl A. Both interleukin 4 and interleukin 13 induce tyrosine phosphorylation of the 140 kDa subunit of the IL-4 receptor. *J. Biol. Chem.* 1995; **270**: 966–970.

99. Markiewicz S, Subtil A, Dautry-Varsat A, Fischer A and De SBG. Detection of three nonsense mutations and one missense mutation in the interleukin 2 receptor gamma chain gene in SCIDX1 that differently affect mRNA processing. *Genomics* 1994; **21**: 291–293.

100. Clark PA, Lester T, Genet S, Jones AM, Hendriks R, Levinsky RJ and Kinnon C. Screening for mutations causing X-linked severe combined immunodeficiency in the IL-2R gamma chain by single strand comformation polymorphism analysis. *Hum. Genet.* 1995; **96**: 427–432.

101. Tassara C, Pepper AE and Puck JM. Intronic point mutation in the IL-2Rγ gene causing X-linked severe combined immunodeficiency. *Hum. Mol. Genet.* 1995; **4**: 1693–1695.

102. Puck JM, Pepper AE, Bedard P-M and Laframboise R. Female germ line mosaicism as the origin of a unique IL-2 receptor gamma chain mutation causing X-linked severe combined immunodeficiency. *J. Clin. Invest.* 1995; **95**: 895–899.

20

T Cell Activation Deficiencies and Disease

Alfredo Corell, Carlos Rodríguez Gallego and Antonio Arnaiz-Villena

IMMUNOLOGÍA, Hospital Universitario '12 de Octubre', Universidad Complutense, Carretera Andalucía, Madrid, Spain

INTRODUCTION

Recognition of foreign antigen by the antigen receptor and accessory molecules on T cells stimulates a range of rapid intracellular signalling events leading to T cell activation and, ultimately, preprogrammed effector functions. In this way, helper T cells are induced to secrete factors that activate macrophages or help B lymphocytes to synthesize specific antibodies, and cytotoxic T cells are induced to secrete factors that lyse infected or neoplastic cells. Therefore, T cell activation encompasses a complex, finely tuned chain of biochemical events that requires the participation of many molecules, both at the cell surface and within the cell. The recognition of T cell activation deficiencies, primary as well as secondary, is growing in human pathology. In this chapter, primary T cell defects, both structural and functional, are described in detail and T cell defects that are associated with other disorders are discussed.

The cellular signal transduction apparatus that is activated upon antigen recognition is complex and involves various enzymes (e.g. kinases, phosphatases) and adaptor molecules which initiatiate new protein–protein interactions and cellular signalling events, leading to the translocation of cytoplasmic transcription factors into the nucleus and activation of nuclear factors culminating in the transcription of new genes, cytokine secretion, T cell proliferation and the initiation of an immune response. A defect in any one of these signalling steps may result in immunodeficiency. The possible mechanisms and molecules known to be involved in this multistep process [1, 2] are summarized in Figure 20.1.

In a small number of human primary immunodeficiencies, careful analysis of T cell phenotype, function, biochemistry and molecular biology has allowed us to diagnose a T cell defect and to define the gene lesion causing the primary protein defect. These we call structural T cell defects. Other primary immunodeficiencies show clear involvement of T lymphocyte activation defects, and although the precise molecular lesions are as yet unknown, no other concomitant disease can be diagnosed. Such immunodeficiencies we term functional T cell defects. The effects of these defects in the normal development and function of the adaptive immune system [3] are depicted in Figure 20.2.

The inheritance status of human primary T cell disorders is heterogeneous: some of them are autosomal recessive, but others are X-linked. There are at least three primary T deficiencies which segregate with the X chromosome (WAS, SCID and hyper IgM syndrome) and the precise locations of the loci responsible for these diseases have been recently identified and are indicated on Figure 20.3. Reports that other B cell deficiencies (XLA) and phagocyte deficiencies (CGD) are also located on the X chromosome have led to the hypothesis that an X-linked gene family is involved in immune system differentiation [4].

EXPLORATION OF T CELL DEFECTS IN THE LABORATORY

A growing list of T cell deficiencies, both primary and secondary, is emerging. However, the rapid advances in T cell biology have allowed the development of a multitude of laboratory tests to assess T cell function in man. Unfortunately, the

Lymphocyte Signalling: Mechanisms, Subversion and Manipulation. Edited by M. M. Harnett and K. P. Rigley © 1997 John Wiley & Sons Ltd.

374

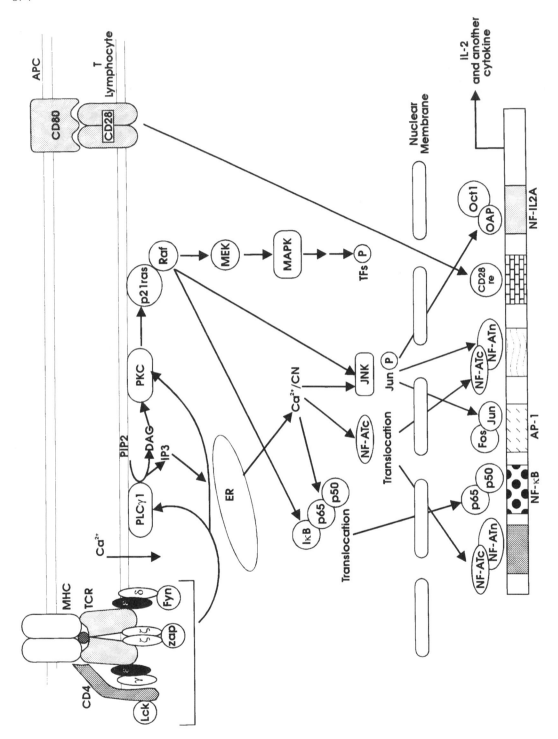

Figure 20.1 Signal transduction events occurring upon T cell activation. PLC, phospholipase-C; ZAP-70, ζ-associated proteins; Lck, Fyn, protein tyrosine kinases; Ras, GTPase, Raf ser/thr kinase, Jun, Fos, transcription factors; DAG, diacylglycerol; PI, phosphatidylinositol; ER, endoplasmic reticulum; JNK, Jun kinase; NF-AT, NF-κB, AP-1, Oct-1, OAP, nuclear activation factors; IkB, NF-κB inhibitor; CD28re, CD28 response element; PKC, protein-kinase C

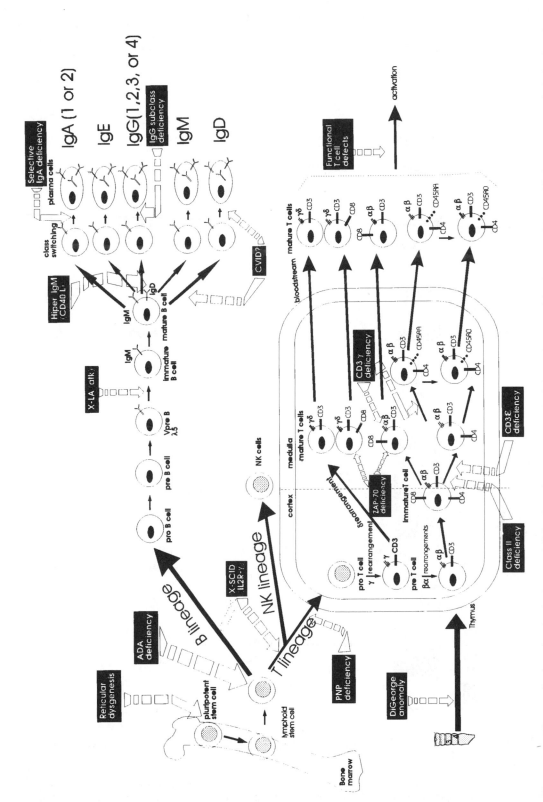

Figure 20.2 The contribution of molecular immunodeficiencies to immune system physiology. The normal development of the specific immune system is depicted, and the steps affected by several immunodeficiencies (in black boxes) are indicated by arrows. ADA, adenosine deaminase; X-SCID, X-linked severe combined immunodeficiency or interleukin 2 receptor (IL-2R) γ chain deficiency; X-LA, X-linked agammaglobulinaemia or tyrosine kinase (atk) agammaglobulinaemia deficiency; Hyper IgM, X-linked immunodeficiency with high IgG or CD40 ligand (CD40L) deficiency; CVID, common variable immunodeficiency; PNP, purine nucleoside phosphorylase

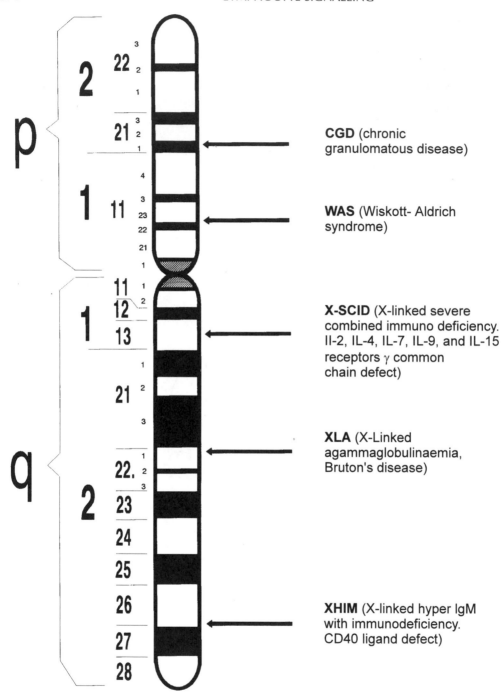

CGD (chronic granulomatous disease)

WAS (Wiskott- Aldrich syndrome)

X-SCID (X-linked severe combined immuno deficiency. Il-2, IL-4, IL-7, IL-9, and IL-15 receptors γ common chain defect)

XLA (X-Linked agammaglobulinaemia, Bruton's disease)

XHIM (X-linked hyper IgM with immunodeficiency. CD40 ligand defect)

Figure 20.3 X-linked immunodeficiencies. The precise location of the genes responsible for these five immunodeficiencies has been characterized in the last 8 years

complexity of some of the tests does not allow their use in all clinical settings. The first step,

obviously, is to ask whether there are any T cells to analyse and if the relative proportions of expected T cell subsets are within normal limits. This is mainly done by flow cytometry using relevant monoclonal antibodies (mAbs). If T cells are indeed present in normal numbers and subsets, the next question is: are they functional?

Phenotypic Analysis

Flow Cytometry

A total leukocyte count indicating the number and percentage of lymphocytes, monocytes and granulocytes is required. Secondly, the number and percentage of lymphocytes expressing a range of T cell surface glycoproteins can by determined by cytofluorometry. The particular combination of molecules borne by a group of cells describes its phenotype. T cells at particular stages of differentiation or with distinct functional capabilities show distinct phenotypes. The determination of cell phenotypes is carried out by using mAbs which bind specific molecules on the T cell surface. A system of classification has been agreed to assign CD (clusters of differentiation) numbers to each molecule. By single colour cytofluorometry [5], the fraction (%) and absolute number of lymphocytes expressing a CD marker can be determined, i.e. CD3 or CD8. However, no information can be obtained about cells expressing two or three CD molecules at a time, i.e. CD3–CD4–CD45RA, or one but not the other (CD8 but not CD3 in NK cells) by this simple method. This can be achieved, however, by two or three colour cytofluorometry, which defines two subsets [3] for each group of three mAbs [6].

In addition to determining the fraction and number of lymphocytes which express a given molecule or group of molecules, flow cytometers with fluorescence detectors can provide information about the relative numbers of molecules per cell. This is indicated by the mean fluorescence in arbitrary units. Thus, while CD8 molecules are expressed in very high numbers on the surface of T cells, termed "bright" expression, in NK cells, CD8 molecules are expressed in lower numbers per cell and termed "dull" expression.

Vβ Usage

mAbs are available which bind to variable regions of $\alpha\beta$ or $\gamma\delta$ T cell receptors. The relative distribution of such variable regions in normal individuals is fairly uniform. In certain diseases, however, dominant representation of a particular region can be observed and denotes an underlying autoimmune or tumoral disorder. This kind of analysis, however, has been quite limited until very recently due to the non-availability of appropriate mAbs and has led to the development of different strategies to study the normal or biased expression of the different TCR variable regions by means of PCR amplification and semi-quantitative analysis of the amplified cDNA [7].

T Cell Function

The above type of assays establish whether there are normal levels of T lymphocytes in a given patient and whether the subsets and molecules that should be there are indeed present. The next question to answer is whether they are functional. Most tests are performed on peripheral blood mononuclear cells (PBMCs), which contain mostly lymphocytes and monocytes. If a defect is found, purified T cells or T cell lines should be derived to confirm the findings. The following assays discriminate defective pathways from a range of perspectives starting with a overall view of the T cell system (skin tests) and finishing with the molecular dissection of the system components (biochemical tests).

Skin Tests

Intradermal skin testing measures delayed type hypersensitivity (DTH) reactions. These reactions estimate T cell functional status by using standard antigens to which most individuals are exposed (Candida, Clostridium, tetani, Corynebacterium diphtheriae, Streptococcus, Tuberculin, Thrichophyton, Proteus). At present, it is the only practical in vivo test of cell-mediated immunity. The percentage of positive reactions to common delayed-hypersensitivity antigens rises progressively from birth and anergy is associated with a wide range of diseases and pathological states. However, the list is so extensive that the isolated finding of skin anergy is of limited diagnostic value.

Proliferation to Lectins, Antigens and Antibodies

Plant lectins such as phytohaemagglutinin (PHA), concanavalin A (Con A) and pokeweed mitogen (PWM), have been used to study lymphocyte function, *in vitro*, for many years. They are still the most widely used and accessible reagents for activating patients' lymphocytes in routine diagnostic work. Their mode of action at a molecular level is not fully understood and to some extent they have been replaced by the use of mAbs to T cell activation antigens [3].

There are several classes of mAbs which are used in proliferation assays. The most widely used are anti-CD3 mAbs. Normal peripheral blood T cells and T cell clones which express the CD3 complex can be activated by anti-CD3 monoclonal antibodies. The technique has been successfully used in attempts to explore the trigger mechanism for the TCR/CD3 receptor complex, and at present is widely used for routine diagnostic studies. Additional activating agents, such as phorbol esters (PMA) or interleukins (IL-2), are often added to optimize the stimulus [8] or to map putative defects: because PMA directly activates PKC, and IL-2 is required after T cell activation, their use allows the definition of early or late T cell activation defects. Other mAbs used to explore more subtle features of T cell function include anti-CD2, anti-CD28, anti-CD26, anti-CD69 and anti-CD43 mAbs. For example, the combination of two CD2 monoclonals (anti-T11$_2$ and anti-T11$_3$) is strongly mitogenic and induces T cell proliferation [9]. Available evidence suggest that CD2-triggered mitogenesis is similar to that via the TCR/CD3 and it sets in motion the same autocrine cycle of increased IL-2R expression, endogenous IL-2 release and T cell proliferation [9]. It is also possible to analyse the CD2 activation pathway by adding suboptimal doses of PMA to non-mitogenic combinations of CD2 monoclonals.

Probably the most studied of the TCR "costimulatory" molecules is CD28. Treatment of mitogen or anti-TCR stimulated T cells with monoclonals directed against CD28 induces a substantial increase in T cell proliferation, primarily due to an increase in IL-2 synthesis. A CD28 response element has been localized on the IL-2 promoter region (Figure 20.1). Anti-CD28 by itself is not mitogenic, but addition of PMA results in a strong proliferation of PBMCs. Likewise, pre-stimulation of PKC by PMA is also required for costimulation via CD26 and CD69 to induce significant lymphocyte proliferation. Indeed, CD69 is not expressed by resting T cells and has to be induced with PMA. The costimulatory activity of CD26 and CD69 can also be observed using suboptimal doses of PHA or anti-CD3 antibodies to up-regulate expression of these activation antigens. In addition, signalling via CD43, which is the major sialoglycoprotein of mature T cells and thymocytes, enhances antigen-specific T cell activation. Several antibodies against CD43 have been shown to have costimulatory effects on T cells and induce cell clustering [10].

Classical antigens do not normally induce strong T cell proliferation *in vitro*, as the number of cells responding to a given antigen is too low to be detected. Also, the responding individual must have been previously exposed to the antigen to obtain a substantial response, and this is not always the case in clinical practice. This problem can be overcome by, for example, inducing significant T cell proliferation in normal donors after vaccination with tetanus toxoid followed by lymphocyte stimulation with the antigen extract. In contrast, the fraction of T cells that respond to alloantigens is quite high (up to 20%) and superantigens, by virtue of their affinity for particular Vβ chains of the TCR and conserved regions of HLA class II molecules, can bind and stimulate significant numbers of T cells irrespective of their specificity. Thus, testing for allogenic or superantigen responses is a good method to assay T cell functions.

Intracellular Calcium Induction [11] and Biochemical Tests

Three of the initial signalling events occurring following TCR/CD3 engagement are protein phosphorylation, phosphatidylinositol 4,5-bisphosphate (PIP$_2$) hydrolysis and Ca^{2+} flux (Figure 20.1): Ligation of the TCR stimulates tyrosine phosphorylation and activation of phospholipase C-γ1, leading to hydrolysis of PIP$_2$ to generate the intracellular messengers, inositol trisphosphate (IP$_3$) and diacylglycerol (DAG). IP$_3$ mobilizes intracellular stores of calcium and DAG is an activator of the ser/thr kinase, protein kinase C (PKC). These events can easily be measured *in vitro* if defects in early signal transduction events are suspected. For example, Ca^{2+} flux can be measured by loading cells with fluorescent molecules that are sensitive to Ca^{2+}–dye concentration (e.g. Indo-1, Fluo-3) and analysing the kinetics of Ca^{2+} mobilization by monitoring the fluorescence intensity of the Ca^{2+}– dye complex in the cell. In addition, TCR-coupled phospholipase C (PLC)-$\gamma 1$ activation can be quantitatively analysed by measuring the levels of IP_3 derived by hydrolysis of radiolabelled-phosphatidylinositol 4,5-bisphosphate (PIP_2).

As stated above, protein phosphorylation of multiple substrates on Ser/Thr or Tyr residues also occurs shortly after T cell activation. Abnormal phosphorylation may thus cause defects in T cell activation and immunodeficiency. Antibodies against phosphorylated tyrosines allow the screening of multiple substrates simultaneously by probing whole cell lysates after electrophoresis. Alternatively, a protein of interest may be immunoprecipitated from cells prelabelled with ^{32}P and its content of labelled phosphorus determined [12]. This analysis defines whether a normal pattern of phosphorylated molecules occurs upon T cell activation and provides primary evidence as to which molecules may be defective. In addition, it is also possible to test for specific protein kinase activities as the kinase activity of various PTKs can be measured by their ability to undergo autophosphorylation, phosphorylate endogeneous co-immunoprecipitated proteins, and phosphorylate exogenous synthetic substrates, *in vitro*. Fyn, Lck, ZAP-70, Csk, PKCs activities can all be tested and such analysis should reveal whether these kinases can be activated following T cell stimulation or if they are constitutively activated in non-activated T cells [13].

Finally, analysis of the normal induction of nuclear activating factors has been significantly facilitated in recent years by the development of the "gel retardation" or "gel mobility shift" assay. This rapid and simple technique is based on the separation of free DNA from DNA/protein complexes due to the differences in their electrophoretic mobilities in native polyacrylamide gels. Such analysis reveals whether activation factors such as NF-κB, NF-AT, AP-1, CD28re, etc., are properly synthesized, activated and/or translocated to the nucleus upon T cell activation [13].

Lymphokine Synthesis

Enzyme-linked immunosorbent assays (ELISA) and radio-immunoassays have been developed to measure IL-2 and most other known cytokines in recent years and have been used in the laboratory to test whether the appropriate induction of such cytokines occurs in T cells (or other cell lineages) upon stimulation with different mitogenic cocktails [14]. Alternatively, appropriate induction of such cytokines can now be tested by the semi-quantitative analysis of specific mRNAs amplification.

T Cell Death Analysis

Apoptosis is a form of programmed cell death shown to play a key role in the regulation of normal development and oncogenesis. Its hallmark biochemical feature is endonuclease activation, giving rise to internucleosomal DNA fragmentation. There are also characteristic morphologic changes, including chromatin condensation, nuclear fragmentation and shrinkage, and the formation of dense chromatin masses (apoptotic bodies). In apoptosis, nuclear changes are observed first, in contrast to the changes seen in necrosis, which usually begin with cell membrane damage. Because DNA fragmentation in apoptosis is internucleosomal, a characteristic "DNA ladder" can be visualized on gel electrophoresis as a series of fragments that are multiples of 180 to 200 bp. In contrast, the random, DNA cleavage in necrosis leads to a DNA smear on gel electrophoresis. There are a wide variety of assays to test apoptosis in T cells including the conventional DNA fragmentation assay, flow cytometric studies based on DNA content, or cell death detection by ELISA [15]. This kind of test is useful when abnormally high counts of T lymphocytes are observed in the absence of diagnosis of a lymphoproliferative syndrome as described in the recently characterized fas deficiency.

STRUCTURAL CELL DEFECTS AFFECTING T CELL ACTIVATION

MHC Class I Deficiency (TAP2 Deficiency)

The defective expression of class I molecules has been called "bare lymphocyte syndrome" and is characterized by abnormal HLA class I expression while HLA class II expression is normal [16]. Recently, a case of two brothers with HLA class I deficiency resulting from mutations in the transporter associated with antigen processing (TAP) protein was reported [17]. The peptide transporter is an heterodimer comprising the products of the genes TAP1 and TAP2, which are located in the MHC class II region. Peptides generated by proteolytic degradation in the cytosol are imported into the endoplasmic reticulum by the transporter for their association with class I molecules. In mutant cell lines, lacking the TAP transporter, the $\beta2$ microglobulins/class I heavy chain complexes do not acquire peptides and become inherently unstable: as a result they are inefficiently transported through the Golgi compartment and consequently few MHC class I molecules are expressed on the cell surface.

Primary Defect

While HLA serotyping found patients to be homozygous for class II expression, no HLA class I molecules were detected on the TAP⁻PBMCs. Also, cytofluorometric analysis revealed that class I expression in PBMC and EBV-transformed B cells derived from patients was 1–3% of that found in control cells. However, CD1a expression on APCs was normal in patients. Interestingly, class I mRNA was found to be present in normal amounts but isoelectric focusing revealed that most of the class I heavy chains remained unsialylated, suggesting protein instability or poor transport via the Golgi compartment. Consistent with a trafficking defect, TAP1 but not TAP2 protein was detected by protein immunoblot analysis in such cell extracts. These results suggested that the TAP2 protein is either missing or truncated and, indeed, a premature stop codon at amino acid 253 was found in the TAP2 gene in these patients. The finding that heterozygous brothers were normal

demonstrated that the TAP2 deficiency is a recessive disease.

Immunological and Clinical Features

Although class I expression defects may be lethal early, in some cases, such as the patients with TAP, deficiency may manifested later in life. Both patients with TAP deficiency suffered from chronic bacterial sinobronchial infections, but no viral infections were recorded. However, antibody titres in serum show that the patients had been infected by several viruses (herpes, measles, varicella, mumps, cytomegalovirus). Flow cytometry showed that while these patients had normal to elevated percentages of $CD4^+CD8^-$ T cells and NK cells, their $CD8^+CD4^-$ T cells were reduced but still present: 11% in one patient and 2% in the other. In the former case (16-year-old patient), high numbers of $CD4^+CD8^+$ double-positive T cells (10%) and $\gamma\delta$T cells (33%, one-third $CD8^+$) were also found. Functional analysis revealed a normal proliferative response in MLC tests suggesting that their $CD4^+$ T cells were normal. In addition, unlike the situation reported for TAP1 knockout mice, allogenic cytotoxicity activity from $CD8^+$ $\alpha\beta$T cells was observed in PBMC from the patient with highest numbers of $CD8^+$ T cells.

Several features are shared between TAP deficiency in humans and TAP⁻ [18] or $\beta2M^-$[19] deficient mice. Firstly, the number of $\alpha\beta$ $CD8^+$ T cells is low in mice and humans. However, $CD8^+$ T cells may appear in the spleen of $\beta2M^-$ mice after viral infections or rejection of skin grafts. Moreover, the eldest of the two brothers with TAP2 deficiency had a significant number (11%) of $CD8^+\alpha\beta$ T cells. Secondly, NK activity was found to be abnormal both in the $\beta2M$-deficient mouse and the TAP2-patients although, in the former, it cannot be assumed that the NK response is completely deficient. The fact that complete absence of NK cells seems to lead to severe viral infections suggests that NK cells may be partially functional in these patients. However, these observations regarding abnormal NK responses are in agreement with recent studies indicating an important role for class I molecules in regulating NK cells' activity [20]. Finally, the finding that number of $\gamma\delta$ T cells (including $\gamma\delta$ $CD8^+$ T cells) was normal or elevated in both the

TAP$^-$ patients and β2M$^-$ deficient mice suggested that CD1a, which was normally expressed in TAP$^-$ individuals, might be involved in the selection of $\gamma\delta$ T cells.

In conclusion, class I deficiency due to TAP2 mutations seems to be a progressive disease, depending on the clinical status following exposure to viral infections. Under conditions of mild infections, antibody responses or CD4$^+$ T cell mediated cytotoxicity may arise as a compensatory mechanism. However, severe viral infections might be fatal.

MHC Class II Deficiencies

HLA class II deficiency is an autosomal recessive disease characterized by low numbers of CD4$^+$ peripheral T cells, and humoral and cellular immunodeficiency [16, 21, 22].

Primary defect

HLA class II deficiency is a disease of gene regulation involving transacting regulatory factors located outside of chromosome 6 [23]. The mRNA of both the α and β chains of HLA-DR, DP and DQ cannot be detected in PBMC from patients [24]. Although clinically homogeneous, the disease is genetically heterogeneous, with three complementation groups described, A, B and C [25]. Patients from groups B and C exhibit a defect in the binding of a specific protein complex, RFX, to the X box motif of MHC class II promoters, while patients from group A (Figure 20.4) bind RFX normally [26]. Genetic complementation analysis of cell lines by cDNA libraries in expression vectors has revealed that patients from complementation group A have mutations in the HLA class II transactivator gene CIITA [27], located on chromosome 16 [28]. CIITA is required for both constitutive and inducible class II expression and is itself induced by IFN-γ. Patients from group C are mutated in the gene that encodes the large subunit of the heteroduplex, RFX5 [29, 30]. Although the gene mutated in group B patients has not been cloned yet, it is tempting to suggest that it encodes the other subunit of RFX5, since purified RFX heterodimers can correct MHC class II transcription *in vitro* in extracts from group B as well as group C patients [29, 30, 31].

Clinical and Immunological Features

All patients with MHC class II deficiency suffered from severe bacterial, fungal and viral infections that occurred within the first year of life. Protracted diarrhoea, malabsorption with osteoporosis and severe failure to thrive are also present. Some patients develop autoimmune cytopenias (anaemia, neutropenia and/or thrombopenia). Since the disease is associated with a high rate of early mortality, bone marrow transplantation (BMT) is considered the only treatment [3, 16, 21].

In class II deficient patients, HLA class II antigen expression is abnormal in all cells which normally express such antigens. Moreover, HLA class I expression is also reduced in some, but not all, patients. In all patients studied, however, correction of MHC class I expression was obtained by *in vitro* treatment of cells with IFN-γ whereas no correction of class II expression was observed [32]. In these patients, the number of peripheral CD4$^+$ T cells is reduced (5–33%), although patients with normal numbers of CD4$^+$ T cells are also reported. Other T cell markers (CD2, CD5, CD7, CD45RA and CD45RO) are expressed normally on T cells from such patients. Moreover, the proliferative response of PBMC to lectins, anti-CD3 mAbs, anti-CD2 mAbs and allogenic cells was generally normal. In contrast, the proliferative response to recall antigens (tetanus toxoid, diphtheria toxoid, PPD or *Candida albicans*) was generally absent. Regarding humoral immunity, while most of the patients are hypogammaglobulinaemic, normal immunoglobulin levels, and even elevated IgM levels, have been observed. However, whereas the synthesis of antibodies following vaccination with protein antigens was never observed, isohaemagglutinins were found to be normal in 50% of the cases. Moreover, some patients had antibodies to viruses or *Candida albicans* when chronically infected. Abnormal antigen presentation in MHC class II deficient patients accounts for these impaired cellular and humoral immune response.

As stated above, the immunodeficiency due to lack of MHC class II expression is a consequence of defective antigen presentation. T cells do not seem to be primarily affected *in vitro* in their functional capacity (as measured by proliferative responses) except in the collaboration with B cells

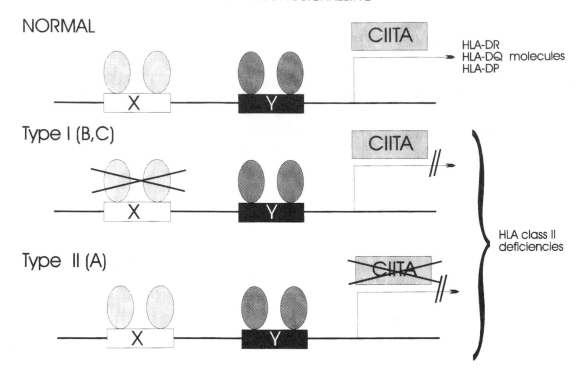

Figure 20.4 Regulation of HLA class II genes in normal individuals and in HLA class II deficiency patients. In type I patients (complementation groups B and C) no factors bind to the X box. In type II patients (complementation group A), mutations in CIITA (class II transactivator) prevent the transcription of HLA class II genes. In both cases, the expression of HLA-DR-DQ and -DP is precluded. Other types of defects may exist

required for eliciting a normal humoral response against antigens. Both T and B cells were found to be abnormal in the presence of normal class II positive APC. Moreover, purified T cells from patients were unable to cooperate *in vitro* with normal B cells and monocytes in the synthesis of specific antibodies. Despite the fact that MHC class II expression appears to be needed for the selection of CD4$^+$ T cells, these cells can be observed, sometimes in substantial numbers, in HLA class II deficient patients [33].

CD40 Ligand Defect (Hyper IgM Syndrome)

Primary Defect

X-linked immunodeficiency with hyper IgM (XHIM) was originally thought to be a primary B cell defect. However, the defect is now known to

be due to mutations in the gene that encodes the ligand for CD40 (CD40L, gp39), assigned to Xq26.3–27.1 [34]. CD40L is a tumour necrosis factor (TNF) family member expressed in activated, but not resting, T cells. Its counterstructure, CD40, belongs to the nerve growth factor receptor (NGF)/TNF receptor family, and is expressed on late pre-B cells in bone marrow, mature B cells and certain accessory cells [35].

Clinical and Immunological Features

XHIM is a rare disorder characterized by very low to absent IgG, IgA and IgE, and normal to increased IgM and IgD serum levels. Patients with XHIM show an increased susceptibility to bacterial infections with symptoms coinciding with the decline of maternally acquired antibodies. Cryptosporidial diarrhoea, *Pneumocystic carinii* pneumonia, neutropenia and autoimmune diseases are also prevalent manifestations of this disease. Patients have normal numbers of circulating B lymphocytes but antibody responses to repeated immunization with T-cell-dependent antigens is restricted to the IgM isotype [36].

CD40 is an important molecule in the differentiation, activation and control of isotype class switching in B lymphocytes. Crosslinking CD40 promotes B cell proliferation, prevents apoptosis of germinal centre B cells, and promotes immunoglobulin class switching [34, 35, 37]. Most muta-

tions described in XHIM patients are point mutations in the extracellular domain of CD40L which prevent the binding to CD40 as assessed by fluorescent labelling with CD40-Ig, a soluble receptor fusion protein. B cells from XHIM patients were capable of proliferating and producing IgE, IgM, IgG and IgA [34, 38] when cultured with anti-CD40 mAbs or non-mutated soluble CD40L plus IL-4 or IL-10, In summary the B cell impairment observed in XHIM patients (reviewed in Chapter 18 of this volume) arises as a consequence of a

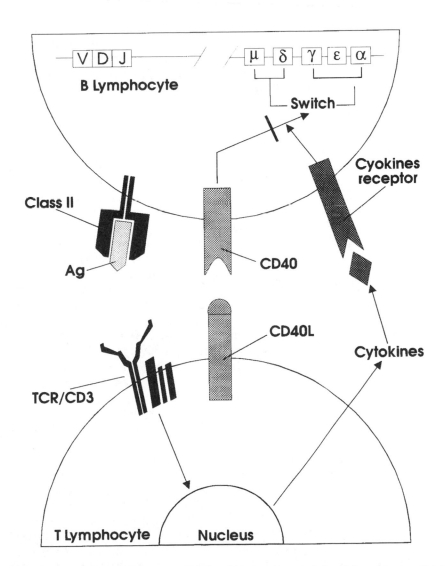

Figure 20.5 Cell interactions and signals during T cell induced Ig isotype switch by B lymphocytes. Cytokines and CD40 engagement are both required. Defects in CD40L prevent signalling through CD40, as in the X-linked hyper IgM syndrome

defective CD40–CD40L interaction, which, in conjunction with cytokines, is necessary for B cell differentiation and isotype switching to IgG, IgA and IgE [Figure 20.5].

A primary T cell defect in XHIM had been also suspected because of the recurrence of opportunistic infections (*P. carinii*, *Cryptosporidium*, toxoplasmosis) associated with T cell abnormalities. Moreover, Mayer *et al.* [39] had reported that B cells from XHIM patients can be successfully induced to produce IgG and IgA when cocultured with T lymphoblasts from patients with Sezary-like syndrome. The suspected T cell defect can now be explained as it is clear that the signalling which occurs during T–B cell collaboration is reciprocal. When B cells present antigens (Ag) to T cells, engagement of MHC class II on B cells induces expression of B7/BB1 (CD80) and B7-2. Recognition of class II-Ag by T cells leads to activation of the T cell and expression of CD40L. The interaction of CD80 and B7-2 on B cells and CD28 and CTLA-4 on T cells allows peripheral T cells to divide and produce cytokines required for T cell differentiation and immunoglobulin class switching. Furthermore, activated B cells expressing CD80 and B7-2 induce T cells to express CD40L and activated T cells expressing CD40L induce resting B cells to express CD80 [34, 35, 37, 40].

Leukocyte Adhesion Deficiency (LAD; CD18 Deficiency)

The disease results from partial or total lack in surface membrane expression of three leukocyte surface glycoprotein heterodimers which constitute an integrin receptor subfamily (β2 integrins), i.e. CD11a/CD18 (LFA-1), CD11b/CD18 (Mac-1) and CD11c/CD18 (p150, 95). Most leukocyte adhesive functions are dependent on the expression and function of the β2 integrins. Their role is essential in leukocyte binding to and migration through the endothelial layer in inflamed tissues.

Primary Defect

Leukocyte adhesion deficiency (LAD) is an autosomal recessive disease due to mutations within the common β2 (CD18) subunit gene, located on chromosome 21q22 [41]. Analysis of LAD CD18 alleles has identified aberrant splicing events, missense mutations and deletions [42]. In contrast, there are no reported cases of structural mutations in the CD11 subunits.

Clinical and Immunological Features

LAD is a rare disease. The most common presentation is due to recurrent pyrogenic bacterial infections of the skin, mucous membranes or deep tissues with impaired pus formation. Recurrent otitis media, severe gingivitis, pharyngitis, stomatitis and perirectal abscesses are common. In addition, fungal infections of soft tissues and delayed umbilical cord separation can frequently be observed. Neutrophilia is invariably seen even during infection-free periods [42–44]. Increased susceptibility to infection is directly related to the inability of granulocytes and monocytes from these patients to migrate to sites of tissue infection and to exhibit their normal functions for a host defence against pathogens. Laboratory tests *in vitro* reveal profound defects in adhesion-related leukocyte functions. Among the defects in phagocytic cells are abnormal binding to complement iC3b, phagocytosis, particle-induced superoxide generation, chemotaxis, adhesion to vascular endothelium or antibody-dependent cellular cytotoxicity. The severity of the disease is directly related to the degree of the deficiency. Patients lacking surface expression of CD11/CD18 often die at an early age from sepsis. Patients with partial deficiency (10 to 20% of normal levels of CD11/CD28) have a milder form of the disease and usually survive into adulthood.

Defects in B and T lymphocyte function *in vitro* have also been demonstrated in these patients: these include antigen-, mitogen-or alloantigen-induced proliferation, natural killing, antibody-dependent killing and T-cell-dependent antibody production. They are more profound at low concentrations of the antigen or during primary stimulation. However, at higher concentrations of stimulus or during secondary stimulation, many of these functions become normal. The normal skin testing with antigens and the lack of susceptibility to viral infections suggest that the *in vivo* function of T lymphocytes is normal. However, it has been shown that the transfer into neonatal mice of human LFA-1$^-$ lymphocytes did not protect them against herpes virus, whereas transfer of LFA-1[+] lymphocytes did. In addition, it has been observed that patients with the severe phenotype of LAD did not reject HLA-incompatible bone marrow transplants. It is possible that CD11a/CD18 interaction with its ligands CD54 (ICAM-1), ICAM-2 and ICAM-3 on APCs may not be compensated by other adhesion molecules that also bind ICAM-1 such as CD43. However, the T cell adhesion pathway involving CD2 and CD58 (LFA-3) molecules [45] or other cell adhesion molecules, such as the CDw49d/CD29 (VLA-4) heterodimer [46,47], may compensate for the LFA-1 deficiency [45] in these patients,

CD3 Defects: γ and ε Deficiencies

In 1986 we described a familial TCR/CD3 expression defect affecting two out of four brothers [48,49]. The first biochemical evidence suggested that the CD3γ component was missing or abnormal in T cells of both siblings [50,51]. Genetic analysis then showed that the defect was due to the confluence of two independent point mutations in the CD3γ genes (Figure 20.6b) of the affected individuals which prevented synthesis of the protein [52]. In spite of sharing the same primary defect, the clinical features of the two patients were disparate. Thus, one of them died at the age of 31 months of severe combined immunodeficiency with autoimmune features (failure to thrive, intractable diarrhoea, autoimmune haemolytic anaemia and viral pneumonia), while the other one is still healthy at the age of 15 [53]; immune system redundancy is almost certainly exemplified in this case. Besides the TCR/CD3 complex expression defect, both individuals also shared a selective IgG2 deficiency and a low response to polysaccharide antigens [54a]; also, the presence of autoimmunity [48,49] provides the best documented example of the unexplained but frequent association between autoimmunity and a molecular (CD3γ) defect [54b].

Phenotypic analyses of such peripheral T cells showed a very low number of cytotoxic (CD8[+]) cells (around five-fold less than normal individuals) and of helper/virgin (CD4[+] CD45RA[+]) cells (around ten-fold less than normal) [55]. In contrast, the number of helper/memory (CD4[+] CD45RO[+] or CD4[+] CD29[+]) T cells was normal. Moreover, other lymphocyte subsets (B, NK cells) were within normal values [55]. These observed abnormalities prompted us to propose a model of asymmetrical interactions between the TCR/CD3 complex and its coreceptors [55]. In this model, CD8 and CD45RA would approach the complex through the CD3γ molecule (or the corresponding isoform) while CD4 and CD45RO would associate through another site of the TCR/CD3 complex (we proposed the homologous CD3δ chain or the corresponding isoform). The absence of CD3γ could prevent the interaction with CD8 and CD45RA and the subsequent selection of these T lymphocytes. The presence of a small number of these cells in the CD3γ deficient individual might be explained by several mechanisms such as extrathymic origin, high-affinity TCR, etc. [56]. The analyses of the few CD8[+] T cells detected in CD3γ deficiency showed a profound functional impairment both *in vitro* and *in vivo* [55] suggesting that the model can also operate in the periphery during antigen recognition.

A T cell line derived from the CD3γ deficient individual showed a normal response to specific antibodies against the defective TCR/CD3 complex with very low production of IL-2 [51]. The responses to protein antigens *in vivo* were also normal [52]. These results suggest that all CD4[+] T cells selected in the absence of CD3γ belong both by phenotypic and functional criteria to the CD45RO subset, and strongly support the hypothesis that CD45RA[+] and CD45RO[+] CD4[+] T cells are independent populations. In spite of the functional disparities found between cells expressing one or other isoform, no functional differences are known between the isoforms. Moreover, they share the same transmembrane and cytoplasmic domains but differ in the extracellular region [57].

Another human TCR/CD3 expression defect has been described [58]. It was clinically associated with a mild immunodeficiency exhibiting a normal number of lymphocytes and immunoglobulin levels. The defect has been ascribed to the confluence of two point mutations in the CD3ε gene (Figure 20.6C) of the affected boy [59]. However, some amounts of normal CD3ε protein could be synthesized by his T lymphocytes giving rise to the expression of low levels of the whole TCR/CD3 complex. Functional analyses *in vitro* showed an activation defect through both the CD2 and the TCR/CD3 complex when using monoclonal anti-

bodies, but a normal response to antigens [60]. These results, together with the mild clinical symp-

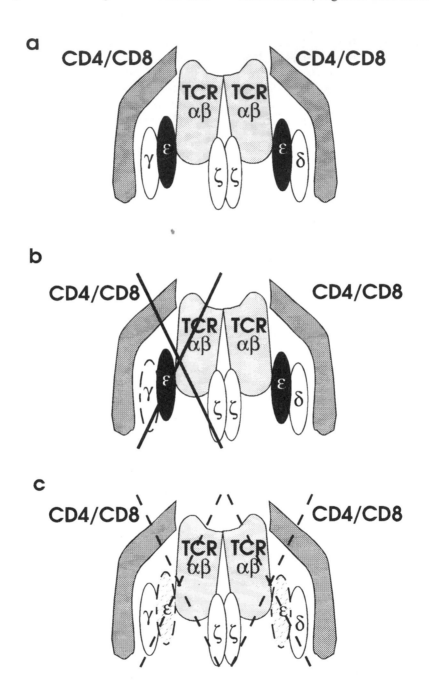

Figure 20.6 Hypothetical models of the T cell receptor–CD3 complex found in immunodeficiencies: (a) T lymphocytes expressing a mixture of the various kinds of T cell receptor–CD3 complexes; (b) the lack of CD3γ, as was found in two patients, prevents the expression of only one type of T cell receptor–CD3 complex; (c) the lack of CD3ε, as was found in one patient, partially prevents the expression of both complexes

toms found in the affected individual suggest that the low number of receptors expressed are functional and sufficient to face the major antigenic challenges *in vivo*. Moreover, as the number of the two main T cell subsets (CD4$^+$ and CD8$^+$) were in the normal range [60], there were no apparent T cell differentiation defects, suggesting that the TCR/CD3 up-regulation occurring in the late stages of normal T cell maturation [61,62] is not crucial for the final outcome of differentiation.

ZAP-70 Deficiency

Several defects in the gene encoding the CD3-zeta-associated PTK, ZAP-70, have been reported [63–65]. Such defects are associated with selective T cell immunodeficiencies characterized by recurrent and persistent viral, fungal or protozoal infections. Interestingly, lack of ZAP-70 prevents the selection of CD8$^+$, but not CD4$^+$, T cells in the thymic medulla and in peripheral blood, demonstrating that this PTK, unlike p56^{lck} and p59^{fyn} [66], is involved in one of the late steps of T cell development (probably associated with the positive selection process). Moreover, although CD4$^+$ T cells appear to develop normally, mature peripheral CD4$^+$ cells show depressed tyrosine phosphorylation and Ca^{2+} flux responses, and are unable to produce IL-2 after stimulation through the TCR/CD3 complex. These findings indicate that the ZAP-70 PTK is involved in the earliest intracellular events associated with TCR/CD3-mediated signal transduction leading to IL-2 production. As yet, the authors have not studied the prevalence of CD45RA/RO cells within this CD4$^+$ subset; thus, a comparison with the previously described CD3γ deficiency cannot be made. However, the fact that both CD3γ and ZAP-70 deficiencies cause a selective defect in the selection of CD8$^+$ T cells suggests that the underlying mechanism could be similar.

Interleukin Common γ Chain Deficiency (X-SCID)

Primary Defect

X-linked severe combined immunodeficiency (X-SCID) accounts for approximately half of all cases of SCID. The gene responsible for X-SCID maps to Xcenq13 and was reported to code primarily for the γ chain of the IL-2 receptor (IL-2Rγ) [67]: the extracellular domain of the γc chain is implicated in the binding of the IL-2 to the IL-2R and its intracellular domain associates with the tyrosine kinase JAK 3 [68]. However, further studies revealed that IL-2Rγ is not only a component of the IL-2 receptor but also of the receptors for IL-4, IL-7, IL-9 and IL-15 (Figure 20.7), and hence is known as the common cytokine receptor γ chain (γc) [69,70]. Thus, mutations in the γc gene affect at least five cytokine systems, explaining the severe immunodeficiency observed in X-SCID when compared with IL-2 deficient patients or with IL-2 knockout mice [71]. At present several patients with mutations in the γc chain have been reported [67, 72, 73].

Clinical and Immunological Features

X-SCID is characterized by severe and persistent infections from early life which are due to a profound impairment of both cellular and humoral immune function. Virtually any type of microorganism may be involved. The majority of patients present with oral candidiasis, pneumonia, protracted diarrhoea and failure to thrive as early as two to three months of age. In the absence of BMT, X-SCID is fatal usually within the first two years of life [3,74,75].

Laboratory findings reveal a complete absence or diminished numbers of immature and mature T cells and NK cells, and normal or elevated numbers of B cells. Any T cells that are detected are usually of maternal origin, although generally these do not cause graft-versus-host disease (GVHD). The lack of T cells can probably be explained by the fact that TCRβ chain rearrangement is arrested in children with X-SCID [76]: while the initial step in TCRβ rearrangement (Dβ–Jβ recombination) is readily detected, Vβ to DJβ rearrangement is undetectable. In addition, proliferation in response to mitogens is almost always completely absent in patients with X-SCID, even those T cells of maternal origin are difficult to activate *in vitro*. Consistent with this, serum antibody level determinations made beyond the perinatal period reveal very low concentrations of IgG and IgA. However, in approximately 50% of the patients IgM levels are in the normal range. Despite the fact that the X-SCID B cells can coop-

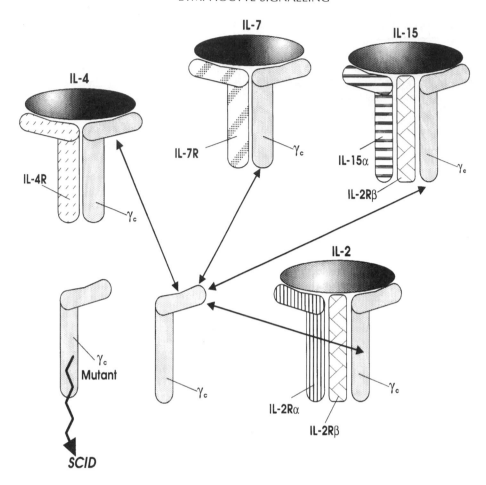

Figure 20.7 Hypothetical structure of the interleukin receptors which share the γc chain. The γ-common (γc) chain is part of the active form of at least five different interleukin receptors. Recently, the γc chain has been involved in the IL-9 receptor. The lack of this chain due to mutations or deletions is responsible for an X-linked severe combined immunodeficiency (X-SCID)

erate *in vivo* with HLA compatible donor T cells following BMT (although antibody responsiveness usually takes more than one year), a primary B cell defect is also suspected in these patients. Thus, although X-SCID B cells express the cell surface markers characteristic of B cells, including surface IgM and IgD, none of them expresses IgG or IgA. However, they show some phenotypic differences when compared with normal B cells [77]. Functional impairment (proliferative response, IgM production) is also observed in X-SCID B cells, *in vitro*, although some of these defects can be reverted in the presence of normal T cells. However, X chromosome inactivation patterns from mothers of children with SCID provide further

support for intrinsic defects in T/NK and B cells [78a]. Defective B cell maturation due to the absence of T cell signals could account for the defective X-SCID B cells. Moreover, differences in the functionality of B cells may arise as a consequence of the severity imposed by different mutations at the γc locus. Indeed, several families have been described that show milder allelic variants of X-SCID [72,73,78a]. Some of these patients can even present with normal numbers of T and NK cells in the newborn period, although these cells are functionally impaired.

In most patients with the typical form of X-SCID, genetic defects in the extracellular region of the γc chain which affect high affinity IL-2 binding, are detected. Other defects have been reported in the cytoplasmic domain of the γc chain leading to disruption of IL-2/IL-2R endocytosis and/or signal transduction (for review, see Chapter 19 of this volume). Finally, one patient has been described with a splice defect: although alternative splicing mechanisms preserved a functional mRNA in 20% of spliced products, a milder form of the disease is still present. Thus, genetic analysis of the γc gene in X-SCID patients has provided new information about the role of the γc chain and associated cytokine receptors in the development and functionality of the lymphoid system. Furthermore, the finding that γc is shared by several cytokine receptors has reinforced our understanding of cytokine pleiotropy and redundancy. Recently, the gene responsible for some autosomal recessive forms of SCID has been reported: mutations in the Jak-3 protein kinase (associated with the γc-chain-containing cytokine receptors) have been found in several SCID patients [78b].

Adenosine Deaminase Deficiency

Primary Defect

In approximately 20% of children with SCID, the disease is caused by deficiency of the enzyme, adenosine deaminase (ADA). ADA is an enzyme of the purine salvage pathway that catalyses the irreversible deamination of adenosine (Ado) and deoxyadenosine (dAdo) to inosine and deoxyinosine respectively (Figure 20.8). These latter compounds are either salvaged back to adenosine or other purines by the activity of the enzymes purine nucleoside phosphorylase (PNP) and hypoxanthine guanine phosphoribosiltransferase (HGPRT), or further metabolized to uric acid and excreted. In ADA deficiency, Ado and dAdo accumulate in the plasma. In addition dAdo within cells is phosphorylated to dATP and accumulates in enormous concentrations. The gene for ADA has been assigned to the long arm of chromosome 20. Multiple genetic defects have been reported and phenotype/genotype correlations have been made in some cases [79].

Clinical and Immunological Features

The clinical phenotype of ADA deficiency comprises a spectrum of diseases ranging from complete lack of ADA activity correlating with classic fulminant neonatal onset SCID, to apparently normal immune function. Mid-spectrum cases comprise patients with slightly delayed onset SCID which is accompanied by some retention of humoral immunity and those with a much latter onset of SCID (even in adulthood) and progressive disease [3,79,80]. Complete lack of ADA activity results in a profound T and B lymphopenia and absence of non-maternally derived immunoglobulins. Patients suffer from overwhelming fungal, viral and bacterial infections with failure to thrive and diarrhoea. In many cases there are markedly elevated IgE and/or hypereosinophilia and autoimmunity symptoms. Although the pathology of this disease is mainly limited to the immune system, same non-lymphoid organs are also affected. At present BMT is the treatment of choice. Enzyme replacement and gene therapy are being assessed.

Pathophysiology

There are several proposed mechanisms whereby the absence of ADA results in toxicity primarily to cells of the immune system [81]. dATP inhibits the activity of the ribonucleotide reductasc (RR) necessary for deoxynucleotide synthesis and, thus, inhibits DNA formation. dAdo causes chromosome breaks *in vitro* and defective mRNA synthesis. In addition, there is a secondary inactivation of the enzyme S-adenosylhomocysteine hydrolase (SAH) involved in methylation reactions, thus inhibiting protein and DNA synthesis. Moreover, some experimental systems suggest that the accumulation of dAdo inhibits PLC-γ-mediated hydrolysis of phosphoinositides [82]. In addition, it has been reported that ADA interacts with the cell surface of Jurkat T cells through the extracellular domain of CD26 [83]: however, the role of this association and the relevance for T cell pathology is poorly understood.

Purine Nucleoside Phosphorylase (PNP) Deficiency

Primary Defect

PNP is an enzyme of the purine salvage pathway catalysing the step adjacent to that catalysed by

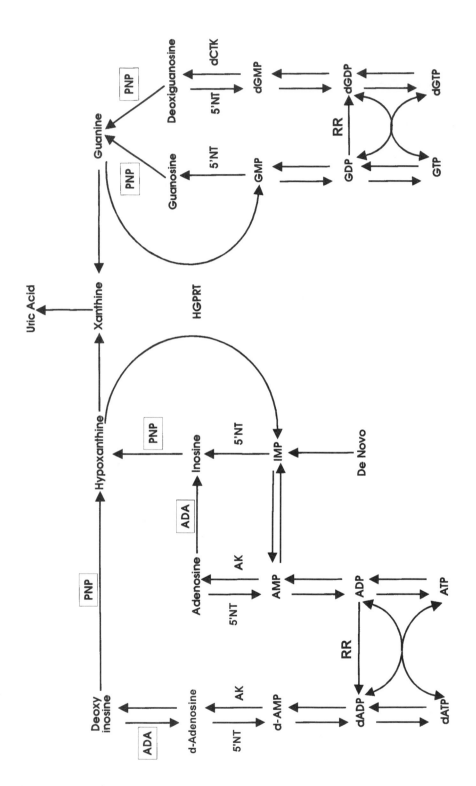

Figure 20.8 Purine metabolic pathways. ADA and PNP sites of action are indicated. In PNP deficiency, inosine, guanosine, deoxyinosine and deoxyguanosine replace uric acid as the end product. In ADA deficiency, adenosine and deoxyadenosine accumulate. dCTK, deoxycytidine kinase; AK, adenosine kinase; 5'NT, 5'nucleotidase; HGPRT, hypoxanthine–guanine phosphoribosyltransferase; RR, ribonucleotide reductase

ADA: PNP mediates the hydrolysis of inosine and deoxyinosine to hypoxanthine and the conversion of guanosine and deoxyguanosine to guanine (Figure 20.8). In PNP deficiency this salvage pathway is not active due to the absence of substrate and, as a consequence, *de novo* purine synthesis takes place to maintain purine levels. Since deoxyguanosine cannot be degraded to guanine, the enzyme deoxycytidine kinase (restricted to thymus and PBMC) converts the deoxyguanosine to dGMP. As phosphorylated nucleotides cannot diffuse out of the cell, lymphoid tissue therefore accumulates dGTP. The mechanisms proposed to account for the toxicity and immunodeficiency in the PNP defect are similar to those proposed for ADA deficiency (see page 389) and are likely to involve inhibition of the enzymes RR and SAH. In addition, inhibition of the terminal deoxinucleotidyl transferase (TdT) and alterations in the intracellular levels of GTP, and thus of cGMP, may also affect T cell differentiation and function. Due to the high mortality rate associated with this deficiency, the thymus is the most affected tissue.

PNP deficiency accounts for approximately 4% of patients with SCID: the PNP gene has been assigned to human chromosome 14q13, and the pattern of inheritance is autosomal recessive. Mutations in this gene have been described as the basis for PNP deficiency [3, 84] clinical and immunological features.

Clinical and Immunological Features

All patients reported so far suffer from recurrent infections of bacterial, viral and fungal origin. Failure to thrive and diarrhoea are usually present. In addition, neurological and autoimmune problems, the most common being haemolytic anaemia, are frequent. One characteristic feature of these patients is that the levels of uric acid are always very low [84].

This deficiency affects T cells more than B cells. T cell numbers are very low, and T cell function is impaired both *in vivo* and *in vitro*. However, as T cell impairment is usually progressive with age, one cannot rule out PNP deficiency when early T cell function is normal. In contrast, B cell function seems to be normal or increased and B cell numbers are usually within the normal range. However, several patients have been reported with poor B cell function, and some of them exhibited progressive decline of B cell function.

Wiskott–Aldrich Syndrome

Primary Defect

Wiskott–Aldrich syndrome (WAS) is an X-linked recessive disease characterized by immunodeficiency, thrombocytopenia and eczema. Molecular genetic studies of WAS patients have localized the WAS locus to the pericentric region of the X chromosome at Xp11.22–Xp11.23. The gene responsible for WAS (WASP) has recently been cloned [85], and several patients with WASP mutations have been reported [85–87].

Clinical and Immunological Features

WAS is recognized by its characteristic presentation of immunodeficiency, thrombocytopenia and eczema [3, 88–90]. However, WAS has been particularly difficult to define in terms of clinical status and the precise immune defect. For example, patients with X-linked thrombocytopenia, but no immunodeficiency or eczemas, have now been shown to have allelic variants of WASP [86, 87]. Recently Ochs *et al.* [87] have established five clinical scores for WAS patients, ranging from score 1 for thrombocytopenia and small platelets and occasionally immunodeficiency, to score 5 which shows thrombocytopenia, eczema, immunodeficiency, severe infections and autoimmune features. This heterogeneity might reflect different mutations in the WASP gene.

Immunodeficiency is generally a feature in WAS (except in some patients of score 1) and is typically a progressive immunodeficiency with clinical and immunological status decreasing with age. Bacterial infections are frequent in WAS patients, and thus they may have recurrent bouts of otitis, sinusitis and pneumonia. Severe periodontal disease also occurs. Meningitis and overwhelming bacterial septicaemia are frequent and may be fatal. Less common, although still potentially lethal, are fulminant viral infection with herpes simplex, varicella or cytomegalovirus. Another complication of WAS is an increased risk of malignancies which can clearly contribute to early mortality.

Carriers of the WAS gene non-randomly inactivate the mutated X chromosome in T and B lymphocytes, granulocytes, monocytes and erythroid cells, suggesting that cells of the most peripheral blood lineages express the gene defect and are selected against at an early stage of haematopoietic development [69]. Indeed, non-random X-inactivation can be detected as early as the granulocyte-macrophage colony-forming-unit (GM-CFU) indicating that WASP mutations are detrimental to early haematopoietic cells [91].

Laboratory studies in WAS patients reveal abnormalities in both the cellular and humoral arms of the immune system. For example, with respect to the thrombocytopenia, circulating platelets are decreased in number and size, and the bleeding time is prolonged. Splenectomy seems to increase both platelet number and size, sometimes even into the normal range. One of the most consistent and striking findings, however, is a characteristic absence (or low circulating levels) of isohaemagglutinins accompanied by low levels of circulating IgM. In contrast, IgA and IgE levels are frequently increased. Another general feature of these patients is their inability to generate anti-carbohydrate antibodies (generally IgM) following vaccination. However, WAS patients have normal levels of IgG2, an isotype which is also thought to contain high levels of anti-polysaccharide activity.

Lymphocyte counts, usually normal at birth, tend to decline with age. The decrease is due to a fall in T populations, mainly of the $CD4^+$ T cells. Functional analysis in WAS patients revealed aberrant delayed-type hypersensitivity and cutaneous anergy to a variety of antigens as well as impaired proliferative responses to anti-CD3 antibodies, lectins and MLR. Moreover, when functional studies were carried out in IL-2 dependent T cell lines and HVS-transformed T cells, impairment of T cell responses, such as proliferation in response to anti-CD3 antibodies, IL-2 production and up-regulation of CD69 and CD28 antigens, was observed [92, 93]. In contrast, early T cell activation events following stimulation with anti-CD3 (CA^{2+} mobilization, tyrosine phosphorylation and GTP binding of the CD3γ chain) were shown to be normal in WAS T cell lines [92]. Interestingly, other stimuli such as allospecific antigens or PMA plus ionomycin elicited normal proliferative responses.

Molecular studies in WAS patients revealed abnormal expression and molecular weights of the O-glycosylated proteins, CD43 and gpIb in lymphocytes and platelets resulting from aberrant O-glycosylation of these molecules. As a result of these abnormalities one constant feature of most WAS patients, even those with more benign forms of the disease, is an impaired proliferative response to periodate ($NaIO_4$) [94]. This oxidizing agent normally induces extensive T blastogenesis by interaction with terminal sialic acid residues on membrane glycoproteins. In addition to the CD43 abnormalities, the expression of other surface molecules such as CD75, CD76, CD24, CD37, CD6 and CD7 was also reported to be altered in WAS cells [93, 95]. As the precise defect has not been identified, it is not possible to exclude a regulatory defect affecting several surface molecules and the overall cellular cytoarchitecture. Such abnormalities might account for the impaired function in WAS lymphocytes as well as for defective T–B cell (and APC) "communication". However, other possible defects, such as an enzyme implicated in glycosylation reactions, cannot be ruled out.

Fas Deficiency

The Fas cell surface antigen is a type I membrane protein belonging to the TNF/NGF receptor family. Its counter-receptor, the FAS ligand (FASL) is a type II membrane protein homologous to members of the TNF family. The human FAS and FASL genes have been mapped to chromosome 10q [96] and 1q23.3 [97] respectively. The importance of Fas was initially indicated by mAbs raised against FAS which were shown to have cell killing activity: it is now clear that FASL cross-links FAS resulting in transduction of the cell death signal. However, whereas TNF kills cells by necrosis, FAS triggers membrane blebbing, cleavage of genomic DNA into nucleosomal fragments and cellular shrinkage, responses bearing the hallmark of apoptosis [98].

The Fas defect in human has been reported to have similar clinical and immunological status to that reported for MRL/*lpr* and MRL/*gld* immunodeficient mice [99–101]. Activated T cells from patients were shown to lack surface expression of the FAS antigen, and anti-Fas (APO-1)-induced apoptosis was absent. Gene sequence analysis revealed that this defect was due to a 290 bp deletion encompassing the terminal part of the intracellular domain-encoding exon of the Fas gene. Lymphocytosis, hydrops fetalis with hepatosplenomegaly, lymphadenopathy and lung infiltrates were observed in these patients. In addition patients presented with hypergammaglobulinaemia with IgG monoclonality and autoimmune thrombocytopenia. Also, as in the mouse model, a high percentage (70%) of their T cells were shown to be CD4CD8 double negative. Two siblings with similar, although less severe, clinical and immunological findings, were also reported. Only the oldest developed autoimmune alterations (neutropenia, anaemia and thrombocytopenia) and a restricted βT cell repertoire. FAS expression was normal in these patients but FAS-mediated T and B apoptosis was impaired. A 2 bp deletion at nucleotide 1005, leading to a stop codon in the cytoplasmic domain was found to be inherited from the mother whereas no defect was detected in the gene inherited from the father. However, a slight defect in FAS-mediated apoptosis with no clinical manifestations was detected in the father. Thymic selection involves apoptosis of those T cells that are negatively selected. Even in the periphery, mature T cells undergo selection: those that recognize self-antigen are first activated, then die by apoptosis (peripheral clonal deletion). In vivo and in vitro results suggest the involvement of the FAS system in the clonal deletion of autoreactive T cells in the periphery. Thus, FAS and FASL mutations prevent this process, and double-negative T cells may be the result of the abnormal peripheral clonal deletion.

Defects Resulting in DNA Recombination and/or Repair Alterations

Rag-1, Rag-2, DNA-PK Anomalies

The large number of genes required to encode the Ig and TCR molecules are generated combinatorially in a process known as V(D)J recombination. Several transacting factors are strongly implicated as a component of V(D)J recombination: endonucleases, TdT, ligases, recombinant activating genes

1 and 2 (RAG1 and RAG2), the DNA-dependent protein kinase complex (DNA-PK), etc. [102]. RAG1 and RAG2 knockout mice lack mature B and T cells due to their inability to initiate V(D)J rearrangement, conferring a non-leaky SCID phenotype to these mice [103]. In contrast, in the "classical" murine SCID, the recombinase machinery is able to initiate V(D)J rearrangement, but coding segments fail to join. As a consequence SCID mice also lack mature B and T lymphocytes. By the age of 12 months, most SCID mice show "leakiness" in their lymphocyte count due to mono-oligoclonal revertants of B and T cell precursors that exhibit an apparently normal V(D)J rearrangement [104]. However, SCID cell lines from different organs exhibit a profound hypersensitivity to DNA-damaging agents and radiation that can cause double-strand breaks. Thus such immunodeficiencies combine defects in site-specific V(D)J recombination and DNA repair [105]. This link between DNA recombination and DNA repair has been illustrated by analysis of chinese hamster ovary (CHO) mutant cell lines that have defects in double-strand DNA break repair and in V(D)J recombination: for example, one of the cell lines (V3) containing mutations in the gene that codes the DNA-PK large catalytic subunit (DNA-PKcs) has the SCID mouse-like phenotype [106].

In contrast to the inbred mouse strains, human SCID is a very heterogeneous disease. Approximately 20–25% of patients with SCID of unknown origin are characterized by the absence of T and B lymphocytes, although functional NK cells can be detected. This disease is inherited as autosomal recessive [107]. Two different groups have reported that recombination events in some T⁻B⁻ SCID patients resemble rearrangement patterns observed in RAG1/2 deficient mice or in SCID mice [103, 108].

Omenn's Syndrome

Nowadays, it is thought that Omenn's syndrome (OS) may correspond to a leaky form of SCID as observed in SCID mice. This hypothesis is further supported by the simultaneous occurrence of OS and SCID in the same family and the fact that OS, as well as SCID T-B-patients show an increased radiosensitivity [109, 110]. Omenn's syndrome is an autosomal recessive, fatal immunodeficiency, characterized by a diffuse erythematous skin eruption, lymphadenopathy, hepatosplenomegaly, diarrhoea, failure to thrive and repeated infections. Lymphocytosis of T cell origin, and hypereosinophilia, frequently with hyper IgE, are the main laboratory findings. The peripheral blood T cells that exist are usually activated (CD25 and HLA-DR are up-regulated) and, furthermore, recent studies have shown oligoclonal expansion of CD⁻CD8⁻ double-negative (DN) T cells. Interestingly, DN T cells from one patient and CD4+CD45RO cells from another patient were shown to have an Th2-like cytokine profile [111, 112]. During a therapeutic trial with IFN-γ in the second patient, an improvement of the immunologic status was observed both *in vivo* and *in vitro* [112]. Thus, the syndrome may result from impaired selection and activation of some residual clones, presumably to epithelial antigens, that might infiltrate skin and gut. In addition, the Th2 phenotype would explain the hyper IgE (IL-4) and hypereosinophilia (IL-5) observed in this syndrome.

DNA Ligase I

A patient has been reported with mutations in the DNA ligase I gene [113]. This patient, a female, had clinical symptoms of retarded growth, sunsensitive facial erythema and severe immunodeficiency. Moreover, cultured cells from this patient were hypersensitive to a wide range of DNA-damaging agents including alkylating agents, ionizing radiation and ultraviolet light. However the V(D)J recombination pathway was found not to be perturbed [114].

Ataxia–Telangiectasia

Ataxia–telangiectasia (A–T) is an autosomal recessive disorder characterized by a progressive cerebellar ataxia with Purkinje cell degeneration, oculocutaneous telangiectasia, immunodeficiency and cellular radiosensitivity [115, 116a]. At least five complementation groups have been described and genetic analyses have localized the major A–T gene to a 500 kb region of 11q.22–23. Recently a single gene, ATM, that is mutated in this disorder has been identified by positional cloning [116b]. A–T cells are abnormally sensitive to ionizing

radiation and a hallmark of A–T is DNA fragility with abnormal chromosome breaks and rearrangements in cultured cells from A–T patients. Whereas in fibroblasts, chromosomal alterations are random, chromosomal translocation breakpoints observed both in B and T lymphocytes involve mainly TCR and Ig loci. Thus, immunodeficiency of IgA. IgE and IgG_2, reduced T cell function and sometimes reduced T cell counts are observed in A–T patients. A–T patients show an increased incidence of malignancies, even in heterozygotes. This may reflect the failure of A–T cells to induce up-regulation of p53 and the GADD45 gene, tumour suppressor products thought to control the arrest of the cell cycle, in response to ionizing radiation [117]. Growth arrest generally occurs in the G_1 phase following DNA damage, however, A–T cells show a foreshortened G1/S delay and an accumulation of cells in G2/M.

FUNCTIONAL T CELL DEFECTS

Signal Transduction Defects

There are several primary TCR/CD3 signal transduction defects in which the molecular lesions have not yet been characterized. However, by using different pharmacological activators and biochemical analyses *in vitro*, it is possible to localize the defect as proximal or distal to the TCR/CD3 complex [13, 118–121].

Gene Induction Defects

Lymphocyte activation deficiencies that do not reflect signal transduction defects but can be reverted, *in vitro*, by the addition of exogenous IL-2, are classified as gene induction defects in which the defective induction of IL-2 (or other cytokines) may underlie the immunodeficiency [122, 123]. For example, recent data from studies on IL-2 knockout mice [70] and cytokine pleiotropy and redundancy [68, 69] (see also page 000 on X-SCID defects) suggest that a defect in IL-2 production might not fully account for the clinical and immunological impairment observed in IL-2 deficient patients. Consistent with this hypothesis, an immunodeficiency presenting with abnormal IL-2, IL-3, IL-4, IL-5 and IFN-γ induction, has been suggested to be due to a primary defect in the transcription factor, NF-AT [124, 125].

OTHER PRIMARY SYNDROMES AFFECTING T CELL FUNCTION

Papillon–Lefèvre Syndrome (PLS)

Papillon–Lefèvre syndrome (PLS) is an autosomal recessive inheritable form of hyperkeratosis palmoplantaris with premature destruction of the periodontium of the deciduous and permanent teeth and an increased susceptibility to bacterial infections, which has been observed in about 20–25% of PLS patients. Parental consanguinity has been found in about 33% of the familial cases [126]. Confluence of both dermal and oral effects clearly differentiates this syndrome from patients with other immunodeficiencies such as leukocyte adhesion deficiency which has similar manifestations. Some controversial immune function test results have also been reported, including decreased neutrophil phagocytosis as well as both impaired and normal responses to T and B cell mitogens [127, 128]. Moreover, IgG plasma cells have been found in oral PLS lesions, together with evidence for T lymphocyte involvement [129]. Thus, as PLS shares some clinical symptomology with well-characterized adhesion molecule defects like LAD, this finding, together with the fact that T cells may be implicated in PLS pathogeny, led Góngora et al. [130] to study cell adhesion molecules in PLS lymphocytes and other hematopoietic populations. A decrease in the absolute number and percentage of PBLs bearing CD2, CD29 or CD45RO molecules was recorded. Interestingly, a low expression of a number of major adhesion molecules (CD18, CD11a, CD29, CD45RO and CD2) was also observed, mainly within the CD3+ population. The possibility of tissue sequestering of T cells with high LFA-1 (CD11a) expression was excluded since this PLS patient showed a stable phenotype during a 4 year follow-up period and cells with the corresponding phenotype were not recorded in patient tissue lesions. Molecular and biochemical analyses did not show evident structural abnormalities in any of the affected molecules. It is suggested from these results that the memory T cell subset (CD45RO and high density of CD2, LFA-1 and CD29 molecules) is diminished in this patient and partially reduced in a sibling. As some of these surface proteins are necessary for activated T lymphocytes to migrate through vascular endothelium, a relative lack of memory T cells may prevent sufficient T cells reaching inflamed tissues (i.e. periodontium) to defend against invasion by microorganisms. Interestingly, retinoids can be used to ameliorate the symptoms of PLS patients [131]: as retinoids alter the naive-to-memory T cell phenotype exchange [132], this suggests that the genetic defect in PLS may somehow be quantitative and related to the metabolism of retinoic acid and its receptors.

DiGeorge Syndrome

DiGeorge syndrome (DGS) is defined by absence or hypoplasia of the thymus and parathyroid together with abnormalities of the major blood vessels and facial abnormalities [133]. The tissues affected in DiGeorge anomaly are those developed from the pharyngeal pouches. However, although there is no single diagnostic feature which is uniformly present and easily ascertained in every patient, in general, the numbers of T cells, B cells and their proliferative responses to mitogens are the best discriminators for diagnosis. While humoral immunity may be normal or altered, the majority of patients have some impairment of antibody formation. On the other hand, NK cells are not affected. However, it should be noted that the immunological test results from DGS patients are very heterogeneous and patients with a minimal thymic defect have even been reported. Moreover, sequential studies of both T cell and B cell immunity are necessary, since spontaneous deterioration with time has been described. Consistent with these variable immunological defects, the aetiology of DGS is presumed to be heterogeneous as there are reported cases of autosomal dominant, autosomal recessive and X-linked modes of inheritance [134]. However, molecular techniques, especially "in situ" hybridization, have shown that most of DGS patients have chromosomal abnormalities, mainly deletions at 22q11 [135, 136].

Common Variable and IgA Immunodeficiencies

Common variable immunodeficiency (CVID) encompasses a heterogeneous group of disorders characterized by hypogammaglobulinaemia, defective antibody production, low to normal number of circulating B cells with an apparent block in the differentiations of these B cells to plasma cells, including an inability to switch isotypes in some patients. As a consequence, recurrent bacterial, mainly sinopulmonar, infections are present. The failure of B cells to mature into immunoglobulin-secreting cells in response to a variety of stimuli led to the assumption that an intrinsic B cell defect was the main pathogenic abnormality in most CVID patients [137]. In recent years, however, several reports have shown that circulating T cells in the majority of CVID patients display abnormalities in subset distribution or functional capacity [3, 138]. For example, several studies have reported reduced percentages of naive CD4$^+$ T cells and normal to elevated levels of memory CD4$^+$ T cells, reduced CD4/CD8 ratios, and an expanded subset of CD8$^+$CD57$^+$ and CD8$^+$DR$^+$ cells. The CD8$^+$CD57$^+$ subset was implicated in the suppression of Ig production, T cell proliferation and B cell differentiation. Reported functional T cell abnormalities include reduced *in vitro* proliferative responses to antigens, lectins, anti-CD3 and anti-CD2 antibodies and reduced cytokine production, in particular decreased IL-2 secretion. However, normal lymphocyte subsets and T cell function have also been extensively reported. Thus, abnormal T cell–B cell collaboration was suspected to be the cause for CVID and, indeed, Farrington *et al.* [139] have shown a reduction of the CD40L (gp39) mRNA and protein expression in 13 out of 39 patients investigated. However, no DNA sequence mutation of the CD40 ligand was observed in two of these patients and thus the genetic basis of CVID is still not known. At present, CVID and IgA deficiency (IgAD), the two most common primary immunodeficiencies, are thought to represent the extreme poles of a disease spectrum, a hypothesis consistent with the fact that patients who initially have IgAD, were later diagnosed as CVID sufferers. In addition, several studies have shown a linkage between CVID/IgAD

and certain HLA class II and class III markers [140–142]. Moreover, additional genes within the Ig heavy chain gene region have also been implicated in this disease [142]. Thus, it is likely that both disorders (CVID/IgAD) are polygenic and that exogenous factors may, in genetically susceptible individuals, determine the degree of expression of Ig genes.

SECONDARY IMMUNODEFICIENCIES WITH IMPAIRED T CELL ACTIVATION

Secondary (or acquired) immunodeficiencies are a heterogeneous group of disorders in which the impairment of the immune response is not the primary defect but the result of several pathologic conditions such as viral infections, atopy, cancer, etc. [3]. An extensive review of these defects is beyond the scope of this chapter so they will only be briefly described here. Moreover, defects secondary to HIV and parasite infection will be discussed in full in Chapters 21 and 22 of this volume, respectively.

T Cell Defects Secondary to Viral Infections

Infections by at least six different viruses have been associated with T cell activation deficiencies: human immunodeficiency virus (HIV), Epstein–Barr virus (EBV), cytomegalovirus (CMV), Human T cell leukaemia virus type 1 (HTLV-I), human herpes viruses 6 and 7 (HHV-6 and HHV-7) and herpes simplex virus (HSV).

HIV Infection

The most important alteration caused by HIV infection is the depletion of CD4$^+$ T cells. Several mechanisms have been proposed as being responsible for this depletion, including a direct cytopathic effect of the virus and activation-induced-programmed cell death or apoptosis [143, 144]. However, impairment of cell-mediated immunity is observed even before the decline of CD4$^+$ T lymphocytes begins. This defect was localized proximal to the TCR/CD3 complex as no Ca^{2+}

increases could be detected whereas other T cell activation pathways, such as that of CD2 and CD28, were unaffected (for review, see Chapter 21 of this volume). It has been proposed that the viral proteins gp120 and p14 may be involved in the functional attrition of CD4$^+$ T cells by binding to CD4 and CD26 [145], respectively. In addition, autoimmune disorders [146] and a Th1 to Th2 switch [147] are thought to play an important role in the clinical status and immune dysfunction of HIV$^+$ patients.

EBV Infection

EBV infection can cause a variety of clinical conditions with infectious mononucleosis being the most common. However, the acute infection causes a generalized impairment of cell-mediated immunity. Although the molecular basis of this immunosuppression is unknown, T lymphocyte anergy has been shown to be caused by a selective and proximal TCR/CD3 signal transduction defect resulting in low proliferation and IL-2 production [148]. It is believed that a viral homologue of IL-10 may be involved in EBV-induced immunosuppression [149, 150].

CMV Infection

CMV causes damage in the host immune response when infection occurs in immunocompromised or immunologically immature hosts. The T cell anergy associated with CMV infection is restricted to the TCR/CD3 activation pathway [151]. CMV encodes a glycoprotein homologue to MHC class I molecules [152] and, thus, the reduction in MHC class I expression observed in lymphocytes from CMV-infected patients may account for some of the immunological impairment observed *in vivo* and *in vitro* following CMV infection.

HTLV-I Infection

HTLV-I infection causes profound morphological, phenotypical and functional changes in T lymphocytes, both *in vivo* and *in vitro*, including a loss of expression of the TCR/CD3 complex due to a selection inhibition of CD3-δ,γ and ϵ genes [153]. Moreover, at least one viral protein, Tax, has been shown to be involved in the transactivation of

several host genes, including the IL-2R and various cytokine genes [154].

HHV-6 and HHV-7

HHV-6 has been reported to infect CD4$^+$ and CD8$^+$ T cells and following HHV-6 infection, there is a progressive decline in the TCR/CD3 complex expression which results in an impairment of T cell function, *in vitro* [155, 156]. In contrast, following infection of T cells with HHV-7, there is a pronounced loss of CD4 expression but only a slight decline in CD3 expression.

HSV Infection

HSV infection can result in transient suppression of several cellular immune responses during acute episodes of infection. The mechanism by which HSV may mediate immune dysfunction include enhanced activity of suppressor T cells and soluble suppressor factors, decrease in cytokine production and inhibition of cytotoxic effector function [157].

As has been shown above, certain viral infections (particularly by retroviruses and herpes viruses) are associated with selective defects of T cell activation. The molecular mechanism in each case is far from established, but it is likely that signalling or cytokine homologues produced by the virus could underlie some of these defects [158, 159]. In addition, viral-induced apoptosis of T cells has been implicated in several infections, both *in vivo* and *in vitro* [143, 160].

T Cells Defects Secondary to Fungal Infections

Several species of fungi including *Candida albicans, Pityrosporum orbiculare, Trychophytom, Blastomyces dermatitidis* or *Coccidioides immitis* share the property of being able to suppress cell-mediated immunity [3]. This immunosuppressive activity has been postulated to be mediated by soluble factors and, indeed, the suppressive effects of mannans and their oligosaccharide fragments in cell-mediated immunity have been demonstrated, both *in vivo* and *in vitro*. Several investigators have proposed that mannans or mannan metabolites act upon monocytes or suppressor T lymphocytes and, in addition, interference with cytokine activities, lymphocyte–monocyte interactions, and lymphocyte homing mechanisms have also been proposed [161].

T Cell Defects Secondary to Helminth Infections

Helminths (cestodes, trematodes and nematodes) are multicellular parasites. Although they do not replicate within their definitive host, the organisms are long-lived in relation to the life-span of the host. The continued survival of the adult forms for long periods in the infected host is suggested to be due to mechanisms developed by helminths for evading the human immune response. In fact a state of general unresponsiveness to mitogens as well as specific anergy to parasite antigens are associated with certain helminth infections (for review, see Chapter 22 of this volume). Several mechanisms have been proposed to account for this immunosuppression including parasite-derived suppressor factors, Th1/Th2 imbalances and suppressor T cells [3, 162, 163].

T Cell Defects Secondary to Protozoal Infections

Trypanosoma cruzi and T. brucei

Infection with *Trypanosoma cruzi* and *T. brucei* gives rise to an immunosuppressive state whose characteristics may vary depending on the species of trypanosome involved ([3]; see also Chapter 22 of this volume). A general consequence of infection, however, is a defect in the production of IL-2 and the expression of IL-2R by activated T cells [164]. In addition, filtrates from *T. cruzi* suspensions inhibit T cell proliferation and expression of IL-2R and the TCR/CD3 complex on the surface of activated T cells, *in vitro* [165]. In addition, defects in the immune response have also been attributed to suppressor T cells, immunosuppressive autoantibodies or the suppressive action of macrophages.

Leishmania

Resistance to Leishmania infection, both in humans and in the BALB/C mouse model, appears to be associated with the expression of T cells that produce IFN-α, IL-2 and positive skin tests to the parasite (Th1). In contrast, susceptibility is associated with Th2 responses (e.g. IL-4, IL-10) and hypergammaglobulinaemia [3, 166]. In humans, a defect in APC function may also play an important role in the state of immune unresponsiveness.

T Cell Defects Secondary to Bacterial Infections

Certain bacterial infections, particularly by mycobacteria, cause both specific and generalized immunosuppression in man. Bacterial suppressor factors, the induction of suppressor T (CD8$^+$) cells or macrophages, modulation of certain T cell surface molecules such as CD2 [167] or suppressor effects of mycobacterial mannans [168] have all been implicated in the general immunosuppression following infection.

T Cell Defects Secondary to Other Primary Immune Systems' Perturbations

Cancer

Certain tumour cells cause T cell immunodeficiencies thus allowing them to escape immune surveillance. For example, Hodgkin's disease (HD) is associated with a profound impairment in the CD2 activation pathway (the CD3 pathway is less affected) and soluble inhibitory factors in HD serum, which are able to bind CD2 molecules, are thought to account for this immunosuppression [169]. In addition, Mizoguchi *et al.* [170] have shown that TCR/CD3 complexes from tumour-bearing mice only contained low amounts of CD3γ and lacked ζ chains: in such cells ζ was found to be replaced by the homologous FcERγ chain and these authors suggested that similar defects may occur in human patients.

Atopy

Several lines of evidence suggest that atopy is related to an impairment or imbalance of T lymphocyte reactivity. Indeed, a decreased T cell proliferative response to antigens, anti-CD3 mAbs and anti-CD2 mAbs associated with low IL-2 production has been reported [171–173]. Costimulation with anti-CD28 mAbs, however, was found to restore both the IL-2 and the proliferative response to such cells triggered via CD3 [173]. Moreover, allergen-specific T cell clones from atopic individuals have typically shown a Th2 cytokine profile, presenting with low levels of IL-2 and IFN-γ production. Furthermore, as two Th2 cytokines (IL-4, IL-10) suppress Th1-mediated responses, an imbalance in Th1/Th2 subsets may underlie the cellular immune defect of atopy. As the CD28 counter-receptors, B7-1 and B7-2, differentially activate the Th1/Th2 developmental pathway [174], costimulation with anti-CD28 mAbs may overcome the CD3 activation defects by bypassing any alterations in the expression of CD28 ligands on APC of atopic patients.

Autoimmunity

Several autoimmune disorders are associated with defective T lymphocyte function. It is not known if these defects are primary or secondary because an autoimmune "hyper-response" HLA haplotype (B8/DR3) is associated with similar T cells defects [3, 175, 176]. In some instances, aberrant costimulation by APC [174] or defective peripheral clonal deletion may allow the selection of autoreactive T lymphocytes and, indeed, several primary immunodeficiencies have associated autoimmune features. In the case of primary T cell deficiencies, defective signalling can allow the selection of hyper-responsive, autoimmune, T cell clones and this abnormal selection may occur in the periphery. Indeed, CD4$^-$ CD8$^-$ (DN) T cells have been observed in patients with systemic lupus erythematous, Omenn's syndrome and FAS deficiency, diseases associated with an abnormal selection of T cells that, moreover, are thought to have an autoimmune origin.

T Cells' Defects Secondary to Ageing and Malnutrition

Ageing is associated with impaired immune responses and increased morbidity resulting from infection. Defective proliferation and IL-2 produc-

tion, *in vitro*, have frequently been reported for T cells obtained from ageing donors and an impairment in early events of T cell activation may be responsible for these observed defects [177, 178].

Correction of nutritional abnormalities has been shown to improve immunity in the elderly [179,180]: this is hardly surprising as proteins, lipids, carbohydrates, vitamins and minerals are essential for cell maintenance and function. When activated, cells of the immune system are very metabolically active, and a deficiency or excess of specific nutrients can influence cellular metabolism, altering immune regulation. Thus in protein-energy malnutrition (PEM) a profound impairment, both in number and function, of T cells is observed [181]. Deficiency of vitamins (vitamin A, E, B$_6$, folic acid) and trace elements such as iron, zinc, manganese or selenium [132,182] generally suppresses T cell mediated immune function. In addition, dietary modulation of polyunsaturated fatty acids has also been related to secondary T cell activation deficiencies both *in vivo* and *in vitro* [183].

Acknowledgements

The authors thank Agustín Madroño Sánchez for his excellent work in designing the figures. This work was financed in part by Fondo de Investigaciones Sanitarias, Ministerio de Sanidad, Spain.

REFERENCES

1. Chan AC, Irving BA and Weiss A. New insights into T-cell antigen receptor structure and signal transduction. *Curr. Opin. Immunol.* 1992; **4**: 246–251.
2. Damjanovich S, Szöllosi J and Trón L. Transmembrane signalling in T cells. *Immunol. Today* 1992; **13**: 12–15.
3. Regueiro JR, Rodríguez-Gallego C and Arnaiz-Villena A. *Human T lymphocyte Activation Deficiencies*. Austin, USA: RG Landes Company, 1994.
4. de Saint Basile G. and Fischer A. X-linked immunodeficiencies: clues to genes involved in T-and B-cell differentiation. *Immunol. Today* 1991; **12**: 456–61.
5. Horam PV, Slezak SE and Poste G. Improved flow cytometric analysis of leucocyte subsets: simultaneous identification of five cell subsets using two-colour immunofluorescence. *Proc. Natl. Acad. Sci. USA* 1986; **83**: 8361–8365.
6. Vuillier F, Scott-Algara D and Digliero G. Extensive analysis of lymphocyte subsets in normal subjects by

three-colour immunofluorescence. *Nouv. Rev. Fr. Hematol.* 1991; **33**: 31–38.

7. Goldman AS, Palkowetz KH, Rudloff HE, Brooks EG and Schmalstieg FC. Repertoire of Valfa and Vbeta regions of T cell antigen receptors on CD4+ and CD8+ peripheral blood T cells in a novel X-linked combined immunodeficiency disease. *Eur. J. Immunol.* 1992; **22**: 1103–1106.

8. Davis L and Lipsky P. Signals involved in T cell activation. Distinct roles of accessory cells, phorbol esters and inter-lukin 1 in activation and cell cycle progression of resting T lymphocytes. *J. Immunol.* 1986; **136**: 3588.

9. Meuer SC, Hussey RE, Fabbi M *et al.* An alternative pathway of T cell activation, a functional role for the 50 kD T11 sheep erythrocyte receptor protein. *Cell* 1984; **36**: 897.

10. Park JK, Rosenstein YJ, Remold-O'Donnell *et al.* Enhancement of T cell activation by the CD43 molecule whose expression is defective in Wiskott–Aldrich syndrome. *Nature* 1991; **350**: 707–709.

11. Berridge MJ. Inositol trisphosphate and calcium signalling. *Nature* 1993; **361**: 315–325.

12. Perlmutter RM, Levin SD, Appleby MW, Anderson SJ and Alberola-Ila J. Regulation of lymphocyte function by protein phosphorylation. *Ann. Rev. Immunol.* 1993; **11**: 451–499.

13. Le Deist F, Hivroz C, Partiseti M, Thomas C and Buc HA. Oleastro. A primary T-cell immunodeficiency associated with defective transmembrane Calcium Influx. *Blood* 1995 (in press).

14. Schandené L, Fester A, Mascart-Lemone F, Crusiaux A, Gérard C and Marchant A. T helper type 2-like cells and therapeutic effects of interferon-gamma in combined immunodeficiency with hypereosinophilia (Omenn's syndrome). *Eur. J. Immunol.* 1993; **23**: 56–60.

15. Perandones CE, Illera VA, Peckham D, Stunz LL and Ashman RF. Regulation of apoptosis in vitro in mature murine spleen T cells. *J. Immunol.* 1993; **151**: 3521–3529.

16. Griscelli C, Lisowska-Grospierre B and Mach B. Combined immunodeficiency with defective expression in MHC class II genes. *Immunodef. Rev.* 1989; **1**: 135–153.

17. De la Salle H, Hanau D, Fricker D, Urlacher A, Kelly A, Salamero J *et al.* Homozygous human TAP peptide transporter mutation in HLA class I deficiency. *Science* 1994; **265**: 237–241.

18. Van Kaer L, Ashton-Rickardt PG, Ploegh HL and Tonegawa S. TAP mutant mice are deficient in antigen presentation, surface class I molecules and CD4−CD8+ T cells. *Cell* 1992; **71**: 1205.

19. Raulet DH. MHC class I-deficient mice. *Adv. Immunol.* 1994; **55**: 381–421.

20. Moretta L, Ciccone E, Poggi A, Mingari MC and Moretta A. Ontogeny, specific functions and receptors of human natural killer cells. *Immunol. Let.* 1994; **40**: 83–88.

21. Griscelli C, Lisowska-Grospierre B and Mach B. Combined immunodeficiency with defective expression in MHC class II genes. In: Rosen FS and Seligman M (eds) *Immunodeficiencies.* Chur, Switzerland: Harwood Academic, 1993: 141–154.

22. Mach B. MHC class II regulation—Lessons from a disease. *N. Engl. J. Med.* 1995; **332**: 120–125.

23. Glimcher LH and Kara CJ. Sequences and factors: a guide to MHC class-II transcription. *Ann. Rev. Immunol.* 1992; **10**: 13–49.

24. Hume CR and Lee JS. Congenital immunodeficiencies associated with absence of HLA class II antigens on lymphocytes results from distinct mutations in trans-acting factors. *Hum. Immunol.* 1989; **26**: 288–309.

25. Benichou B and Strominger JL. Class II-antigen-negative patients and mutnt B-cell lines represent at least three, and probably four, distinct genetic defects defined by complementation analysis. *Proc. Natl. Acad. Sci. USA* 1991; **88**: 4285–4288.

26. Reith W, Satola S, Herrero-Sanchez C, Amaldi I, Lisowska-Grospieree B, Griscelli C *et al.* Congenital immunodeficiency with a regulatory defect in MHC class II gene expression lacks a specific HLA-DR promoter binding protein. *Cell* 1988; **53**: 897–906.

27. Steimle V, Otten LA, Zufferey M and Mach B. Complementation cloning of an MHC class II transactivator mutated in hereditary MHC class II deficiency (or Bare lymphocyte syndrome). *Cell* 1993; **75**: 135–146.

28. Steimle V, Siegrist CA, Mottet A, Lisowska-Grospierre B and Mach B. Regulation of MHC class II expression by IFN-γ is mediated by the transactivator gene CIITA. *Science* 1994; **265**: 106–109.

29. Mach BD. MHC class II deficiency, a disease of gene regulation. Workshop on immunodeficiencies of genetic origin, Instituto Juan March, Madrid, Spain 1995.

30. Durand B, Kobr M, Reith W and Mach B. Functional complementation of MHC class II regulatory mutants by the purified X box binding protein RFX. *Mol. Cell. Biol.* 1994; **14**: 6839–6847.

31. Wolf HM, Hauber I, Gulle H, Thon V, Eggenbauer H, Fischer MB *et al.* Twin boys with major histocompatibility complex class II deficiency but inducible immune responses. *N. Engl. J. Med.* 1995; **332**: 86–90.

32. Lisowska-Grospierre B, Charro DJ and De Préval. A defect in the regulation of major histocompatibility complex class II gene expression in HLA-DR negative lymphocytes from patients with combined immunodeficiency syndrome. *J. Clin. Invest.* 1985; **26**: 381–385.

33. Demotz S, Grey HM and Sette A. The minimal number of Class II MHC–antigen complexes needed for T cell activation. *Science* 1990; **249**: 1028–1030.

34. Callard RE, Armitage RJ, Fanslow WC and Spriggs MK. CD40 ligand and its role in X-linked hyper-IgM syndrome. *Immnol. Today* 1993; **14**: 559–564.

35. Clark EA and Ledbetter JA. How B and T cells talk to each other. *Nature* 1994; **367**: 425–428.

36. Notarangelo LD, Duse M and Ugazio AG. Immunodeficiency with hyper-IGM. *Immunodef. Rev.* 1992; **3**: 101–121.

37. Laman JD, Claassen E and Noelle RJ. Immunodeficiency due to a faulty interaction between T cells and B cells, *Curr. Opin. Immunol.* 1994; **6**: 636–641.

38. Durandy A, Schiff C, Bonnefoy JY, Forveille M, Rousset F and Mazzei G. Induction by anti-CD40 antibody or soluble CD40 ligand and cytokines of IgG, IgA and IgE production by B cells from patients with X-linked hyper IgM syndrome. *Eur. J. Immunol.* 1993; **23**: 2294–2299.

39. Mayer L, Kwan SP, Thompson C, Ko HS, Chiorazzi N and Waldmann T. Evidence for a defect in switching T cells in patients with immunodeficiency and hyperimmunoglobulinemia M. *N. Engl. J. Med.* 1986; **314**: 409–413.

40. Geha RF. CD40 activation and immunodeficiency. Workshop on immunodeficiencies of genetic origin. Instituto Juan March, Madrid, Spain 1995.

41. Corbí AL, Larson RS, Kishimoto TK, Springer TA and Morton CC. Chromosomal location of the genes encoding the leukocyte adhesion receptors LFA-1, Mac-1 and p150, 95: identification of a gene cluster involved in cell adhesion. *J. Exp. Med.* 1988; **167**: 1597–1607.

42. Arnout MA and Michisita M. Genetic abnormalities in leukocyte adhesion molecule deficiency. In: Gupta S and Griscelli C (eds) *New Concepts in Immunodeficiency Disease*. Chichester: John Wiley, 1993: 191–202.

43. Fischer A, Lisowska-Grospierre B, Anderson DC and Springer TA. Leukocyte adhesion deficiencies: molecular basis and functional consequences. *Immunodef. Rev.* 1988; **1**: 39–54.

44. Arnout MA. Leukocyte adhesion molecules deficiency: its structural basis, pathophysiology and implications for modulating the inflamatory response. *Immunol. Rev.* 1990; **114**: 145–180.

45. Van der Wiel van kemenade E, Te Velde AA, De Boer AJ, Weening RS, Fischer A and Melief CJ. Both LFA-1 positive and deficient T cell clones require the CD2/LFA-3 interaction for specific cytolitic activation. *Eur. J. Immunol.* 1992; **22**: 1467–1475.

46. Vennegoor CJ, va der Wiel van Kemenade E, Huijbens RJ, Sanchez-Madrid F, Melief CJ and Figdor CG. Role of LFA-1 and VLA-4 in the adhesion of cloned normal and LFA-1 (CD11/CD18)-deficient T cells to cultured endothelial cells. Indication for a new adhesion pathway. *J. Immunol.* 1992; **148**: 1093–1101.

47. Zocchi MR and Poggi A. NCAM and lymphocyte adhesion in leukocyte adhesion deficiency (LAD) syndrome. *Immunol. Today* 1993; **14**: 94–95.

48. Regueiro JR, Arnaiz-Villena A, Ortiz de Landázuri M, Martín-Villa JM, Vicario JL, Pascual-Ruiz V *et al.* Familial defect of CD3 (T3) expression by T cells associated with rare gut epithelial cell autoantibodies. *Lancet* 1986; **1**: 1274–1275.

49. Regueiro JR, López-Botet M, Ortiz de Landázuri M *et al.* An in vivo functional system lacking polyclonal T-cell surface expression of the CD3/TCR (WT31) complex. *Scand. J. Immunol.* 1987; **26**: 699–708.

50. Arnaiz-Villena A, Pérez-Aciego P, Ballestin C, Sotelo T, Pérez-Seoane C, Martín-Villa JM *et al.* Biochemical basis of a novel T lymphocyte receptor immunodeficiency by immunohistochemistry: a possible CD3γ abnormality. *Lab. Invest.* 1991; **64**: 675–681.

51. Pérez-Aciego P, Alarcón B, Arnaiz-Villena A, Terhorst C, Timón M, Segurado O *et al.* Expression and function of a novel T lymphocyte receptor complex lacking CD3γ. *J. Exp. Med.* 1991; **174**: 319–326.

52. Arnaiz-Villena A, Timón M, Corell A, Pérez-Aciego P, Martin-Villa JM and Regueiro JR. Primary immunodeficiency caused by mutations in the gen encoding the CD3γ

53. Alarcón B, Therhorst C, Arnaiz-Villena A, Pérez-Aciego P and Regueiro JR. Congenital T-cell receptor immunodeficiencies in man. *Immunodef. Rev.* 1990; **2**: 1–16.

54a. Regueiro JR, Pérez-Aciego P, Aparicio P, Martínez AC, Morales P and Arnaiz-Villena A. Low IgG2 and polysaccharide response in a T cell receptor expression defect. *Eur. J. Immunol.* 1990; **20**: 2411–2416.

54b. Martín-Villa JM, Regueiro JR, De Juan MD, Pérez-Aciego P, Pérez-Blas M, Manzanares J, Varela G and Arnaiz-Villena A. T-lymphocyte dysfunctions occurring together with apical gut epithelial cell autoantibodies. *Gastroenterology* 1991; **101**: 390–397.

55. Timón M, Arnaiz-Villena A, Rodríguez-Gallego C, Pérez-Aciego P, Pacheco A and Regueiro JR. Selective disbalances of peripheral blood T lymphocyte subsets in human CD3γ deficiency. *Eur. J. Immunol.* 1993; **23**: 1440–1444.

56. Regueiro JR, Timón M, Pérez-Aciego P, Corell A, Matin-Villa JM, Rodríguez-Gallego C *et al.* From pathology to physiology of the human T-lymphocyte receptor. *Scand. J. Immunol.* 1992; **36**: 363–369.

57. Volarevic S, Niklinska BB, Burns CM, June CH, Weisman AM and Ashwell JD. Regulation of TCR signalling by CD45 lacking transmembrane and extracellular domains. *Science* 1993; **260**: 541–544.

58. Thoenes G, Le Deist F, Fischer A, Griscelli C and Lisowska-Grospierre B. Immunodeficiency associated with defective expression of the T-cell receptor-CD3 complex. *N. Engl. J. Med.* 1990; **322**: 1399.

59. Soudais C, de Villarty J-P, Le Deist F, Fischer A and Lisowska-Grospierre B. Independent mutations of the human CD3-E gene resulting in a T cell receptor/CD3 complex immunodeficiency. *Nat. Genet.* 1993; **3**: 77–81.

60. Le Deist F, Thoenes G, Corado J, Lisowska-Grospierre B and Fischer A. Immunodeficiency with low expression of the T cell receptor/CD3 complex. Effect on T lymphocyte activation. *Eur. J. Immunol.* 1991; **21**: 1641–1647.

61. Nossal GJV. Negative selection of lymphocytes. *Cell* 1994; **76**: 229–239.

62. Von Boehmer H. Positive selection of lymphocytes. *Cell* 1994; **76**: 219–228.

63. Arpala E, Shahar M, Dadi H, Cohen A and Roifman CM. Defective T cell receptor signaling and CD8[+] thymic selection in humans lacking ZAP-70 kinase. *Cell* 1994; **76**: 947–958.

64. Elder ME, Lin D, Clever J, Chan AC, Hope TJ and Weiss A. Human severe combined immunodeficiency due to a defect in ZAP-70, a T cell tyrosine kinase. *Science* 1994; **264**: 1596–1599.

65. Chan A, Kadlecek TA, Elder ME, Filipovich AH, Kuo W-L and Iwashima M. ZAP-70 deficiency in an autosomal recessive form of severe combined immunodeficiency. *Science* 1994; **264**: 1599–1601.

66. Penninger JM, Wallace VA, Kishihara K and Mak T. The role of p56[lck] and p59[fyn] tyrosine kinases and CD45 tyrosine phosphatase in T cell development and clonal selection. *Immunol. Rev.* 1993; **135**: 183–214.

67. Noguchi M, Yi H, Rosenblatt HM, Filipovich AH, Adelstein S and Modi WS. Interleukin-2 receptor γ chain

mutation results in X-linked severe combined immunodeficiency in humans. *Cell* 1993; **73**: 147–157.

68. Russell SM, Johnston JA, Noguchi M, Kawamura M, Bacon CM, Friedman M *et al.* Interaction of IL-2Rβ and γc chains with JAK1 and JAK3: implications for XSCID and XCID. *Science* 1994; **266**: 1042–1045.

69. Leonard WJ, Noguchi M, Russel SM and McBride OW. The molecular basis of X-linked severe combined immunodeficiency: the role of the interleukin-2 receptor γ chain as a common γ chain, γc. *Immunol. Rev.* 1994; **138**: 61–86.

70. Leonard WJ. The defective gene in X-linked severe combined immunodeficiency encodes a shared interleukin receptor subunit: implications for cytokine pleiotropy and redundancy. *Curr. Opin. Immunol.* 1994; **6**: 631–635.

71. Kündig TM, Schorle H, Bachmann MF, Hengartner H, Zinkernagel R and Horak I. Immune responses in interleukin-2-deficient mice. *Science* 1993; **262**: 1059–1061.

72. Ishii N, Asao H, Kimura Y, Takeshita T, Nakamura M and Tsuchiya S. Impairment of ligand binding and growth signalling of mutant IL-2 receptor-γ-chains in patients with X-linked severe combined immunodeficiency. *J. Immunol.* 1994; **153**: 1310–1317.

73. DiSanto JP, Dautry-Varsat A, Certain S, Fischer A and de Saint Basile G. Interleukin-2 (IL-2) receptor γ chain mutations in X-linked severe combined immunodeficiency disease result in the loss of high affinity IL-2 receptor binding. *Eur. J. Immunol.* 1994; **24**: 475–479.

74. Conley ME. X-linked severe combined immunodeficiency. *Clin. Immunol. Immunopathol.* 1991; **61S**: 94–99.

75. Conley ME. X-linked severe combined immunodeficiency. In: Gupta S, Griscelli C (eds) *New Concepts in Immunodeficiency Diseases.* Chichester: John Wiley, 1993: 159–176.

76. Sleasman JW, Harville TO, White GB, George JF, Barrett DJ and Goodenow MM. Arrested rearrangement of TCR Vβ genes in thymocytes from children with X-linked severe combined immunodeficiency disease. *J. Immunol.* 1994; **153**: 442–448.

77. Gougeon M-L, Drean G, Le Deist F, Dousseau M, Fevrier M and Diu A. Human severe combined immunodeficiency disease. Phenotypic and functional characteristics of peripheral blood lymphocytes. *J. Immunol.* 1990; **145**: 2873–2879.

78a. de Saint Basile G, DiSanto J, Dautry-Varsat A, Le Deist M, Cavazzana-Calvo S and Hacein-Bey S. X-SCID caused by IL2Rγ mutations. Workshop on immunodeficiencies of genetic origin. Instituto Juan March, Madrid, Spain, 1995.

78b. Macchi P, Villa A, Giliani S, Sacco MG, Frattini A, *et al.* Mutations of Jak-3 gene in patients with autosomal severe combined immune deficiency (SCID). *Nature* 1995; **377**: 65–68.

79. Hirschhorn R. Adenosine deaminase deficiency. *Immunodef. Rev.* 1990; **2**: 175–198.

80. Hirschhorn R. Overview of biochemical abnormalities and molecular genetics of adenosine deaminase deficiency. *Pediatr. Res.* 1993; **33S**: 35–41.

81. Kredich NM and Hershfield MS. Immunodeficiency disease associated with adenosine deaminase deficiency and purine nucleoside phosphorylase deficiency. In: Scriver CM, Beaudet AI, Sly W and Valle D (eds) *The Metabolic Basis of Inherited Disease.* New York: McGraw-Hill 1989: 1091–1108.

82. Buc HA, Moncion A, Hamet M, Houllier A-M, Thuilier L, Perignon J-L *et al.* Influence of adenosine deaminase inhibition on the phosphoinositide turnover in the initial stages of human T cell activation. *Eur. J. Immunol.* 1990; **20**: 611–615.

83. Kameoka J, Tanaka T, Nohima Y, Schlossman SF and Morimoto C. Direct association of adenosine deaminase with T cell activation antigen CD26. *Science* 1993; **261**: 466–469.

84. Markett ML. Purine nucleoside phosphorylase deficiency. *Immunodef. Rev.* 1991; **3**: 45–81.

85. Derry JMJ, Ochs HD and Francke U. Isolation of a novel gene mutated from Wiskott–Aldrich syndrome. *Cell* 1994; **78**: 635–644.

86. Villa A, Notarangelo L, Macchi P, Mantuano E, Cavagni G, Brugnoni D *et al.* X-linked thrombocytopenia and Wiskott–Aldrich syndrome are allelic diseases with mutations in the WASP gene. *Nat. Genet.* 1995; **9**: 414–417.

87. Ochs H, Zhu Q, Zhang M, Derry JMJ, Blaese M and Junker A. Mutations of the WASP gene cause classic Wiskott–Aldrich syndrome and X-linked thrombocytopenia. Workshop on immunodeficiencies of genetic origin, Instituto Juan March, Madrid, Spain, 1995.

88. Peacocke M and Siminovitch KA. Wiskott–Aldrich syndrome: New molecular and biomedical insights. *J. Amer. Acad. Dermatol.* 1992; **27**: 507–519.

89. Rosenstein Y, Park JK, Biere BE and Burakoff S. The Wiskott–Aldrich syndrome: an immunodeficiency associated with defects of the CD43 molecule. In: Gupta S, Griscelli C (eds) *New Concepts in Immunodeficiency Disease.* Chichester: John Wiley, 1993: 249–268.

90. Remold-O'Donnell E and Rosen FS: Sialophorin (CD43) and the Wiskott–Aldrich syndrome. *Immunodef. Rev.* 1991; **2**: 151–174.

91. Mantuano E, Candotti F, Gigliani S, Parolini O, Lusardi M and Zucchi M. Analysis of X-chromosome inactivation in bone marrow precursors from carriers Wiskott–Aldrich syndrome and X-linked severe combined immunodeficiency: evidence that the Wiskott–Aldrich gene is expressed prior to granulocyte-macrophage colony-forming-unit. *Immunodeficiency* 1993; **4**: 271–276.

92. Molina IJ, Kenney DM, Rosen FS and Remold-O'Donell E. T cell lines characterize events in the pathogenesis of the Wiskott–Aldrich syndrome. *J. Exp. Med.* 1992; **176**: 867–874.

93. Molina IJ, Sancho J, Terhorst C, Rosen F and Remold-O'Donell. T cells of patients with the Wiskott–Aldrich syndrome have a restricted defect in proliferative responses. *J. Immunol.* 1993; **151**: 4383–4390.

94. Siminovitch KA, Greer WL, Axelsson B, Rubin L, Novogrodsky A and Peacocke M. Selective impairment of CD43-mediated T cell activation in the Wiskott–Aldrich syndrome. *Immunodeficiency* 1993; **4**: 99–108.

95. Rosen FS. Wiskott–Aldrich syndrome. Workshop on immunodeficiencies of genetic origin. Instituto Juan March, Madrid, Spain, 1995.

96. Itoh N, Yonehara S, Mizushima S-I, Sameshima M, Hase A, Seto Y *et al.* The polypeptide encoded by the cDNA of human cell surface antigen FAS can mediate apoptosis. *Cell* 1991; **66**: 233–243.

97. Takahashi T, Tanaka M, Inazawa J, Abe T, Suda T and Nagata S. Human FAS ligand: gene structure, chromosomal location and species specificity. *Int. Immunol.* 1994; **6**: 1567–1574.

98. Schulze-Osthoff K. The FAS/APO-1 receptor and its deadly ligand. *Trends Cell Biol.* 1994; **4**: 421–426.

99. Nagata S and Suda T. FAS and FAS ligand: *lpr* and *gld* mutations. *Immunol. Today* 1995; **16**: 39–43.

100. Le Deist F, Hivroz C, Partiseti M, Rieux-Laucat F, Debatin KM and Choquet D. T cell activation deficiencies. In: Caragol I, Español T, Fontan G and Matamoros N (eds) *Progress in Immunodeficiency*, Vol. V. Barcelona: Springer-Verlag Ibérica, 1994: 43–47.

101. Villartay J-P, Rieux-Laucat F, Le Deist F, Hivroz C, Chasseval R and Roberts IAG. APO-1/FAS/CD95 mutations in human lymphoproliferative syndrome associated with autoimmunity and defective apoptosis. Workshop on immunodeficiencies of genetic origin, Instituto Juan March, Madrid, Spain, 1995.

102. Lewis SM. The mechanism of V(D)J rejoining: lessons from molecular, immunological and comparative analysis. *Adv. Immunol.* 1994; **56**: 859–870.

103. Schwarz K, Hansen-Hagge TE and Bartram CR. Recombinase deficiency in mouse and man. *Immunodeficiency* 1993; **4**: 249–252.

104. Bosma MJ. B and T cell leakiness in the SCID mouse mutant. *Immunodef. Rev.* 1992; **3**: 261–276.

105. Hendrickson EA, Qin XQ, Bump EA, Schatz DG, Oettinger M and Weaver DT. A link between double-strand break-related repair and V(D)J recombination: the SCID mutation. *Proc Natl Acad Sci USA* 1991; **88**: 4061–4065.

106. Blunt T, Finnie N, Taccioli GE, Smith GCM, Demengeot J and Gottlieb TM. Defective DNA-dependent protein tyrosine kinase activity is linked to V(D)J recombination and DNA repair defects associated with the murine SCID mutation. *Cell* 1995; **80**: 813–823.

107. Fischer A. Severe combined immunodeficiencies. *Immunodef. Rev.* 1992; **3**: 83–100.

108. Abe T, Tsuge I, Kamachi Y, Torii S, Utsumi K and Akahori Y. Evidence for defects in V(D)J rearrangements in patients with severe combined immunodeficiency. *J. Immunol.* 1994; **152**: 5504–5513.

109. Wirt DP, Brooks EG, Baidya S, Klimpel GR, Waldman TA and Goldblum RM. Novel T-lymphocyte population in combined immunodeficiency with features of graft-versus-host disease. *N. Engl. J. Med.* 1989; **321**: 370–374.

110. de Saint Basile G, Le Deist F, de Villatay J-P. Restricted heterogeneity of T lymphocytes in combined immunodeficiency with Omenn's syndrome. *J. Clin. Invest.* 1991; **87**: 1352–1359.

111. Melamed I, Cohen A and Roifman CM. Expansion of CD3$^+$CD4$^-$CD8$^-$ T cell population expressing high levels of IL-5 in Omenn's syndrome. *Clin. Exp. Immunol.* 1994; **95**: 14–21.

112. Schandené L, Ferster A, Mascart-Lemone F, Crusiaux A and Gérard C. T helper type 2-like cells and therapeutic effects of IFN-γ in combined immunodeficiency with hypereosinophilia (Omenn's syndrome). *Eur. J. Immunol.* 1993; **23**: 56–60.

113. Barnes D, Tomkinson AE, Lehmann AR, Webster DB and Lindahl T. Mutations in the DNA ligase I gene of an individual with immunodeficiencies and cellular hypersensitivity to DNA-damaging agents. *Cell* 1992; **69**: 495–503.

114. Petrini JH, Donovan JW, Dimare C and Weaver D. Normal V(D)J coding junction formation in DNA ligase I deficiency syndromes. *J. Immunol.* 1994; **152**: 176–183.

115. Swift M. Genetic aspects of ataxia-telangiectasis. *Immunodef. Rev.* 1990; **2**: 67–81.

116a. Gatti RA. Ataxia-telangiectasia: Genetic studies. In: Gupta S and Griscelli C (eds) *New Concepts in Immunodeficiency Diseases*. Chichester: John Wiley, 1993: 203–229.

116b. Savitsky K, Bar-Shira A, Gilad S, Rotman G, Ziv Y *et al.* A single Ataxia Telangiectasia Gene with a product similar to PI-3-kinase. *Science* 1995; **268**: 1749–1753.

117. Kastan MB, Zhan Q, El-Delry WS, Carrier F, Jacks T, Walsh WV. A mammalian cell cycle checkpoint pathway utilizing p53 and GADD45 is defective in ataxiatelangiectasia. *Cell* 1992; **71**: 587–597.

118. Chatila T, Wong R, Young M, Miller R, Terhorst C and Gega RS. An immunodeficiency characterized by defective signal transduction in T lymphocytes. *N. Engl. J. Med.* 1989; **320**: 696–702.

119. Rijkers G, Scharenberg JGM, Van Dongen JJM, Neijens HJ and Zegers BJM. Abnormal signal transduction in a patient with severe combined immunodeficiency disease. *Pediatr. Res.* 1991; **29**: 306–309.

120. Hivroz C, Le Deist F, Buc HA, Griscelli C and Fischer A. Functional T cell immunodeficiency characterized by defective TCR/CD3 induced tyrosine phosphorylation. *Immunodeficiency* 1993; **4**: 131–132.

121. Rodríguez-Gallego C, Arnaiz-Villena A, Corell A, Manzanares J, Timón M and Pacheco A. Primary T lymphocyte immunodeficiency associated with a selective impairment of CD2, CD3, CD43 (but not CD28)-mediated signal transduction. *Clin. Exp. Immunol.* 1994; **97**: 386–391.

122. DiSanto J, Keever CA, Small TN, Nichols GL, O'Reilly RJ and Flomenberg N. Absence of interleukin 2 production in a severe combined immunodeficiency disease syndrome with T cells. *J. Exp. Med.* 1990; **171**: 1697–1704.

123. Weinberg K and Parkman R. Severe combined immunodeficiency due to a specific defect in the production of interleukin-2. *N. Engl. J. Med.* 1990; **322**: 1718–1723.

124. Chatila T, Castigli E, Pahwa R, Pahwa S, Chirmule N and Oyaizu N. Primary combined immunodeficiency resulting from defective transcription of multiple lymphokine genes. *Proc. Natl. Acad. Sci. USA* 1990; **87**: 10033–10037.

125. Castigli E, Pahwa R, Good R, Geha R and Chatila T. Molecular basis of a multiple lymphokine deficiency in a patient with severe combined immunodeficiency. *Proc. Natl. Acad. Sci. USA* 1993; **90**: 4728–4732.

126. Haneke E. Papillon-Lefèvre syndrome: keratosis palmoplantaris with periodontopathy: report of a case and review of the cases in the literature. *Hum. Genet.* 1979; **51**: 1–35.

127. Djwari D. Deficient phagocytic function in Papillon-Lefèvre syndrome. *Dermatologica* 1978; **156**: 189.

128. Leo Y, Wollner S and Hacham-Zadah S. Immunological study of patients with Papillon-Lefèvre syndrome. *Clin. Exp. Immunol.* 1980; **40**: 407.

129. Celenligil H, Kansu E, Ruacan S and Eratalay K. Papillon-Lefèvre syndrome: characterization of peripheral blood and gingivial lymphocytes with monoclonal antibodies. *J. Clin. Periodontol.* 1991; **13**: 392.

130. Góngora R, Corell A, Regueiro JR, Carasol M, Rodríguez-Gallego C and Paz-Artal E. Peripheral blood reduction of memory CD29$^+$, CD45RO$^+$, and "bright" CD2$^+$ and LFA-1$^+$ T lymphocytes in Papillon-Lefèvre syndrome. *Hum. Immunol.* 1994; **41**: 185–192.

131. Kellum RE. Papillon-Lefèvre syndrome in four siblings treated with etretinate: a nine year evaluation. *Int. J. Dermatol.* 1989; **28**: 605.

132. Semba RD, Muhilal, Ward BJ, Griffin DE, Scott AL and Natadisastra G. Abnormal T-cell subset proportions in vitamin-A-deficient children. *Lancet* 1993; **341**: 5.

133. Hong R. The DiGeorge anomaly. *Immunodef. Rev.* 1991; **3**: 1–14.

134. Lammer EJ and Opitz JM. The DiGeorge anomaly as a developmental field defect. *Am. J. Med. Genet.* 1986; **2(S)**: 113–127.

135. Driscoll DA, Budarf ML and Emanuel BS. A genetic etiology for Di George syndrome: consistent deletions and microdeletions of 22q11. *Am. J. Hum. Genet.* 1992; **50**: 924–933.

136. Carey AH, Kelly D, Halford S, Wadey R, Wilson D and Goodship D. Molecular genetic study of the frequency of monosomy 22q11 in Di George syndrome. *Am. J. Hum. Genet.* 1992; **51**: 964–970.

137. Mayer L and Fu SM, Cunningham-Rundles C and Kunkel HG. Polyclonal immunoglobulin secretion in patients with common variable immunodeficiency using monoclonal B cells factor. *J. Clin. Invest.* 1984; **75**: 2115–2120.

138. Jaffe JS, Eisenstein E, Sneller MC and Strober W. T-cell abnormalities in common variable immunodeficiency. *Pediat. Res.* 1993; **33(S)**: 24–28.

139. Farrington M, Grosmaire LS, Nonoyama S, Fischer SH, Hollenbaugh D and Ledbetter JA. CD40 ligand expression is defective in a subset of patients with common variable immunodeficiency. *Proc. Natl. Acad. Sci. USA* 1994; **91**: 1099–1113.

140. Olerup O, Smith CIE, Björkander J and Hammarström L. Shared HLA class II associated genetic susceptibility and resistance, related to the HLA-DQB1 gene, in IgA deficiency and common variable immunodeficiency. *Proc. Natl. Acad. Sci. USA* 1992; **89**: 10653–10657.

141. Volanakis JE, Zhu ZB, Schaffer FM, Macon KJ, Palermos J and Barger BO. Major histocompatibility class III genes and susceptibility to immunoglobulin A deficiency and common variable immunodeficiency. *J. Clin. Invest.* 1992; **89**: 1914–1922.

142. Olsson PG, Hammarström L, Cox DW and Smith CIE. Involvement of both HLA and Ig heavy chain haplotypes in human IgA deficiency. *Immunogenetics* 1992; **36**: 389–395.

143. Finkel TH and Banda NK. Indirect mechanisms of HIV pathogenesis: how does HIV kill T cells? *Curr. Opin. Immunol.* 1994; **6**: 605–615.

144. Ameisen JC: The programmed cell death theory of AIDS pathogenesis: implications, testable predictions, and confrontation with experimental findings. *Immunodef. Rev.* 1992; **3**: 237–246.

145. Subramanyam M, Gutheil WG, Bachovchin WW and Huber BI. Mechanism of HIV-1 Tat-induced inhibition of antigen-specific T cell responsiveness. *J. Immunol.* 1993; **150**: 2544–2553.

146. Martín-Villa JM, Camblor S, Costa R and Arnaiz-Villena A. Gut epithelial cell autoantibodies in AIDS pathogenesis. *Lancet* 1993; **342**: 380.

147. Romagnani S and Maggi E. Th1 versus Th2 responses in AIDS. *Curr. Opin. Immunol.* 1994; **4**: 616–622.

148. Pérez-Blas M, Regueiro JR, Ruiz-Contreras J and Arnaiz-Villena A. T lymphocyte anergy during acute infectious mononucleosis is restricted to the clonotypic receptor activation pathway. *Clin. Exp. Immunol.* 1992; **89**: 83–88.

149. Hsu DH, de Waal-Malefyt R, Fiorentino DF *et al.* Expression of interleukin-10 activity by Epstein–Barr virus protein BCRF-1. *Science* 1990; **250**: 830–832.

150. Howard M. and Garra A. Biological properties of interleukin 10. *Immunol. Today* 1992; **13**: 198–200.

151. Timón M, Arnaiz-Villena A, Ruiz-Contreras J, Ramos-Mador JT, Pacheco A and Regueiro JR. Selective impairment of T lymphocyte activation through the T cell receptor/CD3 complex after cytomegalovirus infection. *Clin. Exp. Immunol.* 1993; **94**: 38–42.

152. Beck S and Barrel BG. Human cytomegalovirus encodes a glycoprotein homologous to MHC class I antigens. *Nature* 1988; **331**: 269–272.

153. de Waal Malefyt R, Yssel H, Spits H *et al.* Human T cell leukemia virus type I prevents cell surface expression of the T cell receptor through down-regulation of the CD3γ,δ,ϵ and zeta genes. *J. Immunol.* 1990; **145**: 2297–2303.

154. Beraud C, Sun SC, Ganchi P, Ballard DW and Greene WC. Human T-cell leukemia virus type I Tax associates with and is negatively regulated by the NF-kB2 p100 gene product: implications for viral latency. *Mol. Cell. Biol.* 1994; **14**: 1374–1382.

155. Lusso P, Malnati M, De Maria A, Balotta C, Derocco SE and Markham PD. Productive infection of CD4$^+$ and CD8$^+$ human T cell populations and clones by human herpesvirus 6. Transcriptional down-regulation of CD3. *J. Immunol.* 1991; **147**: 685–691.

156. Furukawa M, Yasukawa M, Yakushijin Y and Fujita S. Distinct effects of human herpesvirus 6 and human herpesvirus 7 on surface molecule expression and function of CD4$^+$ T cells. *J. Immunol.* 1994; **152**: 5768–5775.

157. Rinaldo CR and Torpey DJ. Cell-mediated immunity and immunosuppression in herpes simplex virus infection. *Immunodeficiency* 1993; **5**: 33–90.

158. Gooding LR. Virus proteins that counteract host immune defenses. *Cell* 1992; **71**: 5–7.

159. Ahuja SK, Gao J-L and Murphy PM. Chemokine receptors and molecular mimicry. *Immunol Today* 1994; **15**: 281–287.

160. Akbar AN, Borthwick N, Salmon M *et al.* The significance of low bcl-2 expression by CD45R0 T cells in normal individuals and patients with acute viral infections.

The role of apoptosis in T cell memory. *J. Exp. Med.* 1993; **178**: 427–438.

161. Nelson RD, Shibata N, Podzorski RP and Herron MJ. Candida mannan: chemistry, supression of cell-mediated immunity, and possible mechanisms of action. *Clin. Microbiol. Rev.* 1991; **4**: 1–19.

162. Wilson RA. Immunity and immunoregulation in helminth infections. *Curr. Opin. Immunol.* 1993; **5**: 538–847.

163. Sher A and Coffman RL. Regulation of immunity to parasites by T cells and T cell-derived cytokines. *Ann. Rev. Immunol.* 1992; **10**: 385–409.

164. Kierszenbaum F and Sztein MB. *Trypanosoma cruzi* suppresses the expression of the p75 chain of interleukin-2 receptors on the surface of activated helper and cytotoxic human lymphocytes. *Immunology* 1992; **75**: 546–549.

165. Sztein MB and Kierszenbaum F. Suppression by *Trypanosoma cruzi* of T cell receptor expression by activated human lymphocytes. *Immunology* 1992; **77**: 277–283.

166. Reed SG and Scott P. T-cell and cytokine responses in *leishmaniasis*. *Curr. Opin. Immunol.* 1993; **5**: 524–531.

167. Sheela R, Ilangumaran S and Muthukkaruppan V. Flow cytometric analysis of CD2 modulation on human peripheral blood T lymphocytes by Dharmendra preparation of *Mycobacterium leprae*. *Scand. J. Immunol.* 1991; **33**: 203–209.

168. Moreno C, Mehlert A and Lamb J. The inhibitory effects of micobacterial lipoarabinomannan and polysaccharides upon polyclonal and monoclonal human T cell proliferation. *Clin. Exp. Immunol.* 1988; **74**: 206–210.

169. Roux M, Schraven B, Roux A, Gamm H, Mertelsman R and Meuer S. Natural inhibitors of T-cell activation in Hodgkin's disease. *Blood* 1991; **78**: 2365–2371.

170. Mizoguchi H, O'Shea JJ, Longo DL, Loeffler CM, McVicar DW and Ochoa AC. Alterations in signal transduction molecules in T lymphocytes from tumor-bearing mice. *Science* 1992; **258**: 1795–1798.

171. Leung DY and Geha RS. Immunoregulatory abnormalities in atopic dermatitis. *Clin. Rev. Allergy* 1986; **4**: 67–86.

172. Romano MF, Valerio G, Turco MN, Spadaro G, Venuta S and Formisano S. Defect of CD2-and CD3-mediated activation pathways in T cells of atopic patients: role of interleukin 2. *Cell Immunol.* 1992; **139**: 91–97.

173. Romano MF, Turco MC, Stanziola A, Giarrusso PC, Petrella A, Tassone P *et al.* Defect of interleukin-2 production and T cell proliferation in atopic patients: Restoring ability of the CD28-mediated activation pathway. *Cell Immunol.* 1993; **148**: 455–463.

174. Kuchroo V, Das MP, Brown JA, Ranger AM, Zamvil SS and Sobel RA. B7-1 and B7-2 costimulatory molecules activate differentially the Th1/Th2 developmental pathways: Application to autoimmune disease therapy. *Cell* 1995; **80**: 707–718.

175. McCombs CC and Michalski JP. Lymphocyte abnormality associated with HLA-B8 in healthy young adults. *J. Exp. Med.* 1982; **156**: 936.

176. Hashimoto S, McCombs CC and Michlski JP. Mechanism of a lymphocyte abnormality associated with HLA-B8/DR3 in clinically healthy individuals. *Clin. Exp. Immunol.* 1989; **76**: 317–323.

177. Nagel JE, Chopra RK, Chrest FJ *et al.* Decreased proliferation, interleukin-2 synthesis, and interleukin-2 receptor expression are accompanied by decreased mRNA expression in phytohemagglutinin stimulated cells from elderly donors. *J. Clin. Invest.* 1988; **81**: 1096–1102.

178. Al-Rayes H, Pachas W, Mirza N *et al.* IgE regulation and lymphokine patterns in ageing humans. *J. Allergy. Clin. Immunol.* 1992; **90**: 630–636.

179. Chandra RK. Nutritional regulation of immunity and risk of infection in old age. *Immunology* 1989; **67**: 141–147.

180. Kirk SJ, Hurson M and Regan MC. Arginine stimulates wound healing and immune function in the elderly human beings. *Surgery* 1993; **114**: 155–159.

181. Chandra RK. Immunocompetence in protein-energy malnutrition, a historical perspective. *J. Nutr.* 1991; **122**: 597–600.

182. Bendich A and Chandra RK. *Micronutrients and Immune Actions*. New York: New York Academy of Sciences, 1990.

183. Kinsella JE and Lokesh B. Dietary lipids, eicosanoids and the immune system. *Crit. Care Med.* 1990; **18**(S): 94–113

21

HIV and Disruption of T Cell Signalling

Keith E. Nye

Department of Immunology, The Medical College of Saint Bartholomew's Hospital, London, UK

INTRODUCTION

Many questions have still to be answered concerning the immunopathogenesis of human immunodeficiency virus (HIV) infection which leads to the clinical condition known as the acquired immunodeficiency syndrome (AIDS). The virus, either HIV-1 or HIV-2, uses the CD4 molecule as its cell surface receptor. The CD4 antigen is a glycoprotein that interacts with a specific determinant within MHC class II molecules and is thought to act as an accessory or adhesion molecule, stabilizing the MHC class II–foreign antigen–T cell receptor (TCR) interaction. It also cooperates in transmembrane signalling and thus plays a major role in immune recognition and activation. It is found on most thymocytes, two-thirds of the T lymphocyte population and, in humans, on monocytes and macrophages. Dysfunction of $CD4^+$ cells is seen in early HIV infection, while chronic reduction in the numbers of circulating $CD4^+$ lymphocytes is characteristic of AIDS. As these cells have a central role in the immune system it is not surprising that the cells they control will also be abnormal. Thus lack of cytokine production by $CD4^+$ lymphocytes contributes to the defects observed in antigen presenting cells, natural killer cells, cytotoxic T cells and antibody producing B cells. Cytokine production is controlled through various intracellular signalling pathways and disruption of any part of these mechanisms may account for profound changes in the levels of expression of these proteins and therefore of immune function. This chapter describes the disruption of intracellular signalling pathways observed in T lymphocytes obtained from HIV-infected individuals.

PROPOSED MECHANISMS OF AIDS PATHOGENESIS

For over a decade, since the first isolation of the human immunodeficiency virus, scientists have been trying to find a comprehensive hypothesis to explain the diverse signs and symptoms exhibited by individuals infected with the virus. Clinically, the disease presents itself in discrete stages, beginning with an acute influenza-like illness accompanied by a rapid, though transient, fall in $CD4^+$ T cells and viral p24 antigenaemia. Following this brief acute illness is a protracted, variable phase when the patient has normal health and a near normal circulating CD4 lymphocyte count. At this stage circulating anti-HIV antibodies can be detected. In the great majority of cases the asymptomatic period leads on to progressive immunosuppression and lymphocyte dysfunction accompanied by loss of circulating $CD4^+$ T cells, the re-appearance of viral p24 antigenaemia and the onset of opportunistic infections. This phase of symptomatic HIV infection progresses to overt AIDS and death caused by overwhelming infection. It seems unlikely that a single mechanism can explain all of these effects: it is far more likely to be a meld of several mechanisms. Although it has been shown recently [1, 2] that the number of HIV-infected cells in the lymph nodes is very high, even in asymptomatic subjects, the number of peripheral lymphocytes infected is extremely low and it is unlikely that this frequency of infection can induce sufficient direct cytopathic cell killing to account for the immunodeficiency and decline in $CD4^+$ cells characteristic of this disease.

It has been postulated that autoreactivity is another potential mechanism for immunodeficiency and $CD4^+$ cell loss [3, 4], and various mechanisms have been suggested. One hypothesis is based on the chronic antigenic stimulation characteristic of HIV infection causing polyclonal antibody production by B cells, some of which will be autoantibodies. Another theory involves HIV mimicry of MHC, which it is postulated, leads to chronic allogeneic stimulation, inducing autocytotoxicity and autosuppression. A unifying theory was proposed in 1988 by Ascher and Sheppard [5] which they later named the panergic imnesia hypothesis. This hypothesis is based on the concept that the virus delivers a false physiological signal via the CD4 receptor and thus enhances T cell help in an inappropriate manner leading to the clinical signs of HIV disease, lymphadenopathy, autoimmunity and weight-loss. They used the term panergy to represent a condition where pan-activation could lead to anergy. At this time, the mid to late 1980s, CD4 was regarded as merely a cell-adhesion molecule and its role in signal transduction had not been established. However, we and others had been investigating the effect of HIV and its proteins on membrane-associated receptor signalling pathways, in particular the phosphoinositide signalling pathway: the observed effects of the virus on these intracellular signalling pathways were profound and their description forms the rest of this chapter.

HIV INFECTION AND T CELL SIGNALLING

Disruption of lymphocyte signalling may occur at several levels in HIV infection, latency and re-activation. As described above, HIV-1 infection can be divided into discrete clinical stages and these stages are reflected at the cellular level. The initial infection probably takes place when free virion is introduced into the host via infected bodily fluids during sexual intercourse, blood transfusion or intravenous use of narcotic drugs. The envelope glycoprotein (gp120) of the virus binds to the CD4 receptor on the host T cell generating signals responsible for penetration by the virus. Reverse transcription of viral genomic RNA into double-stranded DNA, and integration into the host genome can then occur fairly rapidly. If HIV infects a quiescent cell, the provirus may remain dormant until the host cell is activated resulting in concomitant viral activation and completion of the viral life-cycle. The second level at which signalling pathways are modulated is during T cell, and therefore viral, activation. Finally, HIV can interfere with T cell signalling at the level of transcriptional regulation via nuclear factors which can be regarded as third messengers. Understanding of the key players in these signal transduction pathways may throw light on viral activation and assist in the design of novel therapies against viral activation.

Effect of HIV Infection on Early TCR-coupled Signals

Studies examining the effect of chronic HIV infection on membrane potential and intracellular calcium concentration ($[Ca^{2+}]i$) in the H9 lymphoblastoid cell line [6] demonstrated that the plasma membrane is depolarized in chronically infected cells, rendering it unresponsive to stimulation with either PHA or anti-CD3. In addition, HIV-infected cells show an increased basal $[Ca^{2+}]i$, which suggests that HIV chronically activates T cells. There is, however, little further increase in $[Ca^{2+}]i$ in response to stimulation with PHA or anti-CD3 (Figure 21.1). This calcium flux is due to hydrolysis of phosphatidylinositol 4,5-bisphosphate (PIP_2) to generate the calcium-mobilizing second messenger, IP_3 and these defects in IP_3-dependent calcium flux have been confirmed for the H9 lymphoblastoid cell line and also observed in lymphocytes from normal controls and HIV-infected individuals [7, 8]. One of the most striking observations made in these experiments is a marked change in the kinetics of inositol polyphosphate metabolism in HIV-infected cells. Small, though detectable, elevations in the basal levels of inositol 1,4,5-trisphosphate {Ins (1,4,5) P_3} and inositol 1,3,4,5-tetrakisphosphate {Ins (1,3,4,5) P_4} are present in quiescent HIV-infected T cells, and while the level of Ins (1,4,5) P_3 increases slightly on stimulation with antigen or lectin, the Ins (1,3,4,5) P_4 rises rapidly over the first 30 seconds, and, abnormally, does not return to its elevated basal level for a further 15 minutes. This defective inositol polyphosphate metabolism,

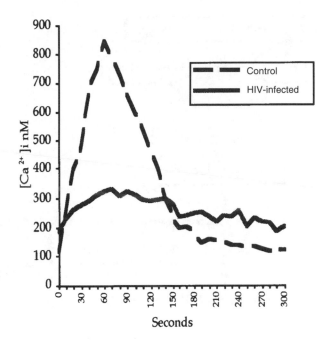

Figure 21.1 Calcium mobilization in HIV-infected lymphocytes. Lymphocytes from either healthy volunteers or HIV-infected individuals, loaded with indo-1 and stimulated with monoclonal anti-CD3 antibody as described in text

although evident in asymptomatic subjects, is much more profound in AIDS patients. The intracellular concentration of inositol 1,4,5-trisphosphate is controlled by the relative activities of the cellular enzymes which catalyse its formation and degradation. Retention of the active isomer in the cell after it has released calcium from the endoplasmic reticulum is detrimental and is, therefore, strictly regulated through one of two pathways. It may either be dephosphorylated directly by the Ins (1,4,5) P_3 / Ins (1,3,4,5) P_4-specific 5-phosphatase and be converted back to inositol or it may be phosphorylated by Ins (1,4,5) P_3-3-kinase to form inositol 1,3,4,5-tetrakisphosphate. The Ins (1,4,5) P_3-3-kinase is activated by the Ca^{2+}/calmodulin complex and there is evidence that the specific 5-phosphatase is regulated by protein kinase C (see Figure 21.2). The HIV-induced defect appears to involve the Ins (1,4,5) P_3/Ins (1,3,4,5) P_4-specific 5-phosphatase.

An increase in the intracellular concentration of ATP has been described in acute viral infection: it is also known that ATP at high physiological concentrations competitively inhibits the activity of Ins

(1,4,5) P_3/Ins (1,3,4,5) P_4-specific 5-phosphatase [9]. We have therefore examined the effect of ATP on this phosphatase and while it is able to attenuate enzyme activity it does not cause the complete inhibition seen in lymphocytes from AIDS patients (see Table 21.1). Thus, although elevated intracellular concentration of ATP may contribute to the signalling defect, it is unable to account for the total dysfunction observed. Indeed, we now have some evidence that a regulatory protein, itself controlled by phosphorylation, may regulate 5-phosphatase activity, in a manner analogous to that described for protein phosphatase-1 [10]. While it is conceivable that HIV or one of its constituent proteins may directly affect the 5-phosphatase, it seems likely, in view of the profound immunological dysfunction observed in HIV disease, that aberrant phosphatase activity may simply be symptomatic of a larger regulatory defect in the infected $CD4^+$ T cell. Currently available evidence suggests that this defect originates at a point proximal to the CD3/TCR/CD4 receptor complex.

One of the enigmas associated with HIV infection is the apparent dichotomy of profound

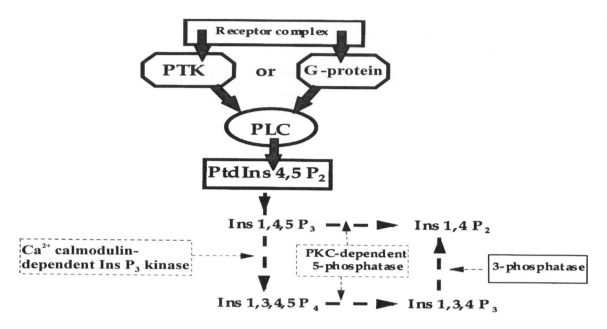

Figure 21.2   Control of the intracellular concentration of Ins 1,4,5 P_3. The dependence on PKC or Ca^{2+}/calmodulin by the 5-phosphatase or 3-kinase respectively, probably provides the negative feedback necessary to limit the duration of the active inositol metabolite in the cell

immunodeficiency stemming from the very low number of lymphocytes shown to be infected in the peripheral blood. The pathogenesis has therefore been attributed to viral proteins such as the envelope protein gp120 and, as described above, a wide range of different, often contradictory, mechanisms have been postulated to explain cell dysfunction and depletion of the $CD4^+$ lymphocyte population. It is well established that the interaction between gp120 and the CD4 molecule is a crucial event in T cell infection and equally well established that the suppression of T cell function emanates from an aberrant signal or signals following receptor ligation. Following the work of Gupta and Vayuvegula [6], several groups investigated the effects of gp120 on T cells [8, 11–13]. Results were variable, partly due to the use of cell lines and partly due to the lack of purity of some of the gp120 preparations. All agreed, however, that the calcium signal, following antigenic stimulation, was attenuated and one paper [12] showed that the CD4 molecule was phosphorylated upon gp120 binding. Although this surface binding was not affected by the addition of H-7, a relatively

Table 21.1   Inhibition of the activity of Ins $(1,4,5)P_3$/Ins$(1,3,4,5)P_4$ specific 5-phosphatase by ATP measured as the production of Ins$(1,3,4)P_3$ (nmol-mg-min). Lysates of cells from healthy controls or HIV infected individuals, each containing 200 μg total protein, incubated with 2 μM[^{3}H]-Ins$(1,3,4,5)P_4$ at 37°C for 10 min

ATP concentration (mM)	Control	Asymptomatic	AIDS
0	0.20 ± 0.012	0.09 ± 0.01	0
2	0.22 ± 0.016	0.07 ± 0.01	0
5	0.14 ± 0.015	0.05 ± 0.01	0
10	0.07 ± 0.010	0.05 ± 0.01	0

non-specific inhibitor of protein kinase C (PKC), viral entry was prevented and it was therefore suggested that PKC and CD4 phosphorylation might play a role in internalization of HIV. Since these experiments were performed we have learned a great deal more of the critical role of CD4-mediated signals in T cell activation and its role as a coreceptor in antigen presentation (Figure 21.3). For example, in the case of the T cell receptor, the CD4 or CD8 surface antigen acts as a coreceptor and other molecules such as CD28 induce separate signals that modify the cellular response to the CD3/TCR/CD4-induced signal. Yet another set of cell surface antigens, the adhesion molecules, increase the avidity of the interaction between the antigen presenting cell and the T lymphocyte, as exemplified by the binding of CD54 (ICAM-1) to the integrins CD11a/CD18 (LFA-1). These molecules, as well as stabilizing the reaction between antigen presenting cell and the T cell, also transduce signals across the membrane. Finally, genetic studies of T cell lines deficient in the protein tyrosine phosphatase CD45, suggest that CD45 is required for signalling through the T cell receptor. The effects of HIV, or its constituent proteins, on these cell surface receptors and their coupling to kinases and phosphatases have been the subject of the majority of recent investigations in the HIV-infected T cell.

Modulation of Signalling by gp120

The greatest enigma and obstacle to our comprehension of the pathology of HIV infection is that we do not understand how the presence of the HIV genome in a minority of CD4$^+$ T cells can lead to the gradual destruction of the entire immune system. Hypotheses have been advanced which propose that the infected cells indirectly impair the function of uninfected cells [3,4,14,15]. There is a body of *in vitro* evidence that binding of the viral envelope glycoproteins gp41, gp120 or the combined gp160, to the cellular CD4 receptor may interfere with the normal function of the CD4$^+$ T cell [11, 16–18]. This may be mediated either at the level of the MHC class II–CD4 interaction, through steric hindrance, or by perturbation of the CD4-dependent signalling pathway.

Early studies of the effect of gp120 on T cell function often presented conflicting results. The main arbiter of function, being the proliferation assay, allowed many points of variation as the cell population chosen for the assay can affect the

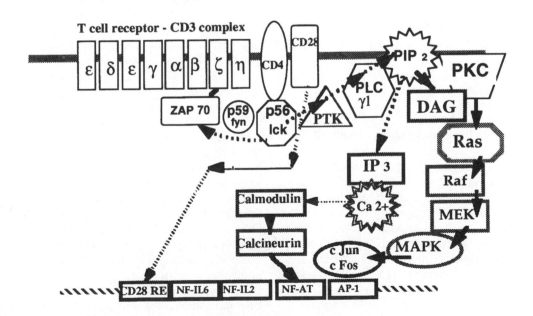

Figure 21.3   The interaction of protein kinases and phosphatases in the regulation of T cells

outcome, as can the particular sample of gp120 employed. For example, the most highly purified recombinant gp120 may contain as much as 5% of spurious protein while gp120 purified from virus is even less pure. The fear is always present that endotoxins, or other impurities capable of modulating T cell function, may be confounding factors and should always be adequately controlled for. This caveat has been heeded in recent experiments and interesting results are adding to our knowledge, not only of HIV pathogenesis, but also of the role of CD4 as an adhesion molecule and coreceptor.

The CD4 molecule binds to a monomorphic region of the MHC class II molecule on the antigen presenting cell and, thus, not only facilitates antigen recognition but also enhances T cell activation by the transmission of a signal across the plasma membrane [19]. The complex formed between the antigen, presented by the class II molecule, and the T cell receptor in association with the CD4 antigen greatly enhances the antigen-specific immune response. The part played by CD4 in this interaction is mediated by the protein tyrosine kinase p56^{lck} which is non-covalently associated with its cytoplasmic tail [20]. If CD4 is crosslinked prior to ligation of the T cell receptor then activation is abrogated, suggesting that CD4 may also have an inhibitory role. This same inhibitory effect may be involved in the dysfunction of CD4$^+$ cells in HIV infection as a consequence of gp120 binding.

Experimental Evidence of HIV–CD4 Interaction in the Modulation of Signalling

In order to investigate the part played by the cytoplasmic domain of the CD4 molecule in HIV infection, CD4-negative human T cell lines were transfected with cDNAs encoding either the full-length molecule or a truncated version lacking most of the cytoplasmic domain [21]. Cells lacking the cytoplasmic domain of the CD4 antigen, and therefore unable to associate with p56^{lck}, allowed greater levels of virus production (6-to 20-fold), as measured by the percentage of cells expressing viral p24 protein and the number of infectious particles shed into the culture supernatant, than

cells bearing the full length molecule. Quantitative PCR showed that internalization and reverse transcription of the virus was similar in both cell types following infection of the cells with HIV. Notably, however, when the antiretroviral agent, azidothymidine (AZT) was added to such cells having comparable amounts of integrated virus, a 50-fold difference was observed in the number of cells expressing viral p24 protein between the two cell groups. While the addition of AZT inhibits *de novo* viral transcription, it does not prevent replication of the virus from integrated proviral DNA in cells infected prior to AZT addition. Viral particles released from these cells will bind to both intact and truncated CD4 molecules leading to their oligermerization. This initiates a signal which negatively regulates the expression of pre-integrated viral DNA. The truncated CD4 molecule is unable to transduce this negative signal as it lacks the cytoplasmic interaction with p56^{lck}, thus allowing enhanced viral replication. A further series of experiments employing a mutant full-length CD4, lacking the p56^{lck} association domain, implicated p56^{lck} in the transduction of the negative signal.

It has been shown that binding of either anti-CD4 monoclonal antibodies or gp120 to CD4 molecules induces p56^{lck} activation and phosphorylation on tyrosine residues of several cellular proteins [8, 20, 22]. We have observed profound changes in the protein tyrosine kinase patterns obtained following stimulation with anti-CD4 antibody of T cells from HIV-infected individuals as compared to cells from a healthy donor (Figure 21.4). Indeed, many of the expected bands were either attenuated or completely missing. The band identified as p56^{lck} diminishes in intensity during the period of stimulation while the band at 32–34 kDa appears to be hyperphosphorylated. One important question raised by this finding is whether the changes are the result of either the innappropriate binding of gp120 to the CD4 molecule or a consequence of altered lymphocyte maturation in the infected lymph nodes. There is, however, recent evidence that T cell receptor antagonists or mutant agonists exhibiting very small changes in peptide sequence, stimulate a distinct pattern of ζ chain phosphorylation and fail to activate the associated ZAP-70 kinase [23]. Moreover, we have recently found that the CD4-

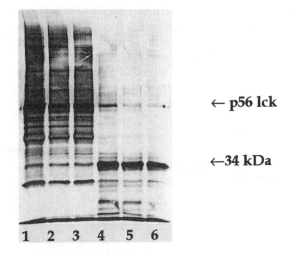

$\leftarrow$ **p56 lck**

$\leftarrow$ **34 kDa**

Figure 21.4 Western blot using anti-phosphotyrosine antibody of normal (lanes 1–3) and HIV-infected T cells (lanes 4–6) following stimulation with anti-CD4 for 2 (1 and 4), 5 (2 and 5) or 15 (3 and 6) minutes

induced ZAP-70 kinase activity observed in normal control peripheral blood lymphocytes is profoundly diminished in virally infected cells. Furthermore, the augmented kinase activity resulting from costimulation with anti-CD3 and anti-CD4 is totally abrogated in infected lymphocytes. In contrast, stimulation by anti-CD3 alone, produces comparable levels of ZAP-70 phosphorylation and kinase activation in infected or unifected PBLs (Guntermann and Nye, unpublished results) Thus, the CD4-dependent enhancement of ZAP-70 kinase activity observed in normal, but not infected cells, indicates the requirement for p56^{lck} in this activation. We hypothesize that in early HIV-1 infection, the activation of p56^{lck} is inhibited by an, as yet, unknown mechanism mediated by the virus or one of its proteins (perhaps by gp120 acting as a mutant MHC II molecule). Failure of p56^{lck} to phosphorylate the immunoreceptor tyrosine-based activation motif (ITAM) within the CD3ζ chains thus prevents recruitment of ZAP-70 to the CD3 receptor complex. Whatever the explanation, it is clear that different signals emanate from receptor ligation in T lymphocytes obtained from HIV-infected subjects. Our present knowledge of the cascade of biochemical events triggered by ligation of the T cell receptor and modulated by coreceptors, which lead eventually to the transcription of numerous genes, is incomplete (see Figure 21.3). Thus studies on cells where

such a pathological condition has modified cell function may direct us to the point of control of that function and, in addition, teach us more about signalling pathways in normal cells.

HIV Infection: Aberrant Cytokine Production and Signalling

The differences in total tyrosine phosphorylation patterns described above may well explain the aberrant production of a number of cytokines seen as a prominent feature of infection with the human immunodeficiency virus. Interleukin-1 (IL-1), interleukin-6 (IL-6), tumour necrosis factor-α (TNF-α) and interferon-γ (IFN-γ) are produced in increased amounts *in vivo*, while the production of interleukin-2 (IL-2) is decreased. Interleukin-2 plays a key role in the development of T cells in the thymus as well as regulating the immune response by enhancing activated T cell growth. Because of the pivotal role of IL-2 and its down-regulation in HIV-1 infection, the signalling pathways responsible for its initial production and thereafter responding to its effects, have been well studied both in this disease and in normal cells (see Figure 21.4).

Further investigation of the inhibitory role of CD4 sheds light on a number of the targets of this negative signal and their role in IL-2 production

[24]. Both recombinant gp120 (rgp120) and anti-CD4 monoclonal antibodies (anti-CD4 mAb) were employed as ligands and the target cells were freshly prepared human peripheral blood mononuclear cells (PBMC). Production of IL-2 was significantly diminished when PBMC were pre-incubated in the presence of either rgp120 or anti-CD4 mAb for 20 minutes prior to stimulation with the phorbol ester PMA and the calcium ionophore ionomycin, relative to the same activated cells that had not been previously exposed to CD4 ligation. Similar results were obtained when anti-CD3 mAb stimulation was substituted for PMA/ionomycin stimulation.

Three transcription factors known to be involved in the regulation of the IL-2 gene, NF-AT, AP-1 and NF-κB, were examined in the context of CD4 mediated inhibition of TCR/CD3 activation. Inhibition of function of these transcription factors, obtained from nuclear extracts of CD4$^+$ T cells, was assessed in gel retardation assays using double-stranded radiolabelled oligonucleotides representing the IL-2 enhancer sites of these transcription factors. Both anti-CD4 mAb

and rgp 120 pretreatment attenuated the bands specific for NF-AT, AP-1 and NF-κB obtained from nuclear extracts of activated CD4$^+$ T cells. In addition, incubation of CD4$^+$ T cells with anti-CD4 mAb, in the absence of TCR/CD3 activation, failed to induce any such bands in the nuclear extracts. The inhibition by rgp120 is concentration-dependent and binding activity is lost in the order NF$-\kappa$B $>$ AP$-1 >$ NF$-$AT. In order to understand the implications of these findings more fully it is necessary to assemble clues from current research in related signalling pathways.

As discussed earlier, it is well established that interaction of the T cell receptor with an appropriately presented antigen leads to the activation of phospholipase C which, in turn, hydrolyses PIP$_2$ (Figure 21.2). This gives rise to the two second messengers, IP$_3$ and DAG, the former binds an intracellular receptor causing the release of stored calcium while the latter activates protein kinase C. The importance of the elevation of calcium concentration in T cell activation was demonstrated more than twenty years ago [25, 26], although one of the immediate targets for

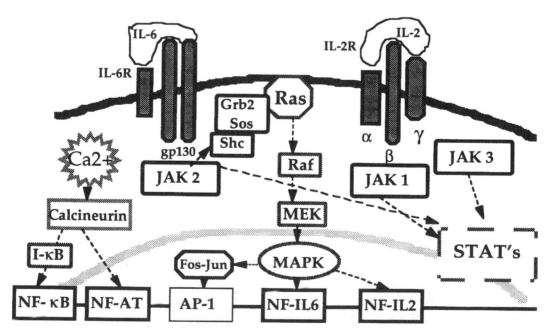

Figure 21.5 Cytokines such as IL-2 and IL-6 may be generated initially as a consequence of monocyte-produced IL-1 binding or of antigen being presented to a cell in association with a MHC molecule. Once secreted, the cytokine can bind to its cognate receptor and act in an autocrine or paracrine manner to increase production of that cytokine

Ca^{2+}, the calcium–calmodulin-dependent phosphatase calcineurin, has only recently been discovered [27]. Two of the nuclear factors described above, NF-κB and AP-1, are thought to be regulated by protein kinase C, while NF-AT is calcium dependent. NF-AT contains a subunit, NF-ATp, present in unstimulated T cells, which forms a complex with Fos and Jun proteins in the nucleus of activated T cells. Purified NF-ATp is a substrate for calcineurin *in vitro* and forms DNA–protein complexes with recombinant Jun homodimers or Jun–Fos heterodimers [28]. The binding of Fos–Fos or Fos–Jun to DNA is an essential prerequisite for formation of the NF-ATp–Jun–Fos–DNA complex which is a key step in the regulation of interleukin-2 gene transcription and therefore is a point of integration between calcium-dependent and protein-kinase-C-dependent pathways in T cell activation and may indeed be the pivotal point of HIV disruption of immune function.

Calcineurin is the target for two clinically important immunosuppressants, cyclosporin A (CsA) and FK506. The phosphatase activity of calcineurin is inhibited when either CsA or FK506 bind to their cognate intracellular receptors, cyclophilin and FK-binding protein (FKBP) respectively, members of a group of proteins known as immunophilins. This inhibition of phosphatase activity correlates with the immunosuppressive effects of these compounds suggesting that calcineurin plays an essential role in regulating transcription of cytokine genes [29]. Surprisingly the drugs do not block synthesis of transcription factors, rather they block the formation of a functional factor [30]. Several experiments have shown that the pre-existing form of nuclear factor of activated T cells (NF-ATp or NF-ATc) and octamer-associated protein (OAP) are the prime functional targets [28, 31, 32]. NF-ATp belongs to a novel family of DNA-binding proteins and is a phosphoprotein of apparent molecular mass 120 kDa. The *N*-terminal third of the protein is extremely rich in proline residues. Approximately 400 amino acids in the centre of the protein make up the DNA-binding region and show weak similarity ($\sim$ 17% identity) to NF-κB. An alternative splicing region at the *C*-terminal gives rise to at least three forms of NF-AT. An increase in intracellular calcium leads to the translocation of NF-ATp to the nucleus where assembly of a functional transcription factor is protein tyrosine kinase and Fos/Jun dependent (see Figures 21.3–21.5)

Cytokine Gene Transcription

The human IL-2 gene promoter possesses at least two purine-rich binding sites for NF-ATp, to which the nuclear factor binds in association with cFos and cJun. This NF-ATp-cFos–cJun complex is able to mediate transcription through the NF-AT binding site of the IL-2 promoter. As previously mentioned, the complex formation is essential, NF-AT does not bind to its specific domain on the IL-2 promoter in the absence of Fos and Jun family proteins. Thus the NF-AT site on the IL-2 promoter is a composite binding site and may explain the effect of glucocorticoids, transforming growth factor-β (TGF-β) and cyclic adenosine monophosphate (cAMP) on IL-2 gene transcription and also the modulatory action of costimulatory signals. There is growing evidence that NF-ATp also regulates the transcription of genes other than IL-2, including IL-3, IL-4, granulocyte-macrophage colony stimulating factor (GM-CSF) and tumour necrosis factor α (TNF-α).

The Effect of CsA and FK506 on Signalling Pathways as a Model for HIV Pathogenesis

The pharmacological suppression induced by either one of two clinically important immunosuppressants, cyclosporin A (CsA) and FK506, may provide useful clues to the mechanisms behind the pathological effects observed following infection by the human immunodeficiency virus (HIV). While cyclosporin and FK506 are chemically distinct molecules, the former is an undecapeptide and the latter a macrolide, their cellular effects are indistinguishable. The cytosolic binding protein for CsA was isolated and described by Handschumacher [33] in 1984. This highly basic protein (pI$\rangle$9.0), subsequently found in many tissues, was named cyclophilin because of its high affinity for the drug. Five years later it was discovered that cyclophilin is actually the enzyme peptidyl prolyl *cis–trans* isomerase (a rotamase). The binding protein for FK506 (FKBP) also possesses rotamase activity, is highly basic and its rotamase activity is inhibited by FK506 binding. Interestingly, FKBP does not share any homology with cyclophilin. When either drug is bound to its specific intracellular receptor the phosphatase activity of calcineurin is inhibited and this correlates with immunosuppressive effect. Thus it appears that calcineurin is the primary target for both of these drugs and that this phosphatase plays an essential role in regulating transcription of cytokine genes.

It is also interesting to note in the context of HIV infection and signalling that human cyclophilins A and B bind specifically to the HIV-1 Gag protein p55gag *in vitro* and that sequences from the capsid domain of p55gag are both essential and sufficient for the virion–cyclophilin association [34, 35]. These same results, most importantly, suggest that the Gag–cyclophilin A interaction is necessary for HIV-1 replication. A single proline mutation in the Gag polyprotein is sufficient to disrupt the Gag–cyclophilin A interaction, the incorporation of cyclophilin A into virions and, therefore, viral replication.

The binding of cyclophilin A to the HIV-1 Gag protein p55gag *in vitro* is inhibited by cyclosporin A and similarly, by SDZ NIM811, a non-immunosuppressive natural analogue of cyclosporin A, but not by the structurally unrelated immunosuppressant, FK506. It is postulated that SDZ NIM811 has its inhibitory effect on HIV-1 transmission by reducing the level of virion-associated cyclophilin A, which is mirrored by a reduction in viral infectivity. Further experiments revealed that the reduction in infectivity mediated by the cyclophilin-binding drugs, involves interaction with virus-producing cells rather than target cells. This observation has two important implications: (i) the possible therapeutic use of a non-immunosuppressive cyclophilin-binding drug in HIV therapy; and (ii) HIV-1-infected cells provide a model to further delineate the pivotal role of this pathway associated with T cell signalling. Only through a complete understanding of such lymphocyte signalling pathways and their disruption in disease may we fully understand pathogenic mechanisms and begin to design rational therapies based on this knowledge.

REFERENCES

1. Pantaleo G, Graziosi C, Demarest JF, Butini L, Montroni M, Fox CH, Orenstein JM, Kotler DP and Fauci AS. HIV infection is active and progressive in lymphoid tissue during the clinically latent stage of disease. *Nature* 1993; **362**: 355–359.

2. Embretson J, Zupanic M, Ribas JL, Burke A, Racz P, Tenner-Racz K and Haase AT. Massive covert infection of helper T lymphocytes and macrophages by HIV during the incubation period of AIDS. *Nature* 1993; **362**: 359–364.

3. Via CS, Morse H and Shearer GM. (Altered immunoregulation and autoimmune aspects of HIV infection: relevant murine models. *Immunol. Today* 1990; **11**: 250–255.

4. Ribera E, Ocana I, Almirante B, Gomez J, Monreal P and Martinez-Vazquez JM. Autoimmune neutropenia and thrombocytopenia associated with development of antibodies to human immunodeficiency virus. *J. Infect.* 1989; **18**: 167–171.

5. Ascher MS and Sheppard HW. AIDS as immune system activation: a model for pathogenesis. *Clin. Exp. Immunol.* 1988; **73**: 165–167.

6. Gupta S and Vayuvegula B. Human immunodeficiency virus-associated changes in signal transduction. *J. Clin. Immunol.* 1987; **7**: 486–489.

7. Nye KE and Pinching AJ. HIV infection of H9 lymphoblastoid cells chronically activates the inositol polyphosphate pathway. *AIDS* 1990; **4**: 41–45.

8. Nye KE, Knox KA and Pinching AJ. Lymphocytes from HIV-infected individuals show aberrant inositol polyphosphate metabolism which reverses after zidovudine therapy. *AIDS* 1991; **5**: 413–417.

9. Shears SB. Kinetic consequences of the inhibition by ATP of the metabolism of inositol 1,4,5-trisphosphate and inositol 1,3,4,5-tetrakisphosphate in liver. Different effects upon the 3- and 5-phosphatases. *Cell. Signal.* 1990; **2**: 191–195.

10. Dent P, Macdougal LK, Mackintosh C, Campbell DG and Cohen P. A myofibrillar protein phosphatase from rabbit skeletal muscle contains the β isoform of protein phosphatase-1 complexed to a regulatory subunit which greatly enhances the dephosphorylation of myosin. *Eur. J. Biochem.* 1992; **210**: 1037–1044.

11. Linette GP, Hartzman RJ, Ledbetter JA and June CH. HIV-1-infected cells show a selective signalling defect after perturbation of CD3/antigen receptor. *Science* 1988; **241**: 573–576.

12. Fields AP, Bednarik DP, Hess A and Stratford-May W. Human immunodeficiency virus induces phosphorylation of its cell surface receptor. *Nature* 1988; **333**: 278–280.

13. Gurley RJ, Ikeuchi K, Byrn RA, Anderson K and Groopman JE. CD4+ lymphocyte function with early human immunodeficiency virus infection. *Proc. Natl. Acad. Sci. USA* 1989; **86**: 1993–1997.

14. Weinhold KJ, Lyerly HK, Stanley Sd, Austin AA, Matthews TJ and Bolognesi DP. HIV-1 gp120-mediated immune suppression and lymphocyte destruction in the absence of viral infection. *J. Immunol.* 1989; **142**: 3091–3097.

15. Habeshaw JA and Dalgleish AG. The relevance of HIV env/CD4 interactions to the pathogenesis of acquired

immune deficiency syndrome. *J. Acq. Immune Defic. Syndr.* 1989; **2**: 457–468.

16. Diamond DC, Sleckman BP, Gregory T, Lasky LA, Greenstein JL and Burakoff SJ. Inhibition of CD4+ T cell function by the HIV envelope protein gp120 *J. Immunol.* 1988; **141**: 3715–3719.

17. Chirmule N, Kalyanaraman V, Oyaizu N and Pakwa S. Inhibitory influences of envelope glycoproteins of HIV-1 on normal immune responses. *J. Acq Immune Defic. Syndr.* 1988; **1**: 425–430.

18. Mittler RS and Hoffman MK. Synergism between HIV gp120 and gp120-specific antibody in blocking human T cell activation. *Science* 1989; **245**: 1380–1382.

19. Weiss A and Littman DR. Signal transduction by lymphocyte antigen receptors. *Cell* 1994; **76**: 263–274.

20. Veilette A, Bookman MA, Horak EM and Bolen JB. The CD4 and CD8 T cell surface antigens are associated with the internal membrane tyrosine-protein kinase p56lck. *Cell* 1988; **55**: 301–305.

21. Tremblay M, Meloche S, Gratton S, Wainberg MA and Sékaly RP. Association of p56lck with the cytoplasmic domain of CD4 modulates HIV-1 expression. *EMBO J.* 1994; **13**: 774–783.

22. Juszczak R, Turchin H, Truneh A, Culp J and Kassis S. Effect of human immunodeficiency virus gp120 glycoprotein on the association of the protein tyrosine kinase p56lck with CD4 in human T lymphocytes. *J. Biol. Chem* 1991; **266**: 11176–11183.

23. Madrenas J, Wange RL, Wang JL, Isakov N, Samelson LE and Germain RN. Phosphorylation without ZAP-70 activation induced by TCR antagonists or partial agonists. *Science* 1995; **267**: 515–518.

24. Jabado N, Le Deist F, Fischer A and Hivroz C. Interaction of HIV gp120 and anti-CD4 antibodies with the CD4 molecule on human CD4+ T cells inhibits the binding activity of NF-AT, NF-κB and AP-1, three nuclear factors regulating interleukin-2 gene enhancer activity. *Eur. J. Immunol.* 1994; **24**: 2646–2652.

25. Allwood G, Asherson GL, Davey MJ and Goodford PJ. The early uptake of radioactive calcium by human lymphocytes treated with phytohaemagglutinin. *Immunology* 1971; **21**: 509–516.

26. Whitney RB and Sutherland RM. Enhanced uptake of calcium by transforming lymphocytes. *Cell. Immunol.* 1972; **5**: 137–142.

27. Clipstone NA and Crabtree GR. Identification of calcineurin as a key signalling enzyme in T-lymphocyte activation. *Nature* 1992; **357**: 695–697.

28. Jugnu J, McCaffrey PG, Miner Z, Kerppola TK, Lambert JN, Verdine GL, Curran T and Rao A. The T cell transcription factor NF-AT p is a substrate for calcineurin and interacts with Fos and Jun. *Nature* 1993; **365**: 352–355.

29. Liu J. FK506 and cyclosporin, molecular probes for studying intracellular signal transduction. *Immunol. Today* 1993; **14**: 290–295.

30. Flanagan WM, Corthésy B, Bram RJ and Crabtree GR. Nuclear association of a T-cell transcription factor blocked by FK506 and cyclosporin A. *Nature* 1991; **352**: 803–807.

31. Granelli-Piperno A. Lymphokine gene expression in vivo is inhibited by cyclosporin A. *J. Exp. Med.* 1990; **171**: 533–544.

32. Mattila PS, Ullman KS and Fiering S. The actions of cyclosporin A and FK506 suggest a novel step in the activation of T lymphocytes. *EMBO J.* 1990; **9**: 4425–4433.

33. Handschumacher R, Harding MW, Rice J and Drugge RJ. Cyclophilin: a specific cytosolic binding protein for cyclosporin A. *Science* 1984; **226**: 544–547.

34. Franke KF, Hui Yuan H and Luban J. Specific incorporation of cyclophilin A into HIV-1 virions. *Nature* 1994; **372**: 359–362.

35. Thali M, Bukovsky A, Kondo E, Rosenwirth B, Walsh CT, Sodroski J and Göttlinger HG Functional association of cyclophilin A with HIV-1 virions. *Nature* 1994; **372**: 363–365.

22

Manipulation of Lymphocyte Signal Transduction Pathways by Eukaryotic Parasites

William Harnett

Department of Immunology, University of Strathclyde, Glasgow, Scotland, UK

INTRODUCTION

Eukaryotic parasitic organisms are responsible for a wide range of human infections throughout the tropics. Several million people die each year as a consequence of infection, and many tens of million more suffer severe and debilitating morbidity [1]. This reflects the fact that methods of control for eukaryotic parasites are largely inadequate. Recently there has been much interest in investigating the biochemistry, molecular biology and interaction with the immune system of eukaryotic parasites in the hope of designing new strategies for control.

Table 22.1 lists the major human pathogens in this "group" of organisms and information relating to their importance. The parasites fall into two categories: (i) protozoa—unicellular organisms which include the causative agents of the most important tropical disease malaria; and (ii) helminths (worm parasites), in particular members of the phyla Nematoda and Platyhelminthes. Generally speaking, the protozoa are associated more with fatalities. Helminths tend to cause morbidity rather than mortality, although infection with schistosomes and, as recently recognized, *Onchocerca volvulus*, can lead to premature death.

INTERACTION OF EUKARYOTIC PARASITES WITH LYMPHOCYTES

Eukaryotic parasitic organisms are no different from any other form of infectious agent, in that they invoke specific immune responses. Clearly, therefore, they interact with lymphocytes. This "normal" antigen-specific association between host and pathogen, is not the topic of this review. Rather, this review is concerned with other ways in which parasites may interact with lymphocytes and which, in general, can be considered to be to the detriment of the host rather than the parasite. These other forms of interaction can be broadly divided into three categories: (i) interaction which causes immunosuppression; (ii) interaction which induces non-specific polyclonal antibody production; and (iii) lymphoproliferation associated with direct parasitism. Each of these will be discussed in some detail and with reference to specific parasites.

INTERACTION WHICH CAUSES IMMUNOSUPPRESSION

Introduction

A characteristic of infection with eukaryotic parasitic organisms is its enduring nature. Parasitic infections in fact are commonly life-long, and indeed there are reports of individual parasitic organisms surviving for more than a decade. Such longevity is remarkable, given that eukaryotic parasites are generally *highly* immunogenic, and dictates that they must have developed extremely effective mechanisms for evading the host immune system. The range of strategies which has evolved is quite remarkable and includes presentation of continuously changing epitopes to the host ("antigenic variation"), molecular mimicry between parasite and host molecules, acquisition of host antigens to act as a disguise to the immune system,

Table 22.1 Major Diseases Caused by Eukaryotic Parasites of Humans

Disease	Causative agent	No. infected (millions)	No. at risk (millions)
Protozoa			
Malaria	*Plasmodium* spp.	276	2100
Sleeping sickness	*Trypanosoma* spp.*	0.2†	50
Chagas' disease	*Trypanosoma cruzi*	17	90
Leishmaniasis	*Leishmania* spp.	12	350
Helminths			
Schistosomiasis	*Schistosoma* spp.	200	60
Lymphatic filariasis	*Wuchereria bancrofti*	119	905
	Brugia malayi		
Onchocerciasis	*Onchocerca volvulus*	18	90

*African.

†Suspected number of new cases per year

and release of enzymes which cleave host effector molecules (e.g. antibodies) (for a review, see reference [2]). Different groups of parasite vary with respect to which of these mechanisms they adopt, but an additional strategy which appears to have virtual universal employment, is the induction of some form of immunosuppression.

The term "immunosuppression" tends to be employed in the field of parasite immunology, when evidence of some form of impaired immune-responsiveness has been observed. Its possible existence had been commented on since the turn of the century, when it was observed that patients harbouring eukaryotic infections often appeared more susceptible to other, e.g. bacterial, infections [3], or were more difficult to vaccinate against pathogens. Furthermore, it has often been noted that animals infected with eukaryotic parasites demonstrate impaired immune responses to heterologous antigens. Parasite-induced "defects" have been described with respect to numerous effector mechanisms of the immune system, but probably most commonly to lymphocyte function. Lymphocyte function has been investigated in a range of ways in these studies, but classically it has involved measuring either antibody responses, or the *in vitro* proliferative response of peripheral blood, spleen or lymph node cells obtained from infected animals or humans (peripheral blood), to parasite antigens and/or mitogens. Using these approaches, "immunosuppression" has been described with respect to all of the major human

eukaryotic parasites and also many parasites either used as models for human parasites, or of economic importance. Arguably it has been most extensively investigated in trypanosomes (both African and South American) and filarial nematodes and these will be discussed in particular detail.

Defects in Lymphocytes During Trypanosome Infections

Trypanosome species represent major pathogens in both South America [4] and Africa [5]. In South America, the causative agent is *Trypanosoma cruzi* and it causes Chagas' disease in humans. In Africa, trypanosomiasis exists as a problem both for humans, in whom it causes sleeping sickness, and for domestic animals. Human disease is caused by *Trypanosoma brucei rhodesiense* and *Trypanosoma brucei gambiense*. A wide range of studies utilizing humans, cattle (for African trypanosomiasis) and as a model, mice, clearly indicate that infection with trypanosomes, particularly acute, is associated with a variety of manifestations of immunosuppression [3,6,7]. These include defects in lymphocyte responsiveness and this phenomenon, as mentioned earlier, has been particularly well studied.

Experimental evidence has in fact exisited for a number of years to support the idea that both T (proliferation studies) and certain B (proliferation studies, antibody production) cell responses are suppressed during trypanosome infections. More recently, attempts have been made to understand the nature of the defects in the lymphocytes. Trypanosome-induced inhibition of lymphocyte proliferation is associated with arrest at the G_0/G_{1a} phase of the cell cycle [8, 9] and reduced levels of mRNA for c-myc, c-fos and IL-2 [10]. Consistent with the latter observation, lymphoid cells from mice infected with trypanosomes have generally been found to release significantly reduced amounts of IL-2 in response to mitogens [11, 12]. A similar effect can be demonstrated *in vitro* when splenocytes from normal mice exposed to PHA are incubated with *T. cruzi*. Studies on IL-2R expression indicate that it is also significantly reduced on mitogen-simulated peripheral blood cells during coculture with both African [9, 13] and South American [14, 15] trypanosomes. With respect to South American trypanosomes, both the proportion of IL-2R$^+$ T cells and the surface density of IL-2R are reduced [7] and the same effect is observed if cells are activated with anti-CD3 or anti-CD2 [16] or indeed with killed *Staphylococcus aureus*, indicating that B cells are also affected [8]. Studies investigating the ability of exogenous IL-2 to restore lymphocyte responsiveness have not produced clear-cut results (reviewed in reference [7]). The critical factor appears to be the level of IL-2R expression: if this is significantly reduced, recovery is absent. Reduced levels of IL-2R protein and mRNA [7] are detected in the cytoplasm of human lymphocytes exposed to *T. cruzi*, indicating that inhibition of proliferation occurs at a stage prior to expression of IL-2R at the plasma membrane. *T cruzi* has also been shown to reduce the expression of other T lymphocyte surface molecules broadly involved in activation, namely the TCR/CD3 complex and CD4 and CD8 [17]. Interestingly, *T.b. rhodesiense* when tested on TCR was found to have no effect [17], suggesting the existence of differences in the mechanism by which African and South American trypanosomes interfere with lymphocyte responses.

The nature of this mechanism is still under investigation but a number of recent studies are providing revealing information. Studies by Tarleton [18] on *T. cruzi* demonstrated a role for a "suppressor" T cell population, which was dependent for its generation/maintenance on prostaglan-

din-producing macrophages, in suppressing IL-2 production. A crucial role for prostaglandins was also shown by Silegham and colleagues [6] who demonstrated that prostaglandin-producing macrophages directly invoked suppression of IL-2 production in lymph node cells from *T. brucei*-infected mice. Suppression of IL-2R expression was also investigated by these workers and found to be independent of prostaglandin synthesis but dependent on a cell population which copurified with macrophages. More recently, a dominant role has been demonstrated for NO released by macrophages in trypanosome-induced inhibition of lymphocyte proliferation [19,20]. This NO synthesis appears to be produced in response to generation of in particular IFN-γ, but also TNF-α [19]. Thus, in spite of inhibiting lymphocyte proliferation, *T.b. brucei* induces lymphocytes to produce IFN-γ *in vitro* [21]. Dramatically elevated levels of this cytokine have also been observed during infection *in vivo* [22–24]. Recently, a factor which is released by *T.b. brucei* and which causes CD8$^+$T cells to release IFN-γ has been characterized [25,26]. This "molecule" is protein in nature and has several molecular weights, the lowest of which is 41–46 kDa. The ability of *T. cruzi* to suppress lymphocyte responses has also recently been shown to be mimicked by a secreted, protease-sensitive, parasite product of molecular weight 30–100 kDa [27]. The exact function of this molecule has yet to be discovered but the observation that administration of recombinant INF-γ to mice prevents immunosuppression caused by *T. cruzi*, suggests that it is unlikely to cause INF-γ release. This observation also emphasizes that, as suggested earlier, immunosuppression invoked by African and South American trypanosomes, in spite of showing some effector similarities, is likely to be mechanistically distinct.

Finally, an intriguing recent observation which also merits mention is that experimental infections with *T. cruzi* can induce programmed T cell death by apoptosis [28]. This can be demonstrated *in vitro* following stimulation of lymphocytes removed from infected animals via TCR/CD3 by for example Con A or appropriate antibodies but not via stimulation of other molecules such as CD69 or Ly-6 A/E. It is observed with CD4$^+$ but not CD8$^+$ T cells. Apoptotic cells were also found following electron microscopic examination of lymphocytes removed from infected donors. Unlike the *in vitro* situation, however, this was also the case with respect to CD8$^+$ cells. The significance of these new data awaits futher exploration but it is pertinent to note at this stage, that the authors observe a linear association between the extent of observed apoptosis and suppression of T cell proliferative responses. Furthermore, this phenomenon may represent more than a *T. cruzi*-specific event among eukaryotic parasites, as it has recently been demonstrated that acute *Plasmodium falciparum* infection is associated with an increased percentage of apoptotic cells [29].

Defects in Lymphocytes During Filarial Infections

Humans act as natural hosts for 7–8 species of filarial nematode, but not surprisingly the ones which have received most study are those which are most pathogenic. These are *Wuchereria bancrofti* and *Brugia malayi* (the "lymphatic filariae") which cause elephantiasis and *Onchocerca volvulus* which causes severe chronic skin lesions and blindness. Not every individual who is infected will demonstrate such major symptoms, however, and indeed the majority of parasitized individuals (particularly with lymphatic filariae) remain apparently disease-free or at least show only mild symptoms. As diseases associated with filarial infection are considered to be associated with a large component of host (immune response)-mediated pathology, it has been suggested that apparently disease-free infected individuals may have depressed immune responses. A number of recent studies generally support this idea in that such individuals—especially those harbouring lymphatic filariae—are generally accepted as having "suppressed" lymphocyte responses relative to uninfected individuals or infected individuals with disease [reviewed in reference [30]). However, it should be borne in mind that pathology can be present to varying degrees during filarial infection, particularly with respect to onchocerciasis, and it may perhaps be more accurate to say that there is some evidence of an inverse correlation between the extent of immunosuppersion and the level of disease.

The role of filarial nematodes as major disease-causing agents in the tropics has dictated that the immunosuppression they invoke has been the subject of much interest. Apparently disease-free individuals harbouring lymphatic filarial parasites have been shown to have impaired DTH, T cell proliferation, cytokine production (IL-2; INF-γ) and in some studies *in vitro* and *in vivo* antibody responses, relative to symptomatic individuals (reviewed in reference [31]). However, it is clearly not the case that the immune response of the asymptomatic individuals is simply non-specifically dampened as these people have 17 times as much anti-parasite IgG$_4$ as their diseased counterparts [32]. There is also evidence of defects in lymphocyte responsiveness to parasite antigens in onchocerciasis, particularly when disease is minimal although the picture is not perhaps quite as clear. Thus, although T cell proliferation has generally been found to be poor, effects on cytokines remain to be clearly established (although there is some evidence of a T$_H$2-type profile predominating) and detectable anti-parasite antibody responses appear to be generally present (reviewed in reference [33]). In relation to this latter point, it is worth noting that studies with animal models of filariasis, indicate that a significant *in vitro* antibody response is possible in the absence of lymphocyte proliferation, suggesting the existence of parasite-induced differential modulation of the two events [34]. Returning to human responses *in vivo*, however, the level of antibody tends to be proportional to the degree of skin pathology and to be particularly high in individuals with sowdah-type onchocerciasis, a form associated with particularly vigorous skin pathology. Furthermore, an inverse correlation has been shown to exist between the level of larval parasites in the skin and several antibody classes/subclasses (reviewed in reference [33]).

The specificity of defects in immune-responsiveness is also unclear with respect to filarial nematodes. Lymphatic filarial parasites generally appear to induce specific defects, but there is evidence from studies on *Onchocerca volvulus* (e.g. on responses to vaccines or mitogens), that a more non-specific effect may in some cases be generated (reviewed in reference [33]). Similar data have been obtained in some studies investigating the immune response to heterologous antigens in animal models of filariasis (see, for example, reference [35]).

How can the lymphocyte hyporesponsiveness observed during filarial infections be explained? The probability exists that more than one explanation may be required, particularly as there is some suggestion that the induction of specific and non-specific immunosuppression may be dependent upon different mechanisms [36]. A very simple explanation for the absence of specific lymphocyte responsiveness seen in many disease-free people infected with parasites is that they simply have a reduced number of filarial antigen-reactive cells. Certainly there is evidence for this [37, 38]. This could arise as a consequence of pre-natal exposure to filarial antigens [39] but it has been known from studies in experimental animals for many years (see, for example, reference [40]) that lymphocyte unresponsiveness starts to appear as patency approaches and follows a period of vigorous responsiveness to parasite antigens. It follows that the parasites must in some way suppress responsiveness during the course of infection. In relation to this, a number of parasite macromolecules or extracts have been described which are able to interfere with lymphocyte responsiveness as measured by proliferation or cytokine production in response to various stimuli [41–45]. In addition, filarial nematodes release eicosanoids, lipid mediators known to inhibit lymphocyte responsiveness (reviewed in reference [46]). Furthermore, several studies demonstrate that successful chemotherapy can reverse the hyporesponsive defects observed in filariasis patients (reviewed in reference [38]).

The nature of the defect underlying the unexpectedly low lymphocyte responses of filariasis patients observed *in vitro* has been investigated in a number of studies, but discrepancies exist in the data obtained. In some cases, a role has been observed for a "suppressor cell", e.g. adherent mononuclear cells or certain T cell populations, in that removal of the offending cells results in reversal of lymphocyte unresponsiveness [47, 48]. Similar findings have been reported from studies on rodent models of filariasis [34, 49]. No evidence of such populations has been noted by other workers investigating their presence in human filariasis however [37, 38]. Studies in which the recuperative powers of exogenous cytokines have been investigated demonstrate a positive effect with IL-2, when using lymphocytes from onchocerciasis patients [44]. In contrast, this result was not observed in a chimpanzee model of onchocerciasis, although IL-2 in combination with IL-4 plus/minus IL-6 was found to be effective, as was IL-4 or IL-6 alone [50]. IL-2 and IL-4 have also been employed in a study on patients with lymphatic filarial parasites, but with little evidence of an ability to rescue responsiveness. Similar results were obtained using anti-IL-4, anti-IL-10, indomethacin, PMA, ionomycin and antibodies to CD26, CD27 and CD28 [51]. Exogenous IL-2 has also been shown to fail to rescue lymphocyte responses in some animal models of lymphatic filariasis including cases in which lack of proliferation has been shown to reflect an inability to produce IL-2 [52].

In terms of individual parasite molecules, those which contain covalently-linked phosphorylcholine (PC) have received most attention. This is in part as a consequence of the known immunomodulatory properties of this structure (see, for example, reference [53]). Excretory-secretory products (ES) of filarial nematodes are particularly well endowed with PC (reviewed in reference [54]) and the appearance of such molecules in the bloodstream (reviewed in references [55, 56]) dictates that they have ample opportunity to interact with the immune system of the parasitized host. A number of studies have shown that PC-ES are more abundant in the bloodstream of hyporesponsive asymptomatic filariasis patients than in responsive symptomatic individuals (reviewed in reference [55]). Although this may simply reflect the fact that the former group generally have greater numbers of parasites, a positive link between PC-ES and hyporesponsiveness appears possible as PC-containing parasite molecules have been shown to interfere with both T [43] and B [45] cell proliferative responses.

The study of Lal and colleagues [43] demonstrates that PC-containing molecules of *B malayi* inhibit phytohaemagglutinin-induced proliferation of human T cells by a mechanism which is dependent upon generation of a "suppressor" T cell population. The work of Harnett and Harnett [45] showed that ES-62, a major PC-ES of the rodent filarial parasite *Acanthocheilonema viteae*, was able to inhibit B cell proliferation induced by using (Fab') fragments of anti-Ig, but not by LPS. This may suggest that PC targets signal transduction through the antigen receptor. Examination of effects on signal transduction pathways associated with ligation of the antigen receptor revealed no effect on generation of inositol phosphates but a general modulation of protein tyrosine kinase activity [57]. Specifically with respect to the latter it was found that sIg-coupling to the protein tyrosine kinases, Syk and Lyn, and to a lesser extent, Blk, but not Fyn or Yes, was inhibited with resultant reduction in signalling through the PI3-kinase and ras/MAPkinase signalling pathways [57]. Similar modulation of TCR/CD3-coupled–PTK-mediated signalling events was also observed in T cells [58]. Furthermore, it has been found that PKC activity [45] was reduced in B cells exposed to the parasite product. A number of signalling molecules involved in B cell activation therefore appear to be affected following exposure to PC. Since a "purified" population of B cells was employed in these studies, it seems that these effects must be due to direct interaction between PC and B cells.

Other "Immunosuppressive" Molecules

The characterization of molecules which promote lymphocyte unresponsiveness is currently a very active area of research. Consequently, new immunosuppressive molecules are continuously being discovered and some additional examples of these are outlined below. At this stage biochemical details of the molecules rather than the molecular nature of the defects they cause is the general rule.

In addition to PC-containing ES, a number of other immunosuppressive molecules have been described for filarial nematodes. These include a heat-stable, low molecular weight (<10 000) ES product released by microfilariae of the bovine parasite *Onchocerca gibsoni*, which inhibits lymphocyte responses to several mitogens [59] and a protease-sensitive high molecular weight ES product released by microfilariae of *B. malayi* which inhibits Con A induced proliferation [41]. Immunosuppressive ES products have also been described with respect to non-filarial nematodes. For example, heat-stable low molecular weight molecules (<26 000) are secreted by adult *Nematospiroides dubius*, a murine parasite, which inhibit proliferation of murine lymphocytes in response to mitogens or other ES products [60] and also heat-

sensitive molecules of generally higher molecular weight which interfere with *in vitro* antibody responses to KLH [61]. With respect to the Platyhelminthes, the major human parasite of this phylum, the schistosome, has been shown to release a low molecular weight (<2000), heat-stable factor which inhibits antigen-or mitogen-induced B and T lymphocyte proliferation [62]. This factor has no effect on IL-2 production or expression of IL-2R or transferrin but it can inhibit IL-2-dependent proliferation [63, 64]. It appears to selectively block proliferation at the G1 transition of the cell cycle. Larvae of *Paragonimus westermani*, a less well known but still medically important platyhelminth causing human paragonimiasis, secrete a neutral thiol protease which when injected into mice suppresses both humoral and cellular immune responses to heterologous antigens [65]. Furthermore the molecule has been shown to reduce expression of MHC and IL-2R on lymphocytes associated with *in vivo* exposure to LPS. The other major parasitic class of Platyhelminthes, the tapeworms, is also associated with immunosuppression and the generation of immunosuppressive products. For example, taeniaestatin, a 19.5 kDa serine protease inhibitor released from larvae of *Taenia taeniaeformis*, inhibits T cell proliferation and IL-2 production [66].

With respect to protozoa, an entirely different category of molecule has been implicated in inhibition of a variety (antigen-mediated and mitogen-induced) of lymphocyte proliferative responses. These are glycosphingolipids extracted from *Leishmania* spp. and the critical component appears to be the carbohydrate moiety [67]. This work therefore complements an earlier report describing a *Leishmania* carbohydrate-rich secreted product which inhibited lymphocyte blast transformation to both parasite antigens and mitogens [68].

INTERACTION WHICH INDUCES NON-SPECIFIC POLYCLONAL ANTIBODY PRODUCTION

Until fairly recently, a characteristic of certain parasitic infections, particularly protozoan, was considered to be a non-specific polyclonal stimulation of B lymphocytes due to parasite products acting as B cell mitogens. This could be most readily measured as elevated antibody levels to non-parasite antigens. Clearly these could not be considered as potentially protective and indeed the idea has been suggested that polyclonal stimulation of B cells may in fact *reduce* the likelihood of protective specific B cell responses arising. More recently, however, this idea of non-specific polyclonal B cell proliferation resulting as a consequence of parasites containing B cell mitogens has begun to be questioned.

There is certainly no doubt that infection with African trypanosomes results in expansion of B cell areas and diminution of T cell areas of lymphoid tissues such as the spleen (reviewed in reference [5]). Furthermore, in trypanosome-infected hosts, IgM production is greatly prolonged and enhanced. Nevertheless, it now seems likely that the sustained production of IgM in trypanosome infections can largely be accounted for by the successive waves of variable antigen types displayed by the parasite. Certainly, the existence of trypanosome antigens acting as B cell mitogens remains to be unequivocally demonstrated. With respect to malaria, it is still accepted that the massive increase in antibody production associated with infection is largely non-specific. However, it appears that this polyclonal stimulation is not due to direct interaction between parasite and B cell, but rather is due to stimulation of T cells which then release cytokines promoting the differentiation of B cells into Ig-secreting plasma cells [69].

Polyclonal stimulation of B lymphocytes by parasite antigens acting as mitogens is probably a questionable idea in the 1990s. The current awareness of cytokine regulation of the immune response offering multiple points of control and stimulation renders it simplistic. It is therefore increasingly likely that examination of "polyclonal B cell responses" will involve investigating factors other than the possibility that parasite products are acting as mitogens.

LYMPHOPROLIFERATION ASSOCIATED WITH DIRECT PARASITISM

Theileria spp. are protozoan parasites some of which are of economic importance due to their ability to infect and subsequently cause death in cattle. The infective sporozoite stage is transmitted to the mammalian host by ticks, whereupon it invades host lymphocytes by a receptor-mediated-like endocytic process. This step in the protozoan's life cycle is of relevance to this review as the parasites subsequently induce blastogenesis and clonal expansion of bovine B (*T. annulata*) and certain subpopulations of T (*T. parva*) lymphocytes. Studies on *T. parva* show that it is able to persist in the continuously dividing cells it produces by synchronization of its cell cycle with that of the host [70]. It is associated with the lymphocyte microtubules [71] and during mitosis becomes incorporated into the mitotic spindle of the lymphocyte [72].

Bovine lymphocytes infected with *T. parva* have many of the characteristics of tumour cells including metastatic behaviour in the parasitized host and the ability to proliferate continuously if placed in culture. *T. parva* infected cells have also been shown to have the property of responding continuously to costimulatory signals. Uninfected T cells are only able to do this after triggering of their antigen receptors. Because it is possible using infected tick salivary glands to parasitize lymphocytes *in vitro* [73], the phenotypic changes associated with parasitism can be studied in a controlled manner. In this respect, it has been demonstrated that infected T cells display quantitative differences in expression of some CD antigens and infected B cells lose the ability to express surface Ig (reviewed in reference [74]). Of even more value, however, is the opportunity which is available to investigate the mechanism of lymphocyte transformation by *Theileria*.

Early studies on the mechanism by which *T. parva* causes lymphocyte proliferation indicated that parasitized cells secreted a substance with T cell growth factor activity [75]. It was subsequently shown that parasitized T and B cell lines investigated, expressed IL-2 mRNA (which was dependent on the presence of the parasite) [76] and that IL-2R was also detectable and constitutively expressed [77]. Furthermore, antibodies against IL-2 [76] and antisense IL-2R RNA [78] were shown to cause partial inhibition of proliferation of cell lines. Constitutive expression of IL-2 and IL-2R is likely to arise ultimately as a consequence of the ability of *T. parva* somehow to activate NF-κB [79]. In relation to this, it has been shown that

T. parva causes changes in the phosphorylation status of protein kinases [80] including Fyn and Lck [81]. Furthermore, it has recently been discovered that *T. parva* is also able to induce at the transcriptional, translational and functional levels, increased expression of bovine casein kinase II [80]. This multifunctional enzyme phosphorylates a number of proteins which modulate cellular proliferation or transformation and hence it has been suggested that it may play a key role in subversion of lymphocyte signal transduction pathways by *Theileria* [see 80].

These recent findings on the effect of *Theileria* on lymphocyte signalling pathways clearly illustrate the progress which is now being made in understanding the molecular events underlying interaction between parasite and host cell. Further data relating to this can be expected in the very near future.

FUTURE PROSPECTS

The existence of an apparent impairment of the immune response during infection with eukaryotic parasites has been known for almost a century. It is only recently, however, that the molecular mechanisms underlying this phenomenon have been subject to investigation. It is now well established that eukaryotic parasites contain, and more importantly release, a variety of molecules which can interupt lymphocyte function either directly or by interaction with other components of the immune system such as macrophages. Although at present the majority of the observed effects associated with these parasite products have been demonstrated *in vitro*, some molecules have been tested, with positive results, *in vivo*. The exact defects in signal transduction events leading to lymphocyte activation which these molecules induce is suddenly an area of parasitology which is receiving a substantial amount of attention. Although elucidation is at an early stage, the tantalizing prospect exists that an understanding of the molecular changes which immunosuppressive parasite molecules promote, may ultimately lead to their consideration as possible therapeutic agents for manipulating the immune system. It is possible for example to speculate that it may be possible to find a molecule which could directly

play a role with respect to controlling lymphocyte dysfunctions such as rheumatoid arthritis or other autoimmune diseases or at least could provide clues for developing new strategies for intervention in these areas. In addition, it is hoped that the complete unravelling of the mechanisms by which *Theileria* "immortalizes" lymphocytes may, by increasing our understanding of lymphocyte signal transduction and the molecular events asociated with immortalization, also be of therapeutic value to lymphoproliferative disorders such as leukaemia and lymphoma.

REFERENCES

1. WHO. Tropical Disease Research. Progress 1975–94. Highlights 1993–4. UNDP/World Bank/WHO. 1995.
2. Dessaint J-PL and Capron A. Survival strategies of parasites in their immunocompetent hosts. In: Warren KS (ed.) *Immunology and Molecular Biology of Parasitic Infections*, 3rd edn. Oxford: Blackwell Scientific Publications, 1993: 87–99.
3. Kierszenbaum F and Sztein MB. Chagas' disease (American trypanosomiasis). In: Kierszenbaum F (ed.) *Impact of Major Parasitic Infections on the Immune System*. London: Academic Press, 1993: 53–86.
4. Takle GB and Snary D. South American trypanosomiasis (Chagas' disease) In: Warren KS (ed.) *Immunology and Molecular Biology of Parasitic Infections*, 3rd edn. Oxford: Blackwell Scientific Publications, 1993: 213–236.
5. Vickerman K, Myler PJ and Stuart KD. African trypanosomiasis. In: Warren KS (ed.) *Immunology and Molecular Biology of Parasitic Infections*, 3rd edn. Oxford: Blackwell Scientific Publications, 1993: 170–212.
6. Silegham M, Flynn JN, Darji A, De Baetselier P and Naessens J. African trypanosomiasis. In: Kierszenbaum F (ed.) *Impact of Major Parasitic Infections on the Immune System*. London: Academic Press, 1993: 1–52.
7. Sztein MB and Kierszenbaum F. Mechanisms of development of immunosuppression during *Trypanosoma* infections. *Parasitol. Today* 1993; **9**: 424–428.
8. Sztein MB and Kierszenbaum F. *Trypanosoma cruzi* suppresses the ability of activated human lymphocytes to enter the cell cycle. *J. Parasitol.* 1991; **77**: 502–550.
9. Sztein MB and Kierszenbaum F. A soluble factor from *Trypanosoma brucei rhodesiense* that prevents progression of activated human T lymphocytes through the cell cycle. *Immunology* 1991; **73**: 180–185.
10. Soong L and Tarleton RL. Selective suppressive effects of *Trypanosoma cruzi* infection on IL-2, c-myc and c-fos gene expression. *J. Immunol.* 1992; **149**: 2095–2102.
11. Harel-Bellan A *et al.* Modulation of T-cell proliferation and interleukin 2 production in mice infected with *Trypanosoma cruzi. PNAS* 1993; **80**: 3466–3469.
12. Mitchell LA, Pearson TW and Gauldie J. Interleukin-1 and interleukin-2 production in resistant and susceptible inbred

mice infected with *Trypanosoma congolese*. *Immunology* 1986; **57**: 291–296.

13. Silegham M, Hamers R and DeBaetselier P. Experimental *Trypanosoma brucei* infections selectively suppress both interleukin 2 production and interleukin 2 receptor expression. *Eur. J. Immunol.* 1987; **17**: 1417–1421.

14. Beltz LA, Sztein MB and Kierszenbaum F. A novel mechanism for *Trypanosoma cruzi*-induced suppression of human lymphocytes. Inhibition of IL-2 receptor expression. *J. Immunol.* 1988; **141**: 289–294.

15. Lopez HM *et al.* Alterations induced by *Trypanosoma cruzi* in activated mouse lymphocytes. *Parasit. Immunol.* 1993; **15**: 273–280.

16. Beltz LA, Kierszenbaum F and Sztein MB. *Trypanosoma cruzi*-induced suppression of human peripheral blood lymphocytes activated via the alternative CD2 pathway. *Infect. Immun.* 1990; **58**: 1114–1116.

17. Sztein MB and Kierszenbaum F. Suppression by *Trypanosoma cruzi* of T-cell receptor expression by activated lymphocytes. *Immunology* 1992; **77**: 277–283.

18. Tarleton RL. *Trypanosoma cruzi*-induced immunosuppression of IL-2 production. II. Evidence for a role for suppressor cells. *J. Immunol.* 1988; **140**: 2769–2773.

19. Schleifer KW and Mansfield JM. Suppressor macrophages in African trypanosomiasis inhibit T cell proliferative responses by nitric oxide and prostaglandins. *J. Immunol.* 1993; **151**: 5492–5503.

20. Mabbott NA, Sutherland IΛ and Sternberg JM. Suppressor macrophages in *Trypanosoma brucei* infection: nitric oxide is related to both suppressive activity and lifespan *in vivo*. *Parasit. Immunol.* 1993; **17**: 143–150.

21. Schleifer KW, Filutowicz H, Schopf LR and Mansfield JM Characterisation of T helper cell responses to the trypanosome variant surface glycoprotein. *J. Immunol.* 1993; **150**: 2910–2919.

22. Bancroft GJ, Sutton CJ, Morris AG and Askonas BA. Production of interferons during experimental African trypanosomiasis. *Clin. Exp. Immunol.* 1983; **52**: 135–143.

23. DeGee ALW, Sonnenfeld G and Mansfield JM. Genetics of resistance to the African trypanosomes. V. Qualitative and quantitative differences in interferon production among susceptible and resistant mouse strains. *J. Immunol.* 1985; **134**: 2723–2726.

24. Kardami E, Pearson TW, Beecroft RP and Fandrich RR. Identification of basic fibroblast growth factor-like proteins in African trypanosomes and *Leishmania*. *Mol. Biochem. Parasitol.* 1992; **51**: 171–178.

25. Olsson T, Bakhiet M, Edlund C, Hojeberg B, Van der Meide P and Kristensson K. Bidirectional activating signals between *Trypanosoma brucei* and T cells: a trypanosome-released factor triggers interferon-γ production that stimulates parasite growth. *Eur. J. Immunol.* 1991; **21**: 2447–2454.

26. Bakhiet M, Olsson T, Edlund C, Hojeberg B, Holmberg K, Lorentzen R and Kristensson K. *Trypanosoma brucei brucei*-derived factor that triggers CD8+ lymphocytes to interferon-γ secretion: purification, characterization and protective effects *in vivo* by treatment with a monoclonal antibody against the factor. *Scand. J. Immunol.* 1993; **37**: 165–178.

27. Kierszenbaum F *et al.* Trypanosomal immunosuppressive factor—a secretion product(s) of *Trypanosoma cruzi* that inhibits proliferation and IL-2 receptor expression by activated human peripheral blood cells, *J. Immunol.* 1990; **144**: 4000–4004.

28. Dos Reis GA, Fonseca MEF and Lopes MF. Programmed T-cell death in experimental Chagas disease. *Parasitol. Today* 1995; **11**: 390–394.

29. Balde AT, Sarthou J-L and Roussilhon C. Acute *Plasmodium falciparum* infection is associated with increased percentages of apoptotic cells. *Immunol. Lett.* 1995; **46**: 59–62.

30. Maizels RM and Lawrence RA. Immunological tolerance—the key feature in human filariasis. *Parasitol. Today* 1991; **7**: 271–276.

31. King CL and Nutman TB. Regulation of the immune response in lymphatic filariasis and onchocerciasis. *Immunol. Today* 1991; **12**: 54.

32. Hussain R, Grugl M and Ottesen EA. IgG antibody subclasses in human filariasis: differential subclass recognition of parasite antigens correlates with different clinical manifestations of infection. *J. Immunol.* 1987; **139**: 2794–2798.

33. Ottesen EA. Immune responsiveness and the pathogenesis of human onchocerciasis. *J. Infect. Dis.* 1995; **171**: 659–671.

34. Prier RC and Lammie PJ. Differential regulation of *in vitro* humoral and cellular immune responsiveness in *Brugia pahangi*-infected jirds. *Infect. Immun.* 1988; **56**: 3052–3057.

35. Zahner H, Sanger I, Chatterjee RK and Seibold G. Altered immune responses (humoral and delayed-type hypersensitivity reactions) to sheep red blood cells in the course of experimental filarial infections (*Litomosoides carinii*, *Brugia malayi*, *Acanthocheilonema viteae*) of *Mastomys natalensis*. *Parasitol. Res.* 1989; **75**: 401–411.

36. Greene BM, Fanning MM and Ellner, JJ Non-specific suppression of antigen-induced lymphocyte blastogenesis in *Onchocerca volvulus* infection in man. *Clin. Exp. Immunol.* 1983; **52**: 259–265.

37. Nutman TB, Kumaraswami V, Pao L, Narayanan PR and Ottesen EA. An analysis of *in vitro* B cell immune responsiveness in human lymphatic filariasis. *J. Immunol.* 1987; **138**: 3954–3959.

38. Nutman TB, Kumaraswami V and Ottesen EA. Parasite-specific anergy in human filariasis. Insights after analysis of parasite antigen-driven lymphokine production. *J. Clin. Invest.* 1987; **79**: 1516–1523.

39. Weil GJ, Hussain R, Kumaraswami V, Tripathy SP, Phillips KS and Ottesen EA. Prenatal allergic sensitization to helminth antigens in offspring of parasite-infected mothers. *J. Clin. Invest.* 1983; **75**: 1124–1129.

40. Weiss N. *Dipetalonema viteae*. *In vitro* blastogenesis of hamster spleen and lymph node cells to phytohemagglutinin and filarial antigens. *Exp. Parasitol.* 1978; **46**: 283–299.

41. Wadee AA, Vickery AC and Piessens WF. Characterisation of immunosuppressive proteins of *Brugia malayi* microfilariae. *Act. Trop.* 1987; **44**: 343–352.

42. Petralanda I, Yarzabal L and Piessens WF. Parasite antigens are present in breast milk of women infected with *Onchocerca volvulus*. *Amer. J. Trop. Med. Hyg.* 1988; **38**: 372–379.

43. Lal RB, Kumaraswami V, Steel C and Nutman TB. Phosphocholine-containing antigens of *Brugia malayi*

nonspecifically suppress lymphocyte function. *Amer. J. Trop. Med. Hyg.* 1990; **42**: 56–64.

44. Elkhalifa MY, Ghalib HW, Dafa'Alla T and Williams JF. Suppression of human lymphocyte responses to specific and non-specific stimuli in human onchocerciasis. *Clin. Exp. Immunol.* 1991; **86**: 433–439.

45. Harnett W and Harnett MM. Inhibition of murine B cell proliferation and down-regulation of protein kinase C levels by a phosphorylcholine-containing filarial excretory–secretory product. *J. Immunol.* 1993; **151**: 4829–4837.

46. Belley A and Chadee K. Eicosanoid production by parasites: from pathogenesis to immunomodulation? *Parasitol. Today* 1995; **11**: 327–334.

47. Piessens WF, Partono F, Hoffman SL, Ratiwayanto S, Piessens PW, Palmieri JR, Koiman I, Dennis DT and Carney WP. Antigen-specific suppressor T lymphocytres in human lymphatic filariasis. *N. Engl. J. Med.* 1982; **307**: 144–148.

48. Ouaissi MA, Dessaint JP, Cornette J, Desmoutis I and Capron A. Induction of non-specific human suppressor cells *in vitro* by defined *Onchocerca volvulus* antigens. *Clin. Exp. Immunol.* 1983; **53**: 634–644.

49. Owhashi M, Horii Y, Ikeda T, Tsukidate S, Fujita K and Nawa Y. Non-specific immune suppression by *CD8+* cells in *Brugia pahangi*-infected rats. *Int. J. Parasitol.* 1990; **20**: 951–956.

50. Luder CGK, Soboslay PT, Prince AM, Greene BM, Lucius R and Schultz-Key H. Experimental onchocerciasis in chimpanzees: cellular responses and antigen recognition after immunisation and challenge with *Onchocerca volvulus* infective third-stage larvae. *Parasitology* 1993; **107**: 87-97.

51. Sartono E, Kruize YCM, Kurniawan A, Maizels RM, van den Elsen PJ, Eggermond MCJA and Yazdanbakhsh M. *In vivo* and *in vitro* modulation of T-cell responses in filariasis. *Parasite* 1994; **1** (suppl.): 17–18.

52. Leiva LE and Lammie PJ. Regulation of parasite antigen-induced T cell growth factor activity and proliferative responsiveness in *Brugia pahangi*-infected jirds. *J. Immunol.* 1989; **142**: 1304–1309.

53. Mitchell GF and Lewers HM. Studies on immune responses to parasite antigens in mice IV. Inhibition of an anti-DNP antibody response with the antigen, DNP-Ficoll containing phosphorylcholine. *Int. Arch. Aller. App. Immunol.* 1976; **52**: 235–240.

54. Harnett W and Parkhouse RME. Structure and function of nematode surface and excretory–secretory products. In: Sood M (ed.) *Perspectives in Nematode Biochemistry and Physiology*. New Delhi: Nadrida Publishing Company, 1995: 207–242.

55. Weil GJ. Parasite antigenemia in lymphatic filariasis. *Exp. Parasitol.* 1990; **71**: 353–356.

56. Harnett W. Molecular approaches to the diagnosis of *Onchocerca volvulus* in man and the insect vector. In: Kennedy, MW (ed.) *Parasitic Nematodes—Antigens, Membranes and Genes*. London: Taylor and Francis, 1991: 195–218.

57. Deehan MR, Harnett MM, Frame MJ and Harnett W. Modulation of antigen receptor-coupling to tyrosine kinase-mediated signalling pathways in B lymphocytes by a filarial nematode excretory–secretory product. Submitted for publication.

58. Harnett MM, Deehan MR, Williams D, and Harnett, W. Modulation of antigen receptor-coupling to signalling pathways in T lymphocytes by a filarial nematode excretory–secretory product. Submitted for publication.

59. Yin Foo D, Nowak M, Copeman B and McCabe M. A low molecular weight immunosuppressive factor produced by *Onchocerca gibsoni*. *Vet. Immunol. Immunopath.* 1983; **4**: 445–451.

60. Monroy FG, Dobson C, Adams JH. Low molecular weight immunosuppressors secreted by adult *Nematospiroides dubius*. *Int. J. Parasitol.* 1989; **19**: 125–127.

61. Pritchard DI, Lawrence CE, Appleby P, Gibb IA, Glover K. Immunosuppressive proteins secreted by the gastrointestinal parasite *Heligmosomoides polygyrus*. *Int. J. Parasitol.* 1994; **24**: 495–500.

62. Dessaint JP, Camus D, Fischer E and Capron A. Inhibition of lymphocyte proliferation by factor(s) produced by *Schistosoma mansoni*. *Eur. J. Immunol.* 1977; **7**: 624–629.

63. Mazingue C, Dessaint JP, Schmitt-Verhulst AM, Cerottini JC and Capron A. Inhibition of cytotoxic T lymphocytes by a schistosome-derived inhibitory factor is independent of an inhibition of the production of interleukin 2. *Int. Arch. Allergy Appl. Immunol.* 1983; **72**: 22–29.

64. Mazingue C, Walker C, Domzig W, Capron A, DeWeck A and Stadler BM. Effect of schistosome-derived inhibitory factor on the cell cycle of T lymphocytes. *Int. Arch. Allergy Appl. Immunol.* 1987; **83**: 12–18.

65. Hamajima F, Yamamoto M, Tsuru S, Yamakami K, Fujino T, Hamajima H and Katsura Y. Immunosuppression by a neutral thiol protease from parasitic helminth larvae in mice. *Parasit. Immunol.* 1994; **16**: 261–273.

66. Leid RW, Suquet CM, Bouwer HGA and Hinrichs DJ. Interleukin inhibition by a parasite proteinase inhibitor, taeniastatin. *J. Immunol.* 1986; **137**: 2700–2702.

67. Giorgio S, Jasiulionis MG, Straus AH, Takahashi HK and Barbieri CL. Inhibition of mouse lymphocyte proliferative response by glycosphingolipids from *Leishmania (L.) amazonensis*. *Exp. Parasitol.* 1992; **75**: 119–125.

68. Londner MV, Frankenburg S, Sluttzky GM and Greenblatt CL. Action of leishmanial excreted factor (EF) on human lymphocyte blast transformation. *Parasit. Immunol.* 1983; **5**: 249–256.

69. Ballet JJ, Jaureguiberry G, Deloron P and Agrapart M. Stimulation of T-lymphocyte-dependent differentiation of activated human B lymphocytes by *Plasmodium falciparum* supernatants. *J. Infect. Dis.* 1987; **155**: 1037–1040.

70. Stagg DA, Chasey D, Young AS, Morzaria SP and Dolan TT. Synchronisation of the division of *Theileria* macroschizonts and their mamalian host cells. *Ann. Trop. Med. Parasitol.* 1980; **74**: 263–265.

71. Shaw MK. *et al.* The entry of *Theileria parva* sporozoites into bovine lymphocytes—evidence for class-1 involvement. *J. Cell. Biol.* 1971; **113**: 87–101.

72. Hullinger L. Cultivation of three species of *Theileria* in lymphoid cells *in vitro*. *J. Protozool.* 1965; **12**: 649–655.

73. Brown CGD, Stagg DAM, Purnell RE, Kanhai GK and Payne RC. Infection and transformation of bovine lymphoid cells *in vitro* by infective particles of *Theileria parva*. *Nature* 1973; **245**: 101–103.

74. ole-Moi Yoi OK. *Theileria parva*: an intracellular proto-zoan parasite that induces reversible lymphocyte transformation. *Exp. Parasitol.* 1989; **69**: 204–210.

75. Baldwin CL and Teale AJ. Alloreactive T-cell clones transformed by *Theileria parva* retain cytolytic activity and antigen specificity. *Eur. J. Immunol.* 1987; **17**: 1859–1862.

76. Heussler VT, Eichhorn M, Reeves R, Magnuson MS, Williams RO and Dobbelaere DA Constitutive IL-2 mRNA expression in lymphocytes infected with the intracellular parasite *Theileria parva*. *J. Immunol.* 1992; **149**: 562–567.

77. Dobbelaere DAE, Prospero PD, Roditi IJ, Kelke C, Eichhorn M, Williams RO, Ahmed JS, Baldwin CL, Clevers H and Morrison WI. Expression of Tac antigen component of bovine interleukin-2 receptor in different leucocyte populations infected with *Theileria parva* or *Theileria annulata*. *Infect. Immun.* 1990; **58**: 3847–3855.

78. Eichhorn M and Dobbelaere DAE. Partial inhibition of *Theileria parva*-infected T-cell proliferation by antisense IL2R α-chain RNA expression. *Res. Immunol.* 1995; **146**: 89–99.

79. Ivanov V, Stein B, Baumann I, Eichhorn M, Dobbelaere DAE, Herrlich P and Williams RO. Infection with the intracellular protozoan parasite *Theileria parva* induces constitutively high levels of NF-κB in bovine T lymphocytes. *Mol. Cell. Biol.* 1989; **11**: 4677–4686.

80. ole-Moi Yoi OK, Brown WC, Iams KP, Nayar A, Tsukamoto T and Macklin MD. Evidence for the induction of casein kinase II in bovine lymphocytes transformed by the intracellular protozoan parasite *Theileria parva*. *EMBO J.* 1993; **12**: 1621–1631.

81. Eichhorn M and Dobbelaere DE. Induction of signal transduction pathways in lymphocytes infected by *Theileria parva*. *Parasitol. Today* 1994; **10**: 469–472.

23

Immunophilin-mediated Inhibition of Lymphocyte Signalling

John E. Kay and David J. Gilfoyle

School of Biological Sciences, University of Sussex, Brighton, UK

IMMUNOSUPPRESSION BY CsA, FK506 AND RAPAMYCIN

Pioneering work in the early 1970s by Dr JF Borel and colleagues at Sandoz Ltd identified cyclosporin A (CsA), a hydrophobic cyclic endecapeptide (Figure 23.1) with some anti-fungal activity isolated from the culture broths of the fungi *Tolypocladium inflatum* and *Cylindrocarpum lucidum*, as a new type of immunosuppressive drug [1]. At therapeutic concentrations CsA showed no general inhibition of mammalian cell proliferation but selectively interfered with an early step in the activation of T lymphocytes by antigen or mitogens. Borel's discovery led to successful clinical trials, and CsA has now been used with considerable success in organ transplantation for more than a decade and has also proved useful in treating some autoimmune conditions [2]. Compared to the drugs previously employed CsA has relatively few serious side effects, the most frequently encountered being nephrotoxicity, although it does have a spectrum of effects on other mammalian cells in experimental systems. A feature common to many of its reported actions is that they involve the inhibition of Ca^{2+}-mediated events.

A decade later FK506 (also known as tacrolimus) was isolated by Fujisawa Pharmaceuticals from the fermentation broth of *Streptomyces tsukubaensis*, a prokaryotic soil microorganism [3, 4]. FK506 is a hydrophobic macrolide, and unrelated in structure to CsA (Figure 23.1). However, it also has some anti-fungal activity, and a spectrum of immunosuppressive effects *in vitro* and *in vivo* that is remarkably similar to that of CsA, except that it is effective at concentrations about two orders of magnitude

lower [5]. It inhibits the same early Ca^{2+}-dependent step in T lymphocyte activation as CsA, and like CsA it selectively blocks the antigen- and mitogen-induced synthesis of IL-2 and other lymphokines. It also has similar effectiveness to CsA in organ transplantation and the same major side effects, although there are some minor side effects that differ [5–7]. It also very frequently mimics the action of CsA in other experimental systems.

Rapamycin, a macrolide produced by *Streptomyces hygroscopicus*, has a strong structural resemblance to FK506 over one-half of its structure (Figure 1). It inhibits the proliferation of some yeasts and some mammalian cells. The initiation of T and B lymphocyte proliferation is especially sensitive, and in mammals *in vivo* rapamycin is strongly immunosuppressive [8, 9]. However, despite its structural similarity to FK506 its mechanism of action is quite distinct. FK506 and CsA inhibit T and B lymphocyte activation only when it is dependent on a Ca^{2+}-mediated intracellular signalling pathway, but rapamycin also blocks the induction of proliferation of T and B cells in response to Ca^{2+}-independent mitogens, such as phorbol esters plus anti-CD28 or LPS (Table 23.1) [10–13]. Unlike FK506 or CsA, rapamycin does not block the induction of IL-2 expression, while IL-2 driven proliferation of activated T cells or T cell lines is blocked by rapamycin but insensitive to FK506 or CsA.

IMMUNOPHILINS AS THE PRIMARY TARGETS OF IMMUNOSUPPRESSIVE DRUGS

The principal intracellular binding protein for CsA is the abundant 17.8 kDa cytoplasmic peptidylpro-

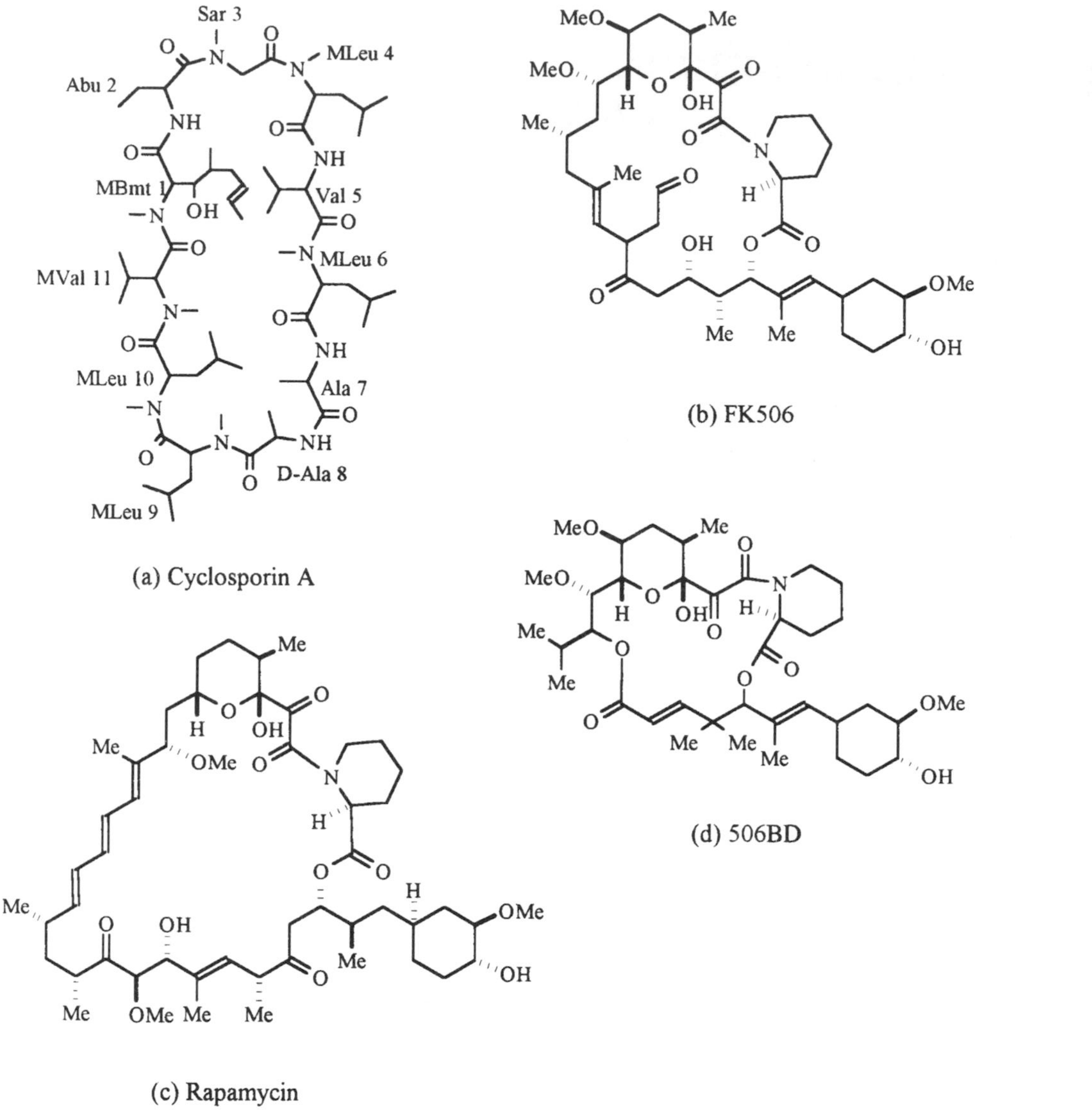

(a) Cyclosporin A

(b) FK506

(c) Rapamycin

(d) 506BD

Figure 23.1 Molecular structures of CsA, FK506, rapamycin and 506BD. CsA, FK506 and rapamycin are naturally produced immunophilin ligands. 506BD is a synthetic macrolide designed to mimic the shared region of FK506 and rapamycin [23]. It inhibits the PPIase activity of FKBP12 but is not immunosuppressive

lyl *cis–trans* isomerase (PPIase) cyclophilin A (CypA) [14–17], while the major binding protein for both FK506 and rapamycin is FKBP12, another abundant cytoplasmic 11.8 kDa PPIase [18, 19]. These PPIases have been termed immunophilins due to their role in immunosuppression, but both CypA and FKBP12 are widely if not universally expressed in mammalian cells and highly conserved homologues are found in many other eukaryotic and prokaryotic organisms. The reaction catalysed by these PPIases or immunophilins is the interconversion of the peptide bonds on the amino side of a proline residue between the *trans* configuration produced by ribo-

Table 23.1   Effects of CsA, FK506 and rapamycin on lymphocyte proliferation in response to mitogens. The strengths of the mitogenic responses observed in response to the different combinations of mitogens and inhibitors are indicated on a scale from (none) to ++++ (strong proliferative response)

	Control	+ CsA	+FK506	+ rapamycin
T cells				
Con A	++++	+	+	+
Anti-CD3	+++	±	±	+
Ionomycin	++	−	−	±
Ionomycin + TPA	++++	+	+	+
Anti-CD28 + TPA	++++	++++	++++	+
Primed T cells				
IL-2	++++	++++	++++	+
B cells				
Anti-Ig	+++	+	+	+
LPS	++++	++++	++++	+

somal protein synthesis and the *cis* configuration in which some proline residues are found in mature proteins. This isomerization is one of the slower steps in protein folding *in vitro* and may be accelerated, at least in the test tube, by PPIases [20, 21].

CsA inhibits the PPIase activity of CypA and both FK506 and rapamycin inhibit the PPIase activity of FKBP12. However, the initial hypothesis that the PPIase-catalysed interconversion of key signalling elements between active and inactive forms might play a critical role in lymphocyte activation has been convincingly disproved, as evidence has accumulated that inhibition of PPIase activity is not a sufficient explanation for the immunosuppressive action. The sub-nanomolar concentrations of FK506 and rapamycin that block T lymphocyte proliferation are too low to inhibit more than a small proportion of the abundant PPIase activity in the cell. In addition FK506 and rapamycin bind competitively to the same FKBP12 site but inhibit lymphocyte activation by quite distinct mechanisms, as noted above, and show reciprocal competitive inhibition of each other's effects [11, 13]. There are also synthetic analogues of CsA [22], FK506 [23, 24] and rapamycin [25] that bind to the immunophilin active site and effectively inhibit its PPIase activity but are not immunosuppressive. Analogues such as 506BD (Figure 23.1) have minimal effects on the proliferation or metabolism of mammalian cells, but all such analogues competitively inhibit the

actions of the immunosuppressive agents that target the same immunophilin.

Perhaps surprisingly in view of their abundance and high degree of sequence conservation, and the occurrence of a variety of natural products that target them, disruption of PPIase genes from yeast or fungi has little effect on their survival or proliferation under a wide variety of environmental conditions [26–30]; indeed, apart from a role of cyclophilins in yeast growth at high temperatures [31, 32], the principal effects are loss of any sensitivity to CsA when the CypA gene is disrupted and loss of sensitivity to FK506 or rapamycin when the FKBP12 gene is inactivated.

CypA and FKBP12 are now recognized as the prototypes of two distinct immunophilin families, different members of which may be expressed in the cytoplasm, endoplasmic reticulum, mitochondria, nucleus and on the cell surface [33, 34]. Each family contains simple PPIases, comprising just the basic PPIase domain plus any signal sequences necessary for targeting to the correct compartment, and also PPIase-domain proteins, in which one or more PPIase domains are found within a larger protein (Table 23.2). A notable feature is that many immunophilins are found in association with other proteins in multi-subunit protein assemblies. CsA binds strongly to all members of the mammalian cyclophilin family studied, and rapamycin binds to all characterized mammalian FKBPs. However, other members of the FKBP family have significantly lower affinity for FK506 than FKBP12 and two family members, the endoplasmic reticulum-retained FKBP13 and the nuclear FKBP25, have markedly decreased affinities for this ligand [46, 47]. A recently identified *Escherichia coli* PPIase, parvulin, appears to represent a third PPIase family, although most members of this family so far identified are prokaryotic [48, 49]. CsA, FK506 and rapamycin do not affect parvulin PPIase activity, and no natural inhibitors have yet been found.

X-ray crystallographic and nuclear magnetic resonance structures have been determined for CypA and FKBP12 with and without bound ligands, and confirm the initial impression from sequence studies that there is no significant homology between the two families [50]. The CypA structure is a β-barrel made up of two antiparallel four stranded β-sheets with α-helices at each end and a variety of exposed loops and turns (Figure 23.2a). CsA and substrate peptides are bound in a long groove on the outer surface of the β-barrel. Amino acids 1–3 and 9–11 of CsA bind into this site, with the methylvaline-11 residue lying in the hydrophobic pocket occupied by the proline residue when a peptide is bound. The remaining CsA residues are exposed to solvent. By contrast the FKBP12 structure is a single five-stranded antiparallel β-sheet twisted to give a hydrophilic convex outer surface and a hydrophobic concave inner

surface (Figure 23.2b). A single α-helix in a long loop passes above the concave surface. The proline binding site, into which the shared sections of the FK506 and rapamycin macrolide rings also fit, lies at one end of the β-sheet and within the hydrophobic core of the FKBP molecule. Site-specific mutagenesis of the CypA or FKBP12 residues that form the proline binding site affects PPIase activity and/or drug sensitivity, but specific changes may have differential effects on the two properties [51–53]. Both the cytoplasmic and periplasmic *E. coli* cyclophilins have a key tryptophan residue in the binding site replaced by phenylalanine, a modification that reduces CsA binding by over 99% without having any corresponding impact on their PPIase activity [54, 55].

It is thus evident that while binding of CsA, FK506 and rapamycin to their target immunophilin is an essential precondition for their immunosuppressive action, it is not in itself sufficient. Immunosuppression requires that the primary drug–immunophilin complex in turn binds to a secondary target, and it is the consequent inactivation of the secondary target protein that blocks lymphocyte activation and thus leads to immunosuppression. For both CsA–cyclophilin and FK506–FKBP complexes the secondary target has been identified as the protein phosphatase calcineurin, while for the rapamycin–FKBP complex

Table 23.2 Characterized human PPIases. The list of human immunophilins below is certainly incomplete. At least two additional FKBP-domain proteins that will doubtless prove to have human homologues have been identified in the mouse. With the exception of CypNK, expressed specifically in natural killer cells, the other immunophilins listed below are widely distributed in mammalian cells

Immunophilin	Amino acid residues	Cellular location	Ref.
CypA	165	Cytoplasm	35
CypB	208*	ER subcompartment—calciosome?	36
CypC	212*	ER & cell membrane	37
CypD	207*	Mitochondria	38
Cyp40	370	Steroid receptor complex	39
CypNK	1403*	Cell membrane, NK cells	40
Cyp358	3224	Nuclear pores	41
FKBP12	108	Cytoplasm	19
FKBP12A	108	Cytoplasm	42
FKBP13	141*	ER—pre-Golgi	43
FKBP25	224	Nucleus	44
FKBP52	459	Steroid receptor complex	45

*Immunophilins containing an *N*-terminal signal sequence.

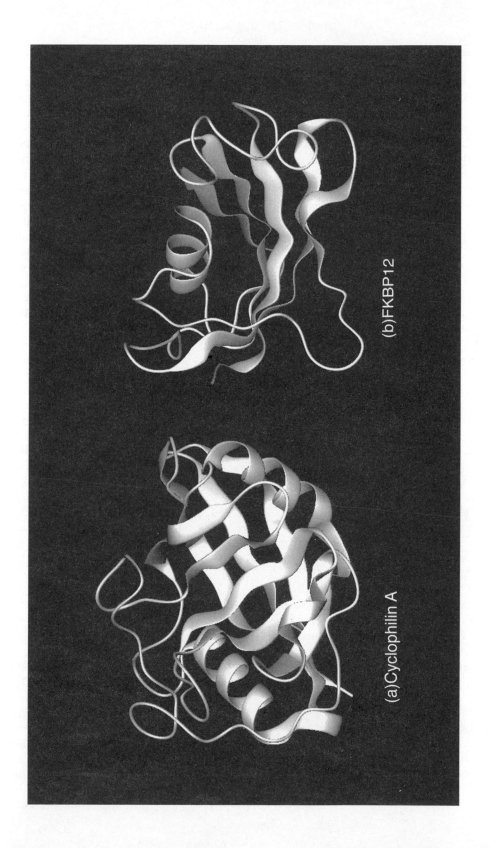

Figure 23.2 Ribbon diagrams of the 3-dimensional structures of (a) CypA and (b) FKBP12. Both immunophilin structures are shown on the same scale, and both are orientated with the PPIase ligand binding site at the front. The strands of the β-sheets and the α-helices are emphasized.

the secondary target is a mammalian homologue of the yeast TOR gene product, a protein that contains a phospholipid kinase homologous domain. In each case the secondary target is a key component of an intracellular signalling pathway. These secondary target proteins are of comparatively low abundance, and their inactivation is due to their ligand-dependent sequestration in complexes with the more abundant and less essential immunophilin.

SEQUESTRATION OF CALCINEURIN BY CsA–CYCLOPHILIN AND FK506–FKBP COMPLEXES

Despite the lack of any evident homology, or even resemblance, between CsA and FK506 or between cyclophilins and FKBPs, it was clear from cell culture and whole animal studies that the principal effects of the two drugs were very similar. The key to understanding this came from studies on retention of proteins present in cell or tissue extracts by CypA-, CypC- and FKBP12-affinity columns in the presence or absence of the corresponding ligands [56, 57]. CypA- and CypC-affinity columns were able to retain the Ca^{2+}-activated serine/threonine protein phosphatase calcineurin (protein phosphatase 2B) in a complex with calmodulin only in the presence of CsA, while FKBP12-affinity columns were able to retain exactly the same proteins in the presence of FK506. Rapamycin could not substitute for FK506, and at high concentrations competitively inhibited the FK506–FKBP12-mediated (but not CsA–CypA-mediated) retention of calcineurin. Addition of CsA or FK506 to T cells inhibited the calcineurin activity of their extracts [58], and both isoforms of calcineurin were affected [59]. The immunosuppressive activity of CsA and FK506 analogues was shown to correlate closely with the ability of their immunophilin complexes to sequester calcineurin [60–62] and overexpression of calcineurin increased the concentration of CsA or FK506 necessary to exert their effects [63, 64]. The immunosuppressive action of ligand concentrations insufficient to inhibit all the PPIase activity is accounted for by the secondary target being much less abundant within the cell [59, 65].

CsA and FK506 complexed with immunophilin homologues can also inactivate calcineurin homologues in plants [66, 67]. In the yeasts *Saccharomyces cerevisiae* and *Schizosaccharomyces pombe* the effects of CsA or FK506 addition are identical to those of inactivation of the yeast calcineurin homologues [68, 69]. Although the *E. coli* periplasmic cyclophilin, which has very low affinity for CsA, also has no detectable effect on mammalian calcineurin in the presence of CsA, both CsA binding and calcineurin sequestration can be restored by a single amino acid substitution [70]. It seems likely that most if not all the biological activities of CsA and FK506 can be explained by calcineurin sequestration.

Mammalian CypA, CypB, CypC and Cyp40 have all been shown to be capable of CsA-dependent calcineurin sequestration in the test tube [56, 57, 71, 39], with CypB the most potent member of the family. When overexpressed in a T cell line CypC was ineffective unless its signal sequence was removed so that the protein was expressed in the cytoplasm [72]. However, CypB was still effective despite its signal sequence, and addition to CypC of the short *C*-terminal CypB peptide (which is believed to direct it to a specific membrane-bound cellular compartment) allowed the modified CypC to sequester calcineurin in the intact cell even when its signal sequence remained.

Of the FKBPs and FKBP-domain proteins studied, only FKBP12 and FKBP12A are able to sequester calcineurin effectively *in vitro* or in the intact cell. FKBP13, FKBP25, FKBP51 and FKBP52 are all very much less active [56, 52, 72–77]. The active FKBPs have a hydrophobic groove on the FKBP–FK506 surface, while in the inactive immunophilins the character of this groove is changed by the presence of charged residues. Site directed mutagenesis to restore the hydrophobic nature of this groove is sufficient to enable FKBP13 to form complexes with FK506 capable of efficient calcineurin sequestration [78, 79]. X-ray crystallographic studies of the bovine FKBP12–FK506–calcineurin complex have confirmed that these residues are directly involved in the interaction [80]. These studies also showed that when the FK506–FKBP12 complex binds to the calcineurin heterodimer it makes contacts with both the phosphatase domain and the B subunit-binding domain of the catalytic calcineurin A subunit and also with the regulatory calcineurin B subunit. Both the FK506 and the FKBP12 participate in the interaction with calcineurin, which is predicted to occur only when Ca^{2+}/calmodulin interaction with the calmodulin-binding domain of calcineurin activates the enzyme, thereby expos-

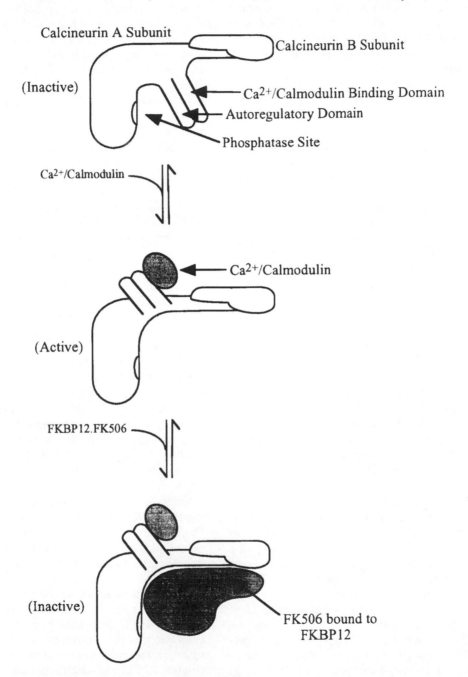

Figure 23.3 Schematic model for the interaction between FK506–FKBP12, calcineurin and calmodulin. The model proposed [80] is that interaction between the FK506–FKBP12 complex and the calcineurin A and B subunits is dependent on the prior interaction between the calcineurin A regulatory domain and calmodulin

ing the FK506–FKBP12 binding site (Figure 23.3). Docking of FK506–FKBP12 at this binding site does not directly inactivate the protein phosphatase catalytic domain but renders it physically inaccessible to large substrates, explaining why FK506–FKBP12 complexes inhibit calcineurin phosphatase activity with large peptide substrates but actually enhance the dephosphorylation of small substrates such as p-nitrophenyl phosphate [56]. The competition between CsA–CypA and FK506–FKBP12 for calcineurin sequestration, and analysis of a number of calcineurin mutations suggests that CsA–CypA and CsA–CypB complexes bind to the same region of calcineurin as FK506–FKBP12, but not to precisely the same residues [56, 81, 82].

Until its identification as a common secondary target for CsA–cyclophilin and FK506–FKBP12 complexes, calcineurin had not been recognized as playing any crucial signalling role in T lymphocyte activation. A principal effect of CsA and FK506 on T lymphocyte activation is inhibition of the early activation of the expression of several of the genes essential for T lymphocytes to proliferate and express their differentiated functions, most evidently the expression of mRNA for IL-2 and other lymphokines [83, 84]. The IL-2 promoter region includes several transcription factor binding sites, some (e.g. NF-AT or NF-IL2A) for factors relatively specific for T cells and others (e.g. NF-κB) more generally expressed. The activation of several such factors is sensitive to inhibition by CsA and FK506 (but not rapamycin). The formation of active NF-AT and NF-IL2A is very strongly inhibited by low concentrations of these inhibitors, but NF-κB activity is less sensitive, with the extent of inhibition correlating with the relative -dependence of the mitogenic stimulus employed. The mechanism of NF-AT activation has been studied in most detail. The active transcription factor found in the nucleus after T lymphocyte activation has two elements: a nuclear component newly synthesized after activation (identified as the widely expressed transcription factor AP-1); and a second component that is present in the cytoplasm of unstimulated T lymphocytes as a phosphorylated precursor, NF–AT$_p$. Expression of AP-1 after activation occurs by a calcineurin-independent pathway, but the dephosphorylation of NF–AT$_p$ is prevented when CsA

or FK506 is added at the time of activation (Figure 23.4). Calcineurin can dephosphorylate NF–AT$_p$ *in vitro* and this appears esential for its translocation into the nucleus. However, NF–AT$_p$ dephosphorylation is not the only role played by calcineurin in T lymphocytes. FK506 and CsA can also down-regulate expression of genes not regulated by NF-AT and block other Ca^{2+} -mediated events such as the induced apoptosis in T cell hybridomas [85, 86] and they can also block degranulation of T cytotoxic cells, a process that does not depend on gene expression [87]. It seems likely that calcineurin has multiple effects in T lymphocyte proliferation and function.

SEQUESTRATION OF THE MAMMALIAN TOR HOMOLOGUE BY RAPAMYCIN–FKBP COMPLEXES

Despite its sharing the same FKBP12 primary target as FK506, rapamycin inhibits T and B lymphocyte activation by a quite distinct mechanism and rapamycin–FKBP12 complexes are unable to sequester calcineurin [56] because the larger exposed domain of rapamycin is unable to fit into the FK506–FKBP12 binding site [80]. The secondary target of rapamycin–FKBP12 was first identified from genetic studies in the yeast *S. cerevisiae*, whose proliferation is also sensitive to rapamycin. Rapamycin-resistant *S. cerevisiae* mutants are readily isolated. Over 90% of these proved to have a recessive mutation inactivating FKBP12, but in a small minority the mutation was dominant, and mapped to one of two different genetic loci encoding the TOR1 and TOR2 proteins (TOR = Target Of Rapamycin) [88]. Inactivation of the more abundant of the two, TOR1, led only to a reduced growth rate, but TOR2 was essential for normal growth.

The TOR proteins encode two large homologous proteins, of 2470 and 2474 amino acids respectively [89–92]. They have at their C-terminal ends functionally equivalent domains with homology to yeast and mammalian phospholipid kinases (Figure 23.5), and although they have not themselves been shown to have either lipid or protein kinase activity, mutation of specific residues predicted to be essential for kinase activity in the putative kinase domain is sufficient to inactivate TOR function [93]. In the dominant rapamycin-resistant mutants a critical serine residue (position 1972 in TOR1 or 1975 in TOR2) is replaced by a

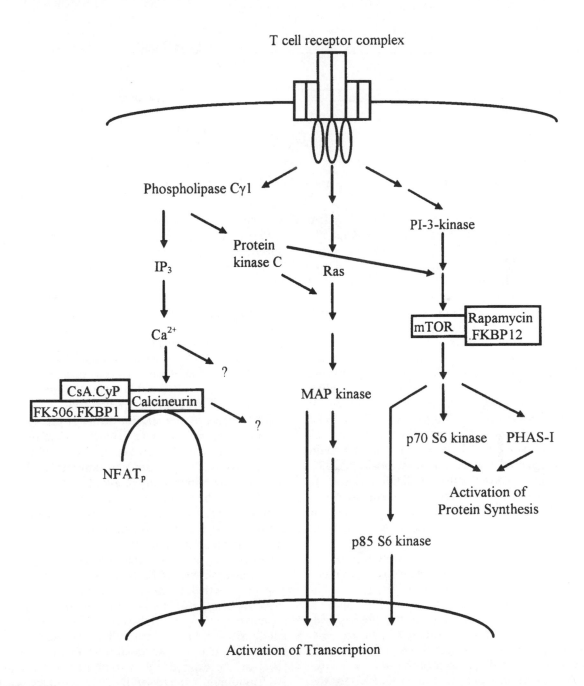

Figure 23.4 Effects of CsA–Cyp, FK506–FKBP12 and rapamycin–FKBP12 complexes on T lymphocyte intracellular signalling pathways. The diagram illustrates the importance of the known ligand–immunophilin secondary targets, calcineurin and mTOR, in T lymphocyte signalling initiated through the T cell receptor complex. Note that none of the ligand–immunophilin complexes affects the activation of the important Ras–MAP kinase pathway. The signalling pathways are necessarily simplified

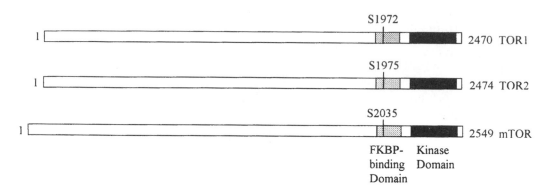

Figure 23.5 Domain structure of *S.cerevisiae* TOR proteins and their mammalian homologue mTOR

larger residue. Using the yeast two-hybrid system or rapamycin-dependent retention on FKBP12 affinity matrices it can be demonstrated that rapamycin–FKBP12 interacts directly with a region of TOR1 or TOR2 surrounding this key serine, and that the interaction is lost if the serine is substituted by a larger alternative (alanine at this position is permissible) [93, 94]. In the *S. cerevisiae* TOR mutants that confer rapamycin resistance the altered protein retains functional activity, but can no longer be sequestered by FKBP12–rapamycin.

Subsequently a 2549 amino acid mammalian homologue of TOR was identified in a number of laboratories, and named FRAP, RAFT1, RATP1 or mTOR [95–98]. This protein also has a *C*-terminal phospholipid kinase homologous domain and a conserved serine residue at position 2035, corresponding to the conserved TOR serine (Figure 23.5). It showed rapamycin-dependent retention on FKBP12-affinity matrices, but was not retained if non-inmmunosuppressive rapamycin analogues were used, or if the FKBP12 was replaced by FKBP25. A fragment around the conserved serine residue was sufficient to ensure binding to rapamycin–FKBP12 using the yeast two-hybrid system, but substitution of the serine by a larger residue again prevented this interaction. When potential FKBP12–rapamycin target proteins were generated using a mouse DNA library, three-quarters of the 28 clones that gave rapamycin-dependent interaction with FKBP12 in the two-hybrid system encoded this mTOR, with the remaining seven encoding six distinct proteins [97]. Further characterization of these additional clones

will be required before it can be concluded that they are additional secondary targets for rapamycin–FKBP12.

Immunoprecipitates of the mammalian TOR homologue from brain or cultured cells have recently been reported to contain phosphatidylinositol-4-kinase activity, although this was not inhibited by addition of rapamycin–FKBP12 [99]. A second study confirmed the presence of trace amounts of this enzyme in immunoprecipitates of recombinant mTOR from Jurkat cell extracts, but found equal amounts of this activity even when the recombinant mTOR was replaced by a kinase-dead variant. However, the immunoprecipitated recombinant mTOR was capable of autophosphorylation, and this protein kinase activity was absent from the kinase-dead variant and was inhibited by rapamycin–FKBP12 [100]. As with yeast TOR, the serine residue essential for interaction with rapamycin–FKBP12 was not needed for kinase activity, but its replacement by a larger residue prevented the inhibition of mTOR protein kinase activity by rapamycin–FKBP12.

The insensitivity of FKBP12-deficient *S. cerevisiae* to growth inhibition by rapamycin indicates that this is the major immunophilin mediating rapamycin action in this yeast. However, rapamycin sensitivity can be restored experimentally by expression in the yeast of either human FKBP12 [101] or the excised FKBP domain of the yeast FKBP-domain protein FKBP46 [29]. Intact yeast FKBP46 is ineffective, probably because of its nucleolar localization. Overexpression of FKBP12 has been shown to enhance sensitivity to rapamycin in murine mast cells while FKBP25 is ineffective [77], but the ability of other FKBPs to interact with mTOR in the presence of rapamycin and thus mediate rapamycin action has not been systematically explored.

Prior to the identification of mTOR as the major secondary target of rapamycin–FKBP12 action, it had already been discovered that addition of rapamycin to lymphoid or non-lymphoid mammalian cells led to the very rapid dephosphorylation of the serine/threonine protein kinase p70 S6 kinase [102,103]. Previous studies had shown that p70 S6 kinase, whose principal identified substrate is the ribosomal protein S6, was phosphorylated and activated via an uncharacterized intracellular signalling pathway in response to a wide variety of proliferative stimuli, including the activation of T lymphocytes. This action of rapamycin implicates its secondary target, mTOR, as a component or modulator of this pathway, and has provided a valuable tool for the elucidation of the role played by the pathway in regulating mammalian cell proliferation. While the details of the pathway are still incompletely understood (Figure 23.4), recent studies suggest that phosphatidylinositide-3-kinase lies upstream from mTOR, and that both the kinase activity of mTOR and its *N*-terminal domain are necessary for p70 S6 kinase phosphorylation [100,104].

Rapamycin-mediated blockade of the mTOR-mediated signalling pathway results in sensitive cells, such as newly activated T or B lymphocytes, failing to express key cell cycle related proteins (or to down-regulate their inhibitors) and thus failing to enter S phase [105–108]. However, other mammalian cells appear much less sensitive, being able to continue to proliferate for several generations or even indefinitely in the presence of high rapamycin concentrations. [109–111].

There now seems little doubt that the effects of rapamycin on cell proliferation are consequent on earlier effects on protein synthesis. Inhibition of protein synthesis in different mammalian cell lines exposed to rapamycin in culture varies (for reasons that are not yet clear) from 20% to 80%, and this inhibition is sufficient to account for the subsequent effects on cell proliferation [112]. Protein synthesis in the rapamycin-sensitive *S. cerevisiae* is strongly inhibited within minutes while in the resistant fission yeast *Schizosaccharomyces pombe*

it is scarcely affected [113]. Strong selective inhibition of the translation of specific mRNA, apparently correlated with the presence of polypyrimidine tracts in their -untranslated region, has been reported [114–116]. The mRNAs affected include those encoding ribosomal proteins and protein synthesis elongation factors, but also growth factors such as IGF-II. However, it is not certain whether these effects on protein synthesis are due to inhibition of S6 phosphorylation or the inactivation of p70 S6 kinase. A rate limiting factor for translational initiation is the mRNA cap-binding protein eIF-4E, whose activity may be controlled by an inhibitor protein PHAS-I. Growth factors can cause the phosphorylation of PHAS-I, leading to it releasing eIF-4E which is then available for protein synthesis. The phosphorylation of PHAS-I in adipocytes and smooth muscle cells has been shown to be prevented by rapamycin and other inhibitors of the rapamycin-sensitive signalling pathway [117,118], but PHAS-I does not seem to be a substrate for p70 S6 kinase, suggesting that mTOR may have more than one method of regulating translation (Figure 23.4). Studies of rapamycin-resistant mutants derived from drug-sensitive cell lines have also shown that effects on protein synthesis may be dissociated from effects on p70 S6 kinase activation [119,120].

WHY ARE CsA, FK506 AND RAPAMYCIN IMMUNOSUPPRESSIVE?

Although the mechanisms of CsA, FK506 and rapamycin action revealed by the studies above offer adequate explanations for the ability of these agents to inhibit lymphocyte activation, the relatively selective immunosuppression observed on their *in vivo* administration remains to be accounted for. The drugs penetrate freely through the body and are rapidly taken up by the tissues, and both the immunophilins that are the primary targets and the intracellular signalling elements that are the secondary targets appear widely distributed in mammalian cells.

Overexpression of immunophilin can increase sensitivity to the corresponding ligands [77] while overexpression of calcineurin may reduce it [63,64], but the hypothesis that lymphocytes might be especially sensitive because they might contain higher levels of immunophilin or lower levels of calcineurin than other cells has not found experimental support. The immunophilins and calcineurin are most highly expressed in the brain but the concentrations of CypA, FKBP12, calcineurin and mTOR in lymphoid cells and tissues appear similar to those in many other mammalian cells. A recent quantitative study of calcineurin distribution in different rat tissues and its sensitivity to inhibition by CsA and FK506 [121] led to the conclusion that both its levels and its drug sensitivity in lymphoid tissues such as spleen and thymus were closely comparable to those in many other tissues, including kidney (in which major side effects are observed in clinical use), lung and muscle (neither tissues in which side effects are commonly experienced). Calcineurin levels were lowest in liver and, in confirmation of previous studies, much higher in brain (Figure 23.6). Drug sensitivity was lower in brain and perhaps also in liver, but closely comparable in the other tissues

studied. It thus seems evident that *in vivo* administration of CsA or FK506 will lead to inhibition of calcineurin activity in many tissues throughout the body, and most if not all of the side effects observed in the clinical use of CsA and FK506 can be ascribed to calcineurin inhibition in other cells. While other explanations cannot yet be altogether excluded, the most credible explanation for the dominance of immunosuppression is that under normal conditions in the mammalian body it is freshly activated lymphocytes that are most dependent on the intracellular signalling pathway(s) mediated by this widely distributed protein phosphatase. In other cells the enzyme may be redundant, due to the presence of alternative protein phosphatases with overlapping specificities. It is relevant that in the yeast *S. cerevisiae* calcineurin is also expressed but not essential for normal vegetative growth. The enzyme becomes essential, and the *S. cerevisiae* sensitive to growth inhibition by CsA or FK506, only during recovery from mating-

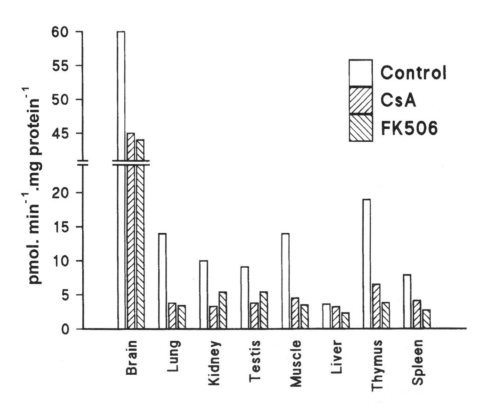

Figure 23.6 Distribution and sensitivity of calcineurin in mammalian tissues. Calcineurin activity in the extracts of a range of mammalian tissues, and its sensitivity to high concentrations of added CsA or FK506, were determined [121]

factor-induced growth arrest [68] or under unusual culture conditions [122]. Similarly with mTOR, essentially complete inhibition of p70 S6 kinase activity by rapamycin, presumably indicating effective sequestration of mTOR, has been found in a wide range of lymphoid and non-lymphoid cells. The greater inhibition of protein synthesis observed in at least some lymphoid cells presumably again reflects greater dependence on this signalling pathway.

It is of course possible that CsA, FK506 or rapamycin may, in association with an immunophilin, sequester other as yet unidentified intracellular proteins, in addition to their currently recognized secondary targets, possibly even proteins specific to lymphocytes. The non-mTOR proteins identified as interacting with rapamycin–FKBP12 in the yeast two-hybrid system may be examples [97]. There are also some minor but significant differences between the side effects of CsA and FK506, such as the hirsutism induced only by CsA, that cannot be ascribed to calcineurin inhibition. However, the mechanisms underlying these specific side effects may not involve immunophilins.

WHY DO MICROORGANISMS PRODUCE IMMUNOSUPPRESSIVE CYCLIC PEPTIDES AND MACROLIDES?

CsA, FK506 and rapamycin are all natural products produced by free-living non-pathogenic soil microorganisms. Their synthesis requires considerable investment—the production of CsA involves at least 40 reaction steps catalysed by a multienzyme complex with an estimated molecular mass of 1400 kDa [123]. As the producing microorganisms themselves contain CsA-and macrolide-sensitive PPIases [124, 125], care also has to be taken to segregate and exclude the inhibitors manufactured. As there is no evident selective advantage to be gained by such microorganisms from the production of immunosuppressive drugs, they are presumably synthesized to retard the growth of competitor microorganisms and their immunosuppressive properties an accident indicative of the remarkable conservation of intracellular signalling components throughout the eukaryotic kingdom.

Prokaryotes do also compete against the producer organisms but are not thought to have homologues of conserved eukaryotic signalling elements such as calcineurin and TOR, so that the significance of both E. coli cyclophilins bearing an amino acid substitution in a highly conserved residue that greatly reduces CsA sensitivity while retaining PPIase activity [54, 55] remains a matter for speculation.

One obvious question is whether the PPIase inhibitors produced by soil microorganisms targeting calcineurin and the TOR protein family for immunophilin sequestration represent the entire repertoire of such mechanisms that have evolved, or whether there are other such ligands with the same basic mechanism but alternative secondary targets awaiting discovery. There are at least suggestions that the full story has not yet unfolded. Both plants and fungi produce cyclic peptides containing pairs of cis-linked proline residues that have been reported to show immunosuppressive activity [126, 127], while Streptomyces hygroscopicus strain F-004081 produces meridamycin, a novel 27-membered macrolide that shares the common element of FK506 and rapamycin and binds to FKBP12 competitively with them to antagonize their effects [128]. Meridamycin is not itself immunosuppressive, but it seems unlikely that FKBP PPIase inhibition alone would justify its production. In addition, the very existence of these natural products suggests a strategy for the design of synthetic bifunctional inhibitors to target designated intracellular proteins for reversible inactivation by sequestration in complexes with immunophilins.

DO IMMUNOPHILINS INTERACT DIRECTLY WITH INTRACELLULAR SIGNALLING COMPONENTS?

A question that has proved easier to ask than to answer is whether the interaction between immunophilins and intracellular signalling elements such as calcineurin and mTOR is entirely an invention of the microorganisms producing cyclosporins and macrolides, or whether they took advantage of a pre-existing physiological association, perhaps mediated by a lower-affinity natural ligand, that might play an important role in regulating these signalling pathways. The conservation over very large evolutionary distances of the ability of immunophilins to interact with calcineurin or TOR homologues in the presence of ligands and the variety of organisms capable of producing such ligands does strongly suggest some natural association. However, a large number of biochemical studies have failed to find evidence for any retention of calcineurin or TOR on immunophilin-affinity matrices or any inhibition of calcineurin activity by immunophilins in the absence of ligand. A single study [129] has reported a weak interaction of yeast CypA and yeast FKBP12 with yeast calcineurin on affinity chromatography, weak inhibition of bovine calcineurin activity by yeast FKBP12 (but not yeast CypA), and a weak interaction between yeast FKBP12 and murine calcineurin A1 using the sensitive yeast two-hybrid system. The ligand-independent interaction with FKBP12 differed from that seen in the presence of FK506 in that it did not depend on the presence of the calcineurin B subunit and was still seen with yeast FKBP12 carrying mutations that interfere with FK506–FKBP12–calcineurin complex formation. However, in another study using the yeast two-hybrid system the interaction between human FKBP12 and calcineurin was completely dependent on FK506 [97].

Whether the PPIase activity of immunophilins represents their main physiological function remains uncertain, especially for the relatively low activity FKBP family [34]. Alternative equally credible hypotheses on the evidence presently available are that they act as chaperones, or as the components of multi-protein assemblies, binding proline residues so as to create a bond in their partner protein intermediate between the *cis* and *trans* conformations. Such an intermediate structure could affect the stability or activity of partner proteins and regulate the activities of such immunophilin-containing multi-subunit assemblies as steroid receptor complexes [130] or the endoplasmic reticulum Ca^{2+}-release channels [131]. While FK506 may affect both steroid receptor properties and Ca^{2+} release into the cytoplasm by dissociating immunophilins from their partners, no evidence of physiological ligands that modulate such immunophilin activities has so far been found.

ACKNOWLEDGEMENTS

We thank Nishith Patel and Victoria Frost for valuable discussions, and the BBSRC Intracellular Signalling Initiative for supporting related work in this laboratory.

REFERENCES

1. Borel JF, Feurer C, Gubler HU and Stahelin H. Biological effects of cyclosporin A: a new antilymphocytic agent. *Agents Actions* 1976; **6**: 468–475.
2. Borel JF. Cyclosporin. *Prog. Allergy* 1986; **38**: 1–465. Thomson AW. *Cyclosporin: Mode of Action and Clinical Applications*. Dordrecht: Kluwer Academic Publishers, 1989.
3. Kino T, Hatanaka H, Hashimoto H, Nishiyama N, Goto T, Okuhara M, Kohsaka M, Aoki H and Imanaka H. FK506, a novel immunosuppressant isolated from a Streptomyces. I. Fermentation, isolation and physicochemical and biological characteristics. *J. Antibiot.* 1987; **40**: 1249–1255.
4. Kino T, Hatanaka H, Miyata S, Inamura N, Nishiyama M, Yajima T, Goto T, Okuhara M, Kohsaka M, Aoki H and Ochiai T. FK506, a novel immunosuppressant isolated from a Streptomyces. II. Immunosuppressive effect of FK506 in vitro. *J. Antibiot.* 1987; **40**: 1256–1265.
5. Peters DH, Fitton A, Plosker GL and Faulds D. Tacrolimus: a review of its pharmacology and therapeutic potential in hepatic and renal transplantation. *Drugs* 1993; **46**: 746–794.
6. European FK506 Multicentre Liver Study Group. Randomised trial comparing tacrolimus (FK506) and cyclosporin in prevention of liver allograft rejection. *Lancet* 1994; **344**: 423–428.
7. McDiarmid SV, Busuttil RW, Ascher NL, Burdick J, D'Alesandro AM, Esquivel C, Kalayoglu M, Klein AS, Marsh JW, Miller CM, Schwartz ME, Shaw BW and So SK. FK506 (tacrolimus) compared with cyclosporine for primary immunosuppression after pediatric liver transplantation. *Transplantation* 1995; **59**: 530–536.
8. Morris RE. Rapamycins: antifungal, antitumour, antiproliferative and immunosuppressive macrolides. *Transplant. Rev.* 1992; **6**: 39–87.
9. Sehgal SN. Immunosuppressive profile of rapamycin. *Ann. NY Acad. Sci.* 1993; **696**: 1–8.
10. Dumont FJ, Staruch MJ, Koprak SL, Melino MR and Sigal, NH. Distinct mechanisms of suppression of murine T cell activation by the related macrolides FK-506 and rapamycin. *J. Immunol.* 1990; **144**: 251–258.
11. Dumont FJ, Melino MR, Staruch MJ, Koprak SL, Fischer, PA and Sigal NH. The immunosuppressive macrolides FK-506 and rapamycin act as reciprocal antagonists in murine T cells. *J. Immunol.* 1990; **144**: 1418–1424.
12. Bierer BE, Mattila PS, Standaert RF, Herzenberg LA, Burakoff SJ, Crabtree G and Schreiber SL. Two distinct signal transmission pathways in T lymphocytes are inhibited by complexes formed between an immunophilin and

either FK506 or rapamycin. *Proc. Natl. Acad. Sci. USA* 1990; **87**: 9231–9235.

13. Kay JE, Kromwel L, Doe SEA and Denyer M. Inhibition of T and B lymphocyte proliferation by rapamycin. *Immunology* 1991; **72**: 544–549.

14. Handschumacher RE, Harding MW, Rice J, Drugge RJ and Speicher DW. Cyclophilin: a specific cytosolic binding protein for cyclosporin A. *Science* 1984; **226**: 544–547.

15. Harding MW, Handschumacher RE and Speicher DW. Isolation and amino acid sequence of cyclophilin. *J. Biol. Chem.* 1986; **261**: 8547–8555.

16. Takahashi N, Hayano T and Suzuki M. Peptidyl-prolyl *cis–trans* isomerase is the cyclosporin A-binding protein cyclophilin. *Nature* 1989; **337**: 473–475.

17. Fischer G, Wittmann-Liebold B, Lang K, Kiefhaber T and Schmid FX. Cyclophilin and peptidyl-prolyl *cis–trans* isomerase are probably identical proteins. *Nature* 1989; **337**: 476–478.

18. Siekierka JJ, Hung SHY, Poe M, Lin CS and Sigal NH. A cytosolic binding protein for the immunosuppressant FK506 has peptidyl-prolyl isomerase activity but is distinct from cyclophilin. *Nature* 1989; **341**: 755–757.

19. Standaert RF, Galat A, Verdine GL and Schreiber SL. Molecular cloning and overexpression of the human FK506-binding protein FKBP. *Nature* 1990; **346**: 671–674.

20. Gething MJ and Sambrook J. Protein folding in the cell. *Nature* 1992; **355**: 33–45.

21. Schmid FX, Mayr LM, Mucke M and Schonbrunner ER. Prolyl isomerases—role in protein folding. *Adv. Prot. Chem.* 1993; **44**: 25–66.

22. Sigal NH, Dumont F, Durette P, Siekierka JJ, Peterson L, Rich DH, Dunlap BE, Staruch MJ, Melino MR, Koprak SL, Williams D, Witzel B and Pisano JM. Is cyclophilin involved in the immunosuppressive and nephrotoxic mechanism of action of cyclosporin A? *J. Exp. Med.* 1991; **173**: 619–628.

23. Bierer BE, Somers PK, Wandless TJ, Burakoff SJ and Schreiber SL. Probing immunosuppressant action with a nonnatural immunophilin ligand. *Science* 1990; **250**: 556–559.

24. Dumont FJ, Staruch MJ, Koprak SL, Siekierka JJ, Lin CS, Harrison R, Sewell T, Kindt VM, Beattie TR, Wyvratt M and Sigal NH. The immunosuppressive and toxic effects of FK506 are mechanistically related—pharmacology of a novel antagonist of FK506 and rapamycin. *J. Exp. Med.* 1992; **176**: 751–760.

25. Ocain TD, Longhi D, Steffan RJ, Caccese RG and Seghal SN. A nonimmunosuppressive triene-modified rapamycin analog is a potent inhibitor of peptidyl prolyl *cis–trans* isomerase. *Biochem. Biophys. Res. Commun.* 1993; **192**: 1340–1346.

26. Tropschug M, Barthelmess IB and Neupert W. Sensitivity to cyclosporin A is mediated by cyclophilin in *Neurospora crassa* and *Saccharomyces cerevisiae*. *Nature*, 1989; **342**: 953–955.

27. Heitman J, Movva NR, Hiestand PC and Hall MN. FK506-binding protein proline rotamase is a target for the immunosuppressant agent FK506 in *Saccharomyces cerevisiae*. *Proc. Natl. Acad. Sci. USA* 1991; **88**: 1948–1952.

28. McLaughlin MM, Bossard MJ, Koser PL, Cafferkey R, Morris RA, Miles LM, Strickler J, Bergsma DJ, Levy MA and Livi GP, The yeast cyclophilin multigene family: purification, cloning and characterization of a new isoform. *Gene* 1992; **111**: 85–92.

29. Benton BM, Zang ZH and Thorner J. A novel FK506-and rapamycin-binding protein (FPR3 gene product) in the yeast *Saccharomyces cerevisiae* is a proline rotamase localized to the nucleolus. *J. Cell. Biol.* 1994; **127**: 623–639.

30. Barthelmess IB and Tropschug M. FK506-binding protein of *Neurospora crassa* (NcFKBP) mediates sensitivity to the immunosuppressant FK506; resistant mutants identify two loci. *Curr. Genet.* 1993; **23**: 54–58.

31. Davis ES, Becker A, Heitman J, Hall MN and Brennan MB. A yeast cyclophilin gene essential for lactate metabolism at high temperature. *Proc. Natl. Acad. Sci. USA* 1992; **89**: 11169–11173.

32. Sykes K, Gething MJ and Sambrook J. Proline isomerases function during heat shock. *Proc. Natl. Acad. Sci. USA* 1993; **90**: 5853–5857.

33. Galat A and Metcalfe SM. Peptidylproline *cis–trans* isomerases. *Prog. Biophys. Molec. Biol.* 1995; **63**: 67–118.

34. Kay JE. Structure–function relationships in the FKBP family of peptidylprolyl *cis–trans* isomerases. *Biochem. J.* 1996; **314**: 361-385

35. Haendler B, Hofer-Warbinek R and Hofer E. Complementary DNA for human T cell cyclophilin. *EMBO J.* 1987; **6**: 947–950.

36. Price ER, Zydowsky LD, Jin M, Baker CH, McKeon FD and Walsh CT. Human cyclophilin B: a second cyclophilin gene encodes a peptidyl-prolyl isomerase with a signal sequence. *Proc. Natl. Acad. Sci. USA* 1991; **88**: 1903–1907.

37. Schneider H, Charara N, Schmitz R, Wehrli S, Mikol V, Zurini MGM, Quesniaux VFJ and Movva NR. Human cyclophilin C: primary structure, tissue distribution, and determination of binding specificity for cyclosporins. *Biochemistry* 1994; **33**: 8218–8224.

38. Bergsma DJ, Eder C, Gross M, Kersten H, Sylvester D, Appelbaum E, Cusimano D, Livi GP, McLaughlin MM, Kasyan K, Porter TG, Silverman C, Dunnington D, Hand A, Pritchett WP, Bossard MJ, Brandt M and Levy MA. The cyclophilin multigene family of peptidyl-prolyl isomerases: characterization of three separate human isoforms. *J. Biol. Chem.* 1991; **266**: 23204–23214.

39. Kieffer LJ, Seng TW, Li W, Osterman DG, Handschumacher RE and Bayney RM. Cyclophilin-40, a protein with homology to the P59 component of the steroid receptor complex. *J. Biol. Chem.* 1993; **268**: 12303–12310.

40. Anderson SK, Gallinger S, Roder J, Frey J, Young HA and Ortaldo JR. A cyclophilin-related protein involved in the function of natural killer cells. *Proc. Natl. Acad. Sci. USA* 1993; **90**: 542–546.

41. Yokoyama N, Hayashi N, Seki T, Pante N, Ohba T, Nishii K, Kuma K, Hayashida T, Miyata T, Aebi U, Fukui M and Nishimoto T. A giant nucleopore protein that binds Ran/TC4. *Nature* 1995; **376**: 184–188.

42. Arakawa H, Nagase H, Hayashi N, Fujiwara T, Ogawa M, Shin S and Nakamura Y. Molecular cloning and expression of a novel human gene that is highly homologous to human FK506-binding protein 12 kDa (hFKBP-12) and character-

ization of two alternatively spliced transcripts. *Biochem. Biophys. Res. Commun.* 1994; **200**: 836–843.

43. Jin YJ, Albers MW, Lane WS, Bierer BE, Schreiber SL and Burakoff SL. Molecular cloning of a membrane associated human FK506-and rapamycin-binding protein, FKBP13. *Proc. Natl. Acad. Sci. USA* 1991; **88**: 6677–6681.

44. Jin YJ, Burakoff SJ and Bierer BE. Molecular cloning of a 25-kDa high affinity rapamycin binding protein, FKBP25. *J Biol Chem*, 1992; **267**: 10942–10945.

45. Peattie DA, Harding MW, Fleming MA, DeCenzo MT, Lippke JA, Livingston DJ and Benasutti M. Expression and characterization of human FKBP52, an immunophilin that associates with the 90-kDa heat shock protein and is a component of steroid receptor complexes. *Proc. Natl. Acad. Sci. USA* 1992; **89**: 10974–10978.

46. Rosborough SL, Fleming M, Nelson PA, Boger J and Harding MW. Identification of FKBP-related proteins with antibodies of predetermined specificity and isolation by FK-506 affinity chromatography. *Transplant. Proc.* 1991; **23**: 2890–2893.

47. Galat A, Lane WS, Standaert RF and Schreiber SL. A rapamycin-selective 25-kDa immunophilin. *Biochemistry*, 1992; **31**: 2427–2434.

48. Rahfeld JU, Rucknagel KP, Schelbert B, Ludwig B, Hacker J, Mann K and Fischer G. Confirmation of the existence of a third family among peptidyl-prolyl *cis/trans* isomerases: amino acid sequence and recombinant production of parvulin. *FEBS Letts.* 1994; **352**: 180–184.

49. Rudd KE, Sofia HJ, Koonin EV, Plunkett G, Lazar S and Rouviere PE. A new family of peptidyl-prolyl isomerases. *Trends Biochem. Sci.* 1995; **20**: 12–14.

50. Braun W, Kallen J, Mikol V, Walkinshaw MD and Wuthrich K. Three dimensional structure and actions of immunosuppressants and their immunophilins. *FASEB J.* 1995; **9**: 63–72.

51. Aldape RA, Futer O, DeCenzo MT, Jarrett BP, Murcko MA and Livingston DJ. Charged surface residues of FKBP12 participate in the formation of the FKBP12–FK506–calcineurin complex. *J. Biol. Chem.* 1992; **267**: 16029–16032.

52. Yang D, Rosen MK and Schreiber SL. A composite FKBP12-FK506 surface that contacts calcineurin. *J. Am. Chem. Soc.* 1993; **115**: 819–820.

53. Futer O, DeCenzo MT, Aldape RA and Livingston DJ. FK506 binding protein mutational analysis. *J. Biol. Chem.* 1995; **270**: 18935–18940.

54. Hayano T, Takahashi N, Kato S, Maki N and Suzuki M. Two distinct forms of peptidylprolyl *cis–trans* isomerase are expressed separately in periplasmic and cytoplasmic compartments of *Escherichia coli* cells. *Biochemistry* 1991; **30**: 3041–3048.

55. Liu J, Chen CM and Walsh CT. Human and *Escherichia coli* cyclophilins: sensitivity to inhibition by the immunosuppressant cyclosporin A correlates with a specific tryptophan residue. *Biochemistry* 1991; **30**: 2306–2310.

56. Liu J, Farmer JD, Lane WS, Friedman J, Weissman I and Schreiber SL. Calcineurin is a common target of cyclophilin-cyclosporin A and FKBP–FK506 complexes. *Cell* 1991; **66**: 807–815.

57. Friedman J and Weissman I. Two cytoplasmic candidates for immunophilin action are revealed by affinity for a new cyclophilin: one in the presence and one in the absence of CsA. *Cell* 1991; **66**: 799–806.

58. Fruman DA, Klee CB, Bierer BE and Burakoff SJ. Calcineurin phosphatase activity in T lymphocytes is inhibited by FK506 and cyclosporin A. *Proc. Natl. Acad. Sci. USA* 1992; **89**: 3686–3690.

59. Mukai H, Kuno T, Chang CD, Lane B, Luly JR and Tanaka C. FKBP12–FK506 complex inhibits phosphatase activity of two mammalian isoforms of calcineurin irrespective of their substrates or activation mechanisms. *J. Biochem.* 1993; **113**: 292–298.

60. Liu J, Albers MW, Wandless TJ, Luan S, Alberg DG, Belshaw PJ, Cohen P, Mackintosh C, Klee CB and Schreiber SL. Inhibition of T cell signalling by immunophilin–ligand complexes correlates with loss of calcineurin phosphatase activity. *Biochemistry* 1992; **31**: 3896–3901.

61. Kawai M, Lane BC, Hsieh GC, Mollison KW, Carter GW and Luly JR. Structure–activity profiles of macrolactam immunosuppressant FK-506 analogues. *FEBS Letts.* 1993; **316**: 107–113.

62. Nelson PA, Akselband Y, Kawamura A, Su M, Tung RD, Rich DH, Kishore V, Rosborough SL, DeCenzo MT, Livingston DJ and Harding MW. Immunosuppressive activity of [MeBm$_2$t]^1, D-diaminobutyryl-8-, and D-diaminopropyl-8-cyclosporin analogues correlates with inhibition of calcineurin phosphatase activity. *J. Immunol.* 1993; **150**: 2139–2147.

63. O'Keefe SJ, Tamura S, Kincaid RL, Tocci MJ and O'Neill EA. FK-506- and CsA-sensitive activation of the interleukin-2 promoter by calcineurin. *Nature* 1992; **357**: 692–694.

64. Clipstone NA and Crabtree GR. Identification of calcineurin as a key signalling enzyme in T lymphocyte activation. *Nature* 1992; **357**: 695–697.

65. Asami M, Kuno T, Mukai H and Tanaka C. Detection of the FK506–FKBP–calcineurin complex by a simple binding assay. *Biochem. Biophys. Res. Commun.* 1993; **192**: 1388–1394.

66. Luan S, Li W, Rusnak F, Assman SM and Schreiber SL. Immunosuppressants implicate protein phosphatase regulation of K$^+$ channels in guard cells. *Proc. Natl. Acad. Sci. USA* 1993; **90**: 2202–2206.

67. Luan S, Lane WS and Schreiber SL. pCyP B: a chloroplast-localized, heat shock-responsive cyclophilin from fava bean. *Plant Cell* 1994; **6**: 885–892.

68. Foor F, Parent SA, Morin N, Dahl AM, Ramadan N, Chrebet G, Bostian KA and Nielsen JB. Calcineurin mediates inhibition by FK506 and cyclosporin of recovery from α-factor arrest in yeast. *Nature* 1992; **360**: 682–684.

69. Yoshida T, Toda T and Yanagida M. A calcineurin-like gene ppbl+ in fission yeast: mutant defects in cytokinesis, cell polarity, mating and spindle pole body positioning. *J. Cell. Sci.* 1994; **107**: 1725–1735.

70. Fejzo J, Etzkorn FA, Clubb RT, Shi Y, Walsh CT and Wagner G. The mutant *Escherichia coli* F112W cyclophilin binds cyclosporin A in nearly identical conformation as human cyclophilin. *Biochemistry* 1994; **33**: 5711–5720.

71. Swanson SKH, Born T, Zydowsky LD, Cho H, Chang HY, Walsh CT and Rusnak F. Cyclosporin-mediated inhibition of bovine calcineurin by cyclophilins A and B. *Proc. Natl. Acad. Sci. USA* 1992; **89**: 3741–3745.

72. Bram R, Hung D, Martin P, Schreiber S and Crabtree G. Identification of the immunophilins capable of mediating inhibition of signal transduction by cyclosporin A and FK506: roles of calcineurin binding and cellular location. *Mol. Cell. Biol.* 1993; **13**: 4760–4769.

73. Wiederrecht G, Hung S, Chan HC, Marcy A, Martin M, Calaycay J, Boulton D, Sigal S, Kincaid RL and Siekierka JJ. Characterization of high molecular wieght FK-506 binding activities reveals a novel FK-506 binding protein as well as a protein complex. *J. Biol. Chem.* 1992; **267**: 21753–21760.

74. Lebeau MC, Myagkikh I, Rouviere-Fourmy N, Baulieu EE and Klee CB. Rabbit FKBP-59/HBI does not inhibit calcineurin activity *in vitro*. *Biochem. Biophys. Res. Commun.* 1994; **203**: 750–755.

75. Sewell TJ, Lam E, Martin MM, Leszyk J, Weidner J, Calaykay J, Griffin P, Williams H, Hung S, Cryan J, Sigal NH and Wiederrecht GJ. Inhibition of calcineurin by a novel FK-506-binding protein. *J. Biol. Chem.* 1994; **269**: 21094–21102.

76. Baughman G, Wiederrecht GJ, Campbell NF, Martin MM and Bourgeois S. FKBP51, a novel T-cell-specific immunophilin capable of calcineurin inhibition. *Mol. Cell. Biol.* 1995; **15**: 4395–4402.

77. Fruman DA, Wood MA, Gjertson CK, Katz HR, Burakoff SJ and Bierer BE. FK506 binding-protein-12 mediates sensitivity to both FK506 and rapamycin in murine mast cells. *Eur. J. Immunol.* 1995; **25**: 563–571.

78. Rosen MK, Yang D, Martin PK and Schreiber SL. Activation of an inactive immunophilin by mutagenesis. *J. Am. Chem. Soc.* 1993; **115**: 821–822.

79. Schultz LW, Martin PK, Liang J, Schreiber SL and Clardy J. Atomic structure of the immunophilin FKBP13–FK506 complex: insights into the complex binding site for calcineurin. *J. Am. Chem. Soc.* 1994; **116**: 3129–3130.

80. Griffith JP, Kim JL, Kim EE, Sintchak MD, Thomson JA, Fitzgibbon MJ, Fleming MA, Caron PR, Hsiao K and Navia MA. X-ray structure of calcineurin inhibited by the immunophilin–immunosuppressant FKBP12–FK506 complex. *Cell* 1995; **82**: 507–522.

81. Milan D, Griffith J, Su M, Price ER and McKeon F. The latch region of calcineurin B is involved in both immunosuppressant–immunophilin complex docking and phosphatase activation. *Cell* 1994; **79**: 437–447.

82. Clipstone NA, Fiorentino DF and Crabtree, GR. Molecular analysis of the interaction of calcineurin with drug–immunophilin complexes. *J. Biol. Chem.* 1994; **269**: 26431–26437.

83. Liu J. FK506 and cyclosporin, molecular probes for studying intracellular signal transduction. *Immunol. Today* 1993; **14**: 290–295.

84. Crabtree GR and Clipstone NA. Signal transmission between the plasma membrane and nucleus of T lymphocytes. *Ann. Rev. Biochem.* 1994; **63**: 1045–1083.

85. Fruman DA, Mather PE, Burakoff SJ and Bierer BE. Correlation of calcineurin phosphatase activity and pro-

86. Yazdanbakhsh K, Choi JW, Li Y, Lau LF and Choi Y. Cyclosporine-A blocks apoptosis by inhibiting the DNA-binding activity of the transcription factor NUR77. *Proc. Natl. Acad. Sci. USA* 1995; **92**: 437–441.

87. Dutz JP, Fruman DA, Burakoff SJ and Bierer, BE. A role for calcineurin in degranulation of murine cytotoxic T lymphocytes. *J. Immunol.* 1993; **150**: 2591–2598.

88. Heitman J, Movva NR and Hall MN. Targets for cell cycle arrest by the immunosuppressant rapamycin in yeast. *Science* 1991; **253**: 905–909.

89. Kunz J, Henriquez R, Schneider U, Deuter-Reinhard M, Movva NR and Hall MN. Target of rapamycin in yeast, TOR2, is an essential phosphatidylinositol kinase homolog required for G1 progression. *Cell* 1993; **73**: 585–596.

90. Cafferkey R, Young PR, McLaughlin MM, Bergsma DJ, Koltin Y, Sathe GM, Faucette L, Eng WK, Johnson RK and Livi GP. Dominant missense mutations in a novel yeast protein related to mammalian phosphatidylinositol 3-kinase and VPS34 abrogate rapamycin cytotoxicity. *Mol. Cell. Biol.* 1993; **13**: 6012–6023.

91. Cafferkey R, McLaughlin MM, Young PR, Johnson RK and Livi GP. Teast TOR (DRR) proteins: amino acid sequence alignment and identification of structural motifs. *Gene* 1994; **141**: 133–136.

92. Helliwell SB, Wagner P, Kunz J, Deuter-Reinhard M, Henriquez R and Hall MN. TOR1 and TOR2 are structurally and functionally similar but not identical phosphatidylinositol kinase homologues in yeast. *Mol. Biol. Cell* 1994; **5**: 105–118.

93. Zheng XF, Fiorentino D, Chen J, Crabtree GR and Schreiber SL. TOR kinase domains are required for two distinct functions, only one of which is inhibited by rapamycin. *Cell* 1995; **82**: 121–130.

94. Stan R, McLaughlin MM, Cafferkey R, Johnson RK, Rosenberg M and Livi GP. Interaction between FKBP12-rapamycin and TOR involves a conserved serine residue. *J. Biol. Chem.* 1994; **269**: 32027–32030.

95. Brown EJ, Albers MW, Shin TB, Ichikawa K, Keith CT, Lane WS and Schreiber SL. A mammalian protein targeted by G1-arresting rapamycin–receptor complex. *Nature* 1994; **369**: 756–758.

96. Sabatini DM, Erdjument-Bromage H, Lui M, Tempst P and Snyder SH. RAFT1: a mammalian protein that binds to FKBP12 in a rapamycin-dependent fashion and is homologous to yeast TORs. *Cell* 1994; **78**: 35–43.

97. Chui MI, Katz H and Berlin V. RAPT1, a mammalian homolog of yeast Tor, interacts with the FKBP12/rapamycin complex. *Proc. Natl. Acad. Sci. USA* 1994; **91**: 12574–12578.

98. Sabers CJ, Martin MM, Brunn GJ, Williams JM, Dumont FJ, Wiederrecht G and Abraham RT. Isolation of a protein target of the FKBP12–rapamycin complex in mammalian cells. *J. Biol. Chem.* 1995; **270**: 815–822.

99. Sabatini DM, Pierchala BA, Barrow RK, Schell MJ and Snyder SH. The rapamycin and FKBP12 target (RAFT) displays phosphatidylinositol 4-kinase activity. *J. Biol. Chem.* 1995; **270**: 20875–20878.

grammed cell death in murine T cell hybridomas. *Eur. J. Immunol.* 1992; **22**: 2513–2517.

100. Brown EJ, Beal PA, Keith CT, Chen J, Shin TB and Schreiber SL. Control of p70 S6 kinase by kinase activity of FRAP *in vivo*. *Nature* 1995; **377**: 441–446.

101. Koltin Y, Faucette L, Bergsma DJ, Levy MA, Cafferkey R, Koser PL, Johnson RK and Livi GP. Rapamycin sensitivity in *Saccharomyces cerevisiae* is mediated by a peptidyl-prolyl *cis–trans* isomerase related to human FK506-binding protein. *Mol. Cell. Biol.* 1991; **11**: 1718–1723.

102. Kuo CJ, Chung J, Fiorentino DF, Flanagan WM, Blenis J and Crabtree GR. Rapamycin selectively inhibits interleukin-2 activation of p70 S6 kinase. *Nature* 1992; **358**: 70–73.

103. Chung J, Kuo CJ, Crabtree GR and Blenis J. Rapamycin–FKBP specifically blocks growth-dependent activation of and signaling by the 70 kd S6 protein kinases. *Cell* 1992; **69**: 1227–1236.

104. Han JW, Pearson RB, Dennis PB, Thomas G. Rapamycin, wortmannin, and the methylxanthine SQ20006 inactivate p70^{s6k} by inducing dephosphorylation of the same subset of sites. *J. Biol. Chem.* 1995; **270**: 21396–21403.

105. Terada N, Lucas JJ, Szepesi A, Franklin RA, Domenico J, Gelfand EW. Rapamycin blocks cell cycle progression of activated T cells prior to events characteristic of the middle to late G1 phase of the cell cycle. *J. Cell. Physiol.* 1993; **154**: 7–15.

106. Morice WG, Brunn GJ, Wiederrecht G, Siekierka JJ and Abraham RT. Rapamycin-induced inhibition of p34^{cdc2} kinase activation is associated with G1/S-phase growth arrest in T lymphocytes. *J. Biol. Chem.* 1993; **268**: 3734–3738.

107. Albers MW, Williams RT, Brown EJ, Tanaka A, Hall FL, Schreiber SL. FKBP-rapamycin inhibits a cyclin-dependent kinase activity and a cyclin D1-cdk association in early G1 of an osteosarcoma cell line. *J. Biol. Chem.* 1993; **268**: 22825–22829.

108. Nourse J, Firpo E, Flanagan WM, Coats S, Polyak K, Lee MH, Massague J, Crabtree GR, Roberts JM. Interleukin-2-mediated elimination of the p27^{kip1} cyclin-dependent kinase inhibitor prevented by rapamycin. *Nature*, 1994; **372**: 570–573.

109. Terada N, Franklin RA, Lucas JJ, Blenis J, Gelfand EW. Failure of rapamycin to block proliferation once resting cells have entered the cell cycle despite inactivation of p70 S6 kinase. *J. Biol. Chem.* 1993; **268**: 12062–12068.

110. Calvo V, Wood M, Gjertson C, Vik T, Bierer BE. Activation of 70-kDa S6 kinase, induced by the cytokines interleukine-3 and erythropoietin and inhibited by rapamycin, is not an absolute requirement for cell proliferation. *Eur. J. Immunol.* 1994; **24**: 2664–2671.

111. Kay, JE, Smith MC, Frost V, Morgan GY. Hypersensitivity to rapamycin of BJAB B lymphoblastoid cells. *Immunology*, 1996; **87**: 390-395

112. Frost V, Smith MC, Morley SJ, Kay JE. Role of p70 S6 kinase in lymphocyte activation. *Biochem. Soc. Trans.* 1996; **24**: 91S.

113. Smith MC, Bowler H, Kay JE. Unpublished data.

114. Jefferies HBJ, Reinhard C, Kozma SC, Thomas G. Rapamycin selectively represses translation of the polypyrimidine tract mRNA family. *Proc. Natl. Acad. Sci. USA* 1994; **91**: 4441–4445.

115. Terada N, Patel HR, Takese K, Kohno K, Nairn AC and Gelfand EW. Rapamycin selectively inhibits translation of mRNAs encoding elongation factors and ribosomal proteins. *Proc. Natl. Acad. Sci. USA* 1994; **91**: 11477–11481.

116. Nielsen FC, Ostergaard L, Nielsen J and Christiansen J. Growth-dependent translation of IGF-II mRNA by a rapamycin-sensitive pathway. *Nature* 1995; **377**: 358–362.

117. Graves LM, Bornfeldt KE, Argast GM, Krebs EG, Kong X, Lin TA and Lawrence JC. cAMP-and rapamycin-sensitive regulation of the association of eukaryotic initiation factor 4E and the translational regulator PHAS-I in aortic smooth muscle cells. *Proc. Natl. Acad. Sci. USA* 1995; **92**: 7222–7226.

118. Lin TA, Kong X, Saltiel AR, Blackshear PJ and Lawrence JC. Control of PHAS-I by insulin in 3T3-L1 adipocytes. Synthesis, degradation and phosphorylation by a rapamycin-sensitive and mitogen-activated protein kinase-independent pathway. *J. Biol. Chem.* 1995; **270**: 18531–18538.

119. Dumont FJ, Altmeyer A, Kastner C, Fischer PA, Lemon KP, Chung J, Blenis J and Staruch MJ. Relationship between multiple biologic effects of rapamycin and the inhibition of pp70S6 protein kinase activity. *J. Immunol.* 1994; **152**: 992–1003.

120. Kay JE, Wellhausen A, Frost V, Morgan GY, Smith MC and Morley SJ. (1996) Rapamycin-resistant human lymphoid cell lines. *Biochem. Soc. Trans.* 1996; **24**: 89S.

121. Su Q, Zhao M, Weber E, Eugster HP and Ryffel B. Distribution and activity of calcineurin in rat tissues. *Eur. J. Biochem.* 1995; **230**: 469–474.

122. Nakamura T, Liu Y, Hirata D, Namba H, Harada S, Hirokawa T and Miyakawa T. Protein phosphatase type 2B (calcineurin)-mediated, FK506-sensitive regulation of intracellular ions in yeast is an important determinant for adaptation to high salt stress conditions. *EMBO J.* 1993; **12**: 4063–4071.

123. Dittmann J, Wenger RM, Kleinkauf H and Lawen A. Mechanism of cyclosporine-A biosynthesis—evidence for synthesis via a single linear undecapeptide precursor. *J. Biol. Chem.* 1994; **269**: 2841–2846.

124. Zocher R, Keller U, Lee C and Hoffmann K. A seventeen kilodaltons peptidyl-prolyl *cis–trans* isomerase of the cyclosporin-producer *Tolypocladium inflatum* is sensitive to cyclosporin A. *J. Antibiot.* 1992; **45**: 265–268.

125. Pahl A and Keller U. FK506-binding proteins from streptomycetes producing immunosuppressive macrolactones of the FK506 type. *J. Bact.* 1992; **174**: 5888–5894.

126. Wieczorek Z, Bengtsson B, Trojnar J and Siemion IZ. The immunosuppressive activity of cyclolinopeptide A. *Peptide Res.* 1991; **4**: 275–283.

127. Siemion IZ, Pedyczak A, Trojnar J, Zimecki M and Wieczorek Z. Immunosuppressive activity of antamanide and some of its analogues. *Peptides* 1992; **13**: 1233–1237.

128. Salituro GM, Zink DL, Dahl A, Nielsen J, Wu E, Huang L, Kastner C, Dumont FJ. Meridamycin: a novel non-immunosuppressive FKBP12 ligand from *Streptomyces hygroscopicus*. *Tetrahdron Lett.* 1995; **36**: 997–1000.

129. Cardenas ME, Hemenway C, Muir RS, Ye R, Fiorentino D and Heitman J. Immunophilins interact with calci-

neurin in the absence of exogenous immunosuppressive ligands. *EMBO J.* 1994; **13**: 5944–5957.

130. Ning YM and Sanchez, ED. Potentiation of glucocorticoid receptor-mediated gene expression by the immunophilin ligands FK506 and rapamycin. *J. Biol. Chem.* 1993; **268**: 6073–6076.

131. Timerman AP, Wiederrecht G, Marcy A and Fleischer S. Characterization of an exchange reaction between soluble FKBP-12 and the FKBP ryanodine receptor complex. *J. Biol. Chem.* 1995; **270**: 2451–2459.

Index

Index compiled by A.C. Purton